EUROPA LEHRMITTEL

EUROPA-TECHNICAL BOOK SERIES
for the Meta[illegible]

Ulrich Fischer
Roland Gomeringer
Max Heinzler
Roland Kilgus
Friedrich Näher
Stefan Oesterle
Heinz Paetzold
Andreas Stephan

Mechanical and Metal Trades Handbook

3rd English edition

Europa-No.: 1910X

VERLAG EUROPA LEHRMITTEL · Nourney, Vollmer GmbH & Co. KG
Düsselberger Straße 23 · 42781 Haan-Gruiten · Germany

Original title:
Tabellenbuch Metall, 45th edition, 2011

Authors:

Ulrich Fischer	Dipl.-Ing. (FH)	Reutlingen
Roland Gomeringer	Dipl.-Gwl.	Meßstetten
Max Heinzler	Dipl.-Ing. (FH)	Wangen im Allgäu
Roland Kilgus	Dipl.-Gwl.	Neckartenzlingen
Friedrich Näher	Dipl.-Ing. (FH)	Balingen
Stefan Oesterle	Dipl.-Ing.	Amtzell
Heinz Paetzold	Dipl.-Ing. (FH)	Mühlacker
Andreas Stephan	Dipl.-Ing. (FH)	Marktoberdorf

Editor:
Ulrich Fischer, Reutlingen

Graphic design:
Design office of Verlag Europa-Lehrmittel, Ostfildern, Germany

Most recent editions of standards and other regulations govern their use.
They can be ordered from Beuth Verlag GmbH, Burggrafenstr. 6, 10787 Berlin, Germany.

The content of the chapter „Program structure of CNC machines according to PAL" (page 412 to 424) complies with the publications of the PAL Prüfungs- und Lehrmittelentwicklungsstelle (Institute for the development of training and testing material) of the IHK Region Stuttgart (Chamber of Commerce and Industry of the Stuttgart region).

English edition: Mechanical and Metal Trades Handbook
3rd edition, 2012

6 5 4 3 2 1

All printings of this edition may be used concurrently in the classroom since they are unchanged, except for some corrections to typographical errors and slight changes in standards.

ISBN 13 978-3-8085-1914-1

Cover design includes a photograph from TESA/Brown & Sharpe, Renens, Switzerland

Translation: Eva Schwarz, 76879 Ottersheim, Germany; www.technische-uebersetzungen-eva-schwarz.de
Typesetting: Satz+Layout Werkstatt Kluth GmbH, D-50374 Erftstadt, Germany; www.slw-kluth.de
Printed by: M.P. Media-Print Informationstechnologie GmbH, D-33100 Paderborn, Germany

Preface

The Mechanical and Metal Trades Handbook is well-suited for shop reference, tooling, machine building, maintenance and as a general book of knowledge. It is also useful for educational purposes, especially in practical work or curricula and continuing education programs.

Target Groups
- Industrial and trade mechanics
- Tool & die makers
- Machinists
- Millwrights
- Draftspersons
- Technical Instructors
- Apprentices in above trade areas
- Practitioners in trades and industry
- Mechanical Engineering students

Notes for the user

The contents of this book include tables and formulae in eight chapters, including Tables of Contents, Subject Index and Standards Index.

The **tables** contain the most important guidelines, designs, types, dimensions and standard values for their subject areas.

Units are not specified in the legends for the **formulae** if several units are possible. However, the calculation examples for each formula use those units normally applied in practice.

The **Table of Contents** in the front of the book is expanded further at the beginning of each chapter in form of a partial Table of Contents.

The **Subject Index** at the end of the book (pages 435–444) is extensive.

The **Standards Index** (pages 425–434) lists all the current standards and regulations cited in the book. In many cases previous standards are also listed to ease the transition from older, more familiar standards to new ones.

Changes in the 3rd edition

In the present edition, we have updated the cited standards and restructured, updated, enhanced or added the following chapters in line with new developments in engineering:

- Fundamentals of technical mathematics
- Strength of materials
- Plastics
- Production management
- Forming
- Welding
- PAL programming system for NC turning and NC milling
- Steel types
- Material testing
- Machining processes
- Injection molding (new)
- GRAFCET

Acknowledgement

Special thanks to Alexander Huter, Vocational Training Specialist – Tool and Die, Ontario, for his input into the English translation of this book. His assistance has been extremely valuable.

November 2012 Authors and publisher

Table of Contents

3 Technical Drawing (TD) 51

4 Materials Science (MS) 111

Standards and other Regulations

Standardization and standards terms

Standardization is the systematic achievement of uniformity of material and non-material objects, such as components, calculation methods, process flows and services for the benefit of the general public.

Standards term	Example	Explanation
Standard	DIN 7157	A standard is the published work of standardization, e.g. the selection of particular fits in DIN 7157.
Part	DIN 30910-2	Standards can comprise several parts associated with each other. The part numbers are appended to the main standard number with hyphens. DIN 30910-2 describes sintered materials for filters for example, whereas Part 3 and 4 deal with sintered materials for bearings and formed parts.
Supplement	DIN 743 Suppl. 1	A supplement contains information for a standard, however no additional specifications. The supplement DIN 743 Suppl. 1, for example, contains application examples of load capacity calculations for shafts and axles described in DIN 743.
Draft	E DIN 743 (2008-10)	Draft standards are made available to the public for examination and commenting. The planned new version of DIN 743 on load-bearing calculations of shafts and axes, for example, has been published since October 2008 as Draft E DIN 743.
Preliminary standard	DIN V 66304 (1991-04)	A preliminary standard contains the results of standardization, which have not been released as a standard because of certain provisos. DIN V 66304, for example, discusses a format for exchange of standard part data for computer-aided design.
Issue date	DIN 76-1 (2004-06)	Date of publication which is made public in the DIN publication guide; this is the date at which time the standard becomes valid. DIN 76-1, which sets undercuts for metric ISO threads has been valid since June 2004 for example.

Types of standards and regulations (selection)

Type	Abbreviation	Explanation	Purpose and contents
International Standards (ISO standards)	ISO	**I**nternational **O**rganization for **S**tandardization, Geneva (O and S are reversed in the abbreviation)	Simplifies the international exchange of goods and services, as well as cooperation in scientific, technical and economic areas.
European Standards (EN standards)	EN	**E**uropean Committee for Standardization (Comité Européen de **N**ormalisation), Brussels	Technical harmonization and the associated reduction of trade barriers for the advancement of the European market and the coalescence of Europe.
German Standards (DIN standards)	DIN	**D**eutsches **I**nstitut für **N**ormung e.V., Berlin (German Institute for Standardization)	National standardization facilitates rationalization, quality assurance, environmental protection and common understanding in economics, technology, science, management and public relations.
	DIN EN	European standard for which the German version has attained the status of a German standard.	
	DIN ISO	German standard for which an international standard has been adopted without change.	
	DIN EN ISO	European standard for which an international standard has been adopted unchanged and the German version has the status of a German standard.	
	DIN VDE	Printed publication of the VDE, which has the status of a German standard.	
VDI Guidelines	VDI	**V**erein **D**eutscher **I**ngenieure e.V., Düsseldorf (Association of German Engineers)	These guidelines give an account of the current state of the art in specific subject areas and contain, for example, concrete procedural guidelines for the performing calculations or designing processes in mechanical or electrical engineering.
VDE printed publications	VDE	**V**erband **D**eutscher **E**lektrotechniker e.V., Frankfurt (Association for Electrical, Electronic & Information Technologies)	
DGQ publications	DGQ	**D**eutsche **G**esellschaft für **Q**ualität e.V., Frankfurt (German Society for Quality)	Recommendations in the area of quality technology.
REFA sheets	REFA	Association for Work Design, Industrial Organization and Corporate Development REFA e.V., Darmstadt	Recommendations in the area of production and work planning.

M

1 Mathematics

Quantity	Symbol	Unit Name	Unit Symbol
Lengths	l	meter	m

Surface area

$$A_s = \pi \cdot d \cdot h + 2 \cdot \frac{\pi \cdot d^2}{4}$$

Lateral surface area

$$A_M = \pi \cdot d \cdot h$$

$$\text{sine} = \frac{\text{opposite side}}{\text{hypotenuse}}$$

$$\text{cosine} = \frac{\text{adjacent side}}{\text{hypotenuse}}$$

$$\text{tangent} = \frac{\text{opposite side}}{\text{adjacent side}}$$

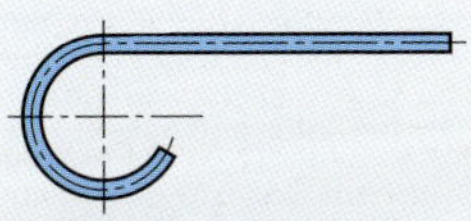

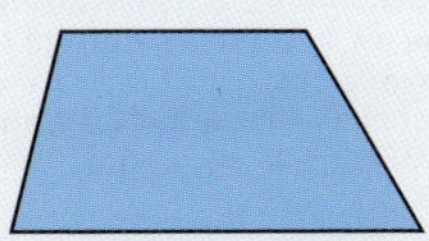

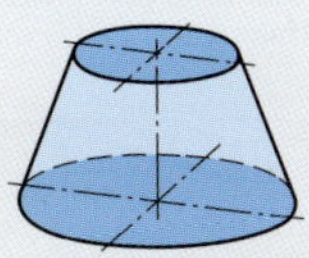

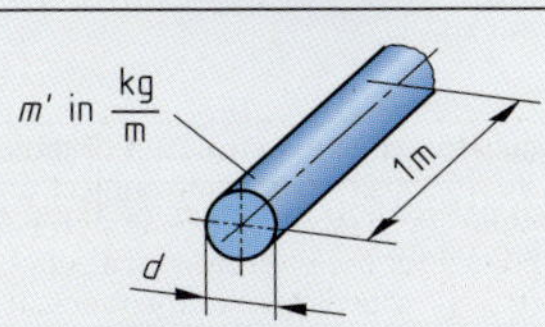

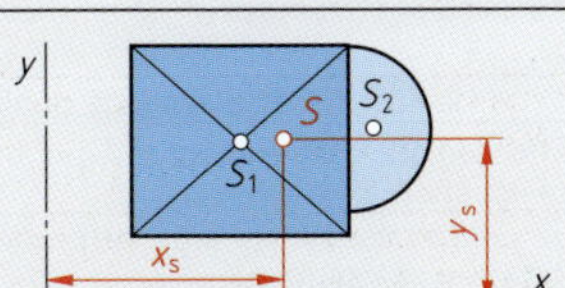

M

Units of measurement

SI[1] Base quantities and base units

cf. DIN 1301-1 (2010-10), -2 (1978-02), -3 (1979-10)

Base quantity	Length	Mass	Time	Electric current	Thermo-dynamic temperature	Amount of substance	Luminous intensity
Base units	meter	kilo-gram	second	ampere	kelvin	mole	candela
Unit symbol	m	kg	s	A	K	mol	cd

[1] The units for measurement are defined in the International System of Units SI (**S**ystème **I**nternational d'Unités). It is based on the seven basic units (SI units), from which other units are derived.

Base quantities, derived quantities and their units

Quantity	Symbol	Unit Name	Unit Symbol	Relationship	Remarks Examples of application
Length, Area, Volume, Angle					
Length	l	**meter**	m	1 m = 10 dm = 100 cm = 1000 mm 1 mm = 1000 µm 1 km = 1000 m	1 inch = 25.4 mm In aviation and nautical applications the following applies: 1 international nautical mile = 1852 m
Area	A, S	square meter are hectare	m^2 a ha	1 m^2 = 10 000 cm^2 = 1 000 000 mm^2 1 a = 100 m^2 1 ha = 100 a = 10 000 m^2 100 ha = 1 km^2	Symbol S only for cross-sectional areas Are and hectare only for land
Volume	V	cubic meter liter	m^3 l, L	1 m^3 = 1000 dm^3 = 1 000 000 cm^3 1 l = 1 L = 1 dm^3 = 10 dl = 0.001 m^3 1 ml = 1 cm^3	Mostly for fluids and gases
Plane angle (angle)	α, β, γ ...	radian degrees minutes seconds	rad ° ′ ″	1 rad = 1 m/m = 57.2957...° = 180°/π 1° = $\frac{\pi}{180}$ rad = 60′ 1′ = 1°/60 = 60″ 1″ = 1′/60 = 1°/3600	1 rad is the angle formed by the intersection of a circle around the center of 1 m radius with an arc of 1 m length. In technical calculations instead of α = 33° 17′ 27.6″, better use is α = 33.291°.
Solid angle	Ω	steradian	sr	1 sr = 1 m^2/m^2	An object whose extension measures 1 rad in one direction and perpendicularly to this also 1 rad, covers a solid angle of 1 sr.
Mechanics					
Mass	m	**kilogram** gram megagram metric ton	kg g Mg t	1 kg = 1000 g 1 g = 1000 mg 1 metric t = 1000 kg = 1 Mg 0.2 g = 1 ct	Mass in the sense of a scale result or a weight is a quantity of the type of mass (unit kg). Mass for precious stones in carat (ct).
Linear mass density	m'	kilogram per meter	kg/m	1 kg/m = 1 g/mm	For calculating the mass of bars, profiles, pipes.
Area mass density	m''	kilogram per square meter	kg/m^2	1 kg/m^2 = 0.1 g/cm^2	To calculate the mass of sheet metal.
Density	ϱ	kilogram per cubic meter	kg/m^3	1000 kg/m^3 = 1 metric t/m^3 = 1 kg/dm^3 = 1 g/cm^3 = 1 g/ml = 1 mg/mm^3	The density is a quantity independent of location.

M

Units of measurement

Quantities and Units (continued)

Quantity	Symbol	Unit Name	Unit Symbol	Relationship	Remarks Examples of application
Mechanics					
Moment of inertia, 2nd Moment of mass	J	kilogram x square meter	$kg \cdot m^2$	The following applies for a homogenous body: $J = \varrho \cdot r^2 \cdot V$	The moment of inertia (2nd moment of mass) is dependent upon the total mass of the body as well as its form and the position of the axis of rotation.
Force Weight	F F_G, G	newton	N	$1\ N = 1 \frac{kg \cdot m}{s^2} = 1 \frac{J}{m}$ $1\ MN = 10^3\ kN = 1\,000\,000\ N$	The force 1 N effects a change in velocity of 1 m/s in 1 s in a 1 kg mass.
Torque Bending mom. Torsional mom.	M M_b T	newton x meter	$N \cdot m$	$1\ N \cdot m = 1 \frac{kg \cdot m^2}{s^2}$	1 N · m is the moment that a force of 1 N effects with a lever arm of 1 m.
Momentum	p	kilogram x meter per second	kg · m/s	1 kg · m/s = 1 N · s	The momentum is the product of the mass times velocity. It has the direction of the velocity.
Pressure Mechanical stress	p σ, τ	pascal newton per square millimeter	Pa N/mm^2	$1\ Pa = 1\ N/m^2 = 0.01\ mbar$ $1\ bar = 100\,000\ N/m^2 = 10\ N/cm^2 = 10^5\ Pa$ $1\ mbar = 1\ hPa$ $1\ N/mm^2 = 10\ bar = 1\ MN/m^2 = 1\ MPa$ $1\ daN/cm^2 = 0.1\ N/mm^2$	Pressure refers to the force per unit area. For gage pressure the symbol p_g is used (DIN 1314). 1 bar = 14.5 psi (pounds per square inch)
Second moment of area	I	meter to the fourth power centimeter to the fourth power	m^4 cm^4	$1\ m^4 = 100\,000\,000\ cm^4$	Previously: Geometrical moment of inertia
Energy, Work, Quantity of heat	E, W	joule	J	$1\ J = 1\ N \cdot m = 1\ W \cdot s = 1\ kg \cdot m^2/s^2$	Joule for all forms of energy, kW · h preferred for electrical energy.
Power Heat flux	P Φ	watt	W	$1\ W = 1\ J/s = 1\ N \cdot m/s = 1\ V \cdot A = 1\ m^2 \cdot kg/s^3$	Power describes the work which is achieved within a specific time.
Time					
Time, Time span, Duration	t	**seconds** minutes hours day year	s min h d a	1 min = 60 s 1 h = 60 min = 3600 s 1 d = 24 h = 86 400 s	3 h means a time span (3 hrs.), 3^h means a point in time (3 o'clock). If points in time are written in mixed form, e.g. $3^h24^m10^s$, the symbol min can be shortened to m.
Frequency	f, ν	hertz	Hz	1 Hz = 1/s	1 Hz ≙ 1 cycle in 1 second.
Rotational speed, Rotational frequency	n	1 per second 1 per minute	1/s 1/min	$1/s = 60/min = 60\ min^{-1}$ $1/min = 1\ min^{-1} = \frac{1}{60\ s}$	The number of revolutions per unit of time gives the revolution frequency, also called rpm.
Velocity	v	meters per second meters per minute kilometers per hour	m/s m/min km/h	1 m/s = 60 m/min = 3.6 km/h $1\ m/min = \frac{1\ m}{60\ s}$ $1\ km/h = \frac{1\ m}{3.6\ s}$	Nautical velocity in knots (kn): 1 kn = 1.852 km/h miles per hour = 1 mile/h = 1 mph 1 mph = 1.60934 km/h
Angular-velocity	ω	1 per second radians per second	1/s rad/s	$\omega = 2\pi \cdot n$	For a rpm of n = 2/s the angular velocity $\omega = 4\pi$/s.
Acceleration	a, g	meters per second squared	m/s^2	$1\ m/s^2 = \frac{1\ m/s}{1\ s}$	Symbol g only for acceleration due to gravity. $g = 9.81\ m/s^2 \approx 10\ m/s^2$

Units of measurement

Quantities and units (continued)

Quantity	Symbol	Unit Name	Unit Symbol	Relationship	Remarks / Examples of application
Electricity and Magnetism					
Electric current Electromotive force Electrical resistance Electrical conductance	I E R G	**ampere** volt ohm siemens	A V Ω S	 1 V = 1 W/1 A = 1 J/C 1 Ω = 1 V/1 A 1 S = 1 A/1 V = 1/Ω	The motion of an electrical charge is called current. The electromotive force is equal to the potential difference between two points in an electric field. The reciprocal of the electrical resistance is called the electrical conductivity.
Specific resistance Conductivity	ϱ $\gamma, \varkappa$	ohm x meter siemens per meter	Ω · m S/m	$10^{-6}\ \Omega \cdot m = 1\ \Omega \cdot mm^2/m$	$\varrho = \frac{1}{\varkappa}$ in $\frac{\Omega \cdot mm^2}{m}$ $\varkappa = \frac{1}{\varrho}$ in $\frac{m}{\Omega \cdot mm^2}$
Frequency	f	hertz	Hz	1 Hz = 1/s 1000 Hz = 1 kHz	Frequency of public electric utility: EU 50 Hz, USA/Canada 60 Hz
Electrical energy	W	joule	J	1 J = 1 W · s = 1 N · m 1 kW · h = 3.6 MJ 1 W · h = 3.6 kJ	In atomic and nuclear physics the unit eV (electron volt) is used.
Phase difference	φ	–	–	for alternating current: $\cos \varphi = \frac{P}{U \cdot I}$	The angle between current and voltage in inductive or capacitive load.
Elect. field strength Elect. charge Elect. capacitance inductance	E Q C L	volts per meter coulomb farad henry	V/m C F H	 1 C = 1 A · 1 s; 1 A · h = 3.6 kC 1 F = 1 C/V 1 H = 1 V · s/A	$E = \frac{F}{Q}$, $C = \frac{Q}{U}$, $Q = I \cdot t$
Power Effective power	P	watt	W	1 W = 1 J/s = 1 N · m/s = 1 V · A	In electrical power engineering: Apparent power S in V · A
Quantity	**Symbol**	**Unit Name**	**Unit Symbol**	**Relationship**	**Remarks / Examples of application**
Electricity and Magnetism					
Thermodynamic temperature Celsius temperature	T, Θ t, ϑ	**kelvin** degrees Celsius	K °C	0 K = – 273.15 °C 0 °C = 273.15 K 0 °C = 32 °F 0 °F = – 17.77 °C	Kelvin (K) and degrees Celsius (°C) are used for temperatures and temperature differences. $t = T - T_0$; T_0 = 273.15 K degrees Fahrenheit (°F): 1.8 °F = 1 °C
Quantity of heat	Q	joule	J	1 J = 1 W · s = 1 N · m 1 kW · h = 3600000 J = 3,6 MJ	1 kcal ≙ 4.1868 kJ
Net calorific value	H_u	joule per kilogram Joule per cubic meter	J/kg J/m^3	1 MJ/kg = 1000000 J/kg 1 MJ/m^3 = 1000000 J/m^3	Thermal energy released per kg fuel minus the heat of vaporization of the water vapor contained in the exhaust gases.

Non-SI units

Length	Area	Volume	Mass	Energy, Power
1 inch (in) = 25.4 mm	1 sq.in = 6.452 cm^2	1 cu.in = 16.39 cm^3	1 oz = 28.35 g	1 PSh = 0.735 kWh
1 foot (ft) = 0.3048 m	1 sq.ft = 9.29 dm^2	1 cu.ft = 28.32 dm^3	1 lb = 453.6 g	1 PS = 735 W
1 yard (yd) = 0.9144 m	1 sq.yd = 0.8361 m^2	1 cu.yd = 764.6 dm^3	1 t = 1000 kg	1 kcal = 4186.8 Ws
1 nautical mile = 1.852 km	1 acre = 4046.856 m^2	1 gallon (US) = 3.785 l	1 short ton = 907.2 kg	1 kcal = 1.166 Wh
1 mile = 1.6093 km	**Pressure**	1 gallon (UK) = 4.546 l	1 Karat = 0.2 g	1 kpm/s = 9.807 W
	1 bar = 14.5 $pound/in^2$	1 barrel = 158.8 l	1 $pound/in^3$ = 27.68 g/cm^3	1 Btu = 1055 Ws
	1 N/mm^2 = 145.038 $pound/in^2$			1 hp = 745.7 W

Formula symbols, Mathematical symbols

Formula symbols

cf. DIN 1304-1 (1994-03)

Formula symbol	Meaning	Formula symbol	Meaning	Formula symbol	Meaning
Length, Area, Volume, Angle					
l	Length	r, R	Radius	α, β, γ	Planar angle
w	Width	d, D	Diameter	Ω	Solid angle
h	Height	A, S	Area, Cross-sectional area	λ	Wave length
s	Linear distance	V	Volume		
Mechanics					
m	Mass	F	Force	G	Shear modulus
m'	Linear mass density	F_W, W	Gravitational force, Weight	μ, f	Coefficient of friction
m''	Area mass density	M	Torque	W	Section modulus
ϱ	Density	T	Torsional moment	I	Second moment of an area
J	Moment of inertia	M_b	Bending moment	W, E	Work, Energy
p	Pressure	σ	Normal stress	W_p, E_p	Potential energy
p_{abs}	Absolute pressure	τ	Shear stress	W_k, E_k	Kinetic energy
p_{amb}	Ambient pressure	ε	Normal strain	P	Power
p_g	Gage pressure	E	Modulus of elasticity	η	Efficiency
Time					
t	Time, Duration	f, ν	Frequency	a	Acceleration
T	Cycle duration	v, u	Velocity	g	Gravitational acceleration
n	Revolution frequency, Speed	ω	Angular velocity	α	Angular acceleration
				$Q, \dot{V}, q_v$	Volumetric flow rate
Electricity					
Q	Electric charge, Quantity of electricity	L	Inductance	X	Reactance
E	Electromotive force	R	Resistance	Z	Impedance
C	Capacitance	ϱ	Specific resistance	φ	Phase difference
I	Electric current	$\gamma, \varkappa$	Electrical conductivity	N	Number of turns
Heat					
T, Θ	Thermodynamic temperature	Q	Heat, Quantity of heat	$\Phi, \dot{Q}$	Heat flow
$\Delta T, \Delta t, \Delta\vartheta$	Temperature difference	λ	Thermal conductivity	a	Thermal diffusivity
t, ϑ	Celsius temperature	α	Heat transition coefficient	c	Specific heat
α_l, α	Coefficient of linear expansion	k	Heat transmission coefficient	H_{net}	Net calorific value
Light, Electromagnetic radiation					
E	Illuminance	f	Focal length	I	Luminous intensity
		n	Refractive index	Q, W	Radiant energy
Acoustics					
p	Acoustic pressure	L_P	Acoustic pressure level	N	Loudness
c	Acoustic velocity	I	Sound intensity	L_N	Loudness level

Mathematical symbols

cf. DIN 1302 (1999-12)

Math. symbol	Spoken	Math. symbol	Spoken	Math. symbol	Spoken
$\approx$	approx. equals, around, about	$\sim$	proportional	log	logarithm (general)
$\hat{=}$	equivalent to	a^n	a to the n-th power, the n-th power of a	lg	common logarithm
...	and so on, etc.	$\sqrt{\ }$	square root of	ln	natural logarithm
∞	infinity	$\sqrt[n]{\ }$	n-th root of	e	Euler number (e = 2.718281…)
$=$	equal to	$\lvert x \rvert$	absolute value of x	sin	sine
$\neq$	not equal to	$\perp$	perpendicular to	cos	cosine
$\stackrel{def}{=}$	is equal to by definition	$\parallel$	is parallel to	tan	tangent
$<$	less than	$\uparrow\uparrow$	parallel in the same direction	cot	cotangent
$\leq$	less than or equal to	$\uparrow\downarrow$	parallel in the opposite direction	(), [], { }	parentheses, brackets open and closed
$>$	greater than	$\sphericalangle$	angle	π	pi (circle constant = 3.14159 …)
$\geq$	greater than or equal to	$\triangle$	triangle		
$+$	plus	$\cong$	congruent to		
$-$	minus	Δx	delta x (difference between two values)	$\overline{AB}$	line segment AB
$\cdot$	times, multiplied by	%	percent, of a hundred	$\overset{\frown}{AB}$	arc AB
$-, /, :, \div$	over, divided by, per, to	‰	per mil, of a thousand	a', a''	a prime, a double prime
Σ	sigma (summation)			a_1, a_2	a sub 1, a sub 2

M

Formulas, Equations, Graphs

Formulas

In most cases, the calculation of physical quantities is done with the help of formulas. They consist of:

- Formula symbols, e.g. v_c for cutting velocity, d for diameter, n for speed
- Operators (calculation rules), e.g. · for multiplication, + for addition, – for subtraction and — (fraction line) for division
- Constants, e.g. π (pi) = 3.14159 ...
- Numbers, e.g. 10, 15 ...

The formula symbols (page 13) are wildcards for quantities. When solving mathematical problems, the known quantities with their units are filled in the formulas. Before or during the calculation process, the units are converted in a way that

- the calculation becomes feasible or
- the result comprises the required unit.

Most quantities and units are standardized (page 10).

The **result** is always a **numerical value** accompanied by a **unit**, e.g. 4.5 m, 15 s

Formula for cutting velocity

$$v_c = \pi \cdot d \cdot n$$

Example:

What is the cutting velocity v_c in m/min for d = 200 mm and n = 630/min?

$$v_c = \pi \cdot d \cdot n = \pi \cdot 200\ \text{mm} \cdot 630\ \frac{1}{\text{min}} = \pi \cdot 200\ \text{mm} \cdot \frac{1\ \text{m}}{1000\ \text{mm}} \cdot 630\ \frac{1}{\text{min}} = \mathbf{395.84\ \frac{m}{min}}$$

Numerical value equations

Numerical value equations or numerical equations are formulas in which the typical conversions of units have already been integrated. The following should be noted when using equations:

The numerical values of the individual quantities may only be used in combination with the designated unit.

- The units are not carried along in the calculation.
- The unit of the quantity to be obtained is predetermined.

Numerical value equation for torque

$$M = \frac{9550 \cdot P}{n}$$

Designated unit		
Designation		Unit
M	Torque	N · m
P	Power	kW
n	Speed	1/min

Example:

What is the torque M of an electrical motor with a driving power of P = 15 kW and a speed of n = 750/min?

$$M = \frac{9550 \cdot P}{n} = \frac{9550 \cdot 15}{750}\ \text{N} \cdot \text{m} = \mathbf{191\ N \cdot m}$$

Equations and graphs

In functional equations, y is the function of x, with x as an independent and y as a dependent variable. The number pairs (x, y) of a value table form a graph in the x-y system of coordinates.

Assigned function

$$y = f(x)$$

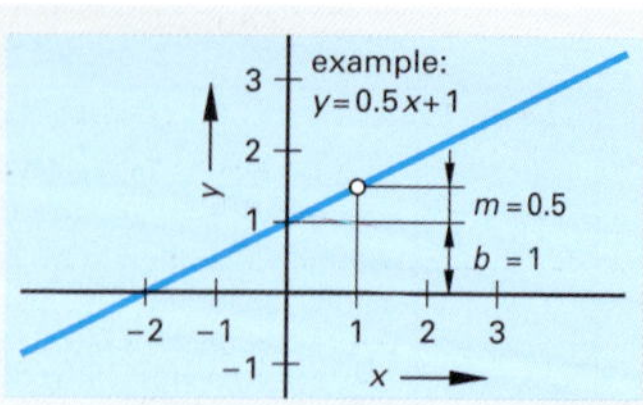

1st example:

$y = 0.5\,x + 1$

x	–2	0	2	3
y	0	1	2	2.5

Linear function

$$y = m \cdot x + b$$

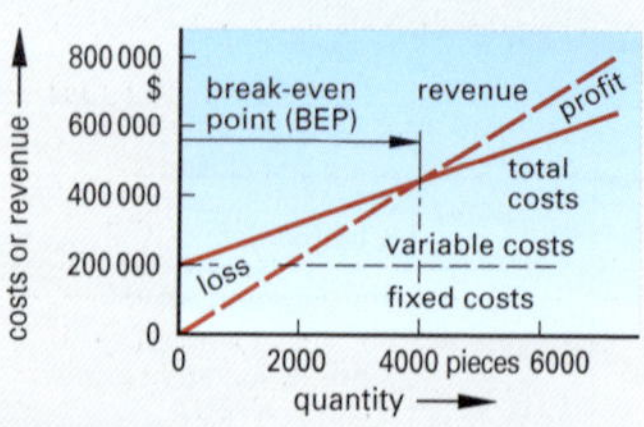

2nd example:

Cost function and revenue function

C_t = 60 \$/piece · Q + 200 000 \$

R = 110 \$/piece · Q

Q	0	4 000	6 000
C_t	200 000	440 000	560 000
R	0	440 000	660 000

C_t total costs → dependent variable
Q quantity → independent variable
C_f fixed costs → y coordinate section
C_v variable costs → gradient of the function
R revenue → dependent variable

Examples:

Cost function

$$C_t = C_V \cdot Q + C_f$$

Revenue function

$$R = R/\text{piece} \cdot Q$$

Transformation of formulas

Transformation of formulas

Formulas and numerical equations are transformed so that the quantity to be obtained stands alone on the left side of the equation. The value of the left side and right side of the formula must not change during the transformation. The following rule applies to all steps of the formula transformation.

Changes applied to the left formula side	=	Changes applied to the right formula side

Formula

$$P = \frac{F \cdot s}{t}$$

left side of the formula = right side of the formula

To be able to trace each step of the transformation, it is useful to mark it to the right next to the formula:

$| \cdot t$ → both sides of the formula are multiplied by t.

$| : F$ → both sides of the formula are divided by F.

Transformations of sums

Example: formula $L = l_1 + l_2$, transformation to find l_2

1 $L = l_1 + l_2$	$\| - l_1$	subtract l_1	3 $L - l_1 = l_2$		invert both sides
2 $L - l_1 = l_1 + l_2 - l_1$		perform subtraction	4 $l_2 = L - l_1$		transformed formula

Transformations of products

Example: formula $A = l \cdot b$, transformation to find l

1 $A = l \cdot b$	$\| : b$	divide by b	3 $\frac{A}{b} = l$		invert both sides
2 $\frac{A}{b} = \frac{l \cdot b}{b}$		cancel b	4 $l = \frac{A}{b}$		transformed formula

Transformations of fractions

Example: formula $n = \frac{l}{l_1 + s}$, transformation to find s

1 $n = \frac{l}{l_1 + s}$	$\| \cdot (l_1 + s)$	multiply by $(l_1 + s)$	4 $n \cdot l_1 - n \cdot l_1 + n \cdot s = l - n \cdot l_1$	$\| : n$	subtract divide by n
2 $n \cdot (l_1 + s) = \frac{l \cdot (l_1 + s)}{(l_1 + s)}$		cancel $(l_1 + s)$ on the right side; solve the term in brackets	5 $\frac{s \cdot n}{n} = \frac{l - n \cdot l_1}{n}$		cancel n
3 $n \cdot l_1 + n \cdot s = l$	$\| - n \cdot l_1$	subtract $- n \cdot l_1$	6 $s = \frac{l - n \cdot l_1}{n}$		transformed formula

Transformations of roots

Example: formula $c = \sqrt{a^2 + b^2}$, transformation to find a

1 $c = \sqrt{a^2 + b^2}$	$\| ()^2$	square equation	4 $a^2 = c^2 - b^2$	$\| \sqrt{\ }$	extract the root
2 $c^2 = a^2 + b^2$	$\| - b^2$	subtract b^2	5 $\sqrt{a^2} = \sqrt{c^2 - b^2}$		simplify the expression
3 $c^2 - b^2 = a^2 + b^2 - b^2$		subtract, invert both sides	6 $a = \sqrt{c^2 - b^2}$		transformed formula

M

Quantities and units

Numerical values and units

Physical quantity

10 mm

Numerical value — Unit

Physical quantities, e.g. 125 mm, consist of a

- **numerical value,** which is determined by measurement or calculation, and a
- **unit,** e.g. m, kg

Units are standardized in accordance with DIN 1301-1 (page 10).

Very large or very small numerical values may be represented in a simplified way as decimal multiples or factors with the help of prefixes, e.g. 0.004 mm = 4 µm.

Decimal multiples or factors of units

cf. DIN 1301-1 (2004-10)

Prefix		Power of ten	Mathematical designation	Examples
Symbol	Name			
T	tera	10^{12}	trillion	12 000 000 000 000 N = 12 · 10^{12} N = 12 TN (teranewtons)
G	giga	10^{9}	billion	45 000 000 000 W = 45 · 10^{9} W = 45 GW (gigawatts)
M	mega	10^{6}	million	8 500 000 V = 8.5 · 10^{6} V = 8.5 MV (megavolts)
k	kilo	10^{3}	thousand	12 600 W = 12.6 · 10^{3} W = 12.6 kW (kilowatts)
h	hecto	10^{2}	hundred	500 l = 5 · 10^{2} l = 5 hl (hectoliters)
da	deca	10^{1}	ten	32 m = 3.2 · 10^{1} m = 3.2 dam (decameters)
–	–	10^{0}	one	1.5 m = 1.5 · 10^{0} m
d	deci	10^{-1}	tenth	0.5 l = 5 · 10^{-1} l = 5 dl (deciliters)
c	centi	10^{-2}	hundredth	0.25 m = 25 · 10^{-2} m = 25 cm (centimeters)
m	milli	10^{-3}	thousandth	0.375 A = 375 · 10^{-3} A = 375 mA (milliamperes)
µ	micro	10^{-6}	millionth	0.000 052 m = 52 · 10^{-6} m = 52 µm (micrometers)
n	nano	10^{-9}	billionth	0.000 000 075 m = 75 · 10^{-9} m = 75 nm (nanometers)
p	pico	10^{-12}	trillionth	0.000 000 000 006 F = 6 · 10^{-12} F = 6 pF (picofarads)

Conversion of units

Calculations with physical units are only possible if these units refer to the same base in this calculation. When solving mathematical problems, units often must be converted to basic units, e. g. mm to m, s to h, mm^2 to m^2. This is done with the help of conversion factors that represent the value 1 (coherent units).

Conversion factors for units (excerpt)

Quantity	Conversion factors, e. g.	Quantity	Conversion factors, e. g.
Length	$1 = \frac{10\ \text{mm}}{1\ \text{cm}} = \frac{1000\ \text{mm}}{1\ \text{m}} = \frac{1\ \text{m}}{1000\ \text{mm}} = \frac{1\ \text{km}}{1000\ \text{m}}$	Time	$1 = \frac{60\ \text{min}}{1\ \text{h}} = \frac{3600\ \text{s}}{1\ \text{h}} = \frac{60\ \text{s}}{1\ \text{min}} = \frac{1\ \text{min}}{60\ \text{s}}$
Area	$1 = \frac{100\ \text{mm}^2}{1\ \text{cm}^2} = \frac{100\ \text{cm}^2}{1\ \text{dm}^2} =$	Angle	$1 = \frac{60'}{1°} = \frac{60''}{1'} = \frac{3600''}{1°} = \frac{1°}{60\ \text{s}}$
Volume	$1 = \frac{1000\ \text{mm}^3}{1\ \text{cm}^3} = \frac{1000\ \text{cm}^3}{1\ \text{dm}^3} =$	Inch	1 inch = 25.4 mm; 1 mm = $\frac{1}{25.4}$ inches

1st example:

Convert volume $V = 3416\ \text{mm}^3$ to cm^3.

Volume V is multiplied by a conversion factor. Its numerator has the unit cm^3 and its denominator the unit mm^3.

$$\boldsymbol{V} = 3416\ mm^3 = \frac{1\ cm^3 \cdot 3416\ mm^3}{1000\ mm^3} = \frac{3416\ cm^3}{1000} = \mathbf{3.416\ cm^3}$$

2nd example:

The angle size specification $\alpha = 42°\,16'$ is to be expressed in degrees (°).

The partial angle 16′ must be converted to degrees (°). The value is multiplied by a conversion factor, the numerator of which has the unit degree (°) and the denominator the unit minute (′).

$$\alpha = 42° + 16' \cdot \frac{1°}{60'} = 42° + \frac{16 \cdot 1°}{60} = 42° + 0.267° = \mathbf{42.267°}$$

M

Calculation with quantities, Percentage and interest calculation

Calculation with quantities

Physical quantities are mathematically treated as products.

- **Adding and subtracting**

  Numerical values that have the same unit are added or subtracted and the unit is carried over to the result.

Example:

$L = l_1 + l_2 - l_3$ mit $l_1 = 124$ mm, $l_2 = 18$ mm, $l_3 = 44$ mm; $L = ?$

$\mathbf{L} = 124\text{ mm} + 18\text{ mm} - 44\text{ mm} = (124 + 18 - 44)\text{ mm} = \mathbf{98\text{ mm}}$

- **Multiplying and dividing**

  The numerical values and the units correspond to the factors of products.

Example:

$F_1 \cdot l_1 = F_2 \cdot l_2$ mit $F_1 = 180$ N, $l_1 = 75$ mm, $l_2 = 105$ mm; $F_2 = ?$

$$\mathbf{F_2} = \frac{F_1 \cdot l_1}{l_2} = \frac{180\text{ N} \cdot 75\text{ mm}}{105\text{ mm}} = 128.57\,\frac{\text{N} \cdot \text{mm}}{\text{mm}} = \mathbf{128.57\text{ N}}$$

- **Multiplying and dividing powers**

  Powers that have the same base are multiplied or divided by adding or subtracting their exponents.

Example:

$W = \frac{A \cdot a^2}{e}$ with $A = 15\text{ cm}^2$, $a = 7.5$ cm, $e = 2.4$ cm; $W = ?$

$$\mathbf{W} = \frac{15\text{ cm}^2 \cdot (7.5\text{ cm})^2}{2.4\text{ cm}} = \frac{15 \cdot 56.25\text{ cm}^{2+2}}{2.4\text{ cm}^1} = 351.56\text{ cm}^{4-1} = \mathbf{351.56\text{ cm}^3}$$

Rules for raising to powers

a base

$m, n \ldots$ exponents

Multiplying powers

$$a^2 \cdot a^3 = a^{2+3}$$

Dividing powers

$$\frac{a^2}{a^3} = a^{2-3}$$

Special cases

$$a^{-2} = \frac{1}{a^2}$$

$$a^1 = a \qquad a^0 = 1$$

Percentage calculation

The **percentage rate** indicates the part of the base value in hundredths.
The **base value** is the value from which the percentage is to be calculated.
The **percent value** is the amount representing the percentage of the base value.

P_r percentage rate, in percent $\quad P_v$ percent value $\quad B_v$ base value

Example:

Weight of raw part: 250 kg (base value); material loss of 2% (percentage rate); material loss in kg = ? (percent value)

$$\mathbf{P_v} = \frac{B_v \cdot P_r}{100\%} = \frac{250\text{ kg} \cdot 2\%}{100\%} = \mathbf{5\text{ kg}}$$

Percent value

$$P_v = \frac{B_v \cdot P_r}{100\%}$$

Interest calculation

P principle
A amount accumulated
I interest
r interest rate per year
t period in days, interest period

1st Example:

$P = \$\,2800.00;\ r = 6\frac{\%}{a};\ t = {}^1/_2\text{ a};\ I = ?$

$$I = \frac{\$\,2800.00 \cdot 6\frac{\%}{a} \cdot 0.5\text{ a}}{100\%} = \$\,84.00$$

2nd Example:

$P = \$\,4800.00;\ r = 5.1\frac{\%}{a};\ t = 50\text{ d};\ I = ?$

$$I = \frac{\$\,4800.00 \cdot 5.1\frac{\%}{a} \cdot 50\text{ d}}{100\% \cdot 360\frac{d}{a}} = \$\,34.00$$

Interest

$$I = \frac{P \cdot r \cdot t}{100\% \cdot 360}$$

1 interest year (1 a) = 360 days (360 d)
360 d = 12 months
1 interest month = 30 days

M

Types of angles, Theorem of intersecting lines, Angles in a triangle, Pythagorean theorem

Types of angles

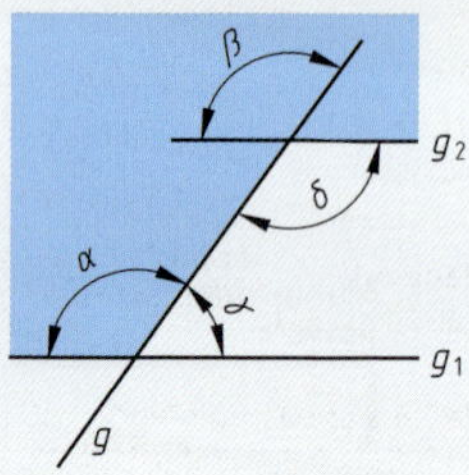

g straight line
g_1, g_2 parallel straight lines
α, β corresponding angles
β, δ opposite angles
α, δ alternate angles
α, γ adjacent angles

If two parallels are intersected by a straight line, there are geometrical interrelationships between the resulting angles.

Corresponding angles

$$\alpha = \beta$$

Opposite angles

$$\beta = \delta$$

Alternate angles

$$\alpha = \delta$$

Adjacent angles

$$\alpha + \gamma = 180°$$

Theorem of intersecting lines

τ_{to} outer torsional stress
τ_{ti} inner torsional stress

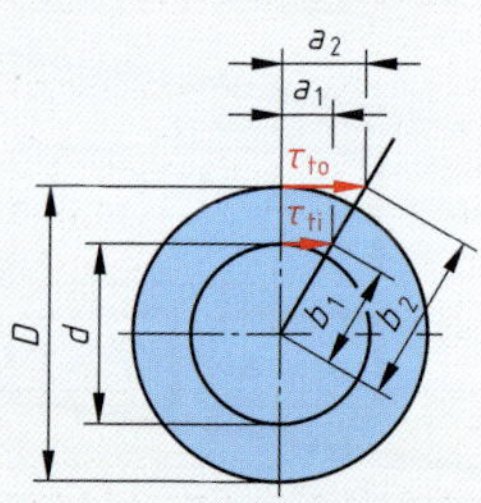

If two intersecting lines are intercepted by a pair of parallels, the resulting segments form equal ratios.

Example:

$D = 40$ mm, $d = 30$ mm,
$\tau_{ta} = 135$ N/mm²; $\tau_{ti} = ?$

$$\frac{\tau_{ti}}{\tau_{to}} = \frac{d}{D} \Rightarrow \tau_{ti} = \frac{\tau_{to} \cdot d}{D}$$

$$= \frac{135\ \text{N/mm}^2 \cdot 30\ \text{mm}}{40\ \text{mm}} = \mathbf{101.25\ N/mm^2}$$

Theorem of intersecting lines

$$\frac{a_1}{a_2} = \frac{b_1}{b_2} = \frac{\frac{d}{2}}{\frac{D}{2}}$$

$$\frac{a_1}{b_1} = \frac{a_2}{b_2} \qquad \frac{b_1}{d} = \frac{b_2}{D}$$

Sum of angles in a triangle

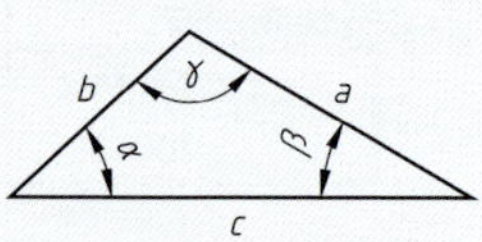

a, b, c sides of the triangle
α, β, γ angles in the triangle

Example:

$\alpha = 21°$, $\beta = 95°$, $\gamma = ?$
$\gamma = 180° - \alpha - \beta = 180° - 21° - 95° = \mathbf{64°}$

Sum of angles in a triangle

$$\alpha + \beta + \gamma = 180°$$

In every triangle, the sum of the interior angles equals 180°.

Pythagorean theorem

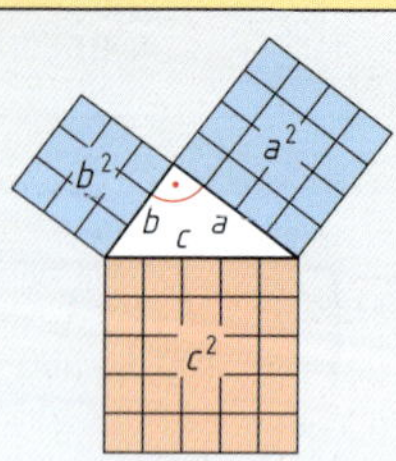

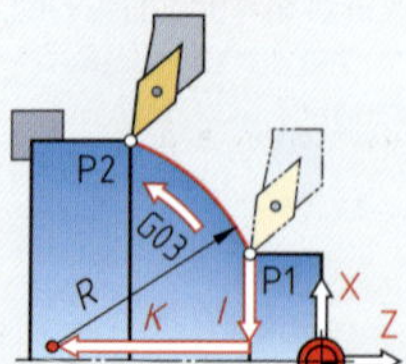

In a **right triangle** the square on the hypotenuse is equal to the sum of the squares on the sides meeting the right angle.

a side
b side
c hypotenuse

1st example:

$c = 35$ mm; $a = 21$ mm; $b = ?$
$\mathbf{b} = \sqrt{c^2 - a^2} = \sqrt{(35\ \text{mm})^2 - (21\ \text{mm})^2} = \mathbf{28\ mm}$

2nd example:

CNC programm with $R = 50$ mm and $I = 25$ mm.
$K = ?$

$c^2 = a^2 + b^2$
$R^2 = I^2 + K^2$
$K = \sqrt{R^2 - I^2} = \sqrt{50^2\ \text{mm}^2 - 25^2\ \text{mm}^2}$
$\mathbf{K = 43.3\ mm}$

Length of the hypotenuse

$$c^2 = a^2 + b^2$$

Square on the hypotenuse

$$c = \sqrt{a^2 + b^2}$$

Length of the sides meeting the right angle

$$a = \sqrt{c^2 - b^2}$$

$$b = \sqrt{c^2 - a^2}$$

M

Functions of triangles

Functions of right triangles (trigonometric functions)

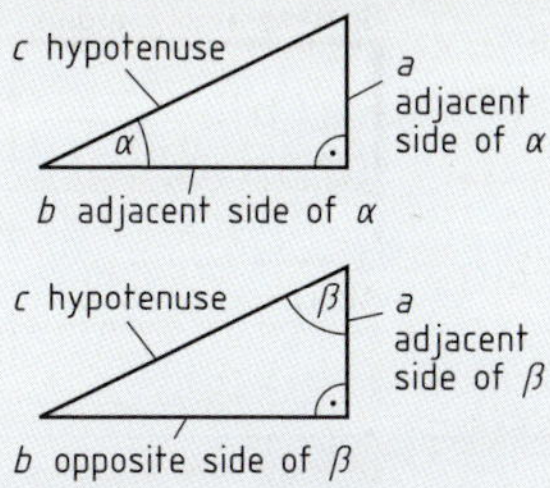

c	hypotenuse (longest side)
a, b	sides, – b is the adjacent side of α – a is the opposite side of α
α, β, γ	angles in the triangle, $\gamma = 90°$
sin	notation of sine
cos	notation of cosine
tan	notation of tangent
$\sin\alpha$	sine of angle α

Trigonometric functions

sine	=	$\dfrac{\textbf{opposite side}}{\textbf{hypotenuse}}$
cosine	=	$\dfrac{\textbf{adjacent side}}{\textbf{hypotenuse}}$
tangent	=	$\dfrac{\textbf{opposite side}}{\textbf{adjacent side}}$

Relations applying to angle α:

$\sin\alpha = \dfrac{a}{c}$	$\cos\alpha = \dfrac{b}{c}$	$\tan\alpha = \dfrac{a}{b}$

Relations applying to angle β:

$\sin\beta = \dfrac{b}{c}$	$\cos\beta = \dfrac{a}{c}$	$\tan\beta = \dfrac{b}{a}$

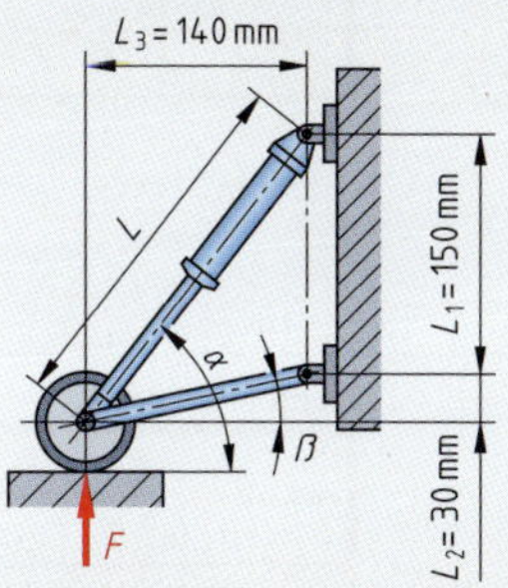

1st example

$L_1 = 150$ mm, $L_2 = 30$ mm, $L_3 = 140$ mm; angle α = ?

$$\tan\alpha = \frac{L_1 + L_2}{L_3} = \frac{180\text{ mm}}{140\text{ mm}} = 1.286$$

Angle $\alpha = 52°$

2nd example

$L_1 = 150$ mm, $L_2 = 30$ mm, $\alpha = 52°$;
Length of the shock absorber L = ?

$$\boldsymbol{L} = \frac{L_1 + L_2}{\sin\alpha} = \frac{180\text{ mm}}{\sin 52°} = \mathbf{228.42\ mm}$$

The calculation of an angle in degrees (°) or as a circular measure (rad) is done with the help of inverse trigonometric functions, e. g. arcsine.

Functions of oblique triangles (law of sines, law of cosines)

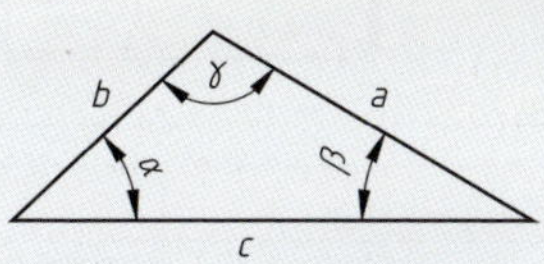

According to the law of sines, the ratios of the sides correspond to the sine of their opposite angles in the triangle. If one side and two angles are known, the other values can be calculated with the help of this function.

Side a	→ opposite angle $\sin\alpha$
Side b	→ opposite angle $\sin\beta$
Hypothenuse c	→ opposite angle $\sin\gamma$

Law of sines

$$a : b : c = \sin\alpha : \sin\beta : \sin\gamma$$

$$\frac{a}{\sin\alpha} = \frac{b}{\sin\beta} = \frac{c}{\sin\gamma}$$

There are many transformation options:

$$a = \frac{b \cdot \sin\alpha}{\sin\beta} = \frac{c \cdot \sin\alpha}{\sin\gamma}$$

$$b = \frac{a \cdot \sin\beta}{\sin\alpha} = \frac{c \cdot \sin\beta}{\sin\gamma}$$

$$c = \frac{a \cdot \sin\gamma}{\sin\alpha} = \frac{b \cdot \sin\gamma}{\sin\beta}$$

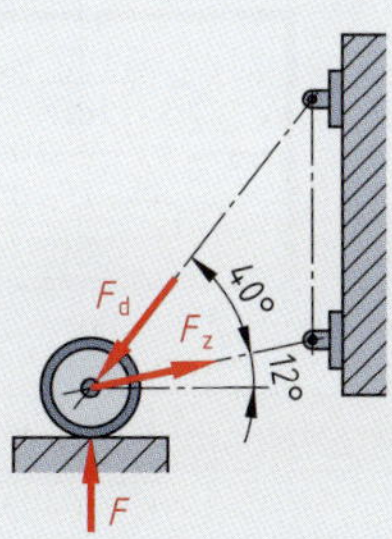

Example:

$F = 800$ N, $\alpha = 40°$, $\beta = 38°$; F_z = ?, F_d = ?

The forces are calculated with the help of the forces diagram.

$$\frac{F}{\sin\alpha} = \frac{F_z}{\sin\beta} \Rightarrow F_z = \frac{F \cdot \sin\beta}{\sin\alpha}$$

$$\boldsymbol{F_z} = \frac{800\text{ N} \cdot \sin 38°}{\sin 40°} = \mathbf{766.24\ N}$$

$$\frac{F}{\sin\alpha} = \frac{F_d}{\sin\gamma} \Rightarrow F_d = \frac{F \cdot \sin\gamma}{\sin\alpha}$$

$$\boldsymbol{F_d} = \frac{800\text{ N} \cdot \sin 102°}{\sin 40°} = \mathbf{1217.38\ N}$$

Diagram of forces

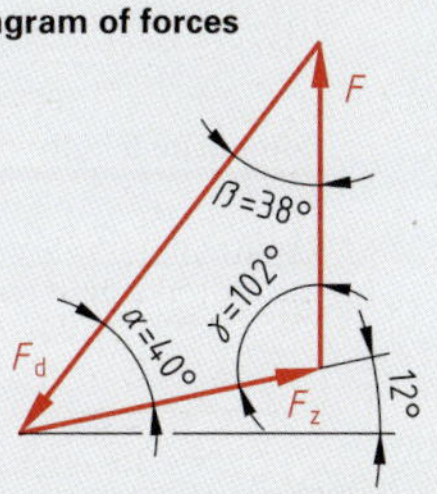

The calculation of an angle in degrees (°) or as a circular measure (rad) is done with the help of inverse trigonometric functions, e. g. arcsine.

Law of cosines

$$a^2 = b^2 + c^2 - 2 \cdot b \cdot c \cdot \cos\alpha$$
$$b^2 = a^2 + c^2 - 2 \cdot a \cdot c \cdot \cos\beta$$
$$c^2 = a^2 + b^2 - 2 \cdot a \cdot b \cdot \cos\gamma$$

Transformation, e. g.

$$\cos\alpha = \frac{b^2 + c^2 - a^2}{2 \cdot b \cdot c}$$

M

Division of lengths, Arc length, Composite length

Sub-dividing lengths

Edge distance = spacing

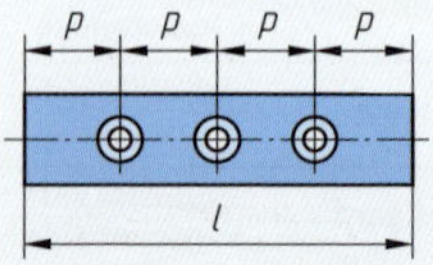

l total length
p spacing
n number of holes

Example:

$l = 2$ m; $n = 24$ holes; $p = ?$

$$\boldsymbol{p} = \frac{l}{n+1} = \frac{2000\text{ mm}}{24+1} = \mathbf{80\,mm}$$

Spacing

$$p = \frac{l}{n+1}$$

Edge distance ≠ spacing

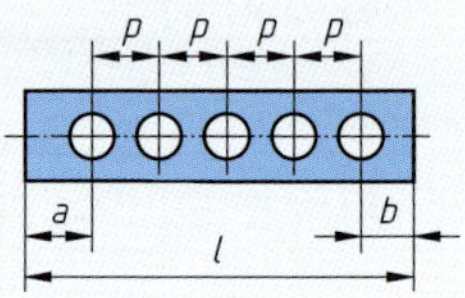

l total length
p spacing
n number of holes
a, b edge distances

Example:

$l = 1950$ mm; $a = 100$ mm; $b = 50$ mm;
$n = 25$ holes; $p = ?$

$$\boldsymbol{p} = \frac{l-(a+b)}{n-1} = \frac{1950\text{ mm} - 150\text{ mm}}{25-1} = \mathbf{75\,mm}$$

Spacing

$$p = \frac{l-(a+b)}{n-1}$$

Subdividing into pieces

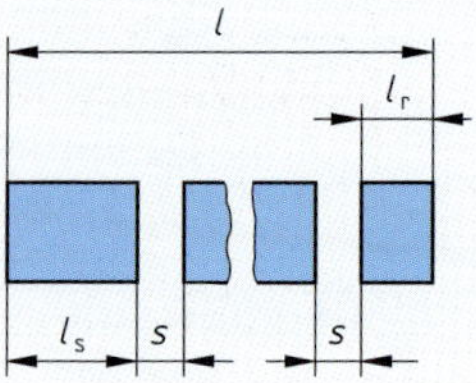

l bar length
z number of pieces
l_s piece length
s saw cutting width
l_r remaining length

Example:

$l = 6000$ mm; $l_s = 230$ mm; $s = 1.2$ mm; $z = ?$; $l_r = ?$

$$\boldsymbol{z} = \frac{l}{l_s+s} = \frac{6000\text{ mm}}{230\text{ mm} + 1.2\text{ mm}} = \mathbf{25.95 = 25\ pieces}$$

$$\boldsymbol{l_r} = l - z \cdot (l_s + s) = 6000\text{ mm} - 25 \cdot (230\text{ mm} + 1.2\text{ mm}) = \mathbf{220\,mm}$$

Number of pieces

$$z = \frac{l}{l_s + s}$$

Remaining length

$$l_r = l - z \cdot (l_s + s)$$

Arc length

Example: Torsion spring

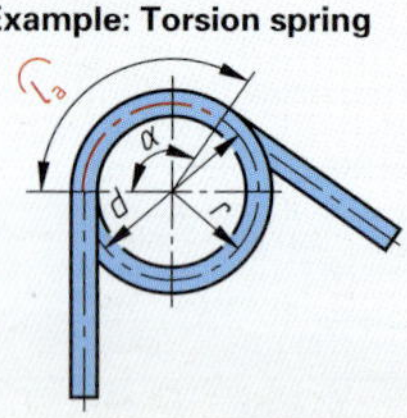

l_a arc length
r radius
α angle at center
d diameter

Example:

$r = 36$ mm; $\alpha = 120°$; $l_a = ?$

$$\boldsymbol{l_a} = \frac{\pi \cdot r \cdot \alpha}{180°} = \frac{\pi \cdot 36\text{ mm} \cdot 120°}{180°} = \mathbf{75.36\,mm}$$

Arc length

$$l_a = \frac{\pi \cdot r \cdot \alpha}{180°}$$

$$l_a = \frac{\pi \cdot d \cdot \alpha}{360°}$$

Composite length

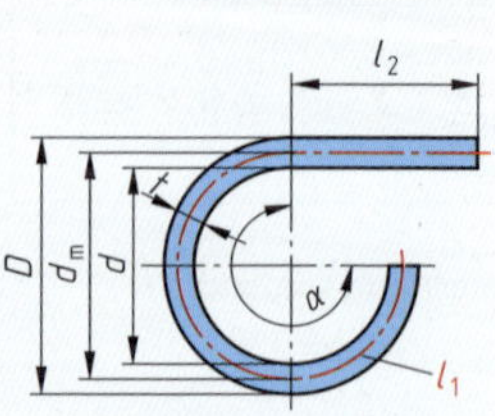

D outside diameter
d_m mean diameter
l_1, l_2 section lengths
α angle at center
d inside diameter
t thickness
L composite length

Example (composite length, picture left):

$D = 360$ mm; $t = 5$ mm; $\alpha = 270°$; $l_2 = 70$ mm;
$d_m = ?$; $L = ?$

$$\boldsymbol{d_m} = D - t = 360\text{ mm} - 5\text{ mm} = \mathbf{355\,mm}$$

$$\boldsymbol{L} = l_1 + l_2 = \frac{\pi \cdot d_m \cdot \alpha}{360} + l_2$$

$$= \frac{\pi \cdot 355\text{ mm} \cdot 270°}{360°} + 70\text{ mm} = \mathbf{906.45\,mm}$$

Composite length

$$L = l_1 + l_2 + \ldots$$

Effective length, Spring wire length, Rough length

M

Effective lengths

Circular ring

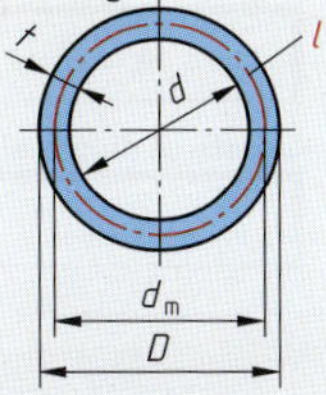

Circular ring sector

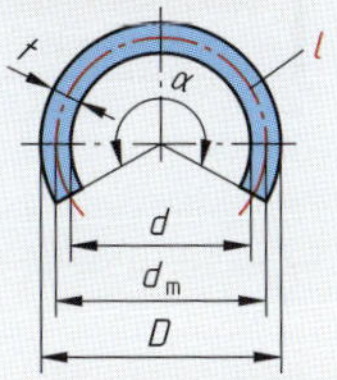

D outside diameter
d inside diameter
d_m mean diameter
t thickness
l effective length
α angle at center

Example (circular ring sector):

$D = 36$ mm; $t = 4$ mm; $\alpha = 240°$; $d_m = ?$; $l = ?$

$d_m = D - t = 36 \text{ mm} - 4 \text{ mm} = 32 \text{ mm}$

$$l = \frac{\pi \cdot d_m \cdot \alpha}{360°} = \frac{\pi \cdot 32 \text{ mm} \cdot 240°}{360°} = \mathbf{67.02\ mm}$$

Effective length of a circular ring

$$l = \pi \cdot d_m$$

Effective length of a circular ring sector

$$l = \frac{\pi \cdot d_m \cdot \alpha}{360°}$$

Mean diameter

$$d_m = D - t$$

$$d_m = d + t$$

Spring wire length

Example: Compression spring

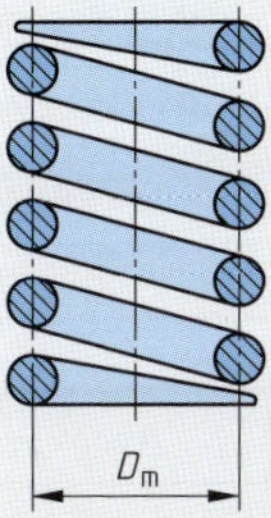

l effective length of the helix
D_m mean coil diameter
i number of active coils

Example:

$D_m = 16$ mm; $i = 8.5$; $l = ?$

$l = \pi \cdot D_m \cdot i + 2 \cdot \pi \cdot D_m$
$= \pi \cdot 16 \text{ mm} \cdot 8.5 + 2 \cdot \pi \cdot 16 \text{ mm} = \mathbf{528\ mm}$

Effective length of the helix

$$l = \pi \cdot D_m \cdot i + 2 \cdot \pi \cdot D_m$$

$$l = \pi \cdot D_m \cdot (i + 2)$$

Rough length of forged parts and pressed parts

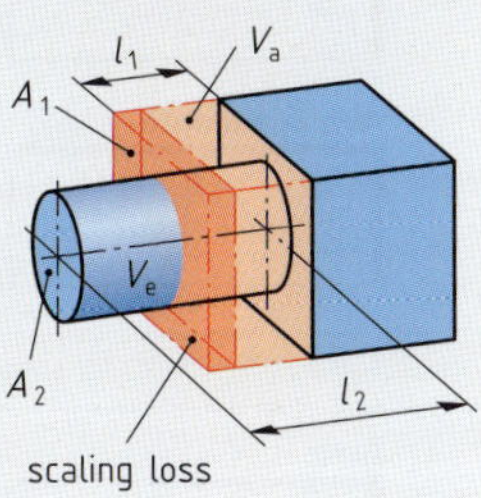

When forming without scaling loss the volume of the rough part is the same as the volume of the finished part. If there is scaling loss or burr formation, this is compensated by a factor that is applied to the volume of the finished piece.

V_a volume of the rough part
V_e volume of the finished part
q addition factor for scaling loss or loss due to burrs
A_1 cross-sectional area of the rough part
A_2 cross-sectional area of the finished part
l_1 initial length of the addition
l_2 length of the solid forged part

Example:

A cylindrical peg d = 24 mm and $l_2 = 60$ mm is pressed onto a flat steel workpiece 50 x 30 mm. The scaling loss is 10 %. What is the initial length l_1 of the forged addition?

$$V_a = V_e \cdot (1 + q)$$

$$A_1 \cdot l_1 = A_2 \cdot l_2 \cdot (1 + q)$$

$$l_1 = \frac{A_2 \cdot l_2 \cdot (1 + q)}{A_1}$$

$$= \frac{\pi \cdot (24 \text{ mm})^2 \cdot 60 \text{ mm} \cdot (1 + 0{,}1)}{4 \cdot 50 \text{ mm} \cdot 30 \text{ mm}} = \mathbf{20\ mm}$$

Volume without scaling loss

$$V_a = V_e$$

Volume with scaling loss

$$V_a = V_e + q \cdot V_e$$

$$V_a = V_e \cdot (1 + q)$$

$$A_1 \cdot l_1 = A_2 \cdot l_2 \cdot (1 + q)$$

Angular areas

Square

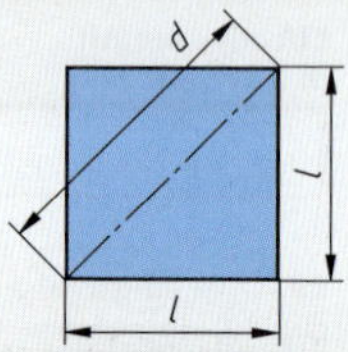

A area
l length of side
d length of diagonal

Example:

$l = 14$ mm; $A = ?$; $d = ?$

$\boldsymbol{A} = l^2 = (14\text{ mm})^2 = \mathbf{196\ mm^2}$

$\boldsymbol{d} = \sqrt{2} \cdot l = \sqrt{2} \cdot 14\text{ mm} = \mathbf{19.8\ mm}$

Area

$$A = l^2$$

Length of diagonal

$$d = \sqrt{2} \cdot l$$

Rhombus (lozenge)

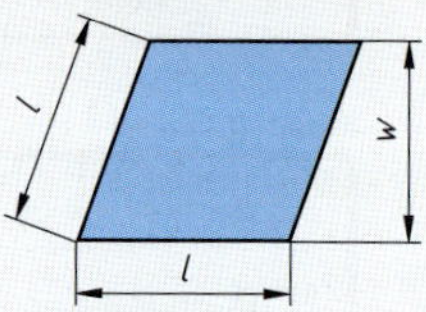

A area
l length of side
w width

Example:

$l = 9$ mm; $w = 8.5$ mm; $A = ?$

$\boldsymbol{A} = l \cdot w = 9\text{ mm} \cdot 8.5\text{ mm} = \mathbf{76.5\ mm^2}$

Area

$$A = l \cdot w$$

Rectangle

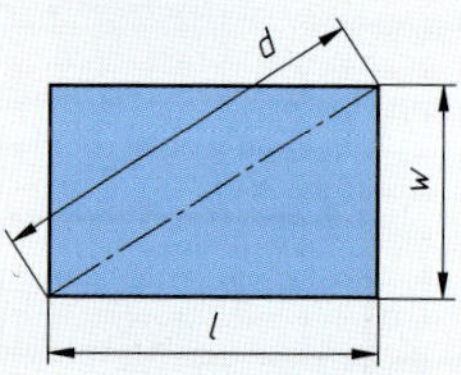

A area
l length
w width
d length of diagonal

Example:

$l = 12$ mm; $w = 11$ mm; $A = ?$; $d = ?$

$\boldsymbol{A} = l \cdot w = 12\text{ mm} \cdot 11\text{ mm} = \mathbf{132\ mm^2}$

$\boldsymbol{d} = \sqrt{l^2 + w^2} = \sqrt{(12\text{ mm})^2 + (11\text{ mm})^2} = \sqrt{265\text{ mm}^2}$
$= \mathbf{16.28\ mm}$

Area

$$A = l \cdot w$$

Length of diagonal

$$d = \sqrt{l^2 + w^2}$$

Rhomboid (parallelogram)

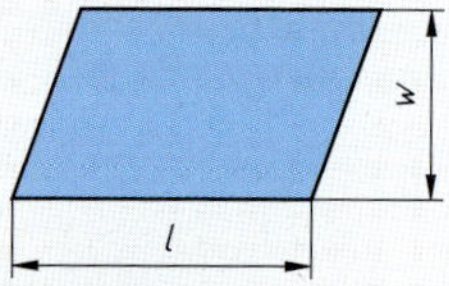

A area
l length
w width

Example:

$l = 36$ mm; $w = 15$ mm; $A = ?$

$\boldsymbol{A} = l \cdot w = 36\text{ mm} \cdot 15\text{ mm} = \mathbf{540\ mm^2}$

Area

$$A = l \cdot w$$

Trapezoid

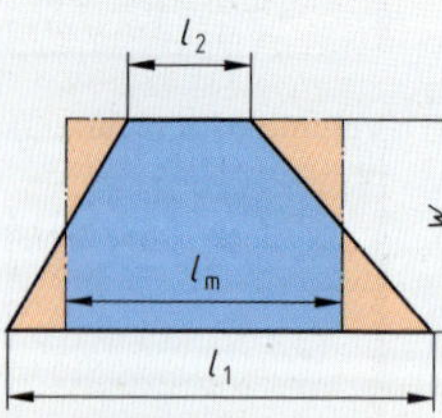

A area
l_1 longer length
l_2 shorter length
l_m average length
w width

Example:

$l_1 = 23$ mm; $l_2 = 20$ mm; $w = 17$ mm; $A = ?$

$\boldsymbol{A} = \frac{l_1 + l_2}{2} \cdot w = \frac{23\text{ mm} + 20\text{ mm}}{2} \cdot 17\text{ mm}$
$= \mathbf{365.5\ mm^2}$

Area

$$A = \frac{l_1 + l_2}{2} \cdot w$$

Average length

$$l_m = \frac{l_1 + l_2}{2}$$

Triangle

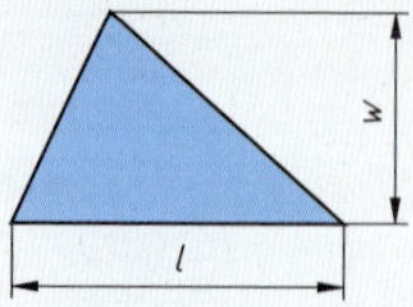

A area
l length of side
w width

Example:

$l_1 = 62$ mm; $w = 29$ mm; $A = ?$

$\boldsymbol{A} = \frac{l_1 \cdot w}{2} = \frac{62\text{ mm} \cdot 29\text{ mm}}{2} = \mathbf{899\ mm^2}$

Area

$$A = \frac{l \cdot w}{2}$$

Triangle, Polygon, Circle

Equilateral triangle

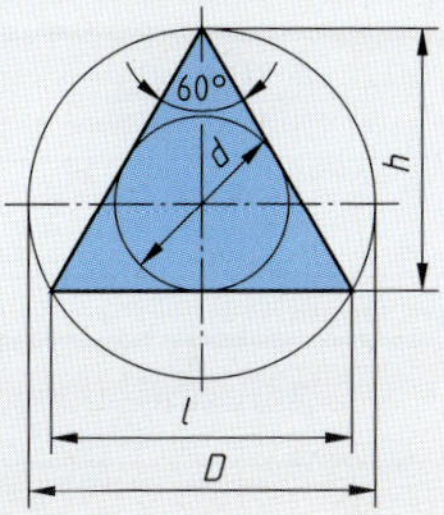

A area
d diameter of inscribed circle
l length of side
h height
D diameter of circumscribed circle

Example:

l = 42 mm; A = ?;

$\mathbf{A} = \frac{1}{4}\cdot\sqrt{3}\cdot l^2 = \frac{1}{4}\cdot\sqrt{3}\cdot(42\,\text{mm})^2 = \mathbf{763.9\,mm^2}$

Diameter of circumscribed circle

$$D = \frac{2}{3}\cdot\sqrt{3}\cdot l = 2\cdot d$$

Area

$$A = \frac{1}{4}\cdot\sqrt{3}\cdot l^2$$

Diameter of inscribed circle

$$d = \frac{1}{3}\cdot\sqrt{3}\cdot l = \frac{D}{2}$$

Triangle height

$$h = \frac{1}{2}\cdot\sqrt{3}\cdot l$$

Regular polygons

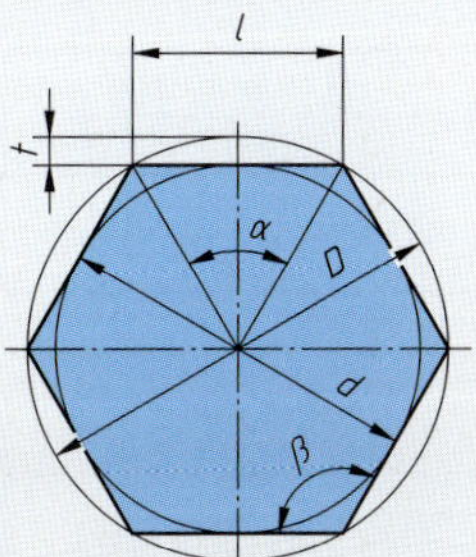

A area
l length of side
D diameter of circumscribed circle
d diameter of inscribed circle
n no. of vertices
α angle at center
β vertex angle
WAF width across flats
t milling depth

Example:

Hexagon with D = 80 mm; l = ?; d = ?; A = ?

$\mathbf{l} = D\cdot\sin\left(\frac{180°}{n}\right) = 80\,\text{mm}\cdot\sin\left(\frac{180°}{6}\right) = \mathbf{40\,mm}$

$\mathbf{d} = \sqrt{D^2 - l^2} = \sqrt{6400\,\text{mm}^2 - 1600\,\text{mm}^2} = \mathbf{69.282\,mm}$

$\mathbf{A} = \frac{n\cdot l\cdot d}{4} = \frac{6\cdot 40\,\text{mm}\cdot 69{,}282\,\text{mm}}{4} = \mathbf{4156.92\,mm^2}$

Diameter of inscribed circle

$$d = \sqrt{D^2 - l^2}$$

Area

$$A = \frac{n\cdot l\cdot d}{4}$$

Diameter of circumscribed circle

$$D = \sqrt{d^2 + l^2}$$

Length of side

$$l = D\cdot\sin\left(\frac{180°}{n}\right)$$

Angle at center

$$\alpha = \frac{360°}{n}$$

Corner angle

$$\beta = 180° - \alpha$$

WAF, D and *t* with squares and hexagons

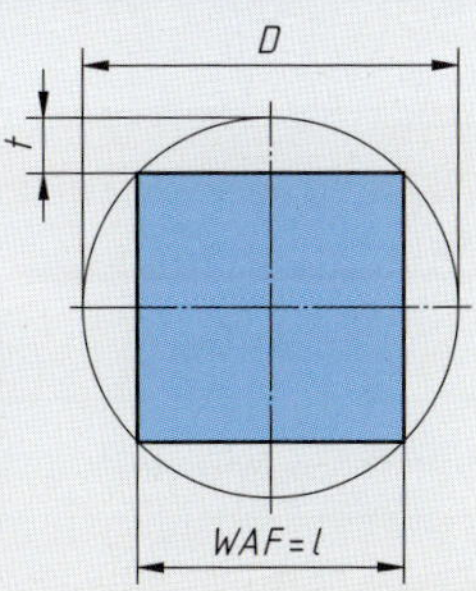

Example:

Hexagon WAF = 24 mm; D = ?; t = ?

$\mathbf{D} = \frac{WAF}{\cos 30°} = \frac{24\,\text{mm}}{0.866} = \mathbf{27.71\,mm}$

$\mathbf{t} = \frac{D - SW}{2} = \frac{27.71\,\text{mm} - 24\,\text{mm}}{2} = \frac{3.71\,\text{mm}}{2} = \mathbf{1.855\,mm}$

Shaft diameter *D*	
Square	$D = \frac{WAF}{\cos 45°}$
Hexagon	$D = \frac{WAF}{\cos 30°}$

Milling depth *t*	
Square and hexagon	$t = \frac{D - WAF}{2}$

Circle

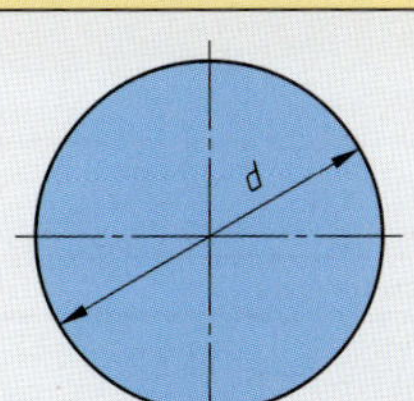

A area
d diameter
C circumference

Example:

d = 60 mm; A = ?; C = ?

$\mathbf{A} = \frac{\pi\cdot d^2}{4} = \frac{\pi\cdot(60\,\text{mm})^2}{4} = \mathbf{2827\,mm^2}$

$\mathbf{C} = \pi\cdot d = \pi\cdot 60\,\text{mm} = \mathbf{188.5\,mm}$

Area

$$A = \frac{\pi\cdot d^2}{4}$$

Circumference

$$C = \pi\cdot d$$

Circular sector, Circular segment, Circular ring, Ellipse

Circular sector

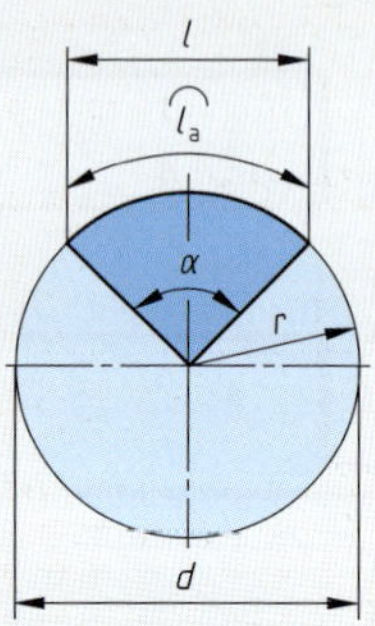

A area
d diameter
l_a arc length
l chord length
r radius
α angle at center

Example:

d = 48 mm; α = 110°; l_a = ?; A = ?

$$l_a = \frac{\pi \cdot r \cdot \alpha}{180^\circ} = \frac{\pi \cdot 24\ \text{mm} \cdot 110^\circ}{180^\circ} = \mathbf{46.1\ mm}$$

$$A = \frac{l_a \cdot r}{2} = \frac{46.1\ \text{mm} \cdot 24\ \text{mm}}{2} = \mathbf{553\ mm^2}$$

Area

$$A = \frac{\pi \cdot d^2}{4} \cdot \frac{\alpha}{360^\circ}$$

$$A = \frac{l_a \cdot r}{2}$$

Chord length

$$l = 2 \cdot r \cdot \sin\frac{\alpha}{2}$$

Arc length

$$l_a = \frac{\pi \cdot r \cdot \alpha}{180^\circ}$$

Circular segment

Circular segment with $\alpha \leq 180°$

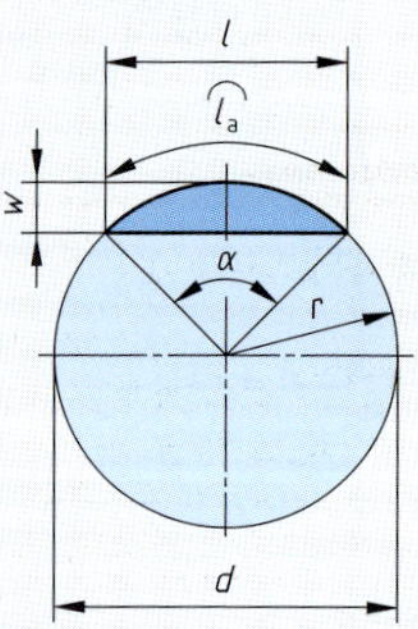

A area
d diameter
l_a arc length
l chord length
w width of segment
r radius
α angle at center

Example:

r = 30 mm; α = 120°; l = ?; w = ?; A = ?

$$l = 2 \cdot r \cdot \sin\frac{\alpha}{2} = 2 \cdot 30\ \text{mm} \cdot \sin\frac{120^\circ}{2} = \mathbf{51.96\ mm}$$

$$w = \frac{l}{2} \cdot \tan\frac{\alpha}{4} = \frac{51.96\ \text{mm}}{2} \cdot \tan\frac{120^\circ}{4} = 14.999\ \text{mm} = \mathbf{15\ mm}$$

$$A = \frac{\pi \cdot d^2}{4} \cdot \frac{\alpha}{360^\circ} - \frac{l \cdot (r - w)}{2}$$

$$= \frac{\pi \cdot (60\ \text{mm})^2}{4} \cdot \frac{120^\circ}{360^\circ} - \frac{51.96\ \text{mm} \cdot (30\ \text{mm} - 15\ \text{mm})}{2}$$

$$= \mathbf{552.8\ mm^2}$$

Radius

$$r = \frac{w}{2} + \frac{l^2}{8 \cdot w}$$

Arc length

$$l_a = \frac{\pi \cdot r \cdot \alpha}{180^\circ}$$

Area

$$A = \frac{\pi \cdot d^2}{4} \cdot \frac{\alpha}{360^\circ} - \frac{l \cdot (r - w)}{2}$$

$$A = \frac{l_a \cdot r - l \cdot (r - w)}{2}$$

Chord length

$$l = 2 \cdot r \cdot \sin\frac{\alpha}{2}$$

$$l = 2 \cdot \sqrt{w \cdot (2 \cdot r - w)}$$

Height of segment

$$w = \frac{l}{2} \cdot \tan\frac{\alpha}{4}$$

$$w = r - \sqrt{r^2 - \frac{l^2}{4}}$$

Circular ring

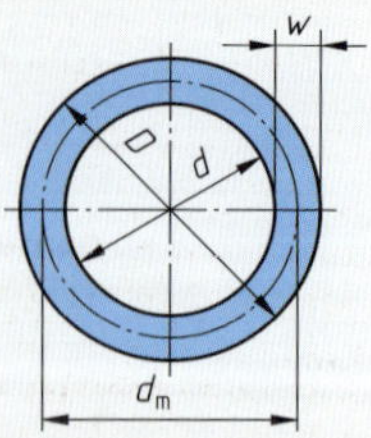

A area
D outside diameter
d inside diameter
d_m mean diameter
w width

Example:

D = 160 mm; d = 125 mm; A = ?

$$A = \frac{\pi}{4} \cdot (D^2 - d^2) = \frac{\pi}{4} \cdot (160^2\ \text{mm}^2 - 125^2\ \text{mm}^2)$$

$$= \mathbf{7834\ mm^2}$$

Area

$$A = \pi \cdot d_m \cdot w$$

$$A = \frac{\pi}{4} \cdot (D^2 - d^2)$$

Ellipse

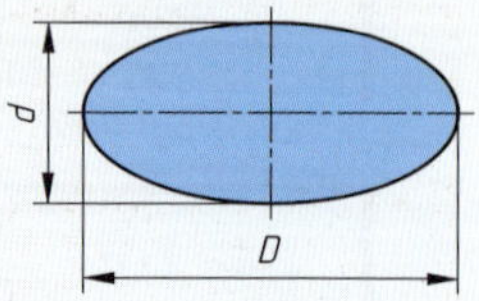

A area
D length
d diameter
C Circumference

Example:

D = 65 mm; d = 20 mm; A = ?

$$A = \frac{\pi \cdot D \cdot d}{4} = \frac{\pi \cdot 65\ \text{mm} \cdot 20\ \text{mm}}{4}$$

$$= \mathbf{1021\ mm^2}$$

Area

$$A = \frac{\pi \cdot D \cdot d}{4}$$

Circumference

$$C \approx \pi \cdot \frac{D + d}{2}$$

M

Cube, Rectangular prism, Cylinder, Hollow cylinder, Pyramid

Cube

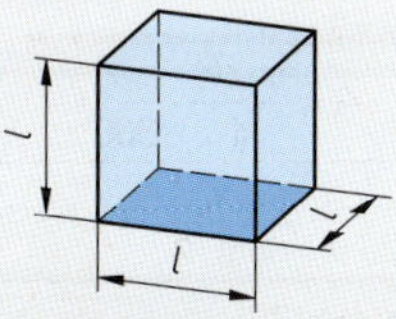

V volume
A_s surface area
l length of side

Example:

$l = 20$ mm; $V = ?$; $A_s = ?$

$\mathbf{V} = l^3 = (20\text{ mm})^3 = \mathbf{8000\ mm^3}$

$\mathbf{A_s} = 6 \cdot l^2 = 6 \cdot (20\text{ mm})^2 = \mathbf{2400\ mm^2}$

Volume

$$V = l^3$$

Surface area

$$A_s = 6 \cdot l^2$$

Rectangular prism

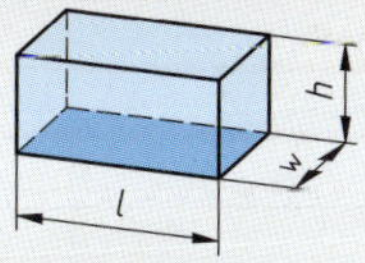

V volume
A_s surface area
l length of side
h height
w width

Example:

$l = 6$ cm; $w = 3$ cm; $h = 2$ cm; $V = ?$

$\mathbf{V} = l \cdot w \cdot h = 6\text{ cm} \cdot 3\text{ cm} \cdot 2\text{ cm} = \mathbf{36\ cm^3}$

Volume

$$V = l \cdot w \cdot h$$

Surface area

$$A_s = 2 \cdot (l \cdot w + l \cdot h + w \cdot h)$$

Cylinder

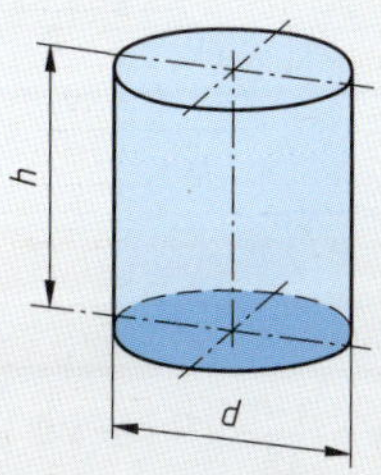

V volume
A_s surface area
A_c cylindrical surface area
d diameter
h height

Example:

$d = 14$ mm; $h = 25$ mm; $V = ?$

$\mathbf{V} = \frac{\pi \cdot d^2}{4} \cdot h$

$= \frac{\pi \cdot (14\text{ mm})^2}{4} \cdot 25\text{ mm}$

$= \mathbf{3848\ mm^3}$

Volume

$$V = \frac{\pi \cdot d^2}{4} \cdot h$$

Surface area

$$A_s = \pi \cdot d \cdot h + 2 \cdot \frac{\pi \cdot d^2}{4}$$

Cylindrical surface area

$$A_c = \pi \cdot d \cdot h$$

Hollow cylinder

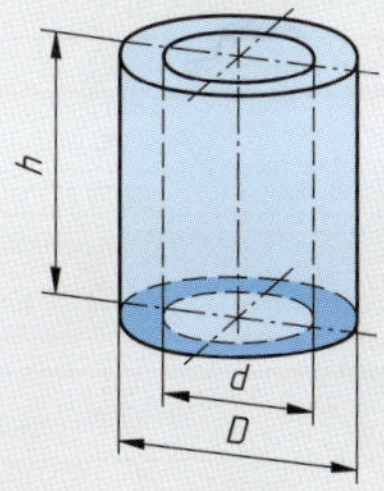

V volume
A_s surface area
D, d diameter
h height

Example:

$D = 42$ mm; $d = 20$ mm; $h = 80$ mm; $V = ?$

$\mathbf{V} = \frac{\pi \cdot h}{4} \cdot (D^2 - d^2)$

$= \frac{\pi \cdot 80\text{ mm}}{4} \cdot (42^2\text{ mm}^2 - 20^2\text{ mm}^2)$

$= \mathbf{85\,703\ mm^3}$

Volume

$$V = \frac{\pi \cdot h}{4} \cdot (D^2 - d^2)$$

Surface area

$$A_s = \pi \cdot (D + d) \cdot \left[\frac{1}{2} \cdot (D - d) + h\right]$$

Pyramid

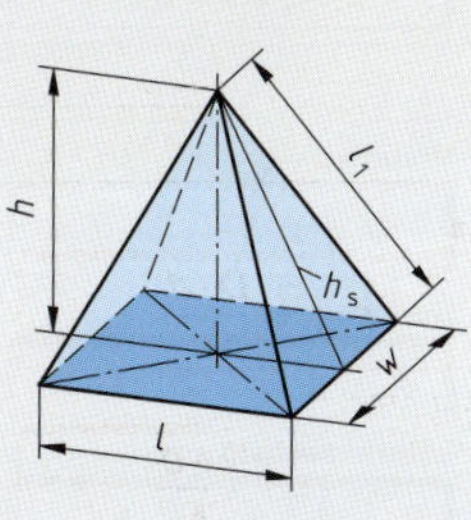

V volume
h height
h_s slant height
l length of base
l_1 edge length
w width of base

Example:

$l = 16$ mm; $w = 21$ mm; $h = 45$ mm; $V = ?$

$\mathbf{V} = \frac{l \cdot w \cdot h}{3} = \frac{16\text{ mm} \cdot 21\text{ mm} \cdot 45\text{ mm}}{3}$

$= \mathbf{5040\ mm^3}$

Volume

$$V = \frac{l \cdot w \cdot h}{3}$$

Edge length

$$l_1 = \sqrt{h_s^2 + \frac{w^2}{4}}$$

Slant height

$$h_s = \sqrt{h^2 + \frac{l^2}{4}}$$

M

Truncated pyramid, Cone, Truncated cone, Sphere, Spherical segment

Truncated pyramid

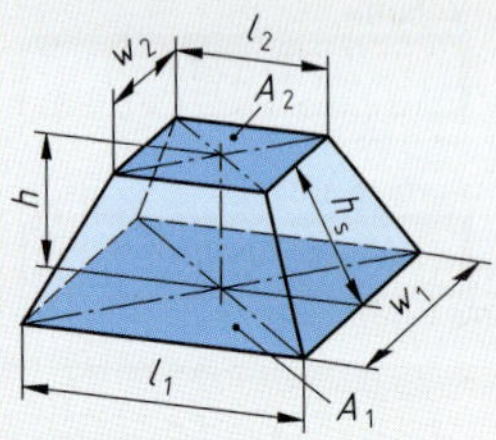

V volume
l_1, l_2 lengths of base
A_1 area of base surface
A_2 top surface
h_s slant height
h height
w_1, w_2 widths

Example:

$l_1 = 40$ mm; $l_2 = 22$ mm; $w_1 = 28$ mm; $w_2 = 15$ mm; $h = 50$ mm; $V = ?$

$$\mathbf{V} = \frac{h}{3} \cdot (A_1 + A_2 + \sqrt{A_1 \cdot A_2})$$
$$= \frac{50\,\text{mm}}{3} \cdot (1120 + 330 + \sqrt{1120 \cdot 330})\,\text{mm}^2$$
$$= \mathbf{34\,299\,mm^3}$$

Volume

$$V = \frac{h}{3} \cdot (A_1 + A_2 + \sqrt{A_1 \cdot A_2})$$

Slant height

$$h_s = \sqrt{h^2 + \left(\frac{l_1 - l_2}{2}\right)^2}$$

Cone

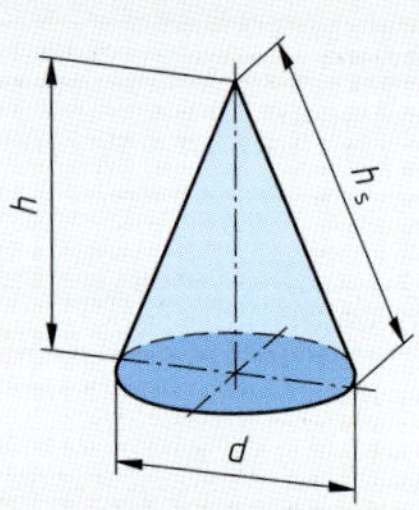

V volume
A_c conical surface area
d diameter
h height
h_s slant height

Example:

$d = 52$ mm; $h = 110$ mm; $V = ?$

$$\mathbf{V} = \frac{\pi \cdot d^2}{4} \cdot \frac{h}{3}$$
$$= \frac{\pi \cdot (52\,\text{mm})^2}{4} \cdot \frac{110\,\text{mm}}{3}$$
$$= \mathbf{77\,870\,mm^3}$$

Volume

$$V = \frac{\pi \cdot d^2}{4} \cdot \frac{h}{3}$$

Conical surface area

$$A_c = \frac{\pi \cdot d \cdot h_s}{2}$$

Slant height

$$h_s = \sqrt{\frac{d^2}{4} + h^2}$$

Truncated cone

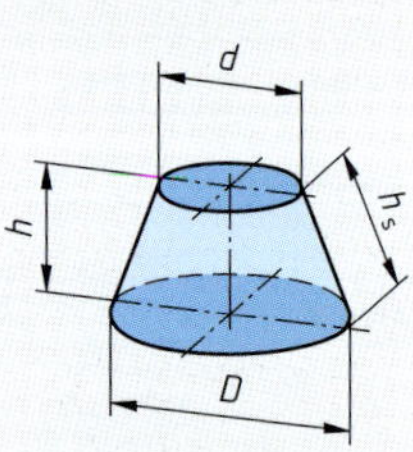

V volume
A_c conical surface area
D diameter of base
d diameter of top
h height
h_s slant height

Example:

$D = 100$ mm; $d = 62$ mm; $h = 80$ mm; $V = ?$

$$\mathbf{V} = \frac{\pi \cdot h}{12} \cdot (D^2 + d^2 + D \cdot d)$$
$$= \frac{\pi \cdot 80\,\text{mm}}{12} \cdot (100^2 + 62^2 + 100 \cdot 62)\,\text{mm}^2$$
$$= \mathbf{419\,800\,mm^3}$$

Volume

$$V = \frac{\pi \cdot h}{12} \cdot (D^2 + d^2 + D \cdot d)$$

Conical surface area

$$A_c = \frac{\pi \cdot h_s}{2} \cdot (D + d)$$

Slant height

$$h_s = \sqrt{h^2 + \left(\frac{D-d}{2}\right)^2}$$

Sphere

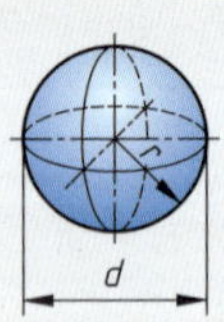

V volume
A_s surface area
d diameter of sphere

Example:

$d = 9$ mm; $V = ?$

$$\mathbf{V} = \frac{\pi \cdot d^3}{6} \cdot \frac{\pi \cdot (9\,\text{mm})^3}{6} = \mathbf{382\,mm^3}$$

Volume

$$V = \frac{\pi \cdot d^3}{6}$$

Surface area

$$A_s = \pi \cdot d^2$$

Spherical segment

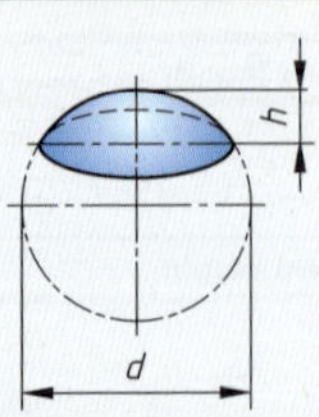

V volume
A_l lateral surface area
A_s surface area
d diameter of sphere
h height

Example:

$d = 8$ mm; $h = 6$ mm; $V = ?$

$$\mathbf{V} = \pi \cdot h^2 \cdot \left(\frac{d}{2} - \frac{h}{3}\right)$$
$$= \pi \cdot 6^2\,\text{mm}^2 \cdot \left(\frac{8\,\text{mm}}{2} - \frac{6\,\text{mm}}{3}\right)$$
$$= \mathbf{226\,mm^3}$$

Volume

$$V = \pi \cdot h^2 \cdot \left(\frac{d}{2} - \frac{h}{3}\right)$$

Surface area

$$A_s = \pi \cdot h \cdot (2 \cdot d - h)$$

Lateral surface area

$$A_l = \pi \cdot d \cdot h$$

Volumes of composite solids, Calculation of mass

Volumes of composite solids

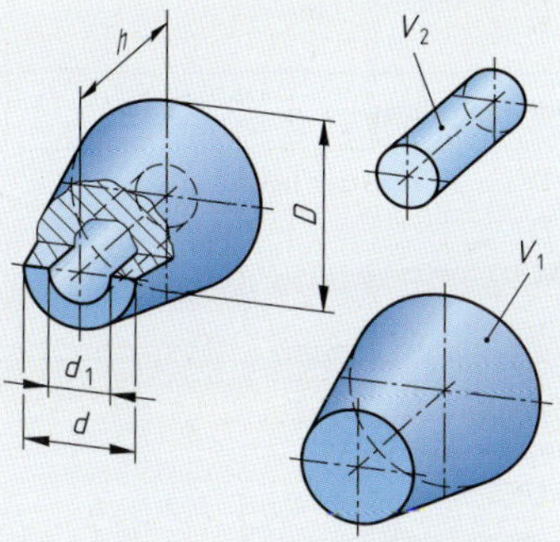

V total volume
V_1, V_2 partial volumes

Total volume

$$V = V_1 + V_2 + \dots - V_3 - V_4$$

Example:

Tapered sleeve; $D = 42$ mm; $d = 26$ mm; $d_1 = 16$ mm; $h = 45$ mm; $V = ?$

$$V_1 = \frac{\pi \cdot h}{12} \cdot (D^2 + d^2 + D \cdot d)$$
$$= \frac{\pi \cdot 45\text{ mm}}{12} \cdot (42^2 + 26^2 + 42 \cdot 26)\text{ mm}^2$$
$$= 41\,610\text{ mm}^3$$
$$V_2 = \frac{\pi \cdot d_1^2}{4} \cdot h = \frac{\pi \cdot 16^2\text{ mm}^2}{4} \cdot 45\text{ mm} = 9048\text{ mm}^3$$
$$\mathbf{V} = V_1 - V_2 = 41\,610\text{ mm}^3 - 9048\text{ mm}^3 = \mathbf{32\,562\text{ mm}^3}$$

Calculation of mass

Mass, general

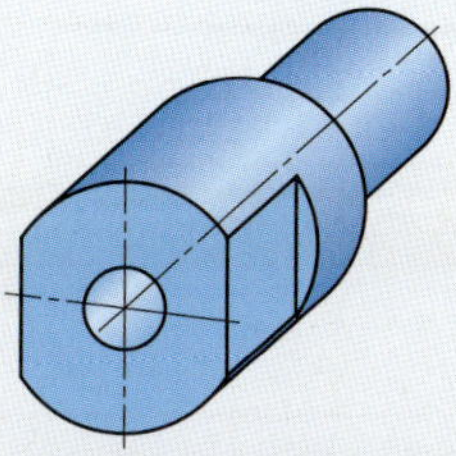

m mass
V volume
ϱ density

Mass

$$m = V \cdot \varrho$$

Example:

Workpiece made of aluminum; $V = 6.4$ dm^3; $\varrho = 2.7$ kg/dm^3; $m = ?$

$$\mathbf{m} = V \cdot \varrho = 6.4\text{ dm}^3 \cdot 2.7\ \frac{\text{kg}}{\text{dm}^3}$$
$$= \mathbf{17.28\text{ kg}}$$

Values for density of solids, liquids and gases: pages 112 and 113

Linear mass density

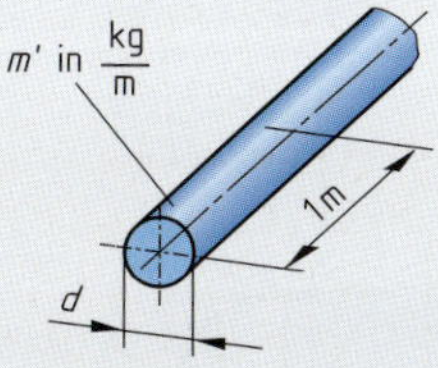

m mass
m' linear mass density
l length

Linear mass density

$$m = m' \cdot l$$

Example:

Steel bar with $d = 15$ mm; $m' = 1.39$ kg/m; $l = 3.86$ m; $m = ?$

$$\mathbf{m} = m' \cdot l = 1.39\ \frac{\text{kg}}{\text{m}} \cdot 3.86\text{ m}$$
$$= \mathbf{5.37\text{ kg}}$$

Application: Calculating the mass of profile sections, pipes, wires, etc. using the table values for m' (page 151)

Area mass density

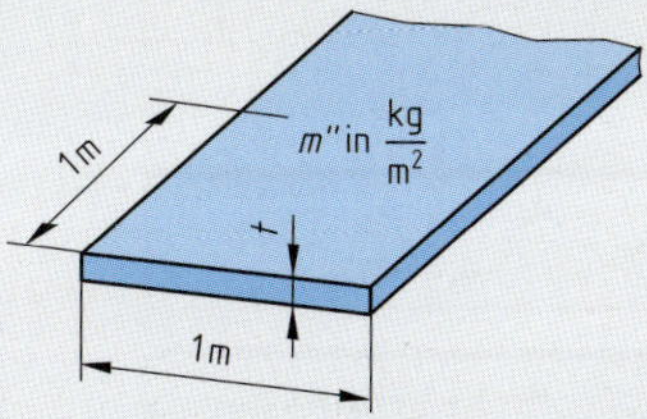

m mass
m'' area mass density
A area

Area mass density

$$m = m'' \cdot A$$

Example:

Steel sheet
$t = 1.5$ mm; $m'' = 11.8$ kg/m^2;
$A = 7.5$ m^2; $m = ?$

$$\mathbf{m} = m'' \cdot A = 11.8\ \frac{\text{kg}}{\text{m}^2} \cdot 7.5\text{ m}^2$$
$$= \mathbf{88.5\text{ kg}}$$

Application: Calculating the mass of sheet metal, foils, coatings, etc using the table values for m'' (page 151)

M

Centroids of Lines and Plane Areas

Centroids of lines

l, l_1, l_2 lengths of the lines C, C_1, C_2 centroids of the lines
x_c, x_1, x_2 horizontal distances of the line centroids from the y-axis
y_c, y_1, y_2 vertical distances of the line centroids from the x-axis

Line segment

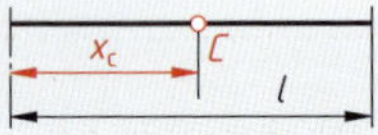

$$x_c = \frac{l}{2}$$

Circular arc

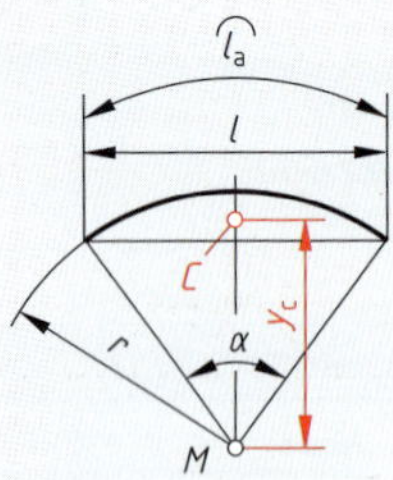

Calculation of l and l_a:
Page 24

General

$$y_c = \frac{r \cdot l}{l_a}$$

$$y_c = \frac{l \cdot 180°}{\pi \cdot \alpha}$$

Semicircular arc

$$y_c \approx 0.6366 \cdot r$$

Quarter circle arc

$$y_c \approx 0.9003 \cdot r$$

Composite continuous lines

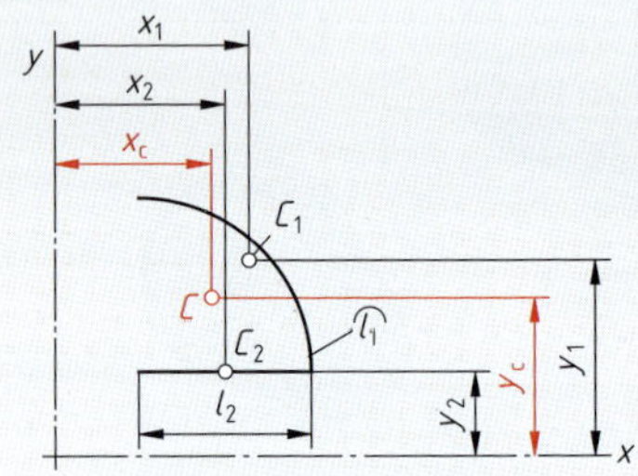

$$x_c = \frac{l_1 \cdot x_1 + l_2 \cdot x_2 + \ldots}{l_1 + l_2 + \ldots}$$

$$y_c = \frac{l_1 \cdot y_1 + l_2 \cdot y_2 + \ldots}{l_1 + l_2 + \ldots}$$

Centroids of plane areas

A, A_1, A_2 areas C, C_1, C_2 centroids of the areas
x_c, x_1, x_2 horizontal distances of the area centroids from the y-axis
y_c, y_1, y_2 vertical distances of the area centroids from the x-axis

Rectangle

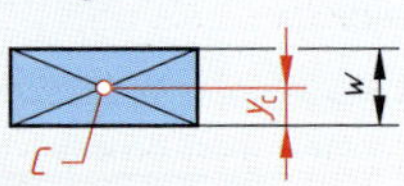

$$y_c = \frac{w}{2}$$

Triangle

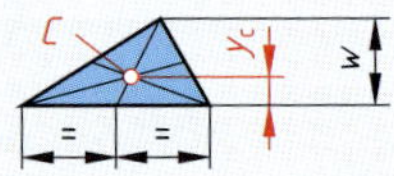

$$y_c = \frac{w}{3}$$

Circular sector

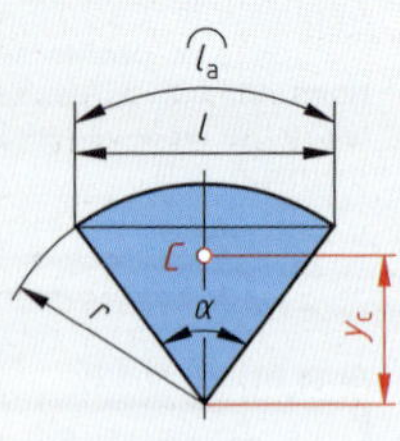

General

$$y_c = \frac{2 \cdot r \cdot l}{3 \cdot l_B}$$

Semi-circle area

$$y_c \approx 0.4244 \cdot r$$

Quarter circle area

$$y_c \approx 0.6002 \cdot r$$

Composite areas

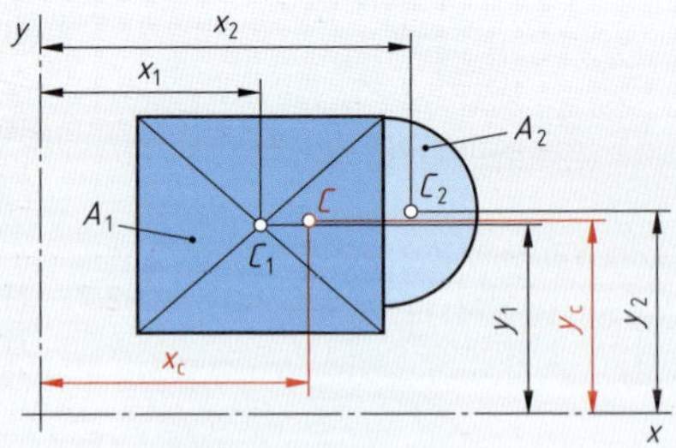

$$x_c = \frac{A_1 \cdot x_1 + A_2 \cdot x_2 + \ldots}{A_1 + A_2 + \ldots}$$

$$y_c = \frac{A_1 \cdot y_1 + A_2 \cdot y_2 + \ldots}{A_1 + A_2 + \ldots}$$

Circular segment

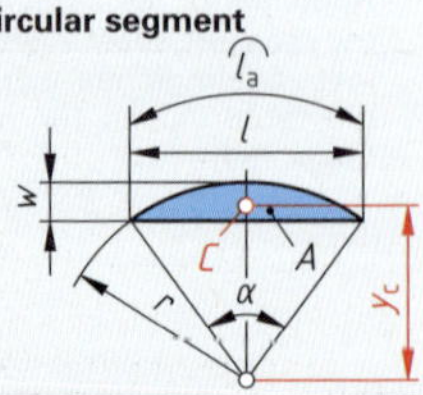

$$y_c = \frac{l^3}{12 \cdot A}$$

2 Physics

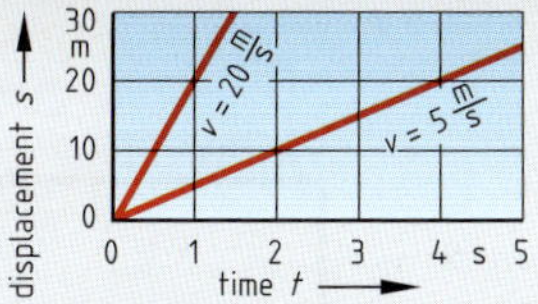
displacement s
30
m
20
10
0
0 1 2 3 4 s 5
time t
v = 20 m/s
v = 5 m/s

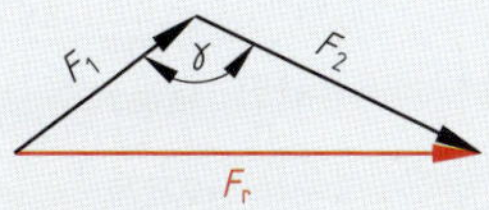
F1
γ
F2
Fr

P

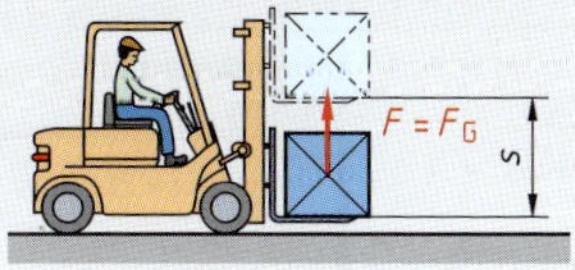
F = FG
s

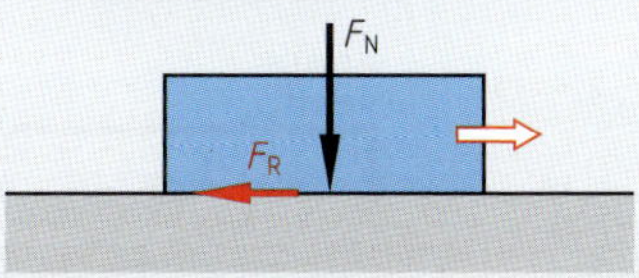
FN
FR

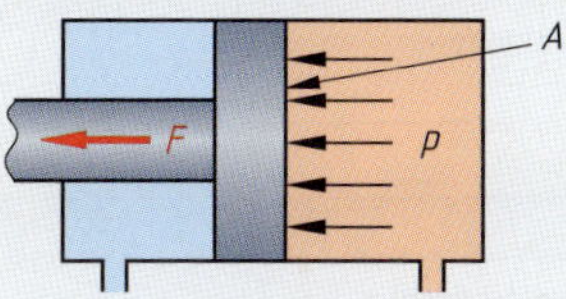
A
F
p

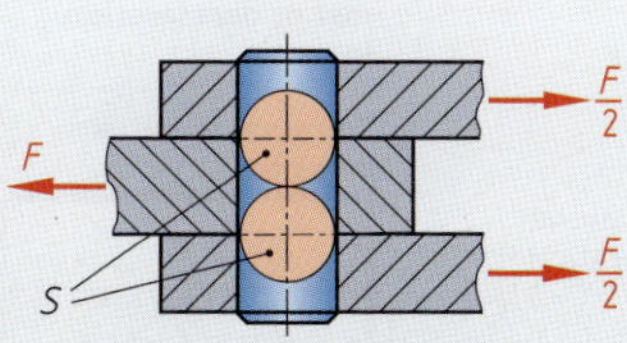
F/2
F
F/2
S

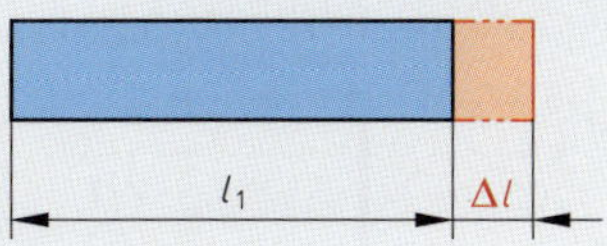
l1
Δl

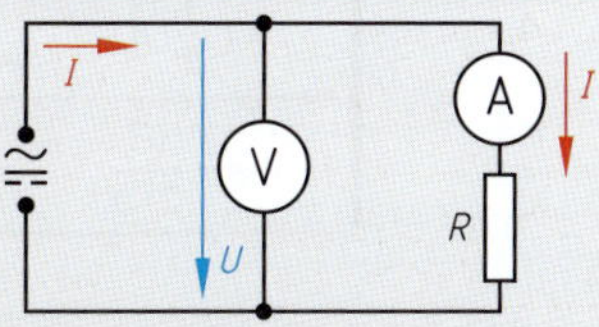
I
A
I
V
R
U

Uniform motion and uniformly accelerated motion

Uniform motion

Linear motion

Displacement-time diagram

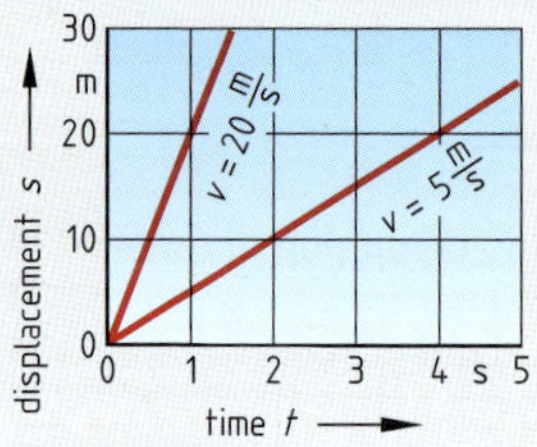

v velocity
t time
s displacement

Example:

v = 48 km/h; s = 12 m; t = ?

Conversion: $48\ \frac{km}{h} = \frac{48\,000\ m}{3600\ s} = 13.33\ \frac{m}{s}$

$\boldsymbol{t} = \frac{s}{v} = \frac{12\ m}{13.33\ m/s} = \mathbf{0.9\ s}$

Velocity

$$v = \frac{s}{t}$$

$1\ \frac{m}{s} = 60\ \frac{m}{min} = 3.6\ \frac{km}{h}$

$1\ \frac{km}{h} = 16.667\ \frac{m}{min} = 0.2778\ \frac{m}{s}$

Circular motion

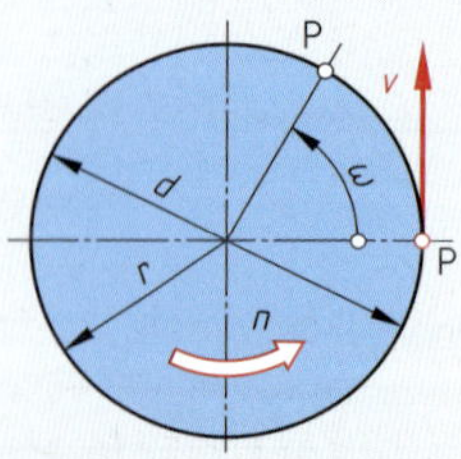

v circumferential velocity, cutting speed
ω angular velocity
n rotational speed
r radius
d diameter

Example:

Pulley, d = 250 mm; n = 1400 min^{-1};
v = ?; ω = ?

Conversion: $n = 1400\ min^{-1} = \frac{1400}{60\ s} = 23.33\ s^{-1}$

$\boldsymbol{v} = \pi \cdot d \cdot n = \pi \cdot 0.25\ m \cdot 23.33\ s^{-1} = \mathbf{18.3\ \frac{m}{s}}$

$\boldsymbol{\omega} = 2 \cdot \pi \cdot n = 2 \cdot \pi \cdot 23.33\ s^{-1} = \mathbf{146.6\ s^{-1}}$

For a cutting speed of a circumferential velocity see page 31.

Circumferential velocity

$$v = \pi \cdot d \cdot n$$

$$v = \omega \cdot r$$

Angular velocity

$$\omega = 2 \cdot \pi \cdot n$$

$\frac{1}{min} = min^{-1} = \frac{1}{60\ s}$

Accelerated and decelerated motion

Linear accelerated motion

Velocity-time diagram

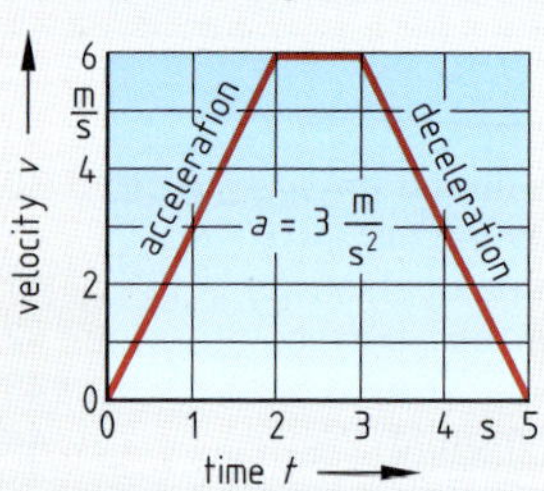

Displacement-time diagram

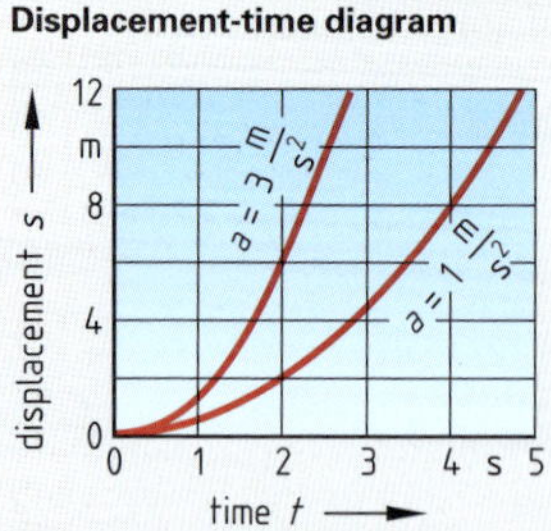

The increase in velocity per second is called **acceleration;** and a decrease is **deceleration.** Free fall is uniformly accelerated motion on which gravitational acceleration *g* is acting.

v terminal velocity (acceleration), or initial velocity (deceleration)
s displacement
a acceleration
t time
g gravitational acceleration

1st example:

Object, free fall from s = 3 m; v = ?

$a = g = 9.81\ \frac{m}{s^2}$

$\boldsymbol{v} = \sqrt{2 \cdot a \cdot s} = \sqrt{2 \cdot 9.81\ m/s^2 \cdot 3\ m} = \mathbf{7.7\ \frac{m}{s}}$

2nd example:

Vehicle, v = 80 km/h; a = 7 m/s^2;
Braking distance s = ?

Conversion: $v = 80\ \frac{km}{h} = \frac{80\,000\ m}{3600\ s} = 22.22\ \frac{m}{s}$

$v = \sqrt{2 \cdot a \cdot s}$

$\boldsymbol{s} = \frac{v^2}{2 \cdot a} = \frac{(22.22\ m/s)^2}{2 \cdot 7\ m/s^2} = \mathbf{35.3\ m}$

The following applies to acceleration from rest or deceleration to rest:

Terminal or initial velocity

$$v = a \cdot t$$

$$v = \sqrt{2 \cdot a \cdot s}$$

Displacement due to acceleration/deceleration

$$s = \frac{1}{2} \cdot v \cdot t$$

$$s = \frac{1}{2} \cdot a \cdot t^2$$

$$s = \frac{v^2}{2 \cdot a}$$

Speeds on machines

Feed rate

Turning

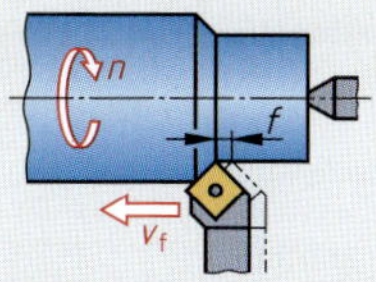

Milling

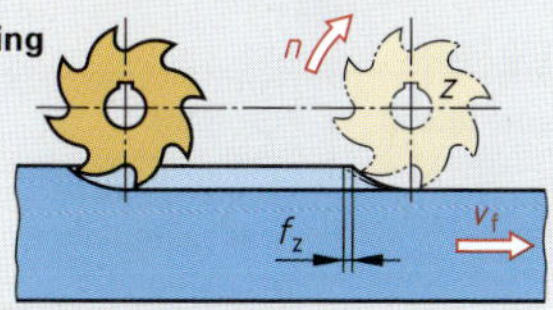

Screw drive

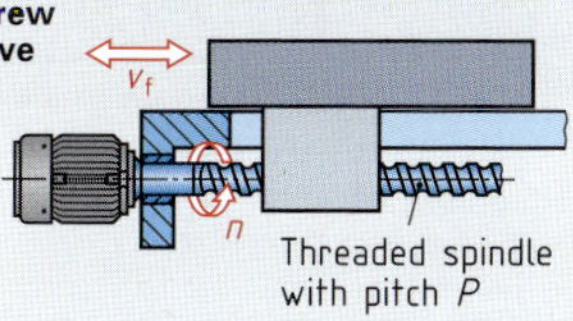

Rack and pinion

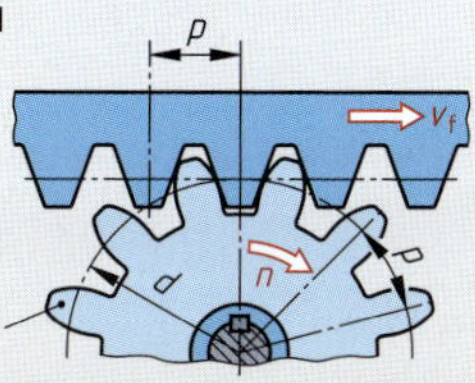

v_f feed rate
n rotational speed
f feed
f_t feed per cutting edge
N number of cutting edges, or number of teeth on the pinion
P thread pitch
p pitch of rack and pinion

1st example:

Plain milling cutter, $N = 8$;
$f_t = 0.2$ mm; $n = 45$/min; $v_f = ?$

$$\mathbf{v_f} = n \cdot f_t \cdot N = 45\,\frac{1}{\text{min}} \cdot 0.2\text{ mm} \cdot 8 = \mathbf{72\,\frac{mm}{min}}$$

2nd example:

Feed drive with threaded spindle,
$P = 5$ mm; $n = 112$/min; $v_f = ?$

$$\mathbf{v_f} = n \cdot P = 112\,\frac{1}{\text{min}} \cdot 5\text{ mm} = \mathbf{560\,\frac{mm}{min}}$$

3rd example:

Feed of rack and pinion,
$n = 80$/min; $d = 75$ mm; $v_f = ?$

$$\mathbf{v_f} = \pi \cdot d \cdot n = \pi \cdot 75\text{ mm} \cdot 80\,\frac{1}{\text{min}}$$
$$= 18\,850\,\frac{\text{mm}}{\text{min}} = \mathbf{18.85\,\frac{m}{min}}$$

Feed rate for drilling, turning

$$v_f = n \cdot f$$

Feed rate for milling

$$v_f = n \cdot f_t \cdot N$$

Feed rate for screw drive

$$v_f = n \cdot P$$

Feed rate for rack and pinion

$$v_f = n \cdot N \cdot p$$

$$v_f = \pi \cdot d \cdot n$$

P

Cutting speed, Circumferential velocity

Cutting speed

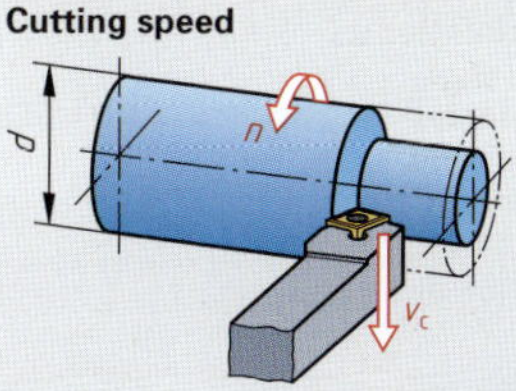

Circumferential velocity

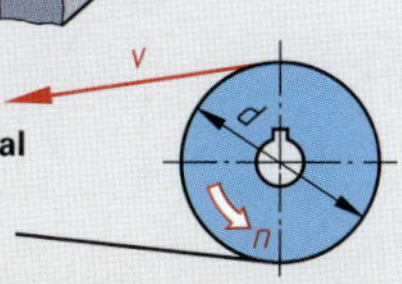

v_c cutting speed
v circumferential velocity
d diameter
n rotational speed

Example:

Turning, $n = 1200$/min; $d = 35$ mm; $v_c = ?$

$$\mathbf{v_c} = \pi \cdot d \cdot n = \pi \cdot 0.035\text{ m} \cdot 1200\,\frac{1}{\text{min}}$$
$$= \mathbf{132\,\frac{m}{min}}$$

Cutting speed

$$v_c = \pi \cdot d \cdot n$$

Circumferential velocity

$$v = \pi \cdot d \cdot n$$

Average speed of crank mechanism

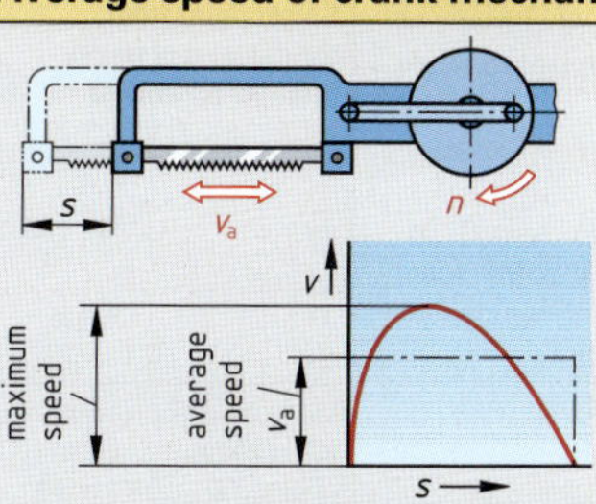

v_a average speed
n number of double strokes
s stroke length

Example:

Power hacksaw,
$s = 280$ mm; $n = 45$/min; $v_a = ?$

$$\mathbf{v_a} = 2 \cdot s \cdot n = 2 \cdot 0.28\text{ m} \cdot 45\,\frac{1}{\text{min}}$$
$$= \mathbf{25.2\,\frac{m}{min}}$$

Average speed

$$v_a = 2 \cdot s \cdot n$$

Representation, addition and resolution of forces

Chosen for the following examples: S_f = 10 N/mm

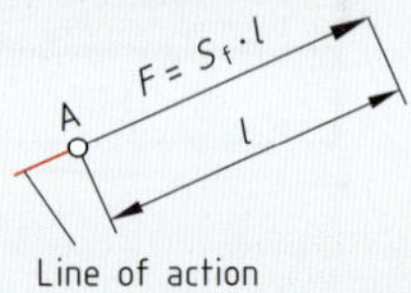

Forces are the cause of changes in motion or deformation of solid bodies.

F_1, F_2, F_i	component forces	l	arrow (vector) length
F_r	resultant	S_f	scale of forces

Forces are represented by arrows (vectors).

The **magnitude** of a force F corresponds to the length of the arrow l.

The **position** of the force is determined by the starting point and the line of action.

The **arrow point** indicates the direction of the force.

Arrow (vector) length

$$l = \frac{F}{S_f}$$

P

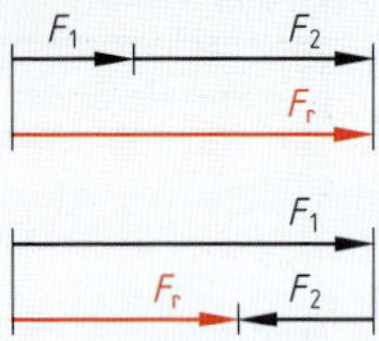

Adding and subtracting of forces on the same line of action

Forces on the same line of action can be added and subtracted algebraically.

Example:

F_1 = 80 N, F_2 = 160 N, same direction; F_r = ?

$\mathbf{F_r} = F_1 + F_2$ = 80 N + 160 N = **240 N**

Resultant
(equivalent force having the same effect as all component forces together)

$$F_r = \Sigma F_i$$

Example: tensioning ropes

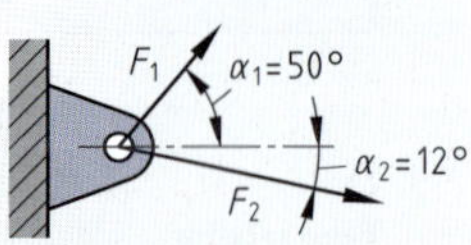

Diagram of forces

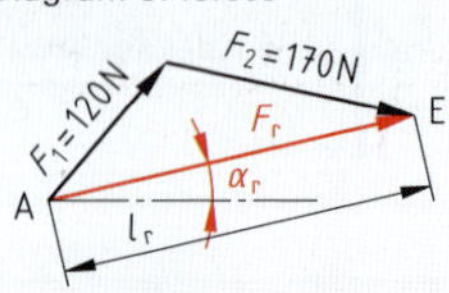

Graphical addition of forces with intersecting lines of action

1. Assemble force vectors from a starting point A to an end point E in any sequence. Make sure that the angle and scale remain unchanged.
2. The resultant F_r runs from point A to point E of the force diagram.
3. Calculate the magnitude of F_r from l_r and S_f and measure the angular position of F_r.

Example:

Tensioning ropes, F_1 = 120 N, F_2 = 170 N; F_r = ?, α_r = ?

Obtained through measuring: l_r = 25 mm, $\boldsymbol{\alpha_r}$ = **13°**

$\mathbf{F_r} = l_r \cdot S_f$ = 25 mm · 10 N/mm = **250 N**

Example: inclined plane

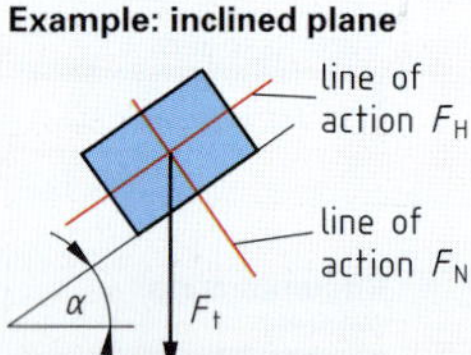

Diagram of forces

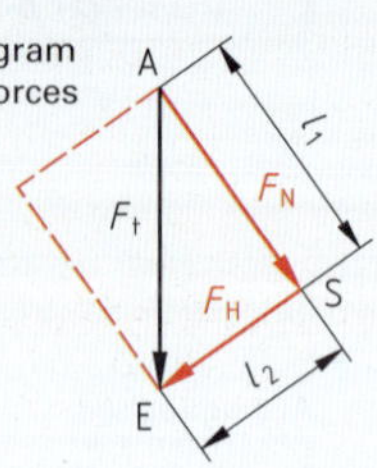

Graphical resolution of a force into two component forces

1. Draw the known force F (= F_t) true to scale and angle and lay the lines of actions of the wanted forces through A and E. They intersect in point S. The line diagram ASE constitutes the diagram of forces.
2. The component forces run from A to S and S to E.
3. Calculate the magnitudes of the component forces from l_1, l_2 and S_f.

Example:

Inclined plane, F_t = 200 N, α = 35°; F_N = ?, F_H = ?
Obtained by measurement: l_1 = 16 mm, l_2 = 11 mm

$\mathbf{F_N} = l_1 \cdot S_f$ = 16 mm · 10 N/mm = **160 N**

$\mathbf{F_H} = l_2 \cdot S_f$ = 11 mm · 10 N/mm = **110 N**

Sketch of the triangle of forces for the calculation (not to scale)

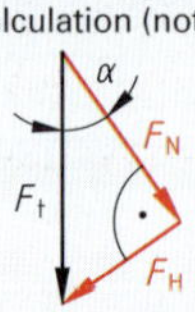

Calculations based on the diagram for forces

The basis of the calculation is the sketch of a force triangle or polygon, which is a force diagram not to scale.

Example:

Inclined plane, F_t = 200 N, α = 35°; F_N = ?, F_H = ?

You can resolve the weight force F_t to F_N and F_H in accordance with the force triangle. The force diagram is right-angled. The trigonometric functions cosine and sine are used for the calculation.

$\mathbf{F_N} = F_t \cdot \cos\alpha$ = 200 N · cos 35° = **163.8 N**

$\mathbf{F_H} = F_t \cdot \sin\alpha$ = 200 N · sin 35° = **114.7 N**

Calculations from the diagram of forces (triangle for forces)

Shape of the force diagram	Required trigonometric function
Force triangle with right angles	Sine Cosine Tangent
Force triangle with oblique angles	Law of sines Law of cosines

Types of forces

Weight

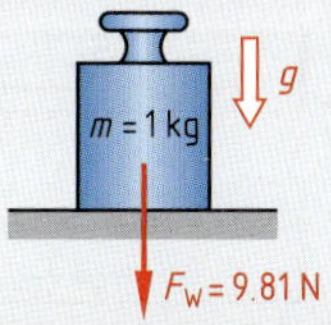

Gravity generates a weight force on a mass.

F_W weight
m mass
g gravitational acceleration

Example:

I-beam, m = 1200 kg; F_W = ?

$\mathbf{F_W} = m \cdot g = 1200\,\text{kg} \cdot 9.81\,\frac{\text{m}}{\text{s}^2} = \mathbf{11\,772\,N}$

Weight

$$F_W = m \cdot g$$

$g = 9.81\,\frac{\text{m}}{\text{s}^2} \approx 10\,\frac{\text{m}}{\text{s}^2}$

Calculation of mass: page 27

P

Forces at acceleration and deceleration

A force is required to accelerate or decelerate a mass.

F acceleration force
m mass
a acceleration

Example:

$m = 50\,\text{kg}$; $a = 3\,\frac{\text{m}}{\text{s}^2}$; $F = ?$

$\mathbf{F} = m \cdot a = 50\,\text{kg} \cdot 3\,\frac{\text{m}}{\text{s}^2} = 150\,\text{kg} \cdot \frac{\text{m}}{\text{s}^2} = \mathbf{150\,N}$

Acceleration force

$$F = m \cdot a$$

$1\,\text{N} = 1\,\text{kg} \cdot \frac{\text{m}}{\text{s}^2}$

Spring force (Hooke's law)

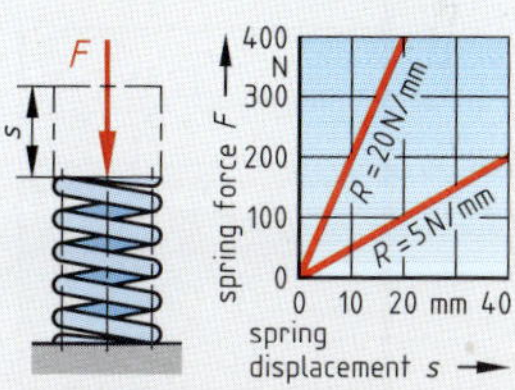

The force and corresponding linear expansion of a spring are proportional within the elastic range.

F spring force
R spring constant
s spring displacement

Example:

Compression spring, R = 8 N/mm; s = 12 mm; F = ?

$\mathbf{F} = R \cdot s = 8\,\frac{\text{N}}{\text{mm}} \cdot 12\,\text{mm} = \mathbf{96\,N}$

Spring force

$$F = R \cdot s$$

Change in spring force

$$\Delta F = R \cdot \Delta s$$

Centrifugal force

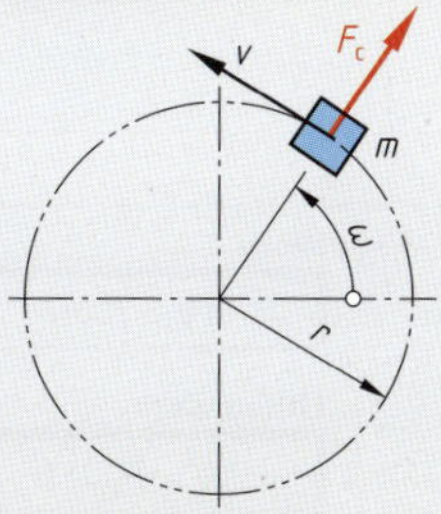

Centrifugal force F_c is generated when a mass is moved along a curvilinear path, e.g. a circle.

F_c centrifugal force
m mass
r radius
ω angular velocity
v circumferential velocity

Example:

Turbine blade, m = 160 g; v = 80 m/s; d = 400 mm; F_c = ?

$\mathbf{F_c} = \frac{m \cdot v^2}{r} = \frac{0.16\,\text{kg} \cdot (80\,\text{m/s})^2}{0.2\,\text{m}} = 5120\,\frac{\text{kg} \cdot \text{m}}{\text{s}^2} = \mathbf{5120\,N}$

Centrifugal force

$$F_c = m \cdot r \cdot \omega^2$$

$$F_c = \frac{m \cdot v^2}{r}$$

Supporting forces

Example of supporting forces

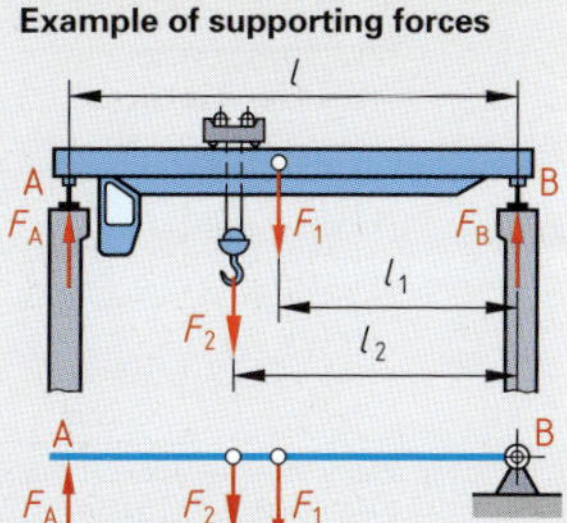

A bearing point is treated as a fulcrum in calculating supporting forces.

F_A, F_B supporting forces
F_1, F_2 forces
l, l_1, l_2 effective lever lengths

Example:

Overhead travelling crane, F_1 = 40 kN; F_2 = 15 kN; l_1 = 6 m; l_2 = 8 m; l = 12 m; F_A = ?

Solution: B is selected as fulcrum point; the supporting force F_A is assumed on a single-ended lever.

$\mathbf{F_A} = \frac{F_1 \cdot l_1 + F_2 \cdot l_2}{l} = \frac{40\,\text{kN} \cdot 6\,\text{m} + 15\,\text{kN} \cdot 8\,\text{m}}{12\,\text{m}} = \mathbf{30\,kN}$

Lever principle

$$\Sigma M_l = \Sigma M_r$$

Supporting force at A

$$F_A = \frac{F_1 \cdot l_1 + F_2 \cdot l_2 \ldots}{l}$$

$$F_A + F_B = {}_1 + F_2 \ldots$$

P

Torque, Mechanical work

Torque and levers

Single-ended lever

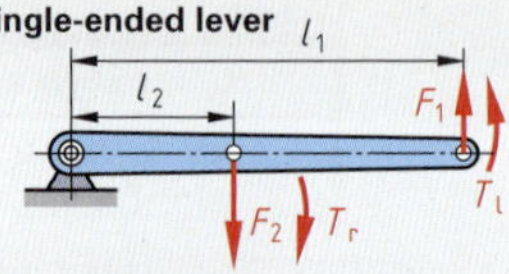

Two-ended lever

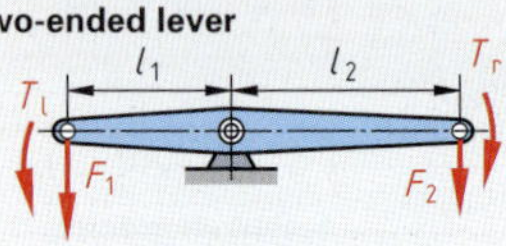

Angular lever

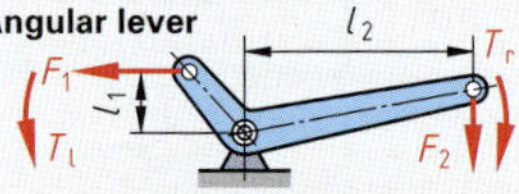

The **effective lever length** is the orthogonal distance between the fulcrum and the line of action of the force. For disk-shaped rotating parts, the lever arm corresponds to the radius r.

T	torque
F	force
l	effective lever length
ΣT_l	sum of all counter-clockwise torques
ΣT_r	sum of all clockwise torques

Example:

Angular lever, $F_1 = 30$ N; $l_1 = 0.15$ m; $l_2 = 0.45$ m; $F_2 = ?$

$$\boldsymbol{F_2} = \frac{F_1 \cdot l_1}{l_2} = \frac{30\ \text{N} \cdot 0.15\ \text{m}}{0.45\ \text{m}} = \mathbf{10\ N}$$

Torque

$$T = F \cdot l$$

Lever principle

$$\Sigma T_l = \Sigma T_r$$

Lever principle with only 2 applying forces

$$F_1 \cdot l_1 = F_2 \cdot l_2$$

Torque in gear drives

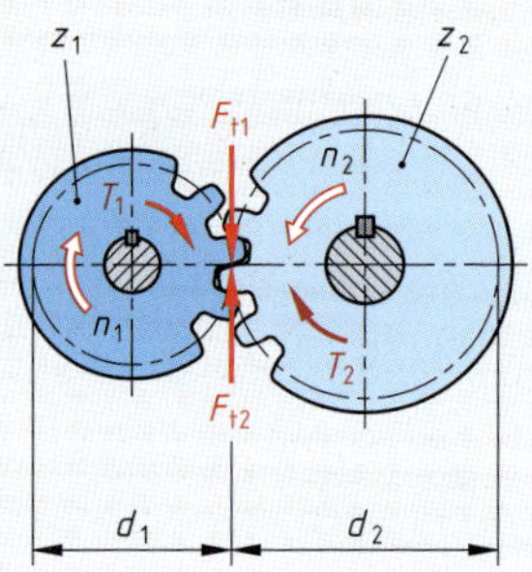

The lever arm of a gear is half of its reference diameter d. Different torques result if two engaging gears have different numbers of teeth.

Driving gear		**Driven gear**	
F_{t1}	tangential force	F_{t2}	tangential force
T_1	torque	T_2	torque
d_1	reference diameter	d_2	reference diameter
z_1	number of teeth	z_2	number of teeth
n_1	speed	n_2	speed

i gear ratio

Example:

Gears, $i = 12$; $T_1 = 60$ N · m; $T_2 = ?$

$\boldsymbol{T_2} = i \cdot T_1 = 12 \cdot 60$ N · m = **720 N · m**

For gear ratios of gear drives, see page 256.

Torques

$$T_1 = \frac{F_{t1} \cdot d_1}{2}$$

$$T_2 = \frac{F_{t2} \cdot d_2}{2}$$

$$T_2 = i \cdot T_1$$

$$\frac{T_2}{T_1} = \frac{z_2}{z_1}$$

$$\frac{T_2}{T_1} = \frac{n_1}{n_2}$$

Mechanical work, lifting work and frictional work

Lifting work

Frictional work

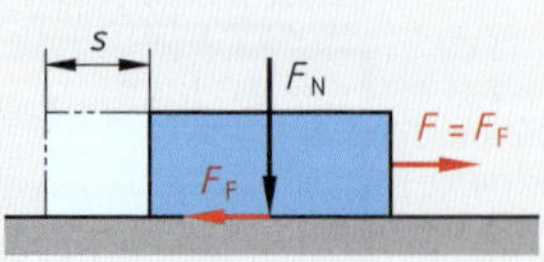

Work is performed when a force acts along a distance.

F	force in direction of travel
F_W	weight
F_F	friction force
F_N	normal force
W	work
s	force distance
s, h	lifting height
μ	friction coefficient

1st example:

Lifting work, $F = 300$ N; $s = 4$ m; $W = ?$

$\boldsymbol{W} = F \cdot s = 300$ N · 4 m = 1200 N · m = **1200 J**

2nd example:

Frictional work, $F_N = 0.8$ kN; $s = 1.2$ m; $\mu = 0.4$; $W = ?$

$\boldsymbol{W} = \mu \cdot F_N \cdot s = 0.4 \cdot 800$ N · 1.2 m = 384 N · m = **384 J**

Work

$$W = F \cdot s$$

Lifting work

$$W = F_W \cdot h$$

Frictional work

$$W = \mu \cdot F_N \cdot s$$

$$1\ \text{J} = 1\ \text{N} \cdot 1\ \text{m} = 1\ \text{W} \cdot \text{s} = 1\ \frac{\text{kg} \cdot \text{m}^2}{\text{s}^2}$$

$1\ \text{kW} \cdot \text{h} = 3.6\ \text{MJ}$

Simple machines and Energy

Block and tackle[1]

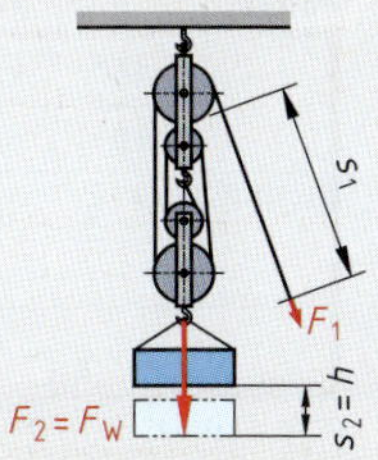

n no. of load-bearing ropes, pulleys

$$F_1 = \frac{F_W}{n}$$

$$s_1 = n \cdot h$$

$$W_2 = F_W \cdot h$$

Inclined plane[1]

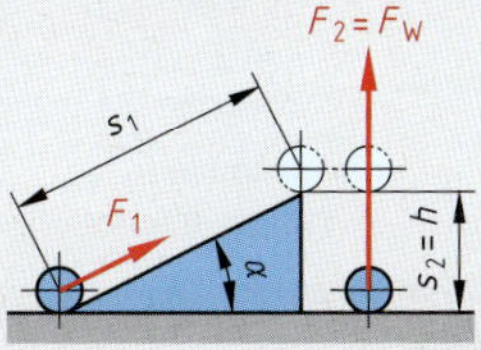

α angle of inclination

$$F_1 \cdot s_1 = F_W \cdot h$$

$$F_1 = F_W \cdot \sin \alpha$$

$$W_2 = F_W \cdot h$$

Wedge[1]

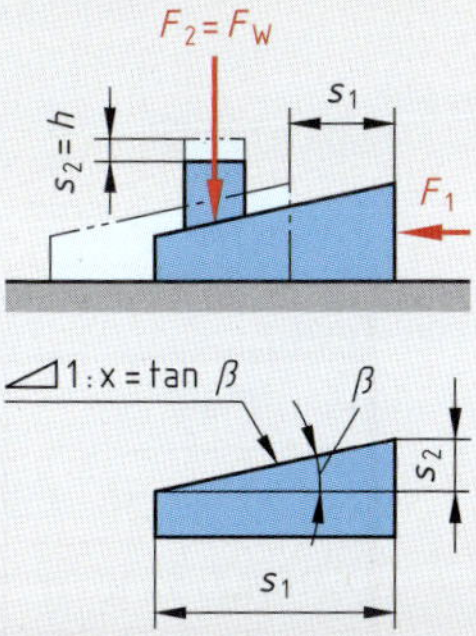

β angle of inclination
$\tan \beta$ inclination

$$F_1 \cdot s_1 = F_2 \cdot h$$

$$F_2 = \frac{F_1}{\tan \beta}$$

$$s_2 = s_1 \cdot \tan \beta$$

$$W_2 = F_2 \cdot h$$

Bolt (angle of motion)[1]

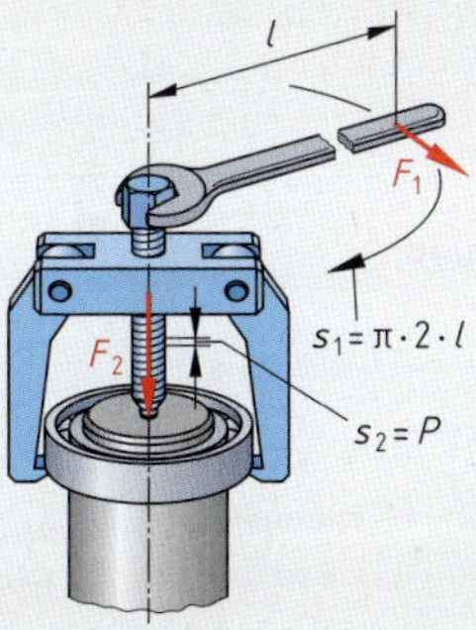

P thread pitch
l lever length

For one full turn

$$F_1 \cdot 2 \cdot \pi \cdot l = F_2 \cdot P$$

$$s_1 = 2 \cdot \pi \cdot l$$

$$W_1 = F_1 \cdot 2 \cdot \pi \cdot l$$

$$W_2 = F_2 \cdot P$$

[1] The formulas apply to a hypothetical frictionless condition, wherein the input work W_1 is equal to the output work W_2, i. e. what is gained in force is lost in displacement.

Potential energy

Gravitational energy

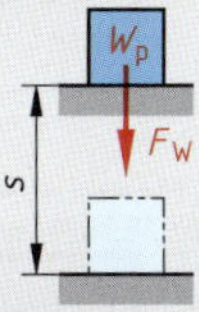

Spring energy

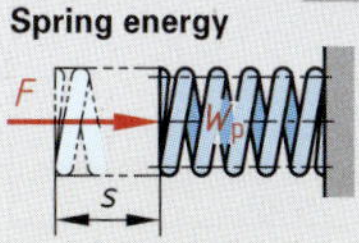

$$R = \frac{F}{s}$$

Potential energy is stored work (gravitational energy, spring energy).

W_p	potential energy	s, h	travel, lift or fall height
F_W	weight		
F	spring force	s	spring displacement
R	spring constant		

Example:

Drop hammer, m = 30 kg; s = 2.6 m; W_p = ?

$$\mathbf{W_p} = F_W \cdot s = 30\,\text{kg} \cdot 9.81\,\frac{\text{m}}{\text{s}^2} \cdot 2.6\,\text{m} = \mathbf{765\,J}$$

Gravitational energy

$$W_p = F_W \cdot s$$

Spring energy

$$W_p = \frac{R \cdot s^2}{2}$$

Kinetic energy

Linear motion

Kinetic energy is energy of motion.

W_k kinetic energy
v velocity
m mass

Example:

Car, m = 1400 kg, v_1 = 50 km/h (13.88 m/s), v_2 = 100 km/h (27.77 m/s); W_{K1} = ?, W_{K2} = ?

$$\mathbf{W_{K1}} = \frac{m \cdot v_1^2}{2} = \frac{1400\,\text{kg} \cdot (13.88\,\text{m/s})^2}{2} = \mathbf{135\,kJ}$$

$$\mathbf{W_{K2}} = \frac{m \cdot v_2^2}{2} = \frac{1400\,\text{kg} \cdot (27.77\,\text{m/s})^2}{2} = \mathbf{540\,kJ}$$

Kinetic energy of linear motion

$$W_k = \frac{m \cdot v^2}{2}$$

P

Power and Efficiency

Power in linear motion

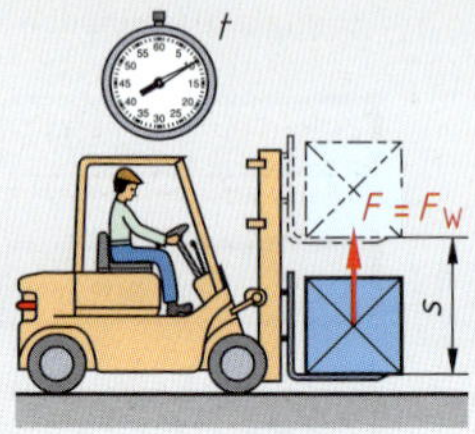

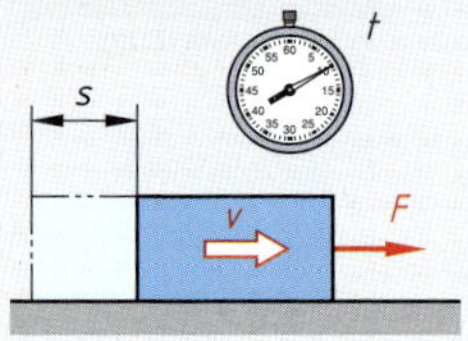

Power is work per unit time.

P	power	s	displacement in the force direction
W	work		
v	velocity	t	time

1st example:

Forklift, $F = 15$ kN; $v = 25$ m/min; $P = ?$

$$\boldsymbol{P} = F \cdot v = 15\,000\ \text{N} \cdot \frac{25\ \text{m}}{60\ \text{s}} = 6250\ \frac{\text{N} \cdot \text{m}}{\text{s}} = 6250\ \text{W} = \mathbf{6.25\ kW}$$

2nd example:

Crane lifts a machine, $m = 1.2$ t; $s = 2.5$ m; $t = 4.5$ s; $P = ?$

$$F_W = m \cdot g = 1200\ \text{kg} \cdot 9.81\ \text{m/s}^2 = 11\,772\ \text{N}$$

$$\boldsymbol{P} = \frac{F_W \cdot s}{t} = \frac{11772\ \text{N} \cdot 2.5\ \text{m}}{4.5\ \text{s}} = 6540\ \text{W} = \mathbf{6.5\ kW}$$

For power in pumps and cylinders see page 402.

Power

$$P = \frac{W}{t}$$

$$P = \frac{F \cdot s}{t}$$

$$P = F \cdot v$$

$$1\ \text{W} = 1\ \frac{\text{J}}{\text{s}} = 1\ \frac{\text{N} \cdot \text{m}}{\text{s}}$$

1 kW = 1.36 PS

Power in circular motion

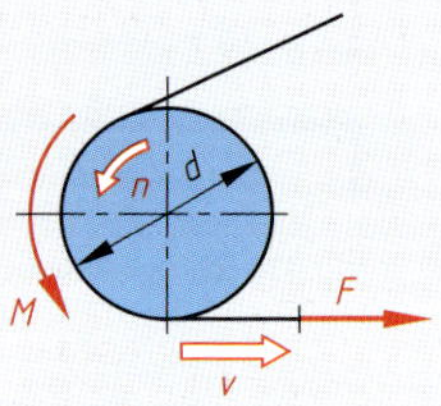

P	power	s	displacement in the force direction
M	torque	t	time
F	tangential force	n	rotational speed
v	velocity	ω	angular velocity

Example:

Belt drive, $F = 1.2$ kN; $d = 200$ mm; $n = 2800$/min; $P = ?$

$$\boldsymbol{P} = F \cdot \pi \cdot d \cdot n$$

$$= 1.2\ \text{kN} \cdot \pi \cdot 0.2\ \text{m} \cdot \frac{2800}{60\ \text{s}} = 35.2\ \frac{\text{kN} \cdot \text{m}}{\text{s}} = \mathbf{35.2\ kW}$$

Numerical equation:
Enter → M in N · m, n in 1/min
Result → P in kW

For cutting power in machine tools see pages 294 and 296.

Power

$$P = F \cdot v$$

$$P = F \cdot \pi \cdot d \cdot n$$

$$P = M \cdot 2 \cdot \pi \cdot n$$

$$P = M \cdot \omega$$

or:

Power

$$P = \frac{M \cdot n}{9550}$$

Efficiency

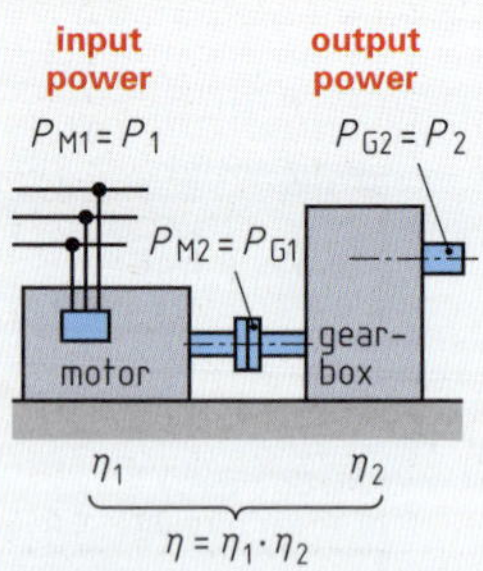

Efficiency refers to the ratio of power or work output to the power or work input.

P_1	input power	P_2	output power
W_1	input work	W_2	output work
η	total efficiency	η_1, η_2	partial efficiencies

Example:

Belt drive, $P_1 = 4$ kW; $P_2 = 3$ kW; $\eta_1 = 85$ %; $\eta = ?$; $\eta_2 = ?$

$$\boldsymbol{\eta} = \frac{P_2}{P_1} = \frac{3\ \text{kW}}{4\ \text{kW}} = \mathbf{0.75}; \qquad \boldsymbol{\eta_2} = \frac{\eta}{\eta_1} = \frac{0.75}{0.85} = \mathbf{0.88}$$

Efficiency

$$\eta = \frac{P_2}{P_1}$$

$$\eta = \frac{W_2}{W_1}$$

Total efficiency

$$\eta = \eta_1 \cdot \eta_2 \cdot \eta_3 \dots$$

Efficiencies η (approximate values)

Brown coal power station	0.32	Gasoline engine	0.27	Screw thread	0.30
Coal power station	0.41	Automobile diesel engine (partial load)	0.24	Pinion gear	0.97
Natural gas power station	0.50	Automobile diesel engine (full load)	0.40	Worm gear, $i = 40$	0.65
Gas turbine	0.38	Large diesel engine (partial load)	0.33	Friction drive	0.80
Steam turbine (high pressure)	0.45	Large diesel engine (full load)	0.55	Chain drive	0.90
Water turbine	0.85	Three phase AC motor	0.85	Wide V-belt drive	0.85
Cogeneration	0.75	Machine tools	0.75	Hydrostatic transmission	0.75

Types of friction, Coefficients of friction

Friction force

Static friction, sliding friction

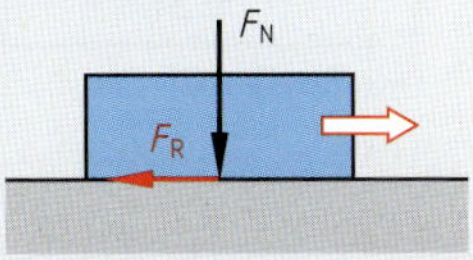

Friction force depends on the normal force FN and

- the type of friction: static, sliding or rolling friction
- the state of lubrication
- the material pairing (combination of materials)

The influences are summarized in the friction coefficient μ, which is assessed by testing.

Friction force for static and sliding friction

$$F_R = \mu \cdot F_N$$

F_N normal force
F_R friction force
μ friction coefficient
f rolling friction coefficient
M_R friction moment
d diameter
r radius

Friction moment

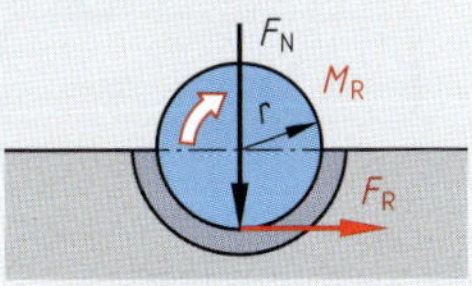

Friction moment

$$M_R = \frac{\mu \cdot F_N \cdot d}{2}$$

$$M_R = F_R \cdot r$$

1st example:

Plain bearing, F_N = 100 N; μ = 0.03; F_R = ?

$\mathbf{F_R} = \mu \cdot F_N = 0.03 \cdot 100\ \text{N} = \mathbf{3\ N}$

2nd example:

Steel shaft in a Cu-Sn plain bearing, μ = 0.05; F_N = 6 kN; d = 160 mm; M_R = ?

$$\mathbf{M_R} = \frac{\mu \cdot F_N \cdot d}{2} = \frac{0.05 \cdot 6000\ \text{N} \cdot 0.16\ \text{m}}{2} = \mathbf{24\ N \cdot m}$$

Rolling friction

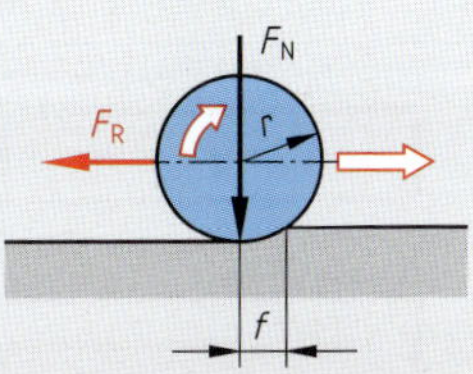

Friction force for rolling friction[1]

$$F_R = \frac{f \cdot F_N}{r}$$

3rd example:

Crane wheel on steel rail, F_N = 45 kN; d = 320 mm; f = 0.5 mm; F_R = ?

$$\mathbf{F_R} = \frac{f \cdot F_N}{r} = \frac{0.5\ \text{mm} \cdot 45\,000\ \text{N}}{160\ \text{mm}} = \mathbf{140.6\ N}$$

[1] caused by elastic deformation between roller body and rolling surface

Coefficients of friction (guideline values)

Material pairing	Example of application	Coefficient of static friction μ dry	Coefficient of static friction μ lubricated	Coefficient of sliding friction μ dry	Coefficient of sliding friction μ lubricated
steel/steel	vise guideway	0.50	0.10	0.40	0.10 – 0.05
steel/cast iron	machine guideway	0.20	0.10	0.18	0.10 – 0.05
steel/Cu-Sn alloy	shaft in solid plain bearing	0.20	0.10	0.10	0.06 – 0.03[2]
steel/Pb-Sn alloy	shaft in multilayer plain bearing	0.15	0.10	0.10	0.05 – 0.03[2]
steel/polyamide	shaft in PA plain bearing	0.30	0.15	0.30	0.12 – 0.03[2]
steel/PTFE	low temperature bearing	0.04	0.04	0.04	0.04[2]
steel/friction lining	shoe brake	0.60	0.30	0.55	0.3 – 0.2
steel/wood	part on an assembly stand	0.55	0.10	0.35	0.05
wood/wood	underlay blocks	0.50	0.20	0.30	0.10
cast iron/Cu-Sn alloy	adjustment gib	0.25	0.16	0.20	0.10
rubber/cast iron	belts on a pulley	0.50	–	0.45	–
rolling element/steel	anti-friction bearing[3], guideway[3]	–	–	–	0.003 – 0.001

[2] The significance of the material pairing decreases with increasing sliding speed and presence of mixed and viscous friction.
[3] Calculation performed in spite of rolling movement, because it is typically similar to calculations of static or sliding friction.

Coefficients of rolling friction (guideline values)[4]

Material pairing	Example of application	Coefficient of rolling friction f in mm
steel/steel	steel wheel on a guide rail	0.5
plastic/concrete	caster wheel on concrete floor	5
rubber/asphalt	car tires on the street	8

[4] Data on coefficients of rolling friction can vary considerably in technical literature.

Types of pressure, Hydraulic power transmission

Pressure

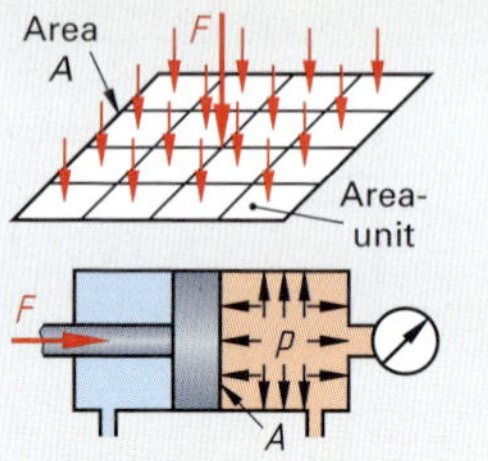

p pressure
F force
A area

Example:

$F = 2$ MN; piston ∅ $d = 400$ mm; $p = ?$

$$\boldsymbol{p} = \frac{F}{A} = \frac{2000000\ \text{N}}{\frac{\pi \cdot (40\ \text{cm})^2}{4}} = 1592\ \frac{\text{N}}{\text{cm}^2} = \mathbf{159.2\ bar}$$

For calculations on hydraulics and pneumatics, see page 402.

Pressure

$$p = \frac{F}{A}$$

Units of pressure

$1\ \text{Pa} = 1\ \frac{\text{N}}{\text{m}^2} = 0.00001\ \text{bar}$

$1\ \text{bar} = 10\ \frac{\text{N}}{\text{cm}^2} = 0.1\ \frac{\text{N}}{\text{mm}^2}$

$1\ \text{mbar} = 100\ \text{Pa} = 1\ \text{hPa}$

Gage pressure, air pressure, absolute pressure

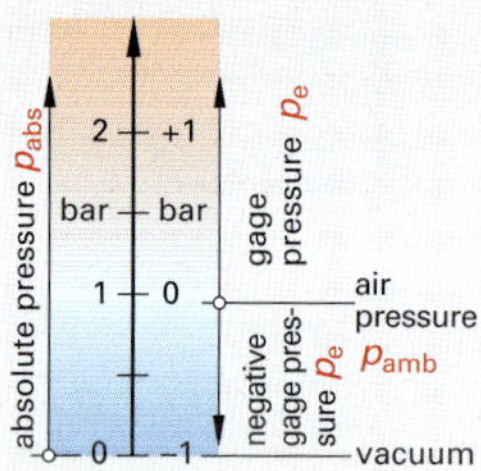

p_e gage pressure (excedens, exceed)
p_{amb} air pressure (ambient)
p_{abs} absolute pressure

The gage pressure is
positive, if $p_{abs} > p_{amb}$ and
negative, if $p_{abs} < p_{amb}$ (vacuum)

Example:

Car tires, $p_e = 2.2$ bar; $p_{amb} = 1$ bar; $p_{abs} = ?$

$\boldsymbol{p_{abs}} = p_e + p_{amb} = 2.2\ \text{bar} + 1\ \text{bar} = \mathbf{3.2\ bar}$

Gage pressure

$$p_e = p_{abs} - p_{amb}$$

$p_{amb} = 1.013\ \text{bar} \approx 1\ \text{bar}$
(standard air pressure)

Gravitational pressure, buoyant force

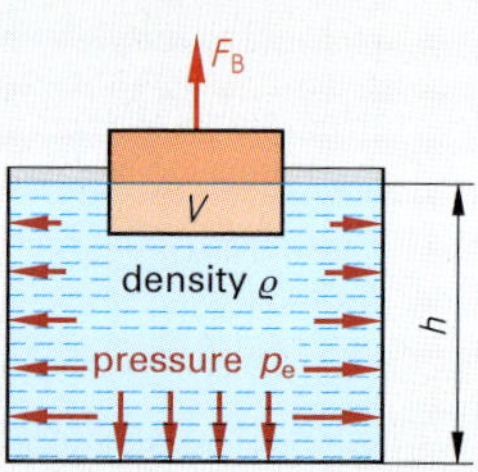

p_e gravitational pressure (inherent pressure)
ϱ density of the liquid
g gravitational acceleration
F_B buoyant force
V displaced volume
h depth of liquid

Example:

What is the gravitational pressure in a water depth of 10 m?

$$\boldsymbol{p_e} = g \cdot \varrho \cdot h = 9.81\ \frac{\text{m}}{\text{s}^2} \cdot 1000\ \frac{\text{kg}}{\text{m}^3} \cdot 10\ \text{m}$$

$$= 98\,100\ \frac{\text{kg}}{\text{m} \cdot \text{s}^2} = 98\,100\ \text{Pa} \approx \mathbf{1\ bar}$$

Gravitational pressure

$$p_e = g \cdot \varrho \cdot h$$

Buoyant force

$$F_B = g \cdot \varrho \cdot V$$

$g = 9.81\ \frac{\text{m}}{\text{s}^2} \approx 10\ \frac{\text{m}}{\text{s}^2}$

For density values, see page 113.

Hydraulic power transmission

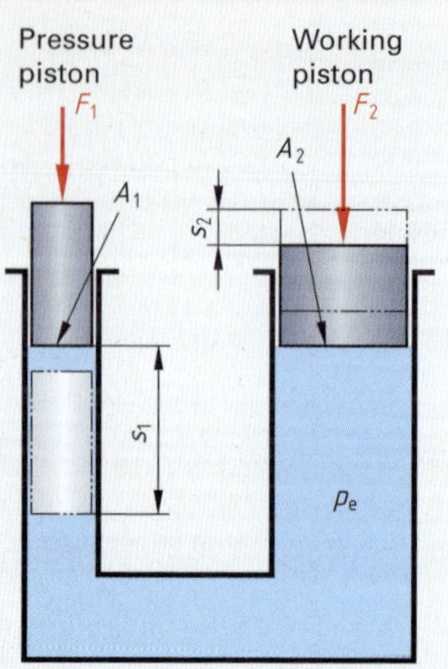

In closed liquid or gas systems, pressure spreads uniformly in all directions.

F_1, F_2 piston forces
A_1, A_2 piston areas
s_1, s_2 piston displacements
i hydraulic transmission ratio
p_e gage pressure

Example:

$F_1 = 200$ N; $A_1 = 5\ \text{cm}^2$; $A_2 = 500\ \text{cm}^2$;
$s_2 = 30$ mm; $F_2 = ?$; $s_1 = ?$; $i = ?$

$$\boldsymbol{F_2} = \frac{F_1 \cdot A_2}{A_1} = \frac{200\ \text{N} \cdot 500\ \text{cm}^2}{5\ \text{cm}^2} = 20\,000\ \text{N} = \mathbf{20\ kN}$$

$$\boldsymbol{s_1} = \frac{s_2 \cdot A_2}{A_1} = \frac{30\ \text{mm} \cdot 500\ \text{cm}^2}{5\ \text{cm}^2} = \mathbf{3000\ mm}$$

$$\boldsymbol{i} = \frac{F_1}{F_2} = \frac{200\ \text{N}}{20\,000\ \text{N}} = \mathbf{\frac{1}{100}}$$

Displaced volume

$$A_1 \cdot s_1 = A_2 \cdot s_2$$

Work on both pistons

$$F_1 \cdot s_1 = F_2 \cdot s_2$$

Relations: forces, areas, displacements

$$\frac{F_2}{F_1} = \frac{A_2}{A_1} = \frac{s_1}{s_2}$$

Transmission ratio

$$i = \frac{F_1}{F_2} = \frac{s_2}{s_1}$$

$$i = \frac{A_1}{A_2}$$

Pressure intensification, Flow velocity, Change of state

Pressure intensification

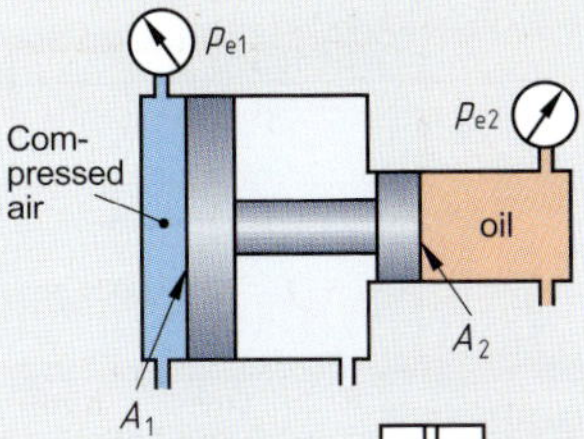

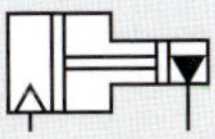

Circuit symbols as per DIN ISO 1219-1

A_1, A_2 piston areas
p_{e1} gage pressure at piston area A_1
p_{e1} gage pressure at piston area A_2
η efficiency of pressure intensifier

Example:

$A_1 = 200\ \text{cm}^2$; $A_2 = 5\ \text{cm}^2$; $\eta = 0.88$;
$p_{e1} = 7\ \text{bar} = 70\ \text{N/cm}^2$; $p_{e2} = ?$

$$\mathbf{p_{e2}} = p_{e1} \cdot \frac{A_1}{A_2} \cdot \eta = 70\ \frac{\text{N}}{\text{cm}^2} \cdot \frac{200\ \text{cm}^2}{5\ \text{cm}^2} \cdot 0.88$$

$$= 2464\ \text{N/cm}^2 = \mathbf{246.4\ bar}$$

Gage pressure

$$p_{e2} = p_{e1} \cdot \frac{A_1}{A_2} \cdot \eta$$

Flow velocity

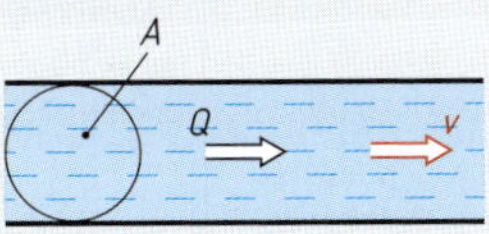

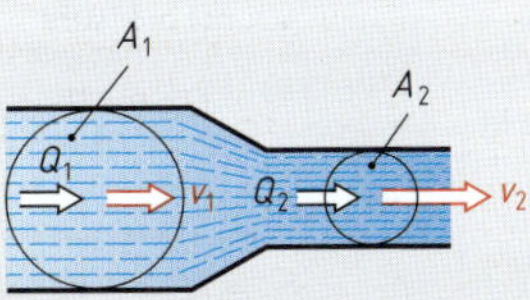

Q, Q_1, Q_2 volume flows
A, A_1, A_2 cross-sectional areas
v, v_1, v_2 flow velocities

Continuity equation

In a pipeline with varying cross-sectional areas, the same volume flow Q passes through each cross section during the time t.

Example:

Pipeline with $A_1 = 19.6\ \text{cm}^2$; $A_2 = 8.04\ \text{cm}^2$ and $Q = 120\ \text{l/min}$; $v_1 = ?$; $v_2 = ?$

$$\mathbf{v_1} = \frac{Q}{A_1} = \frac{120\,000\ \text{cm}^3/\text{min}}{19.6\ \text{cm}^2} = 6122\ \frac{\text{cm}}{\text{min}} = \mathbf{1.02\ \frac{m}{s}}$$

$$\mathbf{v_2} = \frac{v_1 \cdot A_1}{A_2} = \frac{1.02\ \text{m/s} \cdot 19.6\ \text{cm}^2}{8.04\ \text{cm}^2} = \mathbf{2.49\ \frac{m}{s}}$$

Volume flow

$$Q = A \cdot v$$

$$Q_1 = Q_2$$

Ratio of flow velocities

$$\frac{v_1}{v_2} = \frac{A_2}{A_1}$$

Changes of state in gases

Compression

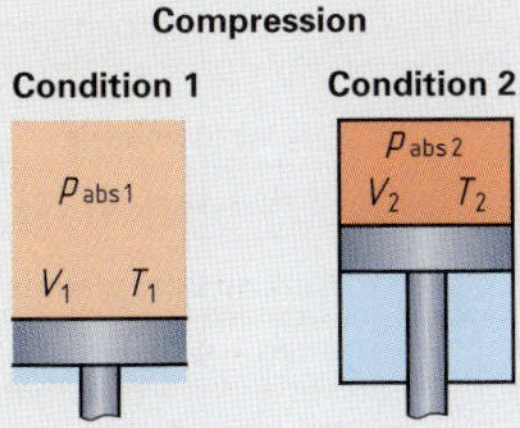

Boyle's Law

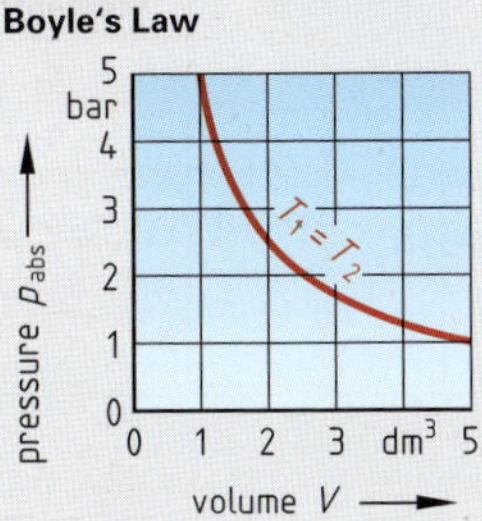

Condition 1

p_{abs1} absolute pressure
V_1 volume
T_1 absolute temperature

Condition 2

p_{abs2} absolute pressure
V_2 volume
T_2 absolute temperature

Example:

A welding-gas cylinder with a volume $V = 20\ \text{dm}^3$ and a filling pressure of 250 bar ($p_{abs1} = 251$ bar) warms up in the sun from initially $t_1 = 15\,°\text{C}$ to $t_2 = 45\,°\text{C}$. How great is the pressure increase Δp in the gas cylinder?

Calculation of absolute temperatures (page 45):

$T_1 = t_1 + 273 = (15 + 273)\ \text{K} = 288\ \text{K}$

$T_2 = t_2 + 273 = (45 + 273)\ \text{K} = 318\ \text{K}$

$$p_{abs2} = \frac{p_{abs1} \cdot T_2}{T_1} = \frac{251\ \text{bar} \cdot 318\,°\text{C}}{288\,°\text{C}}$$

$= 277$ bar

$\mathbf{\Delta p} = p_{abs2} - p_{abs1} = 277\ \text{bar} - 251\ \text{bar}$
$= \mathbf{26\ bar}$

Ideal gas law

$$\frac{p_{abs1} \cdot V_1}{T_1} = \frac{p_{abs2} \cdot V_2}{T_2}$$

Special cases:

with constant temperature

$$p_{abs1} \cdot V_1 = p_{abs2} \cdot V_2$$

with constant volume

$$\frac{p_{abs1}}{T_1} = \frac{p_{abs2}}{T_2}$$

with constant pressure

$$\frac{V_1}{T_1} = \frac{V_2}{T_2}$$

P

Load cases, Types of loading, Material properties, Stress limits

Load cases

static loading	dynamic loading	
stationary	pulsating	alternating
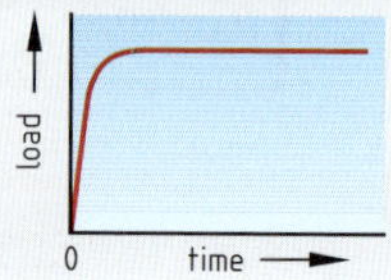	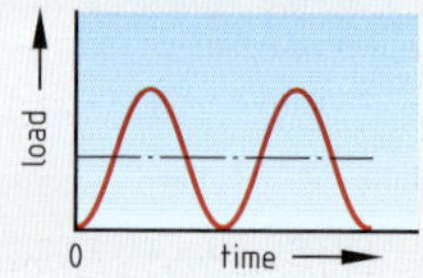	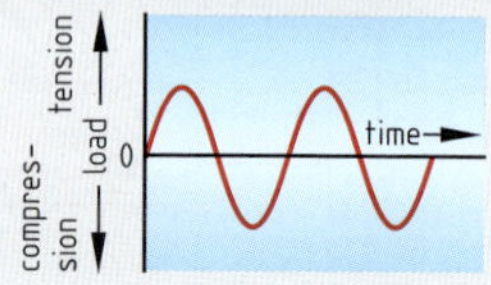
Load case I Magnitude and direction of the load remain the same, e. g. for a weight load hanging on a rope.	**Load case II** The load increases to a maximum value and then falls back to zero, e. g. for crane cables and springs.	**Load case III** The load alternates between a positive and a negative maximum value of equal magnitude, e. g. for rotating axles.

Types of loading, material properties, stress limits

Type of loading	Stress	Elastic deformation	Material properties/stress limits (σ_{limit}) for load case I, material: brittle[1] (e.g. cast iron)	ductile[2] (e.g. steel)	Load case II[3]	Load case III[3]
Tension 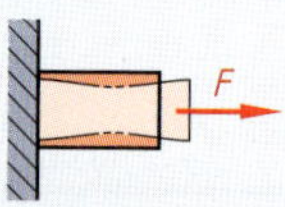	Tensile stress σ_t	Elongation ε Elongation at fracture A	Tensile strength R_m	Yield strength R_e 0.2%-yield point $R_{p0.2}$	Pulsating tensile fatigue strength σ_{tpuls}	Compression-tension fatigue strength σ_{ct}
Compression 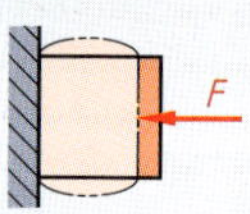	Compression stress σ_c	Compression set ε_c	Compression strength σ_{cB}	Natural compression yield point σ_{cF} 0.2%-offset yield strength $\sigma_{c0.2}$	Pulsating compression fatigue strength σ_{cpuls}	
Bending 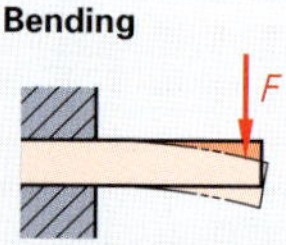	Bending stress σ_b	Deflection f	Bending strength σ_{bB}	Flexural yield strength σ_{bF}	Pulsating bending fatigue strength σ_{bpuls}	Alternating bending fatigue strength σ_{bA}
Shear 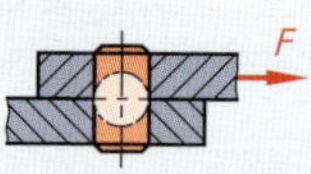	Shear stress τ_s	–	Shear strength τ_{sB}	Shear yield strength τ_{sY}	–	–
Torsion 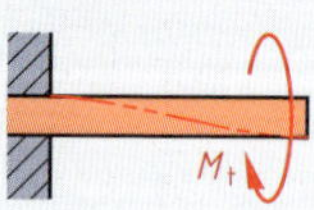	Torsional stress τ_t	Torsion angle φ	Torsional strength τ_{tB}	Torsional yield strength τ_{tF}	Pulsating torsional fatigue strength τ_{tpuls}	Alternating torsional fatigue strength τ_{tA}
Buckling 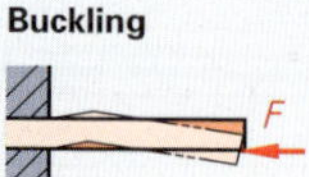	Buckling stress σ_{bu}	–	Buckling strength σ_{buB}	Buckling strength σ_{buB}	–	–

1) Material property, fracture stress limit
2) Material property, plastic yield stress limit
3) Material property, fatigue fracture stress limit (material fatigue)

Strength calculation, Safety factors, Allowable stresses, Modulus of elasticity

Strength calculation, allowable stresses and stress limits for static load

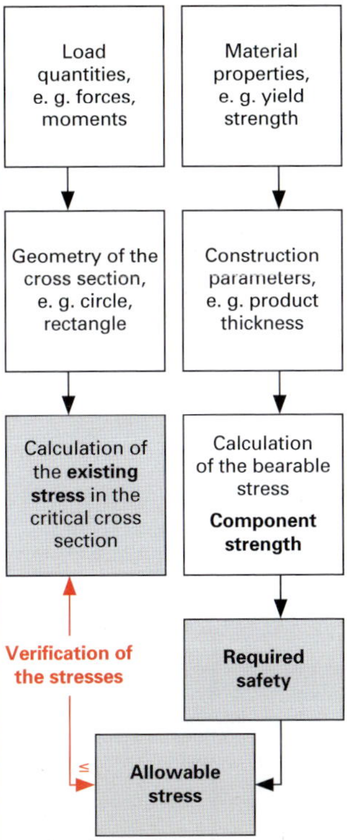

For reasons of safety, components may only be exposed to a portion of the limit stress σ_{lim} that causes permanent deformation, breaking or fatigue failure.

σ_{lim} stress limit depending on the type of load (page 40 and table on the right)
σ_{all} allowable stress
σ_{exis} existing stress
ν safety factor (table on the right)

- **Verification of existing component cross sections**

In the verification, the existing stresses are to be compared to the calculated allowable stresses under consideration of the required safety (safety factor).

- **Pre-dimensioning (estimate calculation of the required component cross section)**

With unknown component thickness, the component strength cannot be calculated accurately. The required component cross section determined in the estimate calculation can be assessed with the help of the guideline values (for material properties and increased safety) in the table below.

Example:

Statically loaded tension rod made of S275; σ_{all} for the pre-dimensioning?, σ_{all} for the verification with d = 25 mm and ν_{req} = 1.5?

Pre-dimens.: $\boldsymbol{\sigma_{all}}$ = **160 N/mm²** (table below)

Verification: $\sigma_{lim} = R_e$ (table on the right) = 265 N/mm² (page 127)

$$\boldsymbol{\sigma_{all}} = \frac{\sigma_{lim}}{\nu} = \frac{265\ \text{N/mm}^2}{1.5} = \mathbf{176\ N/mm^2}$$

Allowable stress (general)

$$\sigma_{all} = \frac{\sigma_{lim}}{\nu}$$

Stress verification

$$\sigma_{exis} \le \sigma_{all}$$

Stress limits σ_{lim} for steel, under	
Tension	R_e; $R_{p0.2}$
Compression	$\sigma_{cV} \approx R_e$
Bending	$\sigma_{bV} \approx 1.2 \cdot R_e$
Torsion	$\tau_{tV} \approx 0.65 \cdot R_e$
Shear	$\tau_{sV} \approx 0.6 \cdot R_e$

Safety factor for general stress verifications in mechanical engineering

Material	Guideline value
Ductile, e. g. steel	$\nu \approx 1.5$
Brittle, e. g. cast iron	$\nu \approx 2.0$

Allowable stresses for pre-dimensioning under static load[1)]

(Guide values with $\nu \approx 1.7$ for steel and steel cast and $\nu \approx 2.1$ for cast iron)

Type of load	Tension	Compression	Bending	Torsion	Shear
Material	**Allowable stress in N/mm²**				
S235	140	140	168	90	85
S275	160	160	190	105	95
E295	175	175	210	115	105
E335	200	200	240	130	120
E360	210	210	250	135	125
C15	260	260	310	170	155
17Cr3	300	300	360	195	180
16MnCr5	375	375	450	245	225
20MnCr5	430	430	515	280	260
18CrNiMo7-6	490	490	590	320	295
C22E	200	200	240	130	120
C45E	290	290	350	190	170

Type of loading	Tension	Compression	Bending	Torsion	Shear
Material	**Allowable stress in N/mm²**				
C60E	340	340	410	220	205
14Cr4	470	470	565	305	280
50CrMo4	530	530	635	345	310
30CrNiMo8	620	620	745	405	370
GE200	115	115	140	75	70
GE240	140	140	170	90	85
GE260	150	150	180	100	90
GE300	175	175	210	115	105
EN-GJS-400	190	245	190	125	125
EN-GJS-500	240	310	240	155	155
EN-GJS-600	285	370	285	185	185
EN-GJS-700	335	435	335	215	215

[1)] Values for normalized structural steels, case hardening steels after case hardening and refining and alloyed quenched and tempered steel. For structural steel engineering, the values specified by DIN 18800 must be applied.

Modulus of elasticity *E* in kN/mm² (average values)

Material	Steel, cast steel	EN-GJL-150	EN-GJL-300	EN-GJS-400	GE200	EN-GJMW-350-4	CuZn40	Aluminum alloy	Titanium alloy
Mod. of *E*	210	85	125	175	210	170	90	70	120

Tensile stress, Compressive stress, Surface pressure

Tensile stress

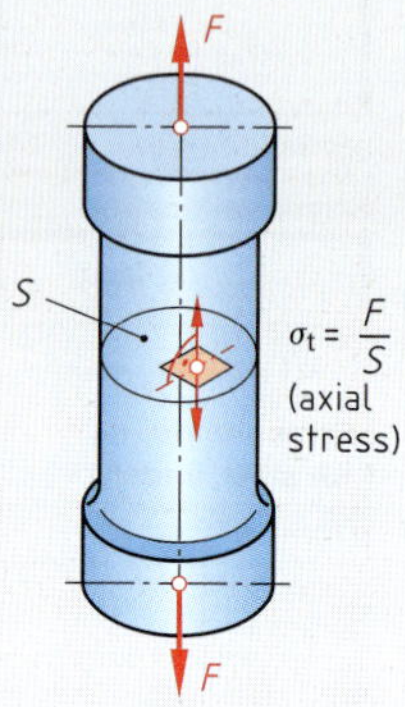

σ_t tensile stress
F tensile force
S cross-sectional area
σ_{tall} allowable tensile stress (guiding values on page 41)
ν safety factor (guiding value ν = 1.5)
S_{req} required cross-sectional area
R_e yield strength

1st example:

Steel wire, d = 3 mm (S = 7.07 mm²), F = 900 N; σ_t = ?

$$\boldsymbol{\sigma_t} = \frac{F}{S} = \frac{900\ \text{N}}{7.07\ \text{mm}^2} = \mathbf{127\ \frac{N}{mm^2}}$$

2nd example:

Pre-dimensioning, bar steel S235, F = 15 kN; S_{req} = ? d = ?

σ_{tall} = 140 N/mm² (page 41)

$$S_{req} = \frac{F}{\sigma_{tall}} = \frac{15\,000\ \text{N}}{140\ \text{N/mm}^2} = 107\ \text{mm}^2 \rightarrow \boldsymbol{d} = \mathbf{12\ mm}$$

Calculation of elastic elongation: see page 191

Tensile stress

$$\sigma_t = \frac{F}{S}$$

Required cross-sectional area

$$S_{req} = \frac{F}{\sigma_{tall}}$$

Allowable tensile stress[1]

$$\sigma_{tall} = \frac{R_e}{\nu}$$

Strength values R_e: see pages 127 to 131

Compressive stress

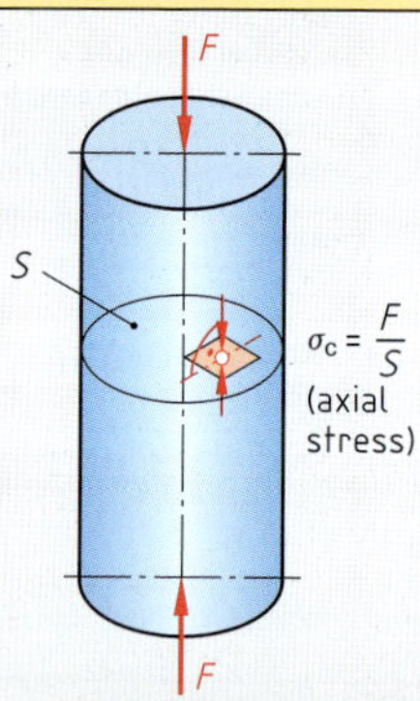

σ_c compressive stress
F compression force
S cross-sectional area
σ_{call} allowable compressive stress (guiding values on page 41)
σ_{cF} natural compression yield point (for steel $\sigma_{cF} \approx R_e$)
ν safety factor (guiding value ν = 1.5)
S_{req} required cross-sectional area
R_e yield strength

Example:

Pre-dimensioning, frame of EN-GJS-400; F = 1200 kN, S_{req} = ?

σ_{call} = 245 N/mm² (see page 41)

$$S_{req} = \frac{F}{\sigma_{call}} = \frac{1\,200\,000\ \text{N}}{245\ \text{N/mm}^2} = 4898\ \text{mm}^2$$

Strength values R_e: see pages 127 to 131

Compressive stress

$$\sigma_c = \frac{F}{S}$$

Required cross-sectional area

$$S_{req} = \frac{F}{\sigma_{call}}$$

Allowable compressive stress[1]

$$\sigma_{call} = \frac{\sigma_{cF}}{\nu}$$

Area compression stress

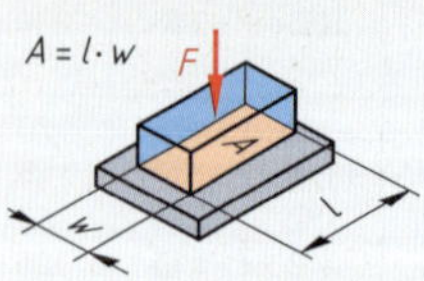

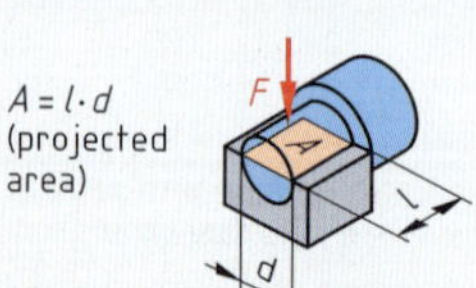

Compressive stresses at the contact surface of two components are referred to as area compression.

F force
p area compression
A contact surface, projected area
A_{req} required contact surface
p_{all} allowable area compression
R_e yield strength

Example:

Two tension straps, each 8 mm thick, are joined with bolts DIN 1445 10h11 x 16 x 30 and loaded with F = 2000 N. p = ?

$$\boldsymbol{p} = \frac{F}{A} = \frac{2000\ \text{N}}{8\ \text{mm} \cdot 10\ \text{mm}} = \mathbf{25\ \frac{N}{mm^2}}$$

Strength values R_e: see pages 127 to 131

Surface pressure

$$p = \frac{F}{A}$$

Required contact surface

$$A_{req} = \frac{F}{p_{all}}$$

Allowable area compression[1][2] (guideline values)

$$p_{all} = \frac{R_e}{1{,}2}$$

[1] The calculation of the allowable stress is only valid for static load on non-brittle materials.
[2] The allowable values specified on page 41 apply to the calculation of machine elements.

Shear, bending and torsion stress

Shear stress

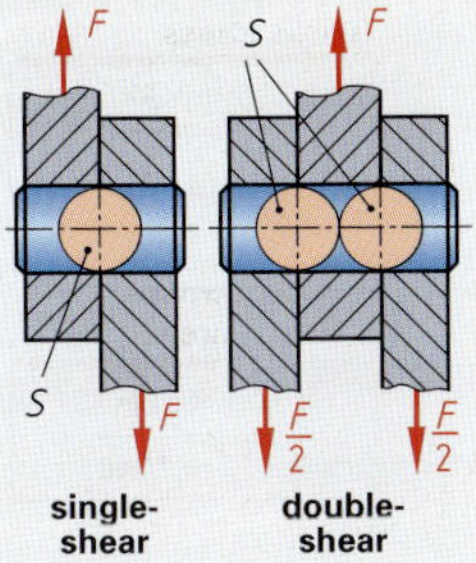

For strength values R_e, see pages 127 to 131.

The cross section under stress must not shear off.

τ_s	shear stress	S_{req}	required cross-sectional area
F	shear force		
S	cross-sect. area	R_e	yield strength

τ_{sall} allowable shear stress (guideline values, page 41)
τ_{sY} shear yield point (for steel $\tau_{sY} \approx 0.6 \cdot R_e$)
ν safety factor (guideline value $\nu = 1.5$)

Example:

Dowel pin ∅ 6 mm ($S = 28.3\ \text{mm}^2$), single-shear, loaded by $F = 2200$ N; $\tau_s = ?$

$$\boldsymbol{\tau_s} = \frac{F}{S} = \frac{2200\ \text{N}}{28.3\ \text{mm}^2} = \mathbf{77.7\ \frac{N}{mm^2}}$$

2) The allowable values specified on page 41 apply to the calculation of machine elements.

Shear stress

$$\tau_s = \frac{F}{S}$$

Required cross-sectional area

$$S_{req} = \frac{F}{\tau_{sall}}$$

Allowable shear stress[1) 2)]

$$\tau_{sall} = \frac{\tau_{sY}}{\nu}$$

Bending stress

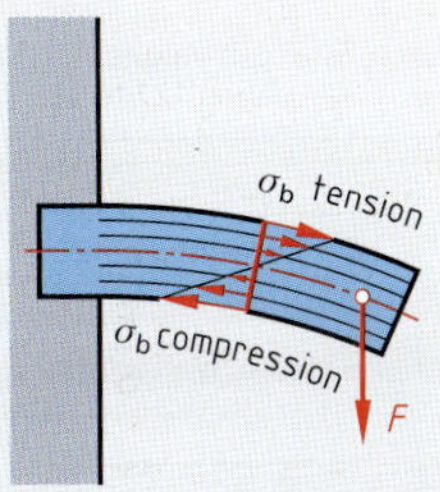

For strength values R_e, see pages 127 to 131.

The maximum tensile or compressive stress in the outer zone of the component is calculated as bending stress.

σ_b	bending stress	R_e	yield strength
M_b	bending moment	F	bending force

W axial section modulus (page 44)
W_{req} required axial section modulus
σ_{all} allowable bending stress (guideline values, p. 41)
σ_{bY} bending yield strength ($\sigma_{bY} \approx 1.2 \cdot R_e$ for steel)
ν safety factor (guideline value $\nu = 1.5$)

Example:

Axle, S275J0, $d = 70$ mm, statically loaded; $\sigma_{ball} = ?$
$R_e = 245\ \text{N/mm}^2$ (page 127)

$$\boldsymbol{\sigma_{ball}} = \frac{\sigma_{bY}}{\nu} = \frac{1.2 \cdot 245\ \text{N/mm}^2}{1.5} = \mathbf{196\ \frac{N}{mm^2}}$$

Bending stress

$$\sigma_b = \frac{M_b}{W}$$

Required section modulus

$$W_{req} = \frac{M_b}{\sigma_{ball}}$$

Allowable bending stress[1)]

$$\sigma_{ball} = \frac{\sigma_{bY}}{\nu}$$

Bending load cases due to a single force acting on a component

Beam restrained at one end

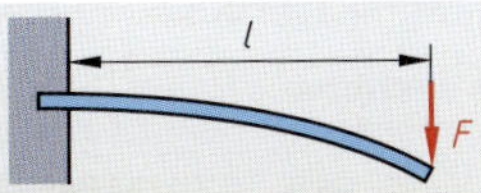

$$M_b = F \cdot l$$

Beam supported by two columns

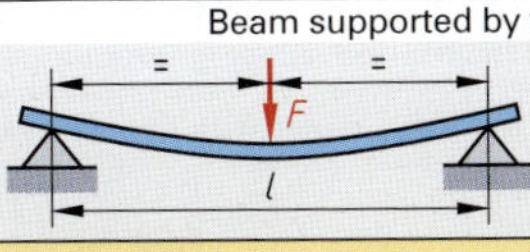

$$M_b = \frac{F \cdot l}{4}$$

Torsional stress

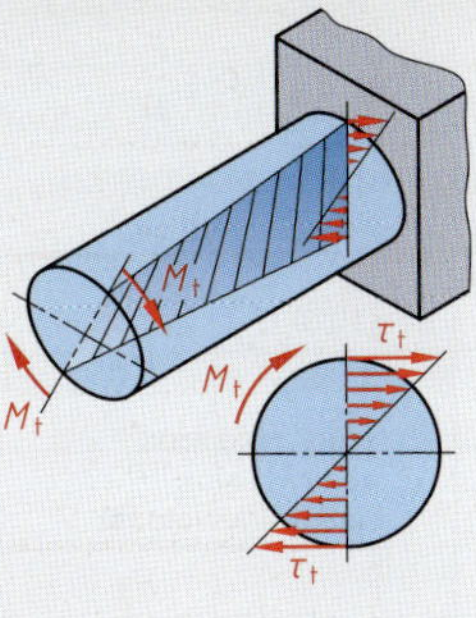

For strength values R_e, see pages 127 to 131.

The maximum stress in the outer zone of the component is calculated as torsional stress.

τ_t	torsional stress	R_e	yield strength

M_t torsional moment
W_p polar section modulus
W_{preq} required polar section modulus
τ_{tall} allowable torsional stress (guideline values, p. 41)
τ_{tY} torsional yield strength ($\tau_{tY} \approx 0.65 \cdot R_e$ for steel)
ν safety factor (guideline value $\nu = 1.5$)

Example:

Shaft, $d = 32$ mm, $M_t = 420$ Nm; $W_p = ?$, $\tau_t = ?$

$$\boldsymbol{W_p} = \frac{\pi \cdot d^3}{16} = \frac{\pi \cdot (32\ \text{mm})^3}{16} = \mathbf{6434\ mm^3}$$

$$\boldsymbol{\tau_t} = \frac{M_t}{W_p} = \frac{420000\ \text{N} \cdot \text{mm}}{6434\ \text{mm}^3} = \mathbf{65.3\ \frac{N}{mm^2}}$$

Torsional stress

$$\tau_t = \frac{M_t}{W_p}$$

Required polar section modulus

$$W_{preq} = \frac{M_t}{\tau_{tall}}$$

Allowable torsional stress[1)]

$$\tau_{tall} = \frac{\tau_{tY}}{\nu}$$

1) The calculation of the allowable stress is only valid for static load on non-brittle materials.

P

Moments of area, Section moduli, Comparison of cross sectional shapes

Moments of area and section moduli[1]

Shape of the cross-section	Bending and Buckling: Second moment of area I	Bending and Buckling: Axial section modulus W	Torsion: Polar section modulus W_p
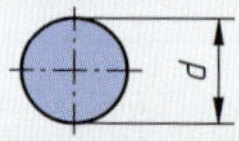	$I = \frac{\pi \cdot d^4}{64}$	$W = \frac{\pi \cdot d^3}{32}$	$W_p = \frac{\pi \cdot d^3}{16}$
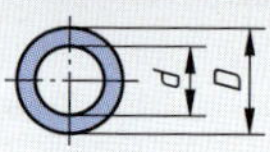	$I = \frac{\pi \cdot (D^4 - d^4)}{64}$	$W = \frac{\pi \cdot (D^4 - d^4)}{32 \cdot D}$	$W_p = \frac{\pi \cdot (D^4 - d^4)}{16 \cdot D}$
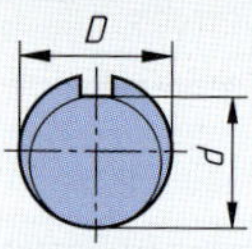	$I = 0.003 \cdot (D + d)^4$	$W = 0.012 \cdot (D + d)^3$	$W_p = 0.2 \cdot d^3$
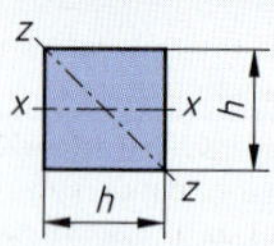	$I_x = I_z = \frac{h^4}{12}$	$W_x = \frac{h^3}{6}$ $W_z = \frac{\sqrt{2} \cdot h^3}{12}$	$W_p = 0.208 \cdot h^3$
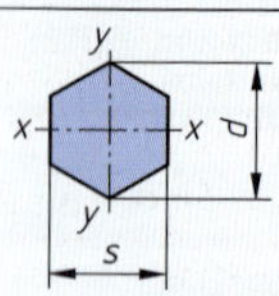	$I_x = I_y = \frac{5 \cdot \sqrt{3} \cdot s^4}{144}$ $I_x = I_y = \frac{5 \cdot \sqrt{3} \cdot d^4}{256}$	$W_x = \frac{5 \cdot s^3}{48} = \frac{5 \cdot \sqrt{3} \cdot d^3}{128}$ $W_y = \frac{5 \cdot s^3}{24 \cdot \sqrt{3}} = \frac{5 \cdot d^3}{64}$	$W_p = 0.188 \cdot s^3$ $W_p = 0.123 \cdot d^3$
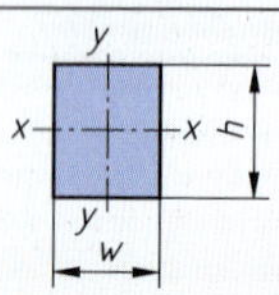	$I_x = \frac{w \cdot h^3}{12}$ $I_y = \frac{h \cdot w^3}{12}$	$W_x = \frac{w \cdot h^2}{6}$ $W_y = \frac{h \cdot w^2}{6}$	–

[1] For second moments of area and axial section moduli for profiles, see pages 145 to 150 and 167 to 169.

Comparison of the load-bearing moments in various shapes of cross sections

Cross section							
	Shape	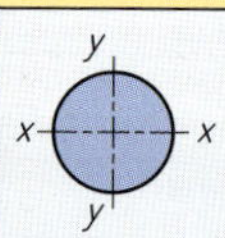	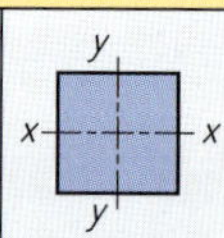	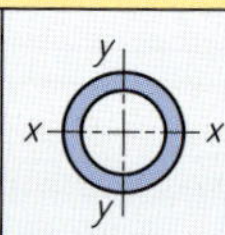	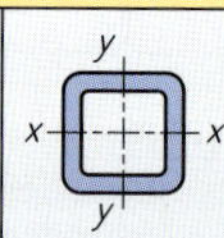	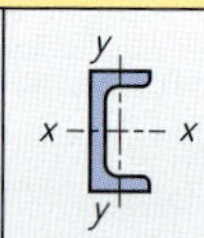	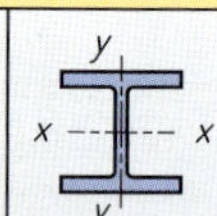
	Standard designation	Round bar EN 10060-100	Square bar EN 10059-100	Pipe EN 10305-100 x 4	Hollow square section EN 10210-2 100x100x6,3	U-channel section DIN 1026-U100	I-beam section DIN 1025-IPB100
Length-related mass	Factor[2] (kg/m)	**1.00** (61.7)	1.27 (78.5)	0.15 (9.4)	0.3 (18.3)	0.17 (10.6)	0.33 (20.4)
Bending W_x	Factor[2] (cm³)	**1.00** (98)	1.7 (167)	0.28 (27.8)	0.69 (67.8)	0.42 (41.2)	0.92 (89.9)
Bending W_y	Factor[2] (cm³)	**1.00** (98)	1.7 (167)	0.28 (27.8)	0.69 (67.8)	0.08 (8.5)	0.34 (33.5)
Torsion W_p	Factor[2] (cm³)	**1.00** (196)	1.06 (208)	0.28 (55.6)	0.56 (110)	–	–
Buckling I_{min}	Factor[2] (cm⁴)	**1.00** (491)	1.7 (833)	0.28 (139)	0.69 (339)	0.06 (29.3)	0.34 (167)

[2] Factor, referenced to round bar EN 10060-100 (cross section in first column of table)

Effects of changes in temperature

Temperature

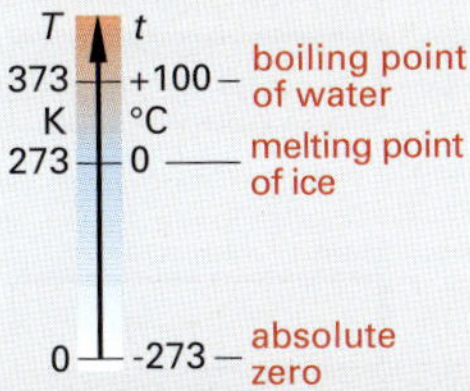

Temperatures are measured in **Kelvin** (K), **degrees Celsius** (Centigrade, °C) or **degrees Fahrenheit** (°F). The Kelvin scale originates at the lowest possible temperature, absolute zero; the origin of the Celsius scale is at the melting point of ice.

T temperature in K (thermodynamic temperature)
t, ϑ temperature in °C
t_F temperature in °F

Example:

$t = 20$ °C; $T = ?$
$T = t + 273 = (20 + 273)$ K $=$ **293 K**

Temperature in Kelvin

$$T = t + 273$$

Temperature in degrees Fahrenheit

$$t_F = 1.8 \cdot t + 32$$

Linear expansion, Change in diameter

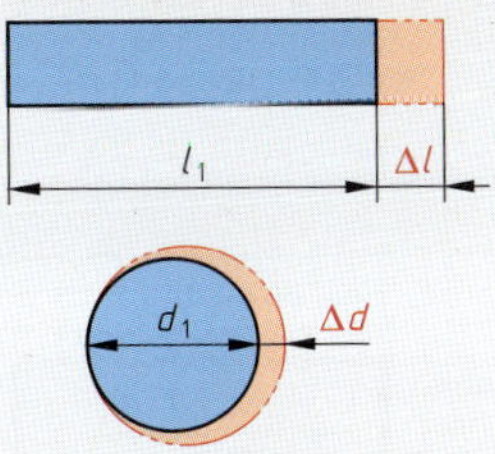

α_l coefficient of linear expansion
Δt, $\Delta\vartheta$ temperature change
Δl linear expansion
Δd change in diameter
l_1 initial length
d_1 initial diameter

Example:

Plate of unalloyed steel, $l_1 = 120$ mm; $\alpha_l = 0.000\,0119 \frac{1}{°C}$
$\Delta t = 550$ °C; $\Delta l = ?$
$\Delta l = \alpha_l \cdot l_1 \cdot \Delta t$
$= 0.000\,0119 \frac{1}{°C} \cdot 120\text{ mm} \cdot 550\text{ °C} =$ **0.785 mm**

Linear expansion

$$\Delta l = \alpha_l \cdot l_1 \cdot \Delta t$$

Change in diameter

$$\Delta d = \alpha_l \cdot d_1 \cdot \Delta t$$

For coefficients of linear expansion see pages 112 and 113

Change in volume

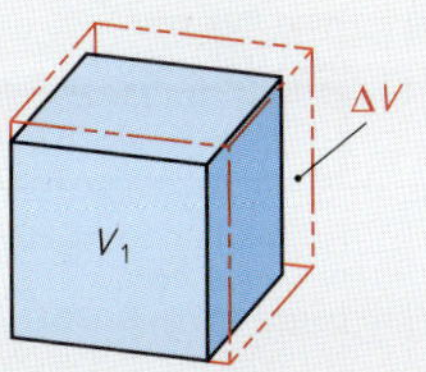

α_V coefficient of volumetric expansion
Δt, $\Delta\vartheta$ temperature change
ΔV change in volume
V_1 initial volume

Example:

Gasoline, $V_1 = 60$ l; $\alpha_V = 0.001 \frac{1}{°C}$; $\Delta t = 32$ °C; $\Delta V = ?$

$\Delta V = \alpha_V \cdot V_1 \cdot \Delta t = 0.001 \frac{1}{°C} \cdot 60\text{ l} \cdot 32\text{ °C} =$ **1.9 l**

Change in volume

$$\Delta V = \alpha_V \cdot V_1 \cdot \Delta t$$

For solids
$\alpha_V = 3 \cdot \alpha_l$

For coefficients of volumetric expansion see page 113.
For volumetric expansion of gases see page 39.

Shrinkage

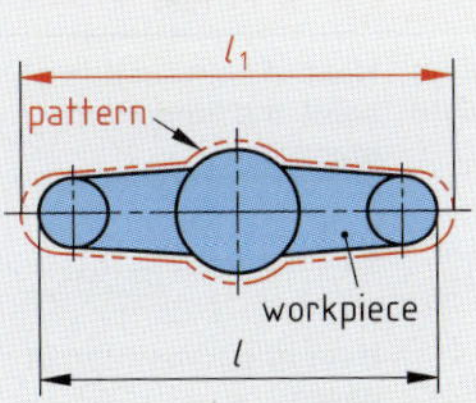

S shrinkage allowance in %
l workpiece length
l_1 pattern length

Example:

Al casting, $l = 680$ mm; $S = 1.2$ %; $l_1 = ?$

$l_1 = \frac{l \cdot 100\,\%}{100\,\% - S} = \frac{680\text{ mm} \cdot 100\,\%}{100\,\% - 1.2\,\%}$
$=$ **688.2 mm**

Pattern length

$$l_1 = \frac{l \cdot 100\,\%}{100\,\% - S}$$

For shrinkage allowances see page 162.

Quantity of heat with changes in temperature

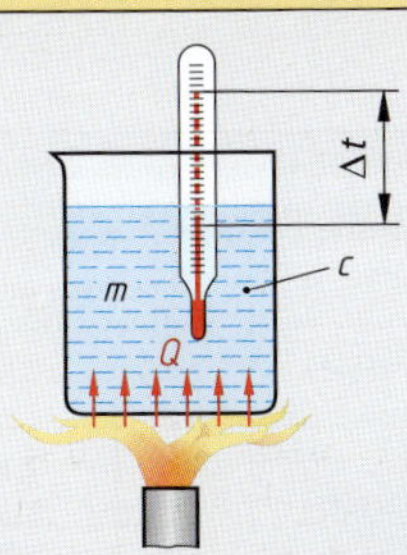

The **specific heat *c*** indicates how much heat is needed to warm up 1 kg of a substance by 1 °C. The same quantity of heat is released again during cooling.

c spec. heat capacity
Δt, $\Delta\vartheta$ temperature change
Q quantity of heat
m mass

Example:

Steel shaft, $m = 2$ kg; $c = 0.48 \frac{\text{kJ}}{\text{kg} \cdot °C}$;
$\Delta t = 800$ °C; $Q = ?$
$Q = c \cdot m \cdot \Delta t = 0.48 \frac{\text{kJ}}{\text{kg} \cdot °C} \cdot 2\text{ kg} \cdot 800\text{ °C} =$ **768 kJ**

Quantity of heat

$$Q = c \cdot m \cdot \Delta t$$

$1\text{ kJ} = \frac{1\text{ kW} \cdot \text{h}}{3600}$

$1\text{ kW} \cdot \text{h} = 3.6$ MJ

For specific heat see pages 112 and 113.

Heat for Melting, Vaporizing, Combustion

Heat of fusion, Heat of vaporization

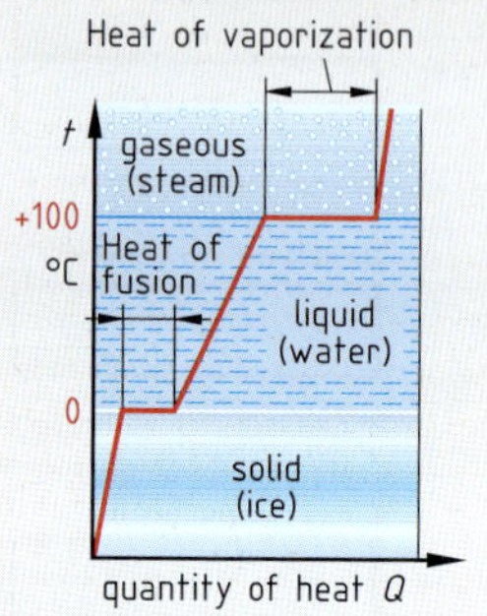

Heat energy is necessary to transform substances from a solid state to a liquid state or from a liquid state to a gaseous state. This is known as the heat of fusion or heat of vaporization.

Q	heat of fusion heat of evaporation	r	specific heat of evaporation
q	specific heat of fusion	m	mass

Example:

Copper, $m = 6.5\ \text{kg};\ q = 213\ \frac{\text{kJ}}{\text{kg}};\ Q = ?$

$\boldsymbol{Q} = q \cdot m = 213\ \frac{\text{kJ}}{\text{kg}} \cdot 6.5\ \text{kg} = 1384.5\ \text{kJ} \approx \mathbf{1.4\ MJ}$

Heat of fusion

$$Q = q \cdot m$$

Heat of vaporization

$$Q = r \cdot m$$

For specific heat of fusion and heat of evaporation see pages 112 and 113.

Heat flux

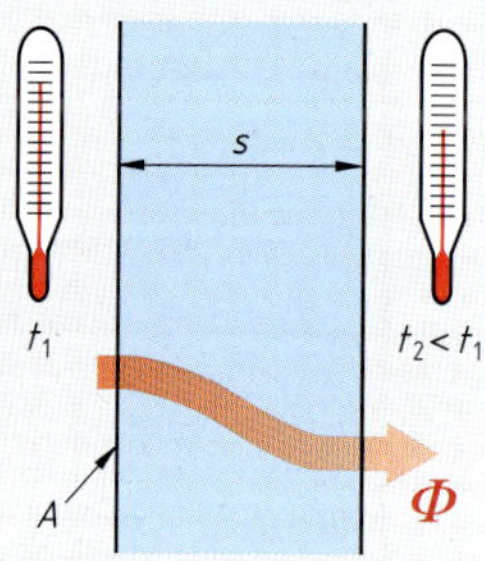

The **heat flux** Φ continually occurs within a substance with movement from higher to lower temperatures.

The **heat transmission coefficient** $\boldsymbol{k}$ also compensates, along with the thermal conductivity of a part, for the heat transmission resistance on the surfaces of the part.

Φ	heat flux	$\Delta t,\ \Delta\vartheta$	temperature difference
λ	thermal conductivity	s	component thickness
k	heat transmission coefficient	A	area of the component

Example:

Heat protection glass, $k = 1.9\ \frac{\text{W}}{\text{m}^2 \cdot {}^\circ\text{C}};\ A = 2.8\ \text{m}^2;$
$\Delta t = 32\ {}^\circ\text{C};\ \Phi = ?$

$\boldsymbol{\Phi} = k \cdot A \cdot \Delta t = 1.9\ \frac{\text{W}}{\text{m}^2 \cdot {}^\circ\text{C}} \cdot 2.8\ \text{m}^2 \cdot 32\ {}^\circ\text{C} = \mathbf{170\ W}$

Heat flux with thermal conduction

$$\Phi = \frac{\lambda \cdot A \cdot \Delta t}{s}$$

Heat flux with heat transmission

$$\Phi = k \cdot A \cdot \Delta t$$

For thermal conductivity values λ see pages 112 and 113. For heat transmission coefficients k see below.

Heat of combustion

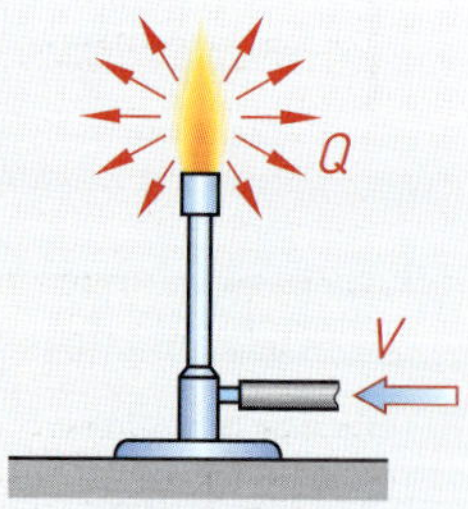

The **net calorific value** $\boldsymbol{H_{net}}$ **(*H*)** of a substance refers to the heat quantity released during the complete combustion of 1 kg or 1 m^3 of that substance.

Q	heat of combustion
H_{net}, H	net calorific value
m	mass of solid and liquid fuels
V	volume of fuel gas

Example:

Natural gas, $V = 3.8\ \text{m}^3;\ H_{net} = 35\ \frac{\text{MJ}}{\text{m}^3};\ Q = ?$

$\boldsymbol{Q} = H_{net} \cdot V = 35\ \frac{\text{MJ}}{\text{m}^3} \cdot 3.8\ \text{m}^3 = \mathbf{133\ MJ}$

Heat of combustion of solid and liquid substances

$$Q = H_{net} \cdot m$$

Heat of combustion of gases

$$Q = H_{net} \cdot V$$

Net calorific value H_{net} (*H*) for fuels						Heat transmission coefficients *k* for construction materials and parts		
Solid fuels	H_{net} MJ/kg	**Liquid fuels**	H_{net} MJ/kg	**Gaseous fuels**	H_{net} MJ/m^3	**Construction elements**	s mm	k $\frac{W}{m^2 \cdot {}^\circ C}$
wood	15–17	alcohol	27	hydrogen	10	outer door, steel	50	5.8
biomass (dry)	14–18	benzene	40	natural gas	34–36	sash window	12	1.3
brown coal	16–20	gasoline	43	acetylene	57	brick wall	365	1.1
coke	30	diesel	41–43	propane	93	intermediate floor	125	3.2
pit coal	30–34	fuel oil	40–43	butane	123	heat insulating board	80	0.39

Quantities and Units, Ohm's Law, Resistance

Electrical quantities and units

Quantity		Unit	
Name	Symbol	Name	Symbol
electrical voltage	E	volt	V
electric current	I	ampere	A
electrical resistance	R	ohm	Ω
electrical conductance	G	Siemens	S
electrical power	P	watt	W

$$1\,\Omega = \frac{1\,\text{V}}{1\,\text{A}}$$

$$1\,\text{W} = 1\,\text{V} \cdot 1\,\text{A}$$

Ohm's Law

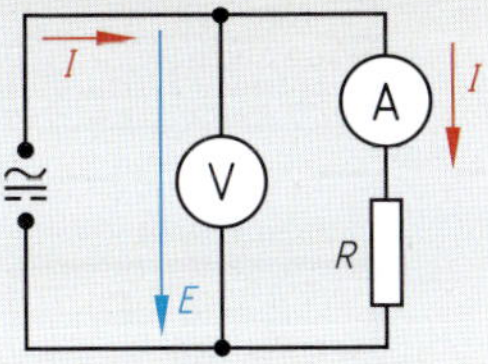

E voltage in V
I electric current in A
R resistance in Ω

Example:

$R = 88\,\Omega$; $E = 230\,\text{V}$; $I = ?$

$\boldsymbol{I} = \frac{E}{R} = \frac{230\,\text{V}}{88\,\Omega} = \mathbf{2.6\,A}$

Electric current

$$I = \frac{E}{R}$$

For circuit symbols see page 375.

Electrical resistance and conductance

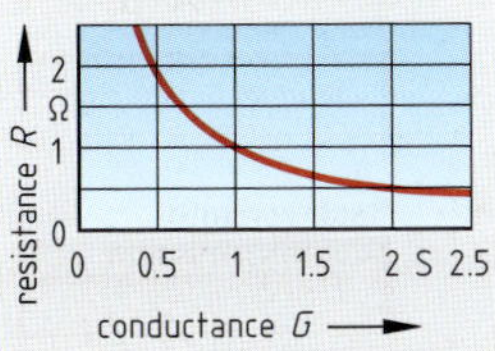

R resistance in Ω
G conductance in S

Example:

$R = 20\,\Omega$; $G = ?$

$\boldsymbol{G} = \frac{1}{R} = \frac{1}{20\,\Omega} = \mathbf{0.05\,S}$

Resistance

$$R = \frac{1}{G}$$

Conductance

$$G = \frac{1}{R}$$

Electrical resistivity, electrical conductivity, conductor resistance

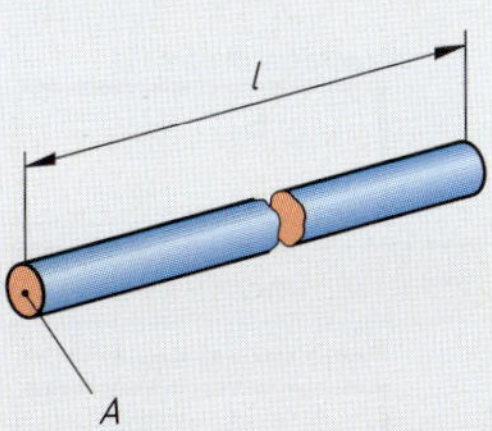

ϱ electrical resistivity in $\Omega \cdot \text{mm}^2/\text{m}$
γ electrical conductivity in $\text{m}/(\Omega \cdot \text{mm}^2)$
R resistance in Ω
A wire cross section in mm^2
l wire length in m

Example:

Copper wire, $l = 100\,\text{m}$;
$A = 1.5\,\text{mm}^2$; $\varrho = 0.0179\,\frac{\Omega \cdot \text{mm}^2}{\text{m}}$; $R = ?$

$\boldsymbol{R} = \frac{\varrho \cdot l}{A} = \frac{0.0179\,\frac{\Omega \cdot \text{mm}^2}{\text{m}} \cdot 100\,\text{m}}{1.5\,\text{mm}^2} = \mathbf{1.19\,\Omega}$

For electrical resistivities, see pages 112 and 113.

Electrical resistivity

$$\varrho = \frac{1}{\gamma}$$

Conductor resistance

$$R = \frac{\varrho \cdot l}{A}$$

Resistance and Temperature

Material	T_k value α in 1/K
aluminum	0.0040
lead	0.0039
gold	0.0037
copper	0.0039
silver	0.0038
tungsten	0.0044
tin	0.0045
zinc	0.0042
graphite	– 0.0013
constantan	± 0.00001

ΔR change in resistance in Ω
R_{20} resistance at 20 °C in Ω
R_t resistance at the temperature t in Ω
α temperature coefficient (T_k value) in 1/K
Δt temperature difference in K

Example:

Resistance of Cu; $R_{20} = 150\,\Omega$; $t = 75\,°\text{C}$; $R_t = ?$

$\boldsymbol{\alpha} = \mathbf{0.0039\,1/K}$; $\boldsymbol{\Delta t} = 75\,°\text{C} - 20\,°\text{C} = 55\,°\text{C} \mathrel{\hat{=}} \mathbf{55\,K}$

$\boldsymbol{R_t} = R_{20} \cdot (1 + \alpha \cdot \Delta t)$

$= 150\,\Omega \cdot (1 + 0.0039\,1/\text{K} \cdot 55\,\text{K}) = \mathbf{182.2\,\Omega}$

Change in resistance

$$\Delta R = \alpha \cdot R_{20} \cdot \Delta t$$

Resistance at temperature *t*

$$R_t = R_{20} + \Delta R$$

$$R_t = R_{20} \cdot (1 + \alpha \cdot \Delta t)$$

P

Current density, Resistor circuits

Current density in wires

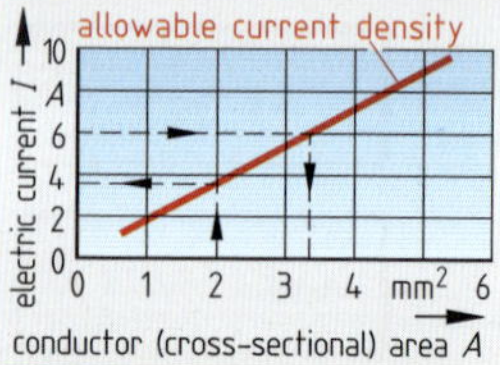

J current density in A/mm²
I electric current in A
A conductor cross section in mm²

Example:

$A = 2.5\ \text{mm}^2;\ I = 4\ \text{A};\ J = ?$

$J = \frac{I}{A} = \frac{4\ \text{A}}{2.5\ \text{mm}^2} = \mathbf{1.6\ \frac{A}{mm^2}}$

Current density

$$J = \frac{I}{A}$$

Voltage drop in wires

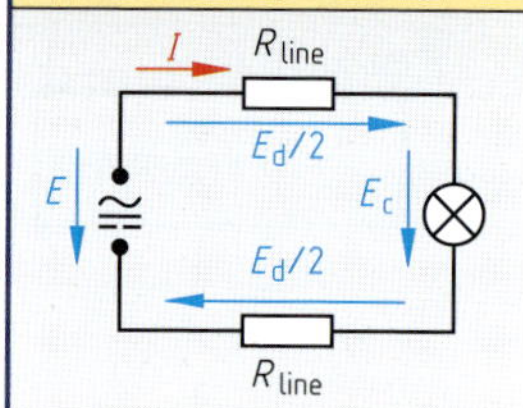

E_d voltage drop in wire in V
E voltage at terminal in V
E_c voltage across load in V
I electric current in A
R_{line} resistance for feed or return line in Ω

Voltage drop

$$E_d = 2 \cdot I \cdot R_{line}$$

Voltage at load

$$E_c = E - E_d$$

Series resistor circuit

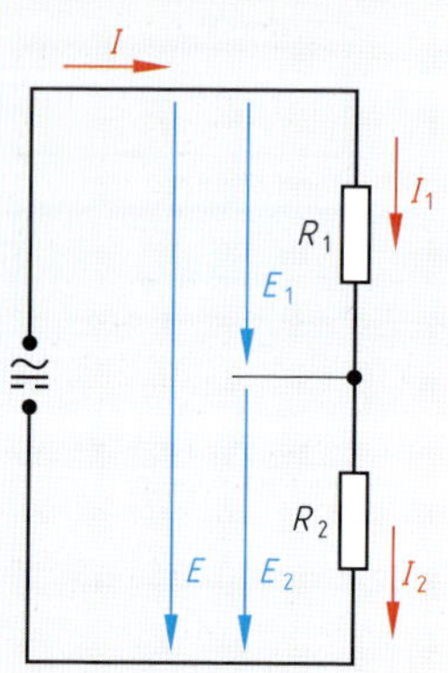

R total resistance, equivalent resistance in Ω
I total current in A
E total voltage in V
R_1, R_2 individual resistances in Ω
I_1, I_2 partial current in A
E_1, E_2 voltage drop across R_1 & R_2 in V

Example:

$R_1 = 10\ \Omega;\ R_2 = 20\ \Omega;\ E = 12\ \text{V};\ R = ?;\ I = ?;$
$E_1 = ?;\ E_2 = ?$

$\boldsymbol{R} = R_1 + R_2 = 10\ \Omega + 20\ \Omega = \mathbf{30\ \Omega}$

$\boldsymbol{I} = \frac{E}{R} = \frac{12\ \text{V}}{30\ \Omega} = \mathbf{0.4\ A}$

$\boldsymbol{E_1} = R_1 \cdot I = 10\ \Omega \cdot 0.4\ \text{A} = \mathbf{4\ V}$

$\boldsymbol{E_2} = R_2 \cdot I = 20\ \Omega \cdot 0.4\ \text{A} = \mathbf{8\ V}$

Total resistance

$$R = R_1 + R_2 + \ldots$$

Total voltage

$$E = E_1 + E_2 + \ldots$$

Total current

$$I = I_1 = I_2 = \ldots$$

Voltage drops

$$\frac{E_1}{E_2} = \frac{R_1}{R_2}$$

Parallel resistor circuit

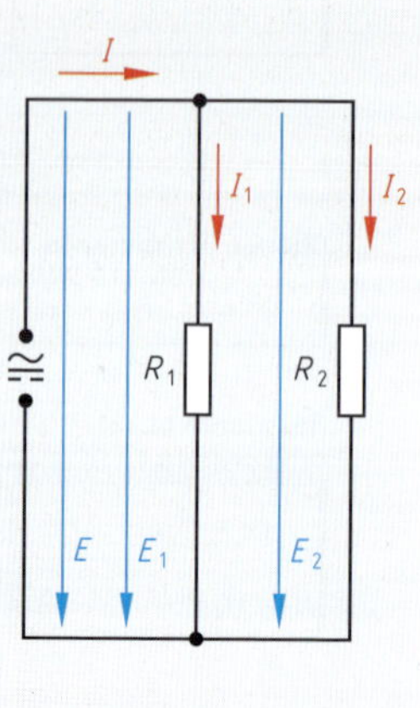

R total resistance, equivalent resistance in Ω
I total current in A
E total voltage in V
R_1, R_2 individual resistances in Ω
I_1, I_2 partial current in A
E_1, E_2 voltage drop across R_1 & R_2 in V

Example:

$R_1 = 15\ \Omega;\ R_2 = 30\ \Omega;\ E = 12\ \text{V};\ R = ?;\ I = ?;$
$I_1 = ?;\ I_2 = ?$

$\boldsymbol{R} = \frac{R_1 \cdot R_2}{R_1 + R_2} = \frac{15\ \Omega \cdot 30\ \Omega}{15\ \Omega + 30\ \Omega} = \mathbf{10\ \Omega}$

$\boldsymbol{I} = \frac{E}{R} = \frac{12\ \text{V}}{10\ \Omega} = \mathbf{1.2\ A}$

$\boldsymbol{I_1} = \frac{E_1}{R_1} = \frac{12\ \text{V}}{15\ \Omega} = \mathbf{0.8\ A};$ $\boldsymbol{I_2} = \frac{E_2}{R_2} = \frac{12\ \text{V}}{30\ \Omega} = \mathbf{0.4\ A}$

1) Use this formula if there are only two parallel resistors in the circuit.

Total resistance

$$\frac{1}{R} = \frac{1}{R_1} + \frac{1}{R_2} + \ldots$$

$$R^{1)} = \frac{R_1 \cdot R_2}{R_1 + R_2}$$

Total voltage

$$E = E_1 = E_2 = \ldots$$

Total current

$$I = I_1 + I_2 + \ldots$$

Partial currents

$$\frac{I_1}{I_2} = \frac{R_2}{R_1}$$

P

Types of current

Direct current (DC; symbol –), DC voltage

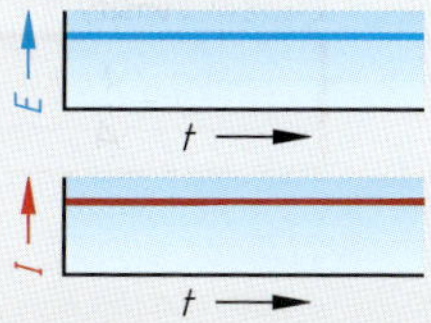

Direct current flows in one direction only and maintains a constant level of current. The voltage is also constant.

I electric current in A
E voltage in V
t time in s

Electric current

$$I = \text{constant}$$

Voltage

$$E = \text{constant}$$

Alternating current (AC); symbol ~), AC voltage

Cycle duration and Frequency

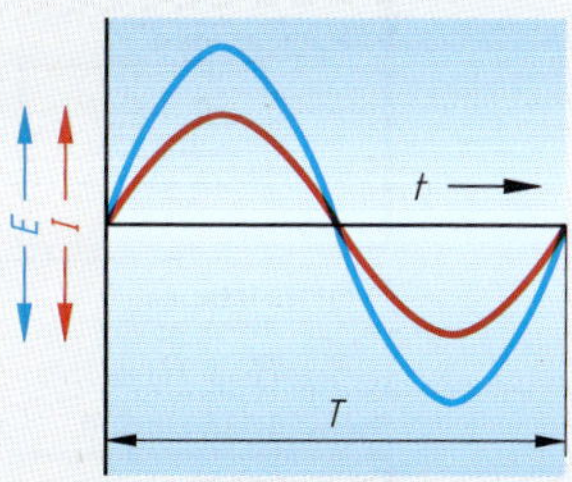

While the voltage is continuously changing in a sinusoidal pattern, the free electrons are also continuously alternating their direction of flow.

f frequency in 1/s, Hz
T period in s
ω angular frequency in 1/s
I electric current in A
E voltage in V
t time in s

Example:

Frequency 50 Hz; $T = ?$

$$\boldsymbol{T} = \frac{1}{50\,\frac{1}{\text{s}}} = \mathbf{0.02\ s}$$

Cycle duration

$$T = \frac{1}{f}$$

Frequency

$$f = \frac{1}{T}$$

Angular frequency

$$\omega = 2 \cdot \pi \cdot f$$

$$\omega = \frac{2 \cdot \pi}{T}$$

1 Hertz = 1 Hz = 1/s = 1 period per second

Maximum value and effective value of current and voltage

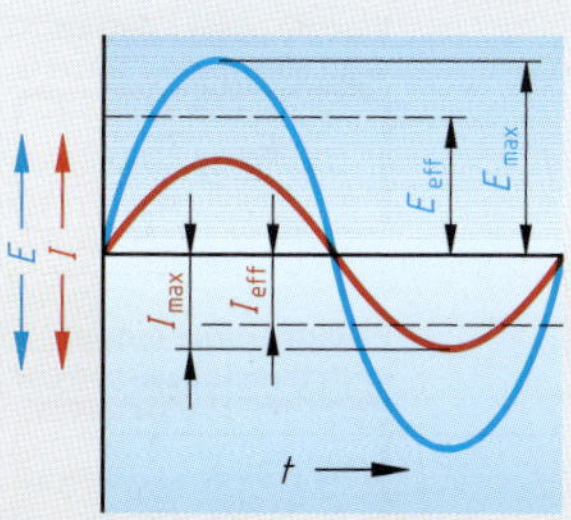

I_{max} maximum value of the electric current in A
I_{eff} effective value of the electric current in A
E_{max} maximum value of the voltage in V
E_{eff} effective value of the voltage in V (voltage that produces the same power as an identical DC voltage across an ohmic resistor).
I electric current in A
E voltage in V
t time in s

Example:

$E_{eff} = 230\ \text{V}$; $E_{max} = ?$

$\boldsymbol{E_{max}} = \sqrt{2} \cdot 230\ \text{V} = \mathbf{325\ V}$

Maximum value of the electric current

$$I_{max} = \sqrt{2} \cdot I_{eff}$$

Maximum value of the voltage

$$E_{max} = \sqrt{2} \cdot E_{eff}$$

Three-phase current

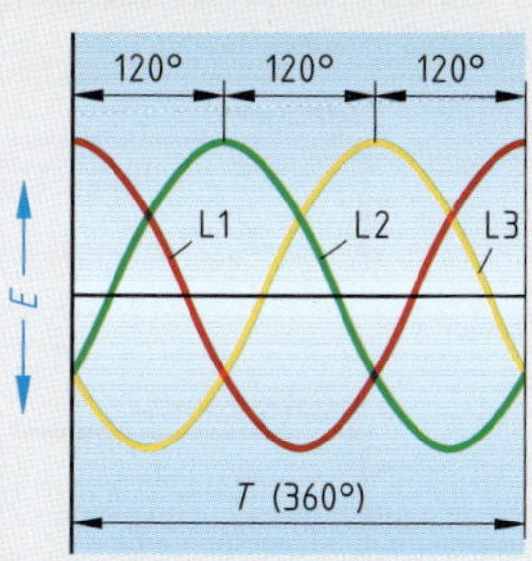

Three-phase current is created from three AC voltages each offset by 120°.

E voltage in V
T period in s
L1 phase 1
L2 phase 2
L3 phase 3

E_{eff} effective voltage between phase wire and neutral wire = 230 V

E_{eff} effective voltage between two phase wires = 400 V

Maximum value of the voltage

$$E_{max} = \sqrt{2} \cdot E_{eff}$$

Electrical Work and Power, Transformers

Electrical work

W electrical work in kW · h
P electrical power in W
t time (power-on time) in h

Example:

Hot plate, $P = 1.8$ kW; $t = 3$ h;
$W = ?$ in kW · h and MJ

$\mathbf{W} = P \cdot t = 1.8 \text{ kW} \cdot 3 \text{ h} = \mathbf{5.4 \text{ kW} \cdot h = 19.44 \text{ MJ}}$

Electrical work

$$W = P \cdot t$$

1 kW · h = 3.6 MJ
= 3 600 000 W · s

Electrical power with non-inductive load[1)]

Direct or alternating current

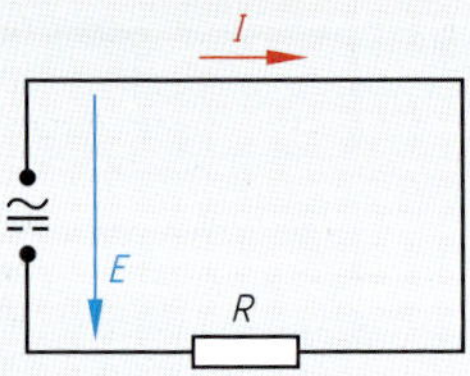

Three-phase current

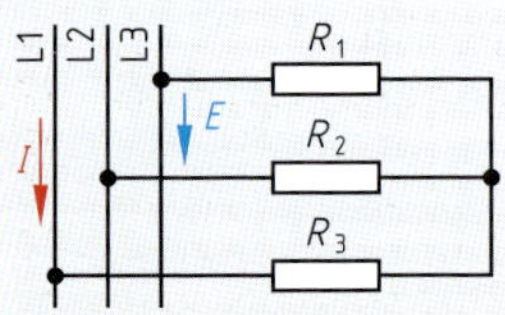

P electrical power in W
E voltage (phase-to-phase voltage) in V
I electric current in A
R resistance in Ω

1st example:

Light bulb, $E = 6$ V; $I = 5$ A; $P = ?$; $R = ?$
$\mathbf{P} = E \cdot I = 6 \text{ V} \cdot 5 \text{A} = \mathbf{30 \text{ W}}$
$\mathbf{R} = \frac{E}{I} = \frac{6 \text{ V}}{5 \text{ A}} = \mathbf{1.2 \; \Omega}$

2nd example:

Annealing furnace, three-phase current,
$E = 400$ V; $P = 12$ kW; $I = ?$
$\mathbf{I} = \frac{P}{\sqrt{3} \cdot E} = \frac{12\,000 \text{ W}}{\sqrt{3} \cdot 400 \text{ V}} = \mathbf{17.3 \text{ A}}$

1) i. e. only with heating devices (ohmic resistors)

Power with direct or alternating current

$$P = E \cdot I$$
$$P = I^2 \cdot R$$
$$P = \frac{E^2}{R}$$

Power with three-phase current

$$P = \sqrt{3} \cdot E \cdot I$$

Electric power output with alternating and three-phase current with reactive or capacitive load component I[2)]

Alternating current

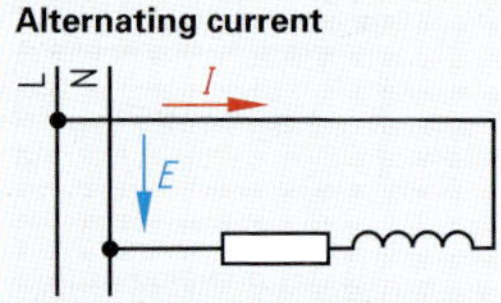

Three-phase current

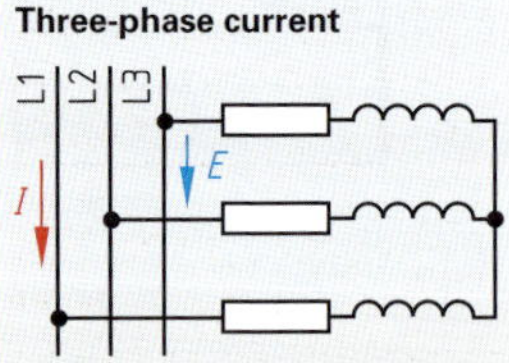

P electrical power output in W
E voltage (phase-to-phase voltage) in V
I electric current in A
cos φ power factor

Example:

Three-phase motor, $E = 400$ V; $I = 2$ A;
$\cos\varphi = 0.85$; $P = ?$

$\mathbf{P} = \sqrt{3} \cdot E \cdot I \cdot \cos\varphi = \sqrt{3} \cdot 400 \text{ V} \cdot 2 \text{ A} \cdot 0.85$
$= 1178 \text{ W} \approx \mathbf{1.2 \text{ kW}}$

2) i. e. in electric motors and generators

Electric power output with alternating current

$$P = E \cdot I \cdot \cos\varphi$$

Electric power output with three-phase current

$$P = \sqrt{3} \cdot E \cdot I \cdot \cos\varphi$$

Transformers

Input side (primary coil) **Output side** (secondary coil)

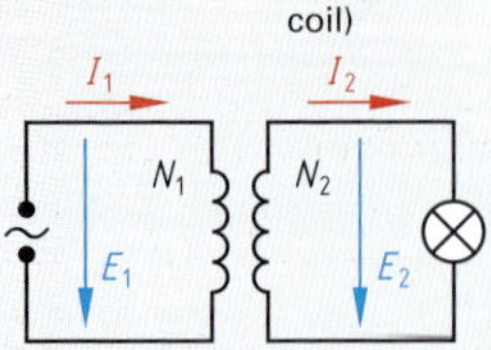

N_1, N_2 number of turns
E_1, E_2 voltages in V
I_1, I_2 current level in A

Example:

$N_1 = 2875$; $N_2 = 100$; $E_1 = 230$ V; $I_1 = 0.25$ A; $E_2 = ?$; $I_2 = ?$

$\mathbf{E_2} = \frac{E_1 \cdot N_2}{N_1} = \frac{230 \text{ V} \cdot 100}{2875} = \mathbf{8 \text{ V}}$

$\mathbf{I_2} = \frac{I_1 \cdot N_1}{N_2} = \frac{0.25 \text{ A} \cdot 2875}{100} = \mathbf{7.2 \text{ A}}$

Voltages

$$\frac{E_1}{E_2} = \frac{N_1}{N_2}$$

Electric current

$$\frac{I_1}{I_2} = \frac{N_2}{N_1}$$

3 Technical drawing

Cartesian coordinate system

cf. DIN 461 (1973-03)

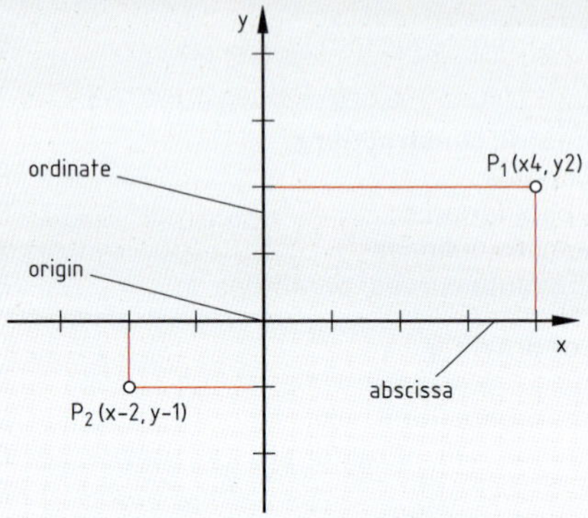

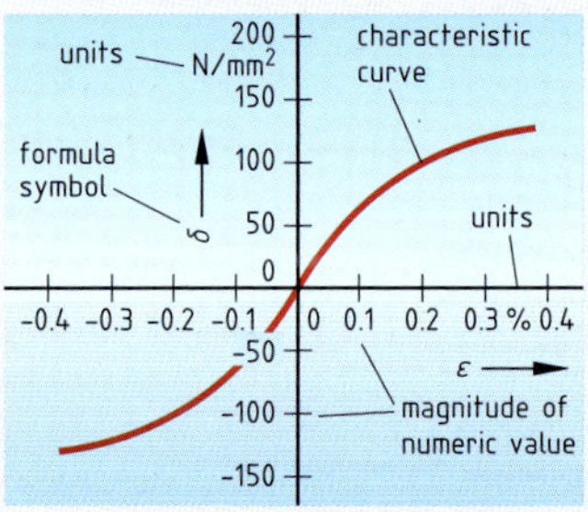

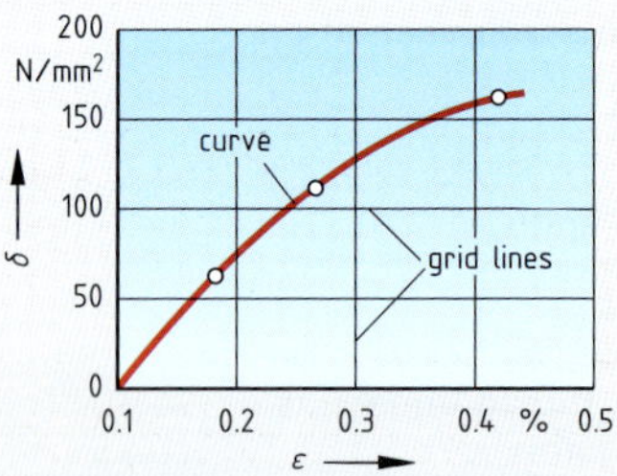

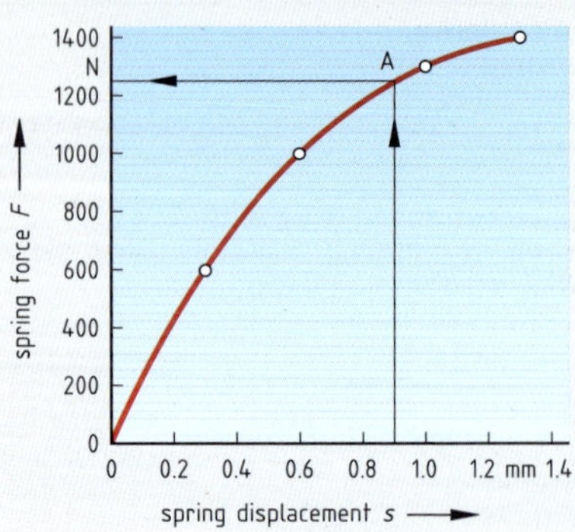

Coordinate axes

- abscissa (horizontal axis; x-axis)
- ordinate (vertical axis; y-axis)

Values to be plotted

- positive: from the origin towards the right, or up
- negative: from the origin towards the left, or down

Marking the positive axis direction with

- arrow heads on the axes, or
- arrows parallel to the axes

Formula symbols are entered in italics on the

- abscissa below the arrow point
- ordinate to the left next to the arrow point

or in front of the arrows parallel to the axes.

Scales are normally linear, but sometimes they are divided logarithmically.

Magnitudes of values. They are placed next to the scale ticks. All negative values have a minus sign.

Value units are placed between the two last positive numbers on the abscissa and ordinate or after the formula symbol.

Grid marks simplify plotting of the values.

Lines (curves) connect the values that have been plotted on the graph.

Line widths. Lines are drawn in the following proportion:
Gridlines : axes : curves = 1 : 2 : 4 .

Graph sections are constructed if values are not to be plotted in each direction from the origin. The origin may also be hidden.

Example (spring characteristic curve):

The following disk spring values are known:

Spring displacement *s* in mm	0	0.3	0.6	1.0	1.3
Spring force *F* in N	0	600	1000	1300	1400

What is the spring force *F* with a spring displacement of $s = 0.9$ mm?

Solution:

The values are plotted on a graph and the points are connected by a curve. A vertical line at $s = 0.9$ mm intersects the curve at point A.

With the help of a horizontal line through A, a spring force of $F \approx 1250$ N is read from the ordinate.

[1] Graphs are used to represent value-based relationships between changing variables.

Polar coordinate systems, Area graphs

Cartesian coordinate system (continued) cf. DIN 461 (1973-03)

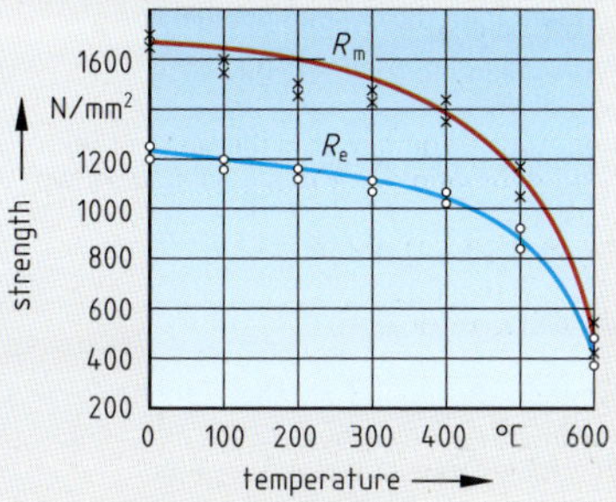

Graphs with multiple curves

When measured values are highly scattered, a different special symbol is used for each curve, e. g: ○, ×, □

Marking the curves

- when the same type of line is used, by using the names or formula symbols of the variables or by using different colors for the curves
- by different types of lines

Polar coordinate system cf. DIN 461 (1973-03)

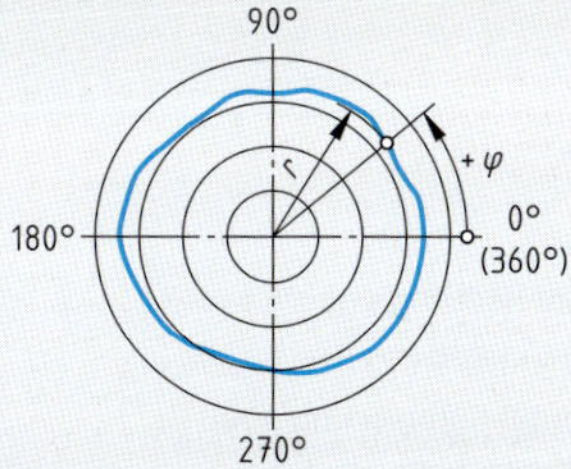

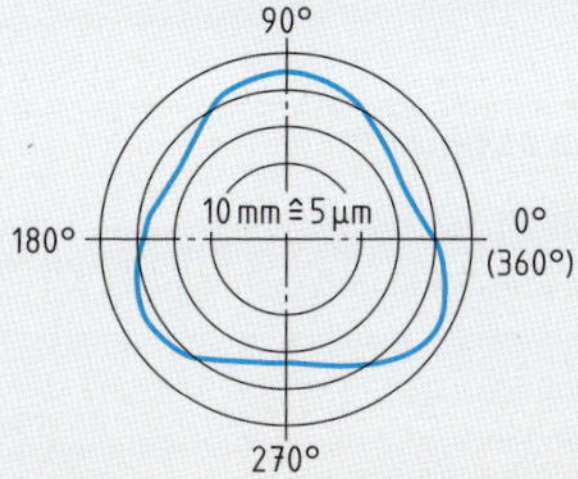

Polar coordinate systems have a 360° division.

Origin (pole). Intersection of horizontal and vertical axis.

Angle layout. The angle 0° is assigned to the horizontal axis to the right of the origin.

Angle position. Positive angles are plotted counter-clockwise.

Radius. The radius corresponds to the magnitude of the value to be plotted. Concentric circles may be drawn about the origin to simplify plotting of the values.

Example:

Using a measuring machine, the roundness of a turned bushing is inspected to see if it lies within the required tolerance.

The out-of-roundness detected was probably caused by clamping the bushing too forcefully in the chuck.

Area graphs

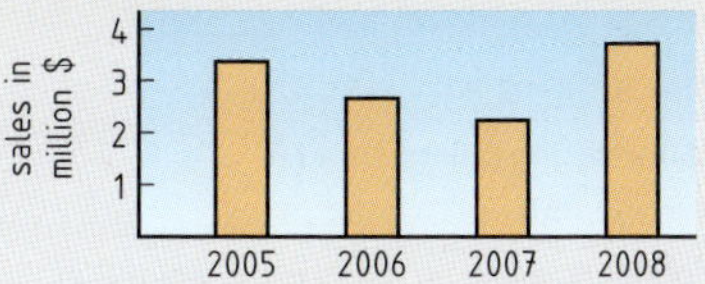

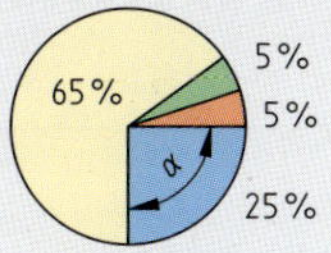

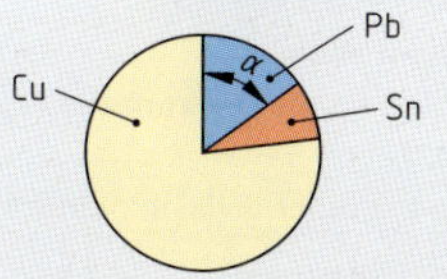

Bar graphs

In bar graphs the quantities to be represented are drawn as horizontal or vertical columns of equal width.

Pie charts

Percent values are normally represented by pie charts. In these the circumference of a circular area corresponds to 100 % (≙ 360°).

Central angle. The percentage x to be plotted determines the corresponding central angle:

$$\alpha = \frac{360° \cdot x\,\%}{100\,\%}$$

Example:

What is the central angle for the percentage of lead in the alloy CuPb15Sn8?

Solution: $\alpha = \frac{360° \cdot 15\,\%}{100\,\%} = 54°$

TD

Line segments, Perpendiculars and Angles

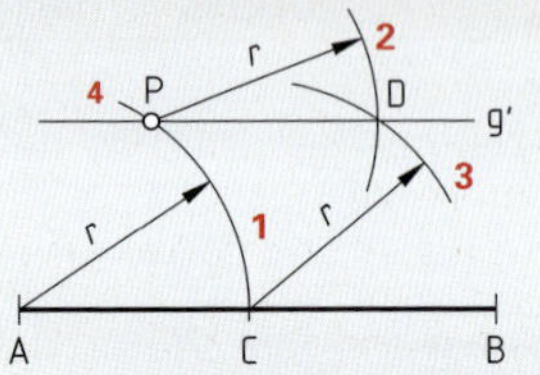

Parallels to a line

Given: Line segment $\overline{AB}$ and point P on the desired parallel line g′

1. Arc with radius *r* about A results in intersecting point C.
2. Arc with radius *r* about P.
3. Arc with radius *r* about C results in intersecting point D.
4. Connecting line segment $\overline{PD}$ is parallel line g′ to $\overline{AB}$.

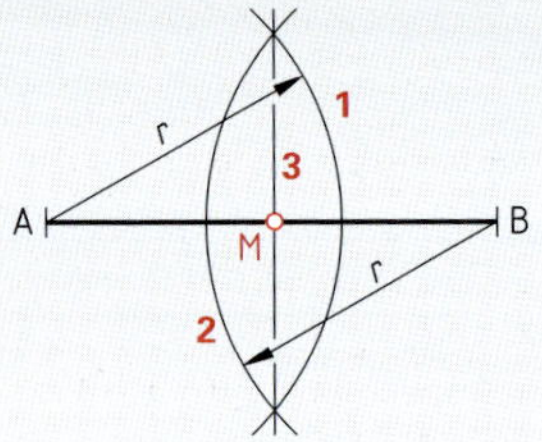

Bisecting a line

Given: Line segment $\overline{AB}$

1. Arc 1 with radius *r* about A; $r > \frac{1}{2}\overline{AB}$.
2. Arc 2 with equal radius *r* about B.
3. The line connecting the intersecting points is the perpendicular bisector or the bisector of line segment $\overline{AB}$.

TD

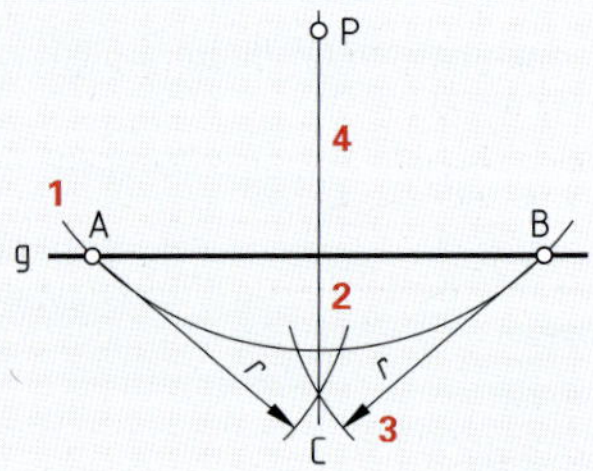

Dropping a perpendicular

Given: Straight line g and point P

1. Any arc 1 about P results in intersecting point A and B.
2. Arc 2 with radius *r* about A; $r > \frac{1}{2}\overline{AB}$.
3. Arc 3 with equal radius *r* about B (intersecting point C).
4. The line joining intersecting point C with P is the desired perpendicular.

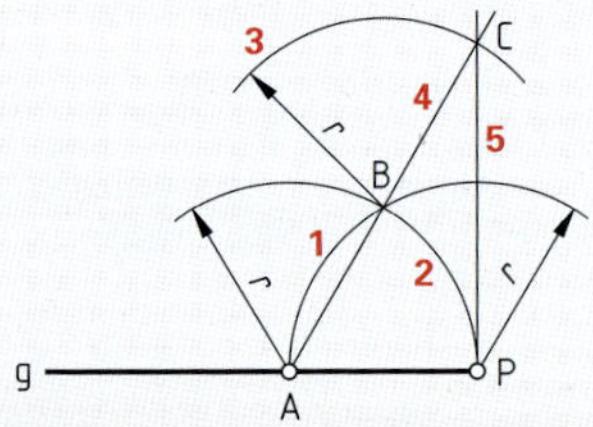

Constructing a vertical line at point P

Given: Straight line g and point P

1. Arc 1 about P with any radius *r* results in intersecting point A.
2. Arc 2 with same radius *r* about point A results in intersecting point B.
3. Arc 3 with equal radius *r* about B.
4. Construct a line from A to B and extend it (to intersecting point C).
5. Construct a line from point C to point P to obtain the vertical at P.

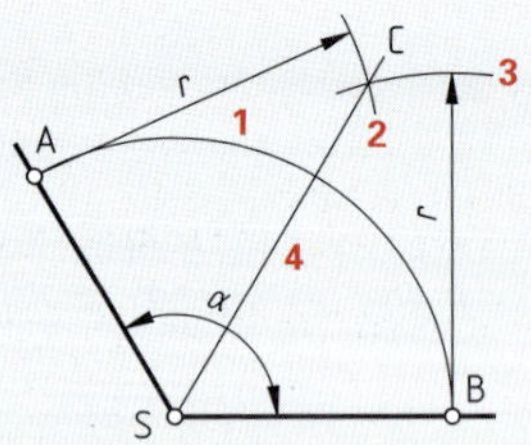

Bisecting an angle

Given: Angle α

1. Any arc 1 about S yields intersecting points A and B.
2. Arc 2 with radius *r* about A; $r > \frac{1}{2}\overline{AB}$.
3. Arc 3 with equal radius *r* about B results in intersecting point C.
4. The line joining intersecting point C with S is the desired bisected angle.

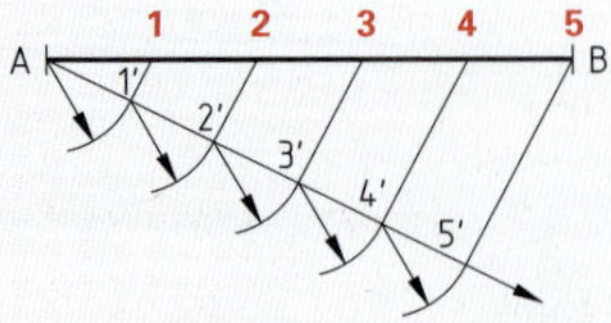

Dividing a line

Given: Line $\overline{AB}$ should be divided into 5 equal parts.

1. Construct a ray from A at any desired angle.
2. Mark 5 equal lengths with a compass on the ray from A.
3. Construct a line from point 5′ to B.
4. Construct parallels to $\overline{5'B}$ through the other division points 1′–4′.

Tangents, Circular arcs, Polygons

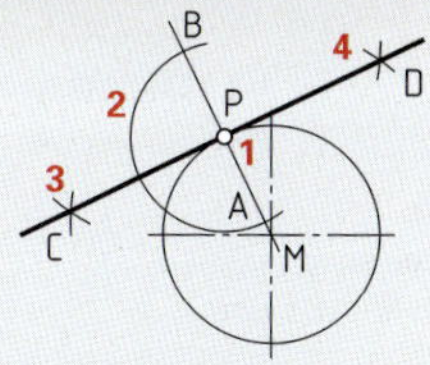

Tangent through point P on a circle

Given: Circle and point P

1. Construct line segment $\overline{MP}$ and extend it.
2. Arc about P gives intersecting points A and B.
3. Arcs about A and B with the same radius yield intersecting points C and D.
4. The line passing through C and D is perpendicular to $\overline{PM}$.

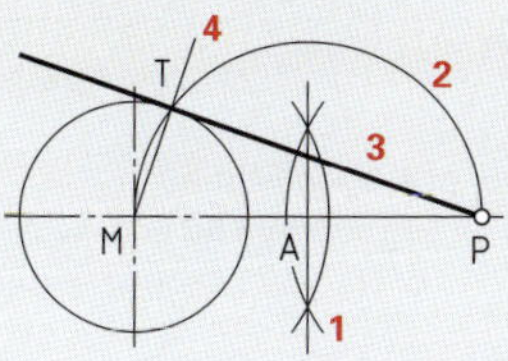

Tangent from a point P to a circle

Given: Circle and point P

1. Bisect $\overline{MP}$. A is the midpoint.
2. Arc about A with radius $r = \overline{AM}$ yields intersecting point P. T is the tangent point.
3. Connect T and P.
4. $\overline{MT}$ is perpendicular to $\overline{PT}$.

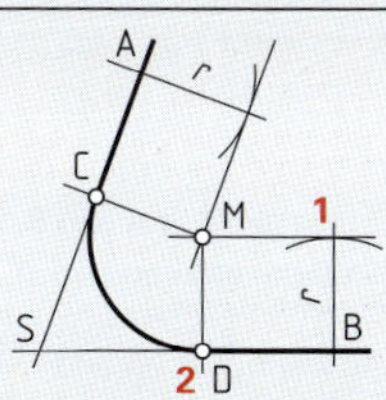

Rounding an angle (arc tangent to two straight lines)

Given: Angle ASB and radius r

1. Construct parallels to $\overline{AS}$ and $\overline{BS}$ of distance r. Their intersection M is the desired center of the circular arc of radius r.
2. The intersection of the perpendiculars from M to the line segments AS and BS are the transition points C and D for the arc.

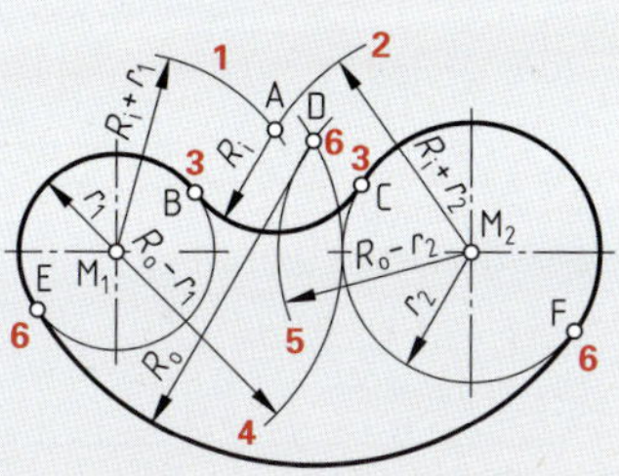

Connecting two circles by arcs

Given: Circle 1 and circle 2; radii R_i and R_o

1. Circle about M_1 with radius $R_i + r_1$.
2. Circle about M_2 with radius $R_i + r_2$ intersects with 1 to yield intersecting point A.
3. Connecting M_1 and M_2 with A yields contact points B and C for the inside radius R_i.
4. Circle about M_1 with radius $R_o - r_1$.
5. Circle about M_2 with radius $R_o - r_2$ combined with step 4 results in the intersecting point D.
6. D connected to M_1 and M_2 and extended gives the contact points E and F for the outside radius R_o.

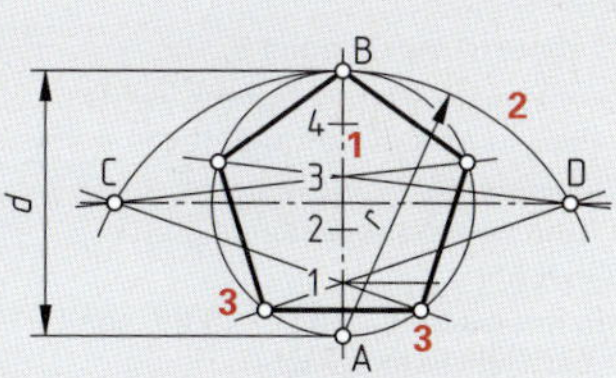

Circumscribed regular polygon (e. g. pentagon)

Given: Circle of diameter d

1. Divide $\overline{AB}$ into 5 equal parts (page 54).
2. An arc centered at A with radius $r = \overline{AB}$ yields points C and D.
3. Construct lines from C and D to 1, 3, etc. (all odd numbers). The intersecting points on the circle yield the desired vertices of the pentagon.

   For **polygons** with an **even number of angles** C and D are connected to 2, 4, 6 etc. (all even numbers).

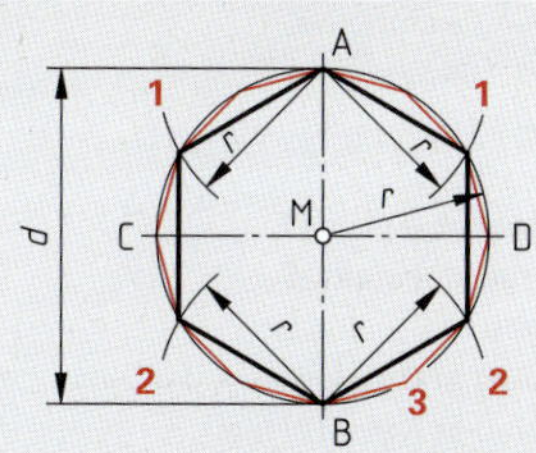

Circumscribed hexagon, dodecagon

Given: Circle of diameter d

1. Arc centered at A with radius $r = \frac{d}{2}$
2. Arc with radius r about B and A.
3. Construct line segments connecting the intersecting points to yield the hexagon.

   For a dodecagon find intermediate points including intersections at C and D.

Inscribed and circumscribed circles for triangles, Circle center point, Ellipse, Spiral

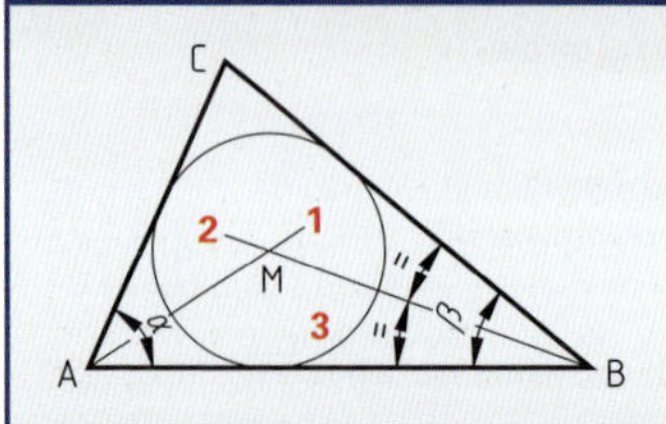

Circle inscribed in a triangle

Given: Triangle A, B, C

1. Bisect angle α.
2. Bisect angle β (intersecting at point M).
3. Inscribed circle about M.

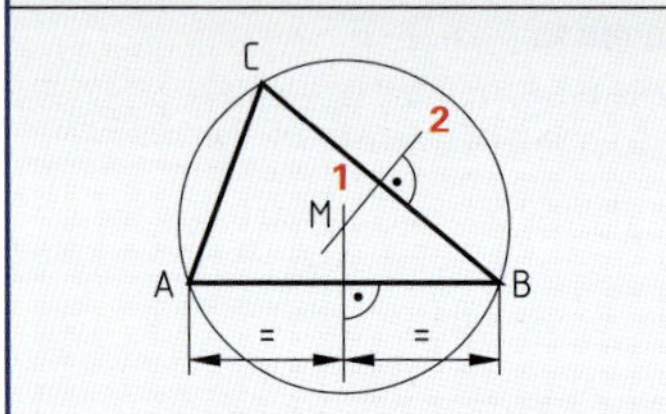

Circle circumscribing a triangle

Given: Triangle A, B, C

1. Construct the perpendicular bisector of line segment $\overline{AB}$.
2. Construct a perpendicular bisector on line segment $\overline{BC}$ (intersecting at point M).
3. Circumscribed circle about M.

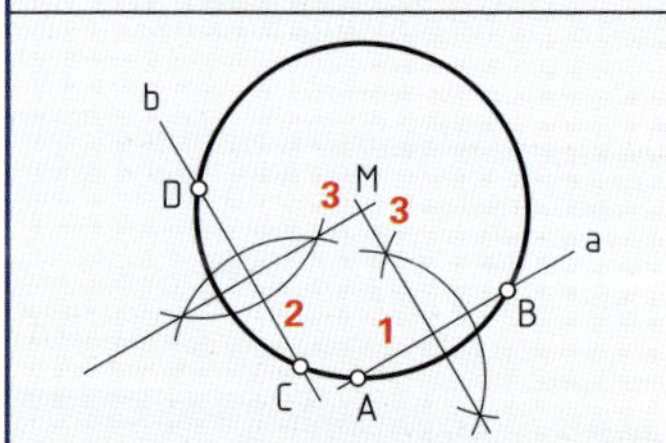

Finding the center of a circle

Given: Circle

1. Choose any straight line a that intersects the circle at A and B.
2. Straight line b (approximately perpendicular to straight line a) intersects circle at C and D.
3. Construct perpendicular bisectors on line segments $\overline{AB}$ and $\overline{CD}$.
4. Intersecting point of the perpendicular bisectors is the center M of the circle.

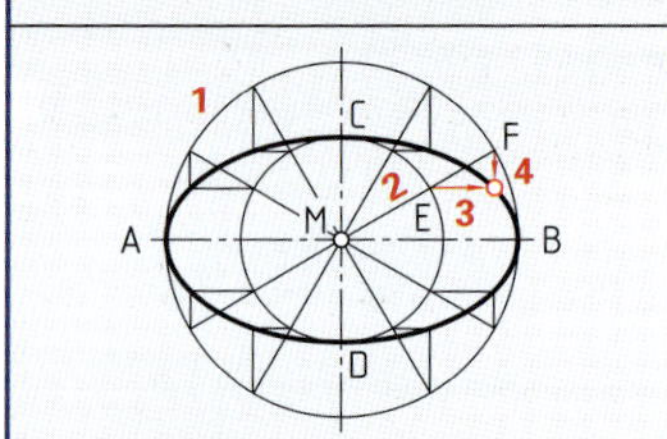

Constructing an ellipse from two circles

Given: Axes $\overline{AB}$ and $\overline{CD}$

1. Two circles about M with diameters $\overline{AB}$ and $\overline{CD}$.
2. Construct several rays through M which intersect both circles (E, F).
3. Construct parallels to the two principle axes $\overline{AB}$ and $\overline{CD}$ through E and F. Intersecting points are points on the ellipse.

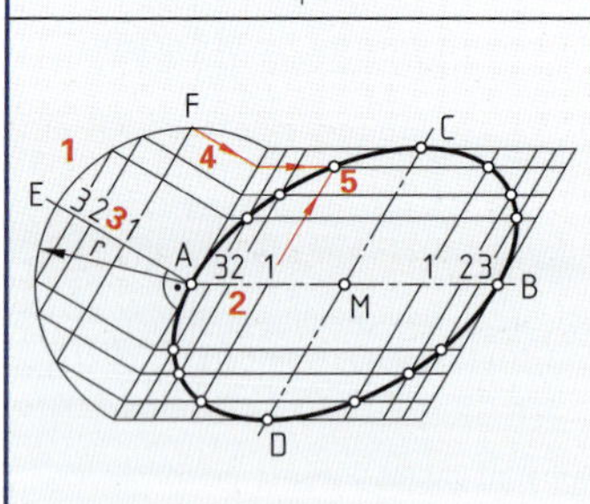

Constructing an ellipse in a parallelogram

Given: Parallelogram with axes $\overline{AB}$ and $\overline{CD}$

1. A semi-circle with radius $r = \overline{MC}$ about A yields point E.
2. Subdividing $\overline{AM}$ (or $\overline{BM}$) into halves, quarters and eighths yields points 1, 2 and 3. Construct parallels to axis CD through these points.
3. Dividing $\overline{EA}$ in halves, quarters and eighths yields points 1, 2 and 3 on the axis AE. Parallels to axis CD through those points give intersecting points F on the circular arc.
4. Construct parallels to $\overline{AE}$ through intersection points F to the semi-circle axis, from there construct parallels to axis AB.
5. Parallel intersection points of matching numbers are points on the ellipse.

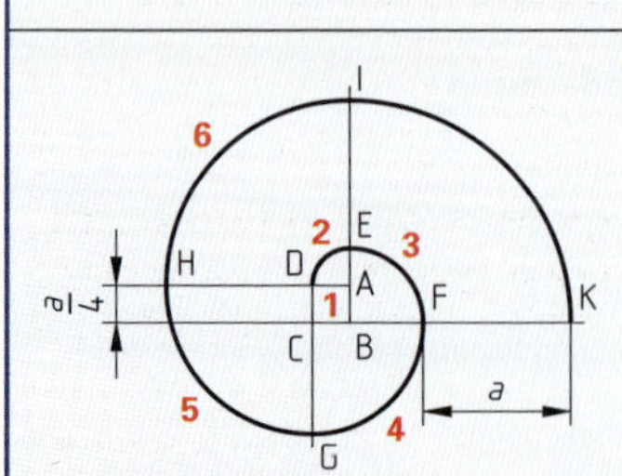

Spiral (approximate construction using a compass)

Given: Rise *a*

1. Construct square ABCD with *a*/4.
2. A quarter circle of radius $\overline{AD}$ centered at A yields E.
3. A quarter circle of radius $\overline{BE}$ centered at B yields F.
4. A quarter circle of radius $\overline{CF}$ centered at C yields G.
5. A quarter circle of radius $\overline{DG}$ centered at D yields H.
6. A quarter circle of radius AH centered at A yields I (etc).

Cycloid, Involute, Parabola, Hyperbola, Helix

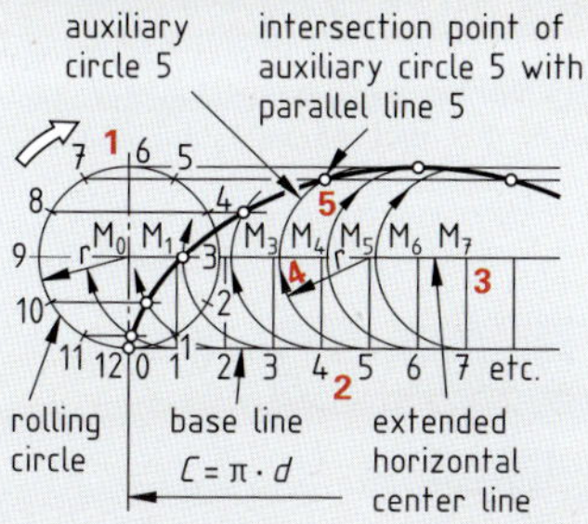

Cycloid

Given: Rolling circle of radius *r*

1. Subdivide the pitch circle into any number of equal sized parts, e. g. 12.
2. Divide the base line (≙ extent of the pitch circle = $\pi \cdot d$) into equal parts, in this case 12.
3. Vertical lines from segment points 1–12 on the base line to the extended vertical center line of the rolling circle yield the midpoints $M_1 - M_{12}$.
4. Construct auxiliary circles about the midpoints $M_1 - M_{12}$ with radius *r*.
5. The intersecting points of these auxiliary circles with the parallels through the points on the rolling circle having the same numbers give the points of the cycloid.

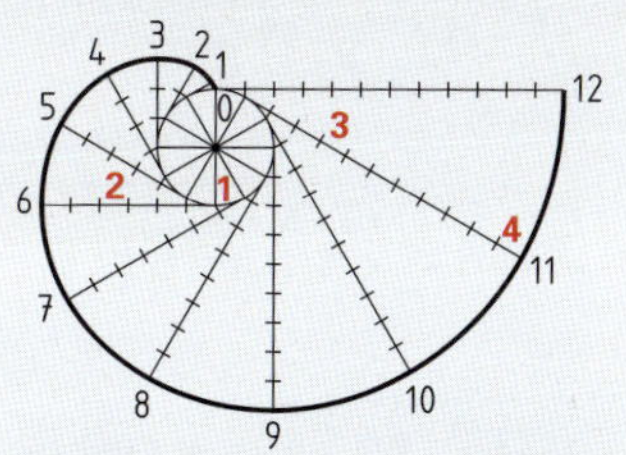

Involute

Given: Circle

1. Subdivide the circle into any desired number of equal sized parts, e. g. 12.
2. Construct tangents to the circle at each section.
3. Mark off the length of the developed circumference on each tangent from its contact point.
4. The curve through the endpoints forms the involute.

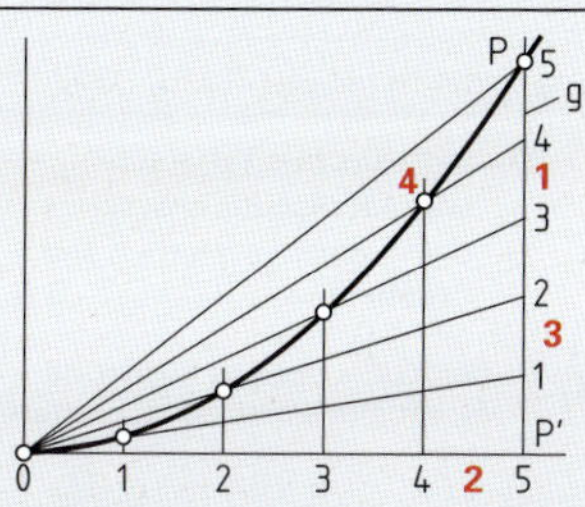

Parabola

Given: Orthogonal parabola axes and parabola point P

1. Parallel g to vertical axis through point P gives P′.
2. Divide distance $\overline{0P'}$ on the horizontal axis into any desired number of parts (e. g. 5) and construct parallels to the vertical axis.
3. Subdivide distance $\overline{PP'}$ into the same number of segments and connect to origin at 0.
4. Intersecting points of the lines with the matching number yield points on the parabola.

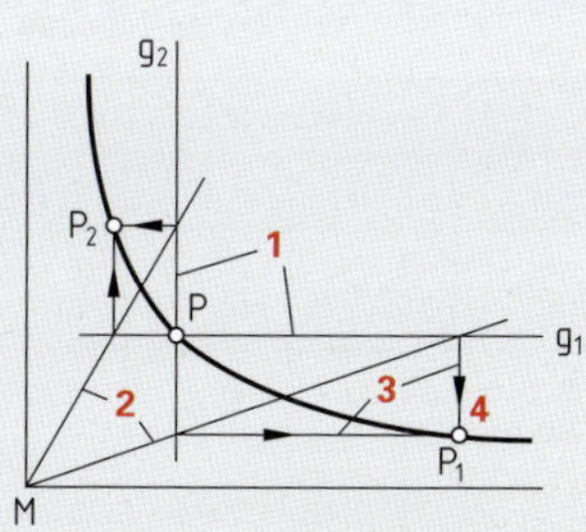

Hyperbola

Given: Orthogonal asymptotes through M and point P on the hyperbola.

1. Construct lines g_1 and g_2 parallel to the asymptotes through point P on the hyperbola.
2. Construct any desired number of rays from M.
3. Construct lines through the intersections of the rays with g_1 and g_2 parallel to the asymptotes.
4. Intersecting points of the parallel lines (P_1, P_2, ...) are points on the hyperbola.

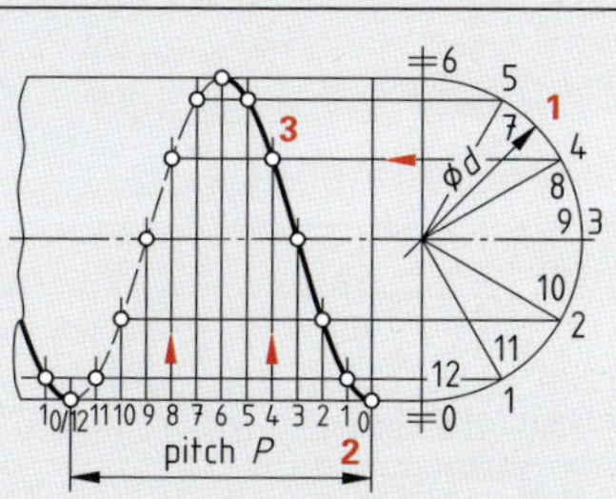

Heliocoidal line (Helix)

Given: Circle of diameter *d* and pitch *P*

1. Divide semicircle into equal sections, e. g. 6.
2. Divide the pitch *P* into twice the number of equal segments, e. g. 12.
3. Extend the same number of horizontal and vertical lines to intersection. The intersecting points yield points on the heliocoidal line.

TD

Fonts

Lettering, fonts

cf. DIN EN ISO 3098-0 (1998-04) and DIN EN ISO 3098-2 (2000-11)

The lettering of technical drawings can be done using type style A (close-spaced) or type style B. Both styles may be drawn vertical (V) or slanted by 15° to the right (I = italics). To ensure good legibility, the distance between the characters should be two line widths. The distance may be reduced to one line width if certain characters are together, e. g. LA, TV, Tr.

Font style B, V (vertical)

ABCDEFGHIJKLMNOPQRSTUVWXYZ

abcdefghijklmnopqrstuvwxyz□

1234567890 IVX[(!?:;"-=+±×·:√ %&)]ϕ

Font style B, I (italic)

ABCDEFGHIJ abcdefghij 1234567890 ϕ□

Font style A, V (vertical)

ABCD efghijk 123456 ϕ□

Font style A, I (italic)

ABCD efghijk 123456 ϕ□

Dimensions

cf. DIN EN ISO 3098-0 (1998-04)

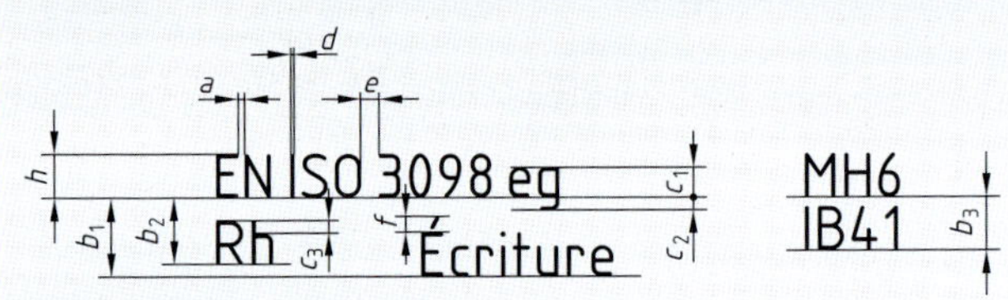

b_1 with diacritic[1] characters
b_2 without diacritic characters
b_3 with upper case letters and numbers

[1] diacritic = used to further differentiate, especially for letters

Character height *h* or height of upper case letters (nominal size) in mm	1.8	2.5	3.5	5	7	10	14	20

Ratio of dimension to character height *h*

cf. DIN EN ISO 3098-3 (1998-04)

Type style	a	b_1	b_2	b_3	c_1	c_2	c_3	d	e	f
A	$\frac{2}{14}h$	$\frac{25}{14}h$	$\frac{21}{14}h$	$\frac{17}{14}h$	$\frac{10}{14}h$	$\frac{4}{14}h$	$\frac{4}{14}h$	$\frac{1}{14}h$	$\frac{6}{14}h$	$\frac{5}{14}h$
B	$\frac{2}{10}h$	$\frac{19}{10}h$	$\frac{15}{10}h$	$\frac{13}{10}h$	$\frac{7}{10}h$	$\frac{3}{10}h$	$\frac{3}{10}h$	$\frac{1}{10}h$	$\frac{6}{10}h$	$\frac{4}{10}h$

Greek Alphabet

cf. DIN EN ISO 3098-3 (2000-11)

Α	α	alpha	Ζ	ζ	zeta	Λ	λ	lambda	Π	π	pi	Φ	φ	phi
Β	β	beta	Η	η	eta	Μ	μ	mu	Ρ	ρ	rho	Χ	χ	chi
Γ	γ	gamma	Θ	ϑ	theta	Ν	ν	nu	Σ	σ	sigma	Ψ	ψ	psi
Δ	δ	delta	Ι	ι	iota	Ξ	ξ	xi	Τ	τ	tau	Ω	ω	omega
Ε	ε	epsilon	Κ	κ	kappa	Ο	ο	omicron	Υ	υ	upsilon			

Roman numerals

I = 1	II = 2	III = 3	IV = 4	V = 5	VI = 6	VII = 7	VIII = 8	IX = 9
X = 10	XX = 20	XXX = 30	XL = 40	L = 50	LX = 60	LXX = 70	LXXX = 80	XC = 90
C = 100	CC = 200	CCC = 300	CD = 400	D = 500	DC = 600	DCC = 700	DCCC = 800	CM = 900
M = 1000	MM = 2000	Examples: MDCLXXXVII = 1687			MCMXCIX = 1999		MMXI = 2011	

Preferred numbers, Radii, Scales

Preferred numbers and series of preferred numbers[1]

cf. DIN 323-1 (1974-08)

R 5	R 10	R 20	R 40	R 5	R 10	R 20	R 40
1.00	1.00	1.00	1.00	4.00	4.00	4.00	4.00
			1.06				4.25
		1.12	1.12			4.50	4.50
			1.18				4.75
	1.25	1.25	1.25		5.00	5.00	5.00
			1.32				5.30
		1.40	1.40			5.60	5.60
			1.50				6.00
1.60	1.60	1.60	1.60	6.30	6.30	6.30	6.30
			1.70				6.70
		1.80	1.80			7.10	7.10
			1.90				7.50
	2.00	2.00	2.00		8.00	8.00	8.00
			2.12				8.50
		2.24	2.24			9.00	9.00
			2.36				9.50
2.50	2.50	2.50	2.50	10.00	10.00	10.00	10.00
			2.65				
		2.80	2.80				
			3.00				
	3.15	3.15	3.15				
			3.35				
		3.55	3.55				
			3.75				

Series	Multiplier
R 5	$q_5 = \sqrt[5]{10} \approx 1.6$
R 10	$q_{10} = \sqrt[10]{10} \approx 1.25$
R 20	$q_{20} = \sqrt[20]{10} \approx 1.12$
R 40	$q_{40} = \sqrt[40]{10} \approx 1.06$

Radii

cf. DIN 250 (2002-04)

						0.2				0.3		**0.4**		0.5		**0.6**		0.8	
1		1.2		**1.6**		2		**2.5**		3		**4**		5		**6**		8	
10		12		**16**	18	**20**	22	**25**	28	**32**	36	**40**	45	**50**	56	**63**	70	**80**	90
100	110	**125**	140	**160**	180	**200**	Values shown in bold font in the table are preferred values.												

Scale factors[2]

cf. DIN ISO 5455 (1979-12)

Actual size	Reduction factors				Enlargement factors		
1 : 1	1 : 2 1 : 5 1 : 10	1 : 20 1 : 50 1 : 100	1 : 200 1 : 500 1 : 1000	1 : 2000 1 : 5000 1 : 10 000	2 : 1 20 : 1	5 : 1 50 : 1	10 : 1

[1] Preferred numbers, e. g. for length dimensions and radii. Their usage prevents arbitrary graduations. In the series of preferred numbers (base series R 5 to R 40), each number of the series is obtained by multiplying the previous number by a constant multiplier for that series. Series 5 (R 5) is preferred over R 10, R 10 over R 20 and R 20 over R 40. The numbers of each series can be multiplied by 10, 100, 1000, etc. or divided by 10, 100, 1000, etc.

[2] For special applications the given enlargement and reduction factors can be expanded by multiplying by whole multiples of 10.

TD

TD

Drawing layout

Paper sizes (ISO)

cf. DIN EN ISO 5457 (2010-11) and DIN EN ISO 216 (2007-12)

Format	A0	A1	A2	A3	A4	A5	A6
Format dimensions[1] in mm	841 x 1189	594 x 841	420 x 594	297 x 420	210 x 297	148 x 210	105 x 148
Drawing area dimensions in mm	821 x 1159	574 x 811	400 x 564	277 x 390	180 x 277	–	–

[1] The height: width aspect ratio of the drawing papers are $1 : \sqrt{2}$ (= 1 : 1.414).

Folding for DIN A4 format

cf. DIN 824 (1981-03)

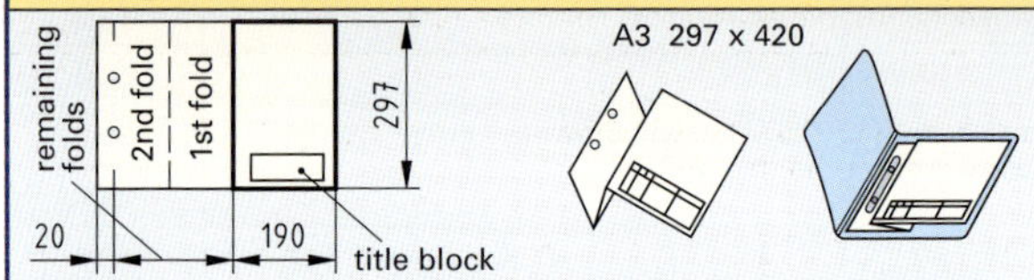

1st fold: Fold right side (190 mm wide) toward the back.

2nd fold: Fold the remainder of the sheet so that the edge of the 1st fold is 20 mm from the left edge of the paper.

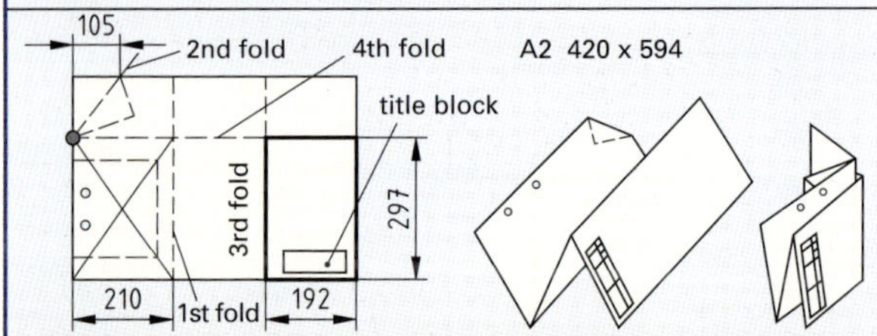

1st fold: Fold the left side (210 mm wide) towards the right.

2nd fold: Fold a triangle of 297 mm height by 105 mm width towards the left.

3rd fold: Fold the right side (192 mm wide) towards the back.

4th fold: Fold the folded packet of 297 mm height toward the back.

Title block

cf. DIN EN ISO 7200 (2004-05), replaces DIN 6771-1

The width of the title block is 180 mm. The sizes of the individual data fields (field widths and heights) are no longer stipulated, in contrast to the previous standard. The table at the bottom of this page has examples of possible field sizes.

Example of a title block:

Resp. dept.	Technical reference	Created by	Approved by	
AB 131 **11**	Susan Miller **12**	Kristin Brown **13**	John Davis **14**	**15**
John Smith Co. **1**	Type of document: Assembly drawing **9**	Document status: released **10**		
	Title, additional title: **2** Circular saw shaft **3** complete with bearing	A225-03300-012 **4**		
		Changes **5**: A · Release date **6**: 2008-01-15 · L. **7**: de · Sheet **8**: 1/3		

Drawing specific callouts, such as scale, projection symbol, tolerances and surface specifications should be indicated on the drawing outside of the title block.

Data fields in the title block

Field no.	Field name	Max. no. of characters	Field name required	Field name optional	Field size (mm) width	Field size (mm) height
1	Owner of the drawing	not specified	yes	–	69	27
2	Title (drawing name)	25	yes	–	60	18
3	Additional title	25	–	yes	60	
4	Drawing number	16	yes	–	51	9
5	Change symbol (drawing version)	2	–	yes	7	
6	Issue date of the drawing	10	yes	–	25	
7	Language identifier (de = German)	4	–	yes	10	
8	Page number and number of pages	4	–	yes	9	
9	Type of document	30	yes	–	60	
10	Document status	20	–	yes	51	
11	Responsible department	10	–	yes	26	
12	Technical reference	20	–	yes	43	
13	Drawing originator	20	yes	–	44	
14	Authorizing person	20	yes	–	43	
15	Classification/key words	not specified	–	yes	24	

Bills of materials, Item numbers

Bill of materials (BOM), parts lists

Bills of materials are indispensable for the information exchange inside and outside a company, e. g. for the assessment of the demand for parts or raw materials. Depending on the purpose of use, various types of BOMs are distinguished, e. g. BOMs for construction, manufacturing and overview of quantities (cf. page 281).

In bill of materials of semi-assembly or assembly drawings (construction bills of materials), all manufactured individual parts (workpieces) and all other parts (e. g. standard and purchased parts) of the assembly or an entire product are listed. Each part is clearly identified, e. g. by specifying the following details:

- item number
- quantity
- unit
- designation
- part or drawing number
- standard code designation
- weight
- comment

The structure of the bill of materials (number of columns) depends on the corporate requirements.

Integrated construction bills of materials

They are placed above the title block of a drawing form (DIN 6771-2, withdrawn).
The parts are listed from bottom to top in the order of their item numbers.

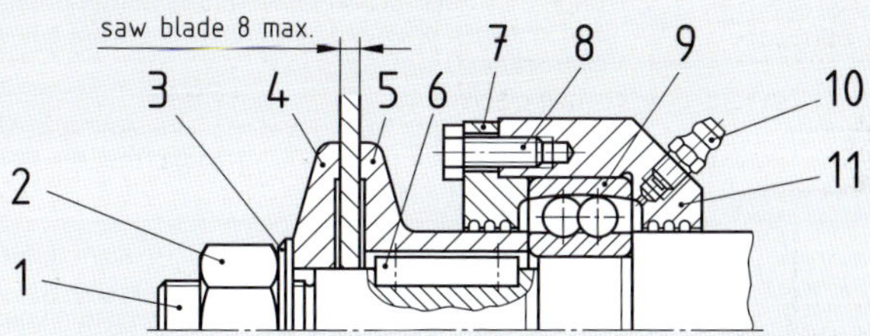

5	1	Retainer	S275JR	
4	1	Blade clamp	S275JR	
3	1	Washer	ISO 7090-20-300 HV	
2	1	Hexagon nut	ISO 8673-M20x1,5-LH-8	
1	1	Shaft	E295	RD45
Item no.	**Quantity/ Unit**	**Designation**	**Material/Standard code designation**	**Comment**

Title block as per DIN EN ISO 7200 (page 60) for a "Circular saw shaft with bearing" (drawing number A226-0096-022)

Separate construction bills of materials

In most cases, the bills of materials are prepared separately from the drawing. In order to ensure that the parts can be allocated to the drawing, the number of the assembly or overview drawing must be specified in the BOM.

Bill of materials				Sheet 1 of 1	
Drawing no. **A226-0096-022**		Designation **Circular saw shaft with bearing**		Date **25 May 2011**	
Item no.	**Part/drawing no.**	**Designation**	**Material**	**Quantity**	**Unit**
01	A226-00972-027	Shaft	E295	1	pc.
02	N701-02064-264	Hexagon nut ISO 8673-M20 x 1.5-LH-8		1	pc.
03	N601-16012-320	Washer ISO 7090-20-300 HV		1	pc.
04	A426-00966-008	Blade clamp	S275JR	1	pc.
05	A526-009761-007	Retainer	S275JR	1	pc.

Item numbers

cf. DIN ISO 6433 (1982-09)

An item number is assigned to each individual part of a semi-assembly or assembly drawing. This number must correspond to the item number in the BOM and the drawing of the individual part. Item numbers are specified as follows (cf. drawing above):

- well structured
- optionally encircled
- with approximately twice the font size of the dimension values
- with reference lines (page 72) to the respective parts

Line types

Lines in mechanical engineering drawings

cf. DIN ISO 128-24 (1999-12)

No.	Name, representation	Examples of application	
01.1	Solid line, thin	• dimension and extension lines • leader and reference lines • root of thread • hatching • position direction of layers (e. g. lamination) • outline of hinged section • short center lines • imaginary intersections from penetrations	• origin circles and dimension line terminators • diagonal crosses to mark plane surfaces • framing details • projection and grid lines • deflection lines on rough and machined parts • marking for repeated details (e. g. root diameter of toothed gear)
	Free-hand line, thin [1]	• preferably hand-drawn representing border of partial or broken views and sections, provided that the border is not a line of symmetry or a center line	
	Break line, thin [1]	• preferably automated drawing representing border of partial or broken views and sections, provided that the border is not a line of symmetry or a center line	
01.2	Solid line, thick	• visible edges and outlines • crests of threads • limit of the usable thread length • cross-section arrow lines • surface structures (e. g. knurls)	• main representations in graphs, edges and flow charts • system lines (steel construction) • mold parting lines in views
02.1	Dashed line, thin	• hidden edges	• hidden contours
02.2	Dashed line, thick	• identifies allowable areas for surface treatment (e. g. heat treatment)	
04.1	Dot-dash line (long dash), thin	• center lines • lines of symmetry	• partial circle in gears • hole circle
04.2	Dot-dash line (long dash), thick	• marking areas of (delimited) required surface treatment (e. g. heat treatment)	• marking section planes
05.1	Two-dot dash-dot line (long dash), thin	• outlines of adjacent parts • final position of movable parts • centroidal axes • contours of the shape • portions in front of the cutting plane • outlines of alternative designs	• contours of finished parts within rough parts • framing special areas or fields • projected tolerance zone

[1] Free-hand and break line types should not be used together in the same drawing.

TD

Lengths of line elements

cf. DIN EN ISO 128-20 (2002-12)

Line element	Line type no.	Length	Line element	Line type no.	Length
long dashes	04.1 and 05.1	$24 \cdot d$	gaps	02.1, 02.2, 04.1, 04.2 and 05.1	$3 \cdot d$
short dashes	02.1 and 02.2	$12 \cdot d$	**Example: Line type 04.2**		
points	04.1, 04.2 and 05.1	$< 0.5 \cdot d$			

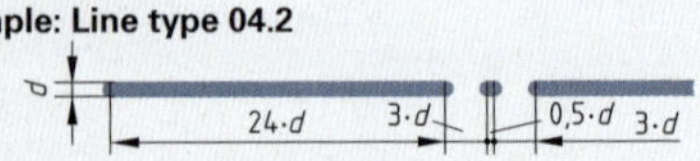

Line types

Line thicknesses and line groups

cf. DIN ISO 128-24 (1999-12)

Line widths. Normally two line types are used in drawings. They are in a ratio of 1 : 2.

Line groups. The line groups are ordered in a ratio of $1 : \sqrt{2}$ ($\approx 1 : 1.4$).

Selection. Line thicknesses and line groups are selected corresponding to the type and size of drawing, as well as to the drawing scale and the requirements of microfilming and/or method of reproduction.

Line group	Associated line thicknesses (dimension in mm) for		
	Thick lines	Thin lines	Dimension and tolerance callouts, graphical symbols
0.25	0.25	0.13	0.18
0.35	0.35	0.18	0.25
0.5	0.5	0.25	0.35
0.7	0.7	0.35	0.5
1	1	0.5	0.7
1.4	1.4	0.7	1
2	2	1	1.4

Examples of lines in technical drawings

cf. DIN ISO 128-24 (1999-12)

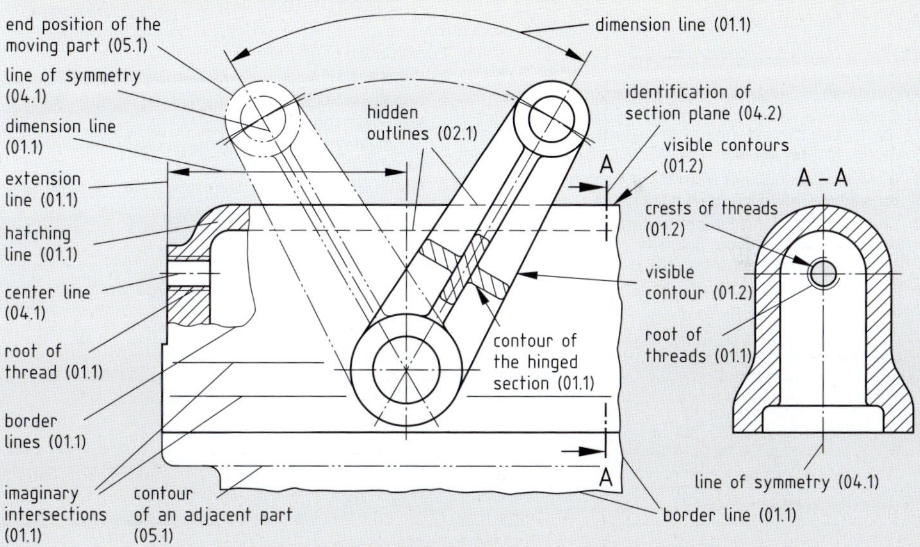

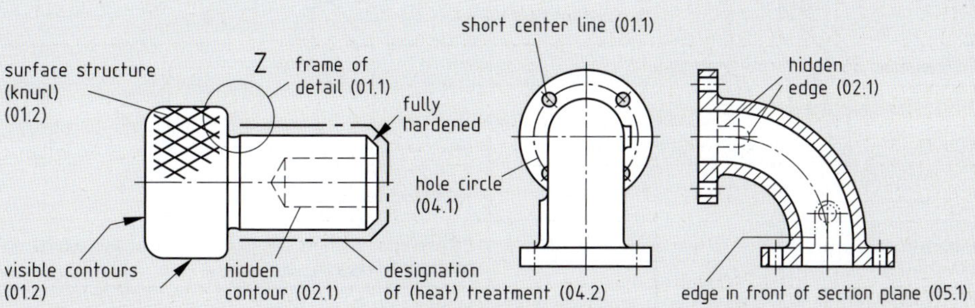

General principles of presentation, Projection methods

General principles of presentation

cf. DIN ISO 128-30 (2002-05) and DIN ISO 5456-2 (1998-04)

Selection of the front view. The view that is selected for the front view is the one which provides the most information regarding shape and dimensions.

Other views. If other views are necessary for clear representation or for complete dimensioning of an object, the following should be observed:

- The selection of the views should be limited to those most necessary.
- Additional views should contain as few hidden edges and contours as possible.

Position of other views. The position of other views is dependent upon the method of projection. For drawings based on the first- and the third-angle projection methods (page 65) the symbol for the projection method must be given in the title block.

Axonometric representation[1]

cf. DIN ISO 5456-3 (1998-04)

Isometric projection

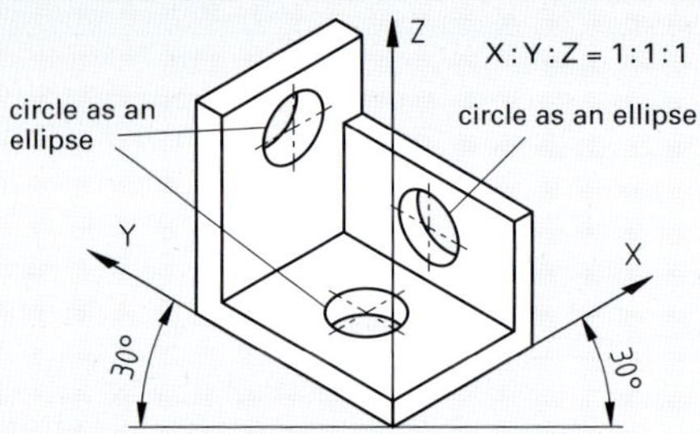

Approximate construction of the ellipse:

1. Construct a rhombus tangential to the hole. Bisect the sides of the rhombus to yield the intersecting points M_1, M_2 and N.
2. Draw connecting lines from M_1 to 1 and from M_2 to 2 to yield the intersecting points 3 and 4.
3. Construct circular arcs with radius *R* about 1 and 2 and with radius *r* about 3 and 4.

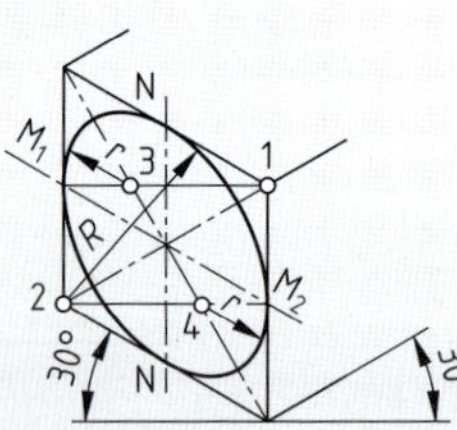

Diametric projection

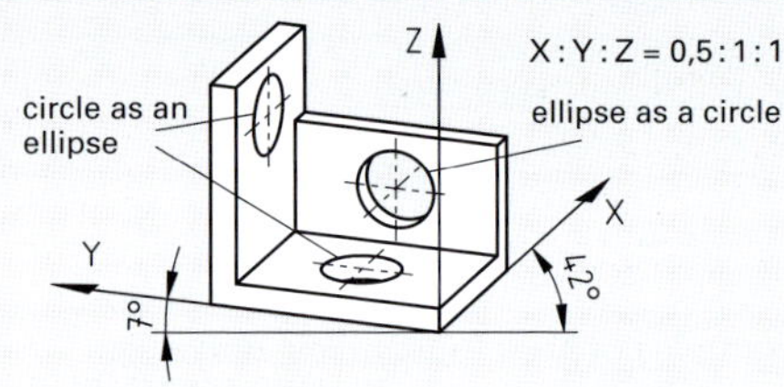

Construction of ellipses:

1. Construct an auxiliary circle with radius $r = d/2$.
2. Subdivide height *d* into any desired number of equal segments and construct grids (1 to 3).
3. Subdivide the diameter of the auxiliary circle into the same number of grids.
4. Transfer the segment lengths *a*, *b* etc. from the auxiliary circle to the rhombus.

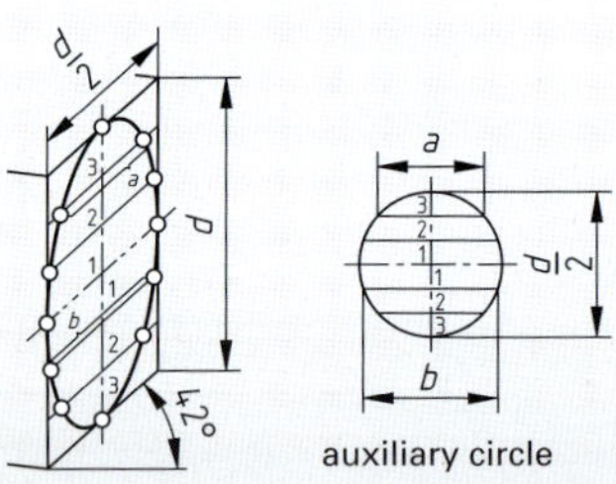

Cavalier projection

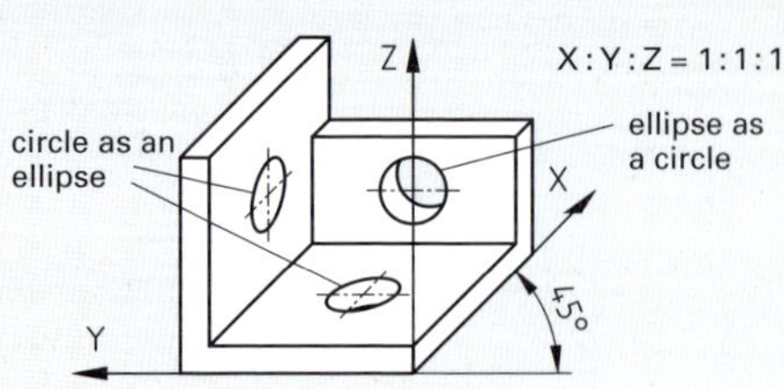

Ellipse construction identical to that on page 56 (ellipse construction in a parallelogram).

Cabinet projection

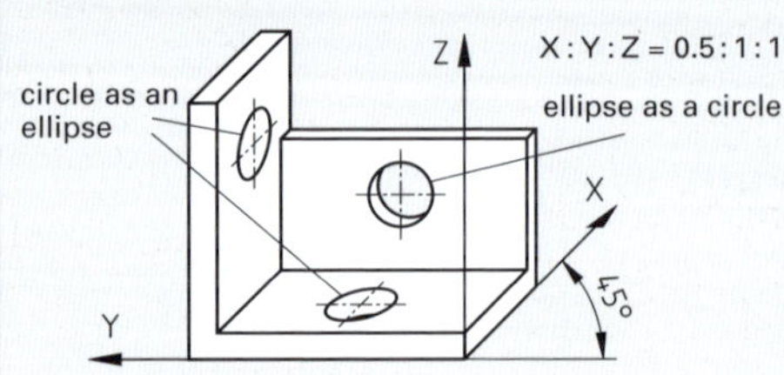

Ellipse construction identical to that of the diametric projection (above).

[1] Axonometric representations: simple, graphical representations.

Projection methods

cf. DIN ISO 128-30 (2002-05) and DIN ISO 5456-2 (1998-04)

Arrow projection method

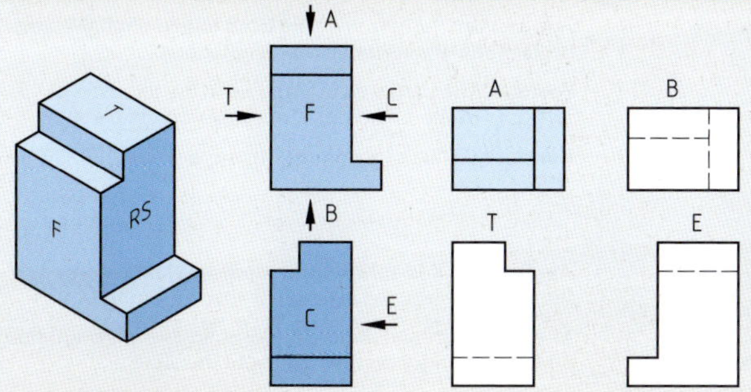

Marking the direction of observation:
- with arrow lines and upper case letters

Marking the views:
- with upper case letters

Locations of the views:
- any location with respect to front view

Layout of upper case letters:
- above the views
- vertical in reading direction
- above or to the right of the arrow lines

First-angle projection

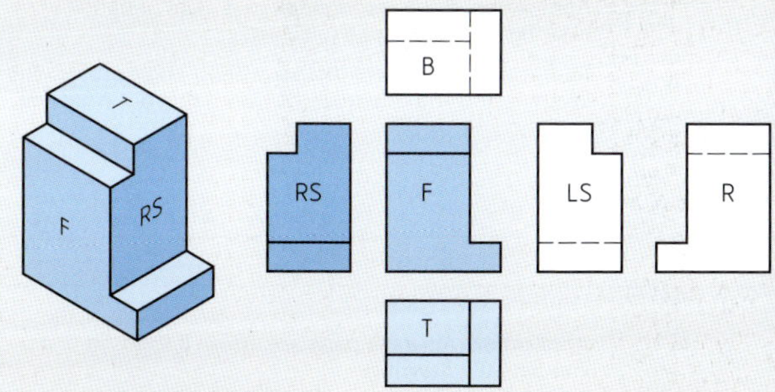

Locations with respect to front view F:		
T	top view	below F
LS	view from the left side	right of F
RS	view from the right side	left of F
B	bottom view	above F
R	rear view	left or right of F

Symbol

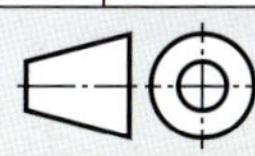

Third-angle projection[1]

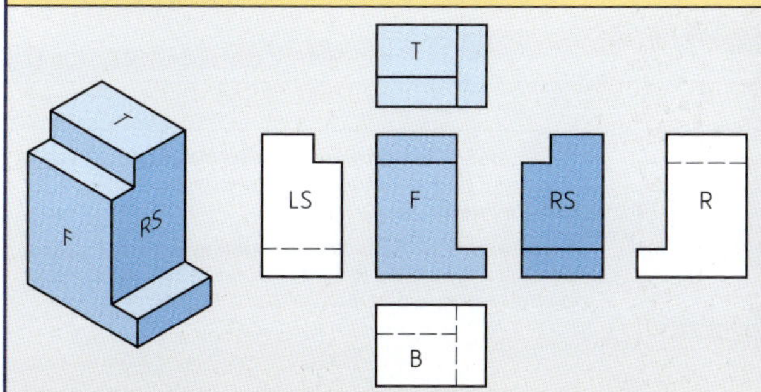

Locations with respect to front view F:		
T	top view	above F
LS	view from the left side	left of F
RS	view from the right side	right of F
B	bottom view	below F
R	rear view	left or right of F

Symbol

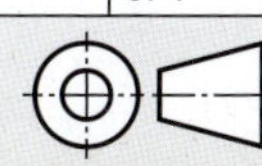

Symbols for projection methods

Symbol[2] for

first-angle projection	**third-angle projection**
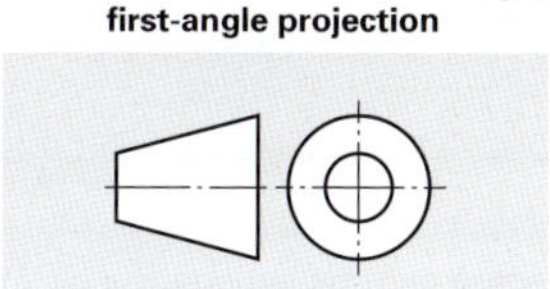	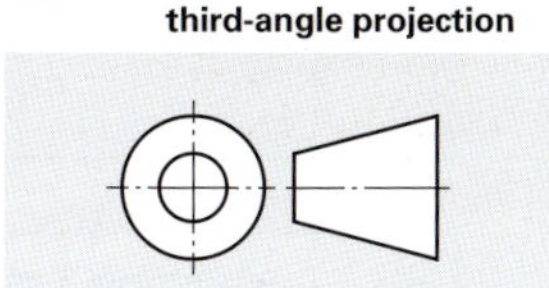

Application in

Germany and most European countries	English speaking countries, e. g. USA/Canada

Symbol for first-angle projection

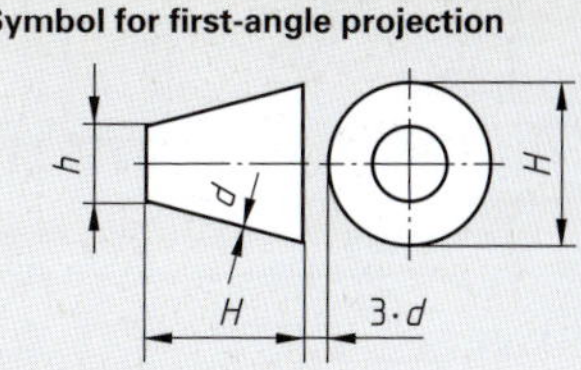

h font height in mm
$H = 2h$
$d = 0.1h$

[1] Second-angle projection is not provided.
[2] The symbol for projection method is included in the drawing layout.

TD

Views

cf. DIN ISO 128-30 and -34 (2002-05)

Partial views

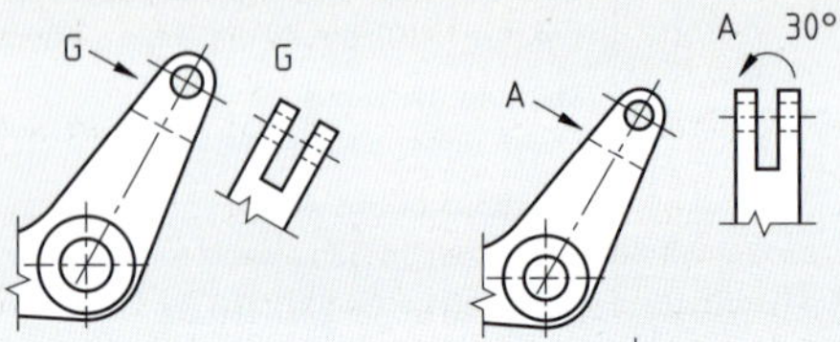

Application. Partial views are used to avoid unfavorable projections or shortened representations.

Position. The partial view is shown in the direction of the arrow or rotated. The angle of rotation must be given.

Boundary. This is identified with a break line.

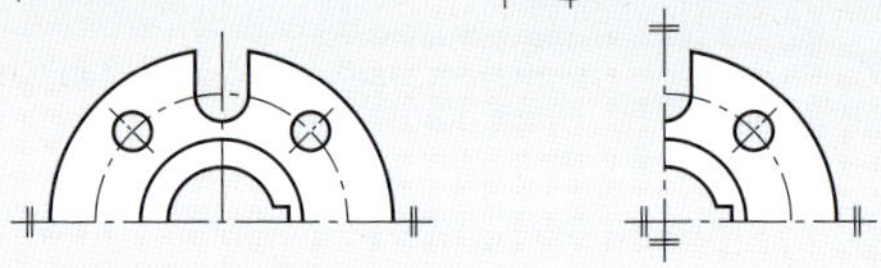

Application. It is sufficient to represent just a portion of the whole object, for example if space is limited.

Marking. With two short parallel solid lines through the line of symmetry on the outside of the view.

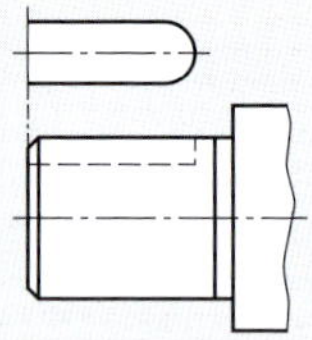

Application. If the representation is clear, a partial view is sufficient instead of a full view.

Representation. The partial view (third-angle projection) is connected with the main view by a thin dot-dash line.

Adjacent parts

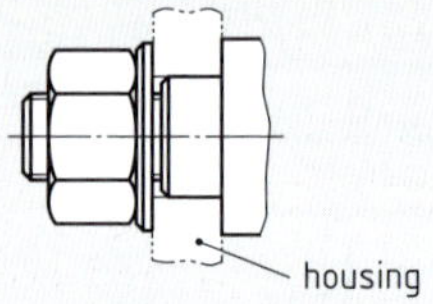

Application. Adjacent parts are drawn if it aids in understanding the drawing.

Representation. This is done with thin two-dot dash-dot lines. Sectioned adjacent parts are not hatched.

Simplified penetrations

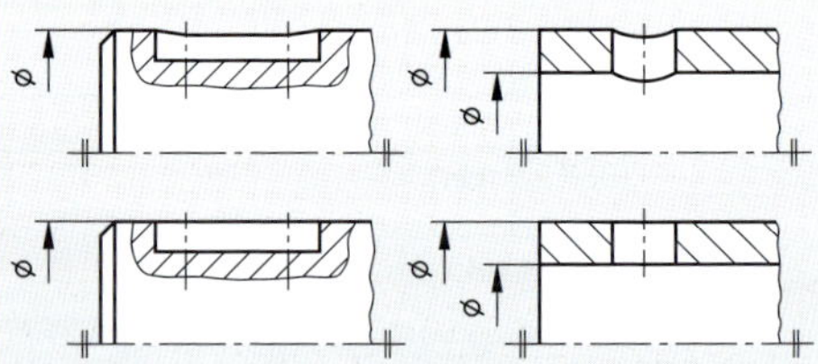

Application. If the drawing remains clearly understandable, rounded penetrating lines may be replaced by straight lines.

Representation. Rounded penetrating lines are drawn with thick solid lines for grooves in shafts and penetrating holes whose diameters significantly differ.

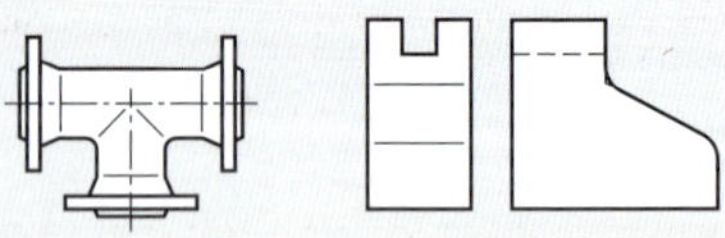

Implied penetrating lines of imaginary intersections and rounded edges are drawn with thin solid lines at the location at which the (circumferential) edge would have been with a sharp edged transition. The thin solid lines do not contact the outline.

Broken views

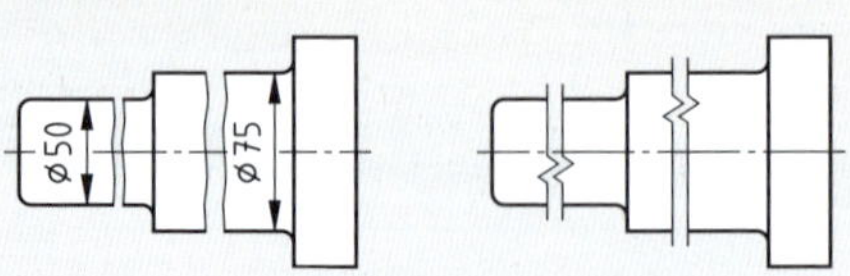

Application. To save space only the important areas of long objects need to be represented.

Representation. The boundary of the remaining parts is shown by free-hand lines or break lines. The parts must be drawn close to each other.

Views

cf. DIN ISO 128-30 and -34 (2002-05)

Repeating geometrical elements

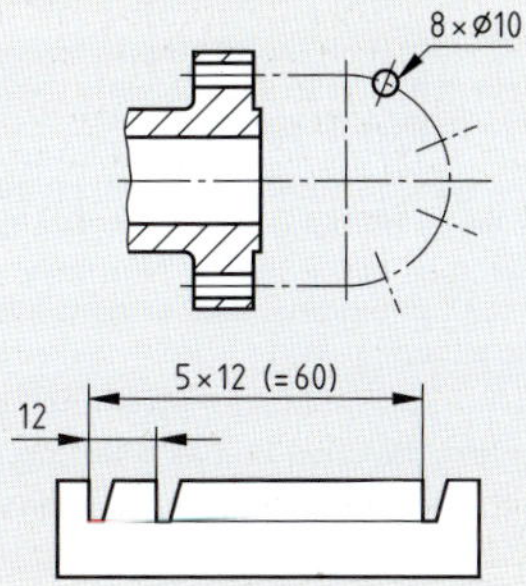

Application. For geometric elements which repeat regularly, the individual element only needs to be drawn once.

Representation. For geometric elements which are not drawn,

- the positions of symmetrical geometric elements are shown with thin dot-dash lines.
- asymmetrical geometric elements of the area in which they are found are drawn with thin solid lines.

The number of repeated elements must be given in the dimensioning.

Parts at a larger scale (details)

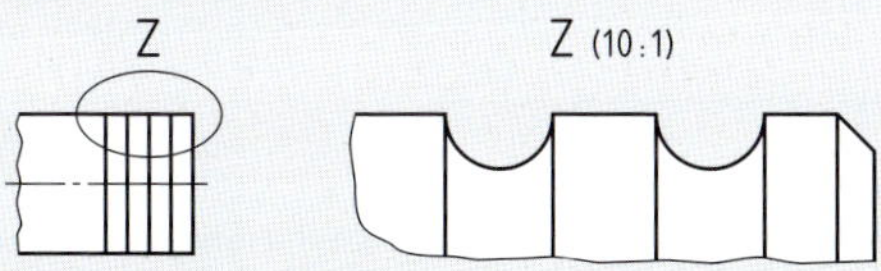

Application. Partial areas of an object which can not be clearly represented may be drawn at a larger scale.

Representation. The partial area is framed with a thin solid line or encircled and marked with a capital letter.

The partial area is represented in an enlarged detail view and is identified with the same capital letter. The enlarged scale is additionally given.

Minimal inclines

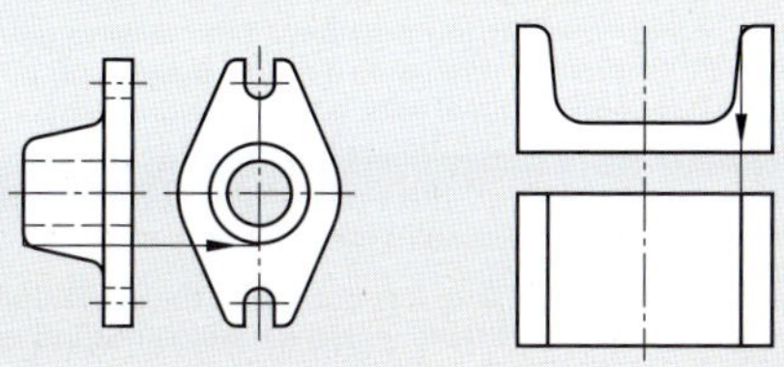

Application. Minimal inclines on slopes, cones or pyramids which cannot be shown clearly, do not have to be drawn in the corresponding projection.

Representation. The edge representing the projection of the smaller dimension is drawn with a thick solid line.

Moving parts

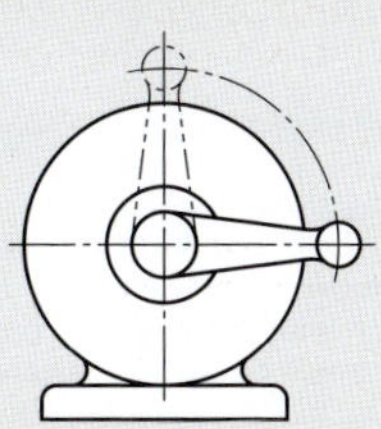

Application. Depicting alternative positions and limits of movement of parts in assembly drawings.

Representation. Parts in alternate positions and limits of movement are drawn with two-dot dash-dot lines.

Surface structures

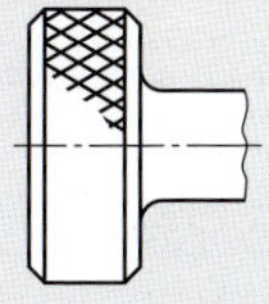

Representation. Structures such as knurls and embossing are represented with thick solid lines. Partial representation of the structure is preferable.

Sectional views

cf. DIN ISO 128-40, -44 and -50 (2002-05)

Section types

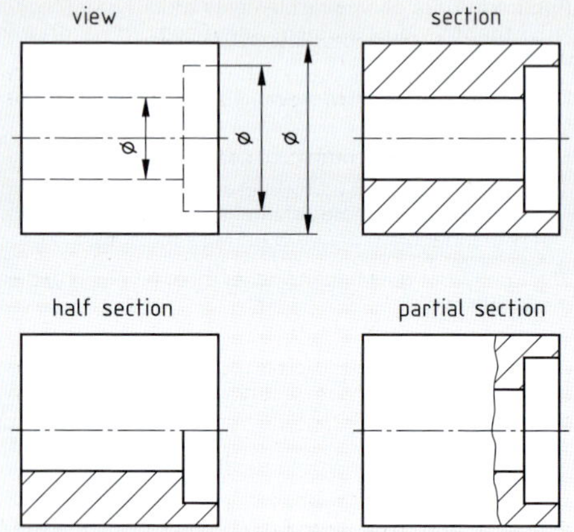

Sectional view. It allows to show clearly the inner shape of an object or individual sections of it.

Section. In a section, the front half of the object is removed to show the internal detail. The section also shows the outlines of the object. Normally, the section is placed in the longitudinal axis or perpendicular to it. However, the sectional plane is freely selectable.

Half section. In a symmetrical object one half is represented as a view, the other half as a section.

Partial section. A partial section shows only a part of the object in section.

Definitions

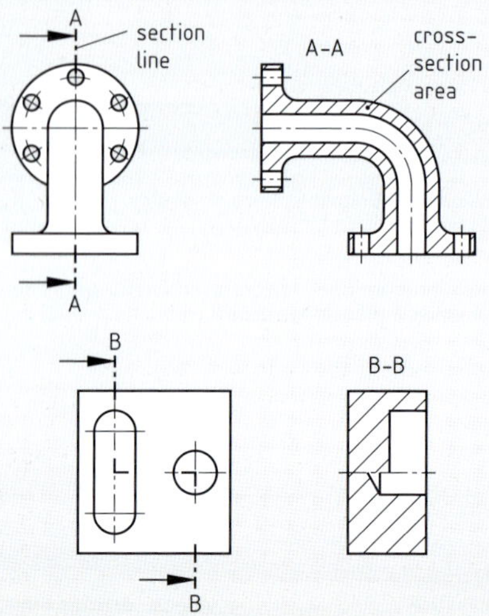

Cutting plane. The cutting plane is the imaginary plane with which the object is sectioned. Complicated objects can also be represented in two or more cutting planes.

Cross-section area. It is formed by the theoretical sectioning of the object. The cross-section area is marked with hatch lines (see below and page 70).

Section line. It marks the position of the cutting plane; for two or more cutting planes it marks the cutting path. The section line is drawn with a thick dot-dash line.

For two or more cutting planes the path of the section line is emphasized on the ends of the corresponding plane using short thick solid lines.

Marking the section line. It is done with the same upper case letters. Arrows drawn with thick solid lines indicate the direction for viewing the cutting plane.

Marking the section. The sectional view is marked with the same upper case reference letters as the section lines.

Hatching of sections

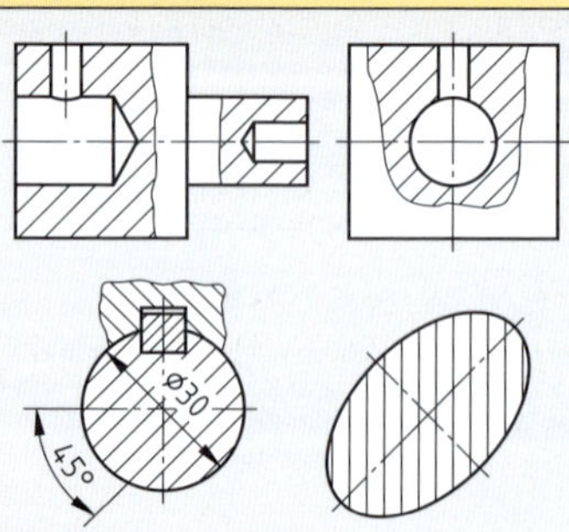

Hatching. The hatching is drawn with parallel solid lines, preferably at an angle of 45° to the centerline or to the main outlines. The hatching is interrupted for lettering.

Hatching is used for

- individual parts – all hatch lines for cross-section areas should be in the same direction and at the same spacing.
- parts adjacent to each other – hatch lines for the different parts should be in different directions or at different spacing.
- large cross-section areas – hatching preferably only near boundaries or edges.

Sectional views

cf. DIN ISO 128-40, -44 and -50 (2002-05)

Special sections

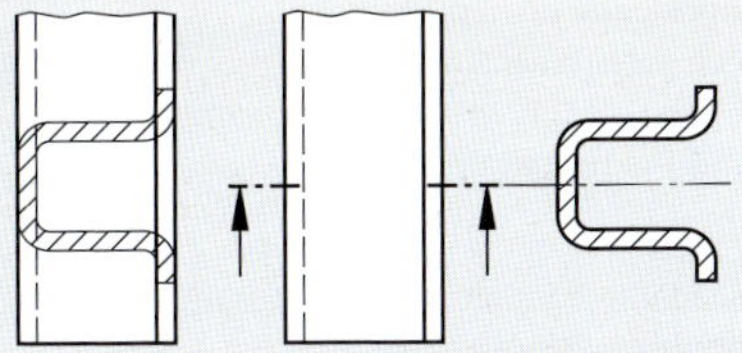

Profile sections. They may be

- drawn rotated in a view (revolved section).

  The contour lines of the section are represented with thin solid lines and are drawn within the interior of the part.
- taken out of a view (removed section).

  The section must be connected with the view by a thin dot-dash line.

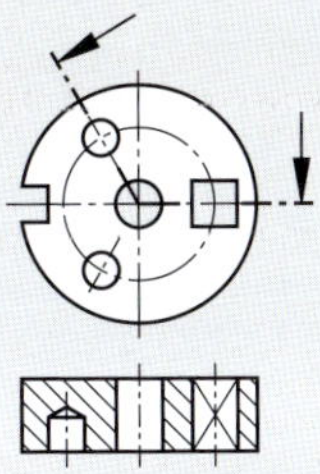

Sections with intersecting planes. If two planes intersect, one cutting plane may be rotated in the projection plane.

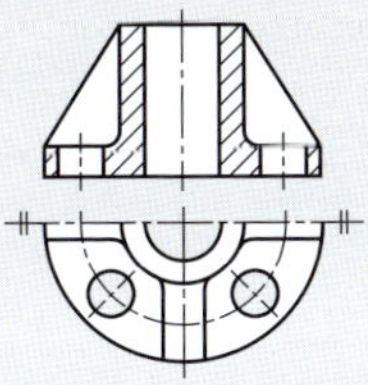

Details of rotated parts. Uniformly arranged details outside of the cross-section area, e .g. holes, may be rotated in the cutting plane.

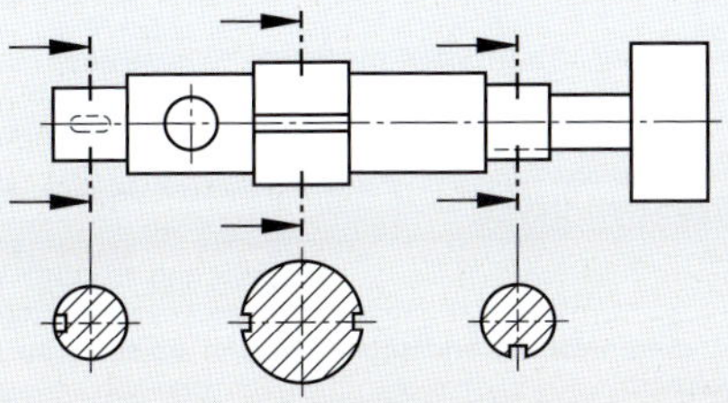

Outlines and edges. Contours and edges lying behind the cutting plane are only drawn if they add clarity to the drawing.

Parts that are not sectioned

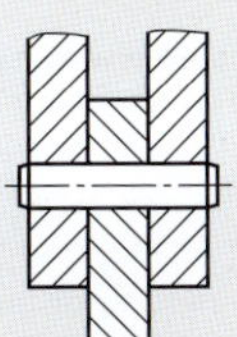

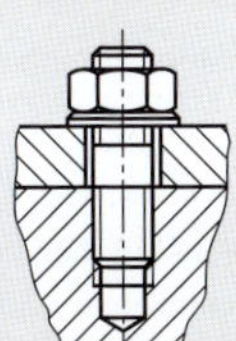

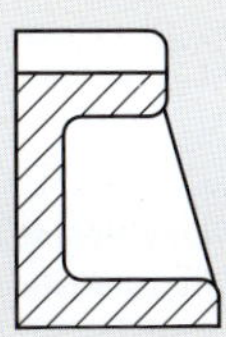

Not sectioned in the lengthwise direction:

- parts that are not hollow, e. g. screws, bolts, pins, shafts
- areas of an individual part which should protrude from the base body, e. g. ribs.

Notes on drawing

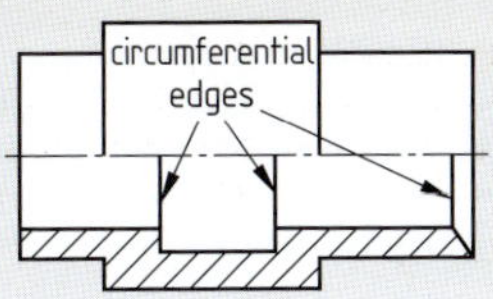

Tool edges

- **Circumferential edges.** Edges exposed by sectioning must be represented.
- **Hidden edges.** In sections the hidden edges are not represented.
- **Edges on the center line.** If an edge falls on a centerline by sectioning, it is represented.

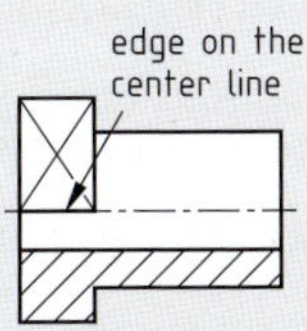

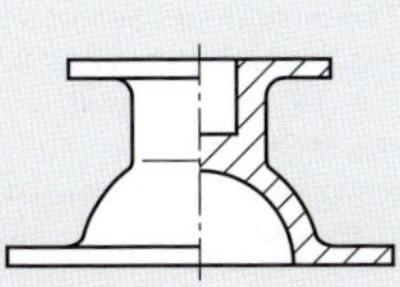

Half-sections in symmetrical objects

Section halves of symmetrical objects are preferably drawn in relation to the center line,

- below, with horizontal center lines
- to the right, for vertical center lines.

TD

Hatching, Systems for entering dimensions

Hatching

cf. DIN ISO 128-50 (2002-05)

Section areas are generally marked with basic hatching without consideration of the material.
Parts whose material should be emphasized can be identified using specific section lining.

Basic hatching (without considering the material)

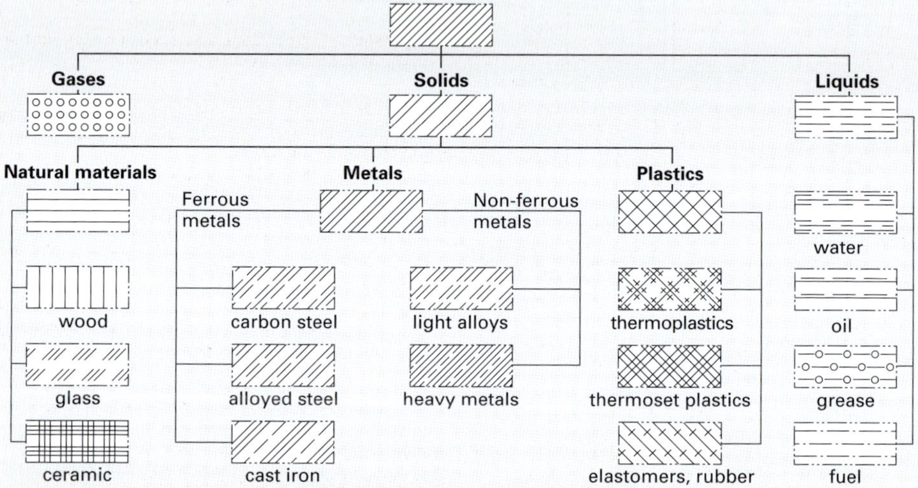

Systems for entering dimensions

cf. DIN 406-10 (1992-12)

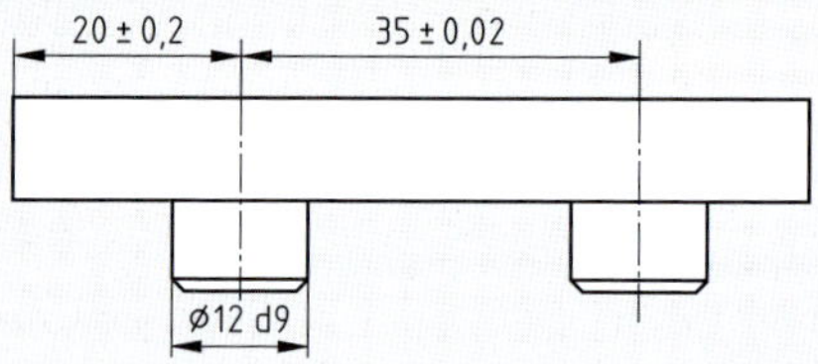

The **dimensioning and tolerancing of objects** can be based on

- function,
- manufacturing or
- inspection.

Several systems of dimensioning may be used within a single drawing.

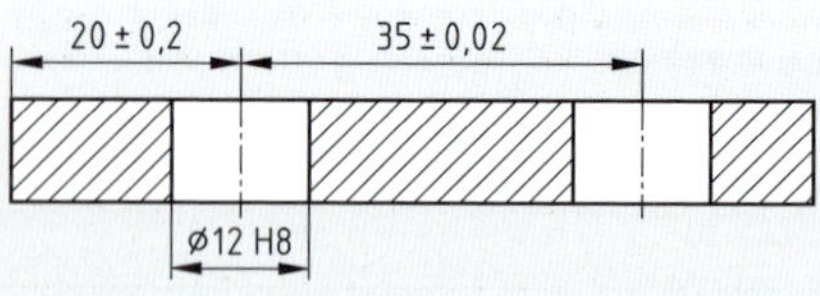

Dimensioning based on function

Characteristic. Selection, entry and tolerancing of the dimensions is done according to design requirements.

The respective manufacturing and inspection procedures are not taken into account.

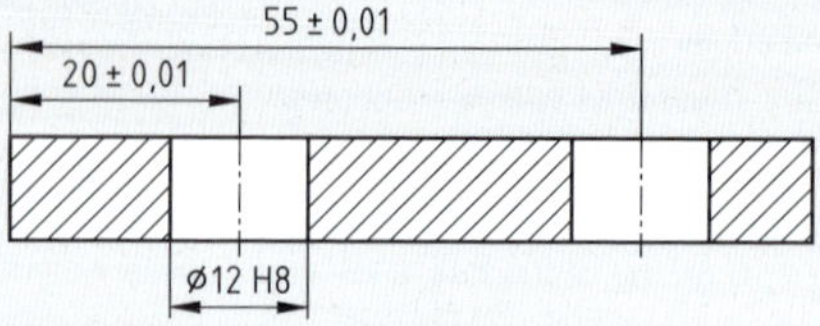

Dimensioning based on manufacturing

Characteristic. Dimensions which are necessary for manufacturing are calculated from functional dimensions.

The dimensioning is based on the intended manufacturing process.

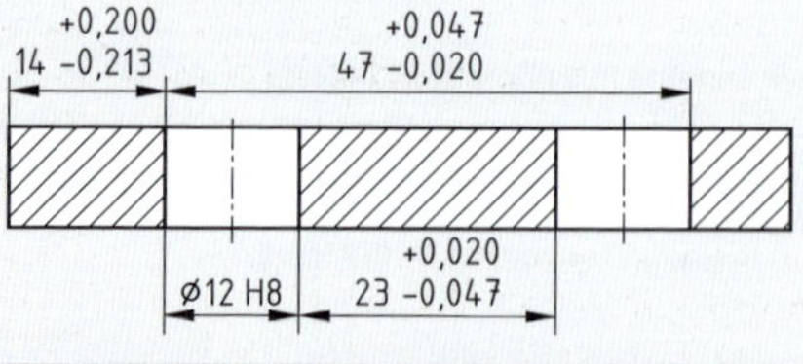

Dimensioning based on inspection

Characteristic. In most cases, dimensions that are required for inspection are calculated with the help of the function-related dimensions.

Dimensioning is based on the intended inspection method.

Dimensioning drawings

Dimension lines, dimension line terminators, extension lines, dimension values

cf. DIN 406-11 (1992-12)

Dimension lines

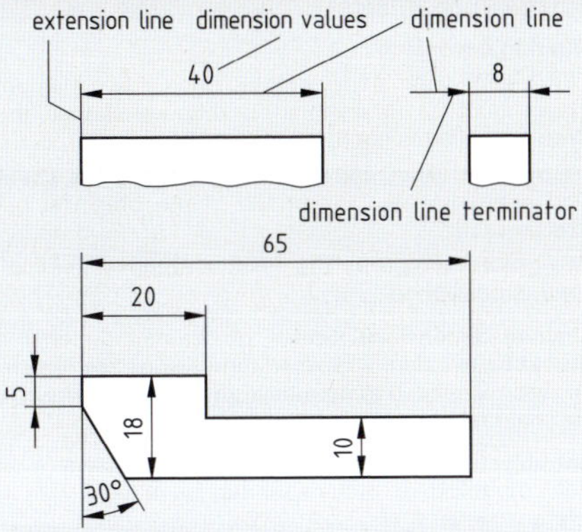

Design. Dimension lines are drawn as thin solid lines.

Entry. Dimension lines are used for:

- length dimensions parallel to the length to be dimensioned
- angle and arc dimensions as a circular arc about the center of the angle or arc.

Limited space. If space is limited, dimension lines may be

- extended to the outside using extension lines
- entered within the object
- drawn to the edges of the part body.

Spacing. Dimension lines should have a minimum distance of

- 10 mm from the edge of bodies and
- 7 mm between each other.

TD

Dimension line terminator

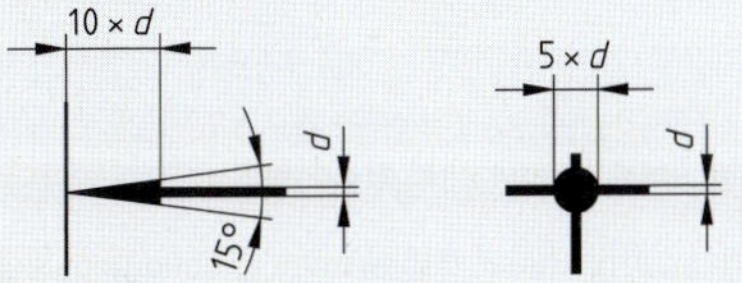

Dimension arrowheads. Generally arrowheads are used to delimit the boundaries of dimension lines.

- arrowhead length: 10 x dimension line width
- angle of lateral side: 15°

Dots. Used if space is limited.

- diameter: 5 x dimension line width

Extension lines

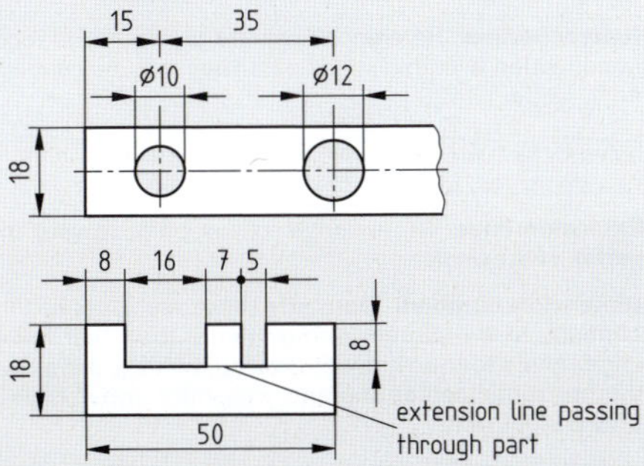

Design. Extension lines are drawn perpendicular to the length to be dimensioned with thin solid lines.

Special features

- **Symmetrical elements.** Centerlines may be used as extension lines within symmetrical elements.
- **Breaks** in extension lines may be used e. g. for entering dimensions.
- **Within a view** the extension lines may be drawn to spatially separate elements of the same or similar shape.
- Extension lines may not be extended **from one view to another view**.

Dimension values

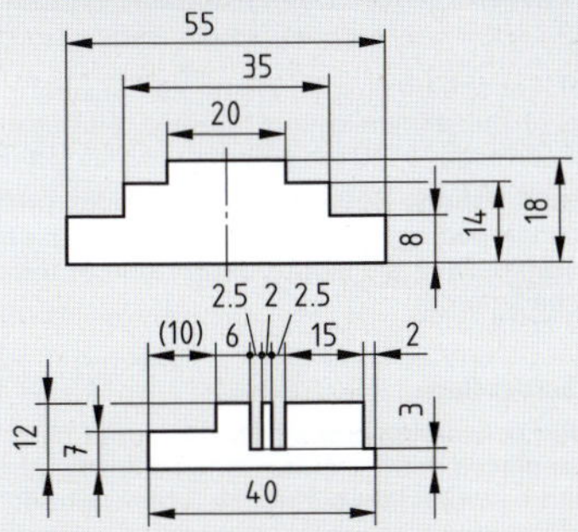

Entry. Dimension values are entered

- in standard lettering according to DIN EN ISO 3098, page 58
- with a minimum font size of 3.5 mm
- above the dimension line
- so that they are legible from below and from the right
- for multiple parallel dimension lines – separated from each other.

Limited space. If there is limited space, the dimension values may be entered

- on a leader line
- over the extension of the dimension line.

Dimensioning drawings

Dimensioning rules, leader and reference lines, angle dimensions, square and width across flats

cf. DIN 406-11 (1992-12) and DIN ISO 128-22 (1999-11)

Dimensioning rules

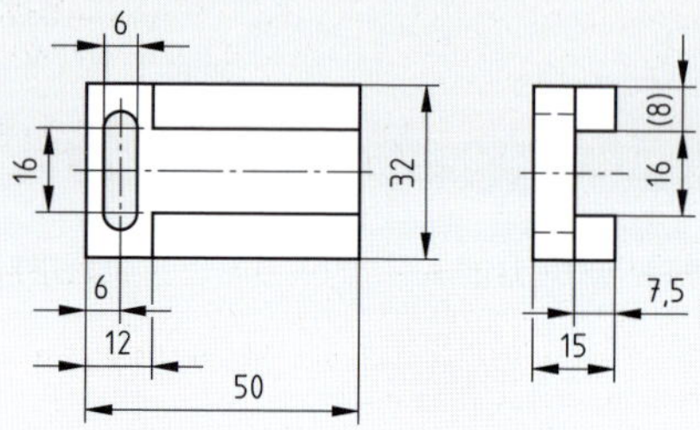

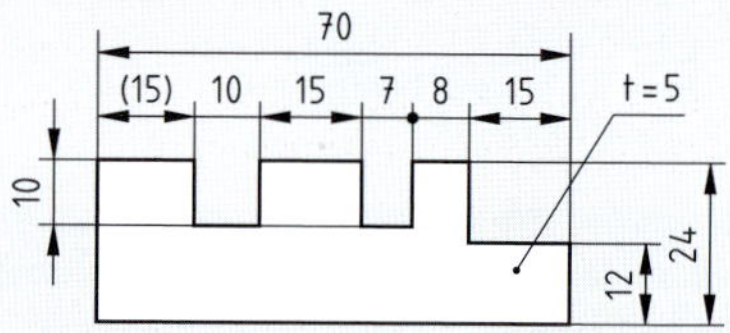

Entering dimensions

- Each dimension is only entered once. If two elements have identical dimensions but different shapes, they must be dimensioned separately.
- If multiple views are drawn, the dimensions should be entered where the shape of the object is best recognized.
- Symmetrical objects. The position of the center line is not dimensioned.

Chained dimensions. Series of chained dimensions should be avoided. If chained dimensions are required for reasons related to manufacturing, one dimension of the chain must be in parentheses.

Flat objects. For flat objects that are only drawn in one view, the thickness dimension may be entered with the reference letter t

- in the view or
- near the view.

TD

Leader and reference lines

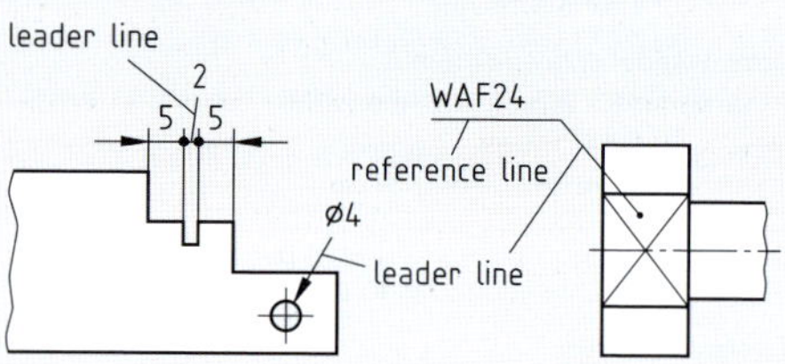

Leader lines. Leader lines are drawn as thin solid lines. They end

- with an arrowhead, if they point to solid body edges or holes.
- with a dot, if they point to a surface.
- without marking, if they point to other lines.

Reference lines. Reference lines are drawn in the reading direction with thin solid lines. They may be connected to leader lines.

Angular dimensions

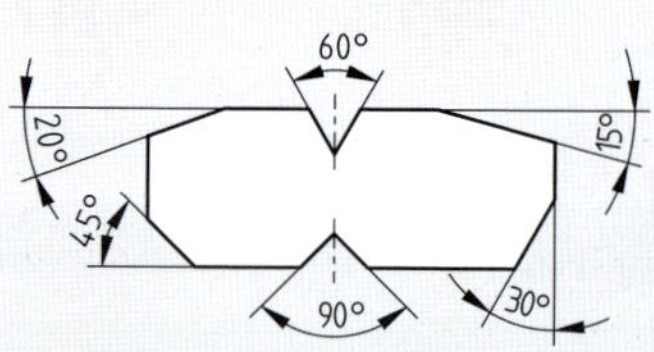

Extension lines. The extension lines point toward the vertex of the angle.

Dimension numbers. Normally these are entered tangentially to the dimensioning line so that their lower edge points to the vertex of the angle if they are above the horizontal center line and with their upper edge if they are below it.

Square, width across flats

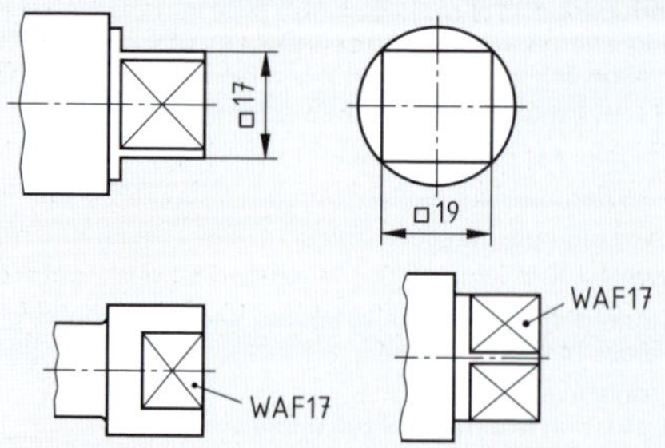

Square

Symbol. For square shaped elements the symbol is set in front of the dimensioning number. The size of the symbol corresponds to the size of the small letters.

Dimension values. Square shapes should preferably be dimensioned in the view in which their shape is recognizable. Only the length of one side of the square should be entered.

Width across flats

Symbol. For widths across flats the upper case letters WAF are placed in front of the dimensioning number, if the width between flats cannot be dimensioned.

Dimensioning drawings

Diameters, radii, spheres, chamfers, inclines, tapers, arc dimensions

cf. DIN 406-11 (1992-12)

Diameter, radius, sphere

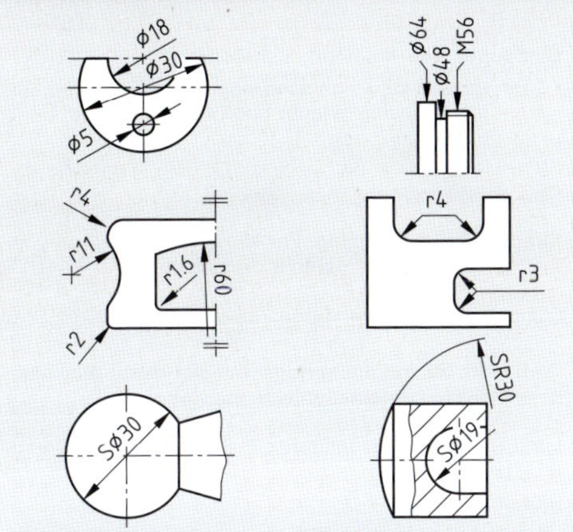

Diameter

Symbol. For all diameters the symbol ∅ is placed before the dimension value. Its overall height corresponds to the height of the dimensioning value.

Limited space. In the case of limited space the dimension references the object feature from the outside.

Radius

Symbol. For radii the lower case letter r is placed before the dimensioning number.

Dimension lines. Dimension lines should be drawn

- from the center of the radius or
- from the direction of the midpoint.

Sphere

Symbol. For spherical object features the capital letter S is placed before the diameter or radius symbol.

TD

Chamfers, countersinks

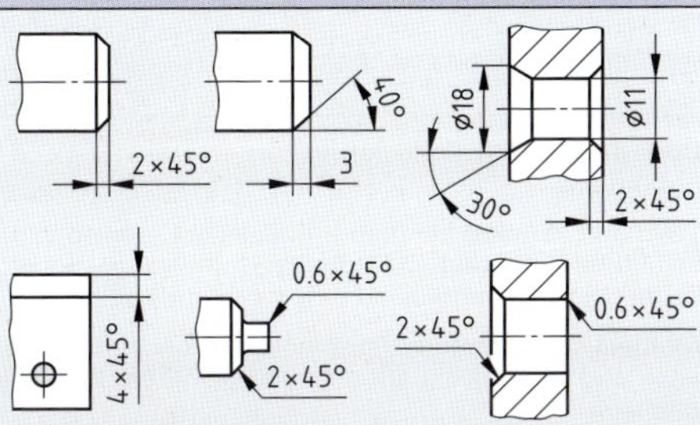

45° chamfers and countersinks of 90° can be simply dimensioned by indicating the angle and the chamfer width. Both drawn and undrawn chamfers may be dimensioned using an extension line.

Other chamfer angles. For chamfers with an angle deviating from 45° the

- angle and the chamfer width or
- the angle and the chamfer diameter

are to be entered.

Inclines, tapers

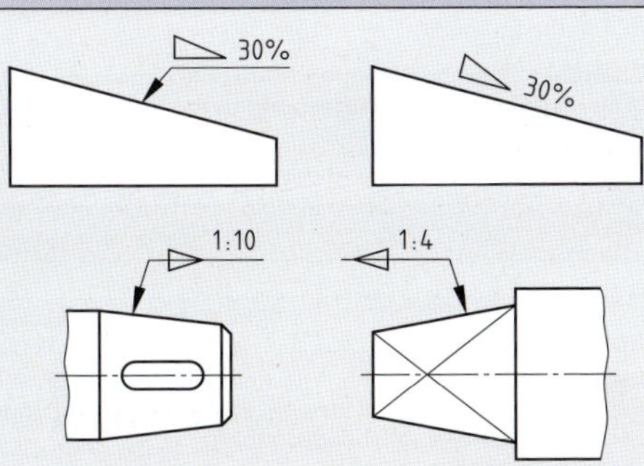

Incline

Symbol. The symbol ⊿ is entered before the dimension values.

Orientation of the symbol. The symbol is oriented so that its incline matches the incline of the object. Preferably the symbol is connected to the inclined surface with a reference line or a leader line.

Taper

Symbol. The symbol ▷ is entered before the dimension values on a reference line.

Orientation of the symbol. The orientation of the symbol must match the direction of the object taper. The reference line of the symbol is connected to the outline of the taper with a leader line.

Arc dimensions

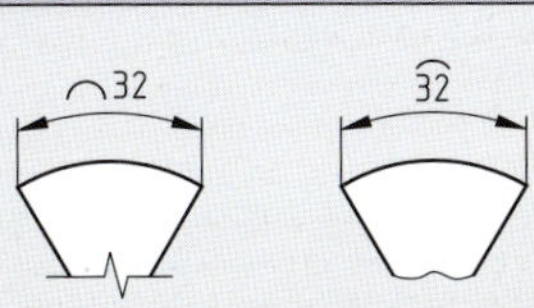

Symbol. The symbol ⌒ is entered before the dimension values. For manual drawing the arc may be labeled with a similar symbol over the dimension value.

Dimensioning drawings

Slots, threads, patterns

cf. DIN 406-11 (1992-12) and DIN ISO 6410-1 (1993-12)

Slots

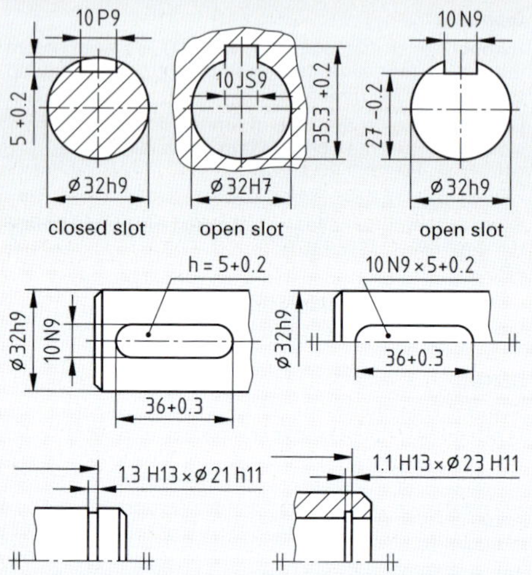

Slot depth. The slot depth is measured

- from the slot side for closed slots
- from the opposing side for open slots.

Simplified dimensioning. For slots represented only in the top view, the slot depth is dimensioned

- with the letter h or
- in combination with the slot width.

With **slots for retaining rings** the slot depth may also be entered in combination with the slot width.

Limit deviations for tolerance classes JS9, N9, P9 and H11: page 105

Slot dimensions

- for wedges see page 238
- for fitted keys see page 239
- for retaining rings see page 265

Threads

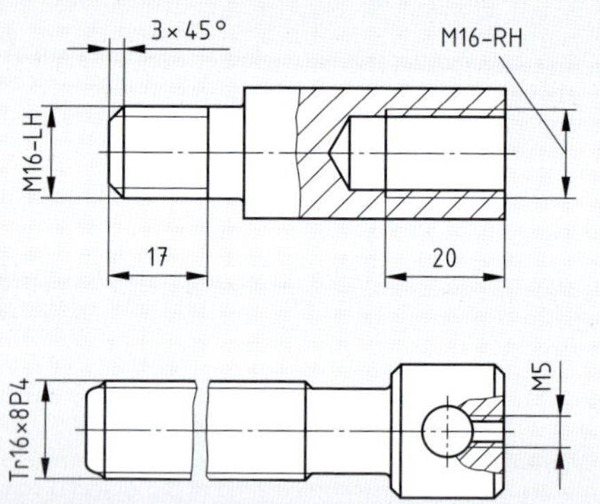

Code designation. Code designators are used for standard threads.

Left hand threads. Left hand threads are marked with LH. If both left hand and right hand threads are found on an object, the right hand threads get the addition RH.

Multiple screw threads. For multiple screw threads the pitch and the spacing are entered behind the nominal diameter.

Length specifications. These give the usable thread length. The depth of the basic hole (page 210) is normally not dimensioned.

Chamfers. Chamfers on threads are only dimensioned if their diameters do not correspond to the thread core or the thread outside diameter.

Radial and linear patterns

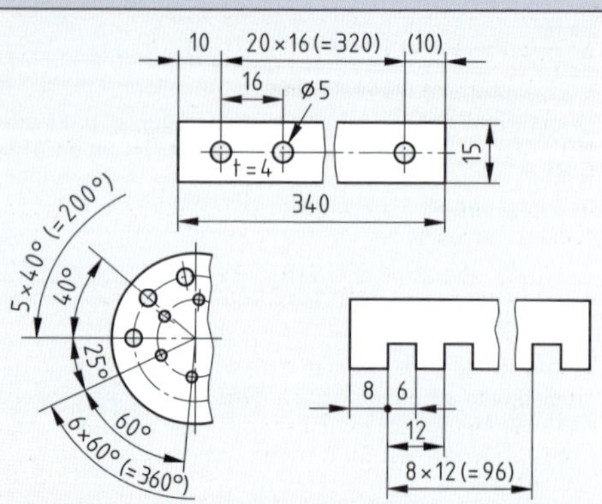

Identical design elements. The following data is given for spacing of identical design elements having the same distance or angle between them

- the number of elements
- the distance between the elements
- the overall length or overall angle (in parentheses).

Dimensioning drawings

Tolerance specifications

cf. DIN 406-12 (1992-12), DIN ISO 2768-1 (1991-06) and DIN ISO 2768-2 (1991-04)

Tolerance specifications using deviations

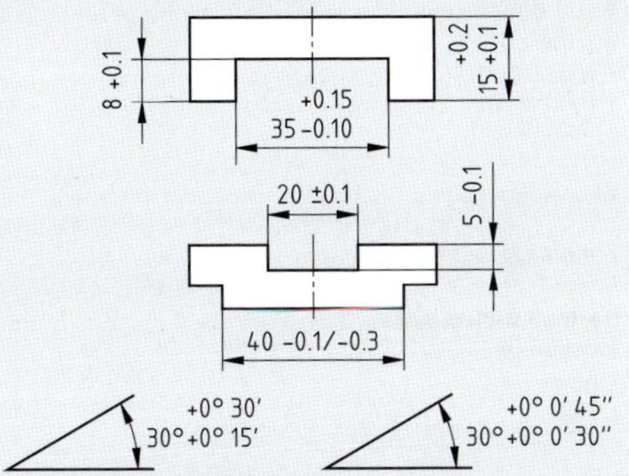

Entry. The deviations are entered

- after the nominal size
- if there are two deviations, the upper deviation is shown above the lower deviation
- for equally large upper and lower deviations by a ± mark before the number value, which is only entered once
- for angle dimensioning with units specified.

TD

Tolerance specifications using tolerance classes

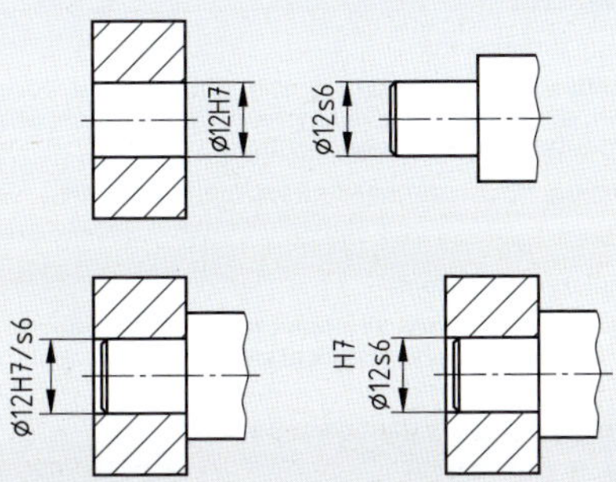

Entry. Tolerance classes are entered for

- single nominal sizes: after the nominal size
- parts shown inserted: the tolerance class of the interior dimension (hole) is before or over the tolerance class of the outer dimension (shaft).

Tolerance specifications for specific areas

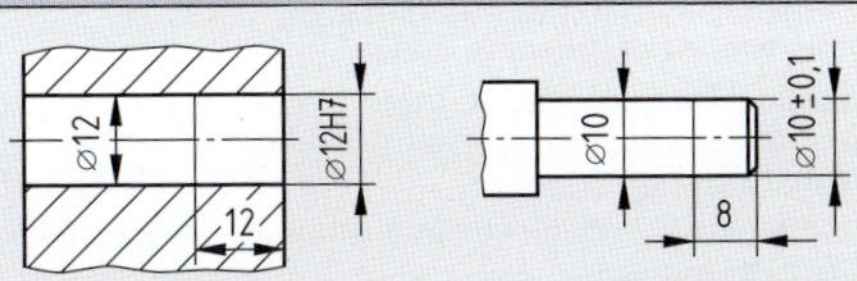

Area of application. The area to which the tolerance applies is bounded by a thin solid line.

Tolerance specifications using general tolerances

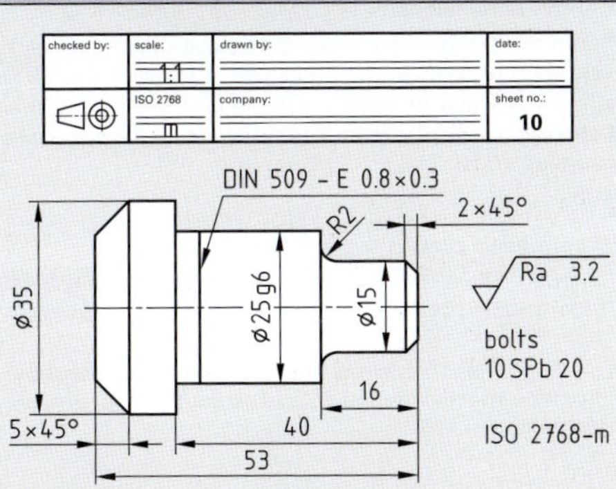

Application. General tolerances are used for

- linear and angular dimensions
- form and position.

They apply to dimensions without individual tolerance entry.

Drawing entry. The note for general tolerances (page 106) can be located:

- near the individual part drawings
- for title blocks according to DIN 6771 (retracted): in the title block.

Entries. Given are:

- the sheet number of the standard
- the tolerance class for linear and angular dimensions
- the tolerance class for form and positional tolerances, as needed.

Dimensioning drawings

Dimensions

cf. DIN 406-10 and -11 (1992-12)

Types of dimensioning

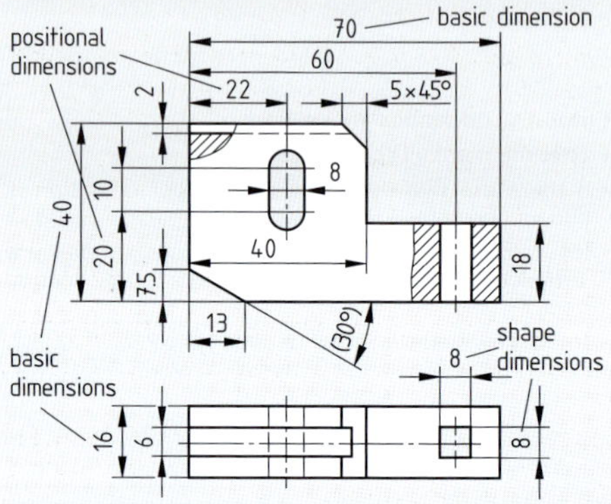

Basic Dimensions. The basic dimensions of an object are the

- total length
- total width
- total height.

Shape dimensions. Shape dimensions establish, e. g. the

- dimensions of slots
- dimensions of shoulders.

Positional dimensions. These are used to specify the location of

- holes
- slots
- elongated holes, etc.

TD

Special dimensions

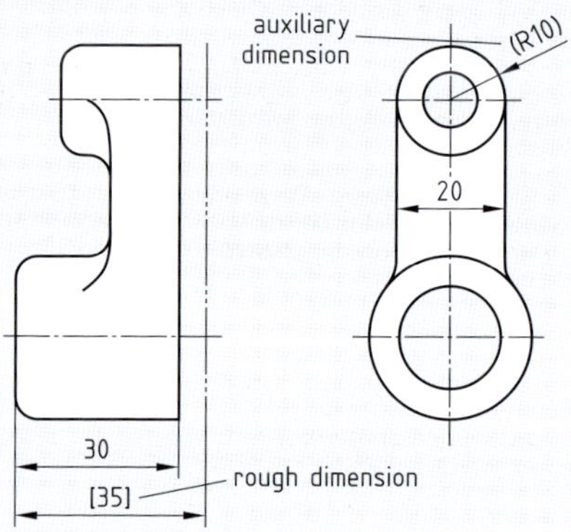

Rough dimensions

Function. Rough dimensions might be used to give information about, for example, the dimensions of cast or forged objects before machining.

Labeling. Rough dimensions are put in brackets.

Auxiliary dimensions

Function. Auxiliary dimensions give additional information. They are not necessary to geometrically define the object.

Labeling. Auxiliary dimensions are

- put in parentheses
- entered without tolerances.

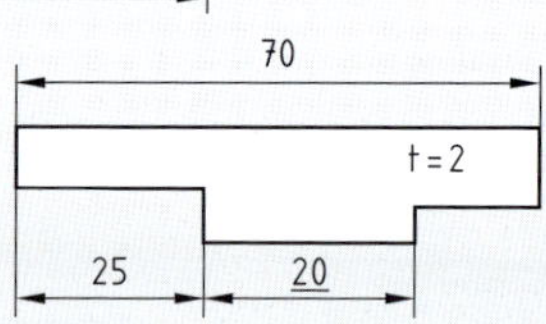

Dimensions not drawn to scale

Labeling. Dimensions not drawn to scale might be used for drawing changes, for example, and they are marked by underlining.

Prohibited are underlined dimensions in computer aided (CAD) drawings.

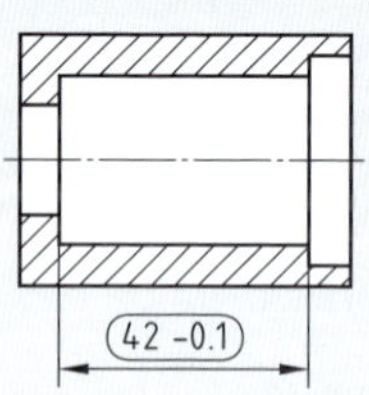

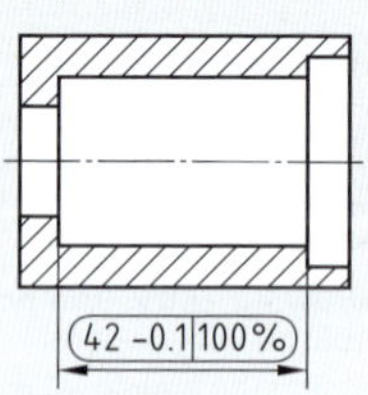

Control dimensions

Function. It should be noted that these dimensions are especially checked by the purchaser. If necessary a 100% check will be performed.

Labeling. Control dimensions are set in frames with rounded ends.

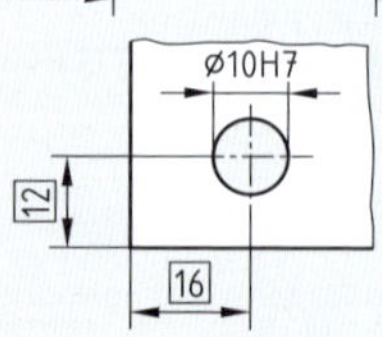

Theoretically precise dimensions

Function. These dimensions give the geometrically ideal (theoretically precise) position of the shape of a design feature.

Labeling. The dimensions are placed in a frame without tolerance specifications and correspond with geometric tolerancing.

Types of dimensioning

Parallel dimensioning, running dimensioning, coordinate dimensioning[1] cf. DIN 406-11 (1992-12)

Stack dimensioning

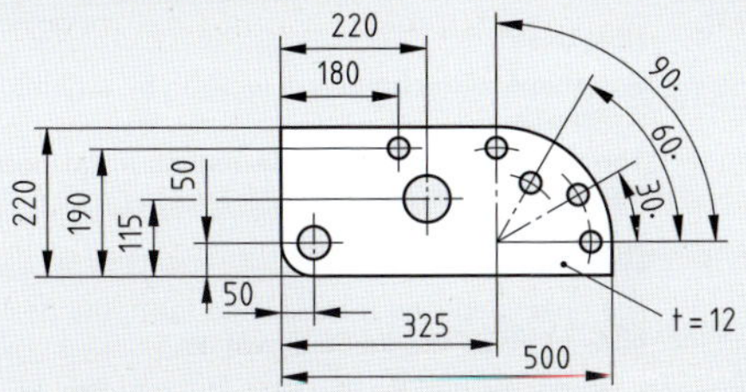

Dimension lines. Several dimension lines are entered together for

- stacked linear dimensions
- concentric angular dimensions.

Running dimensioning

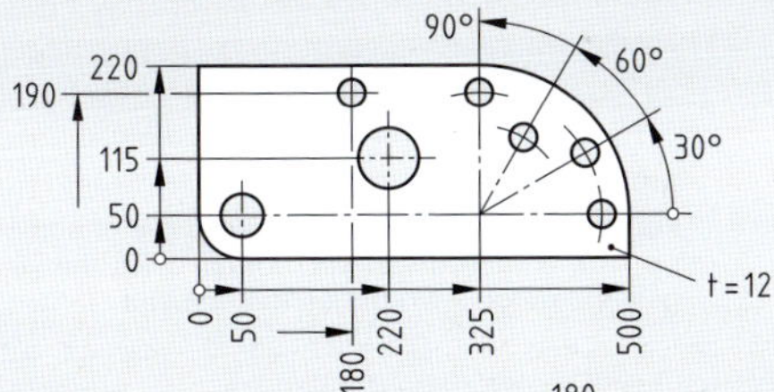

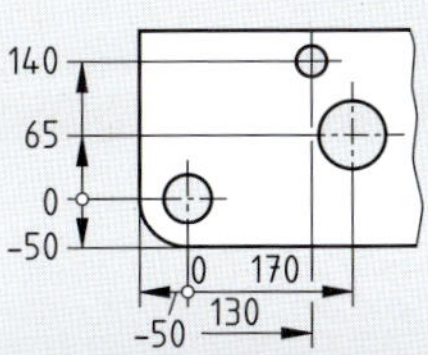

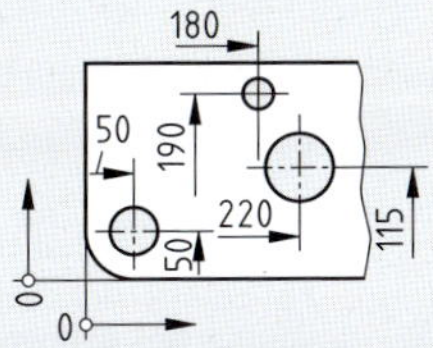

Origin. The dimensions are entered outwards from the origin in each of the three possible directions. The origin is indicated by a small circle.

Dimension lines. The following applies for the entries:

- As a rule only one dimension line is used for each direction.
- If there is limited space two or more dimension lines may be used. The dimension lines may also be shown broken.

Dimensions

- must be provided with a minus sign if they are entered from the origin in the opposite direction.
- may also be entered in the reading direction.

Coordinate dimensioning

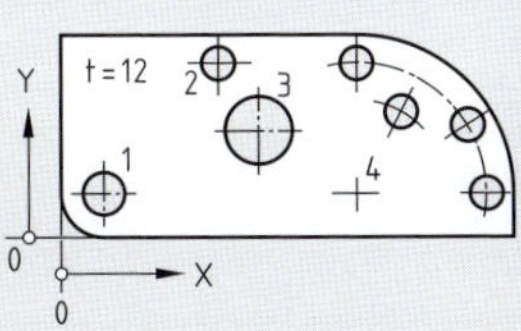

Item	X	Y	*d*
1	50	50	⌀40
2	180	190	⌀30
3	220	115	⌀75
4	325	50	–

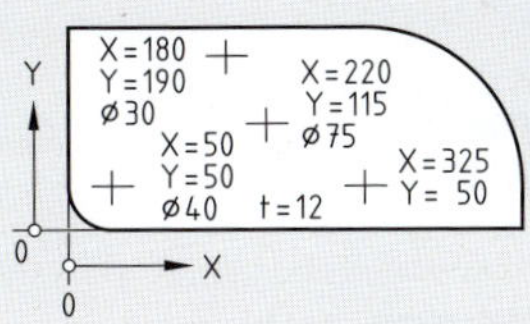

Cartesian coordinates (page 53)

Coordinate values. These are

- entered in tables or
- entered near the coordinate points.

Point of origin. The point of origin

- is entered with a small circle
- can lie at any location of the drawing.

Dimensions. These must be provided with a minus sign if they are entered from the origin in the opposite direction to the positive direction.

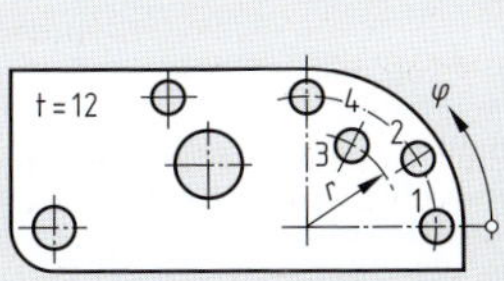

Item	*r*	*φ*	*d*
1	140	0°	⌀30
2	140	30°	⌀30
3	100	60°	⌀30
4	140	90°	⌀30

Polar coordinates (page 53)

Coordinate values. The coordinate values are entered in tables.

[1] Parallel dimensioning, running dimensioning and coordinate dimensioning may be combined with each other.

TD

Simplified presentation in drawings

Simplified representation of holes

cf. DIN 6780 (2000-10)

Hole base, line widths for simplified representation

Full scale representation, full scale dimensioning	Full scale representation, simplified dimensioning	Simplified representation, simplified dimensioning

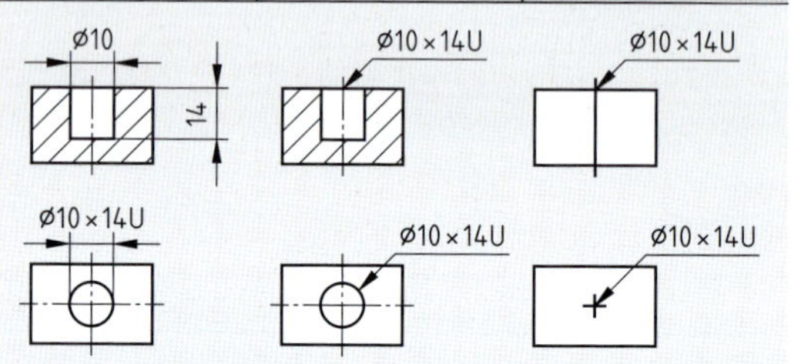

Hole base

The shape of the hole base is given by a symbol if necessary.

The **symbol U** for example means a **flat hole base** (cylindrical end bore).

Line widths

For holes depicted in simplified form, the positions of holes should be drawn as:

- simply the intersecting axes in the top view
- the position of the holes in thick solid lines in parallel axis representation.

TD

Stepped holes, countersinks and chamfers, internal threads

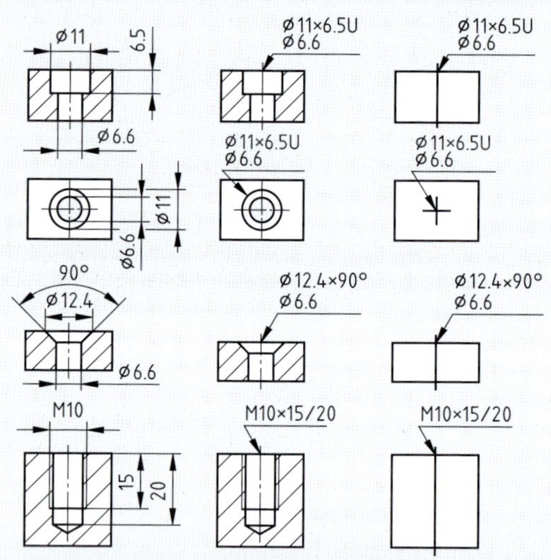

Stepped holes

For holes with two or more steps the dimensions are written under each other. Here the largest diameter is written on the first line.

Countersinks and chamfers

For countersinks and hole chamfers the largest countersink diameter and the countersink angle are given.

Internal threads

The thread length and the hole depth are separated by a slash. Holes without depth specification are drilled through.

Examples

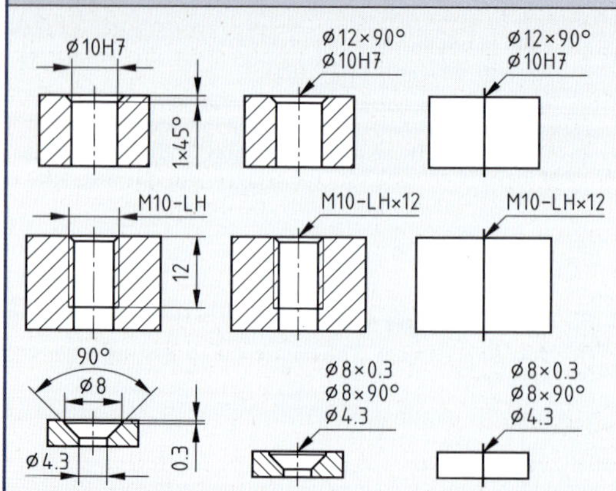

Hole ⌀10H7
Through hole
Chamfer 1 × 45°

Left hand thread M10
Thread length 12 mm
Drilled through core hole

Cylindrical countersink ⌀8
Bore depth 0.3 mm
Through hole ⌀4.3 with
cone shaped counterbore 90°
Countersink diameter ⌀8

Gear types

Representation of gears[1)]

cf. DIN ISO 2203 (1976-06)

Spur gear

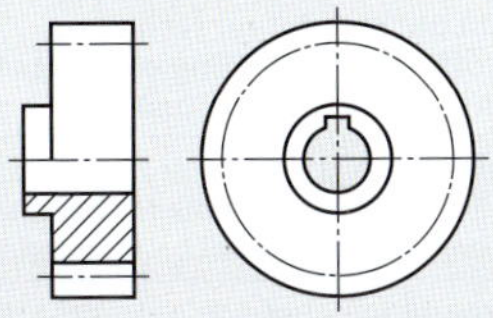

Bevel gear

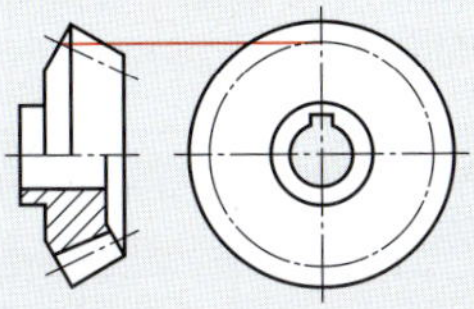

Worm gear

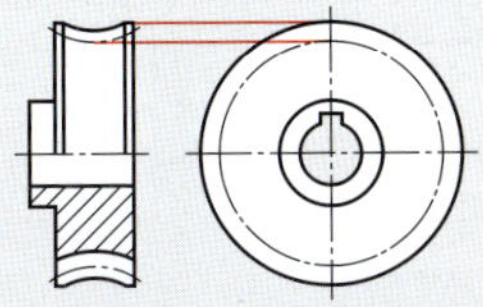

External helical gear

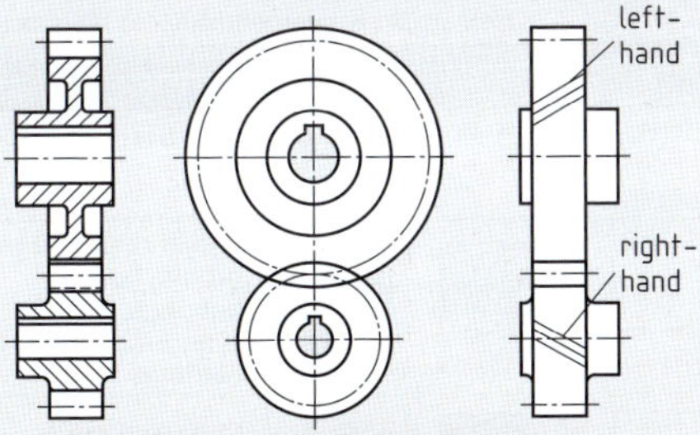

Internal spur gear

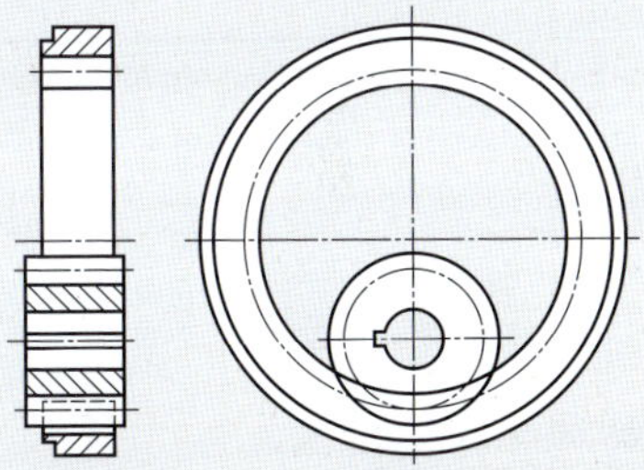

Rack and Pinion

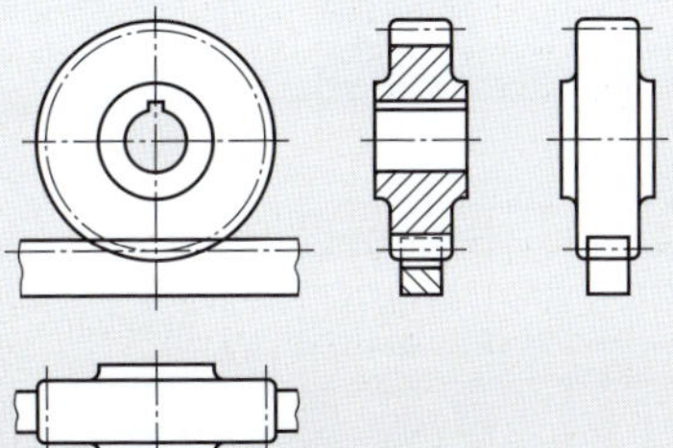

Bevel gear set (shaft angle 90°)

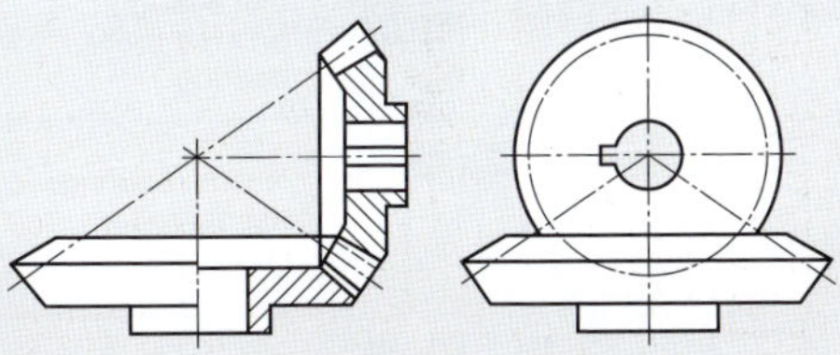

Worm and worm gear

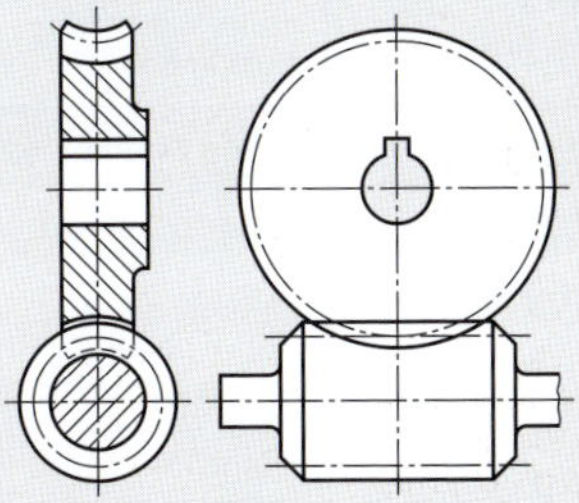

Sprockets

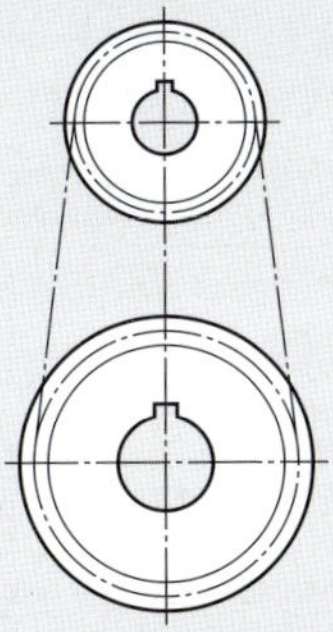

Positive drive belts

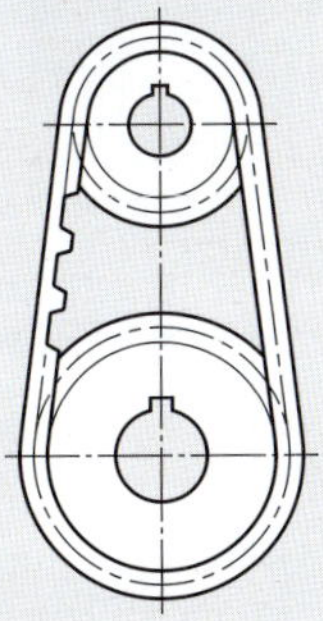

1) Dimensioning of gears on page 97

Roller bearings

Representation of roller bearings

cf. DIN ISO 8826-1 (1990-12) and DIN ISO 8826-2 (1995-10)

Representation			Elements of a detailed simplified representation	
simplified	**graphical**	**explanation**	**element**	**explanation, application**
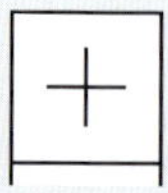	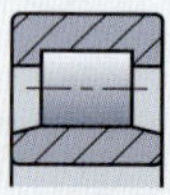	For general purposes a roller bearing is represented as square or rectangular with a free-standing upright cross.		Long, straight line; for representing the axis of the roller bearing elements for bearings that cannot be adjusted.
				Long, curved line; for representing the axis of the roller bearing elements for bearings that can be adjusted (self-aligning bearing).
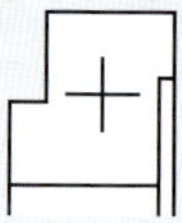	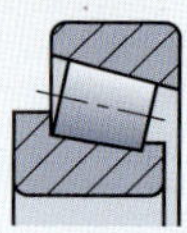	If necessary, the roller bearing can be represented by its outline and a free-standing upright cross.		Short straight line; used to represent the position and number of rows of roller bearing elements.
				Circle; for the representation of roller bearing elements (balls, roller, needle rollers) which are drawn perpendicular to their axis.

Examples of detailed simplified representation of roller bearings

Representation of single-row roller bearings			Representation of double row roller bearings		
detailed simplified	**graphical**	**designation**	**detailed simplified**	**graphical**	**designation**
	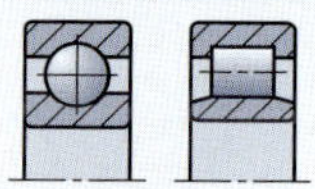	Radial-deep groove ball bearings, cylindrical roller bearings	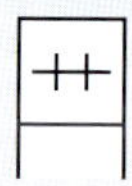	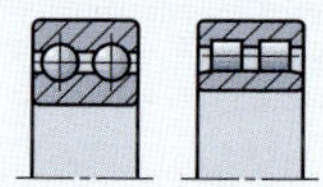	Radial-deep groove ball bearings, cylindrical roller bearings
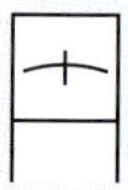	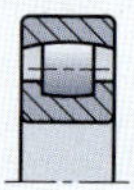	Radial spherical roller bearing (barrel-shaped bearing)	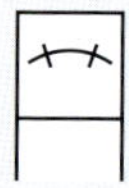	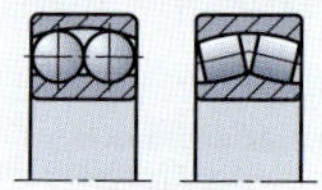	Spherical roller bearing, radial-spherical roller bearing
	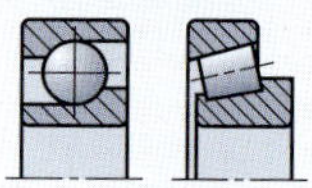	Angular-contact ball bearing, tapered roller bearing		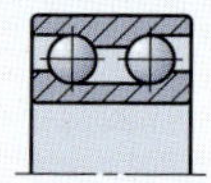	Angular-contact ball bearings
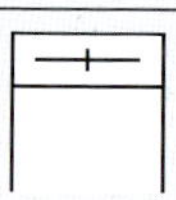	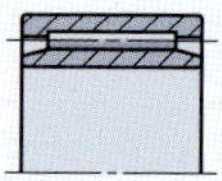	Needle bearing, needle roller assembly	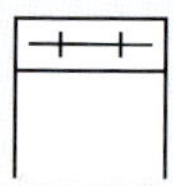	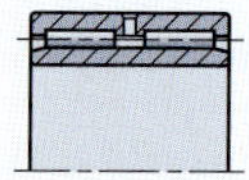	Needle bearing, needle roller assembly
	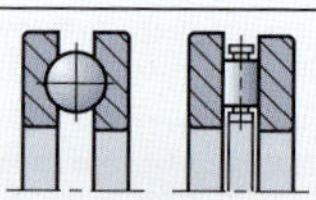	Axial-deep grooved ball bearing, axial-roller bearing	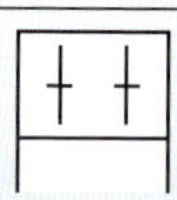	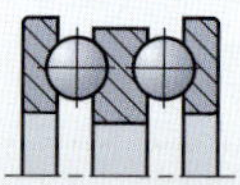	Axial-deep grooved ball bearing, dual action
	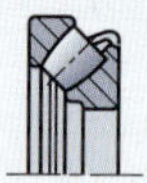	Axial-spherical roller bearing	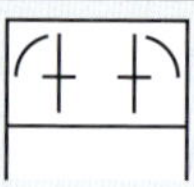	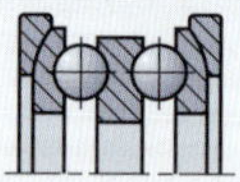	Axial-deep grooved ball bearing with spherical seating, dual action

Combined ball bearings			Representation perpendicular to the rolling element axis	
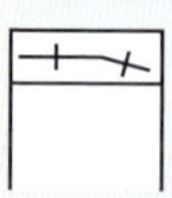	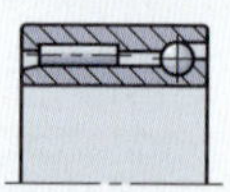	Combined radial-needle bearing with angular-contact ball bearing	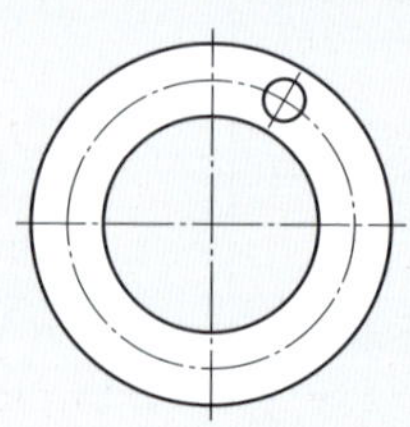	Roller bearing with any desired type of roller element shape (balls, rollers, needles)
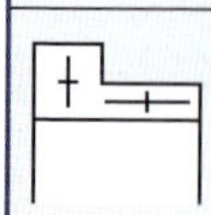	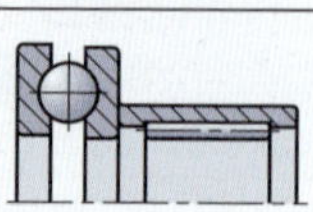	Combined axial-ball bearing with radial needle bearing		

TD

Representation of seals and roller bearings

Simplified representation of seals

cf. DIN ISO 9222-1 (1990-12) and DIN ISO 9222-2 (1991-03)

Representation: simplified	graphical	explanation	Elements of a detailed simplified representation: element	explanation, application
	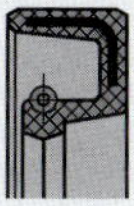	For general purposes a seal is represented by a square or rectangle and a separate diagonal cross-mark. The sealing direction can be given by an arrow.		Long line parallel to the sealing surface; for the fixed (static) sealing element.
				Long diagonal line; for the dynamic sealing element; e.g. the sealing lip. The sealing direction can be given by an arrow.
				Short diagonal line; for dust lip seal, scraper rings.
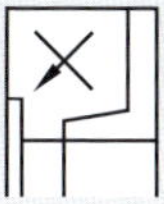	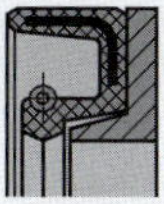	If necessary, the seals can be represented by the outline and a free-standing diagonal cross-mark.		Short lines pointing to the middle of the symbol; for the static parts of U-rings und V-rings, packing.
				Short lines, which point to the middle of the symbol; for the sealing lips of U-rings und V-rings, packing.
				T and U; for non-contact seals.

TD

Examples of detailed simplified representation of seals

Shaft seals and piston rod seals: detailed simplified	graphical	designation for rotation	designation for linear motion	Profile gaskets, packing sets, labyrinth seals: detailed simplified	graphical	detailed simplified	graphical
		Shaft seal without dust lip seal	Rod seal without stripper	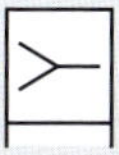			
		Shaft seal with dust lip seal	Rod seal with stripper				
		Shaft seal, dual action	Rod seal, dual action			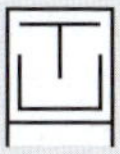	

Examples of simplified representation of seals and roller bearings

Deep grooved roller bearings and radial shaft seal with dust lip seal[1]

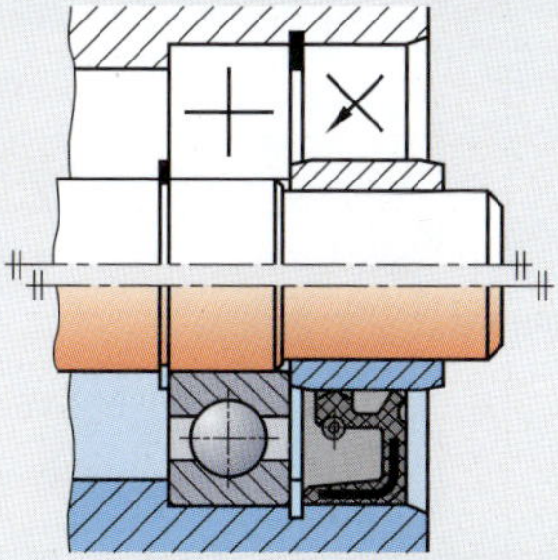

Dual row deep grooved roller bearings and radial shaft seal[2]

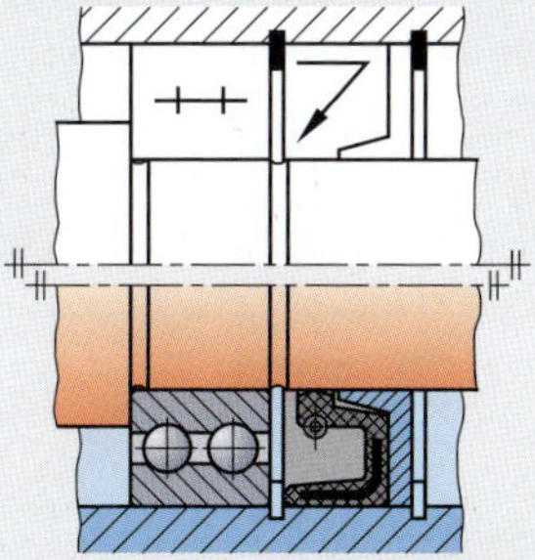

Packing set[2]

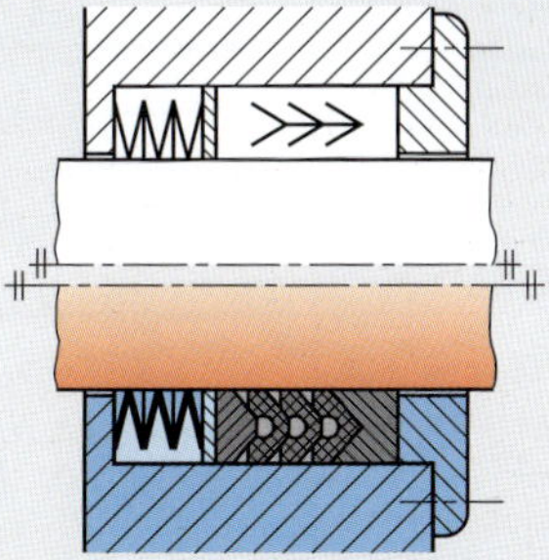

[1] Top half: simplified representation; bottom half: graphical representation.
[2] Top half: detailed simplified representation; bottom half: graphical representation.

TD

Representation of retaining rings, Slots for retaining rings, Springs, Splines and serrations

Representation of retaining rings and slots for retaining rings

	Representation	Assembly dimension	Deviations
Retaining rings for shafts (page 265)	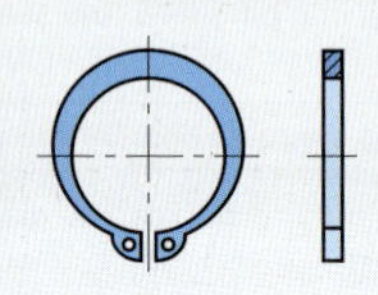	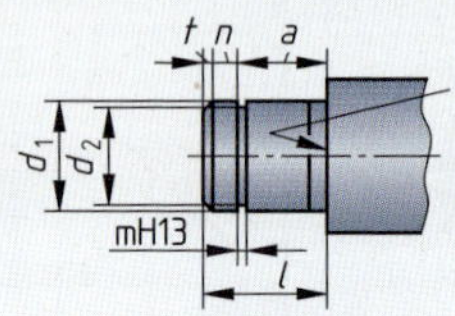 reference plane for dimensioning[1)] a = roller bearing width + retaining ring width	Deviations for d_2: upper deviation: 0 (zero) lower deviation: negative Deviations for a: upper deviation: positive lower deviation: 0 (zero)
Retaining rings for holes (page 265)	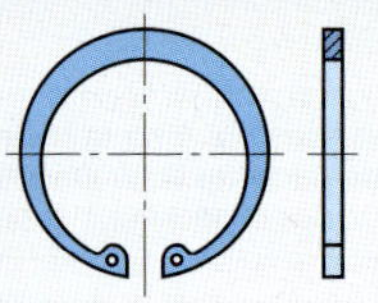	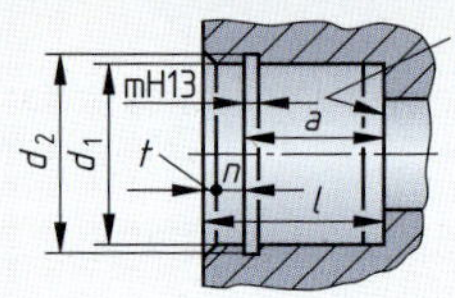 reference plane for dimensioning[1)]	Deviations for d_2: upper deviation: positive lower deviation: 0 (zero) Deviations for a: upper deviation: positive lower deviation: 0 (zero)

[1)] For functional reasons the reference plane for the dimensioning of slots is the locating face of the part to be secured.

Representation of springs

cf. DIN ISO 2162-1 (1994-08)

Name	Representation view	Representation section	Symbol	Name	Representation view	Representation section	Symbol
Cylindrical helical compression spring (round wire)	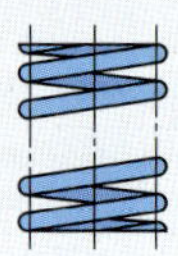	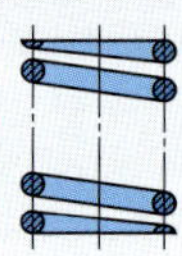	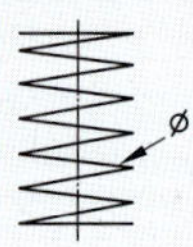	Cylindrical helical tension spring	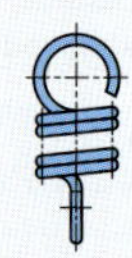	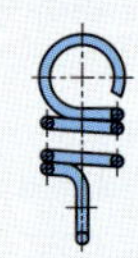	
Cylindrical helical tension spring		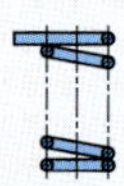		Cylindrical helical compression spring (square wire)	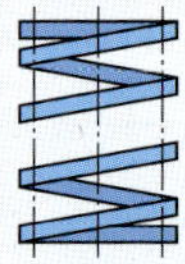	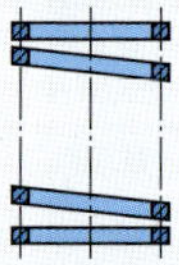	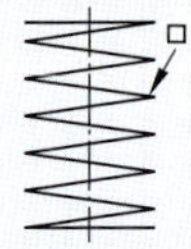
Disk spring (simple)			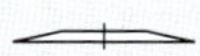	Disk spring assembly (disks layered in alternating directions)	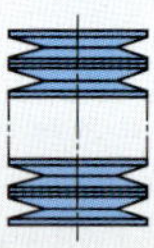	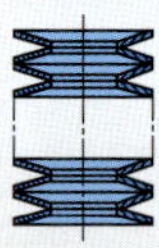	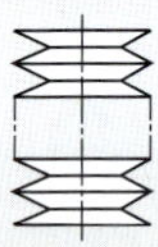
Disk spring assembly (disks layered in the same direction)	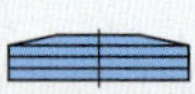	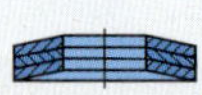	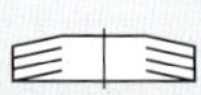				

Representation of splines and serrations

cf. DIN ISO 6413 (1990-03)

	Shaft	Hub	Joint
Splines or spline hubs with straight flanks. Symbol: ⊓	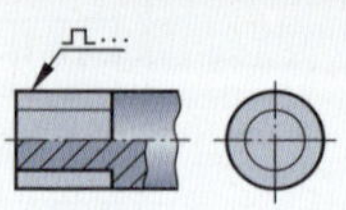	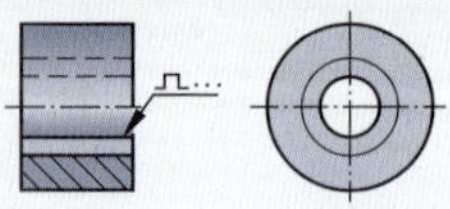	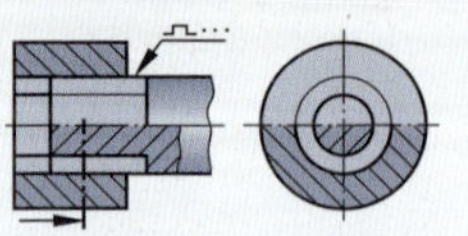
Toothed shafts or toothed hubs with involute splines or serrations. Symbol: ⋀	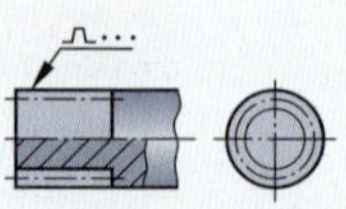	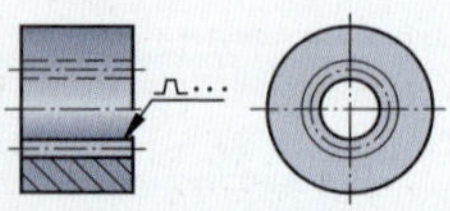	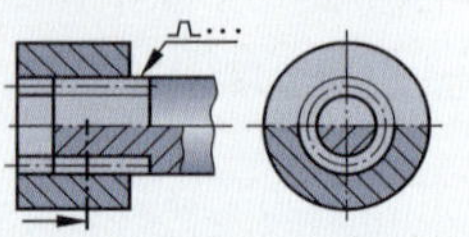
	⇒ **Splines ISO 14-6 x 26 f7 x 30:** Spline profile with straight flanks according to ISO 14, number of splines N = 6, inner diameter d = 26f7, outer diameter D = 30		

Bosses on turned parts, Object corners and edges

Bosses on turned parts
cf. DIN 6785 (1991-11)

Boss dimensions

Example

Drawing entry

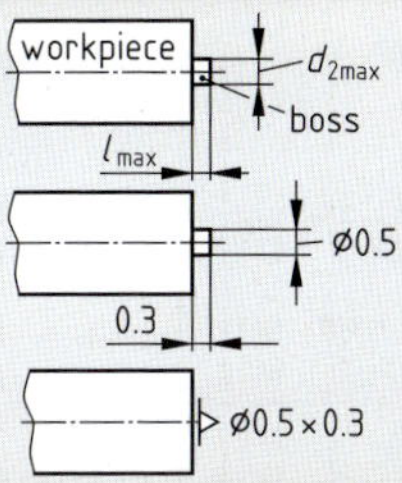

Boss dimensions	Largest diameter of the finished part in mm							
	up to 3	over 3 to 5	over 5 to 8	over 8 to 12	over 12 to 18	over 18 to 26	over 26 to 40	over 40 to 60
$d_{2\,max}$ in mm	0.3	0.5	0.8	1.0	1.5	2.0	2.5	3.5
l_{max} in mm	0.2	0.3	0.5	0.6	0.9	1.2	2.0	3.0

Object corners and edges
cf. DIN ISO 13715 (2000-12)

Edge or corner	Object edge/corner lies in reference to the ideal geometrical form: inside	outside	in area
outer edge 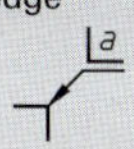	material removal 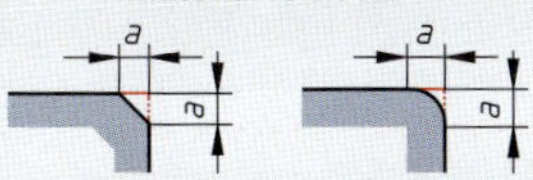	burr 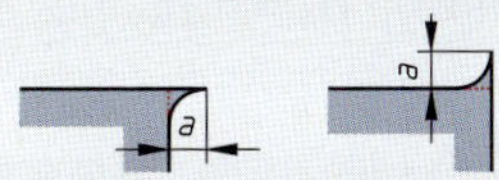	sharp edged 
inner edge 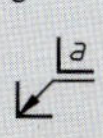	material removal 	transition 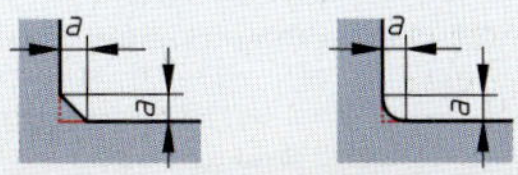	sharp edged 
Dim. *a* (mm)	−0.1; −0.3; −0.5; −1.0; −2.5	+0.1; +0.3; +0.5; +1.0; +2.5	−0.05; −0.02; +0.02; +0.05

Symbol for labeling object edges/corners	Symbol element	Meaning for outer edge	Meaning for inner edge	Burr and material removal direction	outer edge	inner edge
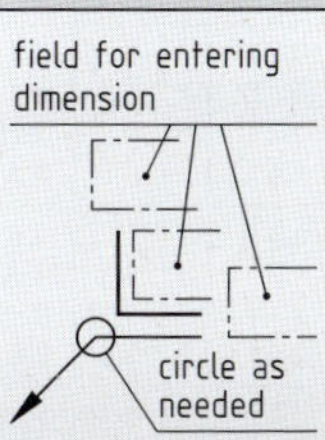	+	Burr allowed, material removal not allowed	Transition allowed, material removal not allowed	Specification allowed for	Burr	Material removal
	–	Removal required, burr not allowed	Removal required, transition not allowed	Example	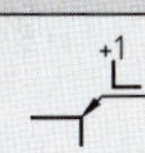	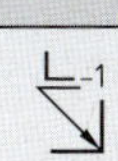
	± [1]	Burr or transition allowed	Material removal or transition allowed	Meaning	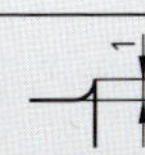	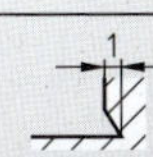

[1] only allowed with a dimension callout

Labeling of object corners and edges

Collective indications

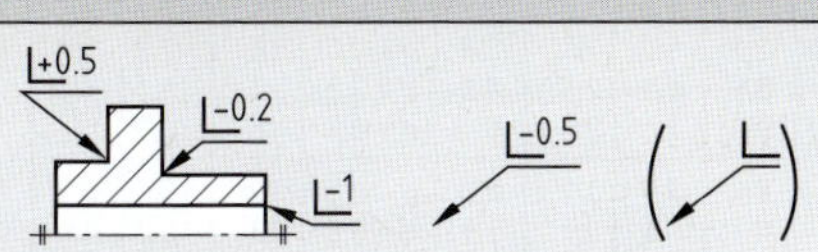

Collective indications apply to all edges for which an edge condition is not given.

Edges for which the collective indication does not apply must be marked in the drawing.

The exceptions are placed after the collective indication in parentheses or indicated by the base symbol.

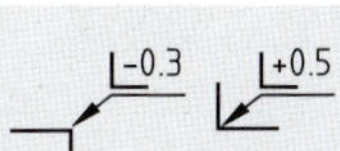

Collective indications which **are only valid for outside or inside edges** are given by the corresponding symbols.

Examples

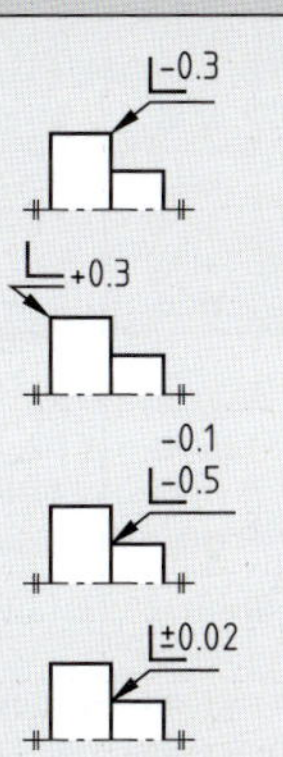

Outside edge without burr. The allowable material removal is between 0 and 0.3 mm.

Outside edge with allowable burr of 0 to 0.3 mm (burr direction specified).

Inside edge with allowable material removal between 0.1 and 0.5 mm (material removal direction not specified).

Inside edge with allowable material removal between 0 and 0.02 mm or allowable transition up to 0.02 mm (sharp edged).

TD

Thread runouts, Thread undercuts

Thread runouts for metric ISO threads

cf. DIN 76-1 (2004-06)

External thread

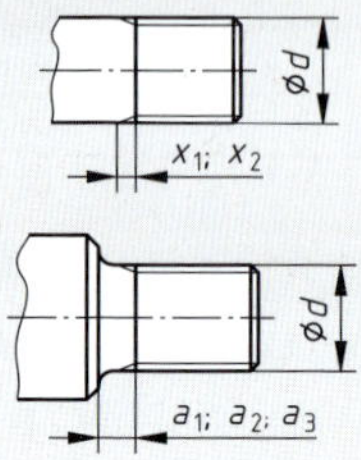

Internal thread

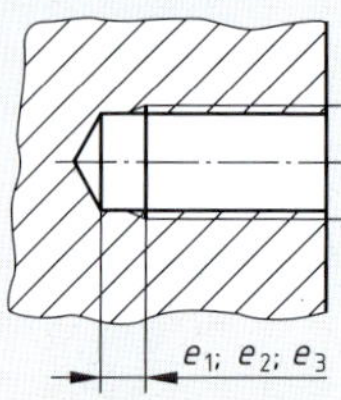

Pitch [1]	ISO standard thread	Thread runout [2]			Pitch [1]	ISO standard thread	Thread runout [2]		
P	d	x_1 max.	a_1 max.	e_1	P	d	x_1 max.	a_1 max.	e_1
0.2	–	0.5	0.6	1.3	1.25	M8	3.2	3.75	6.2
0.25	M1	0.6	0.75	1.5	1.5	M10	3.8	4.5	7.3
0.3	–	0.75	0.9	1.8	1.75	M12	4.3	5.25	8.3
0.35	M1.6	0.9	1.05	2.1	2	M16	5	6	9.3
0.4	M2	1	1.2	2.3	2.5	M20	6.3	7.5	11.2
0.45	M2.5	1.1	1.35	2.6	3	M24	7.5	9	13.1
0.5	M3	1.25	1.5	2.8	3.5	M30	9	10.5	15.2
0.6	–	1.5	1.8	3.4	4	M36	10	12	16.8
0.7	M4	1.75	2.1	3.8	4.5	M42	11	13.5	18.4
0.75	–	1.9	2.25	4	5	M48	12.5	15	20.8
0.8	M5	2	2.4	4.2	5.5	M56	14	16.5	22.4
1	M6	2.5	3	5.1	6	M64	15	18	24

1) For fine threads the dimension of the thread runout is chosen according to the pitch P.

2) As a rule; applies if no other entries are given.
If a shorter thread runout is necessary, this applies:
$x_2 \approx 0.5 \cdot x_1$; $a_2 \approx 0.67 \cdot a_1$; $e_2 \approx 0.625 \cdot e_1$
If a longer thread runout is necessary, this applies:
$a_3 \approx 1.3 \cdot a_1$; $e_3 \approx 1.6 \cdot e_1$

Screw thread undercuts for metric ISO threads

cf. DIN 76-1 (2004-06)

External thread
form A and form B

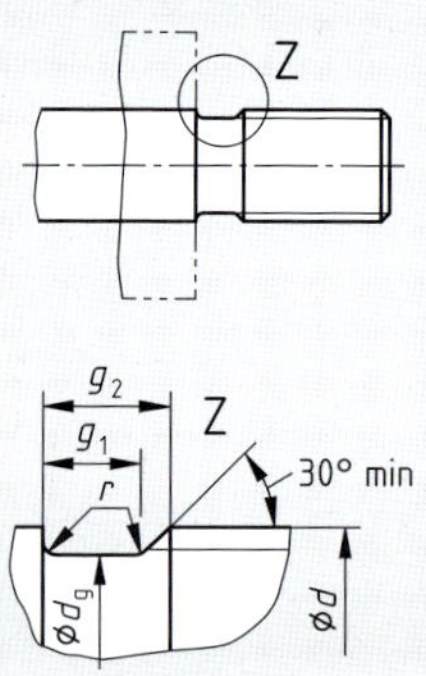

Internal thread
form C and form D

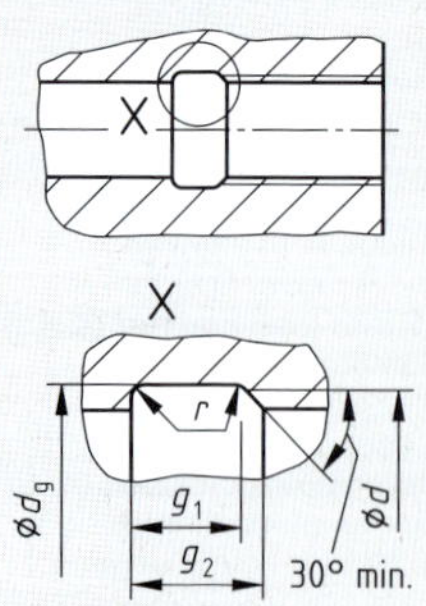

Pitch [1]	ISO standard thread		External threads					Internal threads				
				Form A [2]		Form B [3]			Form C [2]		Form D [3]	
P	d	r	d_g h13	g_1 min.	g_2 max.	g_1 min.	g_2 max.	d_g H13	g_1 min.	g_2 max.	g_1 min.	g_2 max.
0.2	–	0.1	$d - 0.3$	0.45	0.7	0.25	0.5	$d + 0.1$	0.8	1.2	0.5	0.9
0.25	M1	0.12	$d - 0.4$	0.55	0.9	0.25	0.6	$d + 0.1$	1	1.4	0.6	1
0.3	–	0.16	$d - 0.5$	0.6	1.05	0.3	0.75	$d + 0.1$	1.2	1.6	0.75	1.25
0.35	M1.6	0.16	$d - 0.6$	0.7	1.2	0.4	0.9	$d + 0.2$	1.4	1.9	0.9	1.4
0.4	M2	0.2	$d - 0.7$	0.8	1.4	0.5	1	$d + 0.2$	1.6	2.2	1	1.6
0.45	M2.5	0.2	$d - 0.7$	1	1.6	0.5	1.1	$d + 0.2$	1.8	2.4	1.1	1.7
0.5	M3	0.2	$d - 0.8$	1.1	1.75	0.5	1.25	$d + 0.3$	2	2.7	1.25	2
0.6	–	0.4	$d - 1$	1.2	2.1	0.6	1.5	$d + 0.3$	2.4	3.3	1.5	2.4
0.7	M4	0.4	$d - 1.1$	1.5	2.45	0.8	1.75	$d + 0.3$	2.8	3.8	1.75	2.75
0.75	–	0.4	$d - 1.2$	1.6	2.6	0.9	1.9	$d + 0.3$	3	4	1.9	2.9
0.8	M5	0.4	$d - 1.3$	1.7	2.8	0.9	2	$d + 0.3$	3.2	4.2	2	3
1	M6	0.6	$d - 1.6$	2.1	3.5	1.1	2.5	$d + 0.5$	4	5.2	2.5	3.7
1.25	M8	0.6	$d - 2$	2.7	4.4	1.5	3.2	$d + 0.5$	5	6.7	3.2	4.9
1.5	M10	0.8	$d - 2.3$	3.2	5.2	1.8	3.8	$d + 0.5$	6	7.8	3.8	5.6
1.75	M12	1	$d - 2.6$	3.9	6.1	2.1	4.3	$d + 0.5$	7	9.1	4.3	6.4
2	M16	1	$d - 3$	4.5	7	2.5	5	$d + 0.5$	8	10.3	5	7.3
2.5	M20	1.2	$d - 3.6$	5.6	8.7	3.2	6.3	$d + 0.5$	10	13	6.3	9.3
3	M24	1.6	$d - 4.4$	6.7	10.5	3.7	7.5	$d + 0.5$	12	15.2	7.5	10.7
3.5	M30	1.6	$d - 5$	7.7	12	4.7	9	$d + 0.5$	14	17.7	9	12.7
4	M36	2	$d - 5.7$	9	14	5	10	$d + 0.5$	16	20	10	14
4.5	M42	2	$d - 6.4$	10.5	16	5.5	11	$d + 0.5$	18	23	11	16
5	M48	2.5	$d - 7$	11.5	17.5	6.5	12.5	$d + 0.5$	20	26	12.5	18.5
5.5	M56	3.2	$d - 7.7$	12.5	19	7.5	14	$d + 0.5$	22	28	14	20
6	M64	3.2	$d - 8.3$	14	21	8	15	$d + 0.5$	24	30	15	21

⇒ **DIN 76-C:** Screw thread undercut shape C

1) For fine thread screws the dimension of the thread undercut is chosen according to the pitch P.

2) as a rule; always applies if no other entries are made

3) Only in cases where a shorter thread undercut is required.

Representation of threads and screw joints

Representation of threads

cf. DIN ISO 6410-1 (1993-12)

Internal thread

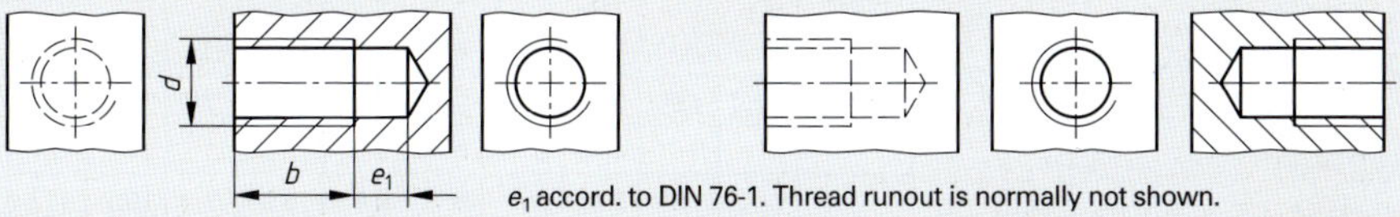

e_1 accord. to DIN 76-1. Thread runout is normally not shown.

Bolt thread

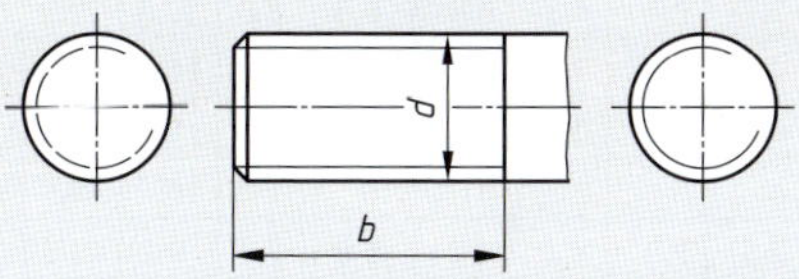

Bolts in internal thread

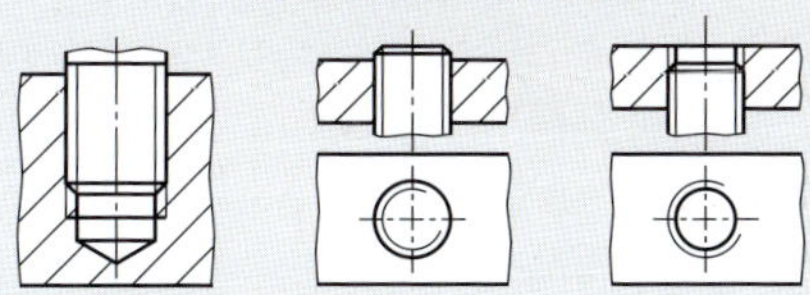

Thread undercut

graphical

symbolic

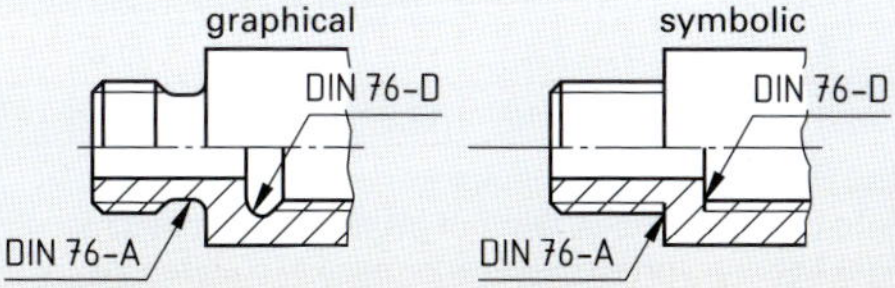

Pipe threads and pipe screw joints

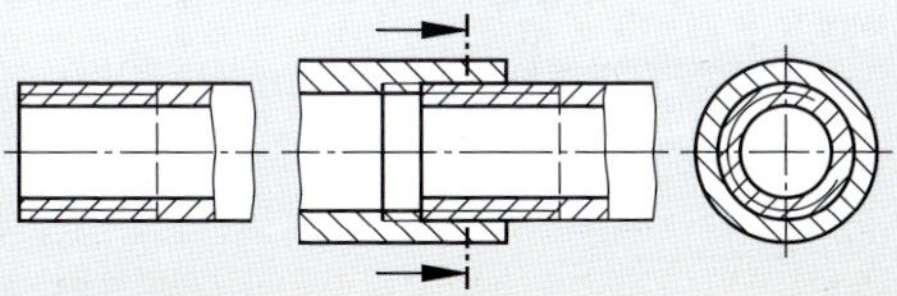

Representation of screw joints

Hexagonal bolt and nut

detailed

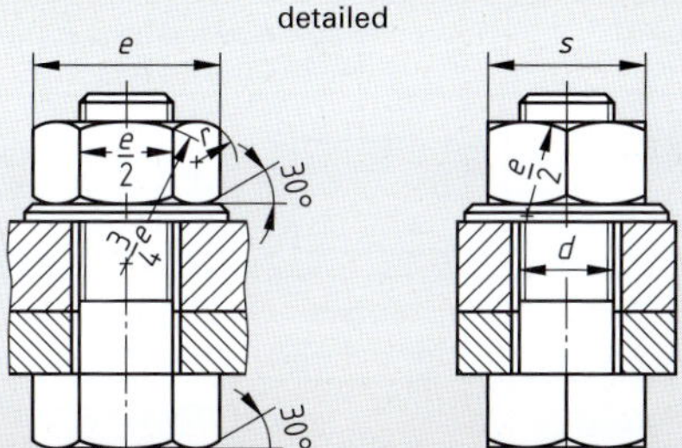

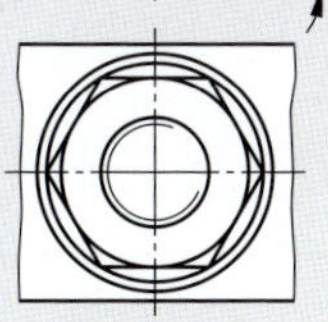

h_1 bolt head hight
h_2 nut height
h_3 washer thickness
e diagonal between corners
s width across flats
d thread nominal ø

simplified

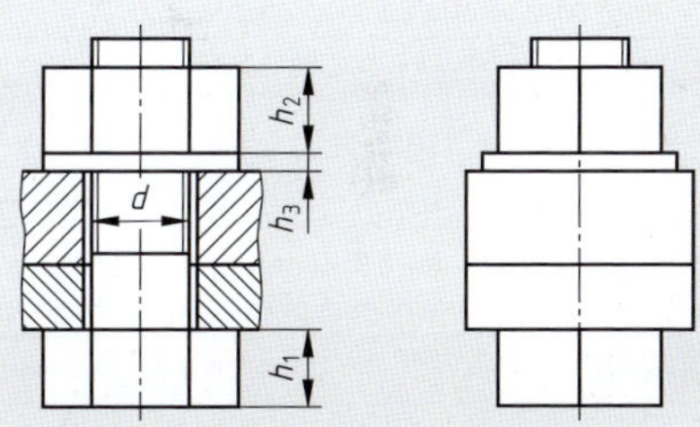

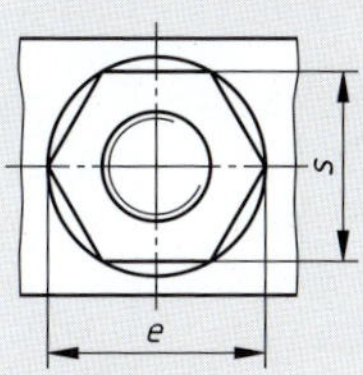

$h_1 \approx 0.7 \cdot d$
$h_2 \approx 0.8 \cdot d$
$h_3 \approx 0.2 \cdot d$
$e \approx 2 \cdot d$
$s \approx 0.87 \cdot e$

Screw joint with cap screw

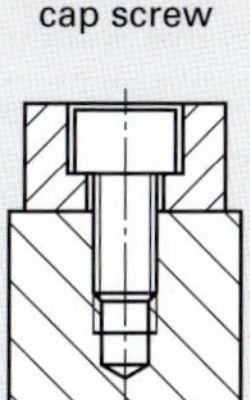

Screw joint with hexagonal screw

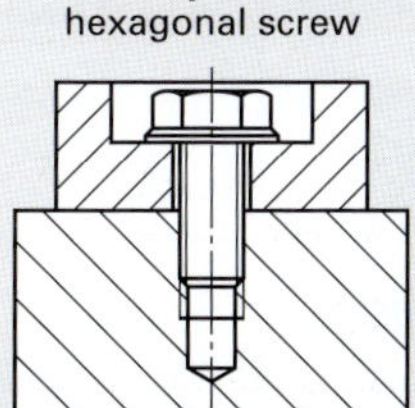

Screw joint with flat head screw

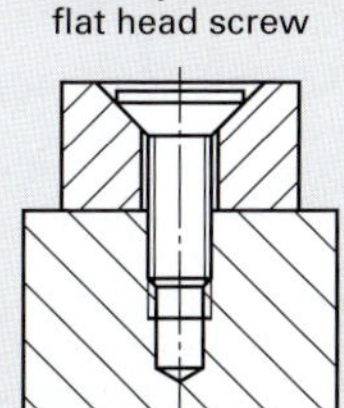

Screw joint with stud

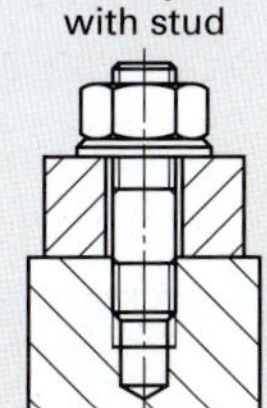

TD

Center holes, Knurls

Center holes

cf. DIN 332-1 (1986-04)

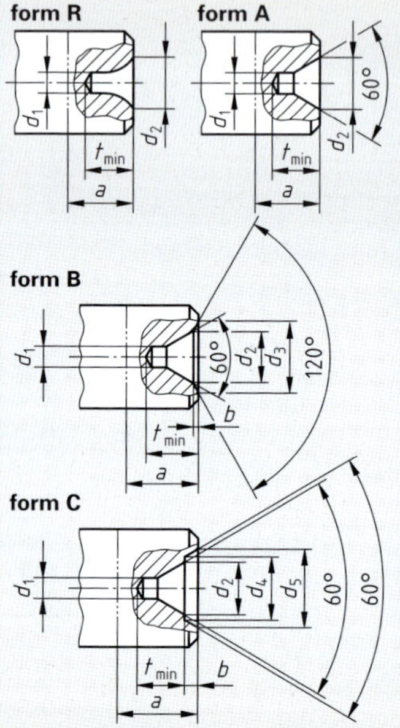

Form		Nominal sizes									
	d_1	1	1.25	1.6	2	2.5	3.15	4	5	6.3	8
	d_2	2.12	2.65	3.35	4.25	5.3	6.7	8.5	10.6	13.2	17
R	t_{min}	1.9	2.3	2.9	3.7	4.6	5.8	7.4	9.2	11.4	14.7
	a	3	4	5	6	7	9	11	14	18	22
A	t_{min}	1.9	2.3	2.9	3.7	4.6	5.9	7.4	9.2	11.5	14.8
	a	3	4	5	6	7	9	11	14	18	22
B	t_{min}	2.2	2.7	3.4	4.3	5.4	6.8	8.6	10.8	12.9	16.4
	a	3.5	4.5	5.5	6.6	8.3	10	12.7	15.6	20	25
	b	0.3	0.4	0.5	0.6	0.8	0.9	1.2	1.6	1.4	1.6
	d_3	3.15	4	5	6.3	8	10	12.5	16	18	22.4
C	t_{min}	1.9	2.3	2.9	3.7	4.6	5.9	7.4	9.2	11.5	14.8
	a	3.5	4.5	5.5	6.6	8.3	10	12.7	15.6	20	25
	b	0.4	0.6	0.7	0.9	0.9	1.1	1.7	1.7	2.3	3
	d_4	4.5	5.3	6.3	7.5	9	11.2	14	18	22.4	28
	d_5	5	6	7.1	8.5	10	12.5	16	20	25	31.5

Form
- **R:** curved bearing surface, without protective countersink
- **A:** straight bearing surface, without protective countersink
- **B:** straight bearing surface, conical protective countersink
- **C:** straight bearing surface, truncated conical protective countersink

Drawing callout for center holes

cf. DIN ISO 6411 (1997-11)

A center hole **is** required on the finished part	A center hole **is allowed** on the finished part	A center hole **may not** be present on the finished part
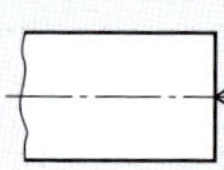	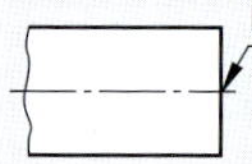	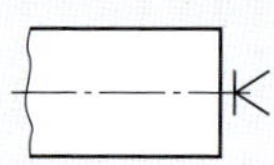

⇒ **< ISO 6411 – A4/8.5:** center hole ISO 6411: a center hole is required on the finished part.
Form and dimensions of the center hole according to DIN 332: form A; d_1 = 4 mm; d_2 = 8.5 mm.

Knurls

cf. DIN 82 (1973-01)

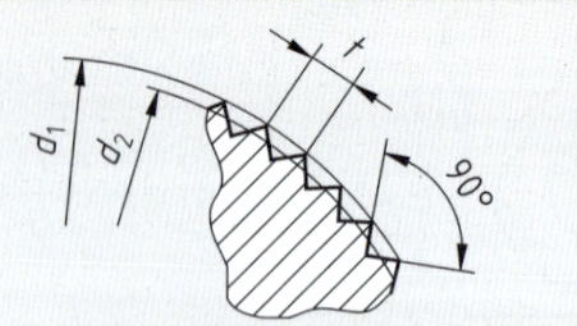

d_1 nominal diameter
d_2 initial diameter
t spacing

Standard spacing values
t: 0.5; 0.6; 0.8; 1.0; 1.2; 1.6 mm

Drawing entry (example):

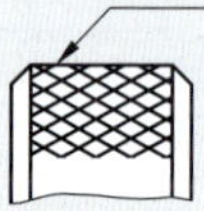

Letter symbol	Representation	Name	Point shape	Initial diameter d_2
RAA	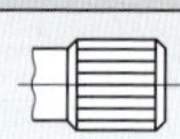	Knurls with axially parallel grooves	–	$d_2 = d_1 - 0.5 \cdot t$
RBR	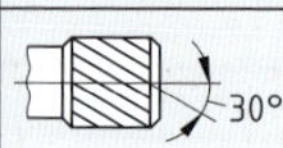	Right-hand knurl	–	$d_2 = d_1 - 0.5 \cdot t$
RBL	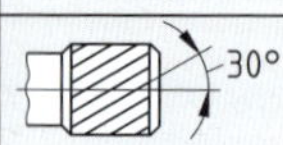	Left-hand knurl	–	$d_2 = d_1 - 0.5 \cdot t$
RGE	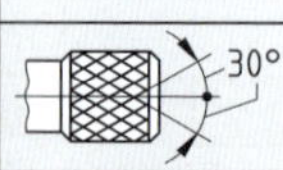	Left-hand/right-hand knurls	raised	$d_2 = d_1 - 0.67 \cdot t$
RGV			recessed	$d_2 = d_1 - 0.33 \cdot t$
RKE	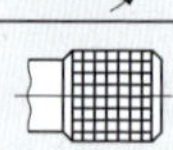	Axial and circumferential knurl	raised	$d_2 = d_1 - 0.67 \cdot t$
RKV			recessed	$d_2 = d_1 - 0.33 \cdot t$

⇒ **DIN 82-RGE 0.8:** Left-hand/right-hand knurls, raised points, t = 0.8 mm

Undercuts

Undercuts[1]

cf. DIN 509 (2006-12)

form E
for cylindrical surface to be further machined

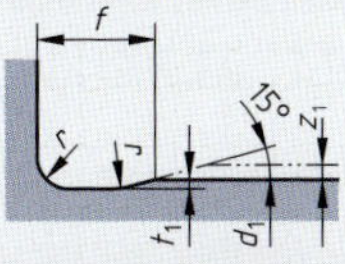

form F
for shoulders and cylindrical surfaces to be further machined

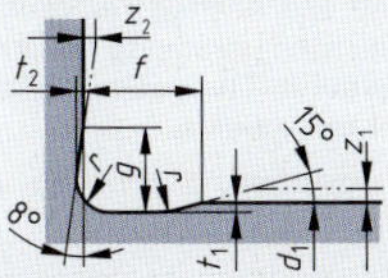

form G
for small transition (for low loading)

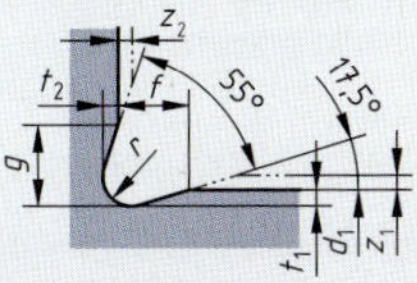

form H
for planar and cylindrical surfaces to be further machined

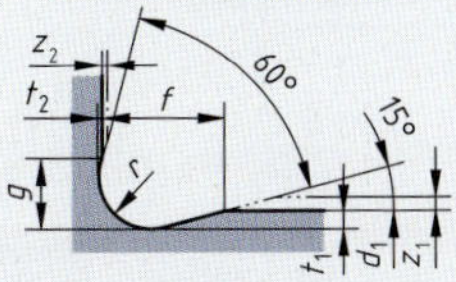

z_1, z_2 = machining allowances

⇒ **Undercut DIN 509 – E 0.8 x 0.3:** form E, radius r = 0.8 mm, undercut depth t_1 = 0.3 mm

Undercut dimensions and countersink dimensions

Form	r[2] ± 0.1		t_1	t_2	f	g	Correlation to diameter d_1[3] for objects with		Minimum dimension a for countersink on the opposing piece[4]				
	Series 1	Series 2	+ 0.1 / 0	+ 0.05 / 0	+ 0.2 / 0		normal loading	increased fatigue strength	Undercut $r \times t_1$	Form E	Form F	Form G	Form H
E and F	–	0.2	0.1	0.1	1	(0.9)	> Ø 1.6–Ø 3	–	0.2 x 0.1	0.2	0	–	–
E and F	0.4	–	0.2	0.1	2	(1.1)	> Ø 3–Ø 18	–	0.4 x 0.2	0.3	0	–	–
E and F	–	0.6	0.2	0.1	2	(1.4)	> Ø 10–Ø 18	–	0.6 x 0.2	0.5	0.15	–	–
E and F	–	0.6	0.3	0.2	2.5	(2.1)	> Ø 18–Ø 80	–	0.6 x 0.3	0.4	0	–	–
E and F	0.8	–	0.3	0.2	2.5	(2.3)	> Ø 18– Ø 80	–	0.8 x 0.3	0.6	0.05	–	–
E and F	–	1	0.2	0.1	2.5	(1.8)	–	> Ø 18–Ø 50	1.0 x 0.2	0.9	0.45	–	–
E and F	–	1	0.4	0.3	4	(3.2)	> Ø 80	–	1.0 x 0.4	0.7	0	–	–
E and F	1.2	–	0.2	0.1	2.5	(2)	–	> Ø 18–Ø 50	1.2 x 0.2	1.1	0.6	–	–
E and F	1.2	–	0.4	0.3	4	(3.4)	> Ø 80	–	1.2 x 0.4	0.9	0.1	–	–
E and F	1.6	–	0.3	0.2	4	(3.1)	–	> Ø 50–Ø 80	1.6 x 0.3	1.4	0.6	–	–
E and F	2.5	–	0.4	0.3	5	(4.8)	–	> Ø 80–Ø 125	2.5 x 0.4	2.2	1.0	–	–
E and F	4	–	0.5	0.3	7	(6.4)	–	> Ø 125	4.0 x 0.5	3.6	2.1	–	–
G	0.4	–	0.2	0.2	(0.9)	(1.1)	> Ø 3–Ø 18	–	0.4 x 0.2	–	–	0	–
H	0.8	–	0.3	0.05	(2.0)	(1.1)	> Ø 18–Ø 80	–	0.8 x 0.3	–	–	–	0.35
H	1.2	–	0.3	0.05	(2.4)	(1.5)	–	> Ø 18–Ø 50	1.2 x 0.3	–	–	–	0.65

[1] All forms of undercut apply to both shafts and holes.
[2] Undercuts with Series 1 radii are preferred.
[3] The correlation to the diameter area does not apply with curved shoulders and thin walled parts. For objects with differing diameters it may be advisable to design all undercuts for all diameters in the same form and size.

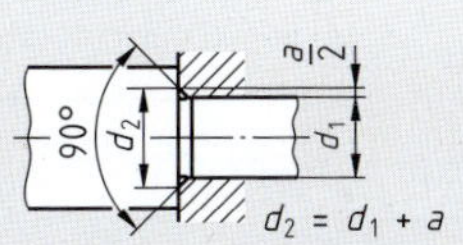

Drawing entry for undercuts

Normally undercuts are represented in drawings as a simplified entry with the designator. However they can also be completely drawn and dimensioned.

Example: Shaft with undercut DIN 509 – F1.2 x 0.2

simplified entry

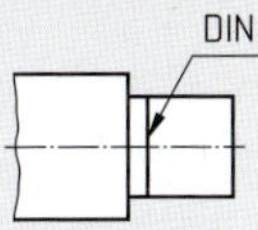

complete entry

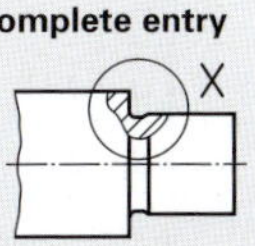

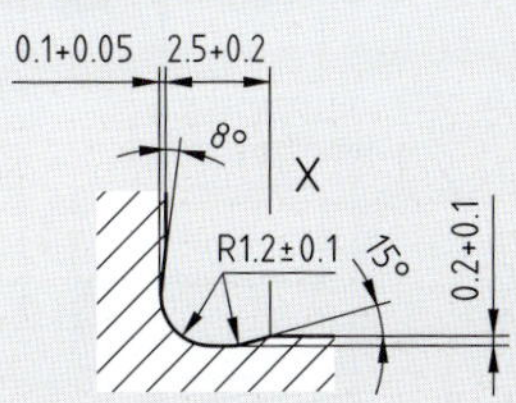

Example: Hole with undercut- DIN 509 – E1.2 x 0.2

simplified entry

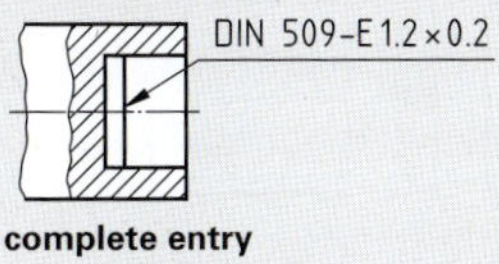

complete entry

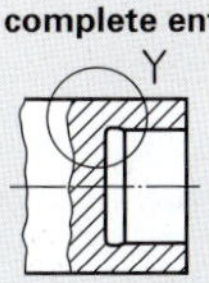

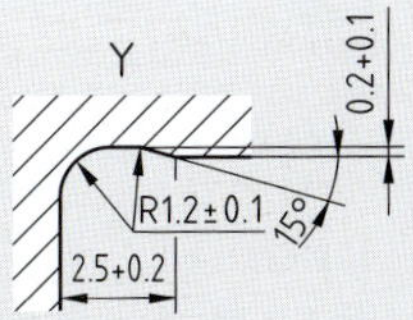

TD

TD

Symbols for Welding and Soldering

Positioning of symbols for welding and soldering in drawings

cf. DIN EN 22553 (1997-03)

Basic terms

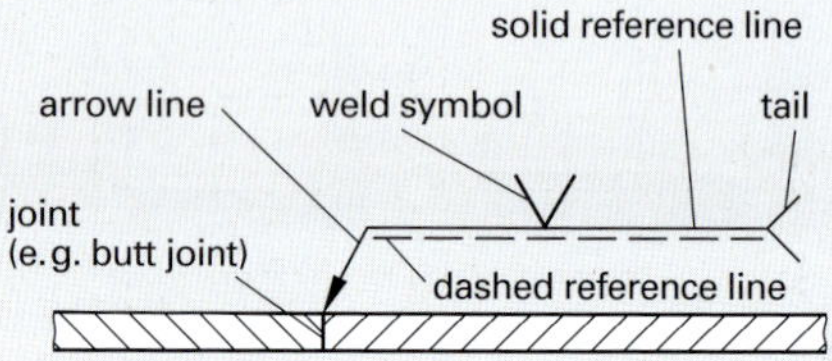

Reference line. This consists of the solid reference line and the dashed reference line. The dashed reference line runs parallel to the solid reference line and above or below it. The dashed reference line is omitted for symmetrical welds.

Arrow line. It connects the solid reference line with the joint.

Tail. Additional entries can be given here as needed for:

- method, process
- evaluation group
- working position
- additional material

Joint. Orientation of the parts to be joined to each other.

Weld information

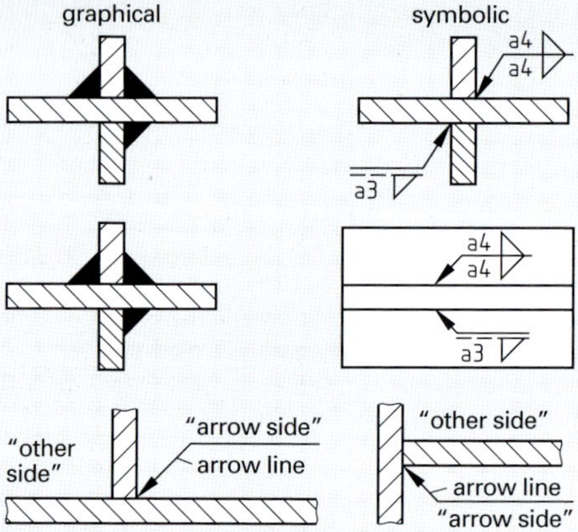

Symbol. The symbol identifies the form of the weld. It is preferably placed normal to the solid reference line, or if necessary on the dashed reference line.

Arrangement of the weld symbol	
position of the weld symbol	position of the weld (weld surface)
solid reference line	"arrow side"
dashed reference line	"other side"

For welds represented in section or view, the position of the symbol must agree with the weld cross section.

Arrow side. The arrow side is that side of the joint to which the arrow line refers.

Other side. The other side of the joint that is opposite the arrow side.

Supplemental and auxiliary symbols

cf. DIN EN 22553 (1997-03)

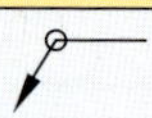	Weld all around		Weld surface hollow (concave)
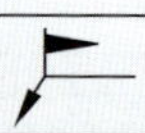	Field weld (weld is made on the construction site)		Weld surface flat (planar)
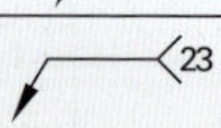	Entry of the welding process in the tail		Weld surface curved (convex)
		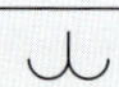	Weld surface notch free

Representation in drawings (basic symbols)

cf. DIN EN 22553 (1997-03)

Weld type/ symbol	Representation graphical	Representation symbolic	Weld type/ symbol	Representation graphical	Representation symbolic
Butt weld ‖	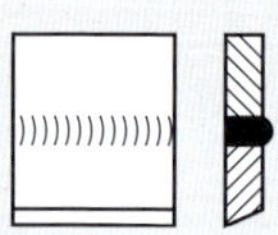	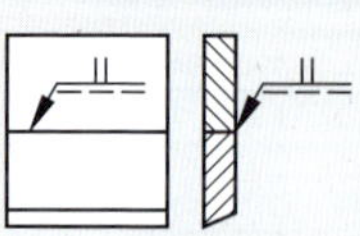	V groove weld V	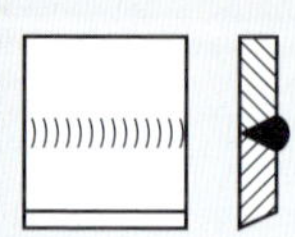	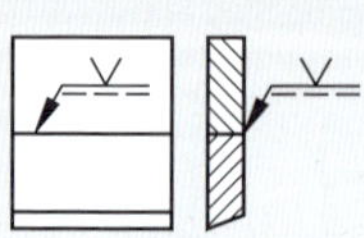
	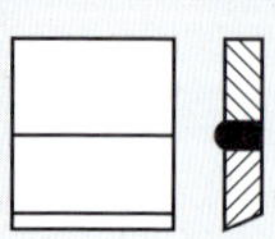	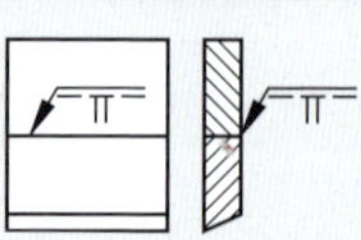		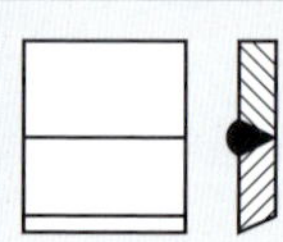	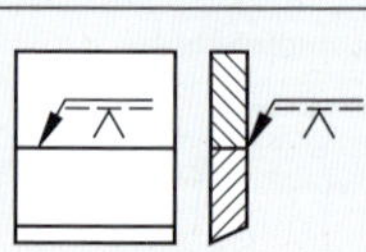

Symbols for Welding and Soldering

Representation in drawings (basic symbols)

cf. DIN EN 22553 (1997-03)

Weld type/ symbol	Representation graphical / symbolic	Weld type/ symbol	Representation graphical / symbolic
Flare-V groove weld	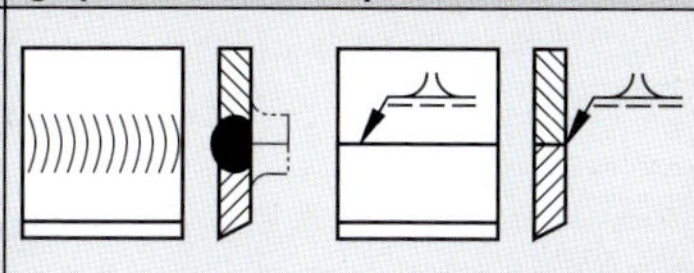	Bevel groove weld	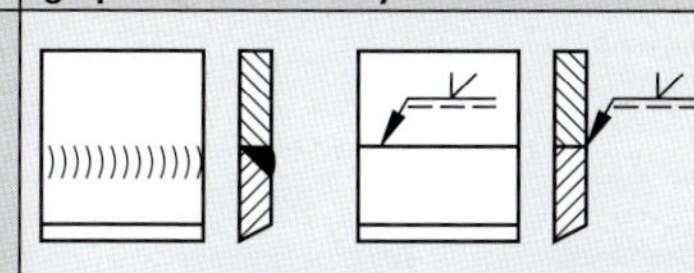
Plug welding	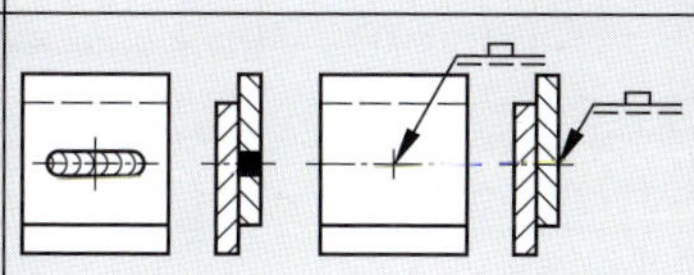		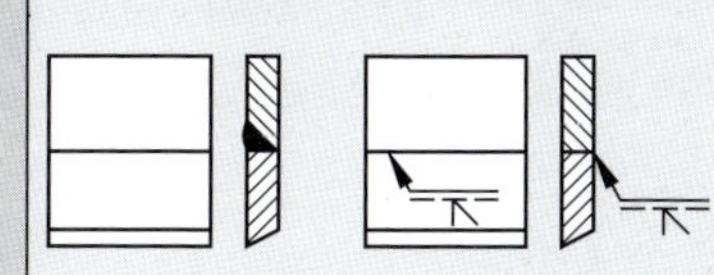
Frontal flush weld III	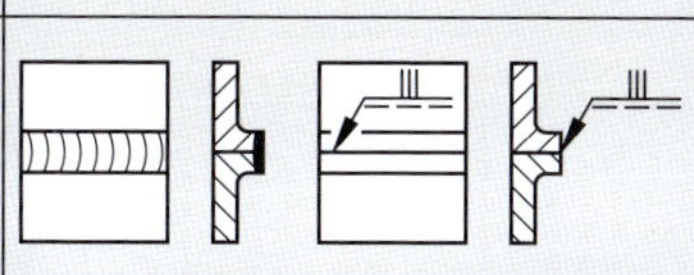	Y-butt weld Y	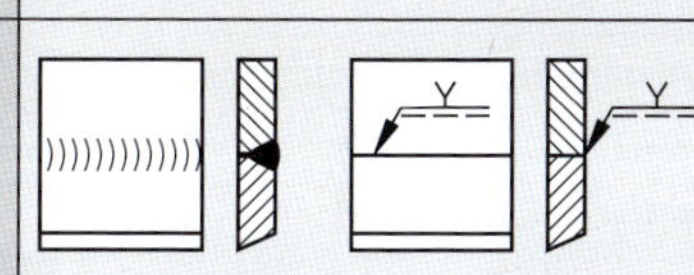
Steep-flanked weld	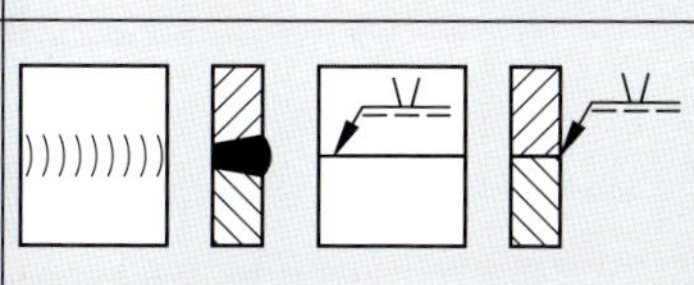	HY-weld	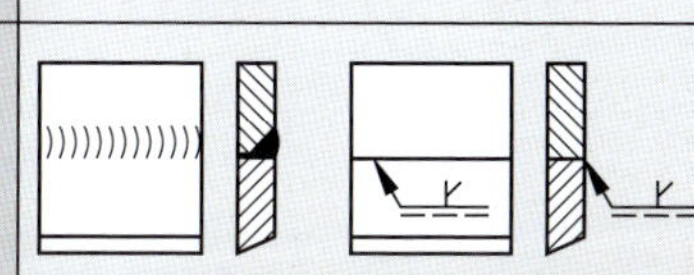
Build-up weld	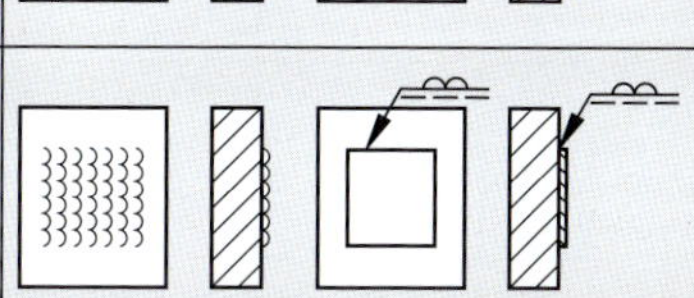	U-groove weld	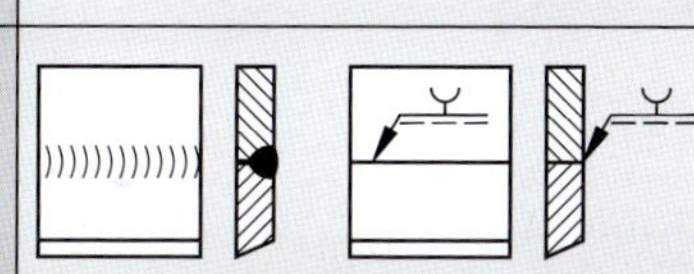
Fold weld	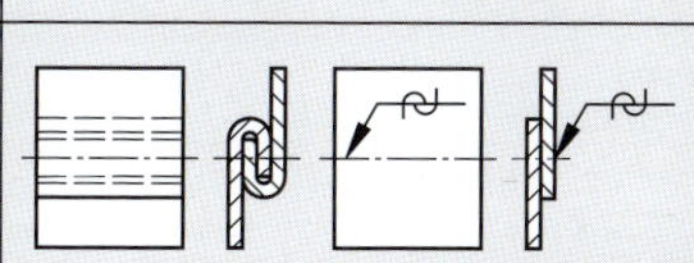	J-groove weld	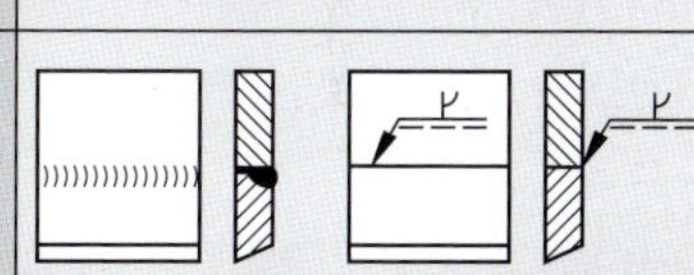
Weld all around	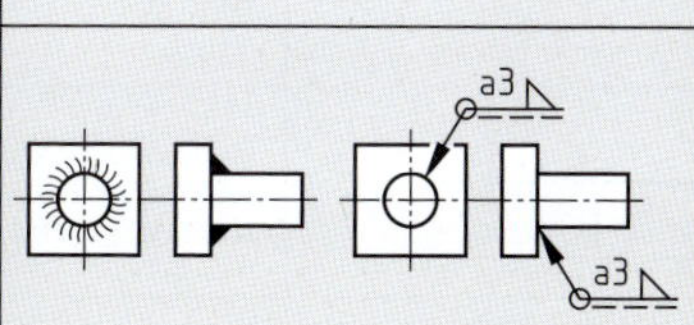	Spot weld O	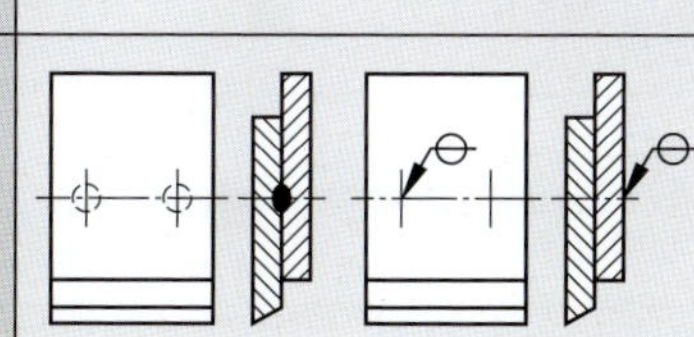
Fillet weld	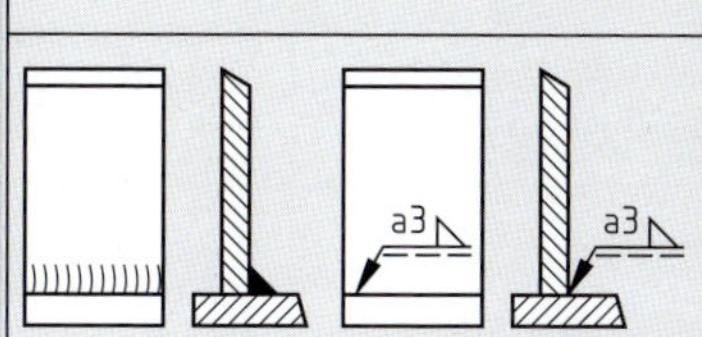	Line weld	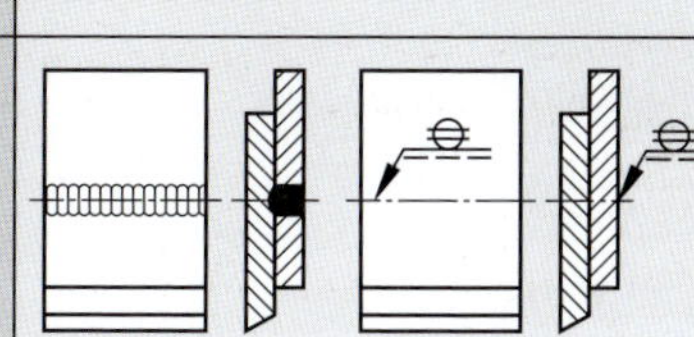
Field weld with 3 mm seam thickness	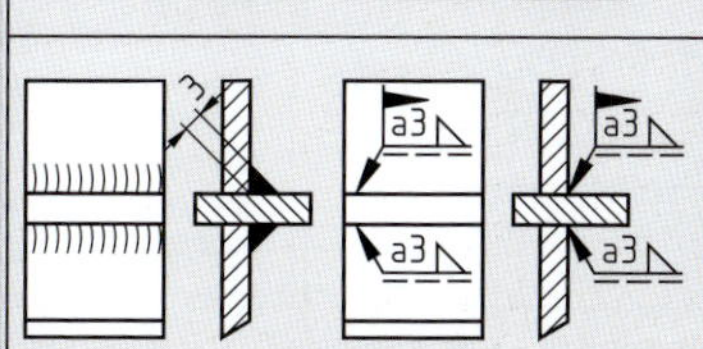	Surface weld =	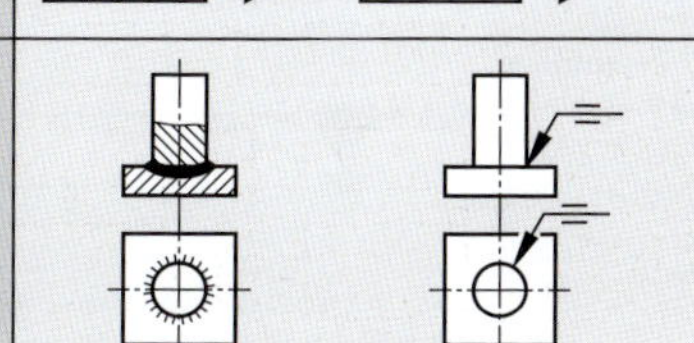

TD

Symbols for Welding and Soldering

Composite symbols for symmetrical welds[1] (examples)

cf. DIN EN 22553 (1997-03)

Weld type	Symbol	Representation	Weld type	Symbol	Representation
D(ouble)-V-weld (X-weld)			D(ouble)-HY-weld		
D(ouble)-bevel weld		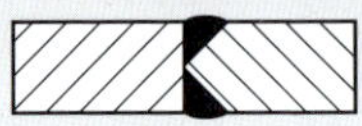	D(ouble)-U-weld		
D(ouble)-Y-weld			[1] The symbols are located symmetrical to the reference line. Example:	graphical 	symbolic 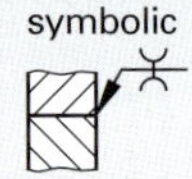

Application examples for auxiliary symbols

cf. DIN EN 22553 (1997-03)

Weld type	Symbol	Representation	Weld type	Symbol	Representation
Flat V-weld			Flat reworked V-weld		
Convex double V-weld		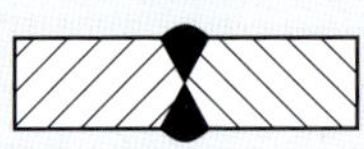	Flat V-weld with flat backing run		
Y-weld with backing run			Hollow fillet weld, weld transfer unnotched		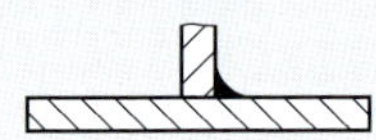

Dimensioning examples

cf. DIN EN 22553 (1997-03)

Weld type	Representation and dimensioning graphical	symbolic	Meaning of the symbolic dimension entry
I-weld (penetrating)	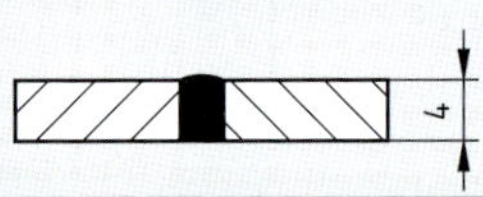	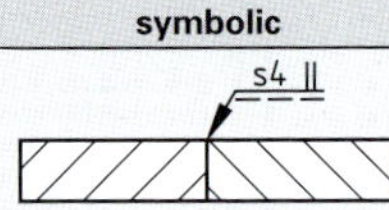	Butt weld, penetrating, weld seam thickness $s = 4$ mm
I-weld (non-penetrating)	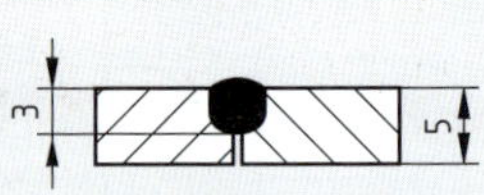	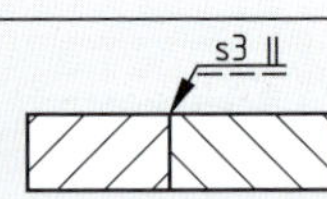	Butt weld, non-penetrating, weld seam thickness $s = 3$ mm, running over the entire object
Flare-V groove weld	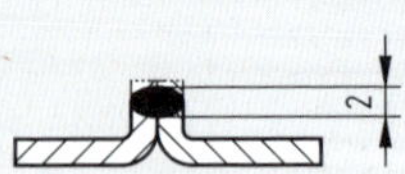	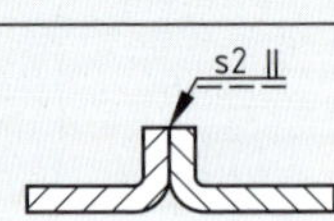	Flare-V groove weld, not completely melted down, weld seam thickness $s = 2$ mm
V-weld (penetrating weld) with backing run	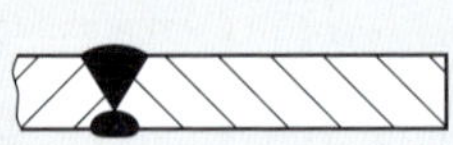	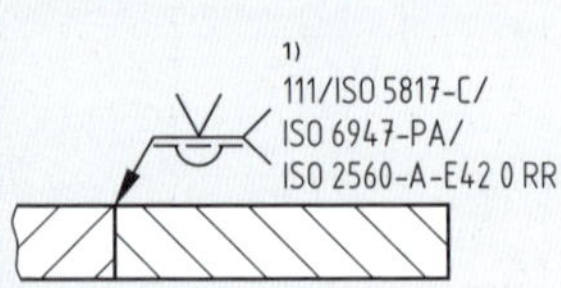	V-weld (penetrating weld) with backing run, fabricated by manual arc welding (code 111 accord. to DIN EN ISO 4063), required evaluation group C accord. to ISO 5817; flat welding position PA accord. to ISO 6947; electrode E 42 0 RR according to ISO 2560-A

[1] Supplementary requirements can be entered in a tail at the end of a reference line.

TD

Symbols for Welding and Soldering, Representation of adhesive, folded and pressed joints

Dimensioning examples (continued)

Weld type	Representation and dimensioning graphical	symbolic	Meaning of the symbolic dimension entry
Fillet weld (continuous)	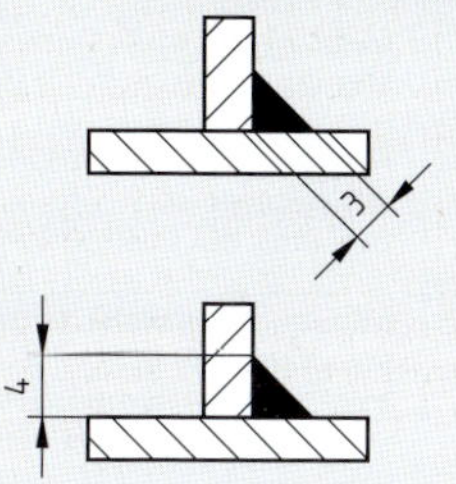	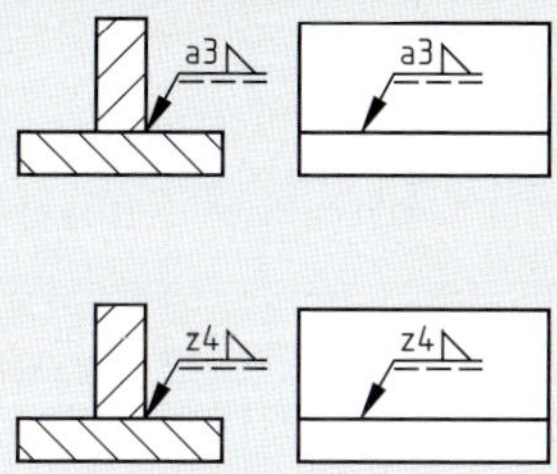	Fillet weld, weld leg thickness a = 3 mm (height of the isosceles triangle) Fillet weld, weld leg thickness z = 4 mm (side length of the isosceles triangle)
Fillet weld (interrupted)	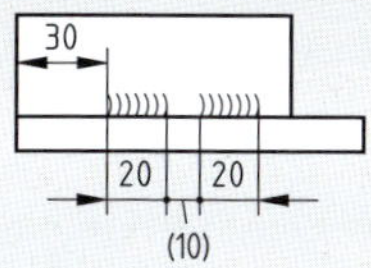	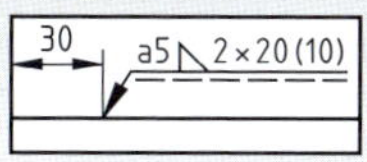	Fillet weld (interrupted), weld leg thickness a = 5 mm; 2 single welds each with l = 20 mm length; weld spacing e = 10 mm, end distance v = 30 mm
Double fillet weld (interrupted)	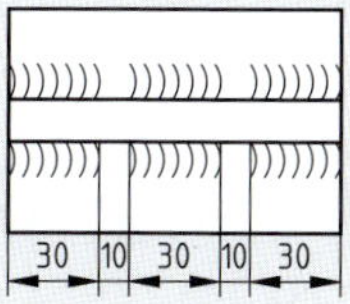	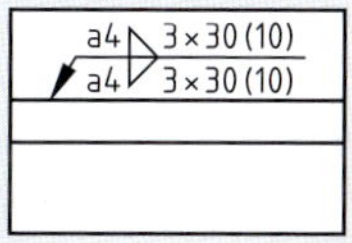	Double fillet weld (interrupted, symmetrical), weld leg thickness a = 4 mm; single weld length l = 30 mm, weld spacing e = 10 mm, without end distance
Double fillet weld (interrupted, staggered)	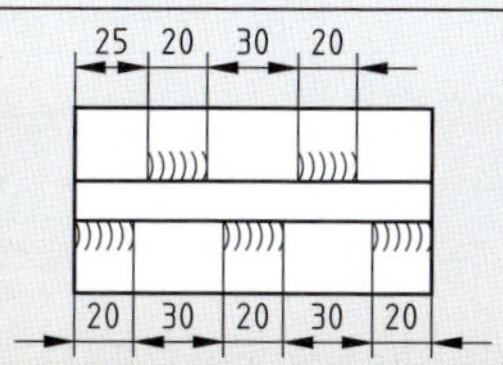	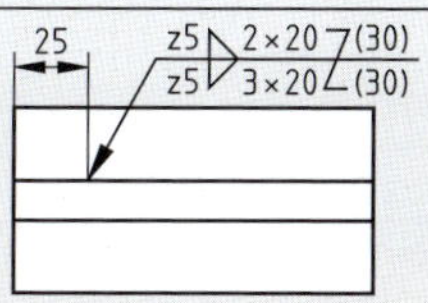	Double fillet weld (interrupted, staggered), weld leg thickness z = 5 mm; single weld length l = 20 mm, weld spacing e = 30 mm, end distance v = 25 mm

Symbolic representation of adhesive, folded and pressed joints (examples)

cf. DIN EN ISO 15785 (2002-12)

Type of joint	Weld type/ symbol	Meaning/ drawing entry	Type of joint	Weld type/ symbol	Meaning/ drawing entry
Adhesive bonded-seams	Surface seam[1)]	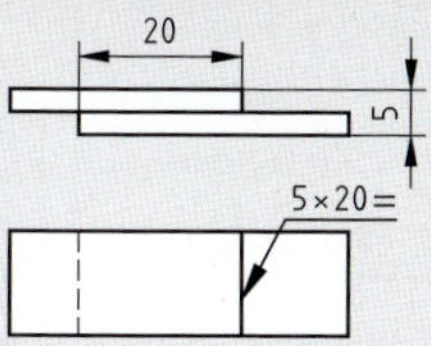	Folded seam	Folded seam 	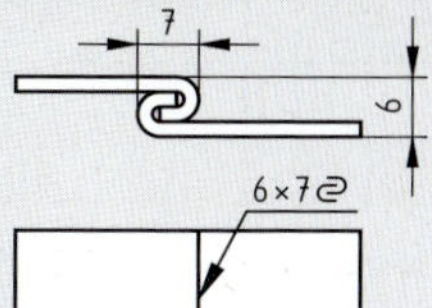
	Slant seam[1)]	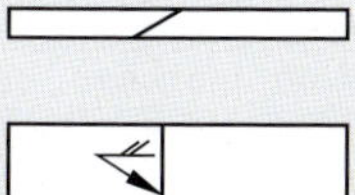	Pressed seam	Pressed seam	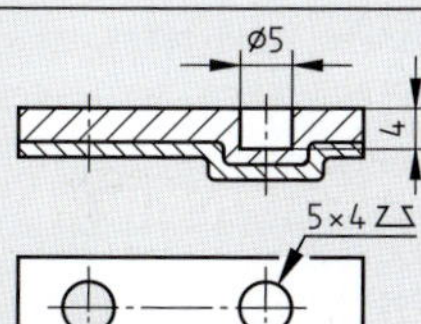

[1)] The adhesive media is not shown for adhesive seams.

TD

Heat treated parts – Hardness specifications

Presentation and indication of heat treated parts on drawings

cf. DIN ISO 15787 (2010-01), replaces DIN 6773

Heat treatment specifications

Term(s) for material condition	Measurable parameters of the material condition			Possible additions
Examples: quenched and tempered hardened hardened and tempered annealed nitrided	hardness value	HRC HV HB	rockwell hardness vickers hardness brinell hardness	**Measuring points.** Entering and dimensioning in the drawing with symbol (⏚).
	Hardening depth	CHD NHD SHD	case hardening depth nitriding depth effective hardening depth	**Heat treatment diagram.** Simplified, usually reduced scale representation of the part near the title block.
	CD CLT		carburizing depth nitride white layer thickness	**Minimum tensile strength or microstructure.** If it is possible to test a part treated in the same batch.
	All entries are made with plus tolerances.			

Identifying areas of the surface to undergo localized heat treatment

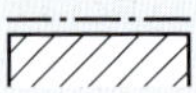
Area must be heat treated.

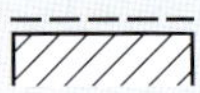
Area may be heat treated.

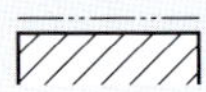
Intermediate area may not be heat treated.

Heat treatment specifications in drawings (examples)

Method	Heat treatment of the entire part – same requirements	Heat treatment of the entire part – different requirements	Heat treatment localized
Quenching and tempering, Hardening, Hardening and tempering	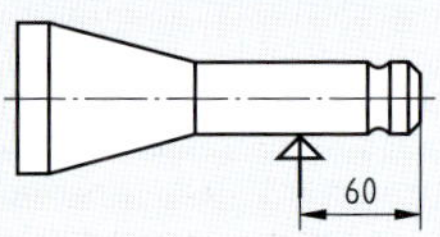 quenched and tempered 350 + 50 HB 2.5/187.5	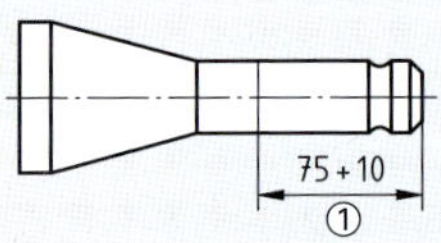 hardened and tempered 58 + 4 HRC ① 40 + 5 HRC	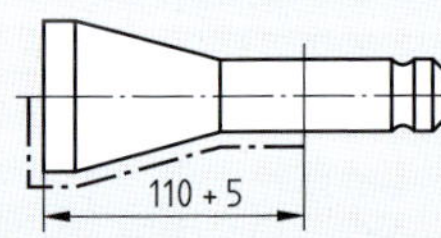 —·— hardened and entire part tempered 60 + 3 HRC
Nitriding, Case hardening	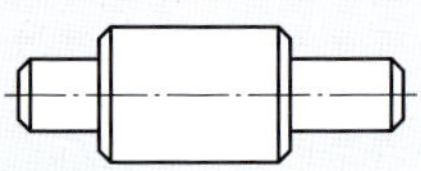 nitrided ≥ 900 HV10 NHD = 0.3 + 0.1	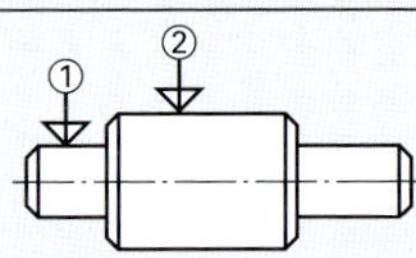 case-hardened and tempered ① 60 + 4 HRC CHD = 0.8 + 0.4 ② ≤ 52 HRC	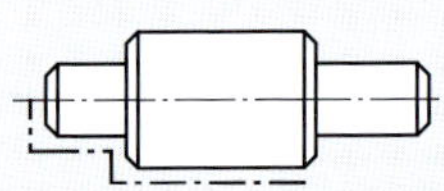 —·— case-hardened and tempered 700 + 100 HV10 CHD = 1.2 + 0.5
Surfaced hardening	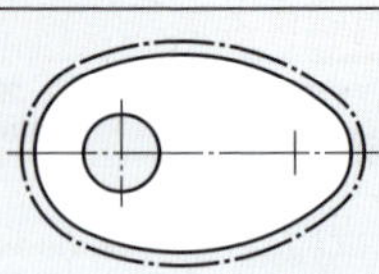 —·— surface hardened 620 + 120 HV50 SHD 500 = 0.8 + 0.8	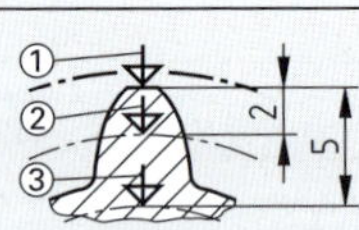 —·— surface hardened and entire part tempered ① 54 + 6 HRC ② ≤ 35 HRC ③ ≤ 30 HRC	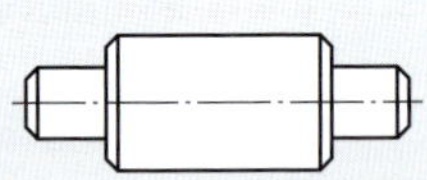 —·— surface hardened and tempered 61 + 4 HRC SHD 600 = 0.8 + 0.8

Hardening depths and tolerances in mm

Case-hardening depth CHD	0.05+0.03	0.1+0.1	0.3+0.2	0.5+0.3	0.8+0.4	1.2+0.5	1.6+0.6
Nitriding depth ND	0.05+0.02	0.1+0.05	0.15+0.02	0.2+0.1	0.25+0.1	0.3+0.1	0.35+0.15
Induction hardening depth SHD	0.2+0.2	0.4+0.4	0.6+0.6	0.8+0.8	1.0+1.0	1.3+1.1	1.6+1.3
Laser/electr. beam hardening depth SHD	0.2+0.1	0.4+0.2	0.6+0.3	0.8+0.4	1.0+0.5	1.3+0.6	1.6+0.8

Control limit hardnesses at the specified hardening depths

Case-hardening depth CHD	550 HV1
Nitriding depth NHD	core hardness + 50 HV0.5
Effective hardening depth SHD	0.8 · minimum surface hardness, calculated in HV

TD

Form deviations and roughness parameters

Form deviations

cf. DIN 4760 (1982-06)

Form deviations are deviations of the actual surface (surfaces ascertainable by measurement) from the geometrically ideal surface, whose standard shape is defined by the drawing.

Degrees of form deviation (Profile section repres. with vertical exaggeration)	**Examples**	**Possible causes**
1st degree: form deviation 	deviation in straightness, roundness	Deflection of the workpiece or the machine during fabrication of the part, malfunction or wear in the guides of the machine tool.
2nd degree: waviness 	waves	Vibrations of the machine, runout or shape deviation of a milling machine during fabrication of the part.
3rd degree: roughness 	grooves	Geometry of the cutting tool, feed or depth of cut of the tool during fabrication of the part.
4th degree: roughness 	scoring, scales, bumps	Sequence of chip formation (e. g. tearing chip), surface deformation due to blasting during fabrication of the part.
5th and 6th degree: roughness Cannot be represented as a simple profile section	matrix structure, lattice structure	Crystallization cycles, matrix changes due to welding or hot working, changes due to chemical effects, e. g. corrosion, etching.

TD

Surface texture profiles and parameters

cf. DIN EN ISO 4287 (2009-11) and DIN EN ISO 4288 (1998-04)

Surface profile	**Parameters**	**Explanations**
Primary profile (act. profile , P profile) 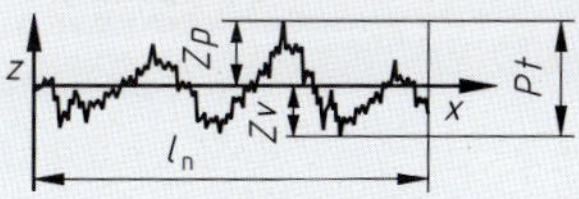	**Total height of the profile *Pt***	The **primary profile** represents the foundation for calculating the parameters of the primary profile and forms the basis for the waviness and roughness profiles. The **total height of the profile *Pt*** is the sum of the height of the highest profile peak *Zp* and the depth of the lowest profile trough *Zv* within the evaluation length l_n.
Waviness profile (W-profile) 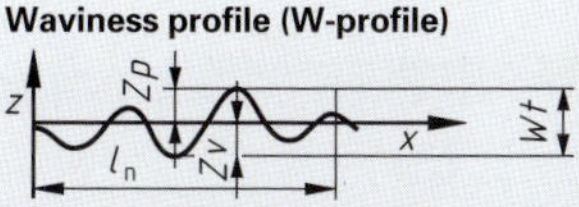	**Total height of the profile *Wt***	The **waviness profile** is obtained by low-pass filtering, i. e. by suppressing the short wavelength components of the profile. The **total height of the profile *Wt*** is the sum of the height of the highest profile peak *Zp* and the depth of the lowest profile trough *Zv* within the evaluation length l_n.
Roughness profile (R-profile) 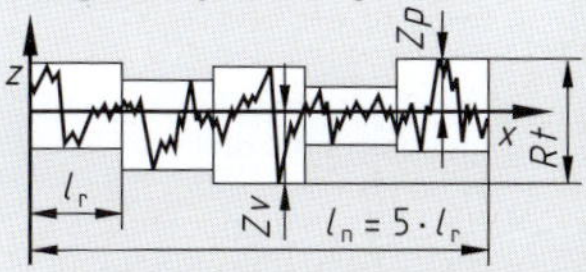	**Total height of the profile *Rt***	The **roughness profile** is obtained by high-pass filtering, i. e. by suppressing the long wavelength components of the profile. The **total height of the profile *Rt*** is the sum of the height of the highest profile peak *Zp* and the depth of the lowest profile trough *Zv* within the evaluation length l_n.
	Rp, Rv	**Height of the highest profile peak *Zp*, depth of the lowest profile trough *Zv*** within the single evaluation length l_r.
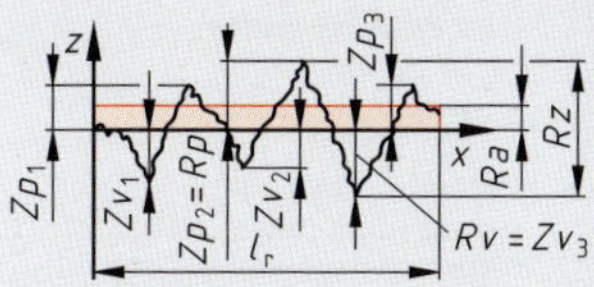	**Highest peak of the profile *Rz* [1]**	The **highest peak of the profile *Rz*** is the sum of the height of the highest profile peak *Zp* and the depth of the lowest profile trough *Zv* within the single evaluation length l_r.
	Arithmetic mean of the profile ordinates *Ra* [1]	The **arithmetic mean of the profile ordinates *Ra*** is the arithmetic mean of all ordinate values *Z(x)* within the single evaluation length l_r.
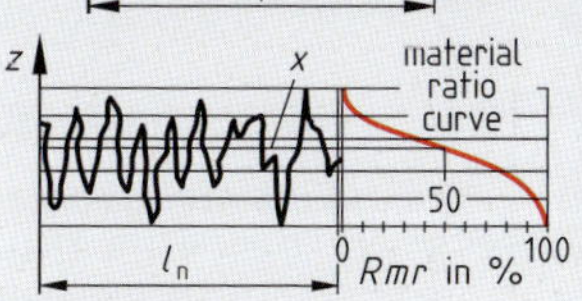	**Material ratio of the profile *Rmr***	The **material ratio of the profile expressed as a percentage, *Rmr*,** is the ratio of the sum of the contributing material lengths at a specified section height to the total evaluation length l_n.
Z(x) height of the profile at any position x; ordinate value l_n evaluation length l_r single evaluation length	**Center line (x-axis) *x***	The **center line (x-axis) *x*** is the line corresponding to the long wavelength profile component which is suppressed by profile filtering.

1) For parameters defined over a single evaluation length, the arithmetic mean of 5 single evaluation lengths to DIN EN ISO 4288 is used for determining the parameters.

Surface testing, Surface indications

Measuring sections for roughness

cf. DIN EN ISO 4288 (1998-04)

Periodic profiles (e. g. turning profiles)	Non-periodic profiles (e. g. grinding and lapping profiles)		Limit wave-length	Single / total evaluation length	Periodic profiles (e. g. turning profiles)	Non-periodic profiles (e. g. grinding and lapping profiles)		Limit wave-length	Single / total evaluation length
Groove width *RSm* mm	*Rz* µm	*Ra* µm	mm	l_r, l_n mm	groove width *RSm* mm	*Rz* µm	*Ra* µm	mm	l_r, l_n mm
> 0.01 – 0.04	up to 0.1	up to 0.02	0.08	0.08/0.4	> 0.13 – 0.4	> 0.5 – 10	> 0.1 – 2	0.8	0.8/4
> 0.04 – 0.13	> 0.1 – 0.5	> 0.02 – 0.1	0.25	0.25/1.25	> 0.4 – 1.3	> 10 – 50	> 2 – 10	2.5	2.5/12.5

Indication of surface finish

cf. DIN EN ISO 1302 (2002-06)

Symbol	Meaning
	All manufacturing processes are allowed.
	Material removal specified, e. g. turning, milling.
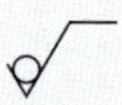	Material removal not allowed or the surface remains in delivered condition.
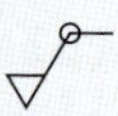	All surfaces around the contour must have the same surface-finish.

Additional marks

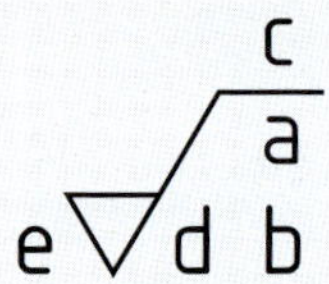

a surface parameter[1] with numerical value in µm, transfer characteristic[2]/individual evaluation length in mm

b secondary surface finish requirement (as described for a)

c manufacturing process

d symbol for the required groove direction (table page 95)

e machining deviation in mm

Examples

Symbol	Meaning
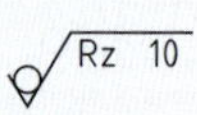	• material removing machining not allowed • *Rz* = 10 µm (upper limit) • standard transfer characteristic[3] • standard evaluation length[4] • "16 % rule"[5]
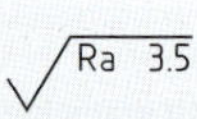	• Machining can be done as desired • standard transfer characteristic[3] • *Ra* = 3.5 µm (upper limit) • standard evaluation length[4] • "16 % rule"[5]
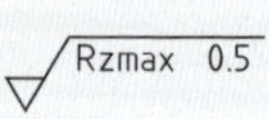	• material removal machining • *Rz* = 0.5 µm (upper limit) • standard transfer characteristic[3] • standard evaluation length[4] • "max. rule"[6]
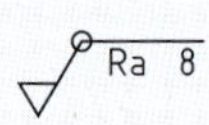	• material removal machining • *Ra* = 8 µm (upper limit) • standard transfer characteristic[3] • standard evaluation length[4] • "16 % rule"[5] • applies all around the contour
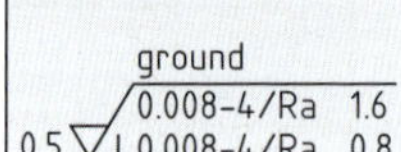	• material removal machining • manufacturing process grinding • *Ra* = 1.6 µm (upper limit) • *Ra* = 0.8 µm (lower limit) • for both *Ra* values: „16 % rule"[5] • transfer characteristic each 0.008 to 4 mm • standard evaluation length[4] • machining deviation 0.5 mm • surface grooves vertical

[1] **surface parameter,** e. g. *Rz*, consists of the profile (here the roughness profile *R*) and the parameters (here: *z*).

[2] **transfer characteristic:** wavelength range between the short wavelength filter λ_s and the long wavelength filter λ_c. The wavelength of the long wavelength filter corresponds to the single evaluation length l_r. If no transfer characteristic is entered, then the standard transfer characteristic applies[3].

[3] **standard transfer characteristic:** the limit wavelength for measurement of the roughness parameters is dependent upon the roughness profile and is taken from tables.

[4] **standard evaluation length** l_n = 5 x single evaluation length l_r.

[5] **"16 % rule":** only 16 % of all measured values may exceed the chosen parameter.

[6] **"max. rule"** ("highest value rule"): no measured value may exceed the specified highest value.

TD

Surface finish symbols

Indication of surface finish

cf. DIN EN ISO 1302 (2002-06)

Symbols for groove direction

Representation of groove direction:

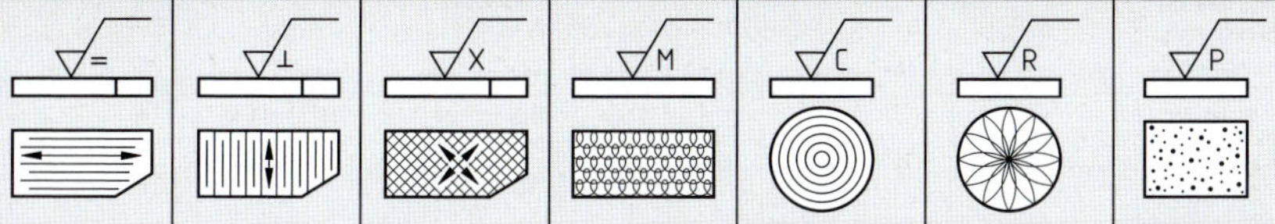

Symbol	=	⊥	X	M	C	R	P
Groove direction	parallel to the projection plane	perpendicular to the projection plane	crossed in two angular directions	multi-directional	approximately concentric to the center	approximately radial to the center	non-grooved surface, non-directional or troughs

Sizes of the symbols

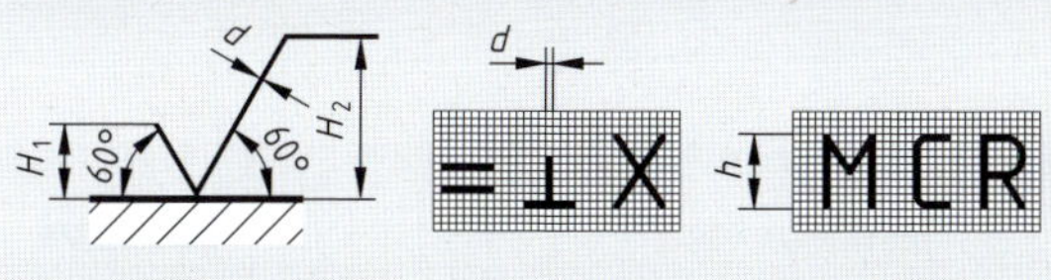

	Letter height h in mm						
	2.5	3.5	5	7	10	14	20
d	0.25	0.35	0.5	0.7	1.0	1.4	2.0
H_1	3.5	5	7	10	14	20	28
H_2	8	11	15	21	30	42	60

Layout of symbols in drawings

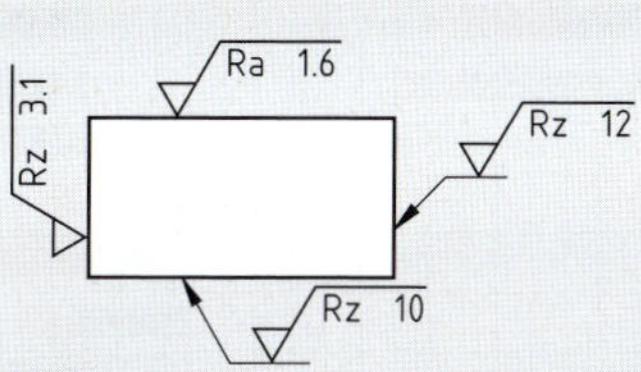

Legibility
from below or from the right

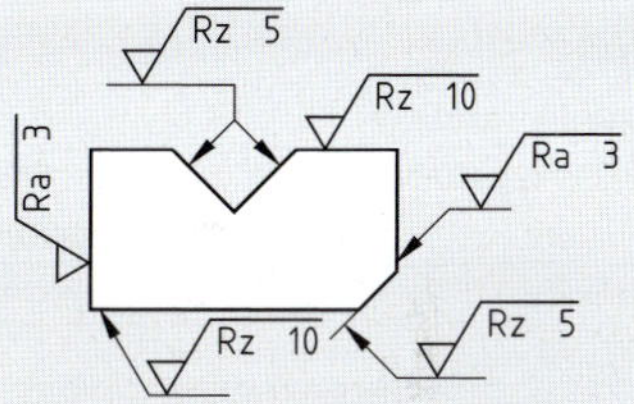

Layout
directly on the surface or with reference and leader lines

Examples of drawing entries

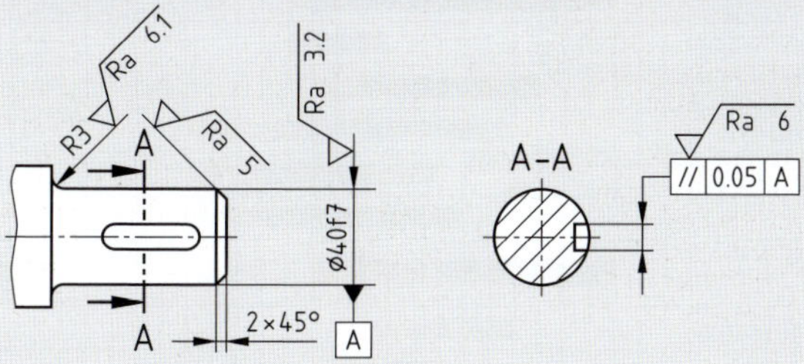

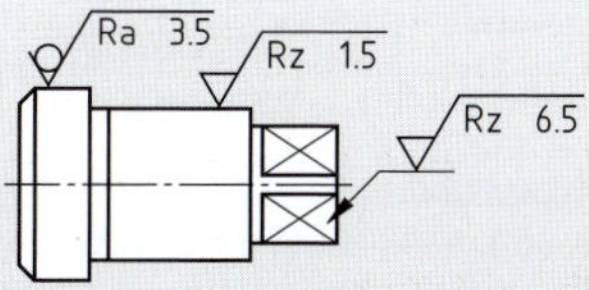

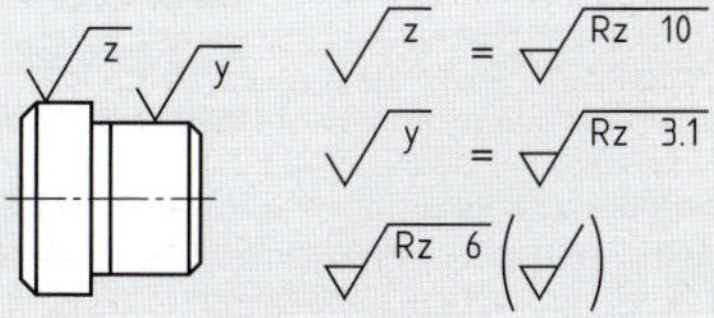

TD

TD

Roughness of surfaces

Recommended assignment of roughness values to ISO tolerance specifications[1)]

Nominal size range from – to mm	5 *Rz*	5 *Ra*	6 *Rz*	6 *Ra*	7 *Rz*	7 *Ra*	8 *Rz*	8 *Ra*	9 *Rz*	9 *Ra*	10 *Rz*	10 *Ra*	11 *Rz*	11 *Ra*
1 – 6	2.5	0.4	4	0.8	6.3	0.8	6.3	1.6	10	1.6	16	3.2	25	6.3
6 – 10							10		16	3.2	25	6.3	40	12.5
10 – 18	4	0.8												
18 – 80			6.3		10	1.6	16	3.2			40		63	
80 – 250	6.3		10	1.6	16		25		25					
250 – 500									40	6.3	63	12.5	100	25

(Recommended values of *Rz* and *Ra* in µm for ISO tolerance grades; values spanning several size ranges are given in the first row of their span.)

Achievable roughness of surfaces[1)]

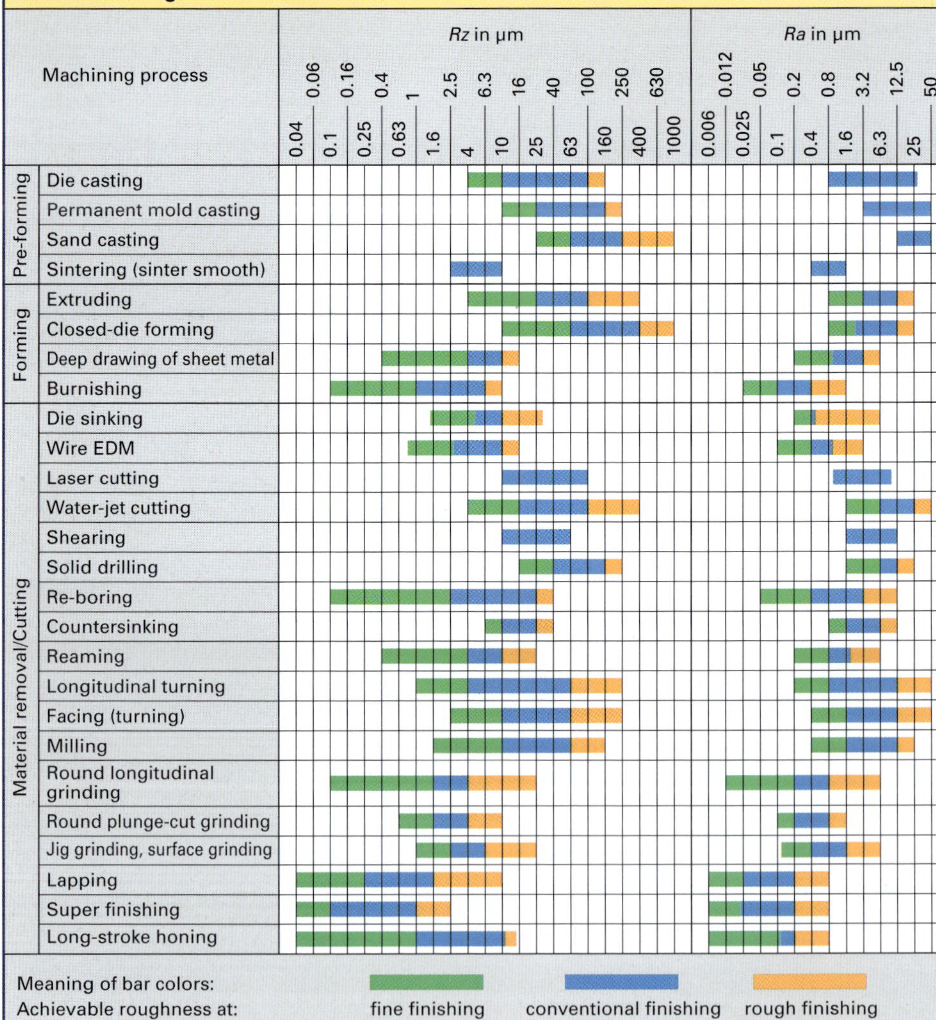

1) Roughness values, unless contained in DIN 4766-1 (withdrawn), are according to the specifications of industry.

Tooth forming quality and dimensioning of gears

Involute spur gearing — cf. DIN 3966-1 (1978-08)

The following geometric specifications are required for spur gears[1]:

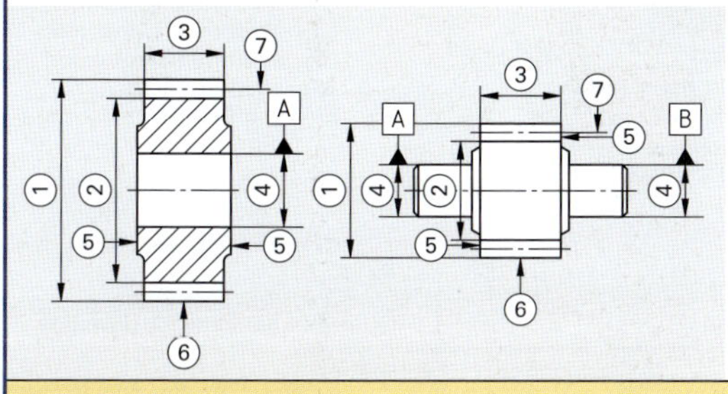

(1) tip diameter d_a with allowances
(2) root diameter d_r, if the tooth height is not specified
(3) tooth width w
(4) reference element
(5) circular runout tolerance and parallelism of the faces
(6) concentricity tolerance
(7) surface quality identification of the tooth flanks as per DIN EN ISO 1302

Straight bevel gearing — cf. DIN 3966-2 (1978-08)

The following geometric specifications are required for bevel gears[1]:

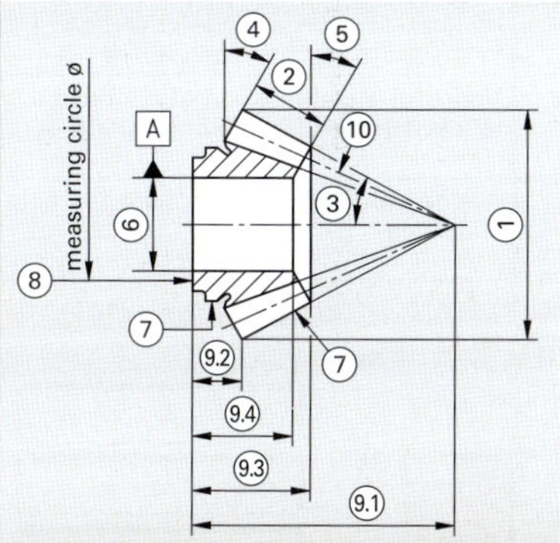

(1) tip diameter d_a with allowances
(2) tooth width w
(3) tip cone angle
(4) complementary angle of the back cone angle
(5) complementary angle of the inner complementary cone angle (if required)
(6) identification of the reference element
(7) concentricity tolerance of the wheel body
(8) circular runout tolerance of the wheel body
(9.1) assembly dimension
(9.2) outer tip circle distance
(9.3) inner tip circle distance
(9.4) distance of auxiliary planes
(10) surface quality identification of the tooth flanks as per DIN EN ISO 1302

Specifications concerning tooth forming — cf. DIN 3966-1 (1978-08), DIN 3966-2 (1978-08)

In addition, the specification of details concerning the tooth forming tool, the adjustment of the gear manufacturing machine and the testing of the tooth forming quality are required for all gearings specified in a table (in a drawing or on a separate sheet).

Straight gearing (outside) for spur gear		
Details		Example
Module	m_n	3
Number of teeth	z	22
Basic rack tooth profile		DIN 867
Tooth forming quality, tolerance zone, test group as per DIN 3961		8 d 25 DIN 3967
Center distance in the housing	a	99±0,05
Mating gear	part number no. of teeth z	25564 44

Meaning of:
8 tooth forming quality from 1 to 12.
d tooth thickness allowance from a to h
25 tooth thickness tolerance from 21 to 30

Tooth forming quality

The tooth forming quality depends on the application (below), the manufacturing process (below) and the gear rotation speed.

Application	Tooth forming quality (1–12)
Gages	2 – 4
Measuring instruments	5 – 7
Motor vehicles	5 – 9
Machine tools	5 – 10
Agricultural machines	6 – 12
Lifting and conveyors	7 – 12

Manufacturing process	Tooth forming quality (1–12)
Honing	1 – 6
Grinding	2 – 8
Shaving	5 – 8
Hobbing, gear planing	5 – 10
Form milling, form planing	7 – 12
Punching, pressing, sintering	8 – 12

[1] Additional dimensions must be specified for the manufacturing of the wheel body, e. g. drilling diameter and groove dimensions.

TD

TD

ISO system of limits and fits

Terms
cf. DIN ISO 286-1 (1990-11)

Hole
- N nominal size
- G_{uH} hole max. dimension
- G_{lH} hole min. dimension
- ES hole upper deviation
- EI hole lower deviation
- T_H dimensional tolerance of hole

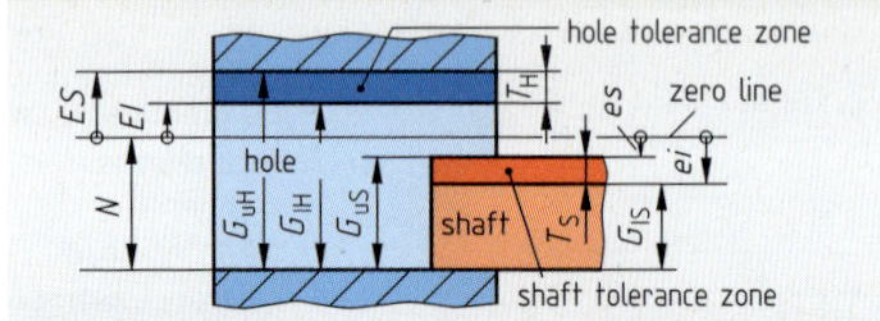

shaft
- N nominal dimension
- G_{uS} shaft max. dimension
- G_{lS} shaft min. dimension
- es shaft upper deviation
- ei shaft lower deviation
- T_S dimensional tolerance of shaft

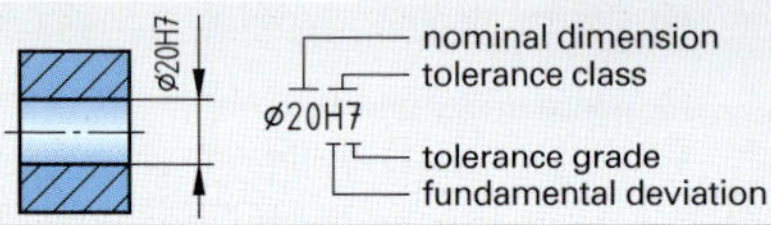

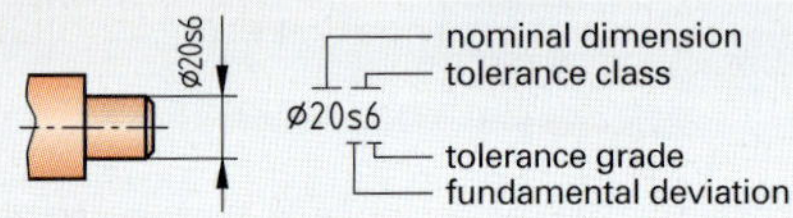

Designation	Explanation	Designation	Explanation
Zero line	It represents the nominal dimension that is referenced by the deviations and tolerances.	**Fundament. tolerance grade**	A group of tolerances assigned to same level of precision, e. g IT7.
Fundamental deviation	The fund. deviation determin. the position of the tolerance zone with resp. to the zero line.	**Tolerance grade**	Number of the fundamental tol. grade, e. g. 7 for the fundamental tolerance grade IT7.
Dimensional tolerance	Difference between the max. and the min. dimension or between the upper and lower deviation.	**Tolerance class**	Name for a combination of a fundamental deviation and a tolerance grade, e. g. H7.
Fundamental tolerance	A tolerance assigned to a fundamental tolerance grade, e. g. IT7 and a nominal dimension range, e. g. 30 to 50 mm.	**Fit**	Planned joining condition between hole and shaft.

Limits, deviations and tolerances
cf. DIN ISO 286-1 (1990-11)

Hole

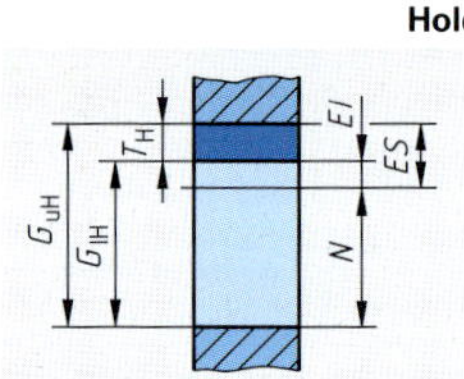

$$G_{uH} = N + ES$$

$$G_{lH} = N + EI$$

$$T_H = ES - EI$$

$$T_H = G_{uH} - G_{lH}$$

Example: Hole ∅ 50 + 0.3/+ 0.1; G_{uH} = ?; T_H = ?

$G_{uH} = N + ES$ = 50 mm + 0.3 mm = 50.30 mm

$T_H = ES - EI$ = 0.3 mm – 0.1 mm = 0.2 mm

Shaft

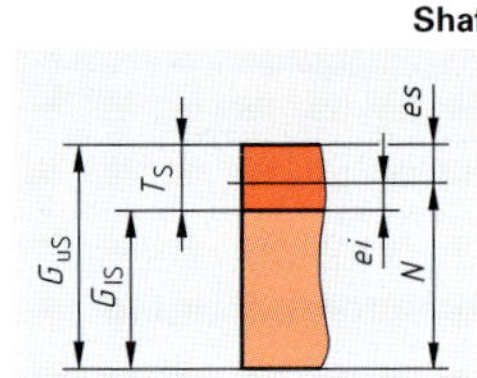

$$G_{uS} = N + es$$

$$G_{lS} = N + ei$$

$$T_S = es - ei$$

$$T_S = G_{uS} - G_{lS}$$

Example: Shaft ∅ 20e8; G_{lS} = ?; T_S = ?

For values for *ei* and *es*, see page 103.

ei = – 73 µm = – 0.073 mm; es = – 40 µm = – 0.040 mm

$G_{lS} = N + ei$ = 20 mm + (– 0.073 mm) = 19.927 mm

$T_S = es - ei$ = – 40 µm – (– 73 µm) = 33 µm

Fits
cf. DIN ISO 286-1 (1990-11)

Clearance fit
- F_{Cmax} max. clearance
- F_{Cmin} min. clearance

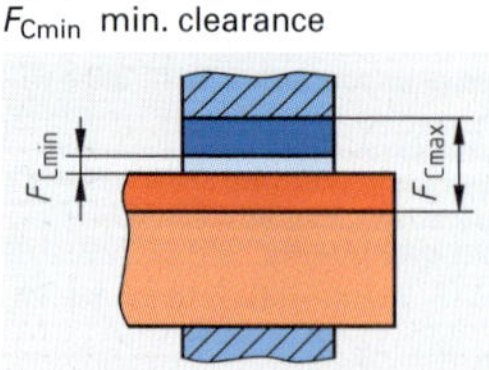

Transition fit
- F_{Cmax} max. clearance
- F_{Imax} max. interference

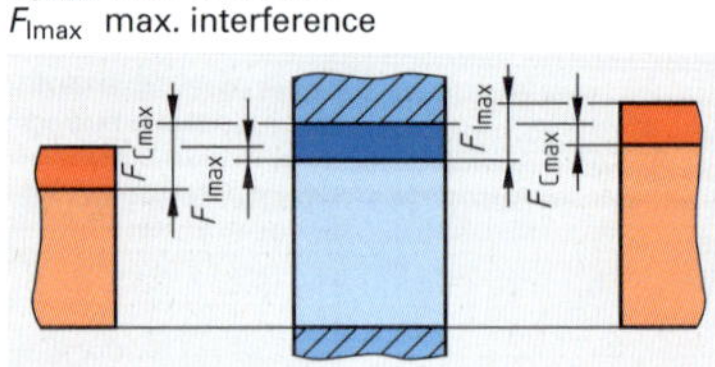

Interference fit
- F_{Imax} max. interference
- F_{Imin} min. interference

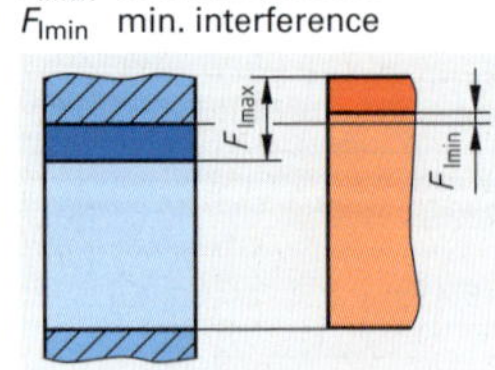

$$F_{Cmin} = G_{lH} - G_{uS}$$

$$F_{Cmax} = G_{uH} - G_{lS}$$

$$F_{Imax} = G_{lH} - G_{uS}$$

$$F_{Imin} = G_{uH} - G_{lS}$$

Example: Fit ∅ 30 H8/f7; F_{Cmax} = ?; F_{Cmin} = ?

For values for *ES*, *EI*, *es*, *ei* see page 103.

$G_{uH} = N + ES$ = 30 mm + 0.033 mm = 30.033 mm

$G_{lH} = N + EI$ = 30 mm + 0 mm = 30.000 mm

$G_{uH} = N + ES$ = 30 mm + (– 0.020 mm) = 29.980 mm

$G_{lH} = N + ES$ = 30 mm + (– 0.041 mm) = 29.959 mm

$F_{Cmax} = G_{uH} - G_{lS}$ = 30.033 mm – 29.959 mm = 0.074 mm

$F_{Cmin} = G_{lH} - G_{uS}$ = 30.000 mm – 29.980 mm = 0.02 mm

ISO system of limits and fits

Fit systems

cf. DIN ISO 286-1 (1990-11)

Fit system: basic hole system (all hole dimensions have the fundamental deviation H)

Fundamental deviations for shafts

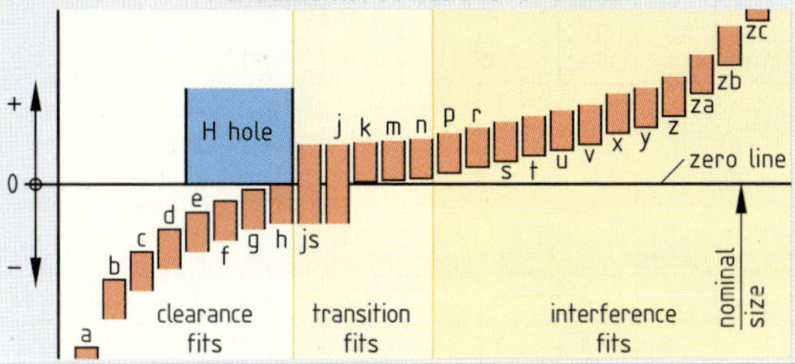

Examples for nominal dimension 25, tolerance grade 7

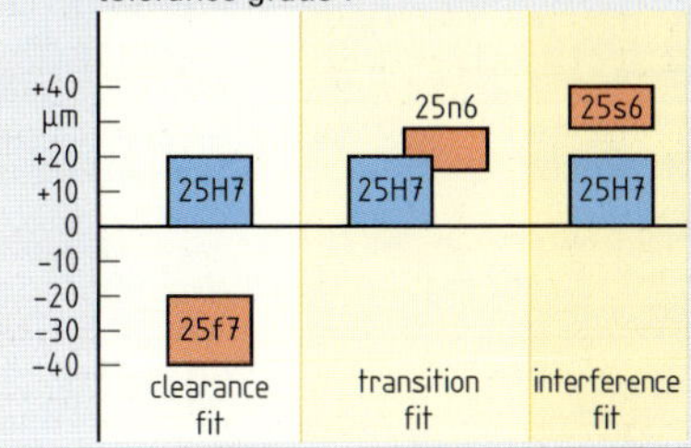

Fit system: basic shaft system (all shaft dimensions have the fundamental deviation h)

Fundamental allowances for holes

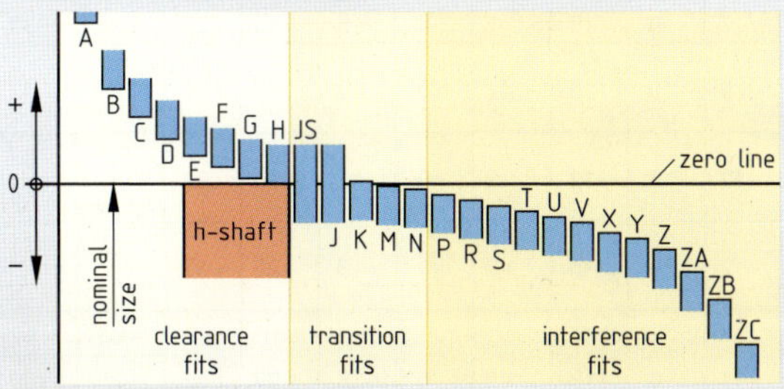

Examples for nominal dimension 25, tolerance grade 6

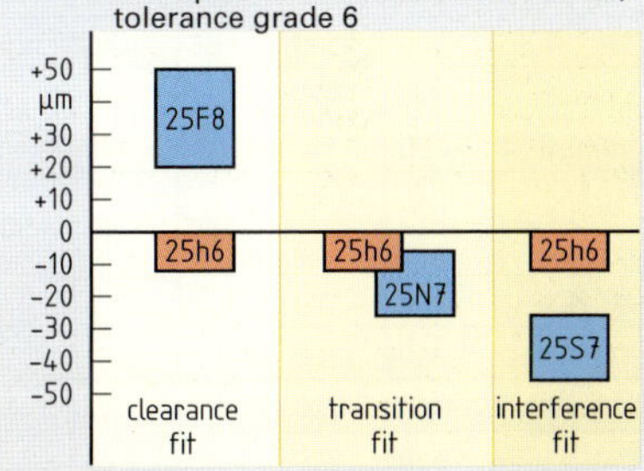

TD

Fundamental tolerances

cf. DIN ISO 286-1 (1990-11)

Nominal dimension range over – to mm	Fundamental tolerance grade																	
	IT1	IT2	IT3	IT4	IT5	IT6	IT7	IT8	IT9	IT10	IT11	IT12	IT13	IT14	IT15	IT16	IT17	IT18
	Fundamental tolerances																	
	µm											mm						
up to 3	0.8	1.2	2	3	4	6	10	14	25	40	60	0.1	0.14	0.25	0.4	0.6	1	1.4
3 – 6	1	1.5	2.5	4	5	8	12	18	30	48	75	0.12	0.18	0.3	0.48	0.75	1.2	1.8
6 – 10	1	1.5	2.5	4	6	9	15	22	36	58	90	0.15	0.22	0.36	0.58	0.9	1.5	2.2
10 – 18	1.2	2	3	5	8	11	18	27	43	70	110	0.18	0.27	0.43	0.7	1.1	1.8	2.7
18 – 30	1.5	2.5	4	6	9	13	21	33	52	84	130	0.21	0.33	0.52	0.84	1.3	2.1	3.3
30 – 50	1.5	2.5	4	7	11	16	25	39	62	100	160	0.25	0.39	0.62	1	1.6	2.5	3.9
50 – 80	2	3	5	8	13	19	30	46	74	120	190	0.3	0.46	0.74	1.2	1.9	3	4.6
80 – 120	2.5	4	6	10	15	22	35	54	87	140	220	0.35	0.54	0.87	1.4	2.2	3.5	5.4
120 – 180	3.5	5	8	12	18	25	40	63	100	160	250	0.4	0.63	1	1.6	2.5	4	6.3
180 – 250	4.5	7	10	14	20	29	46	72	115	185	290	0.46	0.72	1.15	1.85	2.9	4.6	7.2
250 – 315	6	8	12	16	23	32	52	81	130	210	320	0.52	0.81	1.3	2.1	3.2	5.2	8.1
315 – 400	7	9	13	18	25	36	57	89	140	230	360	0.57	0.89	1.4	2.3	3.6	5.7	8.9
400 – 500	8	10	15	20	27	40	63	97	155	250	400	0.63	0.97	1.55	2.5	4	6.3	9.7
500 – 630	9	11	16	22	32	44	70	110	175	280	440	0.7	1.1	1.75	2.8	4.4	7	11
630 – 800	10	13	18	25	36	50	80	125	200	320	500	0.8	1.25	2	3.2	5	8	12.5
800 – 1000	11	15	21	28	40	56	90	140	230	360	560	0.9	1.4	2.3	3.6	5.6	9	14
1000 – 1250	13	18	24	33	47	66	105	165	260	420	660	1.05	1.65	2.6	4.2	6.6	10.5	16.5
1250 – 1600	15	21	29	39	55	78	125	195	310	500	780	1.25	1.95	3.1	5	7.8	12.5	19.5
1600 – 2000	18	25	35	46	65	92	150	230	370	600	920	1.5	2.3	3.7	6	9.2	15	23
2000 – 2500	22	30	41	55	78	110	175	280	440	700	1100	1.75	2.8	4.4	7	11	17.5	28
2500 – 3150	26	36	50	68	96	135	210	330	540	860	1350	2.1	3.3	5.4	8.6	13.5	21	33

The limit deviations of the tolerance grade for the fundamental deviations h, js, H and JS can be derived from the fundamental tolerances: **h:** *es* = 0; *ei* = – IT **js:** *es* = + IT/2; *ei* = – IT/2 **H:** *ES* = + IT; *EI* = 0 **JS:** *ES* = + IT/2; *EI* = – IT/2

ISO fits

Fundamental deviations for shafts (selection)

cf. DIN ISO 286-1 (1990-11)

Fundamental deviations	a	c	d	e	f	g	h	j	k		m	n	p	r	s
Standardized fundamental tolerance grades	IT9 to IT13	IT8 to IT12	IT5 to IT13	IT5 to IT10	IT3 to IT10	IT3 to IT10	IT1 to IT18	IT5 to IT8	IT3 to IT13		IT3 to IT9	IT3 to IT9	IT3 to IT10		
Table applies to ...	all standardized fundamental tolerance grades							IT7	IT4 to IT7	IT8 to IT13	all standardized fundamental tolerance grades				
Nominal dimension over – to mm	Upper deviation *es* in µm							Lower deviation *ei* in µm							
up to 3	–270	–60	–20	–14	–6	–2	0	–4	0	0	+2	+4	+6	+10	+14
3–6		–70	–30	–20	–10	–4	0	–4	+1	0	+4	+8	+12	+15	+19
6–10	–280	–80	–40	–25	–13	–5	0	–5	+1	0	+6	+10	+15	+19	+23
10–18	–290	–95	–50	–32	–16	–6	0	–6	+1	0	+7	+12	+18	+23	+28
18–30	–300	–110	–65	–40	–20	–7	0	–8	+2	0	+8	+15	+22	+28	+35
30–40	–310	–120	–80	–50	–25	–9	0	–10	+2	0	+9	+17	+26	+34	+43
40–50	–320	–130													
50–65	–340	–140	–100	–60	–30	–10	0	–12	+2	0	+11	+20	+32	+41	+53
65–80	–360	–150												+43	+59
80–100	–380	–170	–120	–72	–36	–12	0	–15	+3	0	+13	+23	+37	+51	+71
100–120	–410	–180												+54	+79
120–140	–460	–200	–145	–85	–43	–14	0	–18	+3	0	+15	+27	+43	+63	+92
140–160	–520	–210												+65	+100
160–180	–580	–230												+68	+108
180–200	–660	–240	–170	–100	–50	–15	0	–21	+4	0	+17	+31	+50	+77	+122
200–225	–740	–260												+80	+130
225–250	–820	–280												+84	+140
250–280	–920	–300	–190	–110	–56	–17	0	–26	+4	0	+20	+34	+56	+94	+158
280–315	–1050	–330												+98	+170
315–355	–1200	–360	–210	–125	–62	–18	0	–28	+4	0	+21	+37	+62	+108	+190
355–400	–1350	–400												+114	+208
400–450	–1500	–440	–230	–135	–68	–20	0	–32	+5	0	+23	+40	+68	+126	+232
450–500	–1650	–480												+132	+252

Calculation of limit deviations

Limit deviations for fundamental tolerance grades given in the table row „Table applies to ...“ (above and page 101) can be calculated using tables on this page and page 101 and the formulas below. The values necessary for the fundamental tolerances IT are found in the table on page 99.

Formulas

- for shaft deviations

$ei = es - IT$

$es = ei + IT$

- for hole deviations

$EI = ES - IT$

$ES = EI + IT$

Example 1: Shaft (outside dimension)
∅ 40g5; *es* = ?; *ei* = ?

es (table above) = – 9 µm
IT5 (table page 99) = 11 µm
ei = *es* – IT = – 9 µm – 11 µm = – 20 µm

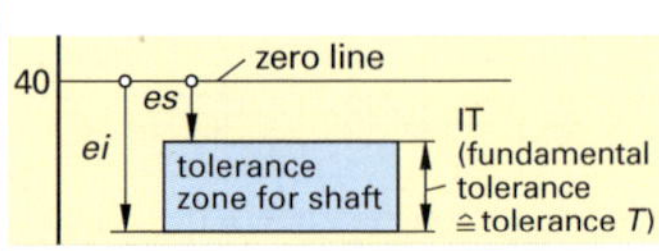

Example 2: Hole (inside dimension)
∅ 100K6; *ES* = ?; *EI* = ?

ES (table page 101) = – 3 µm + Δ
(Value Δ for fundamental tolerance grade IT6 acc. to table, bottom of page 101: 7 µm)
ES = –3 µm + 7 µm = 4 µm
IT6 (table page 99) = 22 µm
EI = *ES* – IT = 4 µm – 22 µm = – 18 µm

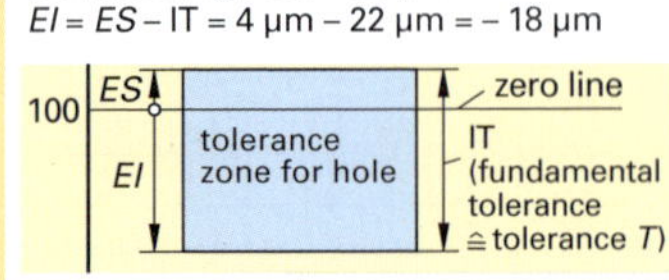

ISO fits

Fundamental deviations for holes (selection)[1]

cf. DIN ISO 286-1 (1990-11)

Fundamental deviations	A	C	D	E	F	G	H	J	K	M	N	P, R, S	P	R	S
Standardized fundamental tolerance grades	IT9 to IT13	IT8 to IT13	IT6 to IT13	IT5 to IT10	IT3 to IT10	IT3 to IT10	IT1 to IT18	IT6 to IT8	IT3 to IT10	IT3 to IT10	IT3 to IT11	IT3 to IT10			
Table applies to ...	all standardized fundamental tolerance grades							IT8	IT3 to IT8			IT3 to IT7	IT 8 to IT10		
Nominal dimension over – to; mm	Lower deviation EI in µm							Upper deviation ES in µm							
up to 3	+270	+60	+20	+14	+6	+2	0	+6	0	−2	−4	Values for the upper deviations ES: the same as fundamental tolerance grade IT8 to IT10, plus Δ	−6	−10	−14
3 – 6		+70	+30	+20	+10	+4	0	+10	−1 + Δ	−4 + Δ	−8 + Δ		−12	−15	−19
6 – 10	+280	+80	+40	+25	+13	+5	0	+12	−1 + Δ	−6 + Δ	−10 + Δ		−15	−19	−23
10 – 18	+290	+95	+50	+32	+16	+6	0	+15	−1 + Δ	−7 + Δ	−12 + Δ		−18	−23	−28
18 – 30	+300	+110	+65	+40	+20	+7	0	+20	−2 + Δ	−8 + Δ	−15 + Δ		−22	−28	−35
30 – 40	+310	+120	+80	+50	+25	+9	0	+24	−2 + Δ	−9 + Δ	−17 + Δ		−26	−34	−43
40 – 50	+320	+130													
50 – 65	+340	+140	+100	+60	+30	+10	0	+28	−2 + Δ	−11 + Δ	−20 + Δ		−32	−41	−53
65 – 80	+360	+150												−43	−59
80 – 100	+380	+170	+120	+72	+36	+12	0	+34	−3 + Δ	−13 + Δ	−23 + Δ		−37	−51	−71
100 – 120	+410	+180												−54	−79
120 – 140	+460	+200	+145	+85	+43	+14	0	+41	−3 + Δ	−15 + Δ	−27 + Δ		−43	−63	−92
140 – 160	+520	+210												−65	−100
160 – 180	+580	+230												−68	−108
180 – 200	+660	+240	+170	+100	+50	+15	0	+47	−4 + Δ	−17 + Δ	−31 + Δ		−50	−77	−122
200 – 225	+740	+260												−80	−130
225 – 250	+820	+280												−84	−140
250 – 280	+920	+300	+190	+110	+56	+17	0	+55	−4 + Δ	−20 + Δ	−34 + Δ		−56	−94	−158
280 – 315	+1050	+330												−98	−170
315 – 355	+1200	+360	+210	+125	+62	+18	0	+60	−4 + Δ	−21 + Δ	−37 + Δ		−62	−108	−190
355 – 400	+1350	+400												−114	−208
400 – 450	+1500	+440	+230	+135	+68	+20	0	+66	−5 + Δ	−23 + Δ	−40 + Δ		−68	−126	−232
450 – 500	+1650	+480												−132	−252

Values for Δ[1] in µm

Fundamental tolerance grade	Nominal dimension over –to in mm: 3 to 6	6 to 10	10 to 18	18 to 30	30 to 50	50 to 80	80 to 120	120 to 180	180 to 250	250 to 315	315 to 400	400 to 500
IT3	1	1	1	1.5	1.5	2	2	3	3	4	4	5
IT4	1.5	1.5	2	2	3	3	4	4	4	4	5	5
IT5	1	2	3	3	4	5	5	6	6	7	7	7
IT6	3	3	3	4	5	6	7	7	9	9	11	13
IT7	4	6	7	8	9	11	13	15	17	20	21	23
IT8	6	7	9	12	14	16	19	23	26	29	32	34

[1] For examples of calculations see page 100.

TD

ISO fits

Basic hole system

cf. DIN ISO 286-2 (1990-11)

Nominal dimension range over – to mm	Limit deviations in µm for tolerance classes[1]															
	for hole	for shafts – Paired with an H6 hole results in a					for hole	for shafts – Paired with an H7 hole results in a								
		clearance, transition, interference fit						clearance fit			transition fit				interference fit	
	H6	h5	j5	k6	n5	r5	**H7**	**f7**	g6	**h6**	j6	k6	m6	**n6**	**r6**	s6
up to 3	+6 0	0 −4	±2	+6 0	+8 +4	+14 +10	+10 0	−6 −16	−2 −8	0 −6	+4 −2	+6 0	+8 +2	+10 +4	+16 +10	+20 +14
3 – 6	+8 0	0 −5	+3 −2	+9 +1	+13 +8	+20 +15	+12 0	−10 −22	−4 −12	0 −8	+6 −2	+9 +1	+12 +4	+16 + 8	+23 +15	+27 +19
6 – 10	+9 0	0 −6	+4 −2	+10 +1	+16 +10	+25 +19	+15 0	−13 −28	−5 −14	0 −9	+7 −2	+10 +1	+15 +6	+19 +10	+28 +19	+32 +23
10 – 14	+11 0	0 −8	+5 −3	+12 +1	+20 +12	+31 +23	+18 0	−16 −34	−6 −17	0 −11	+8 −3	+12 +1	+18 +7	+23 +12	+34 +23	+39 +28
14 – 18																
18 – 24	+13 0	0 −9	+5 −4	+15 +2	+24 +15	+37 +28	+21 0	−20 −41	−7 −20	0 −13	+9 −4	+15 +2	+21 +8	+28 +15	+41 +28	+48 +35
24 – 30																
30 – 40	+16 0	0 −11	+6 −5	+18 +2	+28 +17	+45 +34	+25 0	−25 −50	−9 −25	0 −16	+11 −5	+18 +2	+25 +9	+33 +17	+50 +34	+59 +43
40 – 50																
50 – 65	+19 0	0 −13	+6 −7	+21 +2	+33 +20	+54 +41	+30 0	−30 −60	−10 −29	0 −19	+12 −7	+21 +2	+30 +11	+39 +20	+60 +41	+72 +53
65 – 80						+56 +43									+62 +43	+78 +59
80 – 100	+22 0	0 −15	+6 −9	+25 +3	+38 +23	+66 +51	+35 0	−36 −71	−12 −34	0 −22	+13 −9	+25 +3	+35 +13	+45 +23	+73 +51	+93 +71
100 – 120						+69 +54									+76 +54	+101 +79
120 – 140						+81 +63									+88 +63	+117 +92
140 – 160	+25 0	0 −18	+7 −11	+28 +3	+45 +27	+83 +65	+40 0	−43 −83	−14 −39	0 −25	+14 −11	+28 +3	+40 +15	+52 +27	+90 +65	+125 +100
160 – 180						+86 +68									+93 +68	+133 +108
180 – 200						+97 +77									+106 +77	+151 +122
200 – 225	+29 0	0 −20	+ 7 −13	+33 +4	+51 +31	+100 +80	+46 0	−50 −96	−15 −44	0 −29	+16 −13	+33 +4	+46 +17	+60 +31	+109 +80	+159 +130
225 – 250						+104 +84									+113 +84	+169 +140
250 – 280	+32 0	0 −23	+7 −16	+36 +4	+57 +34	+117 +94	+52 0	−56 −108	−17 −49	0 −32	+16 −16	+36 +4	+52 +20	+66 +34	+126 +94	+190 +158
280 – 315						+121 +98									+130 +98	+202 +170
315 – 355	+36 0	0 −25	+7 −18	+40 +4	+62 +37	+133 +108	+57 0	−62 −119	−18 −54	0 −36	+18 −18	+40 +4	+57 +21	+73 +37	+144 +108	+226 +190
355 – 400						+139 +114									+150 +114	+244 +208
400 – 450	+40 0	0 −27	+7 −20	+45 +5	+67 +40	+153 +126	+63 0	−68 −131	−20 −60	0 −40	+20 −20	+45 +5	+63 +23	+80 +40	+166 +126	+272 +232
450 – 500						+159 +132									+172 +132	+292 +252

[1] The tolerance classes in **bold** print correspond to row 1 in DIN 7157; their use is preferable.

TD

ISO fits

Basic hole system cf. DIN ISO 286-2 (1990-11)

Limit deviations in µm for tolerance classes[1)]

Nominal dimension range over – to mm	for hole	for shafts – Paired with an H8 hole results in a clearance fit				interference fit		for hole	for shafts – Paired with an H11 hole results in a clearance fit					
	H8	d9	e8	**f7**	**h9**	**u8**[2)]	**x8**[2)]	**H11**	a11	c11	d9	d11	h9	h11
up to 3	+14 0	–20 –45	–14 –28	–6 –16	0 –25	+32 +18	+34 +20	+60 0	–270 –330	–60 –120	–20 –45	–20 –80	0 –25	0 –60
3 – 6	+18 0	–30 –60	–20 –38	–10 –22	0 –30	+41 +23	+46 +28	+75 0	–270 –345	–70 –145	–30 –60	–30 –105	0 –30	0 –75
6 – 10	+22 0	–40 –76	–25 –47	–13 –28	0 –36	+50 +28	+56 +34	+90 0	–280 –370	–80 –170	–40 –76	–40 –130	0 –36	0 –90
10 – 14	+27 0	–50 –93	–32 –59	–16 –34	0 –43	+60 +33	+67 +40	+110 0	–290 –400	–95 –205	–50 –93	–50 –160	0 –43	0 –110
14 – 18							+72 +45							
18 – 24	+33 0	–65 –117	–40 –73	–20 –41	0 –52	+74 +41	+87 +54	+130 0	–300 –430	–110 –240	–65 –117	–65 –195	0 –52	0 –130
24 – 30						+81 +48	+97 +64							
30 – 40	+39 0	–80 –142	–50 –89	–25 –50	0 –62	+99 +60	+119 +80	+160 0	–310 –470	–120 –280	–80 –142	–80 –240	0 –62	0 –160
40 – 50						+109 +70	+136 +97		–320 –480	–130 –290				
50 – 65	+46 0	–100 –174	–60 –106	–30 –60	0 –74	+133 +87	+168 +122	+190 0	–340 –530	–140 –330	–100 –174	–100 –290	0 –74	0 –190
65 – 80						+148 +102	+192 +146		–360 –550	–150 –340				
80 – 100	+54 0	–120 –207	–72 –126	–36 –71	0 –87	+178 +124	+232 +178	+220 0	–380 –600	–170 –390	–120 –207	–120 –340	0 –87	0 –220
100 – 120						+198 +144	+264 +210		–410 –630	–180 –400				
120 – 140	+63 0	–145 –245	–85 –148	–43 –83	0 –100	+233 +170	+311 +248	+250 0	–460 –710	–200 –450	–145 –245	–145 –395	0 –100	0 –250
140 – 160						+253 +190	+343 +280		–520 –770	–210 –460				
160 – 180						+273 +210	+373 +310		–580 –830	–230 –480				
180 – 200	+72 0	–170 –285	–100 –172	–50 –96	0 –115	+308 +236	+422 +350	+290 0	–660 –950	–240 –530	–170 –285	–170 –460	0 –115	0 –290
200 – 225						+330 +258	+457 +385		–740 –1030	–260 –550				
225 – 250						+356 +284	+497 +425		–820 –1110	–280 –570				
250 – 280	+81 0	–190 –320	–110 –191	–56 –108	0 –130	+396 +315	+556 +475	+320 0	–920 –1240	–300 –620	–190 –320	–190 –510	0 –130	0 –320
280 – 315						+431 +350	+606 +525		–1050 –1370	–330 –650				
315 – 355	+89 0	–210 –350	–125 –214	–62 –119	0 –140	+479 +390	+679 +590	+360 0	–1200 –1560	–360 –720	–210 –350	–210 –570	0 –140	0 –360
355 – 400						+524 +435	+749 +660		–1350 –1710	–400 –760				
400 – 450	+97 0	–230 –385	–135 –232	–68 –131	0 –155	+587 +490	+837 +740	+400 0	–1500 –1900	–440 –840	–230 –385	–230 –630	0 –155	0 –400
450 – 500						+637 +540	+917 +820		–1650 –2050	–480 –880				

1) The tolerance classes in **bold** print correspond to row 1 in DIN 7157; their use is preferable.
2) DIN 7157 recommends: nominal dimensions up to 24 mm: H8/x8; nominal dimensions over 24 mm: H8/u8.

ISO fits

Basic shaft system

cf. DIN ISO 286-2 (1990-11)

Nominal dimension range over – to mm	Limit deviations in µm for tolerance classes[1]															
	for shafts	for holes – Paired with an h5 shaft results in a					for shafts	for holes – Paired with an h6 shaft results in a								
		clearance fit	transition fit		interference fit			clearance fit			transition fit				interference fit	
	h5	H6	J6	M6	N6	P6	**h6**	**F8**	G7	**H7**	J7	K7	M7	N7	R7	S7
up to 3	0 −4	+6 0	+2 −4	−2 −8	−4 −10	−6 −12	0 −6	+20 +6	+12 +2	+10 0	+4 −6	0 −10	−2 −12	−4 −14	−10 −20	−14 −24
3 – 6	0 −5	+8 0	+5 −3	−1 −9	−5 −13	−9 −17	0 −8	+28 +10	+16 +4	+12 0	+6 −6	+3 −9	0 −12	−4 −16	−11 −23	−15 −27
6 – 10	0 −6	+9 0	+5 −4	−3 −12	−7 −16	−12 −21	0 −9	+35 +13	+20 +5	+15 0	+8 −7	+5 −10	0 −15	−4 −19	−13 −28	−17 −32
10 – 18	0 −8	+11 0	+6 −5	−4 −15	−9 −20	−15 −26	0 −11	+43 +16	+24 +6	+18 0	+10 −8	+6 −12	0 −18	−5 −23	−16 −34	−21 −39
18 – 30	0 −9	+13 0	+8 −5	−4 −17	−11 −24	−18 −31	0 −13	+53 +20	+28 +7	+21 0	+12 −9	+6 −15	0 −21	−7 −28	−20 −41	−27 −48
30 – 40	0 −11	+16 0	+10 −6	−4 −20	−12 −28	−21 −37	0 −16	+64 +25	+34 +9	+25 0	+14 −11	+7 −18	0 −25	−8 −33	−25 −50	−34 −59
40 – 50																
50 – 65	0 −13	+19 0	+13 −6	−5 −24	−14 −33	−26 −45	0 −19	+76 +30	+40 +10	+30 0	+18 −12	+9 −21	0 −30	−9 −39	−30 −60	−42 −72
65 – 80															−32 −62	−48 −78
80 – 100	0 −15	+22 0	+16 −6	−6 −28	−16 −38	−30 −52	0 −22	+90 +36	+47 +12	+35 0	+22 −13	+10 −25	0 −35	−10 −45	−38 −73	−58 −93
100 – 120															−41 −76	−66 −101
120 – 140	0 −18	+25 0	+18 −7	−8 −33	−20 −45	−36 −61	0 −25	+106 +43	+54 +14	+40 0	+26 −14	+12 −28	0 −40	−12 −52	−48 −88	−77 −117
140 – 160															−50 −90	−85 −125
160 – 180															−53 −93	−93 −133
180 – 200	0 −20	+29 0	+22 −7	−8 −37	−22 −51	−41 −70	0 −29	+122 +50	+61 +15	+46 0	+30 −16	+13 −33	0 −46	−14 −60	−60 −106	−105 −151
200 – 225															−63 −109	−113 −159
225 – 250															−67 −113	−123 −169
250 – 280	0 −23	+32 0	+25 −7	−9 −41	−25 −57	−47 −79	0 −32	+137 +56	+69 +17	+52 0	+36 −16	+16 −36	0 −52	−14 −66	−74 −126	−138 −190
280 – 315															−78 −130	−150 −202
315 – 355	0 −25	+36 0	+29 −7	−10 −46	−26 −62	−51 −87	0 −36	+151 +62	+75 +18	+57 0	+39 −18	+17 −40	0 −57	−16 −73	−87 −144	−169 −226
355 – 400															−93 −150	−187 −244
400 – 450	0 −27	+40 0	+33 −7	−10 −50	−27 −67	−55 −95	0 −40	+165 +68	+83 +20	+63 0	+43 −20	+18 −45	0 −63	−17 −80	−103 −166	−209 −272
450 – 500															−109 −172	−229 −292

[1] The tolerance classes in **bold** print correspond to row 1 in DIN 7157; their use is preferable.

ISO fits

Basic shaft system cf. DIN ISO 286-2 (1990-11)

Nominal dimension range over – to mm	Limit deviations in µm for tolerance classes[1]: for shafts	for holes – Pairing with an h9 shaft results in a clearance fit					transition fit			for shafts	for holes – Pairing with an h11 shaft results in a clearance fit			
	h9	**C11**	**D10**	**E9**	**F8**	**H8**	J9/JS9[2]	N9[3]	**P9**	**h11**	A11	**C11**	**D10**	**H11**
up to 3	0 –25	+120 +60	+60 +20	+39 +14	+20 +06	+14 0	+12.5 –12.5	–4 –29	–6 –31	0 –60	+330 +270	+120 +60	+60 +20	+60 0
3 – 6	0 –30	+145 +70	+78 +30	+50 +20	+28 +10	+18 0	+15 –15	0 –30	–12 –42	0 –75	+345 +270	+145 +70	+78 +30	+75 0
6 – 10	0 –36	+170 +80	+98 +40	+61 +25	+35 +13	+22 0	+18 –18	0 –36	–15 –51	0 –90	+370 +280	+170 +80	+98 +40	+90 0
10 – 18	0 –43	+205 +95	+120 +50	+75 +32	+43 +16	+27 0	+21.5 –21.5	0 –43	–18 –61	0 –110	+400 +290	+205 +95	+120 +50	+110 0
18 – 30	0 –52	+240 +110	+149 +65	+92 +40	+53 +20	+33 0	+26 –26	0 –52	–22 –74	0 –130	+430 +300	+240 +110	+149 +65	+130 0
30 – 40	0 –62	+280 +120	+180 +80	+112 +50	+64 +25	+39 0	+31 –31	0 –62	–26 –88	0 –160	+470 +310	+280 +120	+180 +80	+160 0
40 – 50		+290 +130									+480 +320	+290 +130		
50 – 65	0 –74	+330 +140	+220 +100	+134 +60	+76 +30	+46 0	+37 –37	0 –74	–32 –106	0 –190	+530 +340	+330 +140	+220 +100	+190 0
65 – 80		+340 +150									+550 +360	+340 +150		
80 – 100	0 –87	+390 +170	+260 +120	+159 +72	+90 +36	+54 0	+43.5 –43.5	0 –87	–37 –124	0 –220	+600 +380	+390 +170	+260 +120	+220 0
100 – 120		+400 +180									+630 +410	+400 +180		
120 – 140		+450 +200									+710 +460	+450 +200		
140 – 160	0 –100	+460 +210	+305 +145	+185 +85	+106 +43	+63 0	+50 –50	0 –100	–43 –143	0 –250	+770 +520	+460 +210	+305 +145	+250 0
160 – 180		+480 +230									+820 +580	+480 +230		
180 – 200		+530 +240									+950 +660	+530 +240		
200 – 225	0 –115	+550 +260	+355 +170	+215 +100	+122 +50	+72 0	+57.5 –57.5	0 –115	–50 –165	0 –290	+1030 +740	+550 +260	+355 +170	+290 0
225 – 250		+570 +280									+1110 +820	+570 +280		
250 – 280	0 –130	+620 +300	+400 +190	+240 +110	+137 +56	+81 0	+65 –65	0 –130	–56 –186	0 –320	+1240 +920	+620 +300	+400 +190	+320 0
280 – 315		+650 +330									+1370 +1050	+650 +330		
315 – 355	0 –140	+720 +360	+440 +210	+265 +125	+151 +62	+89 0	+70 –70	0 –140	–62 –202	0 –360	+1560 +1200	+720 +360	+440 +210	+360 0
355 – 400		+760 +400									+1710 +1350	+760 +400		
400 – 450	0 –155	+840 +440	+480 +230	+290 +135	+165 +68	+97 0	+77.5 –77.5	0 –155	–68 –223	0 –400	+1900 +1500	+840 +440	+480 +230	+400 0
450 – 500		+880 +480									+2050 +1650	+880 +480		

[1] The tolerance classes in **bold** print correspond to row 1 in DIN 7157; their use is preferable.
[2] The tolerance zones J9/JS9, J10/JS10 etc. are all identical in size and are symmetrical to the zero line.
[3] Tolerance class N9 may not be used for nominal dimensions ≤ 1mm.

TD

General tolerances, Roller bearing fits

General tolerances[1] for linear and angular dimensions

cf. DIN ISO 2768-1 (1991-06)

Tolerance class	Linear dimensions							
	Limit deviations in mm for nominal dimension ranges							
	0.5 to 3	over 3 to 6	over 6 to 30	over 30 to 120	over 120 to 400	over 400 to 1000	over 1000 to 2000	over 2000 to 4000
f (fine)	± 0.05	± 0.05	± 0.1	± 0.15	± 0.2	± 0.3	± 0.5	–
m (medium)	± 0.1	± 0.1	± 0.2	± 0.3	± 0.5	± 0.8	± 1.2	± 2
c (coarse)	± 0.2	± 0.3	± 0.5	± 0.8	± 1.2	± 2	± 3	± 4
v (very coarse)	–	± 0.5	± 1	± 1.5	± 2.5	± 4	± 6	± 8

Tolerance class	Chamfered edges (radii and chamfers)			Angular dimensions				
	Limit deviations in mm for nominal dimension ranges			Limit deviations in degrees and minutes for nominal dimension ranges (shorter angle leg)				
	0.5 to 3	over 3 to 6	6	to 10	over 10 to 50	over 50 to 120	over 120 to 400	400
f (fine)	± 0.2	± 0.5	± 1	± 1°	± 0° 30′	± 0° 20′	± 0° 10′	± 0° 5′
m (medium)								
c (coarse)	± 0.4	± 1	± 2	± 1° 30′	± 1°	± 0° 30′	± 0° 15′	± 0° 10′
v (very coarse)				± 3°	± 2°	± 1°	± 0° 30′	± 0° 20′

General tolerances[1] for form and position

cf. DIN ISO 2768-2 (1991-04)

Tolerance class	Tolerances in mm for straightness and flatness						perpendicularity				symmetry				run
	nominal dimension ranges in mm						nominal dim. ranges in mm (shorter angle leg)				nominal dim. ranges in mm (shorter feature)				
	up to 10	over 10 to 30	over 30 to 100	over 100 to 300	over 300 to 1000	over 1000 to 3000	up to 100	over 100 to 300	over 300 to 1000	over 1000 to 3000	up to 100	over 100 to 300	over 300 to 1000	over 1000 to 3000	
H	0.02	0.05	0.1	0.2	0.3	0.4	0.2	0.3	0.4	0.5	0.5				0.1
K	0.05	0.1	0.2	0.4	0.6	0.8	0.4	0.6	0.8	1	0.6		0.8	1	0.2
L	0.1	0.2	0.4	0.8	1.2	1.6	0.6	1	1.5	2	0.6	1	1.5	2	0.5

1) General tolerances apply to dimensions without individual tolerance entry. Drawing entry page 75.

Tolerances for the installation of roller bearings

cf. DIN 5425-1 (1984-11)

Radial bearing

Inner ring (shaft)

Load case	Fit	Load	Fundamental deviations for shafts[1] with ball bearing	roller bearing
circumferential load 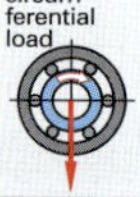	transition or interference fit required	low	h, k	k, m
		medium	j, k, m	k, m, n, p
		high	m, n	n, p, r
point load 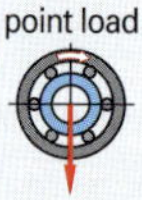	clearance fit allowed	arbitrarily large	j, h, g, f	

Outer ring (housing)

Load case	Fit	Load	Fundamental deviations for housings[1] with ball bearing	roller bearing
point load 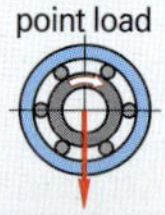	clearance fit allowed	arbitrarily large	J, H, G, F	
circumferential load 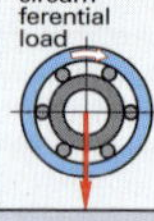	transition or interference fit required	low	J	K
		medium	K, M	M, N
		high	–	N, P

Thrust bearing

Load type	Bearing construction	Shaft washer (shaft) Load case	Fundamental deviat. for shafts[1]	Housing plate (housing) Load case	Fundamental deviations for housing[1]
Combined radial/axial load	angular contact ball bearing spherical roller bearing tapered roller bearing	circumfer. load	j, k, m	point load	H, J
		point load	j	circumfer. load	K, M
Pure axial load	ball bearing roller bearing	–	h, j, k	–	H, G, E

1) Fundamental tolerance grades: for shafts typically IT6, for bores typically IT7. If the smoothness and accuracy of running must satisfy increased requirements, also smaller tolerance grades are specified.

Fit recommendations, possible fits

Fit recommendations[1] cf. DIN 7157 (1966-01)

From row 1	C11/h9, D10/h9, E9/h9, F8/h9, H8/f7, F8/h6, H7/f7, H8/h9, H7/h6, H7/n6, H7/r6, H8/x8 or u8
From row 2	C11/h11, D10/h11, H8/d9, H8/e8, H7/g6, G7/h6, H11/h9, H7/j6, H7/k6, H7/s6

Possible fits (examples) cf. DIN 7157 (1966-01)

Basic hole[2]		Characteristic/application examples	Basic shaft[2]	
Clearance fits				
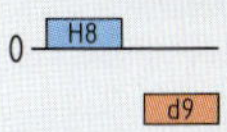	H8/d9	**Loose running fit** Clearance allows for loose fit of mating parts. (i.e. spacer sleeves on shafts)	**D10/h9**	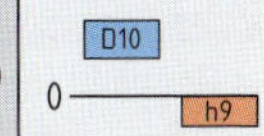
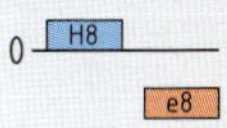	H8/e8	**Free running fit (Medium running fit):** Sufficient clearance is allowed for ease of assembly. (i.e. collar on shaft)	**E9/h9**	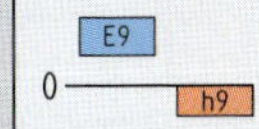
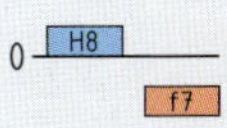	**H8/f7**	**Close running fit:** Clearance allows for parts to be easily assembled by hand while maintaining location accuracy. (i.e. plain bearing of shaft)	**F8/h9**	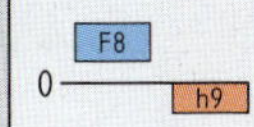
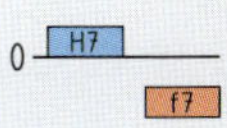	**H7/f7**	**Sliding fit – free:** Clearance allows accurate location and free movement, including turning. (i.e. piston valves in cylinders)	**F8/h6**	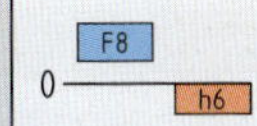
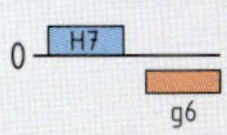	H7/g6	**Sliding fit – constrained:** Clearance allows better locational accuracy while still allowing sliding or turning movement. (i.e. transmission gear on shaft)	G7/h6	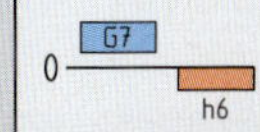
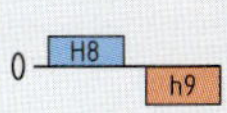	**H8/h9**	**Minimal clearance fit:** Allows locational accuracy and hand force assembly without being a snug fit. (i.e. spacer sleeves)	**H8/h9**	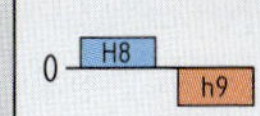
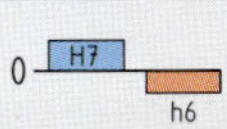	**H7/h6**	**Locational clearance fit:** Allows snug fit of stationary parts that may be assembled by hand force. (i.e. punch in punch holder)	**H7/h6**	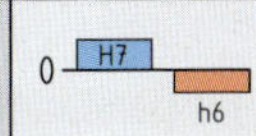
Transition fits				
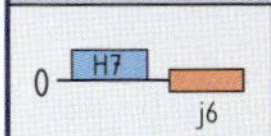	H7/j6	**Locational transition fit – clearance:** For accurate location allowing more clearance than interference. (i.e. gears on shafts)	not specified	
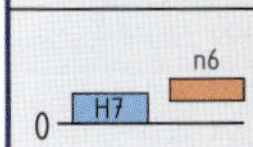	**H7/n6**	**Locational transition fit – interference:** For accurate location where interference is permissible. (i.e. drill bushing in jigs)		
Interference fits				
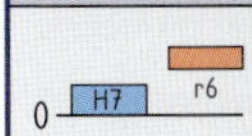	**H7/r6**	**Locational interference fit:** For rigidity and alignment/accurate location without special bore requirements. (i.e. bushings in housings)	not specified	
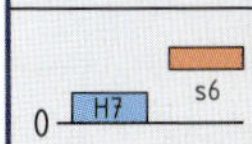	H7/s6	**Medium drive fit:** For ordinary steel parts or shrink fits of light sections. Tightest fit possible for cast iron. (i.e. plain bearing bushings)		
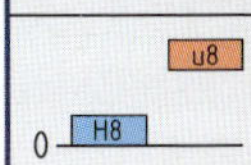	**H8/u8**	**Force fit:** For parts fitting that can withstand high mechanical pressing force or shrink fitting. (i.e. wheel on axle)		
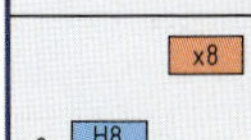	**H8/x8**	**Extreme force fit:** For parts that can only be assembled by stretching or shrinking. (i.e. turbine blade on shaft)		

[1] Deviations from these fit recommendations should only be made in exceptional cases, e. g. installation of roller bearings.
[2] The fits in **bold** print are tolerance combinations according to row 1. Their use is preferred.

TD

Geometric tolerancing

Tolerances of geometry, orientation, location and run-out

cf. DIN EN ISO 1101 (2008-08)

Structure of tolerance specifications

Datum

• **Identification**

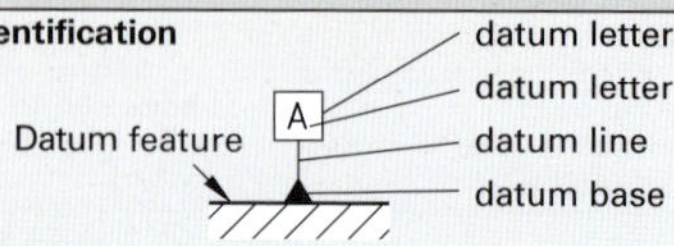

• **Datum** is the

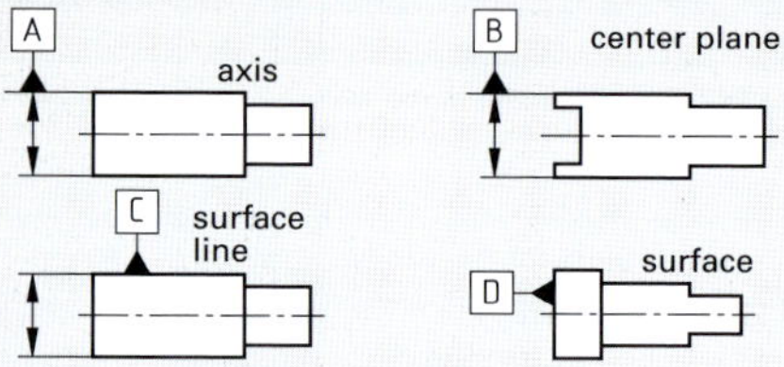

Toleranced element

• **Identification**

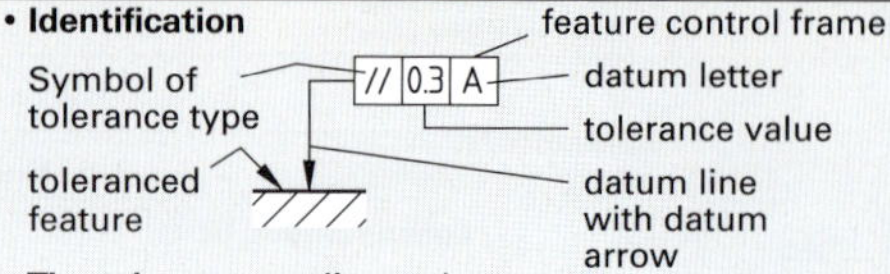

• **The tolerance** applies to the

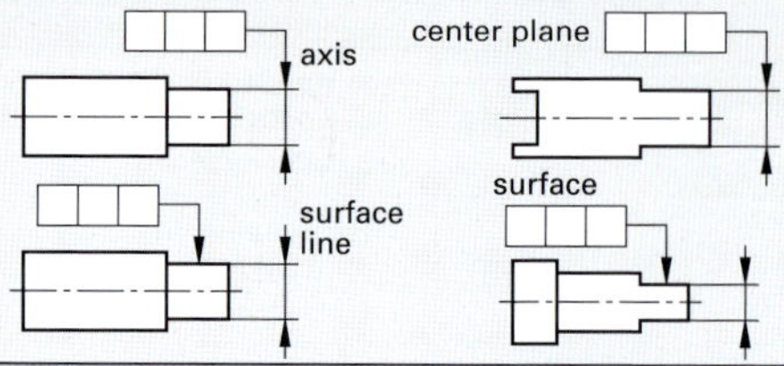

TD

Indications in drawings of datum specifications and toleranced elements

Datum	Simple datum	Common datum	Multiple datum (two or three datums)
Example	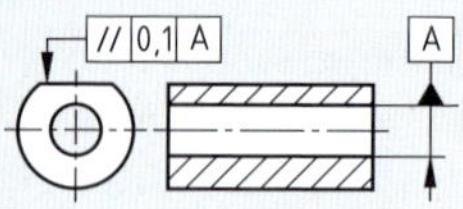	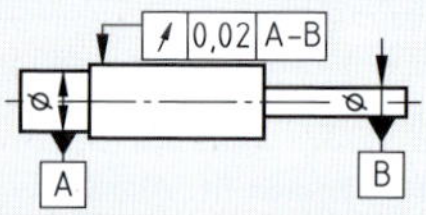	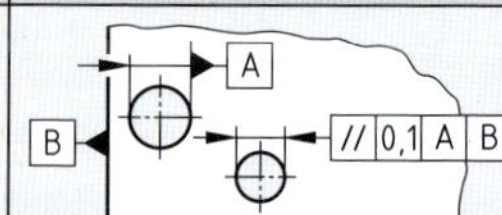
Datum in feature control frame	Individual datum letter	Datum letters separated with hyphens	Order of datum letters according to their importance

Examples

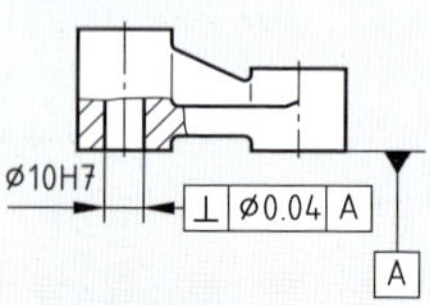

The axis of the hole must run perpendicular (tolerance value 0.04 mm) to the datum surface.

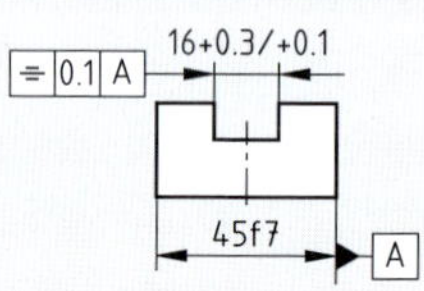

The center plane of the slot must run symmetrically to the center plane of the exterior surface (tolerance value 0.1 mm).

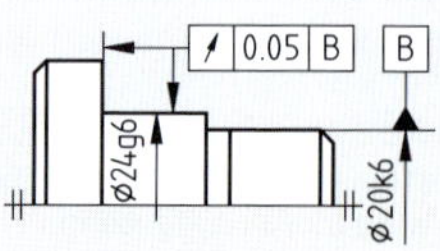

The cylindrical surface Ø24g6 must run true to the axis Ø20k6 and the flat surface must be planar (tolerance value 0.05 mm).

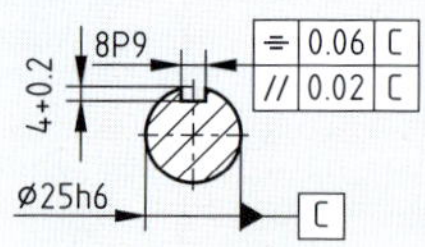

The slot must lie symmetrical (tolerance value 0.06 mm) and parallel (tolerance value 0.02 mm) to the axis Ø25h6.

Indication in drawings

cf. DIN EN ISO 1101 (2008-08)

Geometric characteristic symbols		Representation in drawing (examples)	Explanation	Tolerance zone
Tolerances of form				
—	Straightness	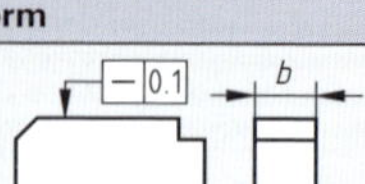	At all points across width *b*, the surface curve must lie between two parallel lines spaced *t* = 0.1 mm apart	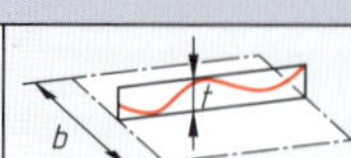
		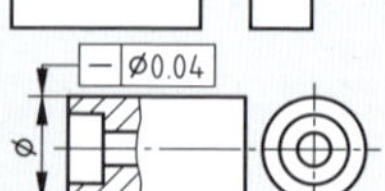	The toleranced axis of the shaft must lie within a cylinder with diameter *t* = 0.04 mm.	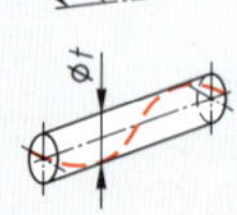
▱	Flatness	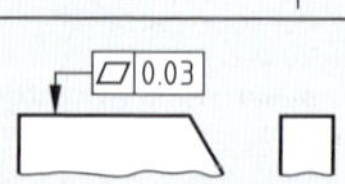	The toleranced surface must be located between two parallel planes spaced apart a distance of *t* = 0.03 mm.	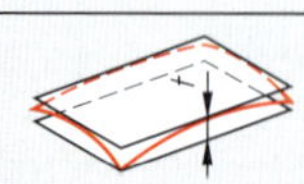

Geometric dimensioning and tolerancing GD & T

Indications in drawings (continued) — cf. DIN EN ISO 1101 (2008-08)

Symbol and toleranced property	Representation in drawing	Explanation	Tolerance zone
Tolerances of form (continued)			
○ Circularity	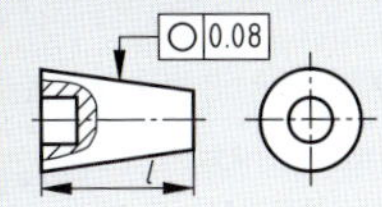	The cone's circumferential line must lie between two concentric circles spaced apart at a distance of $t = 0.08$ mm in each point of the cone length l.	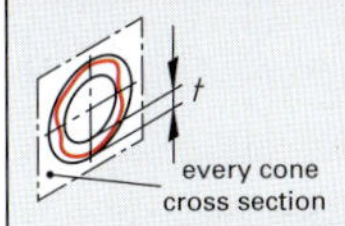
⌭ Cylindricity	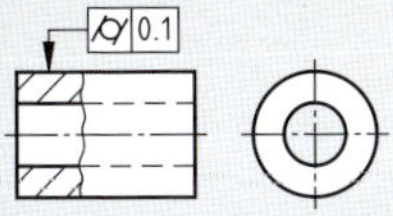	The shell surface of the cylinder must lie between two coaxial cylinders, which are spaced apart at a radial distance of $t = 0.1$ mm.	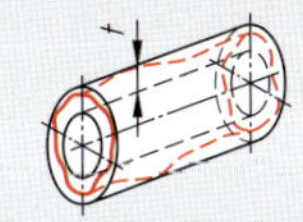
⌒ Profile of line	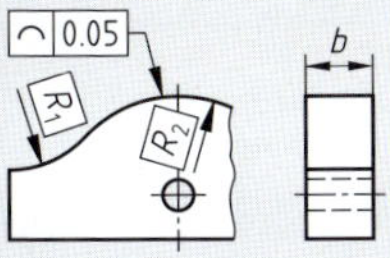	The profile line must lie between two enveloping lines, whose gap is bounded by circles of diameter $t = 0.05$ mm in each point of the object thickness b. The centers of these circles lie on a geometrically ideal line.	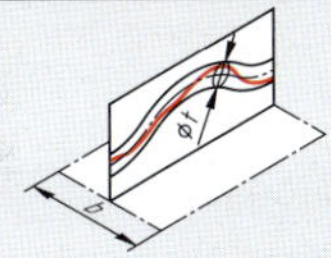
⌓ Profile of surface	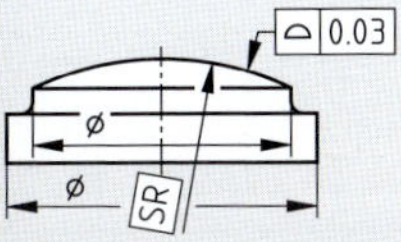	The surface of the sphere must lie between two enveloping surfaces, whose gap $t = 0.3$ mm is created by spheres. The centers of these spheres lie on the geometrically ideal surface.	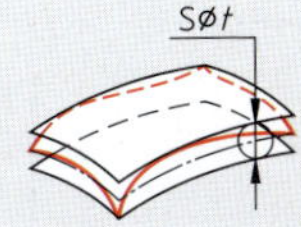
Orientation tolerances			
// Parallelism	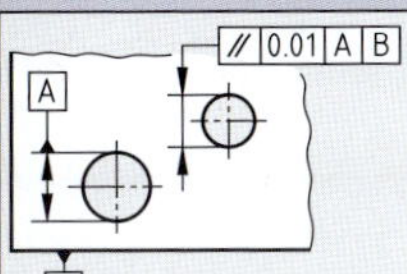	The hole's centerline must lie between two parallel planes spaced apart at a distance of $t = 0.01$ mm. The planes are parallel to datum line A and datum plane B and in line with the defined direction (vertical in this case).	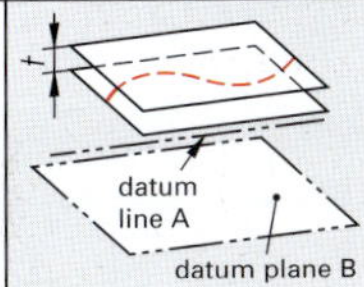
	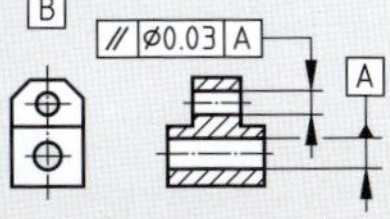	The hole's centerline must lie within a cylinder of diameter $t = 0.03$ mm. The centerline of this cylinder is parallel to datum line (axis) A.	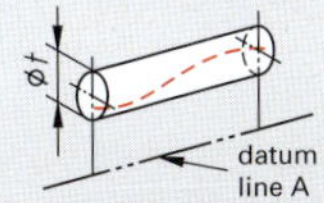
⊥ Perpendicularity	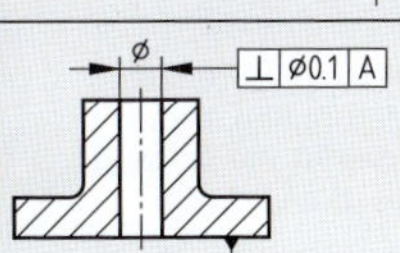	The hole's centerline must lie within a cylinder of diameter $t = 0.1$ mm that is perpendicular to datum plane A.	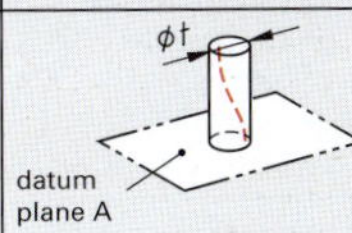
	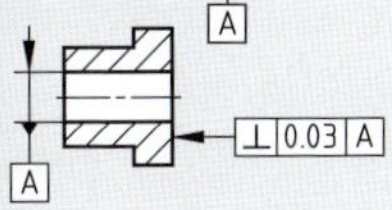	The plane surface must lie between two planes perpendicular to datum line A that are spaced apart at a distance of $t = 0.03$ mm.	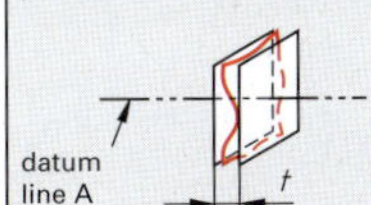
∠ Angularity	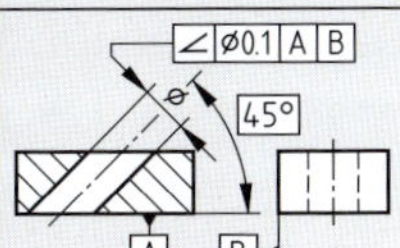	The hole's centerline must lie within a cylinder of diameter $t = 0.1$ mm. The centerline of the cylinder is parallel to datum plane B and inclined at a theoretically exact angle of $\alpha = 45°$ with reference to datum plane A.	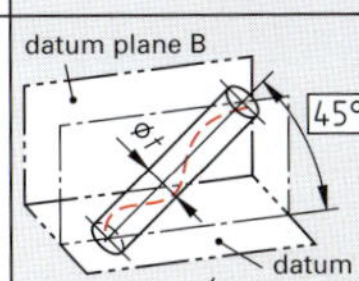
	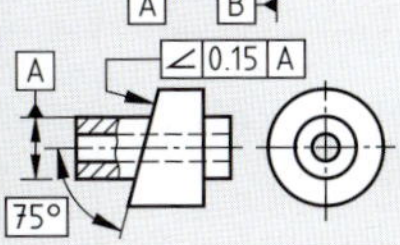	The inclined plane must lie between two parallel planes spaced at a distance of $t = 0.15$ mm that are inclined at a theoretically exact angle of $\alpha = 75°$ with reference to datum line A.	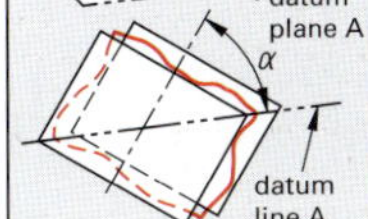

TD

Geometric dimensioning and tolerancing GD & T

Indications in drawings (continued)

cf. DIN EN ISO 1101 (2008-08)

Symbol and toleranced property		Representation in drawing	Explanation	Tolerance zone
Tolerances of location				
⌖	Posi-tion	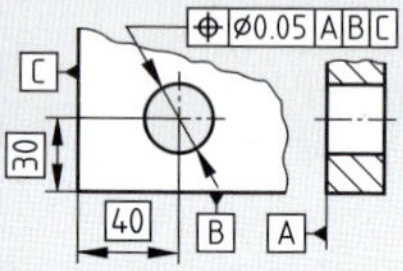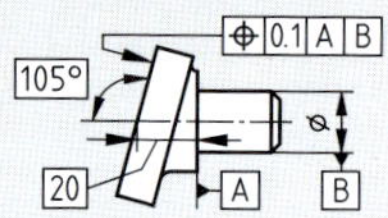	The hole's centerline must lie within a cylinder of diameter t = 0.05 mm. The cylinder's center-line must coincide with the theoretically exact loca-tion of the hole's centerline in regard to the datum planes A, B and C. The surface must lie between two parallel planes spaced apart at a distance of t = 0.1 mm that are symmetrical to the theoretically exact location of the toleranced surface in regard to datum plane A and datum line B.	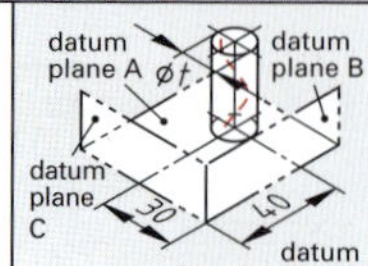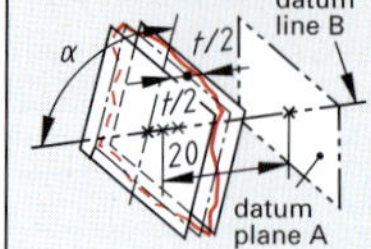
◎	Concen-tricity	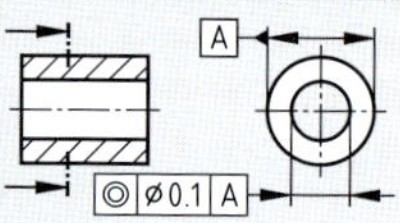	The center of the hole must lie in a circle of dia-meter t = 0.1 mm that is concentric to the datum point A in the cross section.	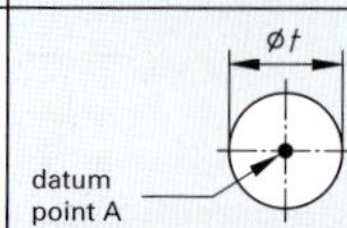
	Coaxi-ality	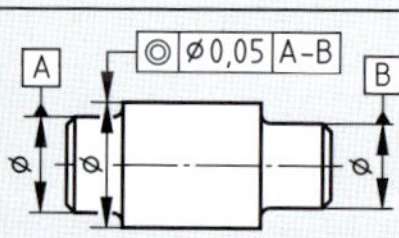	The centerline of all diameters must lie within a cylinder of diameter t = 0.05 mm. The centerline of this cylinder must coincide with the common datum axis A–B.	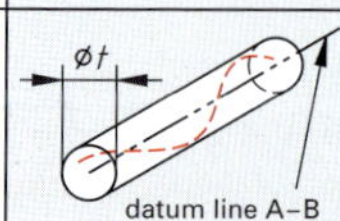
⌯	Sym-metry	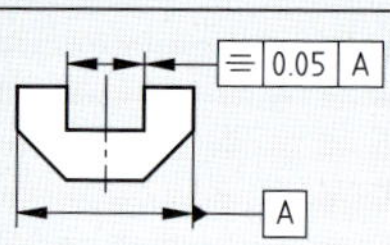	The midplane of the slot must lie between two parallel planes spaced apart at a distance of t = 0.05 mm that are located symmetrical to datum plane A.	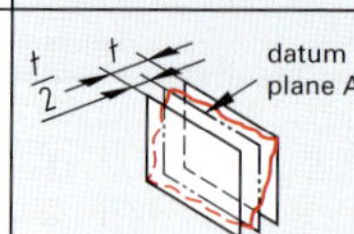
Runout tolerances				
↗	Radial circular runout	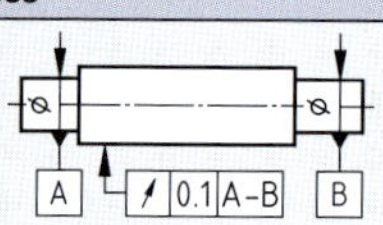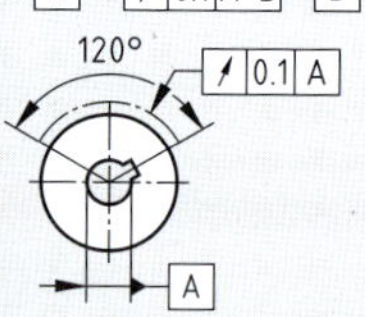	In every cross section, the circumferential line must be perpendicular to the common datum line A–B between two concentric circles in the same plane having a radial distance of t = 0.1 mm. In every cross section, the 120° circumferential line must be perpendicular to datum line A and lie between two concentric circles in the same plane that have a radial distance of t = 0.1 mm.	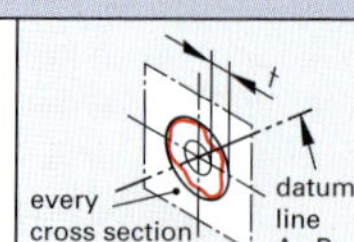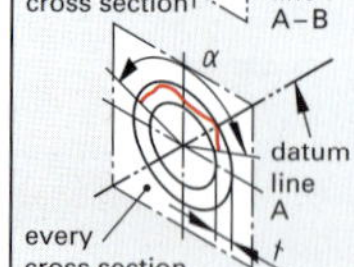
	Axial circular runout	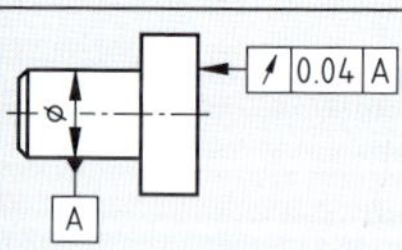	In every diameter, the circumferential line must lie in the plane surface between two circles that have a radial distance of t = 0.04 mm. The centerline of each diameter must coincide with datum line A.	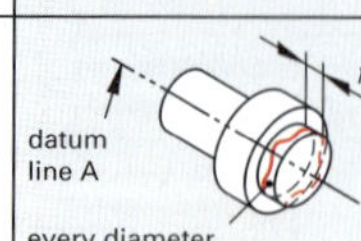
⌰	Total radial runout	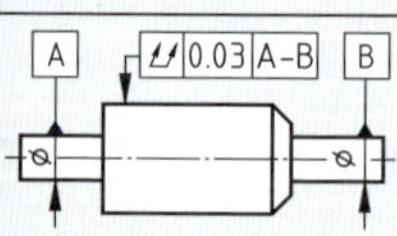	The shell surface must lie between two coaxial cylinders having a radial distance of t = 0.03 mm. The centerlines of these cylinders must coincide with the common datum line A–B.	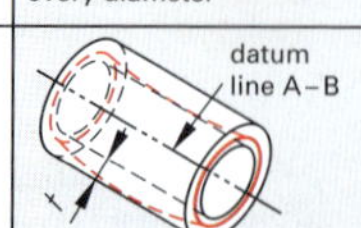
	Total axial runout	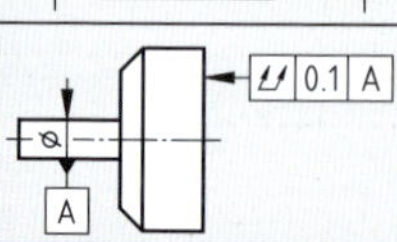	The plane surface must lie between two parallel planes spaced apart at a distance of t = 0.1 mm that are perpendicular to datum line A.	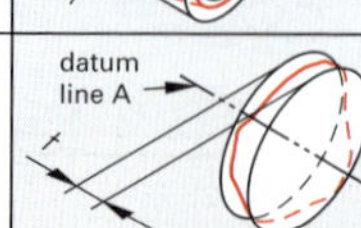

4 Materials science

Tungsten (W)	19.27	3390
Zinc (Zn)	7.13	419.5
Tin (Sn)	7.29	231.9

Unalloyed steels | Alloy steels | Stainless steels

S235	16MnCr5	C60E
31CrMo12	E360	35S20
60WCrV8	X12Cr13	38Si7

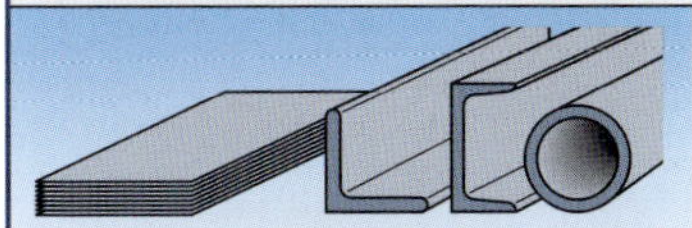

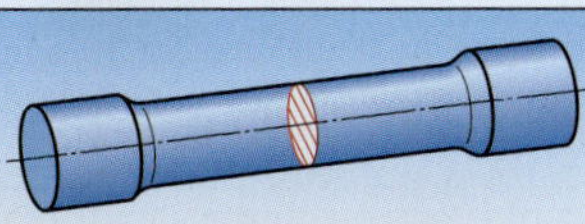

Material characteristics of solids

Solid material

Material	Density ϱ kg/dm^3	Melting temperature at 1.013 bar ϑ °C	Boiling temperature at 1.013 bar ϑ °C	Latent heat of fusion at 1.013 bar q kJ/kg	Thermal conductivity at 20 °C λ W/(m · K)	Mean specific heat at 0–100 °C c kJ/(kg · K)	Specific electrical resistivity at 20 °C ϱ_{20} Ω · mm^2/m	Coefficient of linear expansion 0–100 °C α_l 1/°C or 1/K
Aluminum (Al)	2.7	659	2467	356	204	0.94	0.028	0.0000238
Antimony (Sb)	6.69	630.5	1637	163	22	0.21	0.39	0.0000108
Asbestos	2.1–2.8	≈ 1300	–	–	–	0.81	–	–
Beryllium (Be)	1.85	1280	≈ 3000	–	165	1.02	0.04	0.0000123
Bismuth (Bi)	9.8	271	1560	59	8.1	0.12	1.25	0.0000125
Cadmium (Cd)	8.64	321	765	54	91	0.23	0.077	0.00003
Carbide (K 20)	14.8	> 2000	≈ 4000	–	81.4	0.80	–	0.000005
Carbon (diamond)	3.51	≈ 3550	–	–	–	0.52	–	0.00000118
Cast iron	7.25	1150–1200	2500	125	58	0.50	0.6–1.6	0.0000105
Chromium (Cr)	7.2	1903	2642	134	69	0.46	0.13	0.0000084
Cobalt (Co)	8.9	1493	2880	268	69.1	0.43	0.062	0.0000127
Coke	1.6–1.9	–	–	–	0.18	0.83	–	–
Concrete	1.8–2.2	–	–	–	≈ 1	0.88	–	0.00001
Constantan	8.89	1260	≈ 2400	–	23	0.41	0.49	0.0000152
Copper (Cu)	8.96	1083	≈ 2595	213	384	0.39	0.0179	0.0000168
Cork	0.1–0.3	–	–	–	0.04–0.06	1.7–2.1	–	–
Corundum (Al_2O_3)	3.9–4.0	2050	2700	–	12–23	0.96	–	0.0000065
CuAl alloys	7.4–7.7	1040	2300	–	61	0.44	–	0.0000195
CuSn alloys	7.4–8.9	900	2300	–	46	0.38	0.02–0.03	0.0000175
CuZn alloys	8.4–8.7	900–1000	2300	167	105	0.39	0.05–0.07	0.0000185
Foam rubber	0.06–0.25	–	–	–	0.04–0.06	–	–	–
Glass (quartz glass)	2.4–2.7	520–550[1]	–	–	0.8–1.0	0.83	10^{18}	0.000009
Gold (Au)	19.3	1064	2707	67	310	0.13	0.022	0.0000142
Graphite (C)	2.26	≈ 3550	≈ 4800	–	168	0.71	–	0.0000078
Greases	0.92–0.94	30–175	≈ 300	–	0.21	–	–	–
Gypsum	2.3	1200	–	–	0.45	1.09	–	–
Ice	0.92	0	100	332	2.3	2.09	–	0.000051
Iodine (I)	5.0	113.6	183	62	0.44	0.23	–	–
Iridium (Ir)	22.4	2443	> 4350	135	59	0.13	0.053	0.0000065
Iron oxide (rust)	5.1	1570	–	–	0.58 (pwdr)	0.67	–	–
Iron. pure (Fe)	7.87	1536	3070	276	81	0.47	0.13	0.000012
Lead (Pb)	11.3	327.4	1751	24.3	34.7	0.13	0.208	0.000029
Magnesium (Mg)	1.74	650	1120	195	172	1.04	0.044	0.000026
Magnesium alloy	≈ 1.8	≈ 630	1500	–	46–139	–	–	0.0000245
Manganese (Mn)	7.43	1244	2095	251	21	0.48	0.39	0.000023
Molybdenum (Mo)	10.22	2620	4800	287	145	0.26	0.054	0.0000052
Nickel (Ni)	8.91	1455	2730	306	59	0.45	0.095	0.000013
Niobium (Nb)	8.55	2468	≈ 4800	288	53	0.273	0.217	0.0000071
Phosph., yellow (P)	1.82	44	280	21	–	0.80	–	–
Pit coal	1.35	–	–	–	0.24	1.02	–	–
Platinum (Pt)	21.5	1769	4300	113	70	0.13	0.098	0.000009
Polystyrene	1.05	–	–	–	0.17	1.3	10^{10}	0.00007
Porcelain	2.3–2.5	≈ 1600	–	–	1.6[3]	1.2[3]	10^{12}	0.000004
Quartz, flint (SiO_2)	2.1–2.5	1480	2230	–	9.9	0.8	–	0.000008
Selenium. red (Se)	4.4	220	688	83	0.2	0.33	–	–
Silicon (Si)	2.33	1423	2355	1658	83	0.75	$2.3 \cdot 10^9$	0.0000042
Silicon carbide (SiC)	2.4	disintegrates into C and Si above 3000 °C			9[1]	1.05[1]	–	–
Silver (Ag)	10.5	961.5	2180	105	407	0.23	0.015	0.0000193

[1] transformation temperature [2] cross grain [3] at 800 °C

Material characteristics of solid, liquid and gaseous materials

Solid materials (continued)

Material	Density ϱ kg/dm³	Melting temperature at 1.013 bar ϑ °C	Boiling temperature at 1.013 bar ϑ °C	Latent heat of fusion at 1.013 bar q kJ/kg	Thermal-conductivity at 20 °C λ W/(m · K)	Mean specific heat at 0–100 °C c kJ/(kg · K)	Specific electrical resistivity at 20 °C ϱ_{20} Ω · mm²/m	Coefficient of linear expansion 0–100 °C α_l 1/°C or 1/K
Sodium (Na)	0.97	97.8	890	113	126	1.3	0.04	0.000 071
Steel, unalloyed	7.85	≈ 1500	2500	205	48–58	0.49	0.14–0.18	0.000 011 9
Steel, alloyed	7.9	≈ 1500	–	–	14	0.51	0.7	0.000 016 1
Sulfur (S)	2.07	113	344.6	49	0.2	0.70	–	–
Tantalum (Ta)	16.6	2996	5400	172	54	0.14	0.124	0.000 006 5
Tin (Sn)	7.29	231.9	2687	59	65.7	0.24	0.114	0.000 023
Titanium (Ti)	4.5	1670	3280	88	15.5	0.47	0.42	0.000 008 2
Tungsten (W)	19.27	3390	5500	54	130	0.13	0.055	0.000 004 5
Uranium (U)	19.1	1133	≈ 3800	356	28	0.12	–	–
Vanadium (V)	6.12	1890	≈ 3380	343	31.4	0.50	0.2	–
Wood (air dried)	0.20–0.72	–	–	–	0.06–0.17	2.1–2.9	–	≈ 0.000 04[2]
Zinc (Zn)	7.13	419.5	907	101	113	0.4	0.06	0.000 029

Liquid materials

Material	Density at 20 °C ϱ kg/dm³	Ignition temperature ϑ °C	Freezing or melting temperature at 1.013 bar ϑ °C	Boiling temperature at 1.013 bar ϑ °C	Latent heat of vaporization[2)] r kJ/kg	Thermal-conductivity at 20 °C λ W/(m · K)	Specific heat at 20 °C c kJ/(kg · K)	Coefficient of volume expansion α_V 1/°C or 1/K
Alcohol 95 %	0.81	520	–114	78	854	0.17	2.43	0.001 1
Diesel fuel	0.81–0.85	220	–30	150–360	628	0.15	2.05	0.000 96
Ethyl ether $(C_2H_5)_2O$	0.71	170	–116	35	377	0.13	2.28	0.001 6
Fuel oil EL	≈ 0.83	220	–10	> 175	628	0.14	2.07	0.000 96
Gasoline	0.72–0.75	220	–30 – –50	25–210	419	0.13	2.02	0.001 1
Machine oil	0.91	400	–20	> 300	–	0.13	2.09	0.000 93
Mercury (Hg)	13.5	–	–39	357	285	10	0.14	0.000 18
Petroleum	0.76–0.86	550	–70	> 150	314	0.13	2.16	0.001
Water, distilled	1.00[3)]	–	0	100	2256	0.60	4.18	0.000 18

1) above 1000 °C 2) at boiling temperature and 0.013 bar 3) at 4 °C

Gaseous materials

Material	Density at 0 °C and 1.013 bar ϱ kg/m³	Specific gravity[1)] ϱ/ϱ_L	Melting temperature at 1.013 bar ϑ °C	Boiling temperature at 1.013 bar ϑ °C	Thermal conductivity at 20 °C λ W/(m · K)	Coefficient of thermal conductivity[2)] λ/λ_A	Specific heat at 20 °C and 1,013 bar c_p[3)] kJ/(kg · K)	 c_v[4)] kJ/(kg · K)
Acetylene (C_2H_2)	1.17	0.905	–84	–82	0.021	0.81	1.64	1.33
Air	1.293	1.0	–220	–191	0.026	1.00	1.005	0.716
Ammonia (NH_3)	0.77	0.596	–78	–33	0.024	0.92	2.06	1.56
Butane (C_4H_{10})	2.70	2.088	–135	–0.5	0.016	0.62	–	–
Carbon diox. (CO_2)	1.98	1.531	–57[5)]	–78	0.016	0.62	0.82	0.63
Carbon monox. (CO)	1.25	0.967	–205	–190	0.025	0.96	1.05	0.75
Freon (CF_2Cl_2)	5.51	4.261	–140	–30	0.010	0.39	–	–
Hydrogen (H_2)	0.09	0.07	–259	–253	0.180	6.92	14.24	10.10
Methane (CH_4)	0.72	0.557	–183	–162	0.033	1.27	2.19	1.68
Nitrogen (N_2)	1.25	0.967	–210	–196	0.026	1.00	1.04	0.74
Oxygen (O_2)	1.43	1.106	–219	–183	0.026	1.00	0.91	0.65
Propane (C_3H_8)	2.00	1.547	–190	–43	0.018	0.69	–	–

1) Specific gravity = density of a gas ϱ divided by the density of air ϱ_A.
2) Coefficient of thermal conductivity = the thermal conductivity λ of a gas divided by the thermal conductivity λ_A of air.
3) at constant pressure 4) at constant volume 5) at 5.3 bar

MS

Periodic table of the elements

Atomic number (= proton number) — 11 **Na** — Letter symbols
Sodium — **Element name**; state at 273 K (0 °C) and 1.013 bar:
Relative atomic mass — 22.989

solid:	black print
liquid:	brown print
gaseous:	blue print

Radioactive elements in red, e.g. 222
Synthetic elements in parentheses, e.g. (261)

1) Light metals ϱ ≤ 5 kg/dm³; Heavy non-ferrous metals ϱ > 5 kg/dm³

Period	I A	II A	III B	IV B	V B	VI B	VII B	VIII B	VIII B	VIII B	I B	II B	III A	IV A	V A	VI A	VII A	VIII A
1	1 **H** Hydrogen 1.008																	2 **He** Helium 4.002
2	3 **Li** Lithium 6.941	4 **Be** Beryllium 9.012											5 **B** Boron 10.811	6 **C** Carbon 12.011	7 **N** Nitrogen 14.007	8 **O** Oxygen 15.999	9 **F** Fluorine 18.998	10 **Ne** Neon 20.179
3	11 **Na** Sodium 22.989	12 **Mg** Magnesium 24.305											13 **Al** Aluminum 26.982	14 **Si** Silicon 28.086	15 **P** Phosphorus 30.974	16 **S** Sulfur 32.066	17 **Cl** Chlorine 35.453	18 **Ar** Argon 39.948
4	19 **K** Potassium 39.102	20 **Ca** Calcium 40.078	21 **Sc** Scandium 44.950	22 **Ti** Titanium 47.880	23 **V** Vanadium 50.942	24 **Cr** Chromium 51.996	25 **Mn** Manganese 54.938	26 **Fe** Iron 55.847	27 **Co** Cobalt 58.933	28 **Ni** Nickel 58.690	29 **Cu** Copper 63.546	30 **Zn** Zinc 65.390	31 **Ga** Gallium 69.732	32 **Ge** Germanium 75.590	33 **As** Arsenic 74.922	34 **Se** Selenium 78.960	35 **Br** Bromine 79.904	36 **Kr** Krypton 83.800
5	37 **Rb** Rubidium 85.468	38 **Sr** Strontium 87.620	39 **Y** Yttrium 88.906	40 **Zr** Zirconium 91.224	41 **Nb** Niobium 92.906	42 **Mo** Molybdenum 95.940	43 **Tc** Technetium (98)	44 **Ru** Ruthenium 101.070	45 **Rh** Rhodium 102.906	46 **Pd** Palladium 106.420	47 **Ag** Silver 107.868	48 **Cd** Cadmium 112.410	49 **In** Indium 114.820	50 **Sn** Tin 118.710	51 **Sb** Antimony 121.750	52 **Te** Tellurium 127.600	53 **I** Iodine 126.905	54 **Xe** Xenon 131.290
6	55 **Cs** Cesium 132.905	56 **Ba** Barium 137.340	71 **Lu** Lutecium 174.967	72 **Hf** Hafnium 178.490	73 **Ta** Tantalum 180.948	74 **W** Tungsten 183.850	75 **Re** Ruthenium 186.207	76 **Os** Osmium 190.200	77 **Ir** Iridium 192.200	78 **Pt** Platinum 195.080	79 **Au** Gold 196.967	80 **Hg** Mercury 200.590	81 **Tl** Thallium 204.383	82 **Pb** Lead 207.200	83 **Bi** Bismuth 208.980	84 **Po** Polonium 210	85 **At** Astatine 210	86 **Rn** Radon 222
7	87 **Fr** Francium 223	88 **Ra** Radium 226.025	103 **Lr** Lawrencium (260)	104 **Rf** Rutherfordium* (261)	105 **Ha** Hahnium* (262)	106 **Sg** Seaborgium* (263)	107 **Ns** Nielsbohrium* (264)	108 **Hs** Hassium* (265)	109 **Mt** Meitnerium* (266)									

* **Only name suggestions exist for elements 104 to 109.**
* **Element 104:** also Kurtschatovium (**Ku**) or Dubnium (**Db**);
* **Element 105:** also Joliotium; * **Element 106:** also Unilhexium (**Unh**);
* **Element 107:** also Bohrium (**Bh**) or Unilsptium (**Uns**);
* **Element 108:** also Hahnium (**Hn**) or Uniloctium (**Uno**);
* **Element 109:** also Unilenneadium (**Une**)

Lanthanides 57–71	57 **La** Lanthanum 138.906	58 **Ce** Cerium 140.120	59 **Pr** Praseodymium 140.908	60 **Nd** Neodymium 144.240	61 **Pm** Promethium 145	62 **Sm** Samarium 150.360	63 **Eu** Europium 151.960	64 **Gd** Gadolinium 157.250	65 **Tb** Terbium 158.925	66 **Dy** Dysprosium 162.500	67 **Ho** Holmium 164.930	68 **Er** Erbium 167.260	69 **Tm** Thulium 168.934	70 **Yb** Ytterbium 173.040
Actinides 89–103	89 **Ac** Actinium 227.028	90 **Th** Thorium 232.038	91 **Pa** Protactinium 231.036	92 **U** Uranium 238.029	93 **Np** Neptunium 237	94 **Pu** Plutonium 244	95 **Am** Americium (243)	96 **Cm** Curium (247)	97 **Bk** Berkelium (247)	98 **Cf** Californium (251)	99 **Es** Einsteinium (252)	100 **Fm** Fermium (257)	101 **Md** Mendelevium (258)	102 **No** Nobelium (260)

Nonmetals
Metalloids
Light metals[1]
Heavy metals[1]
Precious metals
Halogens
Noble gases

Chemicals used in metal technology, molecular groups, pH value

Important chemicals used in metal technology

Technical designation	Chemical designation	Formula	Properties	Use
Acetone	Acetone (propanone)	$(CH_3)_2CO$	Colorless, combustible, lightly volatile liquid	Solvent for paint, acetylene and plastics
Acetylene	Acetylene, Ethane	C_2H_2	Highly reactive, colorless gas, highly explosive	Fuel for welding, source material for plastics
Aqueous cleaner	Various surfactants	—COO– —OSO_3– —SO_3–	Various water soluble substances	Solvent, cleaning agent; emulsifying and thickening agent
Carbonic acid	Carbon dioxide	CO_2	Water soluble, non-combustible gas, solidifies at −78 °C	Shielding gas for MAG welding, dry ice as refrigerant
Carbon tetrachloride	Carbon tetrachloride	CCl_4	Colorless, non-combustible liquid, harmful to health	Solvent for fats, oils and paint
Cleaning agent	Organic solvent	C_nH_{2n+2}	Colorless, sometimes lightly combustible liquids	Solvent for fats and oils, cleaning agent
Copper vitriol	Copper sulfate	$CuSO_4$	Blue, water soluble crystal, moderately toxic	Electroplating baths, pest control, for scribing
Corundum	Aluminum oxide	Al_2O_3	Very hard colorless crystal, melting point 2050 °C	Grinding and polishing agent, oxide ceramic materials
Ethyl alcohol	Ethyl alcohol, denatured	C_2H_5OH	Colorless, lightly combustible liquid, boiling point 78 °C	Solvent, cleaning agent, for heating purposes, fuel additive
Hydrochloric acid	Hydrochloric acid	HCl	Colorless, pungent smelling, strong acid	Etching and pickling of metals, manufacture of chemicals
Nitric acid	Nitric acid	HNO_3	Very strong acid, dissolves metals (except precious metals)	Etching and pickling of metals, manufacture of chemicals
Soda	Sodium carbonate	Na_2CO_3	Colorless crystal, slightly water soluble, basic	Degreasing and cleaning baths, water softening
Spirits of ammonia	Ammonium hydroxide	NH_4OH	Colorless, pungent smelling liquid, weak lye	Cleaning agent (fat solvent), neutralization of acids
Sulfuric acid	Sulfuric acid	H_2SO_4	Colorless, oily, odorless liquid, strong acid	Pickling of metals, electroplating baths, storage batteries
Table salt	Sodium chloride	$NaCl$	Colorless, crystalline salt, slightly water soluble	Condiment, for freezing mixtures, for chlorine extraction

Frequently occurring molecular groups

Molecular group		Description	Example	
Designation	Formula		Designation	Formula
Carbide	≡C	Carbon compounds; to some extent very hard	Silicon carbide	SiC
Carbonate	$=CO_3$	Compounds of carbonic acid, addition of heat yields CO_2	Calcium carbonate	$CaCO_3$
Chloride	−Cl	Salts of the hydrochloric acids; usu. dissolve readily in water	Sodium chloride	$NaCl$
Hydroxide	−OH	Hydroxides are produced from metal oxides and water; behave as basics	Calcium hydroxide	$Ca(OH)_2$
Nitrate	$-NO_3$	Salts of the nitric acids; usu. dissolve readily in water	Potassium nitrate	KNO_3
Nitride	≡N	Nitrogen compounds; some of them are very hard	Silicone nitride	SiN
Oxide	=O	Oxygen compounds; most commonly occurring molecular group on earth	Aluminum oxide	Al_2O_3
Sulfate	$=SO_4$	Salts of the sulfuric acids; usu. dissolve readily in water	Copper sulfate	$CuSO_4$
Sulfide	=S	Sulfur compounds; important ores, chip breaker in free cutting steels	Iron(II) sulfide	FeS

pH value

Type of aqueous solution	← increasingly acidic						neutral	increasingly basic →						
pH value	0	1	2	3	4	5	6	7	8	9	10	11	12	13	14
Concentration H^+ in mol/l	10^0	10^{-1}	10^{-2}	10^{-3}	10^{-4}	10^{-5}	10^{-6}	10^{-7}	10^{-8}	10^{-9}	10^{-10}	10^{-11}	10^{-12}	10^{-13}	10^{-14}

MS

Definition and classification of steel

cf. DIN EN 10020 (2000-07)

Steel – Alloy with iron as the main component and a carbon content under 2.0%.

Microstructure – The **microstructural components**, e.g. ferrite, pearlite, carbides, and the **crystalline structure**, e.g. fine grain, coarse grain, bands, determine the **steel properties**, e.g. strength, toughness, workability, machinability, weldability.

Influenced by

Steel production

Composition	Degree of purity	Deoxidation
– carbon content – alloying elements	– non-metallic inclusions – phosphorus and sulfur content	rimmed, semi-killed or killed cast

Subsequent processing

For example:

- **Forming**: rolling, stamping, drawing, bending etc.
- **Heat treatment**: quenching and tempering, surface hardening etc.
- **Annealing**: normalizing, spheroidizing, full annealing etc.
- **Joining**: welding, brazing etc.
- **Coating**: galvanizing etc.

Classification

Unalloyed steels

no alloying element reached the limit value according to **table 1**

Alloy steels

- at least one alloying element reaches the limit value according to **table 1**
- steel types not conforming to the definition for stainless steels

Stainless steels[2]

- chrome content at least 10.5 %
- carbon content maximum 1.2 %

Classification by main characteristics into

- corrosion-resistant steels (pages 133, 134)
- heat resistant steels
- high-temperature steels

Classification[1]

Quality steels	High-grade steels

High-grade steels differ from quality steels due to:

- more careful production
- higher degree of purity
- improved deoxidation
- more exact composition
- improved hardenability

Table 1: Limit values for unalloyed steels

Element	%	Element	%	Element	%
Al	0.30	Mn	1.65	Se	0.10
Bi	0.10	Mo	0.08	Si	0.60
Co	0.30	Nb	0.06	Ti	0.05
Cu	0.40	Ni	0.30	V	0.10
Cr	0.30	Pb	0.40	W	0.30

Main grades

Unalloyed quality steels		Alloy quality steels	
Steel group (excerpt)	Example	Steel group (excerpt)	Example
Unalloyed structural steels	S235JR	Rail steels	R900Mn
Unalloyed steels for quenching & tempering	C45	Magnetic steel sheet and strip	M390-50E
Free cutting steels	10S20	Microalloyed steels with high yield strengths	H400M
Weldable unalloyed fine-grain steels	S275N	Phosphorus alloyed steels with high yield strengths	H180P
Unalloyed press. vessel steels	P235GH		

Unalloyed high-grade steels		Alloy high-grade steels	
Steel group (excerpt)	Example	Steel group (excerpt)	Example
Unalloyed steels for quenching and tempering	C45E	Alloy steels for quenching and tempering	42CrMo4
Unalloyed case hard. steels	C15E	Case hardened alloy steels	16MnCr5
Unalloyed tool steels	C45U	Nitriding steels	34CrAlNi7
Unalloyed steels for flame and induction hardening	C60E	Alloy tool steels	X40Cr14
		High-speed steels	HS6-5-2-5

[1] The main grade "Basic steels" was omitted. All previous basic steels are produced as quality steels.

[2] The stainless steels have their own group. They are alloy steels, so they are not classified as quality or high-grade steels.

Standardization of steel products

The designation of steels and steel products, e. g. sheet metals, bars and pipes, is currently determined by various standards. For a full designation and reliable order specification, the regulations of these standards must be combined.

Example:

Required properties:
- wear-resistant surface
- high core strength
- high durability

Pinion shaft

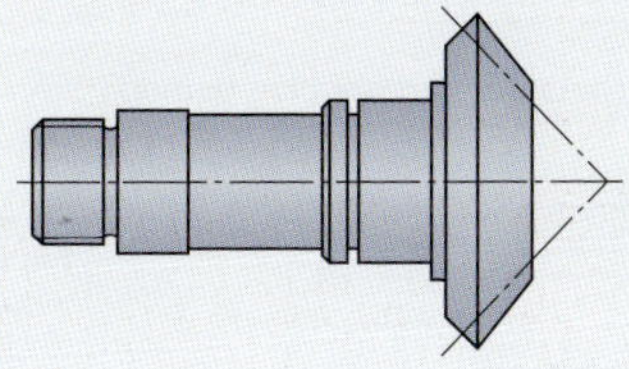

Suitable component manufacturing:
- raw part: long product
- machining and heat treatment

Suitable material groups:
- case hardened steels, page 129
- nitrided steels, page 131
- steels for flame and induction hardening, page 131

Suitable long products[1]**:**
- hot-rolled round bars[1] of steel, page 143
- bright steel, page 144

[1] **Definition of terms for steel products as per DIN EN 10079**

Selected: **Case hardened steels as per DIN EN 10084**

The standard specifies the following details, e.g.

Group	Contents, examples
Definition	Case hardened steel
Classification, designation	Alloyed and unalloyed stainless steels, e.g. C15E, 16MnCr5, 15NiCr13
Ordering details	Quantity, shape of product, code of dimensional standard, dimensions, code name, heat treatment and surface finishing condition
Manufacturing	Killed steel, typical delivery condition
Chemical composition	Carbon contents, alloying elements, non-metallic inclusions
Heat treatment	Temperatures, quenching agent, hardening process and hardness range
Properties	Workability, shearability
Structure	Grain size, inclusions
Steel types	36 different standard case hardened steels
Delivery condition	Soft annealed (+A), treated for hardening range (+TH)
Surface finishing	Hot worked (+HW), hot worked and blast cleaned (+BC)
Workability	Machinable, shearable
Testing	Hardness verification

Selected: **Hot-rolled round bars as per DIN EN 10060**

The standard specifies the following details, e.g.
- designation
- preferred diameter *d*
- length *L* (length types, length ranges)
- limit allowances
- straightness, non-circularity
- measuring rules for parameters

Designation system for steels, code names as per DIN EN 10027-1 (page 119)

The standard specifies the following details, e.g.
- unique designation, notation, determination, classification and structure of the code names
- main code signs and additional signs

Selected: **16MnCr5+A+BC**

16MnCr5	→	main code signs for the chemical composition
+A	→	soft annealed
+BC	→	hot worked and blast cleaned

Designation example:

⇒ **Round bar EN 10060 – 55 x 6000 F steel EN 10084 16MnCr5+A+BC**

Round bar with *d* = 55 mm and *L* = 6000 mm, fixed length (F), of case hardened steel 16 MnCr5, soft-annealed (+A), hot worked and blast cleaned (+BC)

Note: The standardized designations always describe the delivery condition.

The optional additional signs added to the code name or the material number are partly stipulated by the designation system of steels and partly by the steel group standards, e.g. "General structural steels", "Case hardened steels", "Quenched and tempered steels".

MS

Designation of steels using material numbers

Material numbers

cf. DIN EN 10027-2 (1992-09), replaces DIN 17007[1)]

Steel designations (page 119) or material numbers are used to identify and differentiate steels.

Designation of steel (examples):	Designation		Material number (with additional symbol +N)
	42CrMo4+N	or	1.7225+N

The material numbers consist of a 6-character number (five numeric characters and a decimal point). They are better suited for data processing than designations.

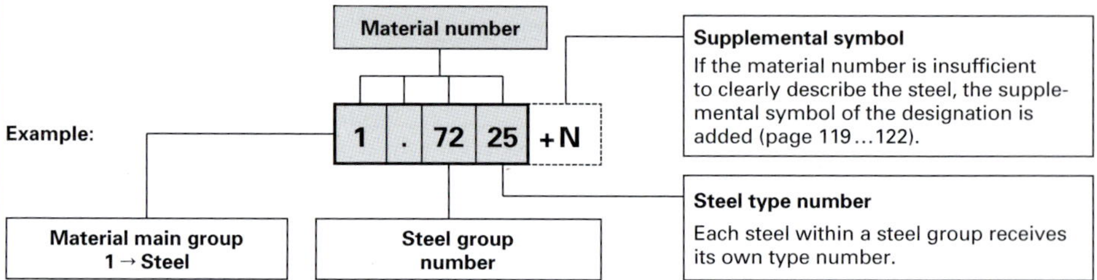

Unalloyed steels

Steel group number	Steel groups[2)]
	Quality steels
01, 91	General structural steels, R_m < 500 N/mm^2
02, 92	Other structural steels not specified for heat treatment with R_m < 500 N/mm^2
03, 93	Steels with C < 0.12 % or R_m < 400 N/mm^2
04, 94	Steels with 0.12 % ≤ C < 0.25 % or 400 N/mm^2 ≤ R_m < 500 N/mm^2
05, 95	Steels with 0.25 % ≤ C < 0.55 % or 500 N/mm^2 ≤ R_m < 700 N/mm^2
06, 96	Steels with C ≥ 0.55 % or R_m ≥ 700 N/mm^2
07, 97	Steels with high phosphorus and sulfur content
	High-grade steels
10	Steels with special physical properties
11	Structural, machine and vessel steels with C < 0.5 %
12	Machine steels with C ≥ 0.5 %
13	Structural, machine and vessel steels with special requirements
15–18	Unalloyed tool steels

Alloy steels

Steel group number	Steel groups
	Quality steels
08, 98	Steels with special physical properties
09, 99	Steels for various areas of application
	High-grade steels
20–28	Alloy tool steels
32	High-speed steels with cobalt
33	High-speed steels without cobalt
35	Roller bearing steels
36, 37	Steels with special magnetic properties
38, 39	Steels with special physical properties
40–45	Stainless steels
	Nickel alloys, chemical resistant, high-temperature
47, 48	Heat resistant steels
49	High-temperature materials
50–84	Structural, machine and vessel steels with various alloy combinations
85	Nitriding steels
87–89	High-strength weldable steels

1) The material numbers remained unchanged with the conversion from DIN 17007 to DIN EN 10027-2.

2) C carbon, R_m tensile strength
Values for tensile strength R_m and for carbon content C are mean values.

Designation system for steels

cf. DIN EN 10027-1 (2005-10)

Designation by application

The codes for steels are composed of main and supplemental symbols. Main symbols reflect the application or chemical composition. Supplemental symbols depend on to the steel or product group.

Example: Pinion shaft

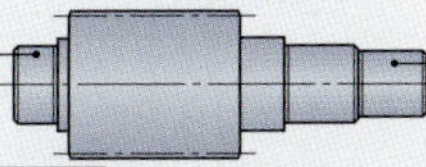

Main symbol | Suppl. symbol

S355JR+AR → Designation → DIN EN 10027-1

Material (examples)

Unalloyed structural steel → Steel group → DIN EN 10025-2

42CrMo4+N → Designation according to the chemical composition (page 121)

Material blank

Hot-rolled round steel bar → DIN EN 10060

Main symbols for the designation by application

Application	Main symbol[1)]		Application	Main symbol[1)]	
Steels for steel construction	**S**	**235**[2)]	Prestressed steels	**Y**	**1770**[3)]
Steels for machine construction	**E**	**360**[2)]	Flat rolled products for cold working	**D**	**X52**[4)]
Steels for pressure vessel construction	**P**	**265**[2)]	Rail steels	**R**	**260**[5)]
Steels for pipes and tubes	**L**	**360**[2)]	Flat products of high-strength steels	**H**	**C400**[6)]
Concrete reinforcing steels	**B**	**500**[2)]	Magnetic steel, sheet and strip	**M**	**400-50**[7)]
Packaging steel, sheet and strip	**T**	**S550**[2)]	To identify **cast steel**, the main symbol is preceded by the letter **G**.		

1) The main symbol is composed of the code letter and a number and may include an additional letter.
2) Yield strength R_e for the smallest product thickness
3) Nominal value for minimum tensile strength R_m
4) As-rolled condition C, D, X followed by two symbols
5) Minimum hardness in accordance with Brinell HBW
6) As-rolled condition C, D, X and minimum yield strength R_e or as-rolled condition CT, DT, XT and minimum tensile strength R_m
7) Maximum magnetic hysteresis loss in W/kg x 100 and nominal thickness x 100 separated by a hyphen

Steels for steel construction

Designation example: **S 235 JR + N**

- S: Code letter for steel construction
- 235: Yield strength R_e for smallest product thickness
- JR + N: Supplemental symbols

Product group (selection)	Standard	Supplemental symbols
Hot-rolled unalloyed structural steels	DIN EN 10025-2	Notch impact energy in J at °C: JR 27 20°; J2 27 −20°; J0 27 0°; K2 40 −20° C special cold workability +AR delivered in as-rolled condition +N normalized
Normalized/normalizing rolled, grain-refined structural steels suitable for welding	DIN EN 10025-3	N normalized or normalizing rolled, notch impact energy values at −20 °C. NL like N, but notch impact energy values at −50 °C
Thermomechanically rolled structural steels suitable for welding	DIN EN 10025-4	M thermomechanically rolled, notch impact energy values at −20 °C ML like M, but with notch impact energy values at −50 °C
Hot-rolled structural steels with higher yield strength in the quenched and tempered state	DIN EN 10025-6	Q quenched and tempered, notch impact energy values at −20 °C QL quenched and tempered, notch impact energy values at −40 °C QL1 quenched and tempered, notch impact energy values at −60 °C
Steels for bright steel products	DIN EN 10277-1, 2	JR, J2, C as per DIN EN 10025-2 (see above) +C drawn; +PL polished +SH peeled; +SL ground
Hot-rolled hollow sections of unalloyed structural steels and grain-refined structural steels	DIN EN 10210-1	JR, J0, J2 and K2 as with DIN EN 10025-2 N, NL as with DIN EN 10025-3 H hollow section

⇒ **S235JR+N:** Steel-construction steel R_e = 235 N/mm², notch impact energy 27 J at 20 °C, normalized (+N)

MS

Designation system for steels

cf. DIN EN 10027-1 (2005-10)

Steels for machine construction

Designation example: **E 355 +AR**

- **E**: Code letter for machine construction
- **355**: Yield strength for the smallest product thickness
- **+AR**: Supplemental symbols

Product group (selection)	Standard	Supplemental symbols
Hot-rolled unalloyed structural steels	DIN EN 10025-2	GC special cold workability +AR delivered in as-rolled condition +N normalized
Steels for bright steel products	DIN EN 10277-1, 2	GC special cold workability +C drawn +PL polished +SH peeled +SL ground
Pipes and tubes, seamless, cold-drawn	DIN EN 10305-1	+A annealed +C bright-drawn/hard +LC brigth-drawn/soft +N normalized +SR bright-drawn and stress relieved
Seamless tubes made of unalloyed and alloyed steel	DIN EN 10297-1	J2 notch impact energy 279 at −20 °C K2 notch impact energy 409 at −20 °C +AR delivered in as-rolled condition +N normalized +QT quenched and tempered

⇒ **E355+AR:** machine construction steel, yield strength R_e = 355 N/mm², delivered in as-rolled condition (+AR)

Flat products for cold working

Designation example: **D C 04 – A – m**

- **D**: Code letter for flat product for cold working
- **C**: Code letter for rolling condition: X rolling condition not specified, C cold-rolled, D hot-rolled
- **04**: Code number for the type of steel, main properties page 139
- **A – m**: Supplemental symbols (product-group specific definition)

Product group (selection)	Standard	Supplemental symbols
Cold-rolled flat products made of soft steels for cold working	DIN EN 10130	**Surface type and finish** A Faults not affecting workability and adhesion of surface coating are permissible. B The better face must be flawless to the extent that the look of quality lacquer finish or coating is not affected. b particularly smooth g smooth m dull r rough
Continuously hot-dip finished strip and sheet made of soft steels for cold working	DIN EN 10346	D hot-dip coating **Coating** (followed by coating mass in g/m², e.g. Z140) +AS aluminum-silicon alloy +AZ aluminum-zinc alloy +Z zinc +ZA zinc-aluminum alloy +ZF zinc-iron alloy
		Coating finish: M small zinc flower with +Z N typical zinc flower with +Z R typical finish with +ZF
		Type of surface: A typical finish B improved finish C best finish

⇒ **DC04 – A – m:** Flat product for cold working (D), cold-rolled (C), steel type 04 (page 139), surface type A, surface finish dull (m)

Flat products made of high-strength steels for cold working

Designation example: **H C 300 – B – g**

- **H**: Code letter for flat product of high-strength steel for cold working
- **C**: Code letter for rolling condition: X rolling condition not specified, C cold-rolled, D hot-rolled
- **300**: 300 yield strength R_e = 300 N/mm²; T500 minimum tensile strength R_m = 500 N/mm²
- **B – g**: Supplemental symbols (product group-specific definition)

Product group (selection)	Standard	Supplemental symbols
Cold-rolled strip and sheet made of micro-alloy steels	DIN EN 10268	B bake-hardening steel Y high-strength I-F steel I isotropic steel P phosphor-alloy steel LA low-alloy/micro-alloy steel **Surface type and finish** for rolling width < 600 mm as with DIN EN 10139 for rolling width ≥ 600 mm as with DIN EN 10130

⇒ **HCT500 – B – g:** Cold-rolled flat product made of high-strength steel (H), cold-rolled (C), minimum tensile strength R_m = 500 N/mm² (T500), surface type B, smooth surface (g)

Designation system for steels

cf. DIN EN 10027-1 (2005-10)

Designation by chemical composition

The main symbols reflect the chemical composition and are created on the basis of four different designation groups. The supplemental symbols depend on the steel group or product group.

Example: Pinion shaft

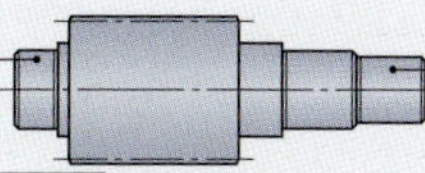

Main symbol | **Suppl. symbol**

42CrMo4+N → Designation → DIN EN 10027-1

Material (examples): Quenched and tempered steel → Steel group → DIN EN 10083-1

S355JR+AR — Designation according to the application (page 119)

Material blank: Hot-rolled round steel bar → DIN EN 10060

Designation groups, examples and application of the main symbols[1)]

Unalloyed steels manganese content < 1% except free-cutting steels	Alloy steels, free-cutting steels unalloyed steels with a manganese content > 1%	Alloy steels average content of individual alloying element above 5%	High-speed steels
C15E	**42CrMo4**	**X12CrNi18-8**	**HS 10-4-3-10** HS: Code letter for high-speed steel 10-4-3-10: Content of alloying elements in percent in the following order W-Mo-V-Co 10 → 10% tungsten (W) 4 → 4% molybdenum (Mo) 3 → 3% vanadium (V) 10 → 10% cobalt (Co)
Application examples: unalloyed case hardened steels, unalloyed quenched and tempered steels, unalloyed tool steels	**Application examples:** free-cutting steels, case hardened alloy steels, quenched and tempered alloy steels, tool alloy steels, spring steels	**Application examples:** **Stainless steels** corrosion-resistant, heat-resistant, high-temperature steels **Tool steels:** cold work steels hot work steels	

1) To identify cast steel, the main symbol is preceded by the letter **G**; to identify powder metallurgical steel, the main symbol is preceded by the letters **PM**.

Unalloyed steels with a manganese content < 1%, except free-cutting steels

Designation example: **C15 E+S+BC**

Main symbols (C15)

C code letter (carbon steel)
15 code number for the carbon content
C_{medium} = 15/100 ≙ 0.15 %

Supplemental symbols (E+S+BC)

Refer to such aspects as special applications, control of the sulphur content, special cold workability, heat treatment states. The definition of the supplemental symbols varies according to the steel group (page 122).

⇒ **C45E+S+BC:** quenched and tempered unalloyed steel, C content 0.45% , prescribed max. sulphur content (E), treated for shearability (+S), blasted (+BC) (supplemental symbols on page 122, quenched & tempered steels)

Alloy steels, free-cutting steels, unalloyed steels with a manganese content > 1%

Designation example: **18CrNiMo7-6 +TH+BC**

Main symbols (18CrNiMo7-6)

18 code number for the carbon content
C_{medium} = 18/100 ≙ 0.18%
Cr, Ni, Mo alloying elements (in the order of their mass portion)
7-6 Alloy contents
Cr_{medium} = 7/4 ≙ 1.75%
Ni_{medium} = 6/4 ≙ 1.5%
Mo = low content

Factors for alloy contents

Alloying elements	Factor
Cr, Co, Mn, Ni, Si, W	4
Al, Be, Cu, Mo, Nb, Pb, Ta, Ti, V, Zr	10
C, Ce, N, P, S	100
B	1000

Supplemental symbols (+TH+BC)

Refer to such aspects as special applications, heat treatment states, quenching stress, surface finish, degree of deformation. The definition of the supplemental symbols varies according to the steel group (page 122).

⇒ **17CrNiMo6-4+TH+BC:** Case hardened alloy steel, C content 0.17% (17), Cr content of 1.5% (6), Ni content 1.0% (4), low Mo content, treated for quenching stress (+TH) and blasted (+BC) (supplemental symbols on page 122, case hardened steels)

MS

Designation system for steels

cf. DIN EN 10027-1 (2005-10)

Steel group/ product group (selection)	Standard	Supplemental symbols
Hot-worked case hardened steels	DIN EN 10084	E prescribed maximum sulphur content R prescribed sulphur content range +H normal hardenability +HH restricted hardness tolerance, upper range +HL restricted hardness tolerance, lower range **Treatment conditions:** +A soft-annealed +S treated for shearability +FP treated for ferrite-pearlite microstructure and quenching stress +U untreated +TH treated for quenching stress **Surface finish:** +BC blasted +HW hot worked +PI pickled
Hot-worked quenched and tempered steels	DIN EN 10083-1 10083-2	E, R as with case hardened steels as per DIN EN 10084 (above) **Treatment conditions** +A soft-annealed +H normal hardenability +N normalized +HL restricted hardness tolerance, lower range +HH restricted hardness tolerance, upper range +QT quenched and tempered +S treated for shearability +U untreated **Surface finish:** +BC blasted +HW hot-worked +P pickled +RM hot-worked and pre-machined
Hot-worked free-cutting steels	DIN EN 10087	Under normal conditions, no supplemental symbols provided (in special cases for direct quenching types: +QT quenched and tempered)
Bright steel products made of case hardened steel, quenched & tempered steel, free-cutting steel	DIN EN 10277-1 10277,3..5	+C cold-drawn +SH peeled +SL ground +PL polished E, R, +A, +FP, +QT as per DIN EN 10083, 10084 (see above)
Seamless steel tubes made of case hardened steels and quenched & tempered steels	DIN EN 10297-1	+A soft-annealed +AR as rolled +N normalized +FP treated for ferrite-pearlite microstructure and quenching stress +QT quenched & tempered +TH treated for quenching stress

MS

⇒ **16MnCr5+A:** case hardened alloy steel, C content 0.16% (16), Mn content 1.25% (5), low Cr content, soft-annealed (+A)

Alloy steels, the content of at least one alloying element is above 5% (without high-speed steels)

Designation example: **X4CrNi18-12 +2D**

Main symbols

X code letter for the designation group
4 code number for medium carbon content
C_{medium} = 4/100 = 0.04%
Cr, Ni main alloying elements (Cr > Ni)
18-12 alloy contents in %
chromium = 18%, nickel = 12%

Supplemental symbols

Specification of heat treatment conditions, the rolling condition, the type of execution, the surface finish.
The definition of the supplemental symbols varies according to the product group.

Steel group/ product group (selection)	Standard	Supplemental symbols (selection)	
		Treatment condition	Type of execution/surface finish
Hot-rolled corrosion-resistant sheets and strips	DIN EN 10088-2	+A annealed +QT quenched & tempered +QT650 quenched & tempered to R_m = 650 N/mm² +AT solution annealed +P precipitation hardened +P1300 precipitation hardened to R_m = 1300 N/mm² +SR stress relieved annealed	+1 hot-rolled products 1U not heat-treated, not descaled 1C heat treated, not descaled 1E heat treated, mechanically descaled 1D heat treated, pickled, smooth 1G ground
Cold-rolled corrosion-resistant sheets and strips	DIN EN 10088-2		+2 cold-rolled products 2C, E, D, G as with hot-rolled products 2B like D but cold-rolled in addition 2R bright-annealed 2Q hardened and tempered, scale-free 2H strain-hardened (with different hardness stages), bright surface

⇒ **X2CrNi18-9+AT+2D:** Alloy steel, C content 0.02% (2), Cr content 18%, Ni content 9%, solution annealed (+AT), cold-rolled (+2), hot-treated, pickled, smooth surface (D)

Overview of Steel Products

Flat and long products, tubes, profiles (selection) — cf. DIN EN 10079 (2007-06)

In the steel mill, semi-finished steels are produced by continuous casting, die-casting, rolling or forging. These products are further processed to flat products, long products or profiles, for instance.

Flat products

Sheet metal

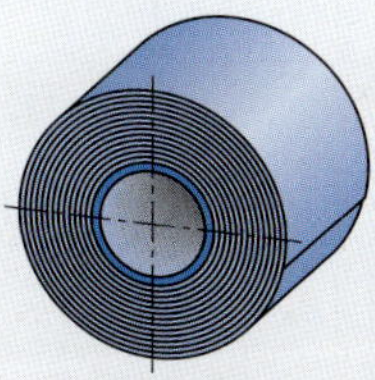

Strip coil

Hot-rolled sheet (page 140)	**Hot-rolled strip** (page 140)
Rectangular or square plates, the edges of the sheets are – uneven and slightly curved in rolled condition – smooth and even in trimmed condition.	Flat products, in coiled condition, the edges of the strip are – uneven and slightly curved in rolled condition – smooth and even in trimmed condition.
Special rolling methods for sheets and strips	
Normalized rolling	Thermomechanical rolling
The degree of reduction and the temperature control during the rolling process produce a structure that corresponds to the „normalized" (+N) condition.	The degree of reduction and the temperature control during the rolling process aim at producing specific steel properties such as a higher yield point, which are not achievable by heat treatment.
Cold-rolled sheets (pages 139, 140)	**Cold-rolled strips** (pages 139, 140)
Rectangular or square plates, smooth surface in various finishings, edges untreated or trimmed, cold-hardened material.	Flat products, in coiled condition, smooth and even surface, edges slightly curved or trimmed, cold-hardened material.
Sheets and strips for packing	
Black plate	**Tin plate**
Sheets and strips of unalloyed steels that were produced by single or double cold rolling (reducing).	Sheets and strips of black plate that were electrolytically tinned on both sides.

Long products

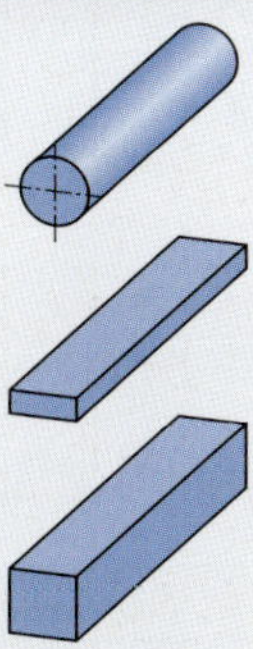

Hot-rolled bars (page 143)	**Wire rod**
Distinction according to the shape of the cross section, e. g. round bars, flat bars, square bars. Surface in rolled condition or descaled/tinted.	Hot rolled and coiled long products with thicknesses $t \leq 5$ mm, smooth surface, cross sections as with the bars.

Bars of bright steel, e.g. round, flat and square bars (page 144)		
Delivery condition		
Cold-drawn	Peeled (round bars)	Ground (round bars)
High shape and dimensional accuracy, high surface quality, cold-hardened border zone, probably decarburized	Machined surface with high shape and dimensional accuracy, high surface quality, no border decarburization	Drawn or peeled bars, either ground or ground and polished, best surface finishing and highest dimensional accuracy

Tubes

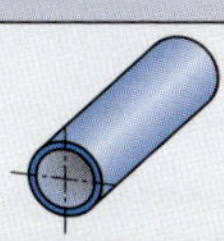

Seamless tubes (page 141)	**Welded tubes**
Production by hot rolling of pre-punched rings, further processing by cold or hot rolling or cold drawing	Circular-shaped flat products with longitudinal or spiral welding seam

Profiles

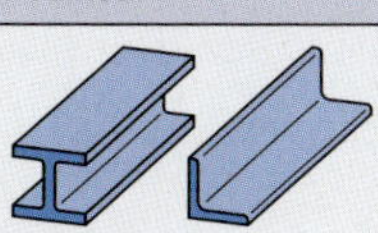

Hot-rolled profiles (pages 145 to 150)	**Cold-rolled profiles**
Designation according to the shape of the cross section, e.g. L-profile, U-profile, I-profile, surface in rolled condition	Designation as with hot-rolled profiles, cold-hardened border zones, fair dimensional stability, smooth surface

MS

Steels – Overview

Subgroups, delivery conditions	Standard	Main characteristics	Areas of application	Product forms[1) S]	B	P	W
Unalloyed structural steels, hot-rolled							page 127
Steels for steel and machine construction	DIN EN 10025-2	• good machinability • weldable, except for S185 • cold and hot workable	Welded constructions in steel and machine construction, simple machine parts	•	•	•	•
Steels for machine construction	DIN EN 10025-2	• machinable • not weldable • cold and hot workable	Machine parts without heat treatment, e. g. by hardening, quenching and tempering	•	•	–	•
Fine-grain steels suitable for welding							page 128
Normalized	DIN EN 10025-3	• weldable • hot workable	Weldments with high toughness, resistance to brittle fracture and aging stability in machine and steel construction	•	•	•	•
Thermomechanically rolled	DIN EN 10025-4	• weldable • not hot workable	Weldments with high toughness, resistance to brittle fracture and aging stability in machine and steel construction	•	•	–	•
Quenched and tempered structural steels with high yield strength							page 128
Alloy steels	DIN EN 10025-6	• weldable • hot workable	High-strength weldments in machine and steel constructions	•	–	–	–
Case hardened steels							page 129
Unalloyed steels	DIN EN 10084	• in spheroidized condition good machinability • hot workable • after surface carburization surface hardenable	Small parts with wear-resistant surface	•	•	–	•
Alloy steels	DIN EN 10084	• in spheroidized condition good machinability • hot workable • after surface carburization surface hardenable	Dynamically stressed parts with wear-resistant surface	•	•	–	•
Quenched and tempered steels							page 130
Unalloyed quality steels	DIN EN 10083-2	• in spheroidized condition good machinability • hot workable • hardenable (uncertain results with unalloyed quality steels)	Parts with high strength, which are not hardened	•	•	–	•
Unalloyed high-grade steels	DIN EN 10083-2	• in spheroidized condition good machinability • hot workable • hardenable (uncertain results with unalloyed quality steels)	Parts with high strength and good toughness	•	•	–	•
Alloy steels	DIN EN 10083-3	• in spheroidized condition good machinability • hot workable • hardenable (uncertain results with unalloyed quality steels)	Highly stressed parts with good toughness	•	•	–	•
Steels for flame and induction hardening							page 131
Unalloyed steels	DIN EN 10083-2,	• in spheroidized condition good machinability • hot workable • directly hardenable; possible to harden individual part areas, e. g. gear tooth face • quenching and tempering of workpieces before hardening	Parts with low core strength but hardening of specific areas	•	•	–	•
Alloy steels	DIN EN 10083-3	• in spheroidized condition good machinability • hot workable • directly hardenable; possible to harden individual part areas, e. g. gear tooth face • quenching and tempering of workpieces before hardening	Larger parts with high core strength and hardening of specific areas	•	•	–	•
Nitriding steels							page 131
Alloy steels	DIN EN 10085	• in spheroidized condition good machinability • hardenable by nitride forming elements, lowest quenching distortion • quenching and tempering of workpieces before nitriding	Parts with increased fatigue strength, parts subject to wear, Parts subjected to temperatures up to 500 °C	•	•	–	•
Spring steels							page 135
Unalloyed and alloy steels	DIN EN 10270 DIN EN 10089	• cold or hot workable • high elastic formability • high fatigue strength	Leaf springs, helical springs, disc springs, torsion bars	–	–	–	•

1) Product forms: S sheets, strips; B bars, e. g. flat, square and round bars; W wires; P profiles, e. g. channels, angles, tees

Steels – Overview

Subgroups, delivery conditions	Standard	Main characteristics	Areas of application	Product forms[1) S	B	P	W
Free cutting steels							page 131
Non-heat-treatable steels	DIN EN 10087	• optimal machinability (short chipping) • non-weldable • might not respond uniformly to heat treatment with case hardening or quench and tempering	Mass produced turned parts with low strength requirements	–	•	–	•
Free cutting case hardened steels	DIN EN 10087		Like unalloyed case hardened steels; better machinability	–	•	–	•
Free cutting quenched and tempered steels	DIN EN 10087		Like unalloyed quenched and tempered steels; better machinability, less fatigue strength	–	•	–	•
Tool steels							page 132
Cold work steels, unalloyed	DIN EN ISO 4957	• in spheroidized condition good machinability • non-cutting cold and hot-workable • full hardening up to max. 10 mm diameter	Low stressed tools for cutting and non-cutting forming at operating temperatures up to 200 °C	•	•	•	•
Cold work steels, alloy	DIN EN ISO 4957	• in spheroidized condition machinable • hot workable • larger case hardening depth, higher strength, more wear-resistant than unalloyed cold work steels	Highly stressed tools for cutting and non-cutting forming at operating temperatures over 200 °C	•	•	–	•
Hot work steels	DIN EN ISO 4957	• in spheroidized condition machinable • hot workable • hardens over the entire cross section	Tools for non-cutting forming at operating temperatures over 200 °C	•	•	–	•
High-speed steels	DIN EN ISO 4957	• in spheroidized condition machinable • hot workable • hardens over the entire cross section	Cutting materials for cutting tools, operating temperatures up to 600 °C, highly stressed forming tools	•	•	–	•
Corrosion resistant steels							pages 133, 134
Ferritic steels	DIN EN 10088-2, DIN EN 10088-3	• machinable • good cold-workable • weldable • heat treatment does not increase strength	Low stressed rust-free parts; parts with high resistance to chlorine induced stress, corrosion cracking	•	•	•	•
Austenitic steels	DIN EN 10088-2, DIN EN 10088-3	• machinable • very good cold workability • weldable • no increase in strength through heat treatment	Non-rusting parts with high corrosion resistance, widest application range of all stainless steels	•	•	•	•
Martensitic steels	DIN EN 10088-2, DIN EN 10088-3	• machinable • in spheroidized condition cold-workable • with low carbon content weldable • heat treatable	Highly stressed non-rusting parts, which can also be quenched and tempered	•	•	•	•

1) Product forms: S sheets, strip; W wires; B bars, e.g. flat, square and round bars; P profiles, e.g. channels, angles, tees

MS

Selecting structural steels, Alloying elements

The scheme below facilitates the selection of structural steels:

Unalloyed steels

Heat treatment, e.g. hardening and tempering: Not intended / Intended

Not intended → Selection by application

Example: unalloyed structural steels (page 127)		
Minimum requirements (examples)	Suitable steel type, quality class	
• strength	S185	High-grade steel
• strength • toughness	E295, E335, E360	High-grade steel
• strength • toughness • weldability	S235JR, S235JO, S275JR, S275JO, S355JR, S355JO	High-grade steel

Additional steel groups (examples)

- cold-rolled flat products from steels with higher strength
- flat products suitable for cold working
- sheets and strips for packing
- electrical sheet
- pressure vessel steels

Intended → Selection by chemical composition

Minimum requirements (examples)	Suitable steel group (examples)	Suitable steel type, quality class	
• thick surface layer • high core strength • high core toughness	Case hardened steels	C10E, C15E, C10R, C15R	Stainless steels
• high strength with great toughness	Quenched and tempered steels	C22E, C35E, C45E, C60E	Stainless steels
• best machinability	Free-cutting steels	10S20, 15SPb20	High-grade steels

Further requirements → Alloyed steels

Steel groups (examples)

- case hardened steels
- quenched and tempered steels
- nitrided steels
- tool steels
- corrosion-resistant steels

Effect of alloying elements

Properties influenced by alloying elements	Alloying elements										
	Cr	Ni	Al	W	V	Co	Mo	Si	Mn	S	P
Tensile strength	●	●	–	●	●	●	●	●	●	–	●
Yield strength	●	●	–	●	●	●	●	●	●	–	●
Impact toughness	○	–	○	–	●	○	●	○	–	○	○
Wear resistance	●	○	–	●	●	●	●	○	○	–	–
Hot workability	○	●	○	○	●	○	●	○	●	○	–
Cold workability	–	–	–	○	–	○	○	○	○	○	○
Machinability	–	○	–	○	–	–	○	○	○	●	●
High-temperature strength	●	●	–	●	●	●	●	●	–	–	–
Corrosion resistance	●	–	–	–	●	–	–	–	–	○	–
Hardening temperature	●	–	–	●	●	–	●	●	○	–	–
Hardenability, temperability	●	●	–	●	●	●	●	●	●	–	–
Nitridability	●	–	●	●	●	–	●	○	●	–	–
Weldability	○	○	●	–	●	–	○	–	○	○	○

● **increase** ○ **decrease** – **no significant effect**

Example: Gears with hard, wear-resistant surfaces, high core strength and high core toughness
Blanks machined by drop forging
Selected steel group: case hardened steels → unalloyed steel C ≤ 0.2%, e.g. C15E,
improved hot workability through Ni, V, Mo, Mn;
suitable steels: 16MnCr5, 20MnCr5 , 16NiCr4 (page 129)

MS

Unalloyed structural steels

Unalloyed structural steels, hot-rolled

cf. DIN EN 10025-2 (2005-04), replaces DIN EN 10025

Steel type Designation	Material number	DO[1]	Notch impact energy at °C	KV J	Tensile strength R_m[2] N/mm²	Yield strength R_e in N/mm² for product thickness in mm ≤ 16	> 16 ≤ 40	> 40 ≤ 63	> 63 ≤ 80	Elongation at fracture A[3] %	Properties, application
Structural and machine construction steels											
S185	1.0035	–	–	–	290–510	185	175	175	175	18	Non-weldable, simple steel constructions
S235JR S235J0 S235J2	1.0038 1.0114 1.0117	FN FN FF	20 0 –20	27	360–510	235	225	215	215	26	Basic machine parts, weldments in steel and machine construction; levers, bolts, axles, shafts
S275JR S275J0 S275J2	1.0044 1.0143 1.0145	FN FN FF	20 0 –20	27	410–560	275	265	255	245	23	
S355JR S355J0 S355J2	1.0045 1.0553 1.0577	FN FN FF	20 0 –20	27	470–630	355	345	335	325	22	Highly stressed weldments in steel, crane and bridge construction
S355K2 S450J0	1.0596 1.0590	FF FF	–20 0	40 27	470–630 550–720	355 450	345 430	335 410	325 390	22 17	
Steels for machine construction											
E295	1.0050	FN	–	–	470–610	295	285	275	265	20	Axles, shafts, bolts
E335	1.0060	FN	–	–	570–710	335	325	315	305	16	Wear parts; pinion gears, worms, spindles
E360	1.0070	FN	–	–	670–830	360	355	345	335	11	

[1] DO Type of deoxidation: – manufacturer's option; FN semi-killed cast steel; FF killed cast steel.
[2] Values apply to product thicknesses from 3 mm to 100 mm.
[3] Values apply to product thicknesses from 3 mm to 40 mm and longitudinal test pieces with $L_0 = 5.65 \cdot \sqrt{S_0}$ (page 191).
The steel types listed in the table are unalloyed quality steels acc. to DIN EN 10020 (page 116).

Technical properties

Weldability

Steels of grade groups JR – J0 – J2– K2 are weldable using all processes.
Increased strength and product thickness also increase the risk of cold cracks.
Steels S185, E295, E335 and E360 are not weldable, because the chemical composition is not specified.

Hot workability

The steels are hot workable. Only products which are ordered and delivered in normalized (+N) or normalizing rolled (+N) condition must meet the requirements of the above table. The treatment condition must be specified at the time of ordering.
Example: S235J0+N or 1.0114+N

Cold workability

The additional C or GC symbol is appended to the designation of a steel type suitable for cold working (edge folding, roll forming, cold-drawing), and these types are also assigned their own material number.

Steel types for cold working

Designation	Material number	Suitable for[1] F	R	C	Designation	Material number	Suitable for[1] F	R	C	Designation	Material number	Suitable for[1] F	R	C
S235JRC S235J0C S235J2C	1.0122 1.0115 1.0119	•	•	•	S275JRC S275J0C S275J2C	1.0128 1.0140 1.0142	•	•	•	S355J0C S355J2C S355K2C	1.0554 1.0579 1.0594	•	•	•
E295GC	1.0533	–	–	•	E335GC	1.0543	–	–	•	E360GC	1.0633	–	–	•

[1] Forming process: F edge folding; R roll forming; C cold drawing; • well-suited – unsuitable

MS

MS

Weldable fine-grain and quenched & tempered structural steels

Weldable fine-grained structural steels, hot-rolled (selection)

cf. DIN EN 10025-3 and DIN EN 10025-4 (2005-04), replaces DIN EN 10113

Steel type		DC[1]	Notch impact energy KV[2] in J at temperatures in °C			Tensile strength R_m N/mm²	Yield strength R_e in N/mm² for nominal thicknesses in mm			Elongation at fracture A %	Properties, application
Designation	Material number		+20	0	−20		≤ 16	> 16 ≤ 40	> 40 ≤ 63		
Unalloyed quality steels											High toughness, brittle fracture and aging resistant; weldments in machinery, crane and bridge construction, automotive manufacturing, conveyors
S275N S275M	1.0490 1.8818	N M	55	47	40	370–510 370–530	275	265	255	24	
S355N S355M	1.0545 1.8823	N M	55	47	40	470–630	355	345	335	22	
Alloy high-grade steels											
S420N S420M	1.8902 1.8825	N M	55	47	40	520–680	420	400	390	19	
S460N S460M	1.8901 1.8827	N M	55	47	40	550–720 540–720	460	440	430	17	

[1] DC Delivery condition: N normalized/normalizing rolled M thermomechanically rolled
[2] Values apply to V-notch longitudinal test pieces.
Assignment of steels: DIN EN 10025-3 → S275N, S355N, S420N, S460N
DIN EN 10025-4 → S275M, S355M, S420M, S460M

Technical properties

Weldability	Hot workability	Cold workability
The steels are weldable. Increased strength and product thickness also increase the risk of cold cracks.	Only steels S275N, S355N, S420N and S480N are hot workable.	Cold-bending or edge folding is guaranteed for nominal thicknesses up to 16 mm, if cold-workability is specified in the order.

Quenched and tempered struc. steels with higher yield strength, hot-rolled (selection)

cf. DIN EN 10025-6 (2009-08), replaces DIN EN 10137-2

Steel type		Notch impact energy KV in J at temperatures in °C			Tensile strength R_m N/mm²	Yield strength R_e in N/mm² for nominal thicknesses in mm			Elongation at fracture A %	Properties, application
Designation[1]	Material number	0	−20	−40		> 3 ≤ 50	> 50 ≤ 100	> 100 ≤ 150		
S460Q S460QL	1.8908 1.8906	40 50	30 40	– 30	550–720	460	440	400	17	High toughness, high resistance to brittle fracture and aging stability; highly stressed weldments in machinery, crane and bridge construction, automotive manufacturing, conveyors
S500Q S500QL	1.8924 1.8909	40 50	30 40	– 30	590–770	500	480	440	17	
S620Q S620QL	1.8914 1.8927	40 50	30 40	– 30	700–890	620	580	560	15	
S890Q S890QL	1.8940 1.8983	40 50	30 40	– 30	940–1100	890	830	–	11	
S960Q S960QL	1.8941 1.8933	40 50	30 40	– 30	980–1150	960	–	–	10	

[1] Q quenched and tempered; QL quenched and tempered, guaranteed minimum values for notched bar impact values to −40 °C

Technical properties

Weldability	Hot workability	Cold workability
The steels are not weldable without limitations. Professional planning of the welding parameters is required. Increased strength and product thickness also increase the risk of cold cracks.	Hot working is not recommended as the material will lose strength in the subsequent heat treatment.	Cold-bending or edge folding is guaranteed for nominal thicknesses up to 16 mm, if cold-workability is specified in the order.

Case hardened steels, unalloyed and alloy

Case hardened steels, hot-rolled (selection)

cf. DIN EN 10084 (2008-06)

Steel type		Hardness HB in delivery condition[2]		Core properties after case hardening[3]			Hardening method[4]		Properties, applications
Designation[1]	Material number	+A	+FP	Tensile strength R_m N/mm²	Yield strength R_e N/mm²	Elong. at fracture A %	D	S	
Unalloyed case hardened steels									
C10E C10R	1.1121 1.1207	131	90–125	49–640	295	16	•	•	Small parts with average stress; levers, pegs, bolts, rollers, spindles, pressed and stamped parts
C15E C15R	1.1141 1.1140	143	103–140	590–780	355	–	•	•	
Alloy case hardened steels									
17Cr3 17CrS3	1.7016 1.7014	174	–	700–900	450	11	•	•	Parts subject to alternating stresses, e.g. in gearbox; gears, bevel and ring gears, driving pinions, shafts, propellershafts
28Cr4 28CrS4	1.7030 1.7036	217	156–207	≥ 700	–	–	•	•	
16MnCr5 16MnCrS5	1.7131 1.7139	207	140–187	780–1080 780–1080	590 590	10 10	○	•	
16NiCr4 16NiCrS4	1.5714 1.5715	217	156–207	≥ 900	–	–	–	•	
18CrMo4 18CrMoS4	1.7243 1.7244	207	140–187	≥ 900	–	–	○	•	
20MoCr3 20MoCrS3	1.7320 1.7319	217	145–185	≥ 900	–	–	•	–	
20MoCr4 20MoCrS4	1.7321 1.7323	207	140–187	880–1180	590	10	•	–	
17CrNi6-6 22CrMoS3-3	1.5918 1.7333	229 217	156–207 152–201	≥ 1100 –	– –	– –	– ○	• •	
15NiCr13 10NiCr5-4	1.5752 1.5805	229 192	166–207 137–187	920–1230 ≥ 900	785 –	10 –	– –	• •	Parts subject to highly alternating stresses, e.g. in gearbox; gears, bevel and ring gears, driving pinion, shafts, propellershafts
20NiCrMo2-2 20NiCrMoS2-2	1.6523 1.6526	212	149–194	780–1080	590	10	•	•	
17NiCrMo6-4 17NiCrMoS6-4 20NiCrMoS6-4	1.6566 1.6569 1.6571	229	149–201 149–201 154–207	≥ 1000 ≥ 1000 ≥ 1100	– – –	– – –	–	•	
20MnCr5 20MnCrS5	1.7147 1.7149	217	152–201	980–1270	685	8	○	•	Parts subject to larger dimensions; pinion shafts, gears, ring gears
18NiCr5-4 14NiCrMo13-4 18CrNiMo7-6	1.5810 1.6657 1.6587	223 241 229	156–207 166–217 159–207	≥ 1100 1030–1390 1060–1320	– – 785	– 10 8	– – –	• • •	

[1] Steel types with added sulfur, e.g. 16MnCrS5, have an improved machinability.

[2] Delivery condition: +A spheroidized; +FP treated for ferrite-pearlite microstructure and hardness range

[3] Strength values are valid for test pieces with 30 mm nominal diameter.

[4] Hardening methods:

D Direct hardening: The workpieces are quenched directly from the carburizing temperature.
S Simple hardening: After carburizing the workpieces are usually left to cool at room temperature. For hardening they are reheated.

• well-suited ○ conditionally suitable – unsuitable

For heat treatment of case hardened steels, see page 154

MS

Quenched and tempered steels, unalloyed and alloy

Quenched and tempered steels, hot-rolled (selection) cf. DIN EN 10083-2 and DIN EN 10083-3

Steel type: Designation	Steel type: Material number	T[1]	Tensile strength R_m in N/mm² > 16 ≤ 40	Tensile strength R_m in N/mm² > 40 ≤ 100	Yield strength R_e in N/mm² > 16 ≤ 40	Yield strength R_e in N/mm² > 40 ≤ 100	Elongation at fracture *EL* in % > 16 ≤ 40	Elongation at fracture *EL* in % > 40 ≤ 100	Properties, applications
Unalloyed quenched and tempered steels[2]									cf. DIN EN 10083-2 (2006-10)
C22E	1.1151	+N +QT	410 470–620	410 –	210 290	210 –	25 22	25 –	Parts subject to lower stresses and small quench and tempering diameters; screws, bolts, axles, shafts, gears
C35 C35E	1.0501 1.1181	+N +QT	520 600–750	520 550–700	270 380	270 320	19 19	19 20	
C45 C45E	1.0503 1.1191	+N +QT	580 650–800	580 630–780	305 430	305 370	16 16	16 17	
C55 C55E	1.0535 1.1203	+N +QT	640 750–900	640 700–850	330 490	330 420	12 14	12 15	
C60 C60E	1.0601 1.1221	+N +QT	670 800–950	670 750–900	340 520	340 450	11 13	11 14	
28Mn6	1.1170	+N +QT	600 700–850	600 650–800	310 490	310 440	18 15	18 16	
Alloy quenched and tempered steels									cf. DIN EN 10083-3 (2007-01)
38Cr2 46Cr2	1.7003 1.7006	+QT	700–850 800–950	600–750 650–800	450 550	350 400	15 14	17 15	Parts subject to higher stresses and larger quenched and tempered diameters; drive shafts, worms, gears
34Cr4 37Cr4	1.7033 1.7034	+QT	800–950 850–1000	700–850 750–900	590 630	460 510	14 13	15 14	
25CrMo4 25CrMoS4	1.7218 1.7213	+QT	800–950	700–850	600	450	14	15	
41Cr4 41CrS4	1.7035 1.7039	+QT	900–1100	800–950	660	560	12	14	Parts subject to high stresses and larger quenched and tempered diameters; shafts, gears, larger forged parts
34CrMo4 34CrMoS4	1.7220 1.7226	+QT	900–1100	800–950	650	550	12	14	
42CrMo4 42CrMoS4	1.7225 1.7227	+QT	1000–1200	900–1100	750	650	11	12	
50CrMo4 51CrV4	1.7228 1.8159	+QT	1000–1200	900–1100	780 800	700	10	12	
30NiCrMo16-6 34CrNiMo6	1.6747 1.6582	+QT	1080–1230 1100–1300	1080–1230 1000–1200	880 900	880 900	10	10 11	Parts subject to highest stresses and large quenched and tempered diameters
36NiCrMo16 30CrNiMo8	1.6773 1.6580	+QT	1250–1450	1100–1300	1050	900	9	10	
20MnB5 30MnB5	1.5530 1.5531	+QT	750–900 800–950	– –	600 650	– –	15 13	– –	
27MnCrB5-2 39MnCrB6-2	1.7182 1.7189	+QT	900–1150 1050–1250	800–1000 1000–1200	750 850	700 800	14 12	15 12	

[1] T treatment condition: +N normalized; +QT quenched and tempered
For unalloyed quenched and tempered steels the treatment conditions +N and +QT also apply to the quality and high-grade steels, for example for C45 and C45E.

[2] Unalloyed quenched and tempered steels C35, C45, C55 and C60 are quality steels, steels C22E, C35E, C45E, C55E and C60E are produced as high-grade steels.

For heat treatment of quenched and tempered steels, see page 155

Nitriding steels, Steels for flame and induction hardening, Free cutting steels

Nitriding steels, hot-rolled (selection)

cf. DIN EN 10085 (2001-07), replaces DIN 17211

Steel type Designation	Material number	Spheroidized hardness HB	Tensile strength[1] R_m N/mm²	Yield strength[1] R_e N/mm²	Elongation at fracture[1] *EL* %	Properties, applications
31CrMo12	1.8515	248	980–1180	785	11	Wear parts up to 250 mm thickness
31CrMoV9	1.8519	248	1000–1200	800	10	Wear parts up to 100 mm thickness
34CrAlMo5-10	1.8507	248	800–1000	600	14	Wear parts up to 80 mm thickness
40CrAlMo7-10	1.8509	248	900–1100	720	13	High-temperature wear parts up to 500 °C
34CrAlNi7-10	1.8550	248	850–1050	650	12	Large parts; piston rods, spindles

[1] Strength values: The values for tensile strength R_m, yield strength R_e and elongation at fracture *EL* apply to material thicknesses from 40 to 100 mm in the quenched and tempered condition.

For heat treatment of nitriding steels, see page 156

Steels for flame and induction hardening, hot-rolled (selection)

cf. DIN EN 10083[1]

Steel type Designation	Material number	Spheroidized hardness HB	T[2]	Tensile strength[2] R_m N/mm²	Yield strength R_e in N/mm² for nominal thicknesses in mm: ≤ 16	> 16 ≤ 40	> 40 ≤ 100	Elongation at fracture *EL* %	Properties, applications
C45E[1]	1.1191	207	+QT	650–800	490	430	370	16	Wear parts with high core strength and good toughness; crank shafts, drive shafts, cam shafts, worms, gears
C60E[1]	1.1221	241	+QT	800–950	580	520	450	13	
37Cr4	1.7034	255	+QT	850–1000	750	630	510	14	
46Cr2	1.7006	255	+QT	800–950	650	550	400	13	
41Cr4	1.7035	255	+QT	900–1100	800	660	560	12	
42CrMo4	1.7225	255	+QT	1000–1200	900	750	650	11	

[1] The previous standard DIN 17212 was withdrawn without replacement. For flame and induction hardenable steels, see quenched and tempered steels DIN EN 10083-3 (page 130). For unalloyed high-quality steels acc. to DIN EN 10083-2, hardness results are only assured if the steels are ordered with austenite grain size ≤ 5.

[2] T treatment condition: +QT quenched and tempered

For heat treatment of steels for flame and induction hardening, see page 155

Free cutting steels, hot-rolled (selection)

cf. DIN EN 10087 (1999-01)

Steel type Designation[1]	Material number	T[2]	For product thicknesses from 16 to 40 mm: Hardness HB	Tensile strength R_m N/mm²	Yield strength R_e N/mm²	Elongation at fracture *EL* %	Properties, applications
11SMn30	1.0715	+U	112–169	380–570	–	–	• Steels unsuitable for heat treatment Small parts subject to low stress; levers, pegs
11SMnPb30	1.0718						
11SMn37	1.0736	+U	112–169	380–570	–	–	
11SMnPb37	1.0737						
10S20	1.0721	+U	107–156	360–530	–	–	• Case hardened steels Wear-resistant small parts; shafts, bolts, pins
10SPb20	1.0722						
15SMn13	1.0725	+U	128–178	430–600	–	–	
35S20	1.0726	+U	154–201	520–680	–	–	• Quenched and tempered steels Larger parts subject to higher stress; spindles, shafts, gears
35SPb20	1.0756	+QT	–	600–750	380	16	
44SMn28	1.0762	+U	187–238	630–800	–	–	
44SMnPb28	1.0763	+QT	–	700–850	420	16	
46S20	1.0727	+U	175–225	590–760	–	–	
46SPb20	1.0757	+QT	–	650–800	430	13	

[1] Steel types with lead additives, e.g. 11SMnPb30, have better machinability.

[2] T treatment condition: +U untreated; +QT quenched and tempered

All free cutting steels are unalloyed quality steels. It is not possible to guarantee a uniform response to case hardening or quench and tempering.

For heat treatment of free cutting steels, see page 156

Cold work steels, Hot work steels, High-speed steels

Tool steels, hot-rolled (selection) cf. DIN EN ISO 4957 (2001-02), replaces DIN 17350

Steel type Designation	Material number	Hardness HB[1] max.	Hardening temperature °C	QM[2]	Tempering temperat. °C	Application examples, properties
Cold work steels, unalloyed						
C45U	1.1730	190	800–830	W	180–300	Non-hardened mounted parts for tools, screwdrivers, chisels, knives
C70U	1.1520	190	790–820	O	180–300	Centering pins, small dies, vise jaws, trimming press
C80U	1.1525	190	780–810	W	180–300	Dies with flat cavities, chisels, cold extruding dies, knives
C105U	1.1545	213	770–800	W	180–300	Simple cutting tools, coining dies, scribers, piercing plugs, twist drills
Cold work steels, alloy						
21MnCr5	1.2162	215	810–840	O	150–180	Complex case hardened press forms for plastics; easily polished
60WCrV8	1.2550	230	880–930	O	180–300	Cutters for steel sheet from 6 to 15 mm, cold punching dies, chisels, center punches
90MnCrV8	1.2842	220	790–820	O	150–250	Die sections, punches, plastic molds, reamers, inspection tools
102Cr6	1.2067	230	820–850	O	100–180	Drills, milling cutters, reamers, small cutting dies, turning centers for lathes
X38CrMo16	1.2316	250	1000–1040	O	650–700	Tools for processing chemically aggressive thermoplastics
40CrMnNiMo8-6-4	1.2738	235	840–870	O	180–220	Plastic molds of all types
45NiCrMo16	1.2767	260	840–870	O, A	160–250	Bending and embossing tools, shearing blades for thick material
X153CrMoV12	1.2379	250	1020–1050	O, A	180–250	Cutting tools sensitive to breaking, milling cutters, broaching tools, shearing blades
X210CrW12	1.2436	255	950–980	O, A	180–250	High-performance cutting tools, broaching tools, stamping tools
Hot work steels						
55NiCrMoV7	1.2714	250	840–870	O	400–650	Plastic molds, small and medium sized dies, hot shearing blades
X37CrMoV5-1	1.2343	235	1020–1050	O, A	550–650	Die casting molds for light alloys, extrusion tools
32CrMoV12-28	1.2365	230	1020–1050	O, A	500–670	Die casting molds for heavy non-ferrous metals, extrusion tools for all metals
X38CrMoV5-3	1.2367	235	1030–1080	O, A	600–700	High-quality dies, highly stressed tools for manufacture of screws
High-speed steels						
HS6-5-2C	1.3343	250	1190–1230	O, A	540–560	Twist drills, reamers, milling cutters, thread cutters, circular saw blades
HS6-5-2-5	1.3243	270	1210–1250	O, A	550–570	Highly stressed twist drills, milling cutters, roughing tools with high toughness
HS10-4-3-10	1.3207	270	1210–1250	O, A	550–570	Lathe tools for automatic machining, high cutting capacity
HS2-9-2	1.3348	250	1190–1230	O, A	540–580	Milling cutters, twist drills and thread cutters, high cutting hardness, high-temp. strength, toughness

[1] Delivery condition: annealed [2] QM Quenching medium; W water; O oil; A air

For designations of tool steels, see pages 121, 122; for heat treatment of tool steels, see page 154

MS

Stainless steels

Corrosion-resistant steels (selection)

cf. DIN EN 10088-2 and 10088-3 (2005-09)

Steel type: Designation	Material number	D[1)] S	D[1)] B	DC[2)]	Thickness *d* mm	Tensile strength R_m N/mm²	Yield strength $R_{p0,2}$ N/mm²	Elongation at fracture *EL* %	Properties, applications
Austenitic steels									
X10CrNi18-8	1.4310	•		C	≤ 8	600–950	250	40	Springs for temperatures up to 300 °C, automotive manufacturing
			•	–	≤ 40	500–750	195	40	
X2CrNi18-9	1.4307	• •		C P	≤ 8 ≤ 75	520–700 500–650	220 200	45	Household containers, chemical and food industry
			•	–	≤ 160	500–700	175	45	
X2CrNiN19-11	1.4306	• •		C P	≤ 8 ≤ 75	520–700 500–700	220 200	45	Equipment and parts exposed to organic and fruit acids
			•	–	≤ 160	460–680	180	45	
X2CrNi18-10	1.4311	• •		C P	≤ 8 ≤ 75	550–750 540–750	290 270	40	Equipment for the dairy and brewery industry, pressure vessels
			•	–	≤ 160	550–760	270	40	
X5CrNi18-10	1.4301	• •		C P	≤ 8 ≤ 75	540–750	230 210	45	Deep-drawn parts in the food industry, easily polished
			•	–	≤ 160	500–700	190	45	
X8CrNiS18-9	1.4305	•		P	≤ 75	500–700	190	35	Parts in the food and dairy industry
			•	–	≤ 160	500–750	190	35	
X6CrNiTi18-10	1.4541	• •		C P	≤ 8 ≤ 75	520–720 500–700	220 200	40	Consumer goods used in the household, parts in the photo industry
			•	–	≤ 160	500–700	190	40	
X4CrNi18-12	1.4303	•		C	≤ 8	500–650	220	45	Chemical industry; bolts, nuts
			•	–	≤ 160	500–700	190	45	
X5CrNiMo17-12-2	1.4401	• •		C P	≤ 8 ≤ 75	530–680 520–670	240 220	40 45	Parts in the paint, oil and textile industry
			•	–	≤ 160	500–700	200	40	
X6CrNiMoTi17-12-2	1.4571	• •		C P	≤ 8 ≤ 75	540–690 520–670	240 220	40	Parts in the textile, synthetic resin and rubber industry
			•	–	≤ 160	500–700	200	40	
X2CrNiMo18-14-3	1.4435	• •		C P	≤ 8 ≤ 75	550–700 520–670	240 220	40 45	Parts with improved chemical resistance for the pulp industry
			•	–	≤ 160	500–700	200	40	
X2CrNiMoN17-13-3	1.4429	• •		C P	≤ 8 ≤ 75	580–780	300 280	35 40	Pressure vessels with increased chemical resistance
			•	–	≤ 160	580–800	280	35	
X2CrNiMoN17-13-5	1.4439	• •		C P	≤ 8 ≤ 75	580–780	290 270	35 40	Resistant to chlorine and higher temperatures; chemical industry
			•	–	≤ 160	580–800	280	35	
X1NiCrMoCu25-20-5	1.4539	• •		C P	≤ 8 ≤ 75	530–730 520–720	240 220	35	Resistant to phosphoric, sulfuric and hydrochloric acids; chemical industry
			•	–	≤ 160	700–800	200	40	

1) D Delivery forms: S sheet, strip → DIN EN 10088-2; B bars, profile → DIN EN 10088-3
2) DC Delivery condition: C cold-rolled strip; P hot-rolled sheet

Stainless steels

Corrosion-resistant steels (continued) cf. DIN EN 10088-2 and 10088-3 (2005-09)

Steel type Designation	Material number	D[1)] S	D[1)] B	DC[2)]	Thickness d mm	Tensile strength R_m N/mm²	Yield strength $R_{p0,2}$ N/mm²	Elongation at fracture EL %	Properties, applications
Ferritic steels									
X2CrNi12	1.4003	•		C P	≤ 8 ≤ 25	450–650	280 250	20 18	Automotive and container manufacturing, conveyors
			•	–	≤ 100	450–600	260	20	
X6Cr13	1.4000	• •		C P	≤ 8 ≤ 25	400–600	240 220	19	Resistant to water and steam; household equipment, fittings
			•	–	≤ 25	400–630	230	20	
X6Cr17	1.4016	• •		C P	≤ 8 ≤ 25	450–600	260 240	20	Good cold workability, able to be polished; flatware, bumpers
			•	–	≤ 100	400–630	240	20	
X2CrTi12	1.4512	•		C	≤ 8	450–650	280	23	Catalytic converters
X6CrMo17-1	1.4113	•		C	≤ 8	450–630	260	18	Automotive manufacturing; trim, hub caps
			•	–	≤ 100	440–660	280	18	
X3CrTi17	1.4510	•		C	≤ 8	450–600	260	20	Welded parts in food industry
X2CrMoTi18-2	1.4521	• •		C P	≤ 8 ≤ 12	420–640 420–620	300 280	20	Bolts, nuts, heaters

1) D Delivery forms: S sheet, strip → DIN EN 10088-2; B bars, profile → DIN EN 10088-3
2) MF Mill finish: C cold-rolled strip; P hot-rolled sheet

Martensitic steels

Steel type Designation	Mat. no.	D[1)] S	D[1)] B	DC[2)]	Thickness d mm	H[3)]	Tensile strength R_m N/mm²	Yield strength $R_{p0,2}$ N/mm²	Elongational fracture EL %	Properties, applications
X12Cr13	1.4006	• •		C P	≤ 8 ≤ 75	A QT650	≤ 600 650–850	– 450	20 12	Resistant to water and steam, food industry
			•	–	≤ 160	QT650	650–850	450	15	
X20Cr13	1.4021	• •		C P	≤ 8 ≤ 75	A QT750	≤ 700 750–950	– 550	15 10	Axles, shafts, pump parts, propellers
			•	–	≤ 160	QT800	800–950	600	12	
X30Cr13	1.4028	• •		C P	≤ 8 ≤ 75	A QT800	≤ 740 800–1000	– 600	15 10	Bolts, nuts, springs, piston rods
			•	–	≤ 160	QT850	850–1000	650	10	
X46Cr13	1.4034	•	 •	C –	≤ 8 ≤ 160	A QT800	≤ 780 850–1000	245 650	12 10	Hardenable; table knives and machine knives
X39CrMo17-1	1.4122	•	 •	C –	≤ 8 ≤ 60	A QT900	≤ 900 900–1100	280 800	12 11	Shafts, spindles, armatures up to 600 °C
X3CrNiMo13-4	1.4313	•		P	≤ 75	QT900	900–1100	800	11	High toughness; pumps, turbine wheels, reactor construction
			• •	–	– ≤ 160	A QT900	≤ 1100 900–1100	320 800	– 12	

1) D Delivery forms: S sheet, strip → DIN EN 10088-2; B bars, profile → DIN EN 10088-3
2) DC Delivery condition: C cold-rolled strip; P hot-rolled sheet
3) H Heat treatment condition: A solution annealed; QT750 → quenched and tempered to minimum tensile strength R_m = 750 N/mm²

Spring steel

Steel wire for springs, patented drawn

cf. DIN EN 10270-1 (2001-12), replaces DIN 17223

Wire type	Minimum tensile strength R_m in N/mm² for the nominal diameter d in mm															
	0.5	0.8	1.0	1.5	2.0	2.5	3.0	3.4	4.0	4.5	5.0	6.0	8.0	10.0	15.0	20.0
SL	–	–	1720	1600	1510	1460	1410	1370	1320	1290	1260	1210	1120	1060	–	–
SM	2200	2050	1980	1850	1740	1690	1630	1590	1530	1500	1460	1400	1310	1240	1110	1020
SH	2480	2310	2330	2090	1970	1900	1840	1790	1740	1690	1660	1590	1490	1410	1270	1160
DM	2200	2050	1980	1850	1740	1690	1630	1590	1530	1500	1460	1400	1310	1240	1110	1020
DH	2480	2310	2230	2090	1970	1900	1840	1790	1740	1690	1660	1590	1490	1410	1270	1160

Wire diameter *d* in mm (selection)

all types, except SL[1]	0.30 – 0.32 – 0.34 – 0.36 – 0.38 – 0.40 – 0.43 – 0.48 – 0.50 – 0.53 – 0.56 – 0.60 – 0.63 – 0.65 – 0.70 – 0.75 – 0.80 – 0.90 – 1.00 – 1.10 – 1.20 – 1.25 – 1.30 – 1.40 – 1.50 – 1.60 – 1.70 – 1.80 – 1.90 – 2.00 – 2.10 – 2.25 – 2.40 – 2.50 – 2.60 – 2.80 – 3.00 – 3.20 – 3.40 – 3.60 – 3.80 – 4.00 – 4.25 – 4.50 – 4.75 – 5.00 – 5.30 – 5.60 – 6.00 – 6.30 – 6.50 – 7.00 – 7.50 – 8.00 – 8.50 – 9.00 – 9.50 – 10.00

[1] Wire type SL is only supplied in diameters d = 1 to 10 mm.

Operating conditions, applications

Wire type	Suitable for springs with:	Applications
SL	Low static loading	Tension springs, compression springs, torsion springs in equipment and machine construction, wire type DH is also suitable for shaped springs.
SM	Moderate static **or**, less often, dynamic loading	
SH	High static **or** low dynamic loading	
DM	Moderate dynamic loading	
DH	High static **or** average dynamic loading	

Wire coatings, delivery forms

Designation	Wire surfaces	Letter symbol	Wire surfaces	Delivery forms
ph	phosphatize	Z	with zinc coating	• in coils or on spools • straightened rods in bundles
cu	copper coated	ZA	with zinc/aluminum coating	

⇒ **Spring wire EN 10270-1 DM 3,4 ph:** Spring type DM, d = 3,4 mm, phosphatized surface (ph)

Hot-rolled steels for quenched and tempered springs

cf. DIN EN 10089 (2003-04), replaces DIN 17221

Steel type: Designation	Steel type: Material number	Hot-rolled Hardness HB	Spheroidized +A Hardness HB	In quenched and tempered condition (+QT)[1]: Tensile strength R_m N/mm²	Yield strength $R_{p\,0,2}$ N/mm²	Elongation at fracture EL %	Properties, applications
38Si7	1.5023	240	217	1300–1600	1150	8	Spring screw locks
46Si7	1.5024	270	248	1400–1700	1250	7	Leaf springs, helical springs
55Cr3	1.7176	> 310	248	1400–1700	1250	3	Larger tension and compression springs
54SiCr6	1.7102	310	248	1450–1750	1300	6	Spring wire
61SiCr7	1.7108	310	248	1550–1850	1400	5.5	Leaf springs, helical springs
51CrV4	1.8159	> 310	248	1400–1700	1200	6	Highly stressed springs

Explanation: [1] Strength values apply to test pieces with d = 10 mm diameter.

⇒ **Round bar EN 10089 – 20 x 8000 – 51CrV4+A:** Bar diameter d = 20 mm, bar length l = 8000 mm, steel type 51CrV4, delivery condition spheroidized (+A)

Wire diameter *d* in mm (selection)	**Delivery forms**
5.0 – 5.5 – 6.0 – 6.5 – 7.0 – 7.5 – 8.0 – 8.5 – 9.0 – 9.5 – 10.0 – 10.5 – 11.0 – 11.5 – 12.0 – 19.0 – 19.5 – 20.0 – 21.0 – 22.0 – 23.0 – 27.0 – 28.0 – 29.0 – 30.0	• directional rods • wire coils

Steels for bright steel products

Bright unalloyed steels (selection)

cf. DIN EN 10277-2 (2008-06)

Steel type: Code name, designation	Steel type: Material number	Thickness range in mm	rolled and peeled (+SH): Hardness HB	rolled and peeled (+SH): Tensile strength R_m N/mm²	cold-drawn (+C): Yield strength $R_{p0.2}$ N/mm²	cold-drawn (+C): Tensile strength R_m N/mm²	cold-drawn (+C): Elongation at fracture A in %	Properties, application
S235JRC	1.0122	> 16 ≤ 40 > 40 ≤ 63	102–140	360–510	260 235	390–730 380–670	10 11	Nonhardenable high-grade steels for general applications, e. g. levers, bolts, axles and shafts under low loads
S355J2C	1.0579	> 16 ≤ 40 > 40 ≤ 63	146–187	470–630	350 335	530–850 500–770	8 9	
E295GC	1.0533	> 16 ≤ 40 > 40 ≤ 63	140–181	470–610	320 300	550–850 520–770	8 9	
E335GC	1.0543	> 16 ≤ 40 > 40 ≤ 63	169–211	570–710	390 340	640–930 620–870	7 8	
C15	1.0401	> 16 ≤ 40 > 40 ≤ 63	98–178	330–600	240 215	380–670 340–600	11 12	High-grade steels, not intended for a heat treatment such as hardening; suitable for parts under higher loads
C35	1.0501	> 16 ≤ 40 > 40 ≤ 63	154–207	520–700	320 300	580–880 520–800	9 9	
C45	1.0503	> 16 ≤ 40 > 40 ≤ 63	172–242	580–820	410 360	650–1000 580–850	8 8	
C60	1.0601	> 16 ≤ 40 > 40 ≤ 63	198–278	670–940	480 –	730–1100 –	6 –	

Available product shapes, surface finishing, notes: Round bars: rolled and peeled (+SH), cold-drawn (+C), ground (+SL), polished (+PL), flat and square bars: rolled and cold-drawn (+C)
Dimensions, allowance limits on page 144

Bright free-cutting steels (selection)

cf. DIN EN 10277-3 (2008-06)

Steel type: Code name, designation	Steel type: Material number	Thickness range in mm	rolled and peeled (+SH): Hardness HB	rolled and peeled (+SH): Tensile strength R_m N/mm²	cold-drawn (+C): Yield strength $R_{p0,2}$ N/mm²	cold-drawn (+C): Tensile strength R_m N/mm²	cold-drawn (+C): Elongation at fracture A in %	Properties, available special product shapes
11SMn30 11SMnPb30	1.0715 1.0718	> 16 ≤ 40 > 40 ≤ 63	112–169	380–570 370–570	375 305	460–710 400–650	8 9	Steels not intended for a heat treatment such as hardening
11SMn37 11SMnPb37	1.0736 1.0737	> 16 ≤ 40 > 40 ≤ 63	112–169	380–570 370–570	375 305	460–710 400–650	8 9	
10S20 10SPb20	1.0721 1.0722	> 16 ≤ 40 > 40 ≤ 63	107–156	360–530	360 295	460–720 410–660	9 10	Steels suitable for case hardening
15SMn13	1.0725	> 16 ≤ 40 > 40 ≤ 63	128–178 128–172	430–600 430–580	390 350	470–770 460–680	8 9	
35S20 35SPb20	1.0726 1.0756	> 16 ≤ 40 > 40 ≤ 63	154–201 154–198	520–680 520–670	360 340	560–800 530–760	8 9	Steels suitable for quenching and tempering; also available as cold-drawn and quenched & tempered (+C+QT) or quenched & tempered and cold-drawn (+QT+C) versions
36SMn14 36SMnPb14	1.0764 1.0765	> 16 ≤ 40 > 40 ≤ 63	166–222 166–219	560–750 560–740	390 360	600–900 580–840	7 8	
44SMn28 44SMnPb28	1.0762 1.0763	> 16 ≤ 40 > 40 ≤ 63	187–242 184–235	630–820 620–790	460 430	660–900 650–870	6 7	
46S20 46SMnPb20	1.0727 1.0757	> 16 ≤ 40 > 40 ≤ 63	175–225 172–216	590–760 580–730	400 380	640–880 610–850	7 8	

Available product shapes, surface finishing, notes: Round bars: rolled and peeled (+SH), cold-drawn (+C), ground (+SL), polished (+PL), flat and square bars: rolled and cold-drawn (+C)
Dimensions, allowance limits on page 144, heat treatment on page 156

Steels for bright steel products

Quenched and tempered bright steels (selection)

cf. DIN EN 10277-5 (2008-06)

Steel type		Thickness range in mm	Mechanical properties in delivery condition					Properties, application
			rolled and peeled (+SH)		quenched & tempered and cold-drawn (+QT+C)			
Code name, designation	Material number		Hardness HB	Tensile strength R_m N/mm²	Yield strength $R_{p0.2}$ N/mm²	Tensile strength R_m N/mm²	Elongation at fracture A in %	
C35E C35R	1.1181 1.1180	> 16 ≤ 40 > 40 ≤ 63	154–207	520–700	455 400	650– 850 570– 770	10 11	Unalloyed quenched and tempered steels for parts under low loads
C45E C45R	1.1191 1.1201	> 16 ≤ 40 > 40 ≤ 63	172–242	580–820	525 455	750– 950 650– 850	9 10	
C60E C60R	1.1221 1.1223	> 16 ≤ 40 > 40 ≤ 63	198–278	670–940	580 545	830–1030 780– 980	7 8	
34CrS4	1.7037	> 16 ≤ 40 > 40 ≤ 63	223	–	580 510	800–1000 700– 900	9 10	Alloyed quenched and tempered steels for parts under higher loads, improved machinability due to sulphur (S)
41CrS4	1.7039	> 16 ≤ 40 > 40 ≤ 63	241	–	670 570	900–1100 800–1000	9 10	
25CrMoS4	1.7213	> 16 ≤ 40 > 40 ≤ 63	212	–	600 520	800–1000 700– 900	10 11	
42CrMoS4	1.7227	> 16 ≤ 40 > 40 ≤ 63	241	–	720 650	1000–1200 900–1100	10 10	
34CrNiMo6	1.6582	> 16 ≤ 40 > 40 ≤ 63	248	–	720 650	1000–1200 1000–1200	9 10	
Available product shapes, surface finishing, notes		Round bars: e.g. rolled and peeled (+SH), cold-drawn (+C), cold-drawn and quenched & tempered (+C+QT) Flat and square bars: e.g. cold-drawn (+C), quenched & tempered and cold-drawn (+QT+C) Dimensions, allowance limits on page 144; heat treatment on page 155						

Case hardened bright steels (selection)

cf. DIN EN 10277-4 (2008-06)

Steel type		Thickness range in mm	Mechanical properties in delivery condition					Properties, application
			rolled and peeled (+SH)		cold-drawn (+C)			
Code name, designation	Material number		Hardness HB	Tensile strength R_m N/mm²	Yield strength $R_{p0.2}$ N/mm²	Tensile strength R_m N/mm²	Elongation at fracture A in %	
C10R	1.1207	> 16 ≤ 40 > 40 ≤ 63	92–163	310–550	250 200	400–700 350–640	12	Unalloyed case hardened steels for small parts under medium load, e. g. bolts, gudgeons, reels
C15R	1.1140	> 16 ≤ 40 > 40 ≤ 63	98–178	330–600	280 240	430–730 380–670	9 11	
C16R	1.1208	> 16 ≤ 40 > 40 ≤ 63	105–184	350–620	300 260	450–750 360–620	11 12	

Alloyed steels			Hardness HB in delivery condition				Properties, application
			+A+SH	+A+C	+FP+SH	+FP+C	
16MnCrS5	1.7139	> 16 ≤ 40 > 40 ≤ 63	207	245 240	140–187	140–240 140–235	Alloyed case hardened steels, improved machinability due to sulfur (S), for parts under higher loads
20MnCrS5	1.7149	> 16 ≤ 40 > 40 ≤ 63	217	255 250	152–201	152–250 152–245	
16NiCrS4	1.5715	> 16 ≤ 40 > 40 ≤ 63	217	255 255	156–207	156–245 156–240	
15NiCr13	1.5752	> 16 ≤ 40 > 40 ≤ 63	255	–	166–217	–	
Available product shapes, surface finishing, notes		Round bars: +A+SH annealed and peeled, +A+C annealed and cold-drawn, +FP+SH treated for ferrite-perlite structure and peeled, +FP+C treated for ferrite-perlite structure and cold-drawn Flat and square bars: +A+C annealed and cold-drawn, +FP+C treated for ferrite- perlite structure and cold-drawn Dimensions, allowance limits on page 144, heat treatment on page 154					

MS

Sheet and strip metal – Classification, overview

Classification according to

Delivery form	
Type	Commercial formats, remarks
Sheet metal for flat products 	Usually rectangular plates in small format: $w \times l$ = 1000 x 2000 mm, med. format: $w \times l$ = 1250 x 2500 mm, large format: $w \times l$ = 1500 x 3000 mm. Sheet thicknesses: s = 0.14 – 250 mm
Coil 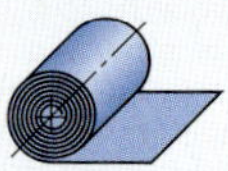	Strip thickness: s = 0.14 – 10 mm appr. strip width w: up to 2000 mm Coil diameter: up to 2400 mm Coil mass: up to 40 t • For feed stock at automatic manufacturing plants or sheet metal blanks for secondary processing

Fabrication process	
Process	Remarks
Hot-rolled	Sheet thicknesses up to approx. 250 mm, surfaces in rolled condition with scales, rolled-in scales, blisters, cracks, sand marks, probably corrected by welding or grinding.
Cold-rolled	Sheet thicknesses up to approx. 10 mm, various surface grades are available, e.g. for high-quality varnishing and surface treatments such as phosphating.
Cold-rolled with surface finishing	• Higher corrosion resistance, due to e.g. galvanizing or organic coating • For decorative purposes, e.g. with plastic or varnish coating • Better workability, e. g. due to textured surfaces

Sheet metal types – Overview (selection)

Main characteristics	Designation, steel types	Standard	Delivery form[1)] Sh	St	thickness range
Cold-rolled sheet and strip					
• cold workable (deep drawing) • weldable • surface paintable	Flat rolled products from soft steels	DIN EN 10130	•	•	0.35–3 mm
	Cold strip from soft steels	DIN EN 10207	–	•	≤ 10 mm
	Flat products with high yield strengths	DIN EN 10268	•	•	≤ 3 mm
	Flat products for enameling	DIN EN 10209	•	•	≤ 3 mm
Cold-rolled sheet and strip with surface finishing					
• higher corrosion resistance • possibly better workability	Hot-dip finished sheet and strip	DIN EN 10346	•	•	≤ 3 mm
	Zinc electroplated flat products from steel for cold working	DIN EN 10152	•	•	0.35–3 mm
	Organically coated flat products from steel	DIN EN 10169-1	•	•	≤ 3 mm
Cold-rolled sheets and strip for packaging					
• corrosion resistant • cold workable • weldable	Black plate for manufacture of tinplate	DIN EN 10205	•	•	0.14–0.49 mm
	Packaging sheet metal from electrolytically tinned or chromed steel	DIN EN 10202	•	•	0.14–0.49 mm
Hot-rolled sheet and strip					
Same properties as the corresponding steel groups (pages 124, 125)	Sheet and strip from unalloyed and alloy steels, e.g. structural steels as per DIN EN 10025-2, fine-grain structural steels as per DIN EN 10025-3, case hardened steels as per DIN EN 10084, quenched and tempered steels as per DIN EN 10083, stainless steels as per DIN EN 10088	DIN EN 10051	•	•	sheet up to 25 mm thickness, strip up to 10 mm thickness
• high yield strength	Sheet metal from structural steels with higher yield strength, quenched and tempered	DIN EN 10025-6	•	–	3–150 mm
• cold workability	Flat products of steel with high yield strength	DIN EN 10149-1	•	•	sheet up to 20 mm thickness

1) Delivery forms: Sh sheet; St strip

Cold-rolled sheet and strip for cold working

Cold-rolled strip and sheet from soft steels

cf. DIN EN 10130 (2007-02)

Steel type: Designation	Steel type: Material number	Type of surface	Tensile strength R_m N/mm²	Yield strength R_e N/mm²	Elongation at fracture EL %	Lack of flow-lines[1]	Properties, Application
DC01	1.0330	A B	270–410	140 280	28	– 3 months	Cold workable, e.g. by deep drawing, weldable, surface paintable; worked sheet parts in automotive, general machine and equipment manufacturing, in the construction industry
DC03	1.0347	A B	270–370	140 240	34	6 months	
DC04	1.0338	A B	270–350	140 210	38	6 months	
DC05	1.0312	A B	270–330	140 180	40	6 months	
DC06	1.0873	A B	270–350	120 180	38	unlimited time	
Delivery forms (standard values)	Sheet thicknesses: 0.25 – 0.35 – 0.4 – 0.5 – 0.6 – 0.7 – 0.8 – 0.9 – 1.0 – 1.2 – 1.5 – 2.0 – 2.5 – 3.0 mm Metal sheet dimensions: 1000 x 2000 mm, 1250 x 2500 mm, 1500 x 3000 mm, 2000 x 6000 mm strip (coils) up to approx. 2000 mm wide						
Explanation	[1] In subsequent non-cutting processes, e.g. deep drawing, no flow lines appear within the given time period. The time period begins at the agreed upon delivery date.						

Type of surface: Designation	Description of the surface	Surface finish: Designation	Finish	Average roughness Ra
A	Defects, e.g. pores, scoring, may not influence the workability and the adhesion of surface coatings.	b g	very smooth smooth	$Ra \leq 0.4$ µm $Ra \leq 0.9$ µm
B	One side of the sheet must be free of defects so that its surface finish will not influence quality painting.	m r	matt rough	0.6 µm $< Ra \leq$ 1.9 µm $Ra >$ 1.6 µm

⇒ **Sheet EN 10130 – DC06 – B – g:** Sheet metal from DC06 material, surface type B, smooth surface

Cold-rolled strip and sheet of high yield steels (selection)

cf. DIN EN 10268 (2006-10)

Steel type: Designation	Steel type: Material number	Tensile strength R_m N/mm²	Yield strength R_e N/mm²	Elongation at fracture EL %	Properties, Application
HC180Y HC220Y HC260Y	1.0922 1.0925 1.0928	340–400 350–420 380–440	180–230 220–270 260–320	36 34 32	Cold workability at high mechanical strength, sophisticated deep-drawn parts
HC180B HC220B HC300B	1.0395 1.0396 1.0444	300–360 320–400 400–480	180–230 220–270 300–360	34 32 26	Good cold workability, increase of the yield strength through heat treatment after the shaping process; exterior parts of the vehicle body
HC180P HC260P HC300P	1.0342 1.0417 1.0448	280–360 360–440 400–480	180–230 280–320 300–360	34 29 26	Good cold workability, high impact resistance and fatigue strength; parts of the body skin, deep-drawn parts
HC260LA HC380LA HC420LA	1.0480 1.0550 1.0556	350–430 440–560 470–590	260–330 380–480 420–520	26 19 17	Good weldability and limited cold workability, good impact resistance and fatigue strength; reinforcing parts of the vehicle body
Forms of delivery, surface finishes	Forms of delivery see DIN EN 10130 (table on top) Surface finishes: The products are available with the surface finish types A and B in accordance with DIN EN 10130. For LA types, e.g. HC380LA, only surface finish type A is available. For rolling width > 600 mm, the surface finishes also comply with DIN EN 10130.				

⇒ **Sheet metal EN 10628 – HC380LA – A – m:** Sheet metal of material HC380LA, surface finish A, matt (m)

MS

MS

Cold-rolled and hot-rolled sheet

Hot-dip-refined strip and sheet metal from soft steels for cold working (selection)

cf. DIN EN 10346 (2009-07) replaces DIN EN 10327

Steel type: Designation	Steel type: Material number	Guarantee for strength values[1]	Tensile strength R_m N/mm²	Yield strength R_e N/mm²	Elongation at fracture *EL* %	Formation of flow lines[2]	Cold working grade
DX51D+Z DX51D+ZF	1.0226+Z 1.0226+ZF	1 month	270 – 500	–	22	Formation of flow lines may occur	machine seamed quality
DX52D+Z DX52D+ZF	1.0350+Z 1.0350+ZF	1 month	270 – 420	140 – 300	26		drawing grade
DX53D+Z DX53D+ZF	1.0355+Z 1.0355+ZF	1 month	270 – 380	140 – 260	30		deep drawing grade
DX54D+Z DX54D+ZF	1.0306+Z 1.0306+ZF	6 months	270 – 350	140 – 220	36 34	6 months free from flow lines[2]	extra deep drawing grade
DX56D+Z DX56D+ZF	1.0322+Z 1.0322+ZF	6 months	270 – 350	120 – 180	39 37		special deep drawing grade
Delivery forms (standard values)	Sheet metal thicknesses: 0.25 – 0.35 – 0.4 – 0.5 – 0.6 – 0.7 – 0.8 – 0.9 – 1.0 – 1.2 – 1.5 – 2.0 – 2.5 – 3.0 mm Sheet metal dimensions: 1000 x 2000 mm, 1250 x 2500 mm, 1500 x 3000 mm, 2000 x 6000 mm strip (coils) up to approx. 2000 mm wide						
Explanation	[1] Values for tensile strength R_m, yield strength R_e and elongation at fracture *EL* are only guaranteed within the given time period. The time period begins at the agreed upon delivery date. [2] In subsequent working, e.g. deep drawing, no flow lines appear within a given period. The time period begins at the agreed upon delivery date.						

Composition and properties of the coating (selection)

Designation	Composition, properties	Designation	Composition, properties
+Z	Coating of pure zinc, shiny flower patterned surface, protection against atmospheric corrosion	+ZF	Abrasion resistant coating of a zinc-iron alloy, uniform matt gray surface, corrosion resistant like +Z

Surface quality and finishing

Designation	Meaning	Surface treatment
A	Flaws like score marks, scratches, pinholes and stripe-shaped marks are allowed.	The surface treatment is indicated by code letters: C → chemically passivated O → oiled P → phosphated S → sealed
B	Improved surface compared to A	
C	Best surface, high-quality painting must be assured on one side of the sheet	

⇒ **Sheet metal EN 10143, 0.5 x 1200 x 2500, steel DIN EN 10346, DX53D+ZF100-B-P:** Sheet metal with limit sizes as per EN 10143, thickness t = 0.5 mm, width w = 1200 mm, length l = 2500 mm, material DX53D, coating of iron-zinc alloy with 100 g/m², improved (B) and oiled (O) surface

Hot-rolled sheet metal and strips

cf. DIN EN 10051 (1997-11)

Materials: Hot-rolled sheet metal and strips according to DIN EN 10051 are produced from steels of various material groups, for example:

Steel group, designation	Standard	Page
Structural steels Case hardened steels Quenched and tempered steels	DIN EN 10025-2 DIN EN 10084 DIN EN 10083-2 – 3	127 129 130
Weldable fine-grain steels Heat-treatable structural steels, high yield strength	DIN EN 10025-3 – 4 DIN EN 10125-6	128 128
Stainless steels Pressure vessel steels	DIN EN 10088-2 – 3 DIN EN 10028	133 –

Properties and applications of the steels are given on the pages for the individual steel.

Delivery forms (standard values): Sheet metal thicknesses: 0.5 – 1.0 – 1.5 – 2.0 – 2.5 – 3.0 – 3.5 – 4.0 – 4.5 – 5.0 – 6.0 – 8.0 – 10.0 – 12.0 – 15.0 – 18.0 – 20.0 – 25.0 mm. Sheet metal and strip dimensions, see DIN EN 10346.

⇒ **Sheet metal EN 10051 – 2.0 x 1200 x 2500 – steel EN 10083-1 – 34Cr4:** sheet metal as per EN 10051, thickness t = 2 mm, width w = 1200 mm, quenched & tempered alloyed steel 34Cr4

Tubes for machine construction, Precision steel tube

Seamless tube for machine construction (selection)

cf. DIN EN 10297-1 (2003-06)

d outside diameter
s wall thickness
S cross-sectional area
m' linear mass density
W_x axial section modulus
I_x axial geometrical moment of inertia

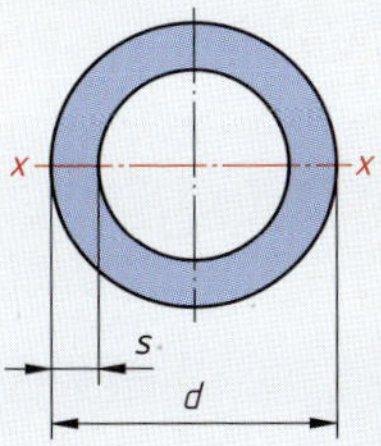

d x *s*	*S* cm²	*m'* kg/m	W_x cm³	I_x cm⁴	*d* x *s*	*S* cm²	*m'* kg/m	W_x cm³	I_x cm⁴
26.9 x 2.3	1.78	1.40	1.01	1.36	54 x 5.0	7.70	6.04	8.64	23.34
26.9 x 2.6	1.98	1.55	1.10	1.48	54 x 8.0	11.56	9.07	11.67	31.50
26.9 x 3.2	2.38	1.87	1.27	1.70	54 x 10.0	13.82	10.85	13.03	35.18
35 x 2.6	2.65	2.08	2.00	3.50	60.3 x 8	13.14	10.31	15.25	45.99
35 x 4.0	3.90	3.06	2.72	4.76	60.3 x 10	15.80	12.40	17.23	51.95
35 x 6.3	5.68	4.46	3.50	6.13	60.3 x 12.5	18.77	14.73	19.00	57.28
40 x 4	4.52	3.55	3.71	7.42	70 x 8	15.58	12.23	21.75	76.12
40 x 5	5.50	4.32	4.30	8.59	70 x 12.5	22.58	17.73	27.92	97.73
40 x 8	8.04	6.31	5.47	10.94	70 x 16	27.14	21.30	30.75	107.6
44.5 x 4	5.09	4.00	4.74	10.54	82.5 x 8	18.72	14.70	31.85	131.4
44.5 x 5	6.20	4.87	5.53	12.29	82.5 x 12.5	27.49	21.58	42.12	173.7
44.5 x 8	9.17	7.20	7.20	16.01	82.5 x 20	39.27	30.83	51.24	211.4
51 x 5	7.23	5.68	7.58	19.34	88.9 x 10	24.79	19.46	44.09	196.0
51 x 8	10.81	8.49	10.13	25.84	88.9 x 16	36.64	28.76	57.40	255.2
51 x 10	12.88	10.11	11.25	28.68	88.9 x 20	43.29	33.98	62.66	278.6

Material, annealing condition	Steel group		Steel type, examples	Annealing condition[1)]
	Machine construction steels	unalloyed	E235, E275, E315	+AR or +N
		alloyed	E355K2, E420J2	+N
	Quenched and tempered steels	unalloyed	C22E, C45E, C60E	+N or +QT
		alloyed	41Cr4, 42CrMo4	+QT
	Case hard. steel, alloyed/unall.		C10E, C15E, 16MnCr5	+A or +N
	Properties and applications of steels, see pages 124 and 125.			

Precision steel tube, cold-drawn seamless (selection)

cf. DIN EN 10305-1 (2010-05)

d outside diameter
s wall thickness
S cross-sectional area
m' linear mass density
W_x axial section modulus
I_x axial geometrical moment of inertia

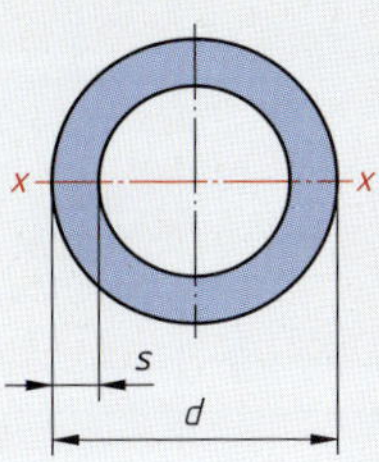

d x *s*	*S* cm²	*m'* kg/m	W_x cm³	I_x cm⁴	*d* x *s*	*S* cm²	*m'* kg/m	W_x cm³	I_x cm⁴
10 x 1	0.28	0.22	0.06	0.03	35 x 3	3.02	2.37	2.23	3.89
10 x 1.5	0.40	0.31	0.07	0.04	35 x 5	4.71	3.70	3.11	5.45
10 x 2	0.50	0.39	0.09	0.04	35 x 8	5.53	4.34	2.53	3.79
12 x 1	0.35	0.27	0.09	0.05	40 x 4	4.52	3.55	3.71	7.42
12 x 1.5	0.49	0.38	0.12	0.07	40 x 5	5.50	4.32	4.30	8.59
12 x 2	0.63	0.49	0.14	0.08	40 x 8	8.04	6.31	5.47	10.94
15 x 2	0.82	0.64	0.24	0.18	50 x 5	7.07	5.55	7.25	18.11
15 x 2.5	0.98	0.77	0.27	0.20	50 x 8	10.56	8.29	9.65	24.12
15 x 3	1.13	0.89	0.29	0.22	50 x 10	12.57	9.87	10.68	26.70
20 x 2.5	1.37	1.08	0.54	0.54	60 x 5	8.64	6.78	10.98	32.94
20 x 4	2.01	1.58	0.68	0.68	60 x 8	13.07	10.26	15.07	45.22
20 x 5	2.36	1.85	0.74	0.74	60 x 10	15.71	12.33	17.02	51.05
25 x 2.5	1.77	1.39	0.91	1.13	70 x 5	10.21	8.01	15.50	54.24
25 x 5	3.14	2.46	1.34	1.67	70 x 10	18.85	14.80	24.91	87.18
25 x 6	3.58	2.81	1.42	1.78	70 x 12	21.87	17.17	27.39	95.88
30 x 3	2.54	1.99	1.56	2.35	80 x 8	18.10	14.21	29.68	118.7
30 x 5	3.93	3.08	2.13	3.19	80 x 10	21.99	17.26	34.36	137.4
30 x 6	4.52	3.55	2.31	3.46	80 x 16	32.17	25.25	43.75	175.0

Materials	Machine steels: E215, E235, E255, E355, E410 Quenched & tempered steels: 26Mn5, C35E, C45E, 26Mo2, 25CrMo4, 42CrMo4 Free-cutting steels: 10S10, 15S10, 18S10, 37S10
Delivery conditions	+C bright-drawn, hard +LC bright-drawn, soft +A annealed +SR bright-drawn, stress-relief annealed +N normalized
Surfaces	Smooth inner and outer surface, roughness limits Ra: Ra ≤ 4 μm for the outer surface of the finishing grades +SR, +A, +N Ra ≤ 4 μm for the outer and inner surface of the finishing grades +C, +LC

Hot-rolled steel profiles

Cross-section	Designation, dimensions	Standard, page	Cross-section	Designation, dimensions	Standard, page
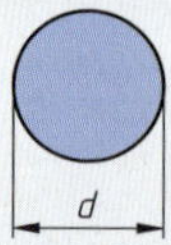	**Round steel bar** *d* = 8–200	DIN EN 10060 page 143	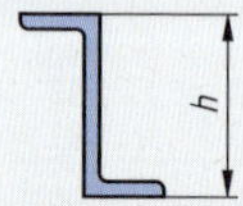	**Z profile steel** *h* = 30–200	DIN 1027
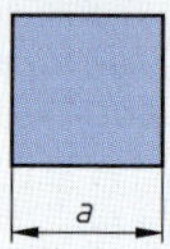	**Square steel bar** *a* = 8–120	DIN EN 10059 page 143	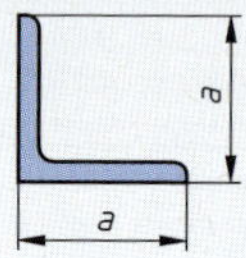	**Equal leg steel angle** *a* = 20–250	DIN EN 10056-1 page 147
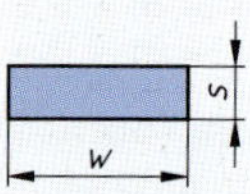	**Flat steel bar** *w* x *s* = 10 x 5 to 150 x 60	DIN EN 10058 page 143	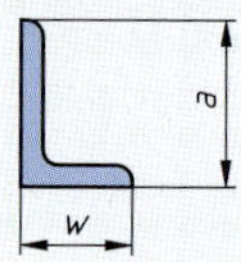	**Unequal leg steel angle** *a* x *w* = 30 x 20 to 200 x 150	DIN EN 10056-1 page 146
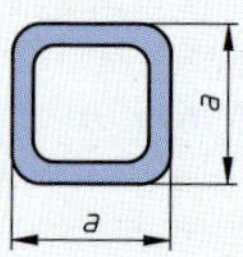	**Square tube** *a* = 40–400	DIN EN 10210-2 page 150	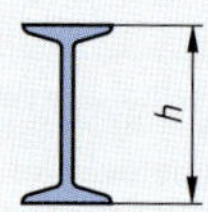	**Narrow I-beam** I series *h* = 80–160	DIN 1025-1
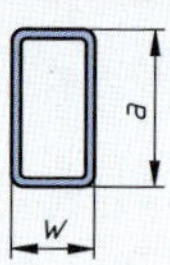	**Rectangular tubes** *a* x *w* = 50 x 25 to 500 x 300	DIN EN 10210-2 page 150	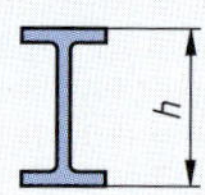	**Medium width I-beam** IPE series *h* = 80–600	DIN 1025-5 page 148
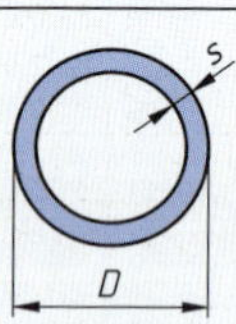	**Circular tube** *D* x *s* = 21.3 x 2.3 to 1219 x 25	DIN EN 10210-1	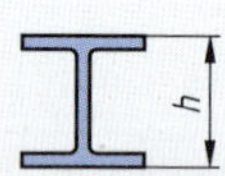	**Wide I-beam** IPB series[1)] *h* = 100–1000	DIN 1025-2 page 149
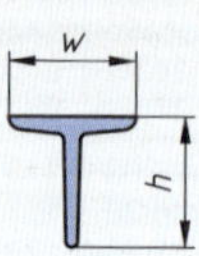	**Equal leg tee** *w* = *h* = 30–140	DIN EN 10055 page 145	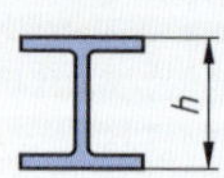	**Wide I-beam** light duty IPBl series[1)] *h* = 100–1000	DIN 1025-3 page 148
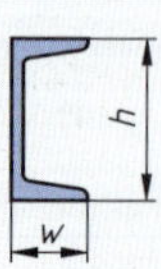	**Steel channel** *h* = 30–400	DIN 1026-1 page 145	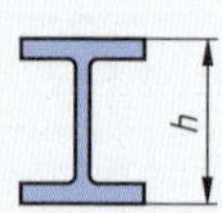	**Wide I-beam** reinforced design IPBv series[1)] *h* = 100–1000	DIN 1025-4 page 149

1) according to EURONORM 53-62: IPB = HE to B, IPBl = HE to A, IPBv = HE to M

MS

Bar steel, hot-rolled

Hot-rolled round bars

cf. DIN EN 10060 (2004-02)

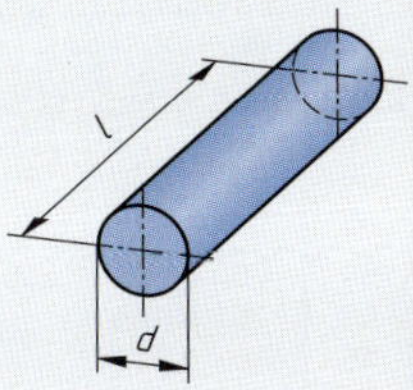

Steel group (selection)	Standard	see page	Surfaces
Unalloyed structural steels	DIN EN 10025-2	127	The surface may show scales, scale inclusions, blisters, cracks and sand marks etc., which might be repaired by grinding.
Free-cutting steels	DIN EN 10087	131	
Case hardened steels	DIN EN 10084	129	
Quenched & tempered steels	DIN EN 10083	130	
Tool steels	DIN EN ISO 4957	132	
Stainless steels	DIN EN 10088	133	

	Nominal diameter *d* in mm								
Nominal sizes	10 12 13	18 19 20	26 27 28	36 38 40	50 52 55	70 73 75	95 100 105	130 135 140	160 165 170
	14 15 16	22 24 25	30 32 35	42 45 48	60 63 65	80 85 90	110 115 120	145 150 155	175 180 200
Lengths *l*	Manufactured lengths (M): *l* = 3000 mm to *l* = 13000 mm Normal lengths (F): *l* = (3000 mm to 13000 mm) ± 100 mm Precision length (E): *l* < 6000 mm ± 25 mm								
⇒	**Round bar EN 10060 – 40 x 5000 E – steel EN 10025-2 – S235JR:** hot-rolled round steel bar, *d* = 40 mm, precision length *l* = 5000 mm ± 25 mm; unalloyed structural steel S235JR								

MS

Hot-rolled flat bars

cf. DIN EN 10058 (2004-02)

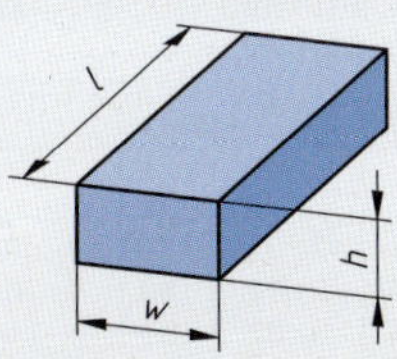

Steels groups, surfaces	See hot-rolled round bars
Lengths *l*	Manufactured lengths (M): *l* = 3000 mm to *l* = 13000 mm Normal lengths (F): *l* = (3000 mm to 13000 mm) ± 100 mm Precision length (E): *l* < 6000 mm ± 25 mm

Nominal sizes: width *w* and height *h* in mm											
w	*h*	*w*	*h*	*w*	*h*	*w*	*h*	*w*	*h*	*w*	*h*
10 12 15	5 5, 6 5–10	16 20 25	5–10 5–15 5–15	30 35 40	5–20 5–20 5–30	45 50 60	5–30 5–30 5–40	70 80 90	5–40 5–60 5–60	100 120 150	5–60 6–60 6–90

Nominal thicknesses *h*: 5, 6, 8, 10, 12, 15, 20, 25, 30, 35, 40, 50, 60, 80 mm

⇒ **Flat bar EN 10058 – 40 x 15 x 3000 M – steel EN 10084 – 20MnCr5:**
hot-rolled flat bar *w* = 40 mm, *h* = 15 mm,
manufactured length *l* = 3000 mm, case hardened 20MnCr5

Hot-rolled square bars

cf. DIN EN 10059 (2004-02)

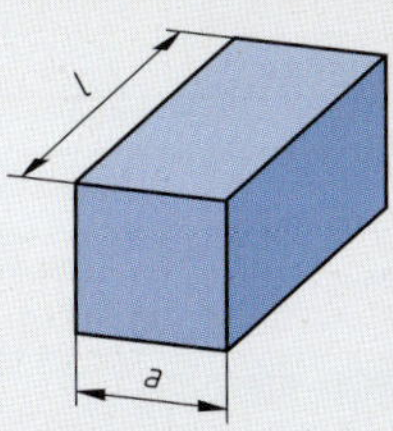

Steel groups, surfaces	See hot-rolled round bars										
Lengths *l*	Manufactured lengths (M): *l* = 3000 mm to *l* = 13000 mm Normal lengths (F): *l* = (3000 mm to 13000 mm) ± 100 mm Precision length (E): *l* < 6000 mm ± 25 mm										
	Side length *a* in mm										
Nominal sizes	8 10 12	13 14 15	16 18 20	22 24 25	26 28 30	32 35 40	45 50 55	60 65 70	75 80 90	100 110 120	130 140 150
⇒	**Square bar EN 10059 – 60 x 5000 F – steel EN 10087 – 35S20:** hot-rolled square bar *a* = 60 mm, normal length *l* = 5000 mm ± 100 mm, material: free-cutting steel 35S20										

Bar steel, bright

Bright round bars

cf. DIN EN 10278 (1999-12)

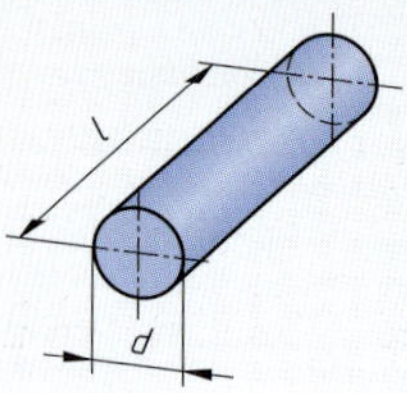

Steel group (selection)	Standard	see page	Delivery condition, surface quality
Steels for general applications	DIN EN 10277-2	136	drawn (+C): smooth, descaled surface peeled (+SH): better surface quality as with (+C), hardly any surface decarbonization or rolling flaws ground (+SL): best surface, best dimensional accuracy
Free-cutting steels	DIN EN 10277-3	136	
Case hardened steels	DIN EN 10277-4	137	
Quenched & tempered steels	DIN EN 10277-5	137	

Allowance limits for diameter *d*	Drawn (+C) h10			Peeled (+SH) h10				Ground (+SL) h9		
Nominal diameter *d* in mm	2.5	6	9.5	16	23	30	42	60	90	150
	3	6.5	10	17	24	32	46	63	100	160
	3.5	7	11	18	25	34	48	65	110	180
	4	7.5	12	19	26	35	50	70	120	200
	4.5	8	13	20	27	36	52	75	125	
	5	8.5	14	21	28	38	55	80	130	
	5.5	9	15	22	29	40	58	85	140	

Lengths *l*	Manufactured lengths: *l* = 3000 mm to *l* = 9000 mm Mill lengths: *l* = 3000 mm or *l* = 6000 mm
⇒	**Round EN 10278 – 25 x mill length 3000 – EN 10277-5 – C45+SH:** round bar *d* = 25 mm, manufactured according to tolerance zone h10, mill length *l* = 3000 mm, material quenched & tempered steel C45, peeled (+SH)

MS

Bright flat bars

cf. DIN EN 10278 (1999-12)

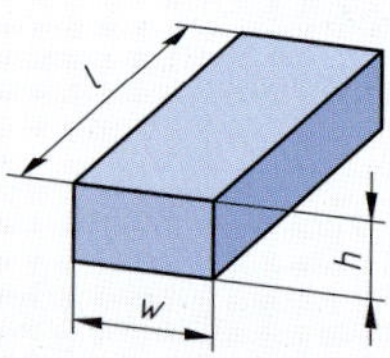

Steel groups, delivery condition		See bright round bars, drawn (+C), smooth, descaled surface	
Allowance limits	Width *w*	≤ 100 mm ⇒ h11	100 mm < *l* ≤ 150 mm ⇒ ± 0.5 mm
	Height *h*	≤ 80 mm ⇒ h11	80 mm < *h* ≤ 100 mm ⇒ h12
Lengths *l*		Manufactured lengths: *l* = 3000 mm to *l* = 9000 mm Mill lengths: *l* = 3000 mm or *l* = 6000 mm	

Nominal sizes: width *w* and height *h* in mm

w	*h*	*w*	*h*	*w*	*h*	*w*	*h*	*w*	*h*	*w*	*h*
5	2–3	12	2–10	18	2–12	28	2–20	45	2–32	70	4–40
6	2–4	14	2–10	20	2–16	32	2–25	50	2–32	80	5–25
8	2–6	16	2–12	22	2–12	36	2–20	56	3–32	90	5–25
10	2–8	18	2–12	25	2–20	40	2–32	63	3–40	100	5–25

Nominal thicknesses *h*: 2, 2,5, 3, 4, 5, 6, 8, 10, 12, 15, 16, 20, 25, 30, 32, 35, 40 mm

⇒ **Flat EN 10278 – 40 x 16 x 5000 – EN 10277-4 – C16R+C:** flat bar *w* = 40 mm, *h* = 16 mm, width and height according to tolerance zone h11, length *l* = 5000 mm, material: case-hardened steel C16R, drawn (+C)

Brights square bars

cf. DIN EN 10278 (1999-12)

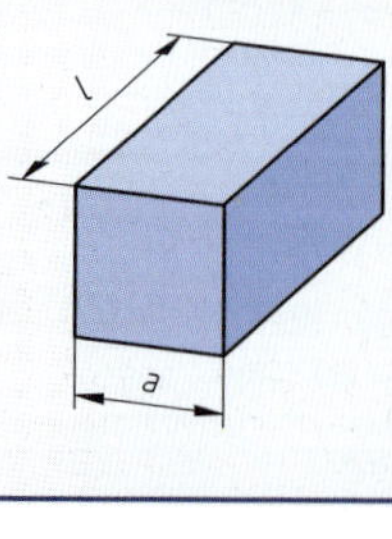

Steel groups, delivery condition	See bright round bars bright, drawn (+C), smooth and descaled surface
Allowance limits	*a* ≤ 80 mm ⇒ h11; *a* > 80 mm ⇒ h12
Lengths *l*	See bright flat bars

Side lengths *a* in mm									
	4	6	9	12	16	22	36	60	80
	4.5	7	10	13	18	25	40	63	100
	5	8	11	14	20	28	45	70	

⇒ **Square bar EN 10278 – 45 x mill length 6000 – EN 10277-3 – 35SPb20+C:** square bar *a* = 45 mm, manufactured according to tolerance zone h11, mill length *l* = 6000 mm, material: free-cutting steel 35SPb20, drawn (+C)

Structural Tee, Steel channel

Equal leg Tee, hot-rolled

cf. DIN EN 10055 (1995-12)

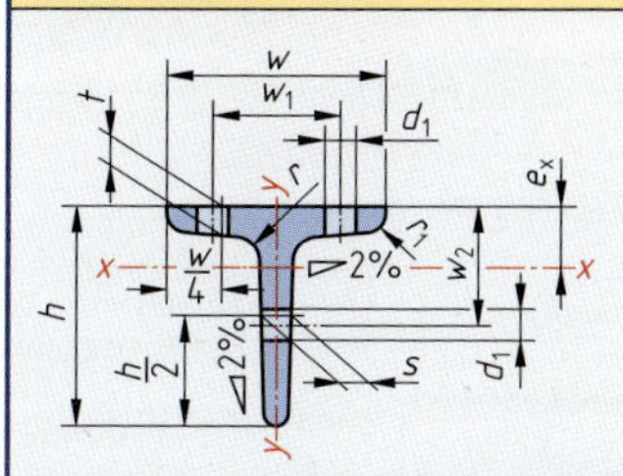

S cross-sectional area
I second moment of inertia
W axial section modulus
m' linear mass density

Material: Unalloyed structural steel DIN EN 10025, e.g. S235JR

Delivery type: Lengths to order with a usual limit deviation of ± 100 mm or a reduced limit deviation ± 50 mm, ± 25 mm, ± 10 mm

$r = s$

$r_1 = \frac{s}{2}$

Designation	Dimensions in mm				Distance of the x axis	For the bending axis x – x		For the bending axis y – y		Tracing dimension accord. to DIN 997		
T	$w = h$	$s = t$	S cm²	m' kg/m	e_x cm	I_x cm⁴	W_x cm³	I_y cm⁴	W_y cm³	w_1 mm	w_2 mm	d_1 mm
30	30	4	2.26	1.77	0.85	1.72	0.80	0.87	0.58	17	17	4.3
35	35	4.5	2.97	2.33	0.99	3.10	1.23	1.04	0.90	19	19	4.3
40	40	5	3.77	2.96	1.12	5.28	1.84	2.58	1.29	21	22	6.4
50	50	6	5.66	4.44	1.39	12.1	3.36	6.06	2.42	30	30	6.4
60	60	7	7.94	6.23	1.66	23.8	5.48	12.2	4.07	34	35	8.4
70	70	8	10.6	8.23	1.94	44.4	8.79	22.1	6.32	38	40	11
80	80	9	13.6	10.7	2.22	73.7	12.8	37.0	9.25	45	45	11
100	100	11	20.9	16.4	2.74	179	24.6	88.3	17.7	60	60	13
120	120	13	29.6	23.2	3.28	366	42.0	179	29.7	70	70	17
140	140	15	39.9	31.3	3.80	660	64.7	330	47.2	80	75	21

⇒ **Tee profile EN 10055 – T50 – S235JR:** Structural steel tee, h = 50 mm, from S235JR

Steel channel, hot-rolled

cf. DIN 1026-1 (2000-03)

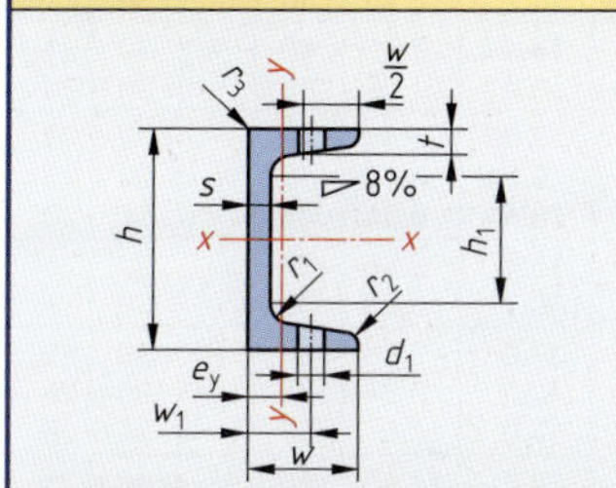

S cross-sectional area
I second moment of inertia
W axial section modulus
m' linear mass density

Material: Unalloyed structural steel DIN EN 10025, e.g. S235J0

Delivery type: Manufactured lengths 3 m to 15 m; normal lengths up to 15 m ± 50 mm; slope angle at $h \leq$ 300 mm: 8%; $h >$ 300 mm: 5%

$r_1 = t$

$r_2 \approx \frac{t}{2}$

$r_3 \leq 0{,}3 \cdot t$

Designation	Dimensions in mm							Distance to the y axis	For the bending axis x – x		For the bending axis y – y		Tracing dimensions DIN 997	
U	h	w	s	t	h_1	S cm²	m' kg/m	e_y cm	I_x cm⁴	W_x cm³	I_y cm⁴	W_y cm³	w_1 mm	d_1 mm
30 x 15	30	15	4	4.5	12	2.21	1.74	0.52	2.53	1.69	0.38	0.39	10	4.3
30	30	33	5	7	10	5.44	4.27	1.31	6.39	4.26	5.33	2.68	20	8.4
40 x 20	40	20	5	5.5	18	3.66	2.87	0.67	7.58	3.97	1.14	0.86	11	6.4
40	40	35	5	7	11	6.21	4.87	1.33	14.1	7.05	6.68	3.08	20	8.4
50 x 25	50	25	5	6	25	4.92	3.86	0.81	16.8	6.73	2.49	1.48	16	8.4
50	50	38	5	7	20	7.12	5.59	1.37	26.4	10.6	9.12	3.75	20	11
60	60	30	6	6	35	6.46	5.07	0.91	31.6	10.5	4.51	2.16	18	8.4
80	80	45	6	8	46	11.0	8.64	1.45	106	26.5	19.4	6.36	25	13
100	100	50	6	8.5	64	13.5	10.6	1.55	206	41.2	29.3	8.49	30	13
120	120	55	7	9	82	17.0	13.4	1.60	364	60.7	43.2	11.1	30	17
160	160	65	7.5	10.5	115	24.0	18.8	1.84	925	116	85.3	18.3	35	21
200	200	75	8.5	11.5	151	32.2	25.3	2.01	1 910	191	148	27.0	40	23
260	260	90	10	14	200	48.3	37.9	2.36	4 820	371	317	47.7	50	25
300	300	100	10	16	232	58.8	46.2	2.70	8 030	535	495	67.8	55	28
350	350	100	14	17.5	276	77.3	60.6	2.40	12 840	734	570	75.0	58	28
400	400	110	14	18	324	91.5	71.8	2.65	20 350	1020	846	102	60	28

⇒ **Channel DIN 1026 – U100 – S235J0:** Steel channel, h = 100 mm, from S235J0

MS

Steel angle

Unequal leg steel angle, hot-rolled (selection)

cf. DIN EN 10056-1 (1998-10)

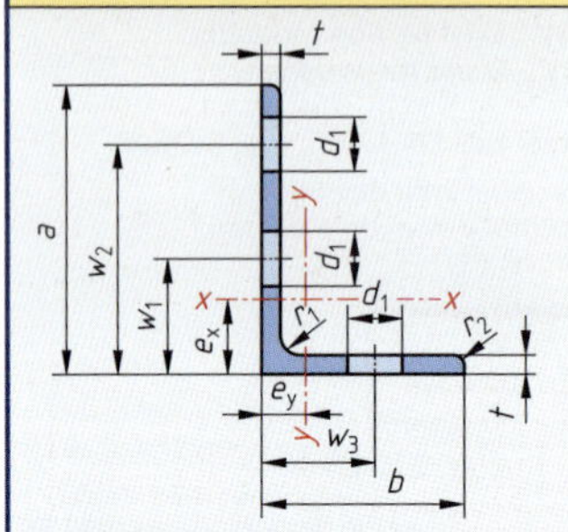

S cross-sectional area
I second moment of inertia
W axial section modulus
m' linear mass density

Material: Unalloyed structural steel DIN EN 10025-2, e.g. S235J0

Delivery type: From 30 x 20 x 3 to 200 x 150 x 15, in manufactured lengths ≥ 6 m < 12 m, normal lengths ≥ 6 m < 12 m ± 100 mm

$r_1 \approx t$

$r_2 \approx \frac{t}{2}$

Designation L	Dimensions in mm a	b	t	S cm²	m′ kg/m	Distances to axes e_x cm	e_y cm	For the bending axis x – x: I_x cm⁴	W_x cm³	y – y: I_y cm⁴	W_y cm³	Tracing dimension accord. to DIN 997 w_1 mm	w_2 mm	w_3 mm	d_1 mm
30 x 20 x 3	30	20	3	1.43	1.12	0.99	0.50	1.25	0.62	0.44	0.29	17	–	12	8.4
30 x 20 x 4	30	20	4	1.86	1.46	1.03	0.54	1.59	0.81	0.55	0.38	17	–	12	8.4
40 x 20 x 4	40	20	4	2.26	1.77	1.47	0.48	3.59	1.42	0.60	0.39	22	–	12	11
40 x 25 x 4	40	25	4	2.46	1.93	1.36	0.62	3.89	1.47	1.16	0.69	22	–	15	11
45 x 30 x 4	45	30	4	2.87	2.25	1.48	0.74	5.78	1.91	2.05	0.91	25	–	17	13
50 x 30 x 5	50	30	5	3.78	2.96	1.73	0.74	9.36	2.86	2.51	1.11	30	–	17	13
60 x 30 x 5	60	30	5	4.28	3.36	2.17	0.68	15.6	4.07	2.63	1.14	35	–	17	17
60 x 40 x 5	60	40	5	4.79	3.76	1.96	0.97	17.2	4.25	6.11	2.02	35	–	22	17
60 x 40 x 6	60	40	6	5.68	4.46	2.00	1.01	20.1	5.03	7.12	2.38	35	–	22	17
65 x 50 x 5	65	50	5	5.54	4.35	1.99	1.25	23.2	5.14	11.9	3.19	35	–	30	21
70 x 50 x 6	70	50	6	6.89	5.41	2.23	1.25	33.4	7.01	14.2	3.78	40	–	30	21
75 x 50 x 6	75	50	6	7.19	5.65	2.44	1.21	40.5	8.01	14.4	3.81	40	–	30	21
75 x 50 x 8	75	50	8	9.41	7.39	2.52	1.29	52.0	10.4	18.4	4.95	40	–	30	23
80 x 40 x 6	80	40	6	6.89	5.41	2.85	0.88	44.9	8.73	7.59	2.44	45	–	22	23
80 x 40 x 8	80	40	8	9.01	7.07	2.94	0.96	57.6	11.4	9.61	3.16	45	–	22	23
80 x 60 x 7	80	60	7	9.38	7.36	2.51	1.52	59.0	10.7	28.4	6.34	45	–	35	23
100 x 50 x 6	100	50	6	8.71	6.84	3.51	1.05	89.9	13.8	15.4	3.89	55	–	30	25
100 x 50 x 8	100	50	8	11.4	8.97	3.60	1.13	116	18.2	19.7	5.08	55	–	30	25
100 x 65 x 7	100	65	7	11.2	8.77	3.23	1.51	113	16.6	37.6	7.53	55	–	35	25
100 x 65 x 8	100	65	8	12.7	9.94	3.27	1.55	127	18.9	42.2	8.54	55	–	35	25
100 x 65 x 10	100	65	10	15.6	12.3	3.36	1.63	154	23.2	51.0	10.5	55	–	35	25
100 x 75 x 8	100	75	8	13.5	10.6	3.10	1.87	133	19.3	64.1	11.4	55	–	40	25
100 x 75 x 10	100	75	10	16.6	13.0	3.19	1.95	162	23.8	77.6	14.0	55	–	40	25
100 x 75 x 12	100	75	12	19.7	15.4	3.27	2.03	189	28.0	90.2	16.5	55	–	40	25
120 x 80 x 8	120	80	8	15.5	12.2	3.83	1.87	226	27.6	80.8	13.2	50	80	45	25
120 x 80 x 10	120	80	10	19.1	15.0	3.92	1.95	276	34.1	98.1	16.2	50	80	45	25
120 x 80 x 12	120	80	12	22.7	17.8	4.00	2.03	323	40.4	114	19.1	50	80	45	25
125 x 75 x 8	125	75	8	15.5	12.2	4.14	1.68	247	29.6	67.6	11.6	50	–	40	25
125 x 75 x 10	125	75	10	19.1	15.0	4.23	1.76	302	36.5	82.1	14.3	50	–	40	25
125 x 75 x 12	125	75	12	22.7	17.8	4.31	1.84	354	43.2	95.5	16.9	50	–	40	25
135 x 65 x 8	135	65	8	15.5	12.2	4.78	1.34	291	33.4	45.2	8.75	50	–	35	25
135 x 65 x 10	135	65	10	19.1	15.0	4.88	1.42	356	41.3	54.7	10.8	50	–	35	25
150 x 75 x 9	150	75	9	19.6	15.4	5.26	1.57	455	46.7	77.9	13.1	60	105	40	28
150 x 75 x 10	150	75	10	21.7	17.0	5.30	1.61	501	51.6	85.6	14.5	60	105	40	28
150 x 75 x 12	150	75	12	25.7	20.2	5.40	1.69	588	61.3	99.6	17.1	60	105	40	28
150 x 75 x 15	150	75	15	31.7	24.8	5.52	1.81	713	75.2	119	21.0	60	105	40	28
150 x 90 x 12	150	90	12	27.5	21.6	5.08	2.12	627	63.3	171	24.8	60	105	50	28
150 x 90 x 15	150	90	15	33.9	26.6	5.21	2.23	761	77.7	205	30.4	60	105	50	28
150 x 100 x 10	150	100	10	24.2	19.0	4.81	2.34	553	54.2	199	25.9	60	105	55	28
150 x 100 x 12	150	100	12	28.7	22.5	4.89	2.42	651	64.4	233	30.7	60	105	55	28
200 x 100 x 10	200	100	10	29.2	23.0	6.93	2.01	1220	93.2	210	26.3	65	150	55	28
200 x 100 x 15	200	100	15	43.0	33.8	7.16	2.22	1758	137	299	38.5	65	150	55	28

⇒ **L EN 10056-1 – 65 x 50 x 5 – S235J0:** Unequal leg steel angle, *a* = 65 mm, *b* = 50 mm, *t* = 5 mm, from S235J0

Steel angle

Equal leg steel angle, hot-rolled (selection)

cf. DIN EN 10056-1 (1998-10)

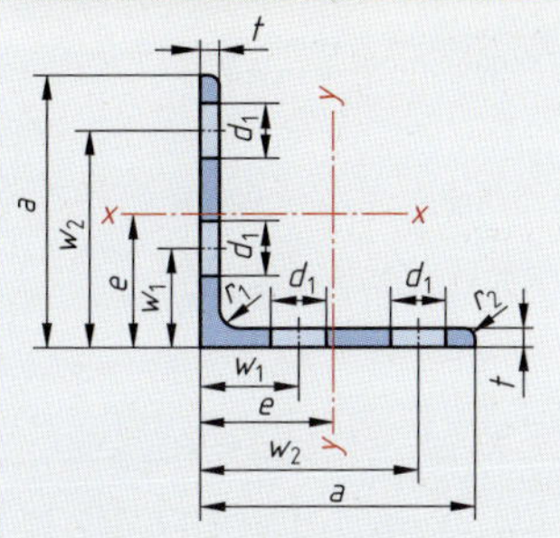

S cross-sectional area
I second moment of inertia
W axial section modulus
m' linear mass density

Material: Unalloyed structural steel DIN EN 10025-2, e.g. S235J0

Delivery type: From 20 x 20 x 3 to 200 x 250 x 35, in manufactured lengths ≥ 6 m < 12 m, normal lengths ≥ 6 m < 12 m ± 100 mm

$r_1 \approx t$

$r_2 \approx \frac{t}{2}$

Designation	Dimensions in mm				Distances to axes	For the bending axis x – x and y – y		Tracing dimension accord. to DIN 997		
L	*a*	*t*	*S* cm²	*m'* kg/m	*e* cm	$I_x = I_y$ cm⁴	$W_x = W_y$ cm³	w_1 mm	w_2 mm	d_1 mm
20 x 20 x 3	20	3	1.12	0.882	0.598	0.39	0.28	12	–	4.3
25 x 25 x 3	25	3	1.42	1.12	0.723	0.80	0.45	15	–	6.4
25 x 25 x 4	25	4	1.85	1.45	0.762	1.02	0.59	15	–	6.5
30 x 30 x 3	30	3	1.74	1.36	0.835	1.40	0.65	17	–	8.4
30 x 30 x 4	30	4	2.27	1.78	0.878	1.80	0.85	17	–	8.4
35 x 35 x 4	35	4	2.67	2.09	1.00	2.95	1.18	18	–	11
40 x 40 x 4	40	4	3.08	2.42	1.12	4.47	1.55	22	–	11
40 x 40 x 5	40	5	3.79	2.97	1.16	5.43	1.91	22	–	11
45 x 45 x 4.5	45	4.5	3.90	3.06	1.25	7.14	2.20	25	–	13
50 x 50 x 4	50	4	3.89	3.06	1.36	8.97	2.46	30	–	13
50 x 50 x 5	50	5	4.80	3.77	1.40	11.0	3.05	30	–	13
50 x 50 x 6	50	6	5.69	4.47	1.45	12.8	3.61	30	–	13
60 x 60 x 5	60	5	5.82	4.57	1.64	19.4	4.45	35	–	17
60 x 60 x 6	60	6	6.91	5.42	1.69	22.8	5.29	35	–	17
60 x 60 x 8	60	8	9.03	7.09	1.77	29.2	6.89	35	–	17
65 x 65 x 7	65	7	8.70	6.83	1.85	33.4	7.18	35	–	21
70 x 70 x 6	70	6	8.13	6.38	1.93	36.9	7.27	40	–	21
70 x 70 x 7	70	7	9.40	7.38	1.97	42.3	8.41	40	–	21
75 x 75 x 6	75	6	8.73	6.85	2.05	45.8	8.41	40	–	23
75 x 75 x 8	75	8	11.4	8.99	2.14	59.1	11.0	40	–	23
80 x 80 x 8	80	8	12.3	9.63	2.26	72.2	12.6	45	–	23
80 x 80 x 10	80	10	15.1	11.9	2.34	87.5	15.4	45	–	23
90 x 90 x 7	90	7	12.2	9.61	2.45	92.6	14.1	50	–	25
90 x 90 x 8	90	8	13.9	10.9	2.50	104	16.1	50	–	25
90 x 90 x 9	90	9	15.5	12.2	2.54	116	17.9	50	–	25
90 x 90 x 10	90	10	17.1	13.4	2.58	127	19.8	50	–	25
100 x 100 x 8	100	8	15.5	12.2	2.74	145	19.9	55	–	25
100 x 100 x 10	100	10	19.2	15.0	2.82	177	24.6	55	–	25
100 x 100 x 12	100	12	22.7	17.8	2.90	207	29.1	55	–	25
120 x 120 x 10	120	10	23.2	18.2	3.31	313	36.0	50	80	25
120 x 120 x 12	120	12	27.5	21.6	3.40	368	42.7	50	80	25
130 x 130 x 12	130	12	30.0	23.6	3.64	472	50.4	50	90	25
150 x 150 x 10	150	10	29.3	23.0	4.03	624	56.9	60	105	28
150 x 150 x 12	150	12	34.8	27.3	4.12	737	67.7	60	105	28
150 x 150 x 15	150	15	43.0	33.8	4.25	898	83.5	60	105	28
160 x 160 x 15	160	15	46.1	36.2	4.49	1100	95.6	60	115	28
180 x 180 x 18	180	18	61.9	48.6	5.10	1870	145	65	135	28
200 x 200 x 16	200	16	61.8	48.5	5.52	2340	162	65	150	28
200 x 200 x 20	200	20	76.3	59.9	5.68	2850	199	65	150	28
200 x 200 x 24	200	24	90.6	71.1	5.84	3330	235	70	150	28
250 x 250 x 28	250	28	133	104	7.24	7700	433	75	150	28

⇒ **L EN 10056-1 – 70 x 70 x 7 – S235J0:** Equal leg steel angle, *a* = 70 mm, *t* = 7 mm, from S235J0

MS

Medium width and wide I-beams

Medium width I-beams (IPE), hot-rolled (selection)

cf. DIN 1025-5 (1994-03)

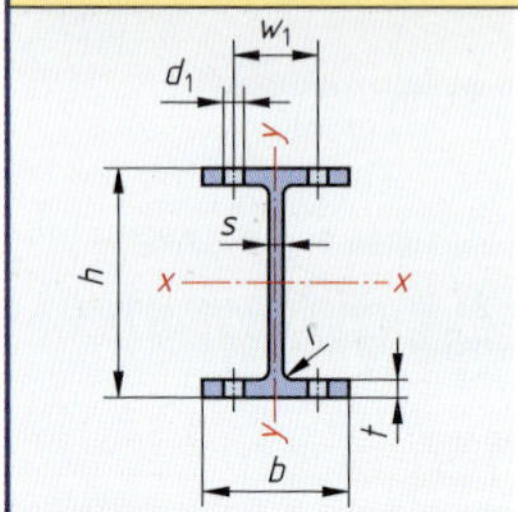

S cross-sectional area
I second moment of inertia
W axial section modulus
m' linear mass density

Material: Unalloyed structural steel DIN EN 10025-2, e.g. S235JR

Delivery type: Standard lengths, 8 m to 16 m ± 50 mm with $h < 300$ mm, 8 m to 18 m ± 50 mm with $h \geq 300$ mm

Designation	Dimensions in mm							For the bending axis				Tracing dimension accord. to DIN 997	
								x – x		y – y			
IPE	*h*	*b*	*s*	*t*	*r*	*S* cm²	*m'* kg/m	I_x cm⁴	W_x cm³	I_y cm⁴	W_y cm³	w_1 mm	d_1 mm
100	100	55	4.1	5.7	7	10.3	8.1	171	34.2	15.9	5.8	30	8.4
120	120	64	4.4	6.3	7	13.2	10.4	318	53.0	27.7	8.7	36	8.4
140	140	73	4.7	6.9	7	16.4	12.9	541	77.3	44.9	12.3	40	11
160	160	82	5.0	7.4	9	20.1	15.8	869	109	68.3	16.7	44	13
180	180	91	5.3	8.0	9	23.9	18.8	1320	146	101	22.2	50	13
200	200	100	5.6	8.5	12	28.5	22.4	1940	194	142	28.5	56	13
240	240	120	6.2	9.8	15	39.1	30.7	3890	324	284	47.3	68	17
270	270	135	6.6	10.2	15	45.9	36.1	5790	429	420	62.2	72	21
300	300	150	7.1	10.7	15	53.8	42.2	8360	557	604	80.5	80	23
360	360	170	8.0	12.7	18	72.7	57.1	16270	904	1040	123	90	25
400	400	180	8.6	13.5	21	84.5	66.3	23130	1160	1320	146	96	28
500	500	200	10.2	16.0	21	116	90.7	48200	1930	2140	214	110	28
600	600	220	12.0	19.0	24	156	122	92080	3070	3390	308	120	28

⇒ **I-profile DIN 1025 – S235JR – IPE 300:** Medium width I-beams with parallel flange surfaces, *h* = 300 mm, from S235JR

MS

Wide I-beams light duty (IPEl), hot-rolled (selection)

cf. DIN 1025-2 (1994-3)

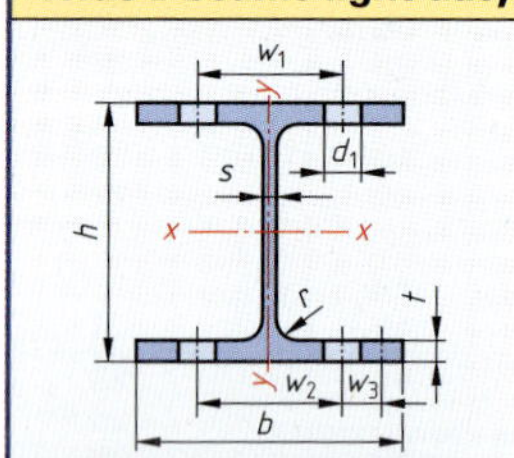

S cross-sectional area
I second moment of inertia
W axial section modulus
m' linear mass density

Material: Unalloyed structural steel DIN EN 10025-2, e.g. S235JR

Delivery type: Standard lengths, 8 m to 16 m ± 50 mm with $h < 300$ mm

$r \approx 3 \cdot s$

Designation	Dimensions in mm						For the bending axis				Tracing dimension accord. to DIN 997			
							x – x		y – y					
IPBl	*h*	*b*	*s*	*t*	*S* cm²	*m'* kg/m	I_x cm⁴	W_x cm³	I_y cm⁴	W_y cm³	w_1	w_2	w_3	d_1
100	96	100	5	8	21.2	16.7	349	72.8	134	26.8	56	–	–	13
120	114	120	5	8	25.3	19.9	606	106	231	38.5	66	–	–	17
140	133	140	5.5	8.5	31.4	24.7	1030	155	389	55.6	76	–	–	21
160	152	160	6	9	38.8	30.4	1670	220	616	76.9	86	–	–	23
180	171	180	6	9.5	45.3	35.5	2510	294	925	103	100	–	–	25
200	190	200	6.5	10	53.8	42.3	3690	389	1340	134	110	–	–	25
240	230	240	7.5	12	76.8	60.3	7760	675	2770	231	–	94	35	25
280	270	280	8	13	97.3	76.4	13670	1010	4760	340	–	110	45	25
320	310	300	9	15.5	124.0	97.6	22930	1480	6990	466	–	120	45	28
400	390	300	11	19	159.0	125.0	45070	2310	8560	571	–	120	45	28
500	490	300	12	23	198.0	155.0	86970	3550	10370	691	–	120	45	28
600	590	300	13	25	226.0	178.0	141200	4790	11270	751	–	120	45	28
800	790	300	15	28	286.0	224.0	303400	7680	12640	843	–	130	40	28

⇒ **I-profile DIN 1025 – S235JR – IPBl 320:** Wide I-beams light duty from S235JR
Designation according to EURONORM 53-62: **HE 320 A**

Wide I-beams

Wide I-beams (IPB), hot-rolled (selection)

cf. DIN 1025-2 (1995-11)

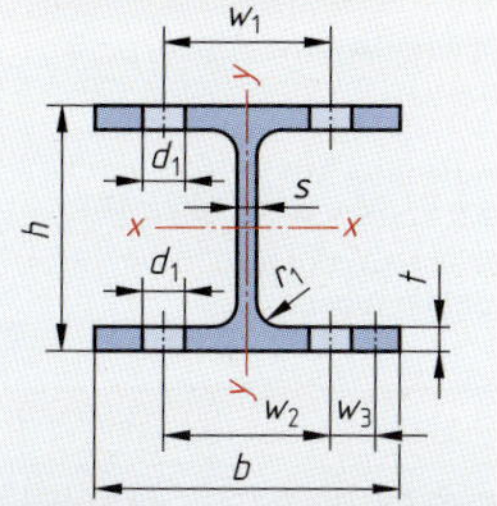

S cross-sectional area
I second moment of inertia
W axial selection modulus
m′ linear mass density

Material: unalloyed structural steel DIN EN 10025-2, e.g. S235JR

Delivery type: standard lengths, 8 m to 16 m ± 50 mm at $h < 300$ mm,
8 m to 18 m ± 50 mm at $h \geq 300$ mm

$r_1 \approx 2 \cdot s$

Designation IPB	Dimensions in mm h	b	s	t	S cm²	m' kg/m	For the bending axis $x-x$: I_x cm⁴	W_x cm³	$y-y$: I_y cm⁴	W_y cm³	Tracing dimension according to DIN 997: w_1 mm	w_2 mm	w_3 mm	d_1 mm
100	100	100	6	10	26.0	20.4	450	89.9	167	33.5	56	–	–	13
120	120	120	6.5	11	34.0	26.7	864	144	318	52.9	66	–	–	17
140	140	140	7	12	43.0	33.7	1510	216	550	78.5	76	–	–	21
160	160	160	8	13	54.3	42.6	2490	311	889	111	86	–	–	23
180	180	180	8.5	14	65.3	51.2	3830	426	1360	151	100	–	–	25
200	200	200	9	15	78.1	61.3	5700	570	2000	200	110	–	–	25
240	240	240	10	17	106	83.2	11260	938	3920	327	–	96	35	25
280	280	280	10.5	18	131	103	19270	1380	6590	471	–	110	45	25
320	320	300	11.5	20.5	161	127	30820	1930	9240	616	–	120	45	28
400	400	300	13.5	24	198	155	57680	2880	10820	721	–	120	45	28
500	500	300	14.5	28	239	187	107200	4290	12620	842	–	120	45	28
600	600	300	15.5	30	270	212	171000	5700	13530	902	–	120	45	28
800	800	300	17.5	33	334	262	359100	8980	14900	994	–	130	40	28

⇒ **I-profile DIN 1025 – S235JR – IPB 240:** Wide I-beam with parallel flange faces, h = 240 mm, made of S235JR, designation according to EURONORM 53-62: **HE 240 B**

MS

Wide I-beams, reinforced version (IPBv) hot-rolled (selection)

cf. DIN 1025-4 (1994-03)

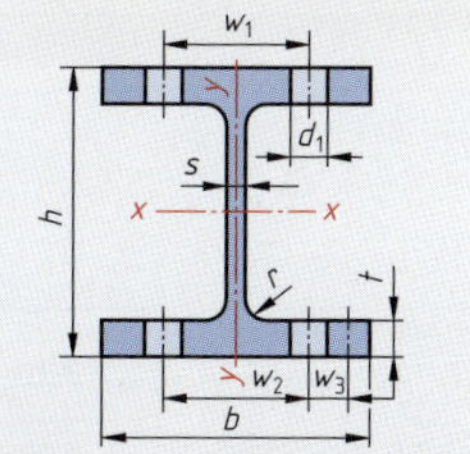

S cross-sectional area
I second moment of inertia
W axial selection modulus
m′ linear mass density

Material: unalloyed structural steel DIN EN 10025-2, e. g. S235JR

Delivery type: standard lengths, 8 m to 16 m ± 50 mm at $h < 300$ mm,
8 m to 16 m ± 50 mm at $h \geq 300$ mm

$r \approx s$

Designation IPBv	Dimensions in mm h	b	s	t	S cm²	m' kg/m	For the bending axis $x-x$: I_x cm⁴	W_x cm³	$y-y$: I_y cm⁴	W_y cm³	Tracing dimension according to DIN 997 in mm: w_1	w_2	w_3	d_1
100	120	106	12	20	53.2	41.8	1140	190	399	75.3	60	–	–	13
120	140	126	12.5	21	66.4	52.1	2020	283	703	112	68	–	–	17
140	160	146	13	22	80.5	63.2	3290	411	1140	157	76	–	–	21
160	180	166	14	23	97.1	76.2	5100	568	1760	212	86	–	–	23
180	200	186	14.5	24	113	88.9	7480	748	2580	277	100	–	–	25
200	220	206	15	25	131	103	10640	967	3650	354	110	–	–	25
240	270	248	18	32	200	157	24290	1800	8150	657	–	100	35	25
280	310	288	18.5	33	240	189	39550	2550	13160	914	–	116	45	25
320	359	309	21	40	312	245	68130	3800	19710	1280	–	126	47	28
400	432	307	21	40	319	250	104100	4820	19340	1260	–	126	47	28
500	524	306	21	40	344	270	161900	6180	19150	1250	–	130	45	28
600	620	305	21	40	364	285	237400	7660	18280	1240	–	130	45	28
800	814	303	21	40	404	317	442600	10870	18630	1230	–	132	42	28

⇒ **I-profile DIN 1025 – S235JR – IPBv 400:** Wide I-beam, reinforced version, made of S235JR, designation according to EURONORM 53-62: **HE 400 M**

Tubes

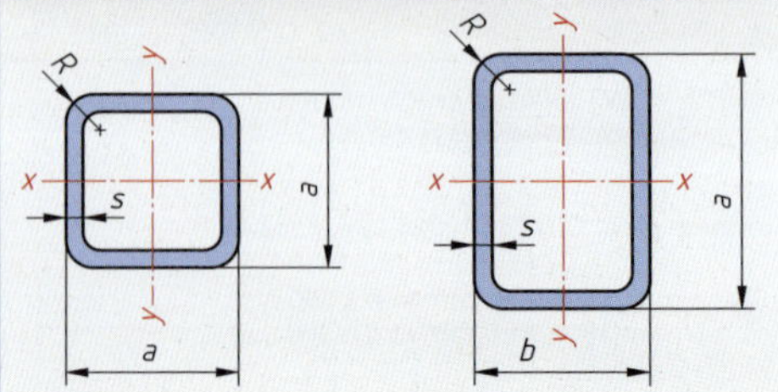

Material: Unalloyed structural steel DIN EN 10025

Delivery type: DIN EN 10210-2
manufactured lengths 4 m to 16 m, profile dimensions $a \times a$ = 20 x 20 to 400 x 400
DIN EN 10219-2
manufactured lengths 4 m to 16 m, profile dimensions $a \times a$ = 20 x 20 to 400 x 400

DIN EN 10210 and DIN EN 10219 also contain circular tubes, along with square and rectangular tubes.

Hot worked square and rectangular tubes

cf. DIN EN 10210-2 (1997-11)

Nominal dimension $a \times a$ $a \times b$ mm	Wall thickness s mm	Linear mass density m' kg/m	Cross section S cm²	Area moments and section moduli for the bending axes $x-x$: I_x cm⁴	W_x cm³	$y-y$: I_y cm⁴	W_y cm³	for torsion: I_p cm⁴	W_p cm³
40 x 40	3.0	3.41	4.34	9.78	4.89	9.78	4.89	15.7	7.10
	4.0	4.39	5.59	11.8	5.91	11.8	5.91	19.5	8.54
50 x 50	2.5	3.68	4.68	17.5	6.99	17.5	6.99	27.5	10.2
	3.0	4.35	5.54	20.2	8.08	20.2	8.08	32.1	11.8
60 x 60	3.0	5.29	6.74	36.2	12.1	36.2	12.1	56.9	17.7
	4.0	6.90	8.79	45.4	15.1	45.4	15.1	72.5	22.0
	5.0	8.42	10.7	53.3	17.8	53.3	17.8	86.4	25.7
50 x 30	3.0	3.41	4.34	13.6	5.43	5.94	3.96	13.5	6.51
	4.0	4.39	5.59	16.5	6.60	7.08	4.72	16.6	7.77
60 x 40	3.0	4.35	5.54	26.5	8.82	13.9	6.95	29.2	11.2
	4.0	5.64	7.19	32.8	10.9	17.0	8.52	36.7	13.7
80 x 40	4.0	6.90	8.79	68.2	17.1	22.2	11.1	55.2	18.9
	5.0	8.42	10.7	80.3	20.1	25.7	12.9	65.1	21.9
	6.0	9.87	12.6	90.5	22.6	28.5	14.2	73.4	24.2
100 x 50	4.0	8.78	11.2	140	27.9	46.2	18.5	113	31.4
	5.0	10.8	13.7	167	33.3	54.3	21.7	135	36.9

⇒ **Tube DIN EN 10210 – 60 x 60 x 5 – S355J0:** Square tube, a = 60 mm, s = 5 mm, made of S355J0

Cold worked, welded, square and rectangular tubes

cf. DIN EN 10219-2 (1997-11)

Nominal dimension $a \times a$ $a \times b$ mm	Wall thickness s mm	Linear mass density m' kg/m	Cross section S cm²	Area moments and section moduli for the bending axes $x-x$: I_x cm⁴	W_x cm³	$y-y$: I_y cm⁴	W_y cm³	for torsion: I_p cm⁴	W_p cm³
30 x 30	2.0	1.68	2.14	2.72	1.81	2.72	1.81	4.54	2.75
	2.5	2.03	2.59	3.16	2.10	3.16	2.10	5.40	3.20
	3.0	2.36	3.01	3.50	2.34	3.50	2.34	6.15	3.58
40 x 40	2.0	2.31	2.94	6.94	3.47	6.94	3.47	11.3	5.23
	2.5	2.82	3.59	8.22	4.11	8.22	4.11	13.6	6.21
	3.0	3.30	4.21	9.32	4.66	9.32	4.66	15.8	7.07
	4.0	4.20	5.35	11.1	5.54	11.1	5.54	19.4	8.48
80 x 80	3.0	7.07	9.01	87.8	22.0	87.8	22.0	140	33.0
	4.0	9.22	11.7	111	27.8	111	27.8	180	41.8
	5.0	11.3	14.4	131	32.9	131	32.9	218	49.7
40 x 20	2.0	1.68	2.14	4.05	2.02	1.34	1.34	3.45	2.36
	2.5	2.03	2.59	4.69	2.35	1.54	1.54	4.06	2.72
	3.0	2.36	3.01	5.21	2.60	1.68	1.68	4.57	3.00
60 x 40	3.0	4.25	5.41	25.4	8.46	13.4	6.72	29.3	11.2
	4.0	5.45	6.95	31.0	10.3	16.3	8.14	36.7	13.7
	5.0	6.56	8.36	35.3	11.8	18.4	9.21	42.8	15.6
80 x 40	3.0	5.19	6.61	52.3	13.1	17.6	8.78	43.9	15.3
	4.0	6.71	8.55	64.8	16.2	21.5	10.7	55.2	18.8
	5.0	8.13	10.4	75.1	18.8	24.6	12.3	65.0	21.7
100 x 40	3.0	6.13	7.81	92.3	18.5	21.7	10.8	59.0	19.4
	4.0	7.97	10.1	116	23.1	26.7	13.3	74.5	24.0
	5.0	9.70	12.4	136	27.1	30.8	15.4	87.9	27.9

⇒ **Tube DIN EN 10219 – 60 x 40 x 4 – S355J0:** Rectangular tube, a = 60 mm, b = 40 mm, s = 4 mm, made of S355J0

Linear mass density and area mass density

Linear mass density[1] (Table values for steel with density ϱ = 7.85 kg/dm³)

d diameter *m'* linear mass density *a* length of side SW widths across flats

Steel wire						Round steel bar					
d mm	*m'* kg/1000 m	*d* mm	*m'* kg/1000 m	*d* mm	*m'* kg/1000 m	*d* mm	*m'* kg/m	*d* mm	*m'* kg/m	*d* mm	*m'* kg/m
0.10	0.062	0.55	1.87	1.1	7.46	3	0.055	18	2.00	60	22.2
0.16	0.158	0.60	2.22	1.2	8.88	4	0.099	20	2.47	70	30.2
0.20	0.247	0.65	2.60	1.3	10.4	5	0.154	25	3.85	80	39.5
0.25	0.385	0.70	3.02	1.4	12.1	6	0.222	30	5.55	100	61.7
0.30	0.555	0.75	3.47	1.5	13.9	8	0.395	35	7.55	120	88.8
0.35	0.755	0.80	3.95	1.6	15.8	10	0.617	40	9.86	140	121
0.40	0.986	0.85	4.45	1.7	17.8	12	0.888	45	12.5	150	139
0.45	1.25	0.90	4.99	1.8	20.0	15	1.39	50	15.4	160	158
0.50	1.54	1.0	6.17	2.0	24.7	16	1.58	55	18.7	200	247

Flat steel bar						Hexagonal steel bar					
a mm	*m'* kg/m	*a* mm	*m'* kg/m	*a* mm	*m'* kg/m	SW mm	*m'* kg/m	SW mm	*m'* kg/m	SW mm	*m'* kg/m
6	0.283	20	3.14	40	12.6	6	0.245	20	2.72	40	10.9
8	0.502	22	3.80	50	19.6	8	0.435	22	3.29	50	17.0
10	0.785	25	4.91	60	28.3	10	0.680	25	4.25	60	24.5
12	1.13	28	6.15	70	38.5	12	0.979	28	5.33	70	33.3
14	1.54	30	7.07	80	50.2	14	1.33	30	6.12	80	43.5
16	2.01	32	8.04	90	63.6	16	1.74	32	6.96	90	55.1
18	2.54	35	9.62	100	78.5	18	2.20	35	8.33	100	68.0

Linear mass density of special profiles

Profile		Page	Profile		Page
Tee	EN 10055	145	Tubes	EN 10210-2	150
Angles, equal legs	EN 10056-1	147	Tubes	EN 10219-2	150
Angles, unequal legs	EN 10056-1	146	Aluminum round bars	DIN 1798	168
Steel channel	DIN1026-1	145	Aluminum square bars	DIN 1796	168
I-beams IPE	DIN 1025-5	148	Aluminum flat bars	DIN 1769	169
I-beams IPB	DIN 1025-2	148	Aluminum round tube	DIN 1795	170
I-beams, narrow	DIN 1025-1	149	Aluminum channel	DIN 9713	170

Area mass density[1] (Table values for steel with density ϱ = 7.85 kg/dm³)

Sheet

s sheet thickness *m''* area mass density

s mm	*m''* kg/m²	*s* mm	*m''* kg/m²	*s* mm	*m''* kg/m²	*s* mm	*m''* kg/m²	*s* mm	*m''* kg/m²	*s* mm	*m''* kg/m²
0.35	2.75	0.70	5.50	1.2	9.42	3.0	23.6	4.75	37.3	10.0	78.5
0.40	3.14	0.80	6.28	1.5	11.8	3.5	27.5	5.0	39.3	12.0	94.2
0.50	3.93	0.90	7.07	2.0	15.7	4.0	31.4	6.0	47.1	14.0	110
0.60	4.71	1.0	7.85	2.5	19.6	4.5	35.3	8.0	62.8	15.0	118

[1] Table values can be calculated for a different material by taking a ratio of its density to the density of steel (7,85 kg/dm³).

Example: Sheet metal with s = 4.0 mm of AlMg3Mn (density 2.66 kg/dm³). From the table: m'' = 31.4 kg/m² for steel.
$AlMg_3Mn$: m'' = 31.4 kg/m² · 2.66 kg/dm³/7.85 kg/dm³ = **10.64 kg/m²**

MS

Iron-carbon phase diagram

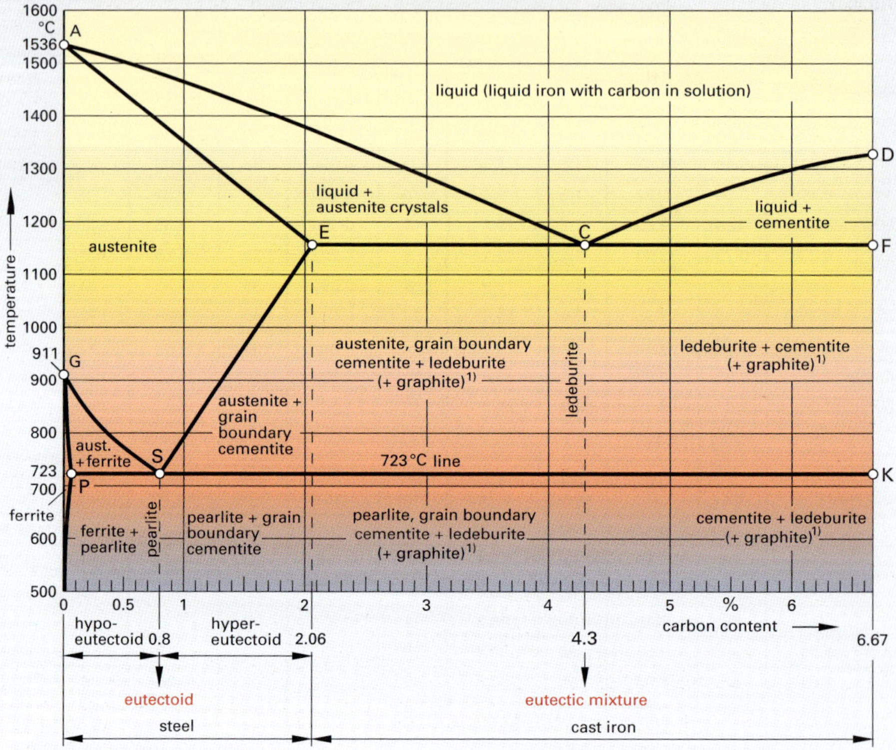

1) For iron types with a C content over 2.06% (cast iron) and additional Si content, a portion of the unalloyed precipitates in the form of graphite.

Heat treatment of steel

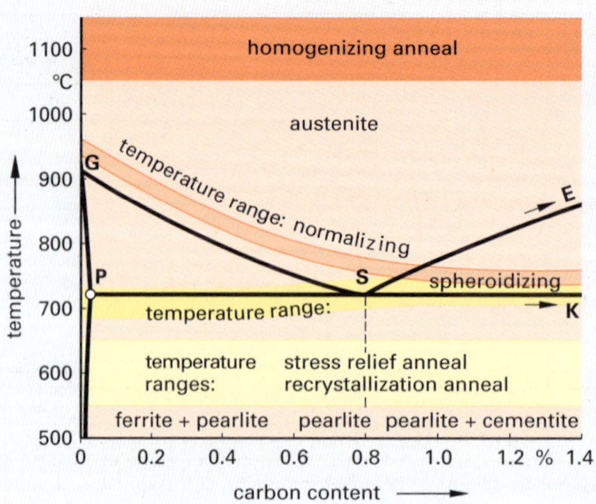

Microstructures of unalloyed steel

Carbon content and crystalline structure
Etchant: 3% nitric acid /alcohol solution
Magnification approx. 500 : 1

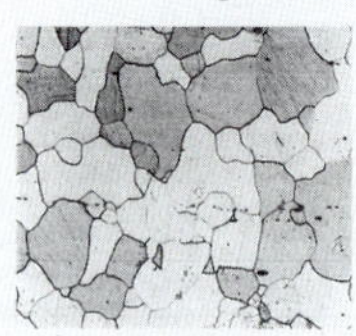

0.1% C
ferrite

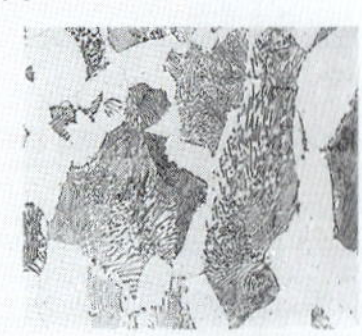

0.45% C
ferrite + pearlite

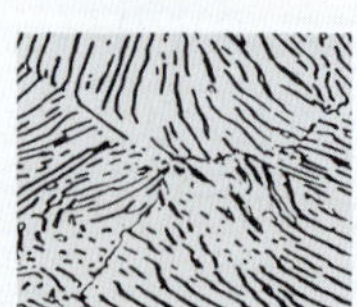

0.8% C
pearlite

1.3% C
pearlite + grain boundary cementite

Heat treatment of steels – Overview

Illustration	Short description	Application, information[1]
Normalizing		
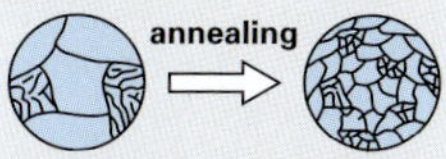	• **Heat** and hold at annealing temperature → structural transformation (austenite) • Controlled **cooling** to room temperature → fine-grained normal structure	To normalize coarse grain structures in rolled, cast, welded and forged products
Spheroidizing		
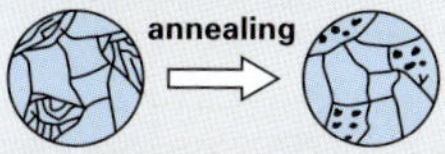	• **Heat** to annealing temperature, hold at temperature or cycle anneal → spheroidizing of the cementite • **Cool down** to room temperature	To improve cold workability, machinability and hardenability; can be used for all steels
Stress relief anneal		
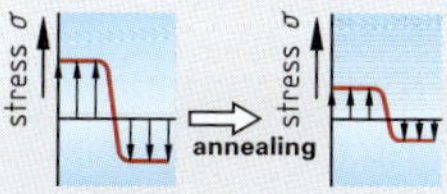	• **Heat** and hold at annealing temperature (below structure transition) → stress relief by plastic deformation of the workpieces • **Cool down** to room temperature	To reduce internal stresses in welded, cast and forged parts; can be used for all steels
Hardening		
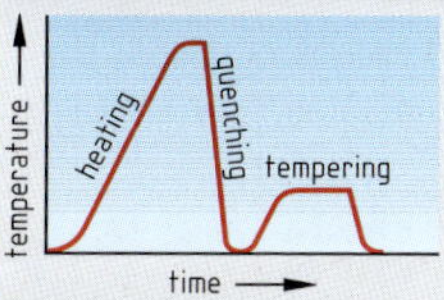	• **Heat** and hold at hardening temperature → structural transformation (austenite) • **Quench** in oil, water, air → hard, brittle, fine structure (martensite) • **Temper** at 100 to 300 °C (212 to 572 °F) → transformation of martensite, higher toughness, working hardness	For parts subject to wear stress, e.g. tools, springs, guideways, press forms; steels suitable for heat treatment with C > 0.3%, e.g. C70U, 102Cr6, C45E, HS6-5-2C, X38CrMoV5-3
Quenching and tempering		
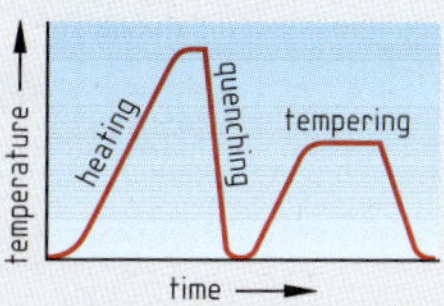	• **Heat** and hold at hardening temperature → structural transformation (austenite) • **Quench** in oil, water, air → hard, brittle, fine-grain structure (martensite), for larger sized parts fine core structure (bainite) • **Temper** at 400 to 650 °C (752 to 1202 °F) → martensite reduction, fine structure, high strength with good toughness	Usually used for dynamically loaded workpieces with high strength and good toughness, e.g. shafts, gears, screws; quenched and tempered steels, see page 130, nitriding steels, see page 131, steels for flame and induction hardening, see page 131, steels for heat-treatable springs, see page 135
Case hardening		
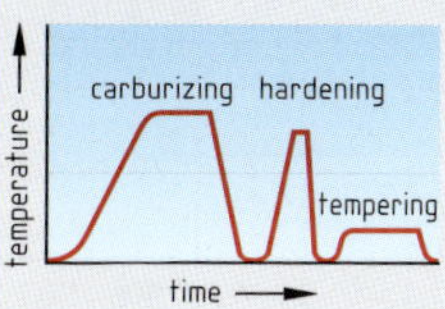	• **Carburize** machined workpieces on the surface layer • **Cool** to room temperature → normal structure (ferrite, pearlite, carbides) • **Harden** (for procedure see hardening) → surface hardening: heat to surface hardening temperature → core hardening: heat to hardening temperature of the core area	For workpieces with wear-resistant surfaces, high fatigue strength and good core strength, e.g. gears, shafts, bolts; **surface hardening:** high wear-resistance, low core strength **core hardening:** high core strength, hard brittle surface; case hardened steels, see page 129, free cutting steels, see page 131
Nitriding		
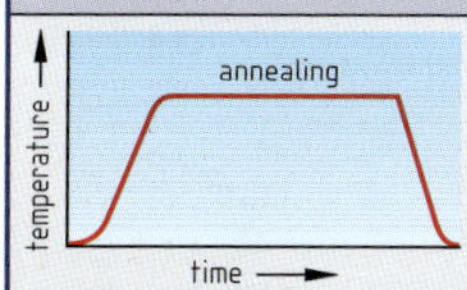	• **Anneal** usually finish-machined workpieces in nitrogen-producing atmospheres → formation of hard, wear-resistant and temperature-resistant nitrides • **Cool** in still air or in nitrogen stream	For workpieces with wear-resistant surfaces, high fatigue strength and good temperature-resistance, e.g. valves, piston rods, spindles; nitriding steels, see page 131

[1] For annealing and tempering temperatures, quenching media and attainable hardness values, see pages 154 to 156.

MS

Tool steels, Case hardened steels

Heat treatment of unalloyed cold work steels

cf. DIN EN ISO 4957 (2001-02)

Steel type: Designation	Steel type: Material number	Hot working temperature °C	Spheroidizing: Temperature °C	Spheroidizing: Hardness HB max.	Hardening: Temperature °C	Hardening: Cooling medium	Hardening: Case harden. depth[1] mm	Hardening: Full harden. up to Ø mm	Surface hardness in HRC ≈ after hardening	after tempering[2] at 100 °C	200 °C	300 °C
C45U	1.1730	1000–800	680–710	207	800–820	water	3.5	15	58	58	54	48
C70U	1.1520			183	790–810		3.0	10	64	63	60	53
C80U	1.1525	1050–800	680–710	192	780–800	water	3.0	10	64	64	60	54
C90U	1.1535	1050–800		207	770–790				64	64	61	54
C105U	1.1545	1000–800		212	770–790				65	64	62	56

[1] For diameters of 30 mm.

[2] The tempering temperature is set according to the application and the desired working hardness. The steels are normally delivered spheroidized.

Heat treatment of alloy cold work steels, hot work steels and high-speed steels

cf. DIN EN ISO 4957 (2001-02)

Steel type: Designation	Steel type: Material number	Hot working temperature °C	Spheroidizing: temperature °C	Spheroidizing: Hardn. HB max.	Hardening: temperature[1] °C	Hardening: cooling medium	Surface hardness in HRC ≈ after hardening	after tempering[2] at 200 °C	300 °C	400 °C	500 °C	550 °C
105V	1.2834	1050–850	710–750	212	780–800	water	68	64	56	48	40	36
X153CrMoV12	1.2379		800–850	255	1010–1030	air	63	61	59	58	58	56
X210CrW12	1.2436	1050–850	800–840	255	96–980	oil	64	62	60	58	56	52
90MnCrV8	1.2842		680–720	229	780–800		65	62	56	50	42	40
102Cr6	1.2067		710–750	223	830–850		65	62	57	50	43	40
60WCrV8	1.2550	1050–850	710–750	229	900–920	oil	62	60	58	53	48	46
X37CrMoV5-1	1.2343	1100–900	750–800	229	1010–1030		53	52	52	53	54	52
HS6-5-2C	1.3343	1100–900	770–840	269	1200–1220	oil, hot bath, air	64	62	62	62	65	65
HS10-4-3-10	1.3207			302	1220–1240		66	61	61	62	66	67
HS2-9-1-8	1.3247			277	1180–1200		66	62	62	61	68	69

[1] The austenitizing time is the holding time at hardening temperature, which is approx. 25 min for cold work steels and approx. 3 min. for high-speed steels. Heating is performed in stages.

[2] High-speed steels are tempered at least twice at 540–570 °C. Holding time at this temperature is at least 60 min.

Heat treatment of case hardened steels

cf. DIN EN 10084 (2008-06)

Steel type[1]: Designation	Steel type[1]: Material number	Carburizing temperature °C	Hardening: Core harden. temperature °C	Hardening: Surf. harden. temperature °C	Tempering °C	Quenching medium	End quench test: Temp. °C	Hardness HRC at distance of: max.[2]	3 mm	5 mm	7 mm
C10E	1.1121	880–980	880–920	780–820	150–200	water	–	–	–	–	–
C15E	1.1141						–	–	–	–	–
17Cr3	1.7016		860–900			oil	880	47	44	40	33
16MnCr5	1.7131						870	47	46	44	41
20MnCr5	1.7147						870	49	49	48	46
20MoCr4	1.7321						910	49	47	44	41
17CrNi6-6	1.5918		830–870				870	47	47	46	45
15NiCr13	1.5752		840–880				880	48	48	48	47
20NiCrMo2-2	1.6523		860–900				920	49	48	45	42
18CrNiMo7-6	1.6587		830–870				860	48	48	48	48

[1] The same values apply to steels with controlled sulfur content, e.g. C10R, 20MnCrS5.

[2] For steels with normal hardenability (+H) at a distance of 1.5 mm from the end face.

Quenched and tempered steels

Heat treatment of unalloyed quenched and tempered steels

cf. DIN EN 10083-2 (2006-10)[1]

Steel types[2] Designation	Material number	Normalizing °C	End quench test °C	Hardness HRC at hardening depth in mm[3] 1	3	5	Quenching and tempering: Hardening[4] °C	Quenching medium	Tempering[5] °C
C22E	1.1151	880 – 940	–	–	–	–	860 – 900	water	550 – 660
C35E[1]	1.1181	860 – 920	870	48 – 58	33 – 55	22 – 49	840 – 880	water or oil	550 – 660
C40E	1.1186	850 – 910	870	51 – 60	35 – 59	25 – 53	830 – 870		
C45E[1]	1.1191	840 – 900	850	55 – 62	37 – 61	28 – 57	820 – 860		
C50E[1]	1.1206	830 – 890	850	56 – 63	44 – 61	31 – 58	810 – 850	oil or water	550 – 660
C55E[1]	1.1203	825 – 885	830	58 – 65	47 – 63	33 – 60	810 – 850		
C60E	1.1221	820 – 880	830	60 – 67	50 – 65	35 – 62	810 – 850		
28Mn6	1.1170	850 – 890	850	45 – 54	42 – 53	37 – 51	840 – 880	water or oil	540 – 680

Heat treatment of quenched and tempered alloy steels (selection)

cf. DIN EN 10083-3 (2007-01)[1]

Steel types[2] Designation	Material number	Surface hardness[6] HRC	End quench test °C	Hardness HRC at hardening depth in mm[3] 1.5	5	15	Quenching and tempering: Hardening[4] °C	Quenching medium	Tempering[5] °C
38Cr2	1.7003	–	850	51 – 59	37 – 54	– 35	830 – 870	oil or water	540 – 680
46Cr2[1]	1.7006	54		54 – 63	40 – 59	22 – 39	820 – 860	oil or water	
34Cr4	1.7033	–	850	49 – 57	45 – 56	27 – 44	830 – 870	water or oil	540 – 680
37Cr4[1]	1.7034	51		51 – 59	48 – 58	31 – 48	825 – 865	oil or water	
41Cr4[1]	1.7035	53		53 – 61	50 – 60	32 – 52	820 – 860	oil or water	
25CrMo4	1.7218	–	850	44 – 52	40 – 51	27 – 41	840 – 900	water or oil	540 – 680
34CrMo4	1.7220	–		49 – 57	48 – 57	34 – 52	830 – 890	oil or water	
42CrMo4[1]	1.7225	53		53 – 61	52 – 61	37 – 58	820 – 880	oil or water	
50CrMo4[1]	1.7228	58	850	58 – 65	57 – 64	48 – 62	820 – 870	oil	540 – 680
51CrV4	1.8159	–		57 – 65	56 – 64	48 – 62	820 – 870	oil	
39NiCrMo3	1.6510	–		52 – 60	50 – 59	43 – 56	820 – 850	oil or water	
34CrNiMo6	1.6582	–	850	50 – 58	50 – 58	48 – 57	830 – 860	oil or water	540 – 660
30CrNiMo8	1.6580	–		48 – 56	48 – 56	46 – 55	830 – 860	oil or water	540 – 660
36NiCrMo16	1.6773	–		50 – 57	48 – 56	47 – 55	865 – 885	air or oil	550 – 650
38MnB5	1.5532	–	850	52 – 60	50 – 59	31 – 47	840 – 880	water/oil	400 – 600
33MnCrB5-2	1.7185	–	880	48 – 57	47 – 57	41 – 54	860 – 900	oil	400 – 600

[1] DIN 17212 „Steels for flame and induction hardening" was withdrawn without replacement. More information about steels for flame and induction hardening on pages 130 and 131 in the section „Quenched and tempered steels".
[2] Identical values apply to the high-grade steels C35 to C60 and steels with controlled sulphur content, such as C35R.
[3] Hardenability requirements: +H normal hardenability
[4] The lower temperature range applies to quenching in water, the higher range to quenching in oil.
[5] The tempering time is 60 minutes minimum.
[6] Minimum surface hardness of the steel after flame or induction hardening.

Hardenability and hardening depth of quenched and tempered steels (scatter bands)

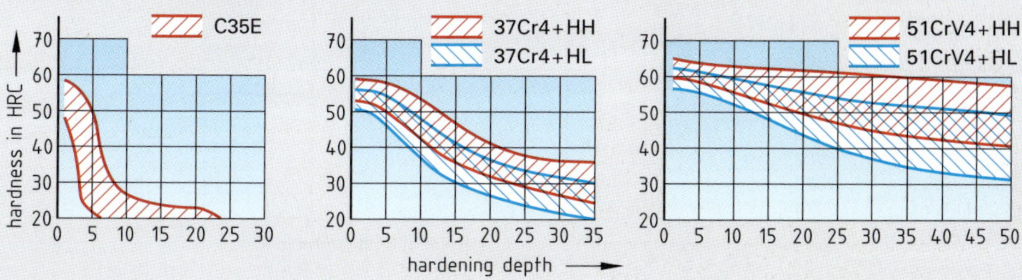

Nitriding steels, Free cutting steels, Aluminum alloys

Heat treatment of nitriding steels

cf. DIN EN 10085 (2001-07)

Steel type: Designation	Steel type: Material number	Heat treatment before nitriding: Spheroid. temperature °C	Quenching and tempering: Hardening Temperature[2] °C	Quenching and tempering: Tempering Quenching medium	Quenching and tempering: Tempering temperature[3)4)] °C	Nitriding treatment[1]: Gas nitriding °C	Nitriding treatment[1]: Nitrocarburizing °C	Nitriding treatment[1]: Hardness[5] HV1
24CrMo13-6	1.8516	650–700	870–970	oil or water	580–700	500–600	570–650	–
31CrMo12	1.8515	650–700	870–930					800
32CrAlMo7-10	1.8505	650–750	870–930					–
31CrMoV9	1.8519	680–720	870–930					800
33CrMoV12-9	1.8522	680–720	870–970					–
34CrAlNi7-10	1.8550	650–700	870–930					950
41CrAlMo7-10	1.8509	650–750	870–930					950
40CrMoV13-9	1.8523	680–720	870–970					–
34CrAlMo5-10	1.8507	650–750	870–930					950

1) The nitriding time is a function of the desired nitriding hardness depth.
2) Austenitizing time at least 0.5 hours.
3) Tempering time at least 1 hour.
4) The tempering temperature should not be less than 50 °C above the nitriding temperature.
5) Hardness of the nitrided surface.

Heat treatment of free cutting steels

cf. DIN EN 10087 (1999-01)

Free cutting case hardened steels

Steel type: Designation	Steel type: Material number	Carburizing temperature °C	Core hardening temperature °C	Surface harden. temperature °C	Quenching medium[1]	Tempering temperature[2] °C
10S20 10SPb20 15SMn13	1.0721 1.0722 1.0725	880–980	880–920	780–820	water, oil, emulsion	150–200

Free cutting quenched and tempered steels

Steel type: Designation	Steel type: Material number	Hardness temperature °C	Quenching medium[1]	Quench. and temp. temperat. °C	Quenched and tempered[3]: R_e N/mm²	Quenched and tempered[3]: R_m N/mm²	Quenched and tempered[3]: A %
35S20 35SPb20	1.0726 1.0756	860–890	water or oil	540–680	430	630–780	15
36SMn14 36SMnPb14	1.0764 1.0765	850–880			460	700–850	14
38SMn28 38SMnPb28	1.0760 1.0761	850–880	oil or water		460		15
44SMn28 44SMnPb28	1.0762 1.0763	840–870			480		16
46S20	1.0757				490		12

1) The choice of quenching medium depends on the shape of the workpiece. 2) Tempering time at least 1 hour.
3) Values apply to diameters $10 < d \leq 16$.

Hardening of aluminum alloys

Alloy EN AW-: Designation	Alloy EN AW-: Material number	Type of age hardening[2]	Solution annealing temperature °C	Artificial aging: temperature °C	Artificial aging: holding time h	Natural aging time days	Age hardened: R_m N/mm²	Age hardened: A %
Al Cu4MgSi	2017	T4	500	100–300	8–24	5–8	390	12
Al Cu4SiMg	2014	T6				–	420	8
Al MgSi	6060	T4	525			5–8	130	15
Al MgSi1MgMn	6082	T6				–	280	6
Al Zn4,5Mg1	7020	T6	470			–	210	12
Al Zn5,5MgCu	7075	T6				–	545	8
Al Si7Mg[1]	42000[1]	T6	525			4	250	1

1) Aluminum casting alloy EN AC-Al Si7Mg or EN AC 42000.
2) T4 solution annealed and naturally aged; T6 solution annealed and artificially aged.

Designation system for cast iron materials

Designations and material numbers

cf. DIN EN 1560 (1997-08)

Cast iron materials are referenced either with a designation or a material number.

Example:

Cast iron with flake graphite, tensile strength R_m = 300 N/mm²

Designation	Material number
EN-GJL-300	EN-JL1050

Material designations

Material designations have up to six characters without spaces, beginning with **EN** (European standard) and **GJ** (cast iron; I iron)

Designation example:

EN	-	GJ	L		-	350		Cast iron with flake graphite
EN	-	GJ	L		-	HB155		Cast iron with flake graphite
EN	-	GJ	S		-	350-22U		Cast iron with spheroidal graphite (ductile Iron)
EN	-	GJ	M	B	-	450-6		Malleable cast iron – blackheart
EN	-	GJ	M	W	-	360-12	W	Malleable cast iron – whiteheart
EN	-	GJ	M		-	HV600(XCr14)		Wear-resistant cast iron
EN	-	GJ	L	A	-	XNiCuCr15-6-2		Austenitic cast iron

Graphite structure (letter)

- L flake graphite
- S spheroidal graphite
- M temper carbon
- V vermicular-graphite
- N no graphite
- Y special structure

Microstructure or macrostructure (letter)

- A austenite
- F ferrite
- P pearlite
- M martensite
- L ledeburite
- Q quenched
- T quenched and tempered
- B not decarburized
- W decarburized

Mechanical properties or chemical composition (numbers/letters)

Mechanical properties

350	minimum tensile strength R_m in N/mm²
350-22	additional elongation at fracture *EL* in %
S	**Test specimen** cast separately
U	cast-on
C	taken from the casting
HB155	max. hardness

Chemical composition

Data are based on steel designations, see pages 121 and 122.

Additional requirements

- D rough casting
- H heat treated casting
- W weldable
- Z additional requirements

Material numbers

Material numbers have seven characters without spaces, beginning with **EN** (European standard) and **J** (iron; I iron)

Designation examples:

EN	-	J	L	2	0 4	7	Cast iron with flake graphite and hardness as characteristic spheroidal
EN	-	J	S	1	0 2	2	graphite casting with cast-on test specimen, characteristic R_m
EN	-	J	M	1	1 3	0	Malleable cast iron without special requirements, characteristic R_m

Graphite structure (letter)

- L flake graphite
- S spheroidal graphite
- M temper carbon
- V vermicular graphite
- N no graphite
- Y special structure

Main characteristic (number)

- 1 tensile strength
- 2 hardness
- 3 chemical composition

Material characteristic number

Every cast iron material is assigned a two-digit number. A higher number indicates a higher strength.

Material requirements (number)

- 0 no special requirements
- 1 separately cast test specimen
- 2 cast-on test specimen
- 3 test specimen taken from the casting
- 4 toughness at room temperature
- 5 toughness at low temperature
- 6 specified weldability
- 7 rough casting
- 8 heat treated casting
- 9 additional requirements

MS

MS

Classification of Cast Iron Materials

Type	Standard	Examples/ material number	Tensile strength R_m N/mm²	Properties	Application examples
Cast iron					
with flake graphite (gray iron)	DIN EN 1561	EN-GJL-150 (GG-15)[1] EN-JL1020	100 to 450	Very good castability, good compression strength, damping capacity, emergency running properties, and good corrosion resistance	For complex workpieces with many contours; very versatile in its applications. Machine frames, gear housings
with spheroidal graphite	DIN EN 1563	EN-GJS-400 (GGG-40)[1] EN-JS1030	350 to 900	Very good castability, high strength even with dynamic loading, surface hardenable	Wear stressed workpieces; clutch parts, fittings, engine/motor construction
with vermicular graphite	ISO 16112	ISO 16112/JV/300	300 to 500	Very good castability, high strength without expensive alloying additions	Automotive parts, engine/motor construction, gear housings
bainitic cast iron	DIN EN 1564	EN-GJS-800-8 EN-JS1100	800 to 1400	Heat treatment and controlled cooling produce bainite and austenite for high strength and good toughness	Highly stressed parts, e.g. wheel hubs, gear rings, ADI castings[2]
wear-resistant castings, white cast iron	DIN EN 12513	EN-GJN-HV350 EN-JN2019	> 1000	Wear-resistant due to martensite and carbides, also alloyed with Cr and Ni	Wear-resistant cast iron, e.g. dressing rolls, dredging shovels, impellers for pumps
Malleable cast iron					
decarburized (whiteheart)	DIN EN 1562	EN-GJMW-350 (GTW-35)[1] EN-JM1010	270 to 570	Decarburization of the surface by tempering. High strength and toughness, ductile	True to shape, thin-walled, impact-loaded parts; levers, brake drums
not decarburized (blackheart)	DIN EN 1562	EN-GJMB-450 (GTS-45)[1] EN-JM1140	300 to 800	Cluster graphite in entire cross-section due to malleablizing. High strength and toughness in larger wall thickness	True to shape, thick walled, impact stressed parts; levers, universal joint yokes
Cast steel					
for general use	DIN EN 10293[3]	GE240 1.0446	380 to 600	Unalloyed and low alloy cast steel for general use	Minimum mechanical values from −10 °C to 300 °C
with improved weldability	DIN EN 10293[4]	G20Mn5 1.6220	430 to 650	Lower carbon content with manganese and microalloy	Welded assembly construction, fine-grain structural steels with larger wall thickness
quenched and tempered cast steel	DIN EN 10293[5]	G30CrMoV6-4 1.7725	500 to 1250	Fine quenched and tempered structure with high toughness	Chains, armor-plating
for pressure vessels	DIN EN 10213	GP280GH 1.0625	420 to 960	Types with high strength and toughness at low and high temperatures	Pressure vessels for hot or cold media, high temperature resistant and tough at low temperatures; rustproof
stainless	DIN EN 10283	GX6CrNi26-7 1.4347	450 to 1100	Resistant to chemical attack and corrosion	Pump impellers in acids, duplex steel
heat-resistant	DIN EN 10295	GX25CrNiSi18-9 1.4825	400 to 550	Resistant to scaling gases	Turbine parts, furnace grates

[1] previous designation [2] ADI → Austempered Ductile Iron
[3] Replaces DIN 1681 [4] Replaces DIN 17182 [5] Replaces DIN 17205

Cast iron with flake graphite, Cast iron with spheroidal graphite

Cast iron with flake graphite (gray iron)

cf. DIN EN 1561 (1997-08)

Tensile strength R_m as identifying characteristic				Hardness HB as identifying characteristic			
Type		Wall thickness	Tensile strength R_m	Type		Wall thickness	Brinell hardness
Designation	Material number	mm	N/mm²	Designation	Material number	mm	HB30
EN-GJL-100	EN-JL1010	5–40	100–200	EN-GJL-HB155	EN-JL2010	40–80	max. 155
EN-GJL-150	EN-JL1020	2.5–300	150–250	EN-GJL-HB175	EN-JL2020	40–80	100–175
EN-GJL-200	EN-JL1030	2.5–300	200–300	EN-GJL-HB195	EN-JL2030	40–80	120–195
EN-GJL-250	EN-JL1040	5–300	250–350	EN-GJL-HB215	EN-JL2040	40–80	145–215
EN-GJL-300	EN-JL1050	10–300	300–400	EN-GJL-HB235	EN-JL2050	40–80	165–235
EN-GJL-350	EN-JL1060	10–300	350–450	EN-GJL-HB255	EN-JL2060	40–80	185–255

⇒ **EN-GJL-100:** Cast iron with flake graphite (gray iron), minimum tensile strength R_m = 100 N/mm²

⇒ **EN-GJL-HB215:** Cast iron with flake graphite (gray iron), maximum Brinell hardness = 215 HB

Properties

Good castability and machinability, vibration damping, corrosion resistance, high compression strength, good sliding properties.

Application examples

Machine frames, bearing housings, plain bearings, pressure-resistant parts, turbine housings.
Hardness as characteristic property provides information on the machinability.

Cast iron with spheroidal (nodular) graphite

cf. DIN EN 1563 (2005-10)

Tensile strength R_m as identifying characteristic

Type		Tensile strength R_m	Yield strength $R_{p\ 0.2}$	Elongation EL	Properties, application examples
Designation	Material number	N/mm²	N/mm²	%	
EN-GJS-350-22-LT[1]	EN-JS1015	350	220	22	Good machinability, low wear resistance; housings
EN-GJS-350-22-RT[2]	EN-JS1014	350	220	22	
EN-GJS-350-22	EN-JS1010	350	220	22	
EN-GJS-400-18-LT[1]	EN-JS1025	400	250	18	
EN-GJS-400-18-RT[2]	EN-JS1024	400	250	18	
EN-GJS-400-18	EN-JS1020	400	250	18	
EN-GJS-400-15	EN-JS1030	400	250	15	
EN-GJS-450-10	EN-JS1040	450	310	10	Good machinability, average wear resistance; fittings, press frames
EN-GJS-500-7	EN-JS1050	500	320	7	
EN-GJS-600-3	EN-JS1060	600	370	3	
EN-GJS-700-2	EN-JS1070	700	420	2	Good surface hardness; gears, steering and clutch parts, chains
EN-GJS-800-2	EN-JS1080	800	480	2	
EN-GJS-900-2	EN-JS1090	900	600	2	

[1] LT for low temperatures [2] RT for room temperature

⇒ **EN-GJS-400-18:** Cast iron with spheroidal (nodular) graphite, minimum tensile strength R_m = 400 N/mm²; elongation at fracture EL = 18%

Hardness HB as identifying characteristic

Type		Tensile strength R_m	Yield strength $R_{p\ 0.2}$	Brinell hardness	Properties, application examples
Designation	Material number	N/mm²	N/mm²	HB	
EN-GJS-HB130	EN-JS2010	350	220	< 160	By specifying hardness values the purchaser can better adapt process parameters to machining of the cast parts. Applications as above.
EN-GJS-HB150	EN-JS2020	400	250	130–175	
EN-GJS-HB155	EN-JS2030	400	250	135–180	
EN-GJS-HB185	EN-JS2040	450	310	160–210	
EN-GJS-HB200	EN-JS2050	500	320	170–230	
EN-GJS-HB230	EN-JS2060	600	370	190–270	
EN-GJS-HB265	EN-JS2070	700	420	225–305	
EN-GJS-HB300	EN-JS2080	800	480	245–335	
EN-GJS-HB330	EN-JS2090	900	600	270–360	

⇒ **EN-GJS-HB130:** Cast iron with spheroidal (nodular) graphite, Brinell hardness HB 130, maximum hardness HB 160

Malleable cast iron, Cast steel

Malleable cast iron[1]

cf. DIN EN 1562 (2006-08)

Type: Designation	Type: Material number	Tensile strength R_m N/mm²	Yield strength $R_{p\,0.2}$ N/mm²	Elongation at fracture EL %	Brinell hardness HB	Properties, application examples
Decarburizing annealed malleable cast iron (whiteheart malleable cast iron)						
EN-GJMW-350-4	EN-JM1010	350	–	4	230	All types have good castability and good machinability. Workpieces with low wall thickness, e.g. levers, chain links
EN-GJMW-400-5	EN-JM1030	400	220	5	220	
EN-GJMW-450-7	EN-JM1040	450	260	7	250	
EN-GJMW-550-4	EN-JM1050	550	340	4	250	
EN-GJMW-360-12	EN-JM1020	360	190	12	200	Especially well suited for welding.

⇒ **EN-GJMW-350-4:** Whiteheart malleable cast iron, R_m = 350 N/mm², EL = 4 %

Type: Designation	Type: Material number	Tensile strength R_m N/mm²	Yield strength $R_{p\,0.2}$ N/mm²	Elongation at fracture EL %	Brinell hardness HB	Properties, application examples
Non-decarburizing annealed malleable iron (blackheart malleable cast iron)						
EN-GJMB-300-6	EN-JM1110	300	–	6	–150	High pressure tightness
EN-GJMB-350-10	EN-JM1130	350	200	10	–150	All types have good castability and good machinability. Workpieces with high wall thickness, e.g. housings, universal joint yokes pistons
EN-GJMB-450-6	EN-JM1140	450	270	6	150–200	
EN-GJMB-500-5	EN-JM1150	500	300	5	165–215	
EN-GJMB-550-4	EN-JM1160	550	340	4	180–230	
EN-GJMB-600-3	EN-JM1170	600	390	3	195–245	
EN-GJMB-650-2	EN-JM1180	650	430	2	210–260	
EN-GJMB-700-2	EN-JM1190	700	530	2	240–290	
EN-GJMB-800-1	EN-JM1200	800	600	1	270–320	

⇒ **EN-GJMB-350-10:** Non-decarburizing annealed malleable cast iron, R_m = 350 N/mm², EL = 10 %

[1] Previous designations: page 158

Cast steel for general applications (selection)

cf. DIN EN 10293 (2005-06)[1]

Type: Designation	Type: Material number	Tensile strength R_m N/mm²	Yield strength $R_{p\,0.2}$ N/mm²	Elongation EL %	Notch impact energy K_v J	Properties, application examples
GE200[2]	1.0420	380–530	200	25	27	For workpieces with average dynamic loading; wheel spiders, levers
GE240[2]	1.0445	450–600	240	22	31	
GE300[2]	1,0558	600–750	300	15	27	
G17Mn5[3]	1.1131	450–600	240	24	70	Improved weldability; composite welded structures
G20Mn5[2]	1.6220	480–620	300	20	60	
GX4CrNiMo16-5-1[3]	1.4405	760–960	540	15	60	
G28Mn6[2]	1.1165	520–670	260	18	27	For workpieces with high dynamic loading; shafts
G10MnMoV6-3[3]	1.5410	600–750	500	18	60	
G34CrMo4[3]	1.7230	620–770	480	10	35	
G32NiCrMo8-5-4[3]	1.6570	850–1000	700	16	50	For corrosion-protected workpieces with high dynamic loading
GX23CrMoV12-1[3]	1.4931	740 000	540	15	27	

[1] DIN 17182 „Steel cast types with improved weldability and toughness“ was withdrawn without replacement.
[2] normalized [3] quenched and tempered

Cast steel for pressure vessels (selection)

cf. DIN EN 10213 (2008-01)

Type: Designation	Type: Material number	Tensile strength[1] R_m N/mm²	Yield strength[1] $R_{p\,0.2}$ N/mm²	Elongation at fracture EL %	Notch impact energy K_v J	Properties, application examples
GP240GH	1.0619	420	240	22	27	For high and low temperatures, e.g. steam turbines, super heated steam armatures, also corrosion resistant
G17CrMo5-5	1.7357	490	315	20	27	
GX8CrNi12	1.4107	540	355	18	45	
GX4CrNiMo16-5-1	1.4405	760	540	15	60	

[1] Values for a wall thickness up to 40 mm

Patterns, Pattern equipment and core boxes

cf. DIN EN 12890 (2000-06)

Materials and grades

Characteristics	Materials: Wood	Materials: Plastic	Materials: Metal
Type of material	Plywood, particle board or sandwich board, hard and soft wood	Epoxy resins or polyurethane with fillers	Cu, Sn, Zn alloys Al alloys Cast iron or steel
Application	Recurring individual pieces and smaller lots, low precision requirements; normally hand molding	Low and high volume production with higher precision requirements; hand and machine molding	Moderate to large volumes with high precision requirements; machine molding
Max. production run for molding	approx. 750	approx. 10 000	approx. 150 000
Quality classes[1]	W1[2], W2, H3	P1[2], P2	M1[2], M2
Surface quality	Sand paper 60–80 grit	Ra = 12.5 µm	Ra = 3.2–6.3 µm

[1] Classification system for the manufacture and use of patterns, pattern equipment and core boxes, according to their application, quality and service life: W wood; P plastic; M metal

[2] best grade

Mold draft for sand casting

Height h mm	Mold draft T in mm: Small draft surfaces: Hand molding: Molding sand clay bonded	Small draft surfaces: Hand molding: Molding sand chem. bonded	Small draft surfaces: Machine molding	Large draft surfaces: Hand molding: Molding sand clay bonded	Large draft surfaces: Hand molding: Molding sand chem. bonded	Large draft surfaces: Machine molding
–30	1.0	1.0	1.0	1.5	1.0	1.0
> 30–80	2.0	2.0	2.0	2.5	2.0	2.0
> 80–180	3.0	2.5	2.5	3.0	3.0	3.0
> 180–250	3.5	3.0	3.0	4.0	4.0	4.0
> 250–1000	+ 1.0 mm each 250 mm					
> 1000–4000	+ 2.0 mm each 1 000 mm					

Paint and color codes on patterns

Surface or partial surface	Cast steel	Nodular cast iron	Gray cast iron	Malleable iron	Heavy metal castings	Light alloy castings
Basic color for areas that should remain unmachined on the casting	blue	purple	red	gray	yellow	green
Areas to be machined on the casting	yellow stripes	yellow stripes	yellow stripes	yellow stripes	red stripes	yellow stripes
Locations of loose parts and their attachments	framed in black					
Locations of chill plates	red	red	blue	red	blue	blue
Core marks	black					
Risers	yellow stripes					

Shrinkage allowances, Dimensional tolerances, Molding and casting methods

Shrinkage allowances

cf. DIN EN 12890 (2000-06)

Cast iron	Shrinkage allowance in %	Other casting materials	Shrinkage allowance in %
with flake graphite	1.0	Cast steel	2.0
with spheroidal graphite, annealed	0.5	Austenitic manganese cast steel	2.3
with spheroidal graphite, not annealed	1.2	Al, Mg, CuZn alloys	1.2
austenitic	2.5	CuSnZn, Zn alloys	1.3
malleable cast iron, decarburizing anneal	1.6	CuSn alloys	1.5
malleable cast iron, no decarburizing anneal	0.5	Cu	1.9

Dimensional tolerances and machining allowances, RMA

cf. DIN ISO 8062 (2008-09)

Examples of tolerance specifications in a drawing:

1. **General tolerances**
   ISO 8062-3 – DCTG 12 – RMA 6 (RMAG H)
   Tolerance grade 12, machining allowance 6 mm
2. Individual tolerances and machining allowances are specified immediately behind the indication of the dimension.

R raw casting – nominal dimension
F dimension after finishing
CTG casting tolerance grade
T total casting tolerance
RMA machining allowance

$$R = F + 2 \cdot RMA + T/2$$

Casting tolerances

Nominal dimensions in mm	Length dimension tolerance *T* in mm for casting tolerance grade *DCTG*															
	1	2	3	4	5	6	7	8	9	10	11	12	13	14	15	16
– 10	0.09	0.13	0.18	0.26	0.36	0.52	0.74	1.0	1.5	2.0	2.8	4.2	–	–	–	–
> 10 – 16	0.10	0.14	0.20	0.28	0.38	0.54	0.78	1.1	1.6	2.2	3.0	4.4	–	–	–	–
> 16 – 25	0.11	0.15	0.22	0.30	0.42	0.58	0.82	1.2	1.7	2.4	3.2	4.6	6	8	10	12
> 25 – 40	0.12	0.17	0.24	0.32	0.46	0.64	0.9	1.3	1.8	2.6	3.6	5	7	9	11	14
> 40 – 63	0.13	0.18	0.26	0.36	0.50	0.70	1.0	1.4	2.0	2.8	4.0	5.6	8	10	12	16
> 63 – 100	0.14	0.20	0.28	0.40	0.56	0.78	1.1	1.6	2.2	3.2	4.4	6	9	11	14	18
> 100 – 160	0.15	0.22	0.30	0.44	0.62	0.88	1.2	1.8	2.5	3.6	5	7	10	12	16	20
> 160 – 250	–	0.24	0.34	0.50	0.70	1.0	1.4	2.0	2.8	4.0	5.6	8	11	14	18	22
> 250 – 400	–	–	0.40	0.56	0.78	1.1	1.6	2.2	3.2	4.4	6.2	9	12	16	20	25
> 400 – 630	–	–	–	0.64	0.90	1.2	1.8	2.6	3.6	5	7	10	14	18	22	28
> 630 – 1000	–	–	–	–	1.0	1.4	2.0	2.8	4	6	8	11	16	20	25	32

Molding and casting methods

Method	Application	Advantages and disadvantages	Casting material	Relative dimensional accuracy[1)] in mm/mm	Achievable roughness *Ra* in µm
Hand molding	large castings, small lots	all sizes, expensive, low dimensional accuracy	GJL, GJS, GS, GJM, Al and Cu alloys	0.00 – 0.10	40 – 320
Machine molding	small to medium sized parts, volume	dimensionally accurate, good surface	GJL, GJS, GS, GJM, Al alloys	0.00 – 0.06	20 – 160
Vacuum molding	medium to large parts, volumes	dimensionally accurate, good surface, high investment costs	GJL, GJS, GS, GJM, Al and Cu alloys	0.00 – 0.08	40 – 160
Shell molding	small parts, large volumes	dimensionally accurate, high mold costs	GJL, GS, Al and Cu alloys	0.00 – 0.06	20 – 160
Investment casting	small parts, large volumes	complex parts, high mold costs	GS, Al alloys	0.00 – 0.04	10 – 80
Die casting	small to medium sized parts, large volumes	dimensionally accurate even with thin walls, fine-grain structure, high investment costs	hot chamber: Zn, Pb, Sn, Mg cold chamber: Cu, Al	0.00 – 0.04	10 – 40

[1)] The ratio of largest relative deviation to the nominal dimension is called the relative dimensional accuracy.

Aluminum, Aluminum alloys – Overview

Alloy group	Material number	Main characteristics	Main areas of application	Product shapes[1] S	B	T
Pure aluminum						page 165
Al (Al content > 99.00%)	AW-1000 to AW-1990 (Series1000)	• very good cold workability • weldable and brazable • difficult for machining • corrosion resistant • anodized for decorative purposes	Containers, conduits and equipment for the food and chemical industry, electrical conductors, reflectors, trims, license plates in automotive manufacturing	•	•	•
Aluminum, wrought aluminum alloys, non-heat treatable (selection)						page 165
AlMn	AW-3000 to AW-3990 (Series 3000)	• cold workable • weldable and solderable • good machinability in work-hardened condition Compared to Series 1000: • higher strength • improved lye resistivity	Roofing, siding, and supporting structures in the construction industry, parts for radiators and air conditioning units in automotive manufacturing, drink and food cans in the packaging industry	•	•	•
AlMg	AW-5000 to AW-5990 (Series 5000)	• good cold workability with high work hardening • limited weldability • good machinability in work-hardened condition and with higher alloy contents • weather and saltwater resistant	Lightweight material for super-structures of commercial vehicles, tank and silo trucks, metal signs, traffic sign, rolling shutters and doors, windows, doors, hardware in the construction industry, machine frames, parts in the construction of jigs and fixtures and mold making	•	•	•
AlMgMn		• good cold workability with high work hardening • good weldability • good cutting machinability • saltwater resistant		•	•	•
Aluminum, wrought aluminum alloys, heat treatable (selection)						page 166
AlMgSi	AW-6000 to AW-6990 (Series 6000)	• good cold and hot workability • corrosion resistant • good weldability • good cutting machinability in heat treated condition	Load-bearing structures in the construction industry, windows, doors, machine beds, hydraulic and pneumatic parts; with Pb, Sn or Bi additions: very good cutting machinable free cutting alloys	•[2]	•[2]	•[2]
AlCuMg	AW-2000 to AW-2990 (Series 2000)	• high-strength values • good high-temperature strength • limited corrosion resistance • limited weldability • good cutting machinability in heat treated condition	Lightweight material in automotive and aircraft construction; with Pb, Sn or Bi additions: very good cutting machinable free cutting alloys	•[2]	•[2]	•[2]
AlZnMgCu	AW-7000 to AW-7990 (Series 7000)	• highest strength of all Al alloys • best corrosion resistance in artificially aged condition • limited weldability • good cutting machinability in heat treated condition	High-strength lightweight material in aircraft industry, machine construction, tools and molds for plastic molding, screws, extruded parts	•	•	•

[1] Product forms: S sheet; B bars; T tubes
[2] Free machining alloys are only delivered as bars or tubes.

MS

Aluminum, wrought aluminum alloys: Designations and material numbers

Designations for aluminum and wrought aluminum alloys
cf. DIN EN 573-2 (1994-12)

The designations apply to wrought products, e.g. sheet, bars, tubes, wires and for wrought parts.

Designation examples:

EN AW - Al 99,98
EN AW - Al Mg1SiCu - H111

EN European standard
AW Aluminum wrought products

Chemical composition, purity

Al 99.98	→	pure aluminum, degree of purity 99,98% Al
Mg1SiCu	→	1% Mg, low percentage of Si and Cu

Material condition (excerpt) cf. DIN EN 515 (1993-12)

Condition	Symbol	Meaning of the symbol	Meaning of the material conditions
manufactured condition	F	Wrought products are manufactured without specifying mechanical limits, e.g. tensile strength, yield strength, elongation at fracture	Wrought products without secondary operations
spheroidized	O O1 O2	Spheroidizing can be replaced by hot working Solution annealed, cooled slowly to room temperature Thermomechanically formed, highest workability	To restore workability after cold working
Work hardened	H12 to H18	Work hardened with the following hardness grades: H12 $\frac{1}{4}$ hard, H14 $\frac{1}{2}$ hard, H16 $\frac{3}{4}$ hard, H18 $\frac{4}{4}$ hard	To assure guaranteed mechanical values, e.g. tensile strength, yield strength
	H111 H112	Annealed with subsequent slight work hardening Slight work hardening	
Heat treated	T1 T2 T3	Solution annealed, stress relieved and naturally age hardened, not redressed Quenched like T1, cold worked and naturally aged Solution heat treated, cold worked and naturally age hardened	To increase in tensile strength, yield strength and hardness, reduction of the cold workability
	T3510 T3511	Solution annealed, stress relieved and naturally aged Like T3510, redressed to hold the limit deviations	
	T4 T4510	Solution annealed, naturally age hardened Solution annealed, stress relieved and naturally age hardened, not redressed	
	T6 T6510	Solution annealed, artificially aged Solution annealed, stress relieved and artificially aged, not redressed	
	T8 T9	Solution annealed, cold worked, artificially aged Solution annealed, artificially aged, cold worked	

Material numbers for aluminum and wrought aluminum alloys
cf. DIN EN 573-1 (2005-2)

Material numbers apply to wrought products, e.g. sheet, bars, tubes, wires and for wrought parts.

Designation examples:

EN AW - 1050A
EN AW - 5154

EN European standard
AW Aluminum wrought products

Indicates that country-specific limits deviate from the original alloy.

Alloy groups

Number	Group	Number	Group
1	pure Al	5	AlMg
2	AlCu	6	AlMgSi
3	AlMn	7	AlZn
4	AlSi	8	other

Alloy modifications

0 → Original alloy
1–9 → Alloys that deviate from the original alloy

Type number

Within an alloy group, e.g. AlMgSi, each type is assigned its own number.

Aluminum, wrought aluminum alloys

Aluminum and wrought aluminum alloys, non-heat treatable (selection)

cf. DIN EN 485-2 (2009-01), DIN EN 754-2, 755-2 (2008-06)

Designation (material-number)[1]	Delivery forms[2] R	Delivery forms[2] S	DC[3]	Material condition[4]	Thickness/ diameter mm	Tensile strength R_m N/mm²	Yield strength $R_{p\,0.2}$ N/mm²	Elong. at fracture *EL* %	Applications, Examples
Al 99.5 (1050A)	•	–	p z z	F, H112 O, H111 H14	≤ 200 ≤ 80 ≤ 40	≥ 60 60–95 100–135	≥ 20 – ≥ 70	25 25 6	Equipment manufacturing, pressure vessels, signs, packaging, trim
	–	•	w	O, H111	0,5–1,4 1,5–2,9 3,0–5,9	65–95 65–95 65–95	≥ 20 ≥ 20 ≥ 20	22 26 29	
Al Mn1 (3103)	•	–	p z z	F, H112 O, H111 H14	≤ 200 ≤ 60 ≤ 10	≥ 95 95–130 130–165	≥ 35 ≥ 35 ≥ 110	25 25 6	Equipment manufacturing, extruded parts, vehicle superstructures, heat exchangers
	–	•	w	O, H111	0.5–1.4 1.5–2.9 3.0–5.9	90–130 90–130 90–130	≥ 35 ≥ 35 ≥ 35	19 21 24	
Al Mn1Cu (3003)	•	–	p z z	F, H112 O, H111 H14	≤ 200 ≤ 80 ≤ 40	≥ 95 95–130 130–165	≥ 35 ≥ 35 ≥ 110	25 25 6	Roofing, facades, load-bearing structures in metal working
	–	•	w	O, H111	0.5–1.4 1.5–2.9 3.0–5.9	95–135 95–135 95–135	≥ 35 ≥ 35 ≥ 35	17 20 23	
Al Mg1 (5005)	•	–	p z z	F, H112 O, H111 H14	≤ 200 ≤ 80 ≤ 40	≥ 100 100–145 ≥ 140	≥ 40 ≥ 40 ≥ 110	18 18 6	Roofing, facades, windows, doors, hardware
	–	•	w	O, H111	0.5–1.49 1.5–2.9 3.0–5.9	100–145 100–145 100–145	≥ 35 ≥ 35 ≥ 35	19 20 22	
Al Mg2Mn0.3 (5251)	•	–	p z z	F, H112 O, H111 H14	≤ 200 ≤ 80 ≤ 30	≥ 160 150–200 200–240	≥ 60 ≥ 60 ≥ 160	16 17 5	Equipment and devices for the food industry
	–	•	w	O, H111	0.5–1.4 1.5–2.9 3.0–5.9	160–200 160–200 160–200	≥ 60 ≥ 60 ≥ 60	14 16 18	
Al Mg3 (5754)	•	–	p z z	F, H112 O, H111 H14	≤ 150 ≤ 80 ≤ 25	≥ 180 180–250 240–290	≥ 80 ≥ 80 ≥ 180	14 16 4	Equipment manufacturing, aircraft industry, body parts, mold making
	–	•	w	O, H111	0.5–1.4 1.5–2.9 3.0–5.9	190–240 190–240 190–240	≥ 80 ≥ 80 ≥ 80	14 16 18	
Al Mg5 (5019)	•	–	p z z	F, H112 O, H111 H14	≤ 200 ≤ 80 ≤ 40	≥ 250 250–320 270–350	≥ 110 ≥ 110 ≥ 180	14 16 8	Optical equipment, packaging
Al Mg3Mn (5454)	•	–	p	F, H112 O, H111	≤ 200	≥ 200 200–275	≥ 85 ≥ 85	10 18	Container construction, including pressure vessels, conduits, tank and silo trucks
	–	•	w	O, H111	0.5–1.4 1.5–2.9 3.0–5.9	215–275 215–275 215–275	≥ 85 ≥ 85 ≥ 85	13 15 17	
Al Mg4.5Mn0.7 (5083)	•	–	p z z	F, H111 O, H111 H12	≤ 200 ≤ 80 ≤ 30	≥ 270 270–350 ≥ 280	≥ 110 ≥ 110 ≥ 200	12 16 6	Mold making and construction of jigs and fix-tures, machine frames

[1] For simplification all designations and material numbers are written without the addition "EN AW-".

[2] Delivery forms: R round bar; S sheet, strip

[3] DC Delivery condition: p extruded; z drawn; w cold-rolled

[4] Material condition, see page 164

Wrought aluminum alloys

Wrought aluminum alloys,
heat treatable (selection)

cf. DIN EN 485-2 (2009-01), DIN EN 754-2, 755-2 (2008-06)

Designation (material-number)[1]	Delivery forms[2] R	Delivery forms[2] S	DC[3]	Material condition[4]	Thickness/ diameter mm	Tensile strength R_m N/mm²	Yield strength $R_{p\,0.2}$ N/mm²	Elong. at fracture *EL* %	Application, Examples
Al Cu4PbMgMn (2007)	•	–	p z z	T4, T4510 T3 T3	≤ 80 ≤ 30 30–80	≥ 370 ≥ 370 ≥ 340	≥ 250 ≥ 240 ≥ 220	8 7 6	Free cutting alloys, also good machinability at high machining outputs, e.g. for turned parts, milled parts
Al Cu4PbMg (2030)	•	–	p z z	T4, T4510 T3 T3	≤ 80 ≤ 30 30–80	≥ 370 ≥ 370 ≥ 340	≥ 250 ≥ 240 ≥ 220	8 7 6	
Al MgSiPb (6012)	•	–	p z z	T5, T6510 T3 T6	≤ 150 ≤ 80 ≤ 80	≥ 310 ≥ 200 ≥ 310	≥ 260 ≥ 100 ≥ 260	8 10 8	
Al Cu4SiMg (2014)	•	–	p z z	O, H111 T3 T4	≤ 200 ≤ 80 ≤ 80	≤ 250 ≥ 380 ≥ 380	≤ 135 ≥ 290 ≥ 220	12 8 12	Parts in hydraulic, pneumatic, automotive and aircraft manufacturing, load-bearing structures in metal manufacturing
	–	•	w	O	0.5–1.4 1.5–2.9 3.0–5.9	≤ 220 ≤ 220 ≤ 220	≤ 140 ≤ 140 ≤ 140	12 13 16	
Al Cu4Mg1 (2024)	•	–	p z z	O, H111 T3 T6	≤ 200 10–80 ≤ 80	≤ 250 ≥ 425 ≥ 425	≤ 150 ≥ 290 ≥ 315	12 9 5	Parts in automotive and aircraft manufacturing, load-bearing structures in metal working
	–	•	w	O	0.5–1.4 1.5–2.9 3.0–5.9	≤ 220 ≤ 220 ≤ 220	≤ 140 ≤ 140 ≤ 140	12 13 13	
Al MgSi (6060)	•	–	p z z	T4 T4 T6	≤ 150 ≤ 80 ≤ 80	≤ 120 ≥ 130 ≥ 215	≤ 60 ≥ 65 ≥ 160	16 15 12	Windows, doors, vehicle superstructures, machine beds, optical equipment
Al Si1MgMn (6082)	•	–	p z z	O, H111 T4 T6	≤ 200 ≤ 80 ≤ 80	≤ 160 ≥ 205 ≥ 310	≤ 110 ≥ 110 ≥ 255	14 14 10	Hardware, parts in mold making and manufacturing of jigs and fixtures, machine beds, equipment in the food industry
	–	•	w	O	0.5–1.4 1.5–2.9 3.0–5.9	≤ 150 ≤ 150 ≤ 150	≤ 85 ≤ 85 ≤ 85	14 16 18	
Al Zn4.5Mg1 (7020)	•	–	p z	T6 T6	≤ 50 ≤ 80	≥ 350 ≥ 350	≥ 290 ≥ 280	10 10	Parts in automotive and aircraft manufacturing, machine beds, superstructures of rail cars
	–	•	w	O	0.5–1.4 1.5–2.9 3.0–5.9	≤ 220 ≤ 220 ≤ 220	≤ 140 ≤ 140 ≤ 140	12 13 15	
Al Zn5Mg3Cu (7022)	•	–	p z	T6, T6510 T6	≤ 80 ≤ 80	≥ 490 ≥ 460	≥ 420 ≥ 380	7 8	Parts in hydraulic, pneumatic and aircraft manufacturing, screws
	–	•	w	T6	3.0–12 12.5–24 25–50	≥ 450 ≥ 450 ≥ 450	≥ 370 ≥ 370 ≥ 370	8 8 7	
Al Zn5.5MgCu (7075)	•	–	p z z	O, H111 T6 T73	≤ 200 ≤ 80 ≤ 80	≤ 275 ≥ 540 ≥ 455	≤ 165 ≥ 485 ≥ 385	10 7 10	Parts in automotive and aircraft manufacturing, mold making and manufacturing of jigs and fixtures, screws
	–	•	w	O	0.4–0.75 0.8–1.45 1.5–2.9	≥ 275 ≥ 275 ≥ 275	≥ 145 ≥ 145 ≥ 145	10 10 10	

[1] For simplification all designations and material numbers are written without the addition "EN AW-".
[2] Delivery forms: R round bar; S sheet, strip
[3] DC Delivery condition: p extruded; z drawn; w cold-rolled
[4] Material condition, see page 164

Aluminum casting alloys

Designation of aluminum castings

cf. DIN EN 1780-1...3 (2003-01), DIN EN 1706 (1998-06)

Aluminum castings are identified by designations or material numbers.

Designation examples:

Designation **EN AC - Al Mg5KF**

- EN European standard; AC Aluminum casting
- Al Mg5 → Chemical composition (table below)
- K → casting method; F → material condition (table below)

Material number **EN AC - 51300KF**

- 51 → Alloy groups (table below)
- 300 → Type number (table below)
- K → casting method; F → material condition (table below)

Chemical composition

Example	Alloy percentage
AlMg5 AlSi6Cu	5% Mg 6% Si, additions of Cu
AlCu4MgTi	4% Cu, additions of Mg and Ti

Alloy groups

No.	Group	No.	Group
21 41	AlCu AlSiMgTi	46 47	AlSi9Cu AlSi(Cu)
42 44	AlSi7Mg AlSi	51 71	AlMg AlZnMg

Type number

Within one alloy group each type has its own number.

Casting method

Letter	Casting method
S	Sand casting
K	Permanent mold casting
D	Die casting
L	Investment casting

Material condition

Letter	Meaning
F O	Casting condition, without subsequent processing Spheroidized
T1 T4	Controlled cooling after pouring, naturally aged Solution annealed and naturally aged
T5 T6	Controlled cooling after pouring, artificially aged Solution annealed and artificially aged

Aluminum casting alloys (selection)

cf. DIN EN 1706 (2010-07)

Designation (material-number)[1]	C_m[2]	M[3]	Strength values in casting condition (F) Hardn. strength HB	Tensile strength R_m N/mm²	Yield $R_{p\,0,2}$ N/mm²	Elongation at fracture EL %	Properties[4] C	P	M	Application
AC-AlMg3 (AC-51100)	S K	F F	50 50	140 150	70 70	3 5	–	–	•	Corrosion resistant, polishable, anodized for decorative purposes; fittings, household appliances, ship building, chemical industry
AC-AlMg5 (AC-51300)	S K	F F	55 60	160 180	90 100	3 4	–	–	•	
AC-AlMg5(Si) (AC-51400)	S K	F F	60 65	160 180	100 110	3 3	–	–	•	
AC-AlSi12 (AC-44100)	S K L	F F F	50 55 50	150 170 150	70 80 80	4 5 4	•	•	○	Resistant to weather influences, for complex, thin-walled and pressure-tight parts; pump and motor housings, cylinder heads, parts in aircraft manufacturing
AC-AlSi7Mg (AC-42000)	S K L	T6 T6 T6	75 90 75	220 260 240	180 220 190	2 1 1	○	•	○	
AC-AlSi12(Cu) (AC-47000)	S K	F F	50 55	150 170	80 90	1 2	•	•	–	
AC-AlCu4Ti (AC-21100)	S K	T6 T6	95 95	300 330	200 220	3 7	–	–	•	Highest strength values, vibration and high temp. resistance; simple castings

[1] For simplification all designations and material numbers are written without "EN", e.g. AC-AlMg3 instead of EN AC-AlMg3 or AC-51000 instead of EN AC-51000.

[2] C_m casting method (table above) [3] M material condition (table above)

[4] C castability, P pressure tightness, M machinability; • very good, ○ good, – conditionally good

Aluminum profiles – Overview, Round bars, Flat bars

Aluminum sections, Overview

Illustration	Fabrication, dimensions	Standard	Illustration	Fabrication, dimensions	Standard
Round bars			**Round tubes**		
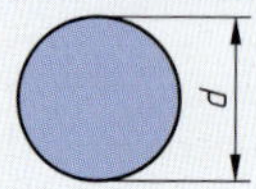	extruded d = 3–100 mm	DIN EN 755-3	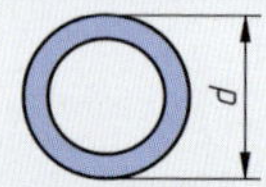	seamless extruded d = 20–250 mm	DIN EN 755-7
	drawn d = 8–320 mm	DIN EN 754-3		cold-drawn seamless d = 3–270 mm	DIN EN 754-7
Square bars			**Square tubes**		
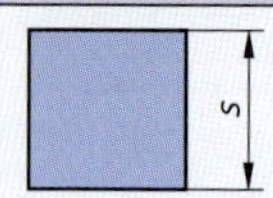	extruded s = 10–220 mm	DIN EN 755-4	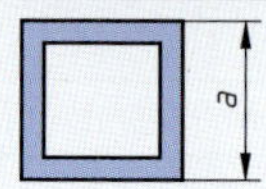	extruded a = 15–100 mm	DIN EN 754-4
	drawn s = 3–100 mm	DIN EN 754-4			
Flat bars			**Flat tubes**		
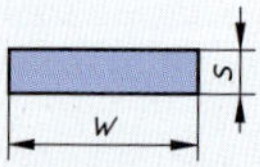	extruded w = 10–600 mm s = 2–240 mm	DIN EN 755-4	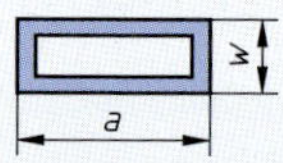	extruded seamless a = 15–250 mm w = 10–100 mm	DIN EN 755-7
	drawn w = 5–200 mm s = 2–60 mm	DIN EN 754-4		cold-drawn seamless a = 15–250 mm w = 10–100 mm	DIN EN 754-7
Sheet and strip			**L profiles**		
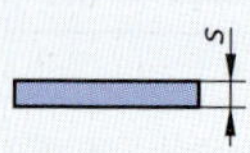	rolled s = 0.4–15 mm	DIN EN 485		sharp corners or round corners h = 10–200 mm	DIN 1771[1)]
Channels			**Tees**		
	sharp corners or round corners h = 10–160 mm	DIN 9713[1)]	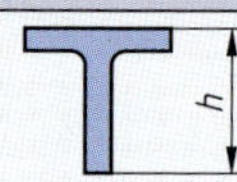	sharp corners or round corners h = 15–100 mm	DIN 9714[1)]

[1)] Standards were withdrawn without replacement.

Round bars, Flat bars, drawn

cf. DIN EN 754-3, 754-4 (2008-06), DIN 1798[1)], DIN 1796[1)]

S cross-sectional area
m' linear mass density
W axial section modulus
I axial moment of inertia

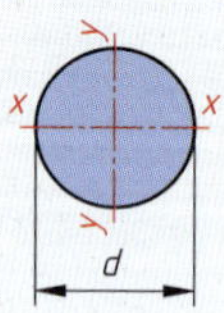

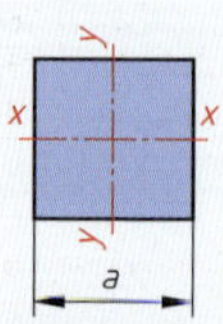

d, a mm	S cm² ○	S cm² □	m' kg/m ○	m' kg/m □	$W_x = W_y$ cm³ ○	$W_x = W_y$ cm³ □	$I_x = I_y$ cm⁴ ○	$I_x = I_y$ cm⁴ □
10	0.79	1.00	0.21	0.27	0.10	0.17	0.05	0.08
12	1.13	1.44	0.31	0.39	0.17	0.29	0.10	0.17
16	2.01	2.56	0.54	0.69	0.40	0.68	0.32	0.55
20	3.14	4.00	0.85	1.08	0.79	1.33	0.79	1.33
25	4.91	6.25	1.33	1.69	1.53	2.60	1.77	3.26
30	7.07	9.00	1.91	2.43	2.65	4.50	3.98	6.75
35	9.62	12.25	2.60	3.31	4.21	7.15	7.37	12.51
40	12.57	16.00	3.40	4.32	6.28	10.68	12.57	21.33
45	15.90	20.25	4.30	5.47	8.95	15.19	20.13	34.17
50	19.64	25.00	5.30	6.75	12.28	20.83	30.69	52.08
55	23.76	30.25	6.42	8.17	16.33	27.73	44.98	76.26
60	28.27	36.00	7.63	9.72	21.21	36.00	63.62	108.00
Materials	Wrought aluminum alloys, see pages 165 and 166.							

[1)] DIN 1796 und DIN 1798 were replaced by DIN EN 754-3 or DIN EN 754-4. The DIN EN standards contain no dimensions. However, dealers continue to offer DIN 1798 and DIN 1796 round and square bars.
○ round bars; □ square bars

Flat bars from aluminum alloys

Flat bars, drawn (selection)

cf. DIN EN 754-5 (2008-06), replaces DIN 1769[1)]

S cross-sectional area
m' linear mass density
e distance to edge
W axial section modulus
I axial moment of inertia

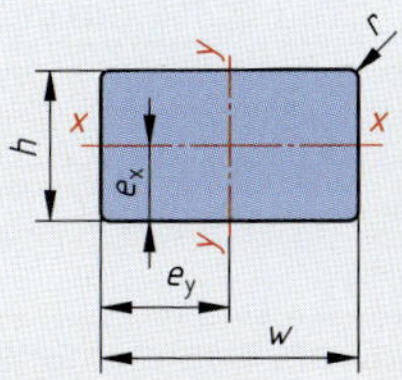

$w \times h$ mm	S cm²	m' kg/m	e_x cm	e_y cm	W_x cm³	I_x cm⁴	W_y cm³	I_y cm⁴
10 × 3	0.30	0.08	0.15	0.5	0.015	0.002	0.05	0.025
10 × 6	0.60	0.16	0.3	0.5	0.060	0.018	0.100	0.050
10 × 8	0.80	0.22	0.4	0.5	0.106	0.042	0.133	0.066
15 × 3	0.45	0.12	0.15	0.75	0.022	0.003	0.112	0.084
15 × 5	0.75	0.24	0.25	0.75	0.063	0.016	0.188	0.141
15 × 8	1.20	0.32	0.4	0.75	0.160	0.064	0.300	0.225
20 × 5	1.00	0.27	0.25	1.0	0.083	0.020	0.333	0.333
20 × 8	1.60	0.43	0.4	1.0	0.213	0.085	0.533	0.533
20 × 10	2.00	0.54	0.5	1.0	0.333	0.166	0.666	0.666
20 × 15	3.00	0.81	0.75	1.0	0.750	0.562	1.000	1.000
25 × 5	1.25	0.34	0.25	1.25	0.104	0.026	0.520	0.651
25 × 8	2.00	0.54	0.4	1.25	0.266	0.106	0.833	1.041
25 × 10	2.50	0.67	0.5	1.25	0.416	0.208	1.041	1.302
25 × 15	3.75	1.01	0.75	1.25	0.937	0.703	1.562	1.953
25 × 20	5.00	1.35	1.0	1.25	1.666	1.666	2.083	2.604
30 × 10	3.00	0.81	0.5	1.5	0.500	0.250	1.500	2.250
30 × 15	4.50	1.22	0.75	1.5	1.125	0.843	2.250	3.375
30 × 20	6.00	1.62	1.0	1.5	2.000	2.000	3.000	4.500
40 × 10	4.00	1.08	0.5	2.0	0.666	0.333	2.666	5.333
40 × 15	6.00	1.62	0.75	2.0	1.500	1.125	4.000	8.000
40 × 20	8.00	2.16	1.0	2.0	2.666	2.666	5.333	10.666
40 × 25	10.00	2.70	1.25	2.0	4.166	5.208	6.666	13.333
40 × 30	12.00	3.24	1.5	2.0	6.000	9.000	8.000	16.000
40 × 35	14.00	3.78	1.75	2.0	8.166	14.291	9.333	18.666
50 × 10	5.00	1.35	0.5	2.5	0.833	0.416	4.166	10.416
50 × 15	7.50	2.03	0.75	2.5	1.875	1.406	6.250	15.625
50 × 20	10.00	2.70	1.0	2.5	3.333	3.333	8.333	20.833
50 × 25	12.50	3.37	1.25	2.5	5.208	6.510	10.416	26.041
50 × 30	15.00	4.05	1.5	2.5	7.500	11.250	12.500	31.250
50 × 35	17.50	4.73	1.75	2.5	10.208	17.864	14.583	36.458
50 × 40	20.00	5.40	2.0	2.5	13.333	26.666	16.666	41.668
60 × 10	6.00	1.62	0.5	3.0	1.000	0.500	6.000	18.000
60 × 15	9.00	2.43	0.75	3.0	2.250	1.687	9.000	27.000
60 × 20	12.00	3.24	1.0	3.0	4.000	4.000	12.000	36.000
60 × 25	15.00	4.05	1.25	3.0	6.250	7.812	15.000	45.000
60 × 30	18.00	4.86	1.5	3.0	9.000	13.500	18.000	54.000
60 × 35	21.00	5.67	1.75	3.0	12.250	21.437	21.000	63.000
60 × 40	24.00	6.48	2.0	3.0	16.000	32.000	24.000	72.000
80 × 10	8.00	2.16	0.5	4.0	1.333	0.666	10.666	42.666
80 × 15	12.00	3.24	0.75	4.0	3.000	2.250	16.000	64.000
80 × 20	16.00	4.52	1.0	4.0	5.333	5.333	21.333	85.333
80 × 25	20.00	5.40	1.25	4.0	8.333	10.416	26.666	106.66
80 × 30	24.00	6.48	1.5	4.0	12.000	18.000	32.000	128.00
80 × 35	28.00	7.56	1.75	4.0	16.333	28.583	37.333	149.33
80 × 40	32.00	8.64	2.0	4.0	21.333	42.666	42.666	170.66
100 × 20	20.00	5.40	1.0	5.0	6.666	6.667	33.333	166.66
100 × 30	30.00	8.10	1.5	5.0	15.000	22.500	50.000	250.00
100 × 40	40.00	10.8	2.0	5.0	26.666	53.333	66.666	333.33

Material Wrought aluminum alloys, see pages 165 and 166.

Edge radii *r*

h mm	r_{max} mm
≤ 10	0.6
> 10–30	1.0
> 30–60	2.0

1) DIN EN 754-5 contains no dimensions. Specialized dealers still offer flat bars in dimensions according to DIN 1769.

Round tubes, Channels from aluminum alloys

Round tubes, cold-drawn seamless (selection)

cf. DIN EN 754-7 (2008-06), replaces DIN 1795[1)]

d outside diameter
s wall thickness
S cross-sectional area
m' linear mass density
W axial section modulus
I axial moment of inertia

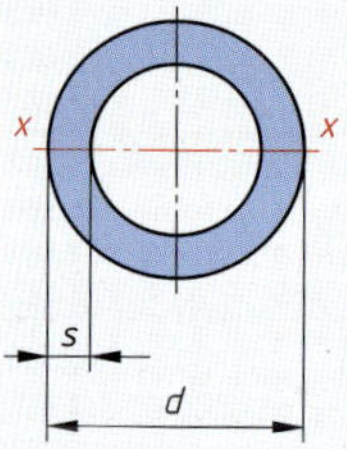

$d \times s$ mm	S cm²	m' kg/m	W_x cm³	I_x cm⁴	$d \times s$ mm	S cm²	m' kg/m	W_x cm³	I_x cm⁴
10 × 1	0.281	0.076	0.058	0.029	35 × 3	3.016	0.814	2.225	3.894
10 × 1.5	0.401	0.108	0.075	0.037	35 × 5	4.712	1.272	3.114	5.449
10 × 2	0.503	0.136	0.085	0.043	35 × 10	7.854	2.121	4.067	7.118
12 × 1	0.346	0.093	0.088	0.053	40 × 3	3.487	0.942	3.003	6.007
12 × 1.5	0.495	0.134	0.116	0.070	40 × 5	5.498	1.484	4.295	8.590
12 × 2	0.628	0.170	0.136	0.082	40 × 10	9.425	2.545	5.890	11.781
16 × 1	0.471	0.127	0.133	0.133	50 × 3	4.430	1.196	4.912	12.281
16 × 2	0.880	0.238	0.220	0.220	50 × 5	7.069	1.909	7.245	18.113
16 × 3	1.225	0.331	0.273	0.273	50 × 10	12.566	3.393	10.681	26.704
20 × 1.5	0.872	0.235	0.375	0.375	55 × 3	4.901	1.323	6.044	16.201
20 × 3	1.602	0.433	0.597	0.597	55 × 5	7.854	2.110	9.014	24.789
20 × 5	2.356	0.636	0.736	0.736	55 × 10	14.137	3.817	13.655	37.552
25 × 2	1.445	0.390	0.770	0.963	60 × 5	8.639	2.333	10.979	32.938
25 × 3	2.073	0.560	1.022	1.278	60 × 10	15.708	4.241	17.017	51.051
25 × 5	3.142	0.848	1.335	1.669	60 × 16	22.117	4.890	20.200	60.600
30 × 2	1.759	0.475	1.155	1.733	70 × 5	10.210	2.757	15.498	54.242
30 × 4	3.267	0.882	1.884	2.826	70 × 10	18.850	5.089	24.908	87.179
30 × 6	4.524	1.220	2.307	3.461	70 × 16	27.143	7.331	30.750	107.62
Material	e.g. aluminum alloys, non-heat treatable, see page 165 aluminum alloys, heat-treatable, see page 166								

[1)] DIN EN 754-7 contains no dimensions. Specialized dealers still offer round tubes in dimensions according to DIN 1795.

Extruded channel sections (selection)

cf. DIN 9713 (1981-09)[1)]

w width
h height
S cross-sectional area
m' linear mass density
W axial section modulus
I axial moment of inertia

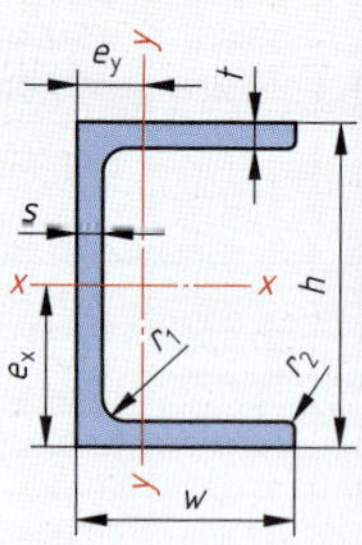

$h \times w \times s \times t$ mm	S cm²	m' kg/m	e_x cm	e_y cm	W_x cm³	I_x cm⁴	W_y cm³	I_y cm⁴
20 × 20 × 3 × 3	1.62	0.437	1.00	0.780	0.945	0.945	0.805	0.628
30 × 30 × 3 × 3	2.52	0.687	1.50	1.10	2.43	3.64	2.06	2.29
35 × 35 × 3 × 3	2.97	0.802	1.75	1.28	3.44	6.02	2.91	3.73
40 × 15 × 3 × 3	1.92	0.518	2.0	0.431	2.04	4.07	0.810	0.349
40 × 20 × 3 × 3	2.25	0.608	2.0	0.610	2.59	5.17	1.30	0.795
40 × 30 × 3 × 3	2.85	0.770	2.0	3.62	7.24	2.49	2.49	2.52
40 × 30 × 4 × 4	3.71	1.00	2.0	1.05	4.49	8.97	3.03	3.17
40 × 40 × 4 × 4	4.51	1.22	2.0	1.49	5.80	11.6	4.80	7.12
40 × 40 × 5 × 5	5.57	1.50	2.0	1.52	6.80	13.6	5.64	8.59
50 × 30 × 3 × 3	3.15	0.851	2.5	0.929	4.88	12.2	2.91	2.70
50 × 30 × 4 × 4	4.91	1.33	2.5	1.38	7.83	19.6	5.65	7.80
50 × 40 × 5 × 5	6.07	1.64	2.5	1.42	9.32	23.3	6.54	9.26
60 × 30 × 4 × 4	4.51	1.22	3.0	0.896	7.90	23.7	4.12	3.69
60 × 40 × 4 × 4	5.31	1.43	3.0	1.29	10.1	30.3	6.35	8.20
60 × 40 × 5 × 5	6.57	1.77	3.0	1.33	12.0	36.0	7.47	9.94
80 × 40 × 6 × 6	8.95	2.42	4.0	1.22	20.6	82.4	10.6	20.6
80 × 45 × 6 × 8	11.2	3.02	4.0	1.57	27.1	108	13.9	21.8
100 × 40 × 6 × 6	10.1	2.74	5.0	1.11	28.3	142	12.5	13.8
100 × 50 × 6 × 9	14.1	3.80	5.0	1.72	43.4	217	19.9	34.3
120 × 55 × 7 × 9	17.2	4.64	6.0	1.74	61.9	295	28.2	49.1
140 × 60 × 4 × 6	12.35	3.35	7.0	1.83	56.4	350	24.7	45.2
Materials	AlMgSi0.5; AlMgSi1; AlZn4.5Mg1							

Rounded edges r_1 and r_2

t mm	r_1 mm	r_2 mm
3, 4	2.5	0.4
5, 6	4	0.6
8, 9	6	0.6

[1)] DIN 9713 was withdrawn without replacement. Specialized dealers still offer channels according to this standard.

Magnesium alloys, Titanium, Titanium alloys

Wrought magnesium alloys (selection) cf. DIN 9715 (1982-08)

Designation	Material-number	Delivery form[1] B	T	D	M[2]	Bar diameter mm	Tensile strength R_m N/mm²	Yield strength $R_{p0,2}$ N/mm²	Elong. at fracture EL %	Properties, application
MgMn2 MgAl3Zn	3.3520 3.5312	•	•	•	F20 F24	≤ 80 ≤ 80	200 240	145 155	15 10	Corrosion resistant, weldable, cold workable; cladding, containers
MgAl6Zn	3.5612	•	•	•	F27	≤ 80	270	195	10	Higher strength, limited weld-ability; lightweight material in automotive, machine and aircraft industry
MgAl8Zn	3.5812	•	•	•	F29 F31	≤ 80 ≤ 80	290 310	205 215	10 6	

[1] Delivery forms: B bars, e.g. round bars; T tubes; D stamped part
[2] M material condition F20 → R_m = 10 · 20 = 200 N/mm²

Magnesium casting alloys (selection) cf. DIN EN 1753 (1997-08)

Designation[1]	Material-number[1]	M[2]	Mate-rial-condi-tion[3]	Hardness HB	Tensile strength R_m N/mm²	Yield strength $R_{p0,2}$ N/mm²	Elong. at fracture EL %	Properties, application
MCMgAl8Zn1	MC21110	S	F T6	50–65 50–65	160 240	90 90	2 8	Very good castability, dynamically loadable, weldable; gear and motor housings
		K K D	F T4 F	50–65 50–65 60–85	160 160 200–250	90 90 140–160	2 8 ≤ 7	
MCMgAl9Zn1	MC21120	S	F T6	55–70 60–90	160 240	90 150	6 2	High-strength, good sliding properties, weldable; automotive and aircraft industry, armatures
		K K D	F T6 F	55–70 60–90 65–85	160 240 200–260	110 150 140–170	2 2 1–6	
MCMgAl6Mn MCMgAl7Mn MCMgAl4Si	MC21230 MC21240 MC21320	D D D	F F F	55–70 60–75 55–80	190–250 200–260 200–250	120–150 130–160 120–150	4–14 3–10 3–12	Fatigue resistant, dynam-ically loadable, high tem-perature resistant, gear and motor housings

[1] For simplification, designations and material numbers are written without the "EN-" prefix, e.g. MCMgAlBZn1 instead of EN-MCMgAl8Zn1.
[2] M casting method: S sand casting; K permanent mold casting; D die casting
[3] Material condition, see designation of aluminum casting alloys, page 167

Titanium, titanium alloys (selection) cf. DIN 17860 (2010-01)

Designation	Material-number	Delivery form[1] S	B	T	Sheet thickness s mm	Hard-ness HB	Tensile-strength R_m N/mm²	Yield strength $R_{p0,2}$ N/mm²	Elong. at fracture EL %	Properties, application
Ti1 Ti2 Ti3	3.7025 3.7035 3.7055	•	•	•	0.4–35	120 150 170	290–410 390–540 460–590	180 250 320	30 22 18	Weldable, solderable, glueable, machinable, cold and hot workable, fatigue resistant, corrosion resistant; weight saving designs in machine construction, electrical engineering, precision engineering, optics and medical technology, chemical industry, food industry, air-craft industry
Ti1Pd Ti2Pd	3.7225 3.7235	•	•	•	0.4–35	120 150	290–410 390–540	180 250	30 22	
TiAl6V6Sn2	3.7175	•	•	•	< 6 6–50	320 320	≥ 1070 ≥ 1000	1000 950	10 8	
TiAl6V4	3.7165	•	•	•	< 6 6 – 100	310 310	≥ 920 ≥ 900	870 830	8 8	
TiAl4Mo4Sn2	3.7185	•	•	•	6 – 65	350	≥ 1050	1050	9	

[1] Delivery forms: S sheet and strip; B bars, e.g. round bars; T tubes

MS

Overview of the heavy non-ferrous metals

Heavy non-ferrous metals have a density ϱ > 5 kg/dm^3. However, in technical literature $\varrho \geq 4.5$ kg/dm^3 is also used as limit for non-ferrous metals.

- Construction materials in machine and plant construction: copper, tin, zinc, nickel, lead and their alloys
- Metals used for alloys: chromium, vanadium, cobalt (for effects of alloying metals, see page 126)
- Precious metals: gold, silver, platinum

Pure metals: Homogeneous structure; low strengths, lesser importance as a construction material; usually used based on material typical properties, e.g. good electrical conductivity.

Heavy non-ferrous metal alloys: Improved properties compared to base metals, such as higher strength, higher hardness, better machinability and corrosion resistance, construction materials for various application. Classified according to **manufacture into wrought alloys and** casting alloys.

Overview of common heavy non-ferrous metals and alloys

Metal, alloy group	Main characteristics	Application examples
Copper (Cu)	High electrical conductivity and thermal conductivity, inhibits bacteria, viruses and molds, corrosion resistant, good appearance, easily recyclable	Pipes in heating and plumbing equipment, cooling and heating coils, electrical wiring, electrical parts, cookware, building facades
CuZn (brass)	Wear-resistant, corrosion-resistant, good hot and cold workability, good machinability, polishable, shiny golden, medium strengths	• Wrought alloys: deep-drawn parts, screws, springs, pipes, instrument parts • Casting alloys: armature housings, plain bearings, precision mechanical parts
CuZnPb	Very good machinability, limited cold workability, very good hot workability	Automatic screw machine parts, precision mechanical parts, fittings, hot-pressed parts
CuZn multi-alloy	Good hot workability, high strengths, wear-resistant, weather-resistant	Armature housings, plain bearings, flanges, valve parts, water housings
CuSn (bronze)	Very corrosion-resistant, good sliding properties, good wear-resistance, strength resulting from cold working is highly variable	• Wrought alloys: hardware, screws, springs, metal hoses • Casting alloys: spindle nuts, worm gears, solid plain bearings
CuAl	High strength and toughness, very corrosion resistant, salt water resistant, heat resistant, highly cavitation resistant	• Wrought alloys: highly stressed lock nuts, ratchet wheels • Casting alloys: armatures in the chemical industry, pump bodies, propellers
CuNi(Zn)	Extremely corrosion resistant, silvery appearance, good machinability, polishable, cold workable	Coins, electrical resistors, heat exchangers, pumps, valves in salt water cooling systems, ship building
Zinc (Zn)	Resistant to atmospheric corrosion	Corrosion protection of steel parts
ZnTi	Good workability, joinable by soft soldering	Roofing, gutters, downspouts
ZnAlCu	Very good castability	Thin walled, finely articulated die castings
Tin (Sn)	Good chemical resistance, non-toxic	Coating of steel sheet
SnPb	Low viscosity	Soft solder
SnSb	Good dry running properties	Small, dimensionally precise die castings, plain bearings with average loading
Nickel (Ni)	Corrosion resistant, high temperature resistant	Corrosion protection layer on steel parts
NiCu	Extremely corrosion resistant and high temp. resist.	Equipment, condensers, heat exchangers
NiCr	Extremely corrosion resistant and very high temperature resistant and nonscaling, e.g. age hardenable	Chemical installations, heating tubes, boiler internals in power plants, gas turbines
Lead (Pb)	Shields against x-ray and gamma rays, corrosion resistant, toxic	Shielding, cable sheathing, tubes for chemical equipment
PbSn	Low viscosity, soft, good dry running properties	Soft solder, sliding sheaths
PbSbSn	Low viscosity, corrosion resistant, good running and sliding properties (low friction)	plain bearings, small, dimensionally precise die castings such as pendulums, parts for measuring equipment, meters

Designation of heavy non-ferrous metals

Designation system (excerpt)

cf. DIN 1700 (1954-07)[1)]

Example: **NiCu30Fe F45**
GD - Sn80Sb

Manufacture, application

E	Electrical material
G	Sand casting
GC	Continuous casting
GD	Die casting
GK	Permanent mold casting
GZ	Centrifugal casting
L	Solder
S	Welding filler alloys

Chemical composition

Example	Comment
NiCu30Fe	Ni-Cu alloy, 30% Cu, trace iron
Sn80Sb	Sn-Sb alloy,, 80% Sn, approx. 20% Sb

Special properties

F45	minimum tensile strength $R_m = 10 \cdot 45\ N/mm^2 = 450\ N/mm^2$
a	age hardened
g	annealed
h	hard
ka	naturally aged
ku	cold worked
ta	partially age hardened
wa	artificially aged
wu	hot worked
zh	drawn hard

1) The standard has been withdrawn. However the material designations are still used in individual standards.

Designation system for copper alloys

cf. DIN EN 1982 (2008-08) and 1173 (2008-08)

Examples: **CuZn31Si - R620**
CuZn38Pb2
CuSn11Pb2 - C - GS

Chemical composition

Example	Meaning
CuZn31Si	Cu alloy, 31% Zn, trace Si
CuZn38Pb2	Cu alloy 38% Zn, 2% Pb
CuSn11Pb2	Cu alloy 11% Sn, 2% Pb

Casting method

GS	Sand casting	GM	Permanent mold casting
GZ	Centrifugal casting	GC	Continuous casting
GP	Die casting		

Product form

C	Material in the form of castings
B	Material in ingot form
	Wrought alloys (without code letter)

Material condition (selection)

Example	Meaning	Example	Meaning
A007	Elongation at fracture $EL = 7\%$	Y450	Yield strength $R_p = 450\ N/mm^2$
D	Drawn, without specified mechanical properties	M	Manufactured condition, without specified mechanical properties
H160	Vickers hardness HV = 160	R620	Minimum tensile strength $R_m = 620\ N/mm^2$

Material numbers for copper and copper alloys

cf. DIN EN 1412 (1995-12)

Example: **C W 024 A**

- C — Copper material
- C Cast material; B Material in ingots; W Wrought material
- 024 — Number between 000 and 999 without specified meaning (sequential number)

Code letters for material groups

Letter	Material group	Letter	Material group
A or B	Copper	H	Copper-nickel alloys
C or D	Copper alloys, percentage of the alloying element < 5%	J K	Copper-zinc alloys Copper-tin alloys
E or F	Copper alloys, percentage of the alloying elements ≥ 5%	L or M N or P	Copper-zinc binary alloys Copper-zinc-lead alloys
G	Copper-aluminum alloys	R or S	Copper-zinc multi-alloys

Material numbers for castings of zinc alloys

cf. DIN EN 12844 (1999-01)

Example: **Z P 04 1 0**

- Z — Zinc alloy
- P — Casting
- 04 — Al content: 04 ≙ 4% aluminum
- 1 — Cu content: 1 ≙ 1% copper
- 0 — Content of the next higher alloying element; 0 = next higher alloying element < 1%

MS

Copper alloys

Wrought copper alloys

Designation, Material number[1]	C[2]	Bars D[3] mm	Hardness HB	Tensile strength R_m N/mm^2	Yield strength $R_{p0,2}$ N/mm^2	Elong. at fracture EL %	Properties, application examples
Copper-zinc alloys							cf. DIN EN 12163 (1998-04)
CuZn28 (CW504L)	R310 R460	4–80 4–10	– –	310 460	120 420	27 –	Very good cold workability, good hot workability, machinable, very easily polished; instrument parts, bushings
	H085 H145	4–80 4–10	85–115 ≥ 145	– –	– –	– –	
CuZn37 (CW508L)	R310 R440	2–80 2–10	– –	310 440	120 400	30 –	Very good cold workability, good hot workability, machinable, very easily polished; deep-drawn parts, screws, springs, press rollers
	H070 H140	4–80 4–10	70–100 ≥ 140	– –	– –	– –	
CuZn40 (CW509L)	R340 H080	2–80	– ≥ 80	340 –	260 –	25 –	Very good hot workability, machinable; rivets, screws
Copper-zinc alloys (multi-alloys)							cf. DIN EN 12163 (1998-04)
CuZn31Si (CW708R)	R460 R530	5–40 5–14	– –	460 530	250 330	22 12	Good cold workability; hot workable, machinable, good sliding properties; sliding parts, bearing bushings, guides
	H115 H140	5–40 5–14	115–145 ≥ 140	– –	– –	– –	
CuZn38Mn1Al (CW716R)	R490 R550	5–40 5–14	– –	490 550	210 280	18 10	Good hot workability, cold workable, machinable, sliding properties, weather resistant; sliding elements, guides
	H120 H150	5–40 5–14	120–150 ≥ 150	– –	– –	– –	
CuZn40Mn2Fe1 (CW723R)	R460 R540	5–40 5–14	– –	460 540	270 320	20 8	Good hot workability, cold workable, machinable, average strength, weather resistant; equipment manufacturing, architecture
	H110 H150	5–40 5–14	110–140 ≥ 150	– –	– –	– –	
Copper-zinc-lead alloys							cf. DIN EN 12164 (2000-09)
CuZn36Pb3 (CW603N)	R340 R550	40–80 2–4	90 150	340 550	160 450	20 –	Excellent machinability, limited cold workability; automatic lathe parts
CuZn38Pb2 (CW608N)	R360 R550	40–80 2–6	90 150	360 550	150 420	25 –	Excellent machinability, good cold and hot workability; screw machine parts
CuZn40Pb2 (CW617N)	R360 R550	40–80 2–4	90 150	360 550	150 420	20 –	Excellent machinability, good hot workability; stamping blanks, gears
Copper-tin alloys							cf. DIN EN 12163 (1998-04)
CuSn6 (CW452K)	R340 R550	2–60 2–6	– –	340 550	230 500	45 –	High chemical resistance, good strength; springs, metal hoses, pipes and bushings for suspension bodies
	H085 H180	2–60 2–6	85–115 ≥ 180	– –	– –	– –	
CuSn8 (CW453K)	R390 R620	2–60 2–6	– –	390 620	260 550	45 –	High chemical resistance, high-strength, good sliding properties; plain bearings, rolled bearing bushings, contact springs
	H090 H185	2–60 2–6	90–120 ≥ 185	– –	– –	– –	
CuSn8P (CW459K)	R390 R620	2–60 2–6	– –	390 620	260 550	45 –	Excellent sliding properties, high wear-resistance, endurance strength; highly stressed plain bearings in automotive and machine manufacturing
	H090 H185	2–60 2–6	90–120 ≥ 185	– –	– –	– –	

[1] Material numbers according to DIN EN 1412, see page 173.

[2] C Material condition according to DIN EN 1173, see page 173. In manufactured condition M all alloys can be delivered up to diameter D = 80 mm.

[3] D Diameter for round bars, width across flats for square bars and hexagonal bars, thickness for flat bars.

Copper and refined zinc alloys

Designation, Material number[1]	C[2]	Bars D[3] mm	Hardness HB	Tensile strength R_m N/mm²	Yield strength $R_{p0.2}$ N/mm²	Elong. at fracture EL %	Properties, application examples
Copper-aluminum alloys							cf. DIN EN 12163 (1998-04)
CuAl10Fe3Mn2 (CW306G)	R590 R690	10–80 10–50	– –	590 690	330 510	12 6	Corrosion-resistant, wear-resistant, fatigue-resistant, high-temperature resistant; screws, shafts, gears, worm gears, valve seats
	H140 H170	10–80 10–50	140–180 ≥ 170	– –	– –	– –	
CuAl10Ni5Fe4 (CW307G)	R680 R740	10–80	– –	680 740	480 530	10 8	Corrosion resistant, wear-resistant, nonscaling, fatigue resistant, high temperature resistant; capacitor bases, control parts for hydraulics
	H170 H200	10–80	170–210 ≥ 200	– –	– –	– –	
Copper-nickel-zinc alloys							cf. DIN EN 12163 (1998-04)
CuNi12Zn24 (CW430J)	R380 R640	2–50 2–4	– –	380 640	270 550	38 –	Very good cold workability, machinable, easily polished; deep-drawn parts, flatware, applied arts, architecture, spring contacts
	H090 H190	2–50 2–4	90–130 ≥ 190	– –	– –	– –	
CuNi18Zn20 (CW409J)	R400 R650	2–50 2–4	– –	400 650	280 580	35 –	Good cold workability, machinable, non-tarnishing, easily polished; membranes, spring contacts, flatware
	H100 H200	2–50 2–4	100–140 ≥ 200	– –	– –	– –	

[1] Material numbers according to DIN EN 1412, see page 173. [2] C Material condition according to DIN EN 1173, see page 173.
[3] *D* Diameter for round bars, width across flats for flat bars and hexagonal bars, thickness for flat bars.

Cast copper alloys

cf. DIN EN 1982 (1998-12)

Designation, Material number[1]	Tensile strength R_m N/mm²	Yield strength $R_{p0,2}$ N/mm²	Elong. at fracture EL %	Hardness HB	Properties, application
CuZn15As-C (CC760S)	160	70	20	45	Excellent soft and hard solderability, salt water resistant; flanges
CuZn32Pb2-C (CC750S)	180	70	12	45	Good machinability, resistant to industrial water up to 90 °C; armatures
CuZn25Al5Mn4Fe-C (CC762S)	750	450	8	180	Very high strength and hardness, good machinability; plain bearings
CuSn12-C (CC483K)	260	140	7	80	High wear-resistance; spindle nuts, worm gears
CuSn11Pb2-C (CC482K)	240	130	5	80	Wear-resistant, good dry running properties; plain bearings
CuSn5Zn5Pb5-C (CC491K)	200	90	13	60	Resistant to seawater, solderable (hard and soft soldering), fittings, enclosures
CuAl10Fe2-C (CC331G)	500	180	18	100	Mechanically stressed parts; levers, housings, bevel gears
CuAl10Fe5Ni5-C (CC333G)	600	250	13	140	Strength and corrosion stressed parts; pumps

1) Material numbers according to DIN EN 1412, see page 173. More cast Cu alloys for plain bearings, see page 257. Strength values apply to separately sand-cast test specimens.

High-grade cast zinc alloys

cf. DIN EN 12844 (1999-01)

ZP3 (ZP0400) ZP5 (ZP0410)	280 330	200 250	10 5	83 92	Very good castability; preferred alloys for die castings
ZP2 (ZP0430) ZP8 (ZP0810)	335 370	270 220	5 8	102 100	Good castability; very good machinability, universally applicable
ZP12 (ZP1110) ZP27 (ZP2720)	400 425	300 300	5 2.5	100 120	Injection, blow, and deep-draw molds for plastics, sheet metal working tools

MS

Composite materials, Ceramic materials

Composite materials

Composite material	Base material[1]	Fiber content %	Density ϱ g/cm³	Tensile strength σ_B N/mm²	Elong. at tear ε_R %	Modulus of elasticity E N/mm²	Service temperature up to °C	Application examples
FRP (Fiberglass reinforced plastic)	EP	60	–	365	3.5	–	–	Shafts, joints, connecting bars, ship hulls, rotor blades
	UP	35	1.5	130	3.5	10800	50	Containers, tanks, pipes, dome lights, body parts
	PA 66	35	1.4	160[2]	5[3]	5000	190	Large-area, stiff housing parts, power plugs
	PC	30	1.42	90[2]	3.5[3]	6000	145	Housings for printers, computers, televisions
	PPS	30	1.56	140	3.5	11200	260	Lamp sockets and coils in electrical equipment
	PAI	30	1.56	205	7	11700	280	Bearings, valve seat rings, seals, piston rings
	PEEK	30	1.44	155	2.2	10300	315	Light construction materials in aerospace applications, metal substitute
CFRP (Carbon fiber reinforced plastic)	PPS	30	1.45	190	2.5	17150	260	Like FRP-PPS
	PAI	30	1.42	205	6	11700	180	Like FRP-PAI
	PEEK	30	1.44	210	1.3	13000	315	Like FRP-PEEK

[1] EP epoxide; UP unsaturated polyester; PA 66 polyamide 66, semi-crystalline; PC polycarbonate; PPS polyphenylene sulfide; PAI polyamideimide; PEEK polyetheretherketone

[2] σ_y yield stress

[3] ε_S elongation at yield stress

Ceramic materials

Material Name	Material Designation	Density ϱ g/cm³	Flexural strength σ_b N/mm²	Modulus of elasticity E N/mm²	Coefficient of linear expansion α 1/K	Properties, application examples
Aluminum silicate	C130	2.5	160	100000	0.000005	Hard, wear-resistant, chemical and heat resistant, high insulating resistance; insulators, catalytic converters, refractory housings
Aluminum oxide	C799	3.7	300	300000	0.000007	Hard, wear-resistant, chemical and heat resistant; ceramic inserts, wire drawing dies, biomedicine
Zirconium oxide	ZrO_2	5.5	800	210000	0.000010	High stability, high strength, heat and chemical resistant, wear-resistant; drawing dies, extrusion dies
Silicon carbide	SiC	3.1	600	440000	0.000005	Hard, wear-resistant, thermal-shock resistance, corrosion-resistant even at high temperatures; abrasives, valves, bearings, combustion chambers
Silicon nitride	Si_3N_4	3.2	900	330000	0.000004	High stability, thermal-shock resistance, high strength; cutting ceramics, guide and runner blades for gas turbines
Aluminum nitride	AlN	3.0	200	300000	0.000005	High thermal conductivity, high electrical insulation property; semiconductors, housings, heatsinks, insulating parts

Sintered metals

Designation system for sintered metals

cf. DIN 30910-1 (1990-10)

Designation example: **Sint - A 1 0 sintered smooth**

- Sint: Sintered metal
- A: Code letters for material class
- 1: 1. 1st number for chemical composition
- 0: 2. 2nd number for further differentiation without systematics
- sintered smooth: Treatment condition

Code letters for material class		
Code letter	**Volume ratio R_x in %**	**Area of application**
AF	< 73	Filters
A	75 ± 2.5	Plain bearings
B	80 ± 2.5	Plain bearings, molded parts with sliding properties
C	85 ± 2.5	plain bearing, molded parts
D	90 ± 2.5	Molded parts
E	94 ± 1.5	Molded parts
F	> 95.5	Sintered forged molded parts

1. 1st number for chemical composition	
Number	**Chemical composition mass fraction in %**
0	**Sintered iron, sint. steel,** Cu < 1% with or without C
1	**Sintered steel,** 1% to 5% Cu, with or without C
2	**Sintered steel,** Cu > 5%, with or without C
3	**Sintered steel,** with or without Cu or C, other alloying elements < 6%, e.g. Ni
4	**Sintered steel,** with or without Cu or C, other alloying elements > 6%, e.g. Ni, Cr
5	**Sintered alloys,** Cu > 60%, e.g. sintered CuSn
6	**Sintered nonferrous heavy metals,** except for no. 5
7	**Sintered light alloys,** e.g. sintered aluminum
8, 9	**Reserved numbers**

Treatment condition

Treatment condition of the material		Treatment condition of the surface	
• sintered	• steam treated	• sintered smooth	• machined
• calibrated	• sintered forged	• calibrated smooth	• surface treated
• heat treated	• isostatically pressed	• sized and coined smooth	

Sintered metals (selection)

vgl. DIN 30910-2, -6 (1990-10), DIN 30910-3 (2004-11), DIN 30910-4 (2010-03)

Designation	Hardness HB_{min}	Tensile strength R_m N/mm²	Chemical composition	Properties, application examples
Sint-AF 40	–	80–200	Sintered steel, Cr 16–19%, Ni 10–14%	Filter parts for gas and liquid filters
Sint-AF 50	–	40–160	Sintered bronze, Sn 9–11%, rem. Cu	
Sint-A 00	> 25	> 60	Sintered iron, C < 0.3%, Cu < 1%	Bearing materials with exceptionally large pore volume for the best emergency running properties; bearing liners, bearing bushings
Sint-A 20	> 30	> 80	Sintered steel, C < 0.3%, Cu 15–25%	
Sint-A 50	> 25	> 70	Sintered bronze, C < 0.2%, Sn 9–1%, rem. Cu	
Sint-A 51	> 20	> 60	Sintered bronze, C 0.2–2%, Sn 9–11%, rem. Cu	
Sint-B 00	> 30	> 80	Sintered iron, C < 0.3%, Cu < 1%	Plain bearings with very good dry running properties, low stressed molded parts
Sint-B 10	> 40	> 150	Sintered steel, C < 0.3%, Cu 1–5%	
Sint-B 50	> 30	> 90	Sintered bronze, C < 0.2%, Sn 9–11%, rem. Cu	
Sint-C 00	> 40	> 120	Sintered iron, C < 0.3%, Cu < 1%	Plain bearings, molded parts with average stress with good sliding properties; auto parts, levers, clutch parts
Sint-C 10	> 55	> 200	Sintered steel, C < 0.3%, Cu 1–1,5%	
Sint-C 40	> 100	> 300	Sintered steel, Cr 16–19%, Ni 10–14%, Mo 2%	
Sint-C 50	> 35	> 140	Sintered bronze, C < 0.2%, Sn 9–11%, rem. Cu	
Sint-D 00	> 50	> 170	Sintered iron, C < 0.3%, Cu < 1%	Molded parts for higher stresses; wear-resistant pump parts, gears, some are corrosion-resistant
Sint-D 10	> 60	> 250	Sintered steel, C < 0.3%, Cu 1–5%	
Sint-D 30	> 80	> 460	Sintered steel, C < 0.3%, Cu 1–5%, Ni 1–5%	
Sint-D 40	> 130	> 400	Sintered steel, Cr 16–19%, Ni 10–14%, Mo 2%	
Sint-E 00	> 60	> 240	Sintered iron, C < 0.3%, Cu < 1%	Molded parts for precision engineering, for household appliances, for the electrical industry
Sint-E 10	> 100	> 340	Sintered steel, C < 0.3%, Cu 1–5%	
Sint-E 73	> 55	> 200	Sintered aluminum Cu 4–6%	
Sint-F 00	> 140	> 600	Sinter forged steel, containing C and Mn	Sealing rings, flanges for muffler systems
Sint-F 31	> 180	> 770	Sinter forged steel, containing C, Ni, Mn, Mo	

MS

MS

Overview of plastics

General properties	Advantages:	Disadvantages:
	• low density • electrically insulating • heat and sound absorbing • decorative surface • economical forming • weather and chemical resistance	• lower strength and heat resistance in comparison to metals • some are combustible • some are nonresistant to solvents • limited material reutilization

Classification	Thermoplastics	Thermosets	Elastomers
Processing	Hot workable Weldable Generally glueable Machinable	Not workable Non-weldable Glueable Machinable	Not workable Non-weldable Glueable Machinable at low temperatures
Fabrication	Injection molding Injection blow molding Extruding	Pressing Transfer molding Injection molding, molding	Pressing Injection molding Extruding
Recycling	Easily recyclable	Not recyclable, possible reuse as filler	Not recyclable

Structure	Temperature behavior

Amorphous thermoplastica

Filamentary macromolecules without cross-linking

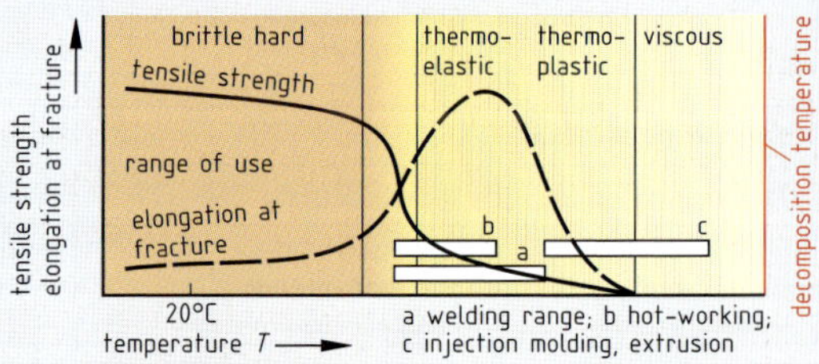

Semi-crystalline thermoplastic

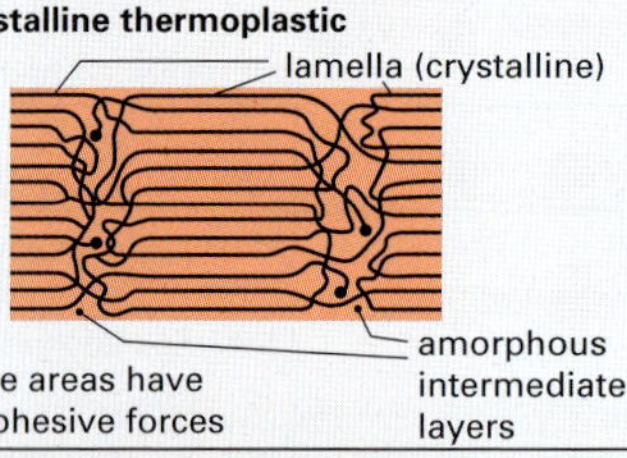

Crystalline areas have greater cohesive forces

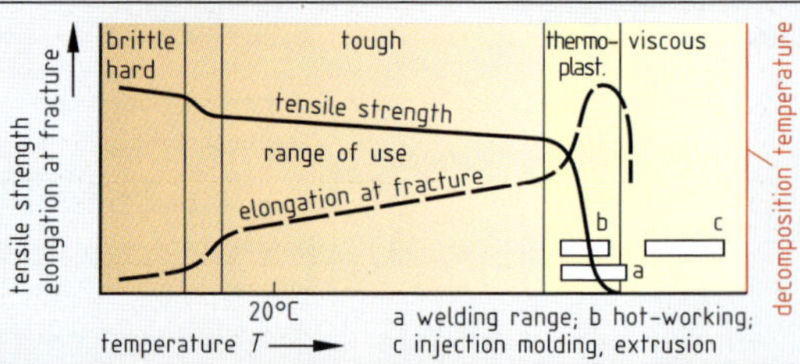

Filamentary thermoset plastics

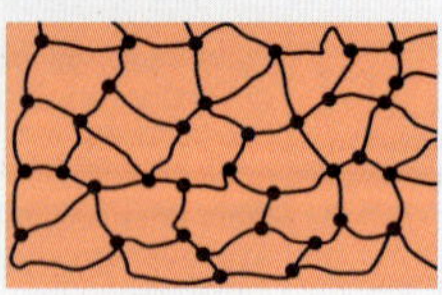

Macromolecules with many cross-links

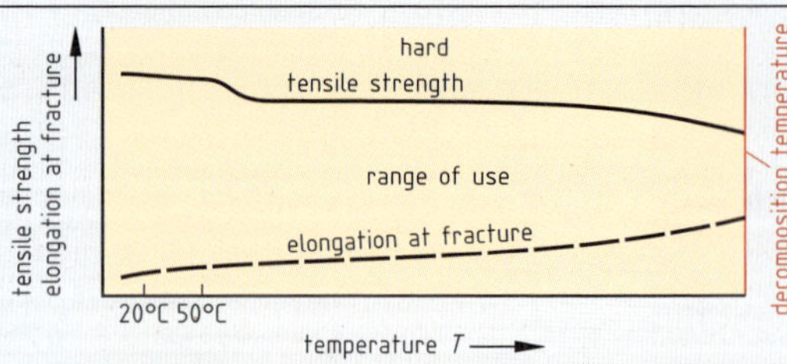

Filamentary elastomers

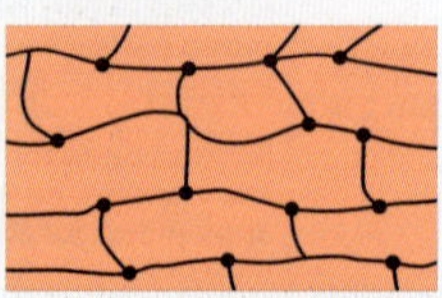

Macromolecules in random condition with few cross-linkages

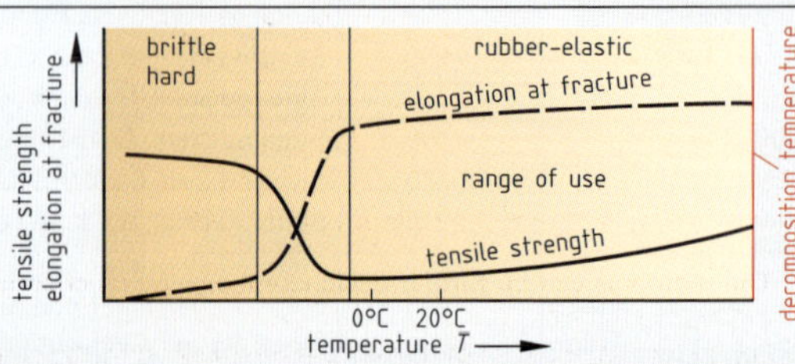

Basic polymers, fillers and reinforcing materials

Designations for basic polymers

cf. DIN EN ISO 1043-1 (2002-06)

Designation	Meaning	Type[1]	Designation	Meaning	Type[1]	Designation	Meaning	Type[1]
ABS	Acrylonitrile butadiene styrene	T	PAK	Polyacrylate	T	PTFE	Polytetrafluoroethylene	T
AMMA	Acrylonitrile-methyl-methacrylate	T	PAN	Polyacrylonitrile	T	PUR	Polyurethane	D
			PB	Polybutene	T	PVAC	Polyvinyl acetate	T
			PBT	Polybutylene terephthalate	T	PVB	Polyvinyl butyral	T
ASA	Acrylonitrile-styrene-acrylate	T	PC	Polycarbonate	T	PVC	Polyvinyl chloride	T
CA	Cellulose acetate	T	PE	Polyethylene	T	PVDC	Polyvinylidene chloride	T
CAB	Cellulose acetate butyrate	T	PEEK	Polyether ether ketone		PVF	Polyvinyl fluoride	T
CF	Cresol-formaldehyde	D	PET	Polyethyleneterephthalate	T	PVFM	Polyvinyl formaldehyde	T
CMC	Carboxymethyl cellulose	MNM	PF	Phenol formaldehyde	D	PVK	Poly-N-vinylcarbazole	T
CN	Cellulose nitrate	MNM	PI	Polyimide	T	SAN	Styrene-acrylonitrile	T
CP	Cellulose propionate	T	PMMA	Polymethylmethacrylate	T	SB	Styrene-butadiene	T
EC	Ethyl cellulose	MNM	POM	Polyoxymethylene; Polyformaldehyde	T	SI	Silicone	D
EP	Epoxide	D				SMS	Styrene-α-methylstyrene	T
EVAC	Ethylene-vinyl acetate	E	PP	Polypropylene	T	UF	Urea-formaldehyde	D
MF	Melamine formaldehyde	D	PPS	Polyphenylene sulfide	T	UP	Unsaturated polyester	D
PA	Polyamide	T	PS	Polystyrene	T	VCE	Vinyl chloride-ethylene	T

[1] MNM modified natural materials; E elastomers; D thermoset plastics; T thermoplastics

Code letters for designation of special properties

cf. DIN EN ISO 1043-1 (2002-06)

CL[1]	Special properties	CL[1]	Special properties	CL[1]	Special properties	CL[1]	Special properties
B	block, brominated	F	flexible; liquid	N	normal; novolak	T	temperature
C	chlorinated; crystalline	H	high; homo	O	oriented	U	ultra; no plasticizers
D	density	I	impact tough	P	plasticized	V	very
E	foamed; elastomer	L	linear, low	R	raised; resol; hard	W	weight
		M	moderate, molecular	S	saturated; sulphonated	X	cross-linked, cross-linkable

⇒ **PVC-P**: Polyvinylchloride, plasticized; **PE-LLD**: Linear Polyethylene low density

[1] code letter

Code letters and abbreviations for fillers and reinforcing materials

cf. DIN EN ISO 1043-2 (2002-04)

Abbreviation for material[1]

Designation	Material	Designation	Material	Designation	Material	Designation	Material
B	Boron	G	Glass	P	Mica	T	Talc
C	Carbon	K	Calcium carbonate	Q	Silicate	W	Wood
D	Aluminum trihydrate	L	Cellulose	R	Aramid	X	not specified
E	Clay	M	Mineral, metal[2]	S	Synthetic materials	Z	other

Abbreviations for shape and structure

Designation	Shape, structure	Designation	Shape, structure	Designation	Shape, structure	Designation	Shape, structure
B	pearls, balls, beads	G	ground stock	N	nonwoven (thin)	VV	veneer
		H	whiskers	P	paper	W	woven
C	chips, shavings	K	knitwear	R	roving	X	not specified
D	powder	L	laminates	S	peelings, flakes	Y	yarn
F	fibers	M	matted, thick	T	spun yarn, cord	Z	other

⇒ **GF**: glass fiber; **CH**: carbon whisker; **MD**: mineral powder

[1] The materials can be further designated, e.g. by its chemical symbol or another symbol from relevant international standards.

[2] For metals (M) the type of metal must be specified by the chemical symbol.

MS

Identification, Distinguishing characteristics

Methods for identifying plastics

Floating test: Solution density in g/cm³	Floating test: Plastics floating
0.9–1.0	PB, PE, PIB, PP
1.0–1.2	ABS, ASA, CAB, CP, PA, PC, PMMA, PS, SAN, SB
1.2–1.5	CA, PBT, PET, POM, PSU, PUR
1.5–1.8	Organically filled molding material
1.8–2.2	PTFE

Solubility in solvents

Thermosets and PTFE are not soluble.

Other thermoplastics are soluble in certain solvents; e.g. PS is soluble in benzene or acetone.

Visual test – Appearance of the specimen is: transparent	cloudy
CA, CAB, CP, EP, PC, PS, PMMA, PVC, SAN	ABS, ASA, PA, PE, POM, PP, PTFE

Touch

Waxy to the touch: PE, PTFE, POM, PP

Behavior when heated

- Thermopl. soften and melt
- Thermosets and elastomers decompose without softening

Burning test

- flame color
- fire behavior
- soot formation
- odor of the smoke

Distinguishing characteristics of plastics

Designation[1)]	Density g/cm³	Burning behavior	Other characteristics
ABS	≈ 1.05	Yellow flame, soots strongly, smells like coal gas	Tough elastic, is not dissolved by carbon tetrachloride, sounds dull
CA	1.31	Yellow, sputtering flame, drips, smells like distilled vinegar and burnt paper	Pleasant to the touch, sounds dull
CAB	1.19	Yellow, sputtering flame, drips burning, smells like rancid butter	Sounds dull
MF	1.50	Very flammable, chars with white edges, smells like ammonia	Very brittle, rattling sound (compare to UF)
PA	≈ 1.10	Blue flame with yellow edges, drips in fibers, smells like burnt horn	Tough elastic, not brittle, sounds dull
PC	1.20	Yellow flame, goes out after flame is removed, soots, smells like phenol	Tough hard, not brittle, rattling sound
PE	0.92	Light flame with blue core, drips off burning, odor like paraffin, smoke hardly visible (compare with PP)	Wax like surface, can be scratched with the fingernail, not brittle, working temperature > 230 °C
PF	1.40	Very flammable, yellow flame, chars, smells like phenol and burnt wood	Very brittle, rattling sound
PMMA	1.18	Luminous flame, fruity odor, crackles, drips	Clear when uncolored, sounds dull
POM	1.42	Bluish flame, drips, smells like formaldehyde	Not brittle, rattling sound
PP	0.91	Light flame with blue core, drips off burning, odor like paraffin, smoke hardly visible (compare with PE)	Cannot mark with fingernail, not brittle
PS	1.05	Yellow flame, soots strongly, smells sweet like coal gas, drips off burning	Brittle, sounds like tinny metal, is dissolved by carbon tetrachloride among others
PTFE	2.20	Nonflammable, strong odor when red hot	Waxy surface
PUR	1.26	Yellow flame, very strong odor	Polyurethane, rubber elastic
	≈ 0.05		Polyurethane foam
PVC-U	1.38	Very flammable, extinguishes after the flame is removed, smells like hydrochloric acid, chars	Rattling sound (U = hard)
PVC-P	1.20–1.35	Can be more flammable than PVC-U, depending on plasticizer, smells like hydrochloric acid, chars	Rubbery flexible, no sound (P = soft)
SAN	1.08	Yellow flame, soots strongly, smells like coal gas, drips off burning	Tough elastic, is not dissolved by carbon tetrachloride
SB	1.05	Yellow flame, soots strongly, smells like coal gas and rubber, drips off burning	Not as brittle as PS, is dissolved by carbon tetrachloride among other things
UF	1.50	Very flammable, chars with white edges, smells like ammonia	Very brittle, rattling sound (compare to MF)
UP	2.00	Luminous flame, chars, soots, smells like styrene, glass fiber residue	Very brittle, rattling sound

1) Compare to page 179

Thermoset plastics

Abbreviation, chemical designation	Trade names (selection)	Look, density[1] g/cm³	Elongation at fracture[2] N/mm²	Impact toughness kJ/mm²	Working temperature[2] °C
PF Phenol formaldehyde	Bakelite, Kerit, Supraplast, Vyncolit, Ridurid	Yellowish brown 1.25	40–90	4.5–5.0	140–150
MF Melamine-formaldehyde resin	Bakelite, Resopal, Hornit	Colorless 1.45	30	6.5–7.0	100–130
UF Urea-formaldehyde resin	Bakelite UF, Resamin, Urecoll	Colorless 1.5	35–55	4.5–7.5	80
UP Unsaturated polyester resin	Palatal, Rütapal, Polylite, Bakelite, Ampal, Resipol	Yellowish, glass clear 1.12–1.27	50–80	5.0–10.0	50
EP Epoxy resin	Epoxy, Rütapox, Araldit, Grilonit, Supraplast, Bakelite	Yellow, opaque 1.15–1.25	55–80	10.0–22.0	80–100

Code, chemical designation	Mechanical properties	Electrical properties	Contact with food, water absorption[2]
PF Phenol formaldehyde	Hard, brittle, strength depends on filler	Satisfactory insulating properties	Not permitted 50–300 mg
MF Melamine-formaldehyde resin	Hard, brittle, less notch-sensitive than UF, scratch resistant, high post-shrinkage	Satisfactory insulating properties, non-tracking	Partly permitted 180–250 mg
UF Urea-formaldehyde resin	Hard, brittle, notch-sensitive	Satisfactory insulating properties	Not permitted 300 mg
UP Unsaturated polyester resin	Brittle to tough, high strength and rigidity, weather-resistant	Fair insulating properties, excellent track resistance	Partly permitted 30–200 mg
EP Epoxy resin	Brittle to tough, high strength and rigidity, weather-resistant	Excellent insulating properties, non-tracking	Recognized as safe to a great extent 10–30 mg

Code, chemical designation	Resistant to	Not resistant to	Suitable for[3] a	m	Application
PF Phenol formaldehyde	Oil, grease, alcohol, benzene, gasoline, water	Strong acids and lyes	++	+	Housings, bearings, handles, pumps, ignition systems, gears, pot and pan handles
MF Melamine-formaldehyde resin	Oil, grease, alcohol, weak acids or lyes	Strong acids and lyes	+	+	Bright-colored electrical products: switches, plugs, terminals; tableware
UF Urea-formaldehyde resin	Solvents, oil, grease	Strong acids and lyes, boiling water	+	+	Bright-colored screw connections; sanitary products; electrical installation material
UP Unsaturated polyester resin	Gasoline, UV light, weathering, mineral lubricants	Mineral acids, acetone, organic acids, strong lyes	+	++	Silos, tanks for fuel oil and beverages, chassis, spoilers, sports boats, relays, tennis rackets
EP Epoxy resin	Thinned acids and lyes, alcohol, gasoline, oil, grease	Strong acids and lyes; acetone	++	+	Cast resins: gages, models; laminates: automotive; molding compounds: precision parts with metal core

1) Unreinforced
2) Depending on the type of reinforcing fibers and the processing method (compression molding or transfer molding)
3) a adhesive bonding, m machining, + well, ++ very well

MS

Thermoplastics

Abbreviation, chemical designation	Trade names (selection)	Density g/cm³, structure	Resistant to	Not resistant to	Working temperature °C
Transparent plastics[1]					
PC Polycarbonate	Makrolon, Lexan, Tecanat, Calibre	1.20–1.24 amorphous	Gasoline, grease, oil, water (< 60 °C)	Lyes, acetone, benzene, water (> 60 °C)	−100 to +115
PET Polyethylene terephthalate	Arnite, Rynite, Valox, Hostadur	1.33–1.38 semi-crystalline	Oil, grease, fuels	Hot water, acetone, concentrated acids and lyes	−20 to +115
PMMA Polymethyl methacrylate	Acrylite, Plexiglas, Plexidur, Perspex	1.19 amorphous	Aqueous acids and lyes, grease, light	Gasoline containing benzene, spirit, nitrocellulose lacquer, concentrated lyes	−40 to +80
PS Polystyrene	Vestyron, Luran, Empera, Styron	1.05 amorphous	Lyes, alcohol, water, non-aging	Gasoline, acetone, UV-sensitive	−20 to +70
SAN Styrene acrylonitrile	Luran, Lustran, Kibisan, Tyril	1.08 amorphous	Gasoline, oil, weak acids and lyes	Acetone, UV-sensitive	+90
Engineering plastics1[)]					
ABS Acrylonitrile butadiene styrene	Lustran, Magnum, Terluran, Tarodur	1.02–1.07 amorphous	Gasoline, mineral oil, grease, water	Concentrated mineral acids, benzene	−30 to +80
CA Cellulose acetate	Tenite, Acetate, Vuscacelle, Cellolux, Dexel	1.26–1.29 amorphous	Grease, oil, gasoline, water, benzene	Strong acids, lyes, alcohol	0 to +70
PA 6 Polyamide 6	Durethan B, Ultramid, Vydyne, Ertalon, Taromid	1.12–1.15 semi-crystalline	Gasoline, oil, grease, weak lyes	Strong lyes, phenols, mineral acids	−40 to +85
PA 66 Polyamide 66	Acromid, Durethan A, Acromit A, Ultramid A	1.12–1.14 semi-crystalline	Gasoline, oil, grease, weak lyes	Strong lyes, phenols, mineral acids	−30 to +95
PE HD Polyethylene, high density	Hostalen, Lupolen, Vestolen	0.94–0.96 semi-crystalline	Water, alcohol, oil, gasoline	Strong oxidants	−50 to +80
POM Polyoxymethylene, polyformaldehyde	Tenac, Delrin, Hostaform, Ultraform	1.41–1.43 crystalline	Gasoline, mineral oil, suds, alcohol	Strong acids, UV-rays, water > 65 °C	−50 to +110
PP Polypropylene	Hostalen, Vestolen, Inspire	0.90–0.92 semi-crystalline	Suds, weak lyes, alcohol	Gasoline, benzene	0 to +110
PVC-P Polyvinyl chloride, soft	Vestolit, Coroplast	1.20–1.35 amorphous	Alcohol, oil, gasoline	Benzene, organic solvents	−20 to +60
PVC-U Polyvinyl chloride, hard	Hostalit, Vestolit, Vinidur	1.37–1.44 amorphous	Gasoline, oil, acids, lyes, alcohol	Benzene, nitric acid	−5 to +60
High-performance plastics[1]					
PEEK Polyether ether ketone	Hostalec, Ketron, Victrex	1.27 amorph. 1.32 semicr.	Most chemicals	UV-rays, concentrated nitric acid	−80 to +250
PI Polyimide	Kinel, Meldin, Vespel	1.43 amorphous	Alcohol, kerosene, dilute acids	Hot water, weathering, acids, lyes	−250 to +240
PPS Polyphenylene sulfide	Techtron, Ryton, Tedur	1.43 semi-crystalline	Concentrated hydrochloric and sulfur acids	Concentrated nitric acid, UV-rays	−50 to +220
PSU Polysulfone	Mindel, Tecason, Ultrason, Udel	1.24 amorphous	Grease, oil, gasoline, alcohol, hot water	Benzene, UV-rays	−50 to +150
PTFE Polytetrafluor ethylene	Teflon, Hostalon, Polyflon	2.14–2.20 semi-crystalline	Nearly all aggressive substances, UV-rays	Alkaline metals	−200 to +260

[1] Common classification in retail

MS

Thermoplastics

Abbre-viation	Yield stress N/mm²	Yield strain %	Suitable for[1) a]	w	m	General properties	Application examples
Transparent plastics							
PC	65	80	+	+	++	Hard, abrasion-resistant, impact-resistant, partly permitted for food	Lenses, spectacle glasses, inspection glasses, tableware, housings, protective glasses, car lamps, CDs, helmets
PET	90	15	+	+	+	Very hard, highly resistant to wear and compression	Packing material, cam disks, gears, sliding bearings, housings, plumbing, magnetic tape
PMMA	60–80	5.5	+	+	++	Good optical properties, hard, brittle, scratch-resistant	Spectacle glasses, scales, tail lights, tubs, housings, control buttons
PS	50	3	++	+	++	Hard, brittle, notch-sensitive	Lamps, combs, tooth brushes, bobbin cores, relays, transparent packing
SAN	60–70	2–3	++	+	+	Stiff, impact-resistant, scratch-proof, hard surface	Transparent housing parts and packing, scale disks, tableware, emergency triangles
Engineering plastics							
ABS	37	4	+	+	++	Highly impact-resistant (also at –40 °C), hard, scratch-proof	Enclosures and control elements for audio and video equipment, radiator shrouds, spoilers, toys
CA	37	–	+	+	+	High strength, tough, impact-resistant, scratch-proof	Handles, ballpoint pens, combs, toys, switch buttons
PA 6	45	> 200	+	+	++	High strength, abrasion-resistant, excellent sliding properties	Gears, sliding bearings, coupling elements, cam disks, motorcycle helmets
PA 66	55	> 100	+	+	++	Harder than PA 6, higher load-bearing capacity, lower water absorption	Roller bearing cages, bearing bushes, screws, oil filters, intake pipes, motorcycle helmets
PE HD	20–30	9	–	+	–	Fracture proof, also under frost conditions, not scratch-resistant, good electrical isolator	Handles, gaskets, seals, fuel tanks, sliding elements, water pipes
POM	65–70	35	–	++	++	Excellent strength and shape stability, tough, abrasion-resistant	Thin-walled precision parts, gears, sliding elements, pump parts, housings
PP	30	8	–	+	+	Like PE HD, however, not frost-resistant	Fan blades, pump housings, spoilers, truck fenders, transformer enclosures, suitcase shells, toys
PVC-P (soft)	17–29	240–350	+	+	–	Soft, flexible, abrasion-resistant, narrow temperature range	Hoses, tubes, pipes, gaskets, cable insulation, toys, suitcases
PVC-U (hard)	50–60	10–50	++	++	+	High strength and hardness, notch-sensitive	Fittings, pipes, tanks, oil and beverage bottles, gutters, cable channels
High-performance plastics							
PEEK	110	20–25	+	+	+	High tensile and bending strength, impact-resistant, notch-sensitive	Base material for printed circuits, replacement for metals, implants, valves
PI	74	8[3)]	+	+	+	High hardness and strength, low toughness, wear-resistant	Gears, components for jet engines, turbine blades, piston rings, sliding bearings
PPS	78[2)]	5[3)]	+	+	++	High strength at high temperatures, low toughness	Often reinforced with fibers, valves, parts for pumps and carburetors, fuel cells, sensors
PSU	80[2)]	10[3)]	+	+	+	High strength, good toughness and heat resistance	Structural parts under high mechanical an thermal loads, daylight projectors
PTFE	20–40[2)]	250–400	–	–	+	High chemical resistance, good electrical insulator	Maintenance-free bearings, piston rings, gaskets, isolators, valves, pumps

[1)] a adhesive bonding, w welding, m machining; [2)] Tensile strength; [3)] Elongation at tear;
++ very well, + well, – not at all or only under particular conditions

MS

Semi-finished products made of thermoplastics

Round bars

cf. DIN 16980 (1987-05)

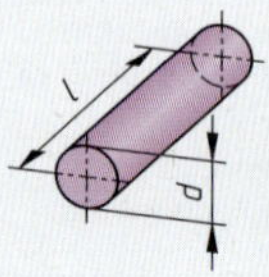

Material	PMMA	PA 6	PP	PA 66 GF 30
Color	Crystal clear	Natural[1], black	Natural, gray	Black
d in mm *l* in mm	15–100 1000	3–300 1000; 3000	3–500 2000	10–200 1000; 3000
Material	**PVC**	**PET**	**PC**	**PE-HD**
Color	Black, white, red, gray	Light gray, natural, black	Natural	Natural, black
d in mm *l* in mm	3–300 1000; 2000; 3000	3–200 1000; 2000; 3000	3–200 1000; 3000	3–500 2000; 1000

Tubes and hollow bars[2]

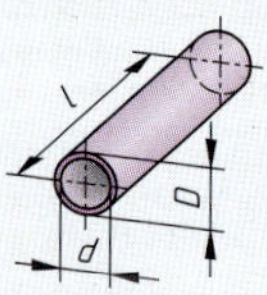

Material	PMMA	PA 6[2]	PA 66[2]	PVC
Color	Transparent, white	Natural, black	Natural	Gray
D x *d* in mm *l* in mm	5 x 3 ... 400 x 390 2000	20 x 10 ... 280 x 200 1000; 2000; 3000	20 x 10 ... 350 x 310 1000; 2000; 3000	6 x 4 ... 200 x 196 5000
Material	**PC**	**PET[2]**	**POM[2]**	**POM GF 25[2]**
Color	Colorless	Natural	Natural, black	Natural, black
D x *d* in mm *l* in mm	10 x 1... 250 x 5 2000	20 x 12 ... 200 x 150 1000; 2000; 3000	50 x 30 ... 200 x 150 1000; 2000	50 x 30 ... 200 x 150 1000; 2000

MS

Flat bars

vgl. DIN 16986 (1989-06)

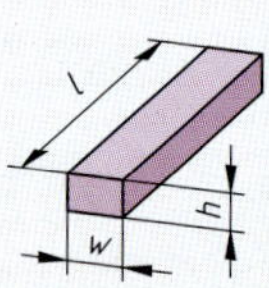

Material	PA 6	PA 6 GF 30	PA 66 PE	PA 12
Color	Natural, black	Natural, black	Natural	Natural, black
l in mm *w* in mm *h* in mm	1000; 2000; 3000 300; 500 5–100	1000; 2000; 3000 300; 500 10–50	1000; 2000; 3000 300; 500 5–100	1000; 2000; 3000 300; 500 5–100
Material	**PET**	**POM**	**POM GF 23**	**POM PTFE**
Color	Natural, black	Natural	Natural, black	Natural, black
l in mm *w* in mm *h* in mm	1000; 2000; 3000 300; 500 5–100	1000; 2000; 3000 300; 500 5–100	1000; 2000; 3000 300; 500 10–50	1000; 2000; 3000 300; 500 10–50

Boards, panels

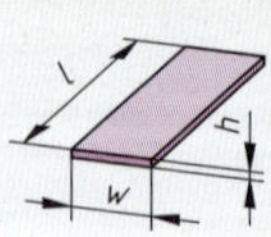

Material	PMMA	PA 6	PET	POM
Color	Transparent	Natural, black	Natural, black	Natural, black
l in mm *w* in mm *h* in mm	2000; 3050 1220; 2030 0.5–100	1000; 2000; 3000 620; 1000 3–100	1000; 2000; 3000 620; 1000 3–100	1000; 2000; 3000 620; 1000 0.5–100

Various PVC profiles

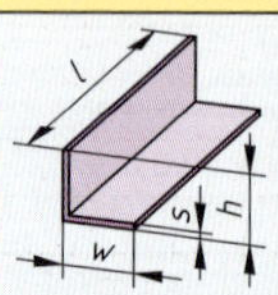

Profile section	U-channel section	T-section	Angle section	Square tube
Color	Gray	Gray	Gray	Gray
l in mm *w* in mm *h* in mm *s* in mm	3000 13–90 15–20 1.5–2.5	3000 30–50 30–50 4; 5	3000 15–90 15–90 2–7	3000 20–120 20–120 1.5–2.5

⇒ **Flat bar DIN 16986 – PA 6 – 20 x 300:** h = 20 mm, b = 300 mm, made of polyamide 6

1) Natural means that no substances were added in order to change the color of the material.
2) In general, hollow bars have a greater wall thickness than tubes.

Elastomers, Foam materials

Elastomers (rubber)

Abbreviation[1]	Designation	Density g/cm³	Tensile strength[2] N/mm²	Elong. at fracture %	Working temperature °C	Properties, application examples
BR	Butadiene rubber	0.94	2 (18)	450	−60 to +90	High abrasion resistance; tires, belts, V-belts
CO	Epichlorhydrin rubber	1.27 −1.36	5 (15)	250	−30 to +120 −10 to +120	Vibration damping, oil and gasoline resistant; seals, heat resistant dampers
CR	Chloroprene rubber	1.25	11 (25)	400	−30 to +110	Oil and acid resistant, very flammable, seals, hoses, V-belts
CSM	Chlorosulfonated polyethylene	1.25	18 (20)	300	−30 to +120	Aging and weather resistant, oil resistant; insulating material, molded goods, films
EPDM	Ethylene-propylene rubber	0.86	4 (25)	500	−50 to +120	Good electrical insulator, not resistant against oil and gasoline; seals, profiles, bumpers, cold water hoses
FKM	Fluoro rubber	1.85	2 (15)	450	−10 to +190	Abrasion resistant, best thermal resistance; aerospace and automotive industries; rotary shaft seals, O-rings
IIR	Isobutene-Isoprene rubber	0.93	5 (21)	600	−30 to +120	Weather and ozone resistant; cable insulation, automotive hoses
IR	Isoprene rubber	0.93	1 (24)	500	−60 to +60	Low resistance to oil, high strength; truck tires, spring elements
NBR	Acrylonitrile-butadiene rubber	1.00	6 (25)	450	−20 to +110	Abrasion resistant, oil and gasoline resistant, electr. conductors, O-rings, hydraulic hoses, rotary shaft seals, axial seal
NR	Natural rubber Isoprene rubber	0.93	22 (27)	600	−60 to +70	Low resistance to oil, high strength; truck tires, spring elements
PUR	Polyurethane rubber	1.25	20 (30)	450	−30 to +100	Elastic, wear-resistant; timing belts, seals, couplings
SIR	Styrene-Isoprene rubber	1.25	1 (8)	250	−80 to +180	Good electr. insulator, water repellant O-rings, spark plug caps, cylinder head and joint sealing
SBR	Styrene-Butadiene rubber	0.94	5 (25)	500	−30 to +80	Low resistance to oil and gasoline; tires, hoses, cable sheathing

[1] cf. DIN ISO 1629 (1992-03) [2] Value in parentheses = with additive or filler reinforced elastomer

Foam materials

cf. DIN 7726 (1982-05)

Foam materials consist of open cells, closed cells or a mixture of closed and open cells. Their raw density is lower than that of the structural substance. A distinction is made between hard, medium hard, soft, elastic, soft elastic and integral foam material.

Stiffness, hardness	Raw material base of the foam material	Cell structure	Density kg/m³	Max. working temperature °C[1]	Thermal conductivity W/(K · m)	Water absorption in 7 days Vol.-%
Hard	Polystyrene	Predominantly closed cell	15 – 30	75 (100)	0.035	2 – 3
	Polyvinylchloride		50 – 130	60 (80)	0.038	< 1
	Polyethersulfone		45 – 55	180 (210)	0.05	15
	Polyurethane		20 – 100	80 (150)	0.021	1 – 4
	Phenolic resin	Open cell	40 – 100	130 (250)	0.025	7 – 10
	Urea-formaldehyde resin		5 – 15	90 (100)	0.03	20
Medium-hard to soft-elastic	Polyethylene	Predominantly closed cell	25 – 40	up to 100	0.036	1 – 2
	Polyvinylchloride		50 – 70	−60 to +50	0.036	1 – 4
	Melamine resin		10.5 – 11.5	up to 150	0.033	approx. 1
	Polyurethane polyester type	Open cell	20 – 45	−40 to +100	0.045	–
	Polyurethane polyether type					

[1] Long-term working temperature, short-term in parentheses

MS

Plastics processing

Injection molding and extrusion

Abbreviation	Injection molding temperature in °C		Injection pressure in bar	Extrusion process temperature in °C	Shrinkage in %	Tolerance group[1] for		
	Substance	Mold				General tolerances	Dimensions with deviations Series 1[2]	Dimensions with deviations Series 2[2]
PE	160–300	20–70	500	190–230	1.5–3.5	150	140	130
PP	170–300	20–100	1200	235–270	0.8–2[3]	150	140	130
PVC, hard	170–210[4]	30–60	1000–1800	170–190	0.2–0.5	130	120	110
PVC, soft	170–200[4]	20–60	300	150–200	1–2.5	–	–	–
PS	180–250	30–60	–	180–220	0.3–0.7	130	120	110
SB	180–250	20–70	–	180–220	0.4–0.7	130	120	110
SAN	200–260	40–80	–	180–200	0.5–0.6	130	120	110
ABS	200–240	40–85	800–1800	180–220	0.4–0.7	130	120	110
PMMA	200–250	50–90	400–1200	180–250	0.3–0.8	130	120	110
PA	210–290	80–120	700–1200	230–275	1–2	130	120	110
POM	180–230[4]	50–120	800–1700	180–220	1–3.5	140	130	120
PC	280–320[4]	80–120	> 800	240–290	0.7–0.8	130	120	110
PF[5]	90–110[4]	170–190	800–2500	–	0.5–1.5[3]	140	130	120
MF[6]	95–110[4]	160–180	1500–2500	–	0.6–1.7[3]	130	120	110
UF[5]	95–110	150–160	1500–2500	–	0.4–0.6	140	130	120

1) See table below 2) Series 1: Can be maintained without special effort, Series 2: Requires high finishing effort
3) Transverse and longitudinal shrinkage may differ 4) With screw injection molding machine
5) With organic filler material 6) With inorganic filler material

Tolerances for plastic molded parts

cf. DIN 16901 (1982-11); withdrawn

Tolerance group from table above	Code-letter[1]	Nominal dimension range over – up to in mm												
		0–1	1–3	3–6	6–10	10–15	15–22	22–30	30–40	40–53	53–70	70–90	90–120	120–160
General tolerances														
150	A	±0.23	±0.25	±0.27	±0.30	±0.34	±0.38	±0.43	±0.49	±0.57	±0.68	±0.81	±0.97	±1.20
	B	±0.13	±0.15	±0.17	±0.20	±0.24	±0.28	±0.33	±0.39	±0.47	±0.58	±0.71	±0.87	±1.10
140	A	±0.20	±0.21	±0.22	±0.24	±0.27	±0.30	±0.34	±0.38	±0.43	±0.50	±0.60	±0.70	±0.85
	B	±0.10	±0.11	±0.12	±0.14	±0.17	±0.20	±0.24	±0.28	±0.33	±0.40	±0.50	±0.60	±0.75
130	A	±0.18	±0.19	±0.20	±0.21	±0.23	±0.25	±0.27	±0.30	±0.34	±0.38	±0.44	±0.51	±0.60
	B	±0.08	±0.09	±0.10	±0.11	±0.13	±0.15	±0.17	±0.20	±0.24	±0.28	±0.34	±0.41	±0.50
Tolerances for dimensions with deviations														
140	A	0.40	0.42	0.44	0.48	0.54	0.60	0.68	0.76	0.86	1.00	1.20	1.40	1.70
	B	0.20	0.22	0.24	0.28	0.34	0.40	0.48	0.56	0.66	0.80	1.00	1.20	1.50
130	A	0.36	0.38	0.40	0.42	0.46	0.50	0.54	0.60	0.68	0.76	0.88	1.02	1.20
	B	0.16	0.18	0.20	0.22	0.26	0.30	0.34	0.40	0.48	0.56	0.68	0.82	1.00
120	A	0.32	0.34	0.36	0.38	0.40	0.42	0.46	0.50	0.54	0.60	0.68	0.78	0.90
	B	0.12	0.14	0.16	0.18	0.20	0.22	0.26	0.30	0.34	0.40	0.48	0.58	0.70
110	A	0.18	0.20	0.22	0.24	0.26	0.28	0.30	0.32	0.36	0.40	0.44	0.50	0.58
	B	0.08	0.10	0.12	0.14	0.16	0.18	0.20	0.22	0.26	0.30	0.34	0.40	0.48

1) A for dimensions which do not depend on mold dimensions; B for dimensions which depend on mold dimensions

Polyblends, Reinforcing fibers, Laminated materials

Polyblends

Polyblends (also known as "blends") are mixtures of different thermoplastics. The special properties of these copolymers result from numerous possible combinations of the properties of the original materials.

Abbreviation	Designation	Components	Special properties	Application examples
S/B	Styrene/butadiene	90% polystyrene, 10% butadiene rubber	Brittle hard, at low temperatures not impact tough	Stacking boxes, fan housings, radio housings
ABS	Acrylonitrile/butadiene/ styrene	90% styrene-acrylonitrile, 10% nitrile rubber	Brittle hard, impact tough even at low temperatures	Telephones, dash-boards, hub caps
PPE + PS	Polyphenylenether + Polystyrene	various compositions; possibly can be reinforced with 30% glass fiber	High hardness, high cold impact toughness to –40 °C, physiologically harmless	Radiator grill, computer parts, medical equipment, solar panels, trims
PC + ABS	Polycarbonate + Acrylnitrile/Butadiene/ Styrene	various compositions	High strength, hardness, toughness, dimensional stability under heat, impact tough, shock-proof	Instrument panels, fenders, office machine housings, lamp housings in motor vehicles
PC + PET	Polycarbonate + Polyethyleneterephthalate	different compositions	Exceptional impact toughness and shock resistance	Motorcycle helmets, automotive parts

Reinforcing fibers

Designation	Density kg/dm³	Tensile strength N/mm²	Elongation at fracture %	Special properties	Application examples
Glass fiber GF	2.52	3400	4.5	Isotropic[1], good strength, high-temp. strength, inexpensive	Body parts, aircraft industry, sailboats
Aramide fibers AF[3]	1.45	3400 – 3800	2.0–4.0	Lightest reinforcing fiber, ductile, fracture tough, strongly anisotropic[1], radar-penetrable	Highly stressed light parts, crash helmets, bulletproof vests
Carbon fiber CF	1.6–2.0	1750 – 5000[2]	0.35–2.1[2]	Extremely anisotropic[1], high-strength, light, corrosion resistant, good electr. conductor	Parts for racing cars, sails for racing yachts, aerospace applications

Thermosets (e.g. UP and EP resins) and thermoplastics with high working temperatures (e.g. PSU, PPS, PEEK, PI) are used as embedding materials (so-called **matrix**).

[1] Isotropic = the same material properties in all directions; anisotropic = material properties in the direction of the fibers are different from those transverse to fibers
[2] Depends significantly on the fiber defect sites occurring during the manufacturing process
[3] Trade name "Kevlar"

Laminated materials[1]

cf. DIN EN 60893 (2004-12)

Resin types		Types of reinforcing materials	
Type of resin	Designation	Abbreviation	Designation
EP	Epoxy resin	CC	Cotton fabric
MF	Melamine (formaldehyde) resin	CP	Cellulose paper
PF	Phenolic (formaldehyde) resin	CR	Combined reinforcing material
UP	Unsaturated polyester resin	GC	Glass fiber fabric
SI	Silicone resin	GM	Fiber glass mat
PI	Polyimide resin	WV	Wood veneer
Nominal thicknesses t in mm	0.4; 0.5; 0.6; 0.8; 1.0; 1.2; 1.5; 2; 2.5; 3; 4; 5; 6; 8; 10; 12; 14; 16; 20; 25; 30; 35; 40; 45; 50; 60; 70; 80; 90; 100		
⇒	**Board IEC 60893 – 3 – 4 – PF CP 201, 10 x 500 x 1000:** Board made of phenolic (formaldehyde) resin/cellulose paper (PF CP 201) according to IEC standard[2] 60893-3-4 with t = 10 mm, w = 500 mm, l = 1000 mm.		

[1] Applications: insulators for electrical equipment, for instance, or bearing liners, rollers and gears for machine construction
[2] IEC = International Electrotechnical Commission (international standard)

MS

Testing of plastics: Tensile properties, Hardness testing

Determination of the tensile properties on plastics

cf. DIN EN ISO 527-1 (1996-04)

Typical stress-strain curves

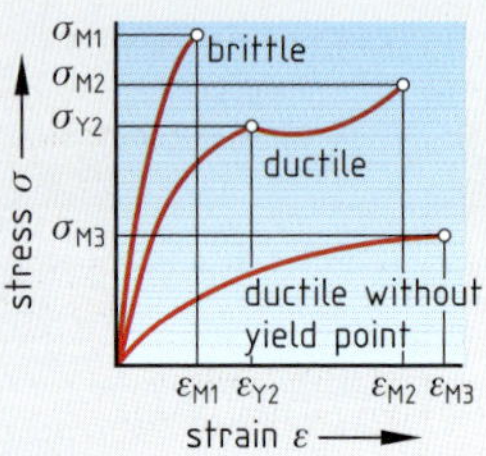

Test specimens

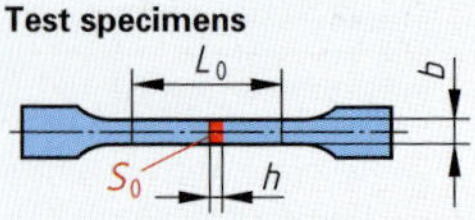

F_M maximum force
F_Y yield stress
ΔL_{FM} change in length with maximum load
ΔL_{FY} change in length with yield strength
L_0 gage length
S_0 initial cross section
σ_M tensile strength
σ_Y yield strength
ε_M maximum elongation
ε_Y yield strain

Test Specimens

For each property, e.g. tensile strength, yield strength, yield strain, at least five test specimens must be tested.

Application
- thermoplastic injection molded and extrusion molding materials
- thermoplastic slabs and films
- thermoset molding materials
- thermoset slabs
- fiber reinforced composite materials, thermoplastic and thermoset plastic

Tensile strength

$$\sigma_M = \frac{F_M}{S_0}$$

Yield strength

$$\sigma_Y = \frac{F_Y}{S_0}$$

Maximum elongation

$$\varepsilon_M = \frac{\Delta L_{FM}}{L_0} \cdot 100\,\%$$

Yield strain

$$\varepsilon_Y = \frac{\Delta L_{FY}}{L_0} \cdot 100\,\%$$

Test speed					Test specimen according to							
						DIN EN ISO 527-2 for molding materials				DIN EN ISO 527-3 for films		
Test speed in mm/min				Toler-ance	Type	1A	1B	5A	5B	2	4	5
					L_0 mm	50 ± 0.5	50 ± 0.5	20 ± 0.5	10 ± 0.2	50 ± 0.5	50 ± 0.5	25 ± 0.25
1	2	5	10	± 20%	h mm	4 ± 0.2	4 ± 0.2	≥ 2	≥ 1	≤ 1	≤ 1	≤ 1
20	50	100	200	± 10%	b mm	10 ± 0.2	10 ± 0.2	4 ± 0.1	2 ± 0.1	10–25	25.4 ± 0.1	6 ± 0.4

⇒ **Tensile test ISO 527-2/1A/50:** Tensile test according to ISO 527-2; specimen type 1A; test speed 50 mm/min

MS

Hardness test on plastics

cf. DIN EN ISO 2039-1 (2003-06)

Ball indentation test

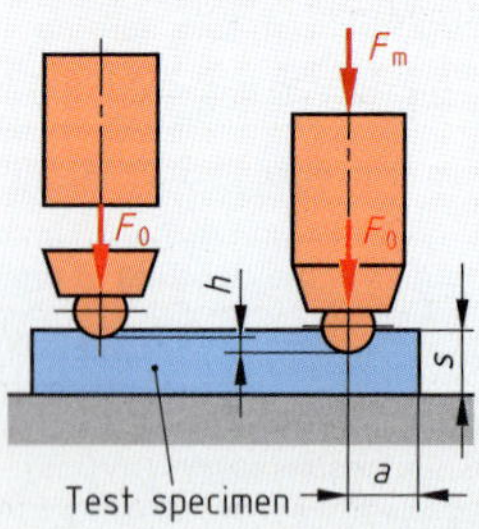

F_0 preload 9.8 N
F_m test load
h depth of penetration
a distance from edge
s specimen thickness

Test Specimens

distance from edge $a \geq 10$ mm, minimum specimen thickness $s \geq 4$ mm

Test load F_m in N	Ball indentation hardness H in N/mm² for indentation depth h in mm: 0.16	0.18	0.20	0.22	0.24	0.26	0.28	0.30	0.32	0.34
49	22	19	16	15	13	12	11	10	9	9
132	59	51	44	39	35	32	30	27	25	24
358	160	137	120	106	96	87	80	74	68	64
961	430	370	320	290	260	234	214	198	184	171

⇒ **Ball indentation hardness ISO 2039-1 H 132:** $H = 31$ N/mm² at $F_m = 132$ N

Hardness test by Shore on plastics

cf. DIN EN ISO 868 (2003-06)

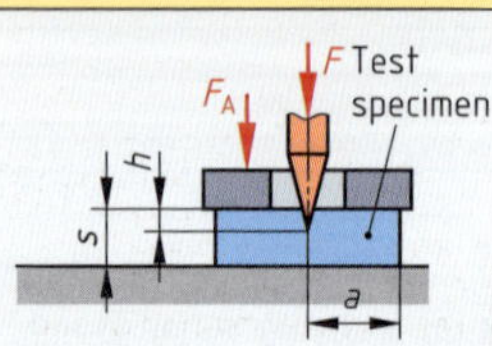

Indenters for

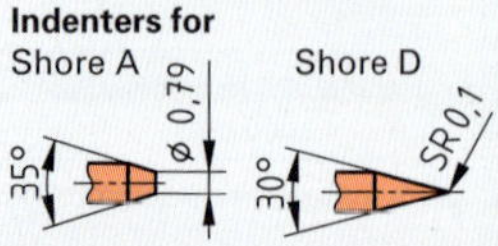

F_A contact pressure in N
F test load
h depth of penetration
a distance from edge
s specimen thickness

Test Specimens

Distance from edge $a \geq 9$ mm, minimum specimen thickness $s \geq 4$ mm

Test conditions for the Shore A and Shore D methods			
Test method	F_{max} in N	F_A in N	Application
A	7.30	10	if Shore hardness with Type D is < 20
D	40.05	50	if Shore hardness with Type A is > 90

⇒ **85 Shore A:** Hardness value 85; test method Shore A

Material testing methods – Overview

Illustration	Process	Application, information
Tensile test		page 191
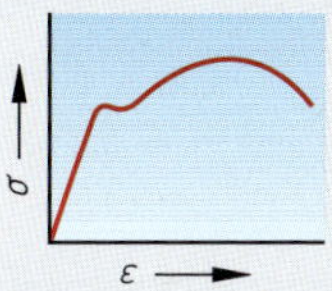	Standard tensile test specimens are pulled to fracture. The changes in tensile force and strain are measured and plotted on a graph. This is converted to a stress-strain curve.	Determination of material characteristic values, for example – calculation of static load strength – prediction of forming behavior – obtaining data for machining processes
Hardness test by Brinell HB		page 193
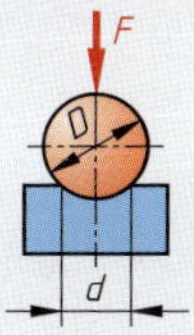	• Indenter ball is loaded with standardized test load *F* – test load depends on ball diameter *D* and on the material group → Degree of loading, see page 193 • Indentation diameter *d* is measured • Hardness is determined based on the test load and the surface area of indentation	Hardness test, e.g. on steels, cast iron materials, non-ferrous metals, which – are not hardened – have a metallic bright testing surface – are softer than 650 HB
Hardness test by Rockwell		page 194
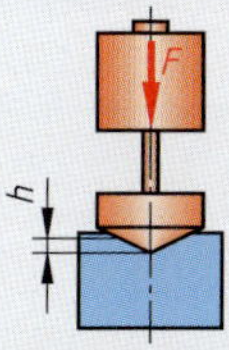	• Indenter (diamond cone, carbide ball) is loaded with minor test load → measurement baseline • Impact with major test load → permanent deformation of the test piece • Removal of the major load • Hardness is displayed directly on the test device and is based on the depth of penetration *h*	Hardness testing by different methods, e.g. on steels and non-ferrous metals, – in soft or hardened condition – with small thicknesses **Methods HRA, HRC:** hardened and high-strength metals **Methods HRB, HRF:** soft steel, non-ferrous metals
Hardness test by Vickers		page 194
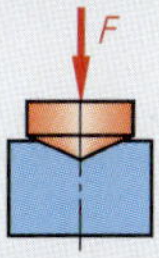	• The diamond pyramid is loaded with variable loads – test load is a function of parameters such as test piece thickness or grain size in matrix structure • The diagonals of the indentation are measured • Hardness is determined based on the test load and surface area of indentation	Universal method for testing – soft and hardened materials – thin layers – individual microstructural components of metals
Hardness test by penetrant testing (Martens hardness)		page 195
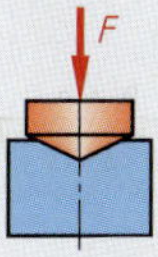	• Diamond pyramid is loaded with variable loads – test load is based on parameters such as test piece thickness or grain size • The load is logged continuously as a function of penetration depth • Martens hardness is determined **during** loading	Method for testing all materials, e.g. – soft and hardened metals – thin layers, also carbide coatings and paint coating – individual microstructure components – ceramic, hard material, etc.
Hardness test by ball penetration test		page 188
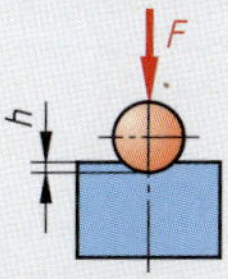	• The test ball is loaded with initial load → measurement baseline • Impact with established test load – test load must produce a penetration depth of 0.15–0.35 mm • The penetration depth is measured after 30 s loading time • Ball indentation hardness is determined	Testing of plastics and hard rubber. Ball indentation hardness provides comparison values for research, development and quality control.

MS

Material testing methods – Overview

Illustration	Process	Applications, notes
Hardness test by Shore		page 188
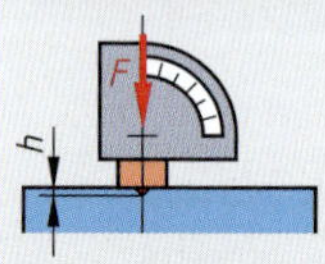	• The testing device (durometer) is pressed on the test piece with contact pressure F • The spring loaded indenter penetrates into the test piece • Working time 15 s • The shore hardness is displ. directly on the device	Control of plastics (elastomers). It is hardly possible to derive any relationships to other material properties from the shore hardness.
Shear test (DIN 50141 withdrawn without replacement)		
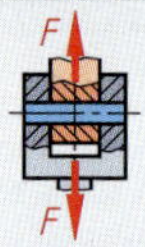	• Cylindrical specimens are loaded in standardized equipment until fractured due to shearing • Breaking strength is determined from the maximum shearing force and cross-sectional area of the test specimen	Used to determine the shear strength τ_{sB}, e.g. – for strength calculations of shear loaded parts, e.g. pins – to predict cutting forces in forming
Notched-bar impact bending test		page 192
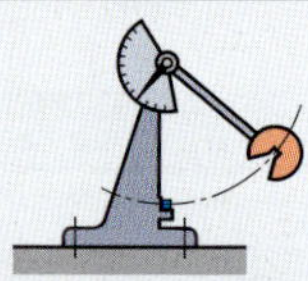	• Notched test specimens are subjected to bending load by pendulum impact and are fractured • Notch impact toughness = energy required to deform and fracture the test specimen	– To test metallic materials for behavior after impact bending loads – To monitor heat treatment results, e.g. with quenching and tempering – To test the temperature behavior of steels
Erichsen cupping test		
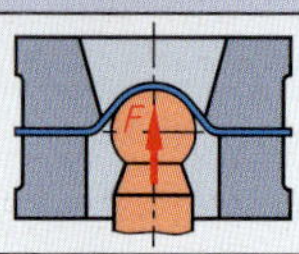	• Sheet metal clamped on all sides is deformed until crack formation by a ball • The deformation depth until crack propagation is a measure of deep drawing capability	– For testing of sheet metal and strip for their deep drawing capability – Evaluation of the sheet surface for changes during cold working
Fatigue test		page 192
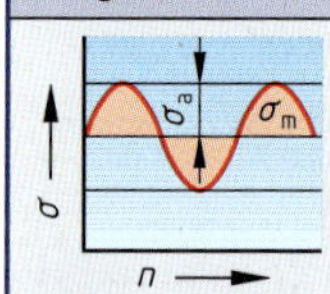	• Cylindrical specimens with polished surface are alternately loaded with constant mean stress σ_m and variable alternating stress amplitude σ_A, until fracture. The graphical representation of the series of tests yields the Wöhler (S-N) curve	Used to determine material properties with dynamic loading, e.g. – fatigue strength, fatigue endurance and fatigue strength under alternating stresses – endurance limit
Ultrasonic testing		
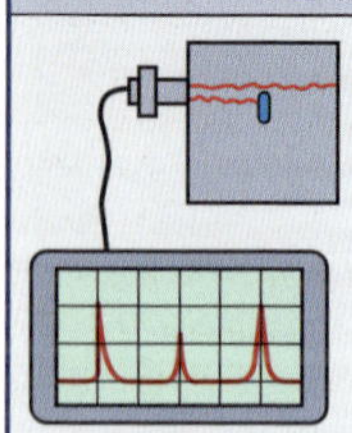	• A transducer sends ultrasonic signals through the workpiece. The waves are reflected by the front wall, the back wall and by defects of a certain size • The screen of the testing device displays the echoes • The test frequency determines the detectable defect size which is limited by the grain size of the test specimen	– Nondestructive testing of parts, e.g. for cracks, cavities, gas holes, inclusions, lack of fusion, differences in microstructure – To determine the type of defect, the size and the location of the defect – To measure wall and layer thicknesses
Metallography		
	Etching metallographic test specimens (microsections) develops the microstructure which can then be observed under the metallographic microscope. Specimen preparation: Removal → avoid structural transformation Embedding → sharp edged microsections Grinding → removal of layers of deformation Polishing → high surface quality Etching → structural development	– To check the crystalline structure – To monitor heat treatments, forming and joining processes – To determine grain distribution and grain size – Defect testing

Tensile test, Tensile test specimens

Tensile test

cf. DIN EN ISO 6892-1 (2009-12), replaces DIN EN 10002-1

Stress-strain diagram with distinct yield point, e.g. for soft steel

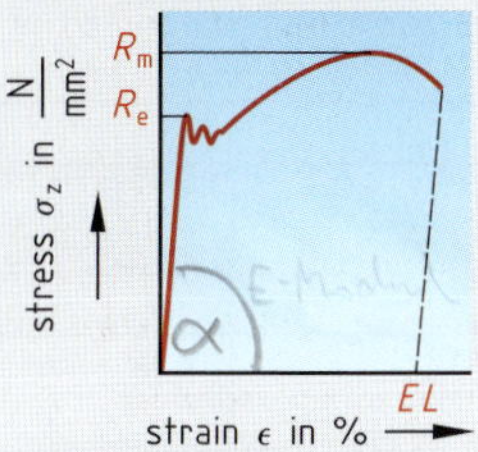

Stress-strain diagram without distinct yield point, e.g. for quenched and tempered steel

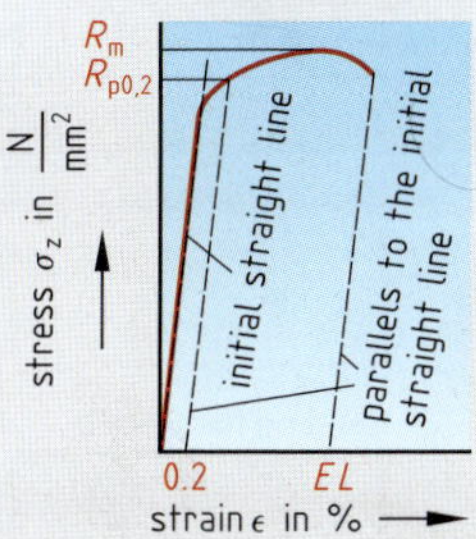

EL	elongation at fracture
F	tensile force
F_m	maximum force
F_e	force at yield strength limit
$F_{p0.2}$	force at yield strength limit at 0.2% strain offset
L_0	initial gage length
L_u	gage length after fracture
d_0	initial diameter of the test specimen
S_0	initial cross section of the test specimen
S_u	smallest test specimen cross section after fracture
ε	normal strain
Z	reduction of area at fracture
σ_z	tensile stress
R_m	tensile strength
R_e	yield strength
$R_{p\,0.2}$	yield strength at 0.2% strain offset
V_s	yield strength ratio

Tensile test specimens

As a rule, round proportional bars with an initial gage length of $L_0 = 5 \cdot d_0$ are used.

Unmachined specimens are allowed with
- uniform cross sections, e.g. for specimens of sheet metal, profiles, wires
- cast test specimens, e.g. of cast iron materials or non-ferrous casting alloys

Elongation at fracture EL

If tensile test specimens are used that contract during the test, the initial gage length L_0 has an effect on the elongation at fracture EL.

Smaller initial gage length L_0 → greater elongation at fracture EL

Yield strength ratio: $V_s = R_e\ (R_{p\,0.2})/R_m$
It provides information about the heat treatment condition of the steels:

normalized	$V_s \approx 0.5–0.7$
quenched & tempered	$V_s \approx 0.7–0.95$

Tensile stress

$$\sigma_z = \frac{F}{S_0}$$

Tensile strength

$$R_m = \frac{F_m}{S_0}$$

Yield strength

$$R_e = \frac{F_e}{S_0}$$

Yield strength at 0.2 % strain offset

$$R_{p0.2} = \frac{F_{p0.2}}{S_0}$$

Normal strain

$$\varepsilon = \frac{L - L_0}{L_0} \cdot 100\%$$

Elongation at fracture

$$EL = \frac{L_u - L_0}{L_0} \cdot 100\%$$

Reduction of area at fraction

$$Z = \frac{S_0 - S_u}{S_0} \cdot 100\%$$

MS

Tensile test specimens

cf. DIN 50125 (2009-07)

Shape B

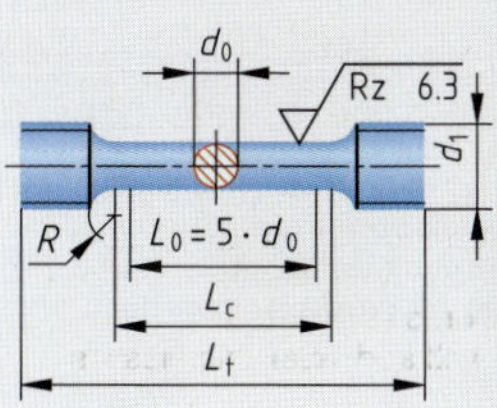

Shape E

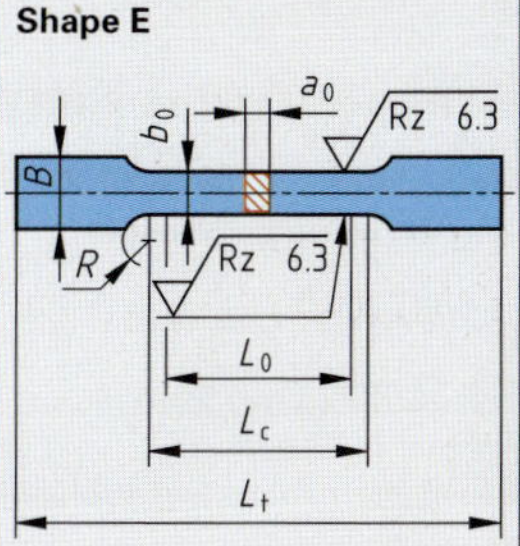

Round tensile test specimens, shapes A and B

		4	5	6	8	10	12	14	Shapes, application
	d_0	4	5	6	8	10	12	14	Shapes, application
	L_0	20	25	30	40	50	60	70	**Shape A:** machined test specimens for clamping in self-locking jaws
	L_c [1]	24	30	36	48	60	72	84	
	R [1]	3	4	5	6	8	9	11	
Shape A	d_1	5	6	8	10	12	15	17	**Shape B:** machined test specimens with threaded heads produce more precise elongation measurements
	L_t [1]	60	74	92	115	138	162	186	
Shape B	d_1	M6	M8	M10	M12	M16	M18	M20	
	L_t [1]	41	51	60	77	97	118	134	

Tensile test specimens, other shapes

	a_0	3	4	5	6	7	8	10	Shapes, application
	b_0	8	10	10	20	22	25	25	Flat specimen for clamping in self-locking jaws
Shape E	L_0	30	35	40	60	70	80	90	
	L_c	38	45	51	77	89	102	114	Tensile test specimens of strips, metal sheets, flat bars and profiles
	R [1]	12	12	12	15	20	20	20	
	L_t [1]	104	120	126	197	222	246	258	
Shape C	Machined round test specimens with shouldered ends								
Shape F	Unmachined sections of round bars								
Explanation	[1] Minimum dimensions								
⇒	**Tensile test specimen DIN 50125 – A10x50:** Shape A, d_0 = 10 mm, L_0 = 50 mm								

Notched-bar impact bending test, Rotary bending test

Notched-bar impact bending test in accordance with Charpy

cf. DIN EN 10045 (1991-04)

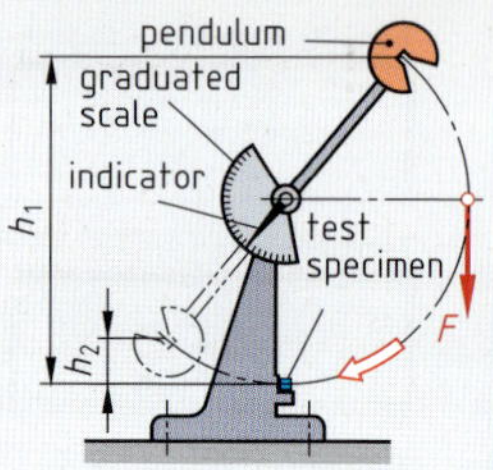

Test specimen cross section

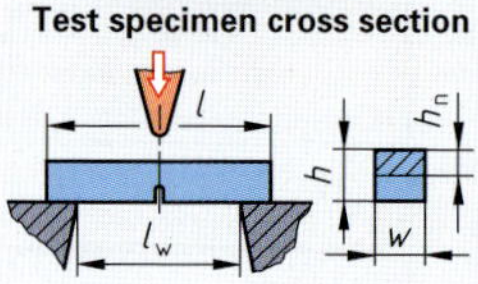

Notch shapes

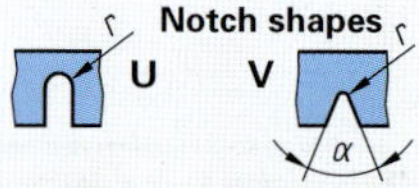

KU Notch impact energy in J, measured on a test specimen with U-notch

KV Notch impact energy in J, measured on a test specimen with V-notch

Test specimen

The test specimen must be completely machined. Fabrication of the test material should alter the material's microstructure as little as possible. No notch should be visible with the naked eye at the notch root, which runs parallel to the notch axis.

Notch impact test specimens

Designation	Notch shape	Test dimension in mm or degree (°)						
		l	l_w	h	w	h_n	r	α
Standard test specimen	U	55	40	10	10	5	1.0	–
Standard test specimen	V	55	40	10	10	8	0.25	45°
DVM test specimen[1]	U	55	40	10	10	7	1.0	–
Explanation	[1] Deutscher Verband für Materialprüfung (German Association for Material Testing)							

⇒

KU = 115 J: Standard test specimen with U-notch, notch impact energy 115 J, work capacity of the pendulum impact tester 300 J

KV150 = 85 J: Standard test specimen with V-notch, notch impact energy 85 J, work capacity of the pendulum impact tester 150 J

Rotary bending test (bending fatigue strength)

cf. DIN 50113 (1982-03)

MS

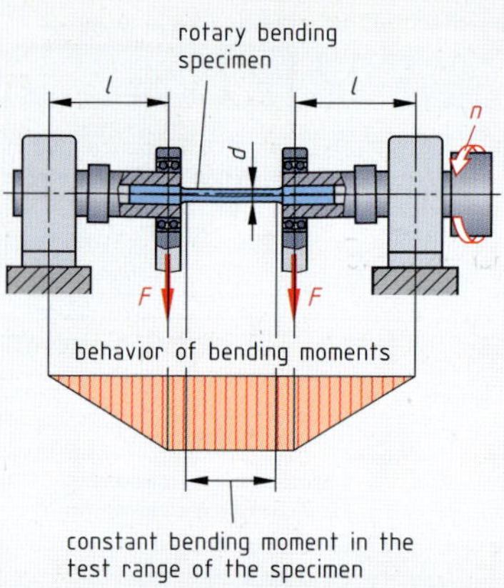

Wöhler (S-N) diagram of steel

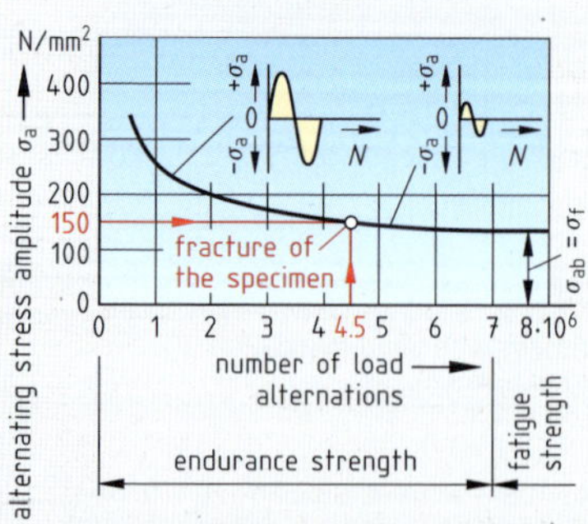

n speed in 1/min

N number of load alternations

d specimen diameter in mm

l distance between bearings

F bending force in N

M bending moment in N · mm

W section modulus in mm³

σ_a alternating stress amplitude in N/mm²

σ_f fatigue strength in N/mm²

σ_{ab} alternate bending strength in N/mm² ($\sigma_{ab} = \sigma_f$)

Section modulus

$$W = \frac{\pi \cdot d^3}{32}$$

Bending force

$$F = \frac{\sigma_a \cdot W}{l}$$

Test specimens

Cylindrical specimens with a ground or polished surface are alternately loaded (strain hardening is to be avoided). Diameter of specimen $d \leq 16$ mm

Test

A bending force F is applied to the test specimens rotating at speeds of n = 3000/min to 12000/min in such a manner that an alternating bending load applies with the preselected stress amplitude σ_a. The steel specimens break within the endurance range after $N < 7 \cdot 10^6$ load alternations. Loads in the fatigue range do not produce any fracture of specimens.

Test results

Fatigue strength σ_f = alternate bending strength σ_{ab}, endurance strength σ_N

Example:

Specimen diameter d = 10 mm, distance between bearings l = 100 mm, selected alternating stress amplitude σ_a = 150 N/mm²

Target: Section modulus W, bending force F to be adjusted

Solution: $W = \frac{\pi \cdot d^3}{32} = \frac{\pi \cdot (10\ \text{mm})^3}{32} = 98.17\ \text{mm}^3$

$$F = \frac{\sigma_a \cdot W}{l} = \frac{150\ \text{N/mm}^2 \cdot 98.17\ \text{mm}^3}{100\ \text{mm}} = \mathbf{147.3\ N}$$

Failure of the specimen at N = 4.5 · 10⁶ load alternations (see Wöhler (S-N) curve)

Hardness test by Brinell

Hardness test by Brinell

cf. DIN EN ISO 6506-1 (2006-03)

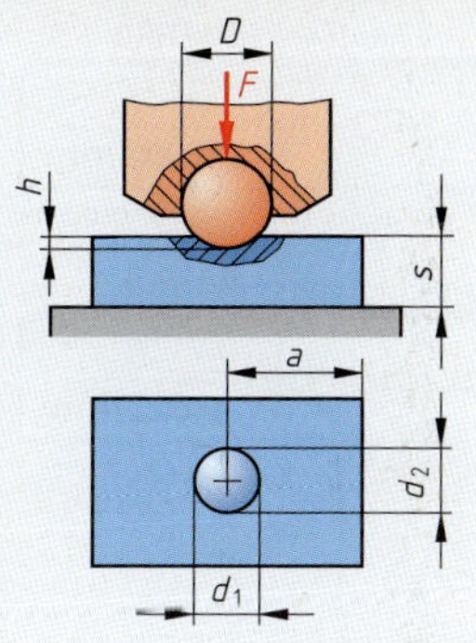

F	test load in N
D	ball diameter in mm
d	diameter of the impression in mm
d_1, d_2	individual measurement values of the impression diameter in mm
h	depth of impression in mm
s	minimum thickness of the test specimen in mm
a	distance from edge in mm

Test conditions

Impression diameter
$0.24 \cdot D \leq d \leq 0.6 \cdot D$

Minimum test specimen thickness $s \geq 8 \cdot h$

Distance from edge $a \geq 3 \cdot d$

Test specimen surface: metallic bright

Impression diameter

$$d = \frac{d_1 + d_2}{2}$$

Brinell hardness

$$\text{HBW} = \frac{0.204 \cdot F}{\pi \cdot D \cdot (D - \sqrt{D^2 - d^2})}$$

Designation examples:

180 HBW 2.5 / 62.5
600 HBW 1 / 30 / 25

Hardness value	Indenter	Ball diameter	Test force F	Impact time
Brinell hardness 180 Brinell hardness 600	W carbide ball	2.5 mm 1 mm	62.5 · 9.80665 N = 612.9 N 30 · 9.80665 N = 294.2 N	Unspecified: 10 to 15 s Value entry: 25 s

Degree of loading, ball diameter, test loads and test materials

Degree of loading $0.102 \cdot F/D^2$	Test load F in N with ball diameter D[1] in mm				Test range	
	1	2.5	5	10	Materials	Brinell hardness HBW
30	294.2	1839	7355	29420	Steel, nickel and titanium alloys Cast iron Copper, copper alloys	≤ 650 ≥ 140 > 200
15	–	–	–	14710	Light metal, light metal alloys	> 35
10	98.07	612.9	2452	9807	Cast iron Light metal, light metal alloys Copper, copper alloys	< 140 > 35 35 – 200
5	49.03	306.5	1226	4903	Copper, copper alloys Light metals, light metal alloys	< 35 35–80
2.5	24.52	153.2	612.9	2452	Light metals, light metal alloys	< 35
1	9.807	61.29	245.2	980.7	Lead, tin	–

[1] Small ball diameters for fine-grained materials, thin specimens or hardness tests in the outer layer. For hardness tests on cast iron, the ball diameter D must be ≥ 2.5 mm. Hardness values are only comparable if the tests were carried out with the same degree of loading.

Minimum thickness s of the specimens

Ball diameter D in mm	Minimum thickness s in mm for impression diameter d[1] in mm																	
	0.25	0.35	0.5	0.6	0.8	1.0	1.2	1.3	1.5	2.0	2.4	3.0	3.5	4.0	4.5	5.0	5.5	6.0
1	0.13	0.25	0.54	0.8														
2			0.23	0.37	0.67	1.07	1.6											
2.5				0.29	0.53	0.83	1.23	1.46	2.0									
5							0.58	0.69	0.92	1.67	2.45	4.0						
10											1.17	1.84	2.53	3.34	4.28	5.36	6.59	8.0

Example: D = 2.5 mm, d = 1.2 mm → minimum specimen thickness s = 1.23 mm

[1] Table fields without thickness indicated lie outside of the test range $0.24 \cdot D \leq d \leq 0.6 \cdot D$

MS

MS

Hardness test by Rockwell, Hardness test by Vickers

Hardness test by Rockwell

cf. DIN EN ISO 6508-1 (2006-03)

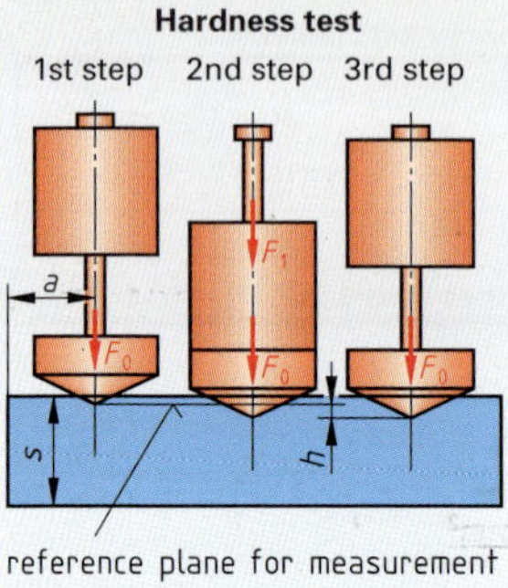

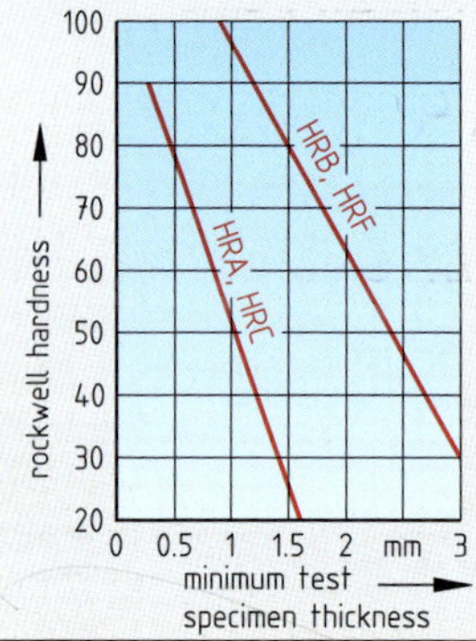

F_0 minor load in N
F_1 major load in N
h permanent indentation depth in mm
s test specimen thickness
a distance from edge

Test conditions

Surface of specimen is ground to Ra = 0.8–1.6 µm. The machining of the specimen must not result in any changes to the microstructure.

Distance from edge $a \geq 1$ mm

Rockwell hardness HRA, HRC

$$\text{HRA, HRC} = 100 - \frac{h}{0.002\ \text{mm}}$$

Rockwell hardness HRB, HRF

$$\text{HRB, HRF} = 130 - \frac{h}{0.002\ \text{mm}}$$

Designation examples:

65 HRC
70 HRBW

Hardness value	Test method	
65 70	HRC Rockwell hardness – C, test with diamond cone	HRBW Rockwell hardness – B, test with carbide ball

Test method, applications (selection)

Method	Indenter	F_0 in N	F_1 in N	Measurement range from – to	Application
HRA	Diamond cone, cone angle 120°	98	490.3	20–88 HRA	Hardened steel, high-strength metals
HRC		98	1373	20–70 HRC	
HRB	Carbide ball (W) 1.5785 mm	98	882.6	20–100 HRB	Soft steel, non-ferrous metals
HRF		98	490.3	60–100 HRF	

Hardness test by Vickers

cf. DIN EN ISO 6507-1 (2006-03)

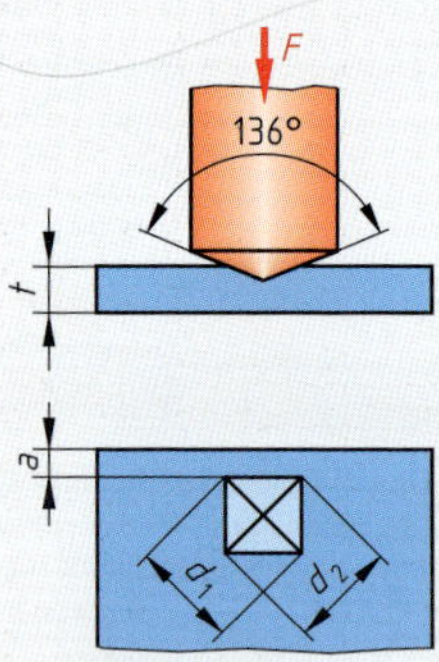

F test load in N
d diagonal of the indentation in mm
s test specimen thickness
a distance from edge

Test conditions

Surface of specimen is ground to Ra = 0.4–0.8 µm. The machining of the specimen must not result in any changes to the microstructure.

Distance from edge $a \geq 2.5 \cdot d$

Diagonal of the impression

$$d = \frac{d_1 + d_2}{2}$$

Vickers hardness

$$\text{HV} = 0.1891 \cdot \frac{F}{d^2}$$

HB ≈ ca. 0,95 · HV

Designation examples:

540 HV 1 / 20
650 HV 5

Hardness value	Test load F	Working time
Vickers hardn. 540 Vickers hardn. 650	1 · 9.80665 N = 9.807 N 5 · 9.80665 N = 49.03 N	Value entry 20 s Unspecified: 10 to 15 s

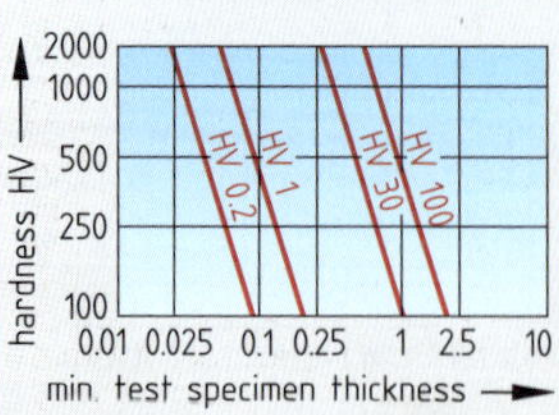

Test conditions and applied loads for the Vickers hardness test

Test condition	HV100	HV50	HV30	HV20	HV10	HV5
Test load in N	980.7	490.3	294.2	196.1	98.07	49.03
Test condition	HV3	HV2	HV1	HV0.5	HV0.3	HV0.2
Test load in N	29.42	19.61	9.807	4.903	2.942	1.961

Martens hardness, Conversion of hardness values

Martens hardness by penetrant testing

cf. DIN EN ISO 14577 (2003-05)

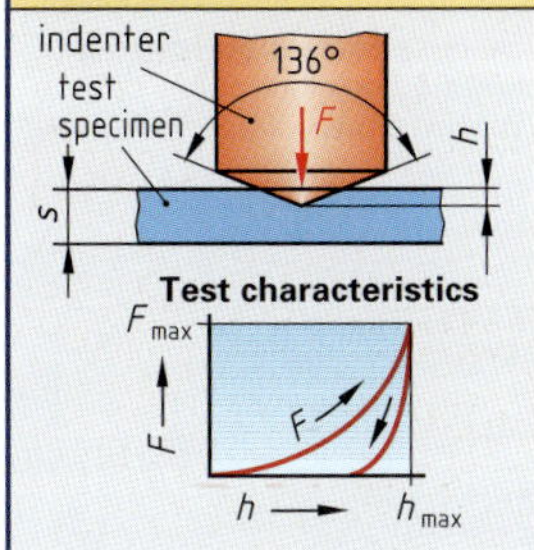

F test load in N
h depth of penetration in mm
s specimen thickness in mm

Test specimen surface			
Material	Average roughness Ra at F		
	0.1 N	2 N	100 N
Aluminum	0.13	0.55	4.00
Steel	0.08	0.30	2.20
Carbide	0.03	0.10	0.80

Martens hardness

$$HM = \frac{F}{26.43 \cdot h^2}$$

Designation: **HM 0.5 / 20 / 20 = 5700 N/mm²**

Test method	Test load F	Test duration	Application of load	Martens hardn. value
Martens hardness	0.5 N	20 s	within 20 s	5700 N/mm²

Test range	Conditions	Applications
Macro range	2 N ≤ F ≤ 30 kN	Universal hardness test, e.g. for all metals, plastics, carbides, ceramic materials; micro and nano ranges: thin layer measurement, microstructure components
Micro range	F < 2 N or H > 0.2 µm	
Nano range	h ≤ 0.2 µm	

Conversion tables for hardness values and tensile strength[1)]

cf. DIN EN ISO 18265 (2004-02)

Tensile strength R_m N/mm²	Vickers hardness HV (F ≧ 98 N)	Brinell hardness HB30	Rockwell hardness HRC	HRA	HRB[2)]	HRF[2)]
255	80	76	–	–	–	–
285	90	86	–	–	48	83
320	100	95	–	–	56	87
350	110	105	–	–	62	91
385	120	114	–	–	67	94
415	130	124	–	–	71	96
450	140	133	–	–	75	99
480	150	143	–	–	79	(101)
510	160	152	–	–	82	(104)
545	170	162	–	–	85	(106)
575	180	171	–	–	87	(107)
610	190	181	–	–	90	(109)
640	200	190	–	–	92	(110)
675	210	199	–	–	94	(111)
705	220	209	–	–	95	(112)
740	230	219	–	–	97	(113)
770	240	228	20	61	98	(114)
800	250	238	22	62	100	(115)
835	260	247	24	62	(101)	–
865	270	257	26	63	(102)	–
900	280	266	27	64	(104)	–
930	290	276	29	65	(105)	–
965	300	285	30	65	–	–
1030	320	304	32	66	–	–
1095	340	323	34	68	–	–

Tensile strength R_m N/mm²	Vickers hardness HV (F ≧ 98 N)	Brinell hardness HB30	Rockwell hardness HRC	HRA
1155	360	342	37	69
1220	380	361	39	70
1290	400	380	41	71
1350	420	399	43	72
1420	440	418	45	73
1485	460	437	46	74
1555	480	456	48	75
1595	490	466	48	75
1665	510	485	50	76
1740	530	504	51	76
1810	550	523	52	77
1880	570	542	54	78
1955	590	561	55	78
2030	610	580	56	79
2105	630	599	57	80
2180	650	618	58	80
–	670	–	59	81
–	690	–	60	81
–	720	–	61	82
–	760	–	63	83
–	800	–	64	83
–	840	–	65	84
–	880	–	66	85
–	920	–	68	85
–	940	–	68	86

1) Applies to unalloyed and low alloy steels and cast steel. Special tables of this standard are to be used for quenched and tempered, cold worked and high-speed steels, as well as for various carbide types. Considerable deviations are to be expected for high-alloyed and/or work-hardened steels.
2) The values in parentheses lie outside of the measurement range.

Corrosion

Electrochemical series of metals

In galvanic corrosion the same processes occur as in electrical elements where the base metals are corroded. The voltage produced between two dissimilar metals under influence of a conducting liquid (electrolyte) can be taken from the standard potentials of the electrochemical series. Standard potential refers to the voltage produced between the electrode material and a platinum electrode immersed in hydrogen.

Passivation (formation of protective layers) alters the voltage between the elements.

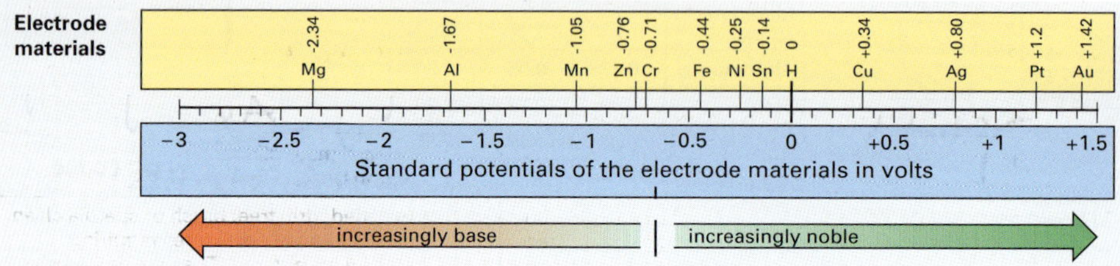

Example: The standard potentials of Cu = +0.34 V and Al = −1.67 V yield a voltage of U = +0.34 V − (−1.67 V) = 2.01 V between Cu and Al.

Corrosion behavior of metallic materials

MS

Materials	Corrosion behavior	Resistance in following environment: Dry ambient air	Country air	Industrial air	Sea air	Salt water
Unalloyed and alloy steels	Only resist corrosion in dry areas	●	◔	◔	○	○
Stainless steels	Resistant, but not against aggressive chemicals	●	●	◐	◐	◐
Aluminum and Al alloys	Resistant, except the Al alloys containing Cu	●	◐	◐	◐	● to ◔
Copper and Cu alloys	Resistant, especially Cu alloys containing Ni	●	●	◐	◐	● to ◔

● resistant ◐ fairly resistant ◔ non-resistant ○ unusable

Corrosion protection

Preparation of metal surfaces before coating

Processing step	Purpose	Process
Mechanical cleaning and creating a good surface for adherence	Removal of mill scale, rust and dirt	Grinding, brushing, blasting with water jet mixed with silica sand
Chemical cleaning and creating an optimal surface finish	Removal of mill scale, rust and grease residues Roughing or smoothing the surface	Etching with acid or lye; degreasing with solvents; chemical or electrochemical polishing

Preventative actions for corrosion protection

Actions	Examples
Select suitable materials	Stainless steel for parts for preparation in the paper production
Observe corrosion protection principles in design	Same material on contact points, insulation layers between the parts, avoiding gaps
Protective layers: • protective oil or lubricant • chemical surface treatment • protective paint	Oiling sliding tracks and measuring tools Phosphatizing, burnishing Lacquer coat, possible after previous phosphatizing
Metallic coatings	Hot-dip galvanizing, galvanic metal plating, e.g. chrome plating
Cathodic corrosion protection	Part to be protected, e.g. a ship propeller, is connected to a sacrificial anode
Anodic oxidation of Al materials	A corrosion-resistant permanent oxide layer is produced on the part, e.g. a rim

Disposal of substances*

Waste management laws

cf. Closed Substance Cycle and Waste Management Act (2001-10)

Important principles of recycling management

- Avoid waste, e.g. by in-house recycling management or a low-waste product design.
- Utilize material waste, e.g. by recovery of raw materials from waste (secondary raw materials).
- Use waste for recovery of energy (energy use), e.g. use as substitute fuel.
- Waste must be recycled properly without adverse effect on the well being of the general public.

The disposal of waste is subject to monitoring by the responsible authorities (usually the administrative district). In particular, wastes hazardous to health, air or water, explosive, and flammable especially need to be monitored.

The waste producer is responsible for proper disposal and documentation of disposal.

Examples of waste requiring special monitoring (hazardous waste) in metal processing industry[1)]

Disposal code	Description of the type of waste	Appearance, description, source	Special instructions, actions
150199D1	Packaging containing hazardous impurities	Barrels, canisters, buckets and cans contain residues of paints, lacquers, solvents, cleaning agents, rust preventatives, rust and silicone removers, spackle, etc.	Emptied, drip free, brush or spatula clean containers are not wastes requiring special monitoring. They are considered as retail packaging. Disposal using the dual system or in metal bins using a waste management company. Bins with dried paint are similar to house-hold commercial waste.
		Spray cans with residual contents	Spray cans should be avoided if possible; they must be disposed as hazardous waste.
160602	Nickel cadmium batteries	Rechargeable batteries, e.g. from drills and screwdrivers, etc.	All batteries containing contaminants are labeled. The dealer must accept their return at no charge. Consumers are required to return them to the dealer or to a public recycling center.
160603	Mercury dry cells	Coin cell batteries, mercury containing monocell batteries	
160604	Alkaline batteries	Non-rechargeable batteries	
060404	Mercury containing waste	Fluorescent lamps	Can be recycled. Return to dealer or to waste disposer. Do not put in glass recycling!
120106	Used machining oils, containing halogens, no emulsion	Water free drilling, turning, grinding and cutting oils, so-called cooling lubricants	Avoid cooling lubricants as much as possible, e.g. by • dry machining • minimum quantity cooling lubrication Separated collection of different cooling lubricants, emulsions, solvents. Inquire with supplier for reprocessing or combustion (energy recycling) options.
120107	Used machining oils, halogen free, no emulsion	Old, water free honing oil	
110	Synthetic machining oils	Cooling lubricants from synthetic oils, e.g. on ester-based	
130202	Non- chlorinated machine, gear and lubricating oils	Used oil and gear oil, hydraulic oil, compressor oil from piston air compressors	Recycling through supplier or a licensed waste disposal service. Used oils of known origin may be recycled by secondary refining or energy recovery. Do not mix with other materials!
150299D1	Vacuumed and filter materials, wipe cloths and protective clothing with hazardous contaminants	For example, used rags, cleaning cloths; brushes contaminated with oil or wax, oil binders, oil and lubricant cans	Option of using a rental service for cleaning cloths.
130505	Other emulsions	Condensation water from compressors	Use compressor oils with de-emulsifying properties; inquire about the option of oil free compressors.
140102	Other halogenated solvents and solvent mixtures	Per (-chloroethane) Tri (-chloroethene) Mixed solvents	Recycling by suppliers and test replacement with aqueous cleaning solution.

1) Regulation governing wastes requiring special monitoring – BestbüAbfV (1999-01), Appendix 1: Wastes listed in the European Waste Catalog (EAK waste) are considered to be especially hazardous. Appendix 2: EAK waste requiring special monitoring as well as waste types not on the EAK list (Letter "D" in Disposal code).

*) According to European Standards

Hazardous materials and material characteristics of hazardous gases

Identification and handling of hazardous materials

cf. EC Directive R 67/548/EEC[1]

Substance	Identification[2]			Substance	Identification[2]		
	Symbol	R-phrases	S-phrases		Symbol	R-phrases	S-phrases
Acetone	F, Xi	11; 36; 66; 67	9; 16; 26	Tetrachlor-ethane ("Per")	Xn; N	40; 51/53	23; 36/37; 61
Acetylene	F+	5; 6; 12	(2); 9; 16; 33	Kerosine	T	45	53; 45
Acrylonitrile	F, T, N	45; 11; 23/24; 25; 37/38; 41; 43; 51/53	9; 16; 45; 53; 61	Phenol	T; C	23/24/25; 34; 48/20/21/22; 68	24/25; 26; 28; 36/37; 39; 45
Ammonia	C; N	34; 50	26; 36/37/39; 61	Phosphoric acid	C	34	23; 45
Arsenic	T; N	23/25; 50/53	20/21; 28; 45; 60; 61	Propane	F+	12	9; 16
Asbestos	T	45; 48/23	53; 45	Mercury	T; N	23; 33; 50/53	7; 45; 60; 61
Gasoline	T	45; 65	53; 45	Hydrochloric acid	C	34; 37	26; 45
Benzene	F; T	45; 46; 11; 36/38; 48/23/ 24/25; 65	53; 45	Oxygen	O	8	17
Lead compounds	T; N	61; 20/22; 33; 62; 50/53	53; 45; 60; 61	Lubricating grease	T	45	53; 45
Chromium compounds	T; N	49; 43; 50/53	53; 45; 60; 61	Lubricating oil	T	45	53; 45
Hydrofluoric acid (HF)	T+; C	26/27/28; 35	7/9; 26; 36/37; 45	Sulphoric acid	C	35	26; 30; 45
Ceramic mineral fibers	T	49; 38	53; 45	Styrene	Xn	10; 20; 36/38	23
Carbon monoxide	F+; T	61; 12; 23; 48/23	53; 45	Turpentine, oil	Xn; N	10; 20/21; 36/38; 43; 51/53; 65	36/37; 46; 61; 62
Fiber glass	Xn	38; 40	35/37	Trichlorethylene (Tri)	T	45; 36/38; 52/53; 67	53; 45; 61
Nicotine	T+; N	25; 27; 51/53	36/37; 45; 61	Hydrogen	F+	12	9; 16; 33

[1] As per Art. 1a of the Regulation on Hazardous Materials applicable in Germany since 31 October 2005

[2] Cf. R-phrases on page 199, S-phrases on page 200, Safety signs on page 342; the slash (/) between the number indicates a combination of R-phrases or S-phrases.

Material characteristics of hazardous gases

Gas	Density ratio to air	Ignition temperature	Lower ignition limit vol.-% gas in air	Upper ignition limit vol.-% gas in air	Additional information
Acetylene	0.91	305 °C	1.5	82	With a pressure p_e > 2 bar self-disintegration and explosion
Argon	1.38	incombustible	–	–	Loss of breath; danger of suffocation
Butane	2.11	365 °C	1.5	8.5	Narcotic effect; suffocating effect
Carbon dioxide	1.53	incombustible	–	–	Liquid CO_2 and dry ice lead to serious frostbyte
Carbon monoxide	0.97	605 °C	12.5	74	Potent blood poison; damage to vision, lungs, liver, kidneys and hearing
Hydrogen	0.07	570 °C	4	75.6	Spontaneous combustion with high escaping speeds; forms explosive mixtures with air, O_2 and Cl
Nitrogen	0.97	incombustible	–	–	Lose of breath in enclosed spaces; danger of suffocation
Oxygen	1.1	incombustible	–	–	Greases and oils react with oxygen explosively; fire-promoting gas
Propane	1.55	470 °C	2.1	9.5	Loss of breath; liquid propane causes damage to skin and eyes

Hazardous substances, R-phrases*

Hazardous substances adversely affect the safety and health of humans and endanger the environment. They must be specially labeled (see page 363). The following R-phrases[1] are standard phrases and point out the special risks when handling a hazardous substance. Special safety data sheets for each hazardous substance contain further extensive information. R-phrases were replaced by Hazard Statements (cf. page 364)

R-phrases: Notes on special risks

cf. RL 67/548/EWG[2] (2004-04)

R-phrases[3]	Meaning	R-phrases[3]	Meaning
R 1	Explosive when dry	R 34	Causes burns
R 2	Risk of explosion by shock, friction, fire, or other sources of ignition	R 35	Causes severe burns
R 3	Extreme risk of explosion by shock, friction, fire, or other sources of ignition	R 36	Irritating to the eyes
R 4	Forms very sensitive explosive metallic compounds	R 37	Irritating to respiratory system
R 5	Heating may cause an explosion	R 38	Irritating to the skin
R 6	Explosive with or without contact with air	R 39	Danger of very serious irreversible effects
R 7	May cause fire	R 40	Limited evidence of a carcinogenic effect
R 8	Contact with combustible material may cause fire	R 41	Risk of serious damage to eyes
R 10	Flammable	R 42	May cause sensitization by inhalation
R 11	Highly flammable	R 43	May cause sensitization by skin contact
R 12	Extremely flammable	R 44	Risk of explosion if heated under confinement
R 13	Extremely flammable liquid gas	R 45	May cause cancer
R 14	Reacts violently with water	R 46	May cause heritable genetic damage
R 15	Contact with water liberates extremely flammable gases	R 48	Danger of serious damage to health by prolonged exposure
R 16	Explosive when mixed with oxidizing substances	R 49	May cause cancer by inhalation
R 17	Spontaneously flammable in air	R 50	Very toxic to aquatic organisms
R 18	In use, may form flammable/explosive vapor-air mixture	R 51	Toxic to aquatic organisms
R 19	May form explosive peroxides	R 52	Harmful to aquatic organisms
R 20	Harmful by inhalation	R 53	May cause long-term adverse effects in the aquatic environment
R 21	Harmful in contact with skin	R 54	Toxic to flora (plants)
R 22	Harmful if swallowed	R 55	Toxic to fauna (animals)
R 23	Toxic by inhalation	R 56	Toxic to soil organisms
R 24	Toxic in contact with skin	R 57	Toxic to bees
R 25	Toxic if swallowed	R 58	May cause long-term adverse effects in the environment
R 26	Very toxic by inhalation	R 59	Dangerous to the ozone layer
R 27	Very toxic in contact with skin	R 60	May impair fertility
R 28	Very toxic if swallowed	R 61	May cause harm to the unborn child
R 29	Contact with water liberates toxic gases	R 62	Possible risk of impaired fertility
R 30	Can become highly flammable in use	R 63	Possible risk of harm to the unborn child
R 31	Contact with acids liberates toxic gases	R 64	May cause harm to breastfed babies
R 32	Contact with acids liberates very toxic gases	R 65	Harmful: May cause lung damage if swallowed
R 33	Danger of cumulative effects	R 66	Repeated exposure may cause skin dryness or cracking
		R 67	Vapors may cause drowsiness and dizziness
		R 68	Possible irreversible damage

[1] R = Risk [2] EU-Directive, Appendix III
[3] Combinations of the risk phrases are possible; e.g. R 23/24: Toxic by inhalation and in contact with skin
*) According to European Standards

MS

Hazardous substances, S-phrases*

The following standardized recommended safety measures (S-phrases)[1] are to be followed while handling hazardous substances and preparations. By complying with them dangers can be avoided or reduced. S-phrases were replaced by Precautionary Statements (cf. page 364).

S-(safety)-phrases: Recommended Safety Measures

cf. RL 67/548/EWG[2] (2004-04)

S-phrase[3]	Meaning
S 1	Keep locked up
S 2	Keep out of the reach of children
S 3	Keep in a cool place
S 4	Keep away from living quarters
S 5	Keep contents under … (appropriate liquid to be specified by the manufacturer)
S 6	Keep contents under … (appropriate linert gas to be specified by the manufacturer)
S 7	Keep container tightly closed
S 8	Keep container dry
S 9	Keep container in a well-ventilated place
S 12	Do not keep the container sealed
S 13	Keep away from food, drink and animal feeding stuffs
S 14	Keep away from ... (incompatible materials to be indicated by the manufacturer)
S 15	Keep away from heat
S 16	Keep away from sources of ignition – no smoking
S 17	Keep away from combustible materials
S 18	Handle and open container with care
S 20	When using do not eat or drink
S 21	When using do not smoke
S 22	Do not breathe dust
S 23	Do not breathe gas/fumes/vapor/spray (appropriate wording to be specified by the manufacturer)
S 24	Avoid contact with skin
S 25	Avoid contact with eyes
S 26	In case of contact with eyes, rinse immediately with plenty of water and seek medical advice
S 27	Take off immediately all contaminated clothing
S 28	After contact with skin, wash immediately with plenty of ... (to be specified by the manufacturer)
S 29	Do not empty into drains
S 30	Never add water to this product
S 33	Take precautionary measures against static discharges
S 35	This material and its container must be disposed of in a safe way
S 36	Wear suitable protective clothing
S 37	Wear suitable gloves
S 38	In case of insufficient ventilation, wear suitable respiratory equipment
S 39	Wear eye/face protection
S 40	To clean the floor and all objects contam. by this material, use ... (to be specif. by the manufacturer)
S 41	In case of fire and/or explosions do not breathe fumes
S 42	During fumigation/spraying wear suitable respiratory equipment (appropriate wording to be specified by the manufacturer)
S 43	In case of fire, use ... (indicate in the space the precise type of fire-fighting equipment if water increases risk, add: 'Never use water')
S 45	In case of accident or if you feel unwell, seek medical advice immediately (show the label where possible)
S 46	If swallowed, seek medical advice immediately and show this container or label
S 47	Keep at temperature not exceeding ... °C (To be specified by the manufacturer)
S 48	Keep wet with ... (appropriate material to be specified by the manufacturer)
S 49	Keep only in the original container
S 50	Do not mix with … (to be specified by the manufacturer)
S 51	Use only in well-ventilated areas
S 52	Not recommended for interior use on large surface areas
S 53	Avoid exposures[4], obtain special instructions before use
S 56	Dispose of this material and its container at hazardous or special waste collection point
S 57	Use appropriate container to avoid[5] environmental contamination
S 59	Refer to manufacturer/supplier for information on recovery/recycling
S 60	This material and its container must be disposed of as hazardous waste
S 61	Avoid release to the environment. Refer to special instructions/safety data sheets
S 62	If swallowed, do not induce vomiting: seek medical advice immediately and show this container or label
S 63	In case of accident by inhalation: move victim to fresh air and keep at rest
S 64	If swallowed, rinse mouth with water (only if the person is conscious)

[1] S = safety [2] EU- Directive, Appendix IV
[3] Combinations of the S phrases are possible; e.g. S 20/21: when using do not eat, drink or smoke.
[4] i.e. do not expose yourself to this hazard [5] Contamination, infestation
*) According to European Standards

5 Machine elements

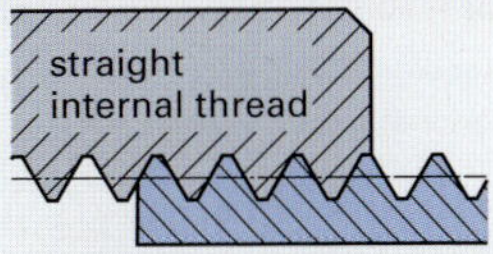

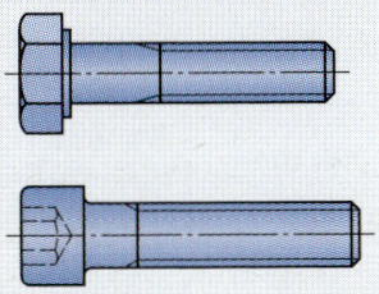

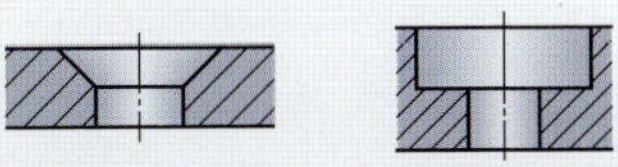

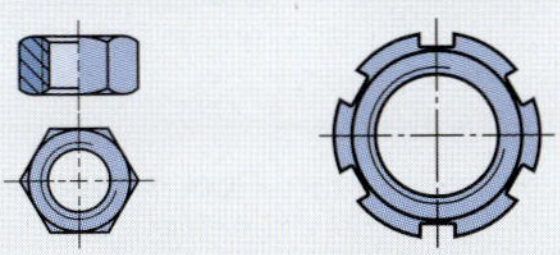

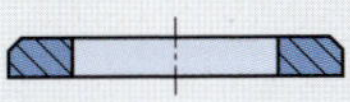

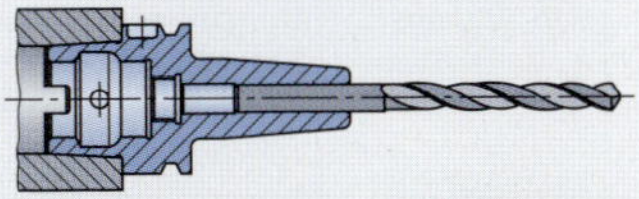

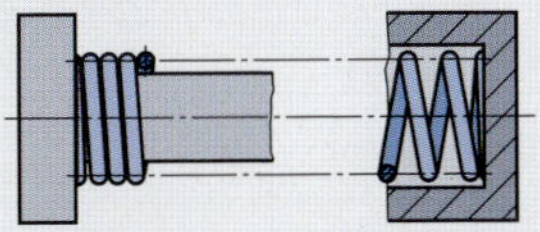

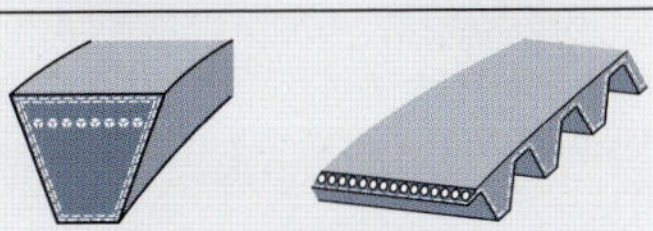

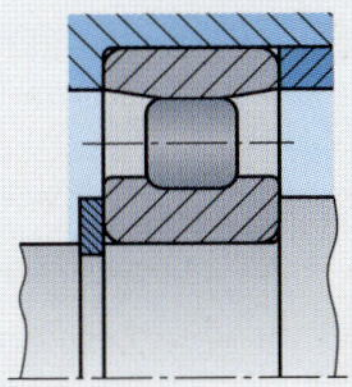

Types of threads, Overview

cf. DIN 202 (1999-11)

Right-hand threads, single-start

Thread name	Thread profile	Code letter	Designation example	Nominal sizes	Application
Metric threads ISO threads		M	DIN 14–M 08	0.3 to 0.9 mm	Clocks, precision mechanisms
			DIN 13–M 30	1 to 68 mm	General purpose (coarse thread)
			DIN 13–M 20 × 1	1 to 1000 mm	General purpose (fine thread)
Metric threads with large clearance			DIN 2510–M 36	12 to 180 mm	Bolts/screws with anti-fatigue shank
Metric straight internal threads			DIN 158–M 30 × 2	6 to 60 mm	Drain plugs and grease nipples
Metric taper external threads	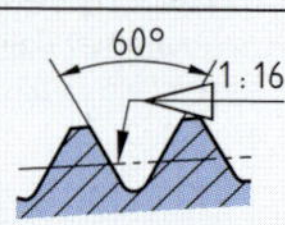	M	DIN 158–M 30 × 2 keg	6 to 60 mm	Drain plugs and grease nipples
Pipe threads, straight	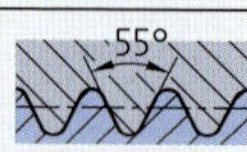	G	DIN ISO 228–G1 $^1/_2$ (internal) DIN ISO 228–G $^1/_2$A (external)	$^1/_8$ to 6 inches	Does not seal on thread
Parallel pipe threads (internal threads)	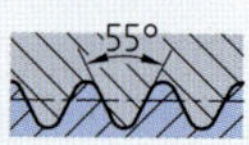	Rp	DIN 2999–Rp $^1/_2$	$^1/_{16}$ to 6 inch	Pipe threads, seals on thread; for threaded pipe, fittings, screwed pipe joints
			DIN 3858–Rp $^1/_8$	$^1/_8$ to 1 $^1/_2$ inch	
Taper pipe threads (external threads)	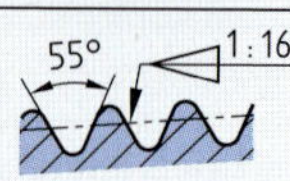	R	DIN 2999–R $^1/_2$	$^1/_{16}$ to 6 inches	
			DIN 3858–R $^1/_8$-1	$^1/_8$ to 1 $^1/_2$ inches	
Metric ISO trapezoidal threads	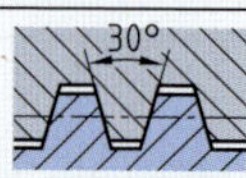	Tr	DIN 103–Tr 40 × 7	8 to 300 mm	General purpose as transmission screw threads
Buttress threads	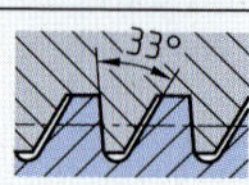	S	DIN 513–S 48 × 8	10 to 640 mm	General purpose as transmission screw threads
Knuckle threads	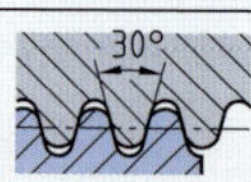	Rd	DIN 405–Rd 40 × $^1/_6$	8 to 200 mm	General purpose
			DIN 20400–Rd 40 × 5	10 to 300 mm	Knuckle threads with large thread overlap
Sheet metal threads	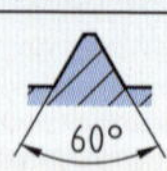	ST	ISO 1478–ST 3,5	1.5 to 9.5 mm	For self-tapping screws

Designation of left-hand and metric multiple start threads

cf. DIN ISO 965-1 (1999-11)

Type of thread	Explanation	Code designation (examples)
Left-hand threads	The code designation "LH" is placed after the complete thread designation (LH = Left-Hand).	M 30–LH Tr 40 × 7–LH
Multiple start right-hand thread	The lead Ph and the pitch P follow the code designation and the thread diameter.	M 16 x Ph 3 P 1,5 or M 16 x Ph 3 P 1,5 (double-start)
Multiple start left-hand thread	"LH" is placed after the thread designation of the multiple start.[1]	M 14 x Ph 6 P 2-LH or M 14 x Ph 6 P 2 (triple-start)-LH

[1] For parts which have right-hand and left-hand threads, "RH" (Right-Hand) is placed after the thread designation of the right-hand thread and "LH" (Left-Hand) after the left-hand thread. The number of starts for multiple-starts is found by: **no. of starts = lead Ph/ pitch P.**

Thread standards of various countries (selection)[1]

Thread name	Thread profile	Code	Thread designation: Example	Thread designation: Meaning	Country[2]
Unified **N**ational **C**oarse Thread	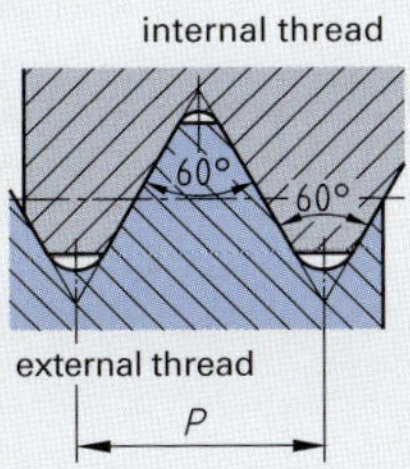	UNC	$^1/_4$–20 UNC–2A	ISO-UNC-thread with $^1/_4$ inch nominal diameter, 20 threads/inch, Class 2A	ARG, AUS, CAN, GBR, IND, JPN, NOR, PAK, SWE and others
Unified **N**ational **F**ine Thread		UNF	$^1/_4$–28 UNC–3A	ISO-UNF threads with $^1/_4$ inch nominal diameter, 28 threads/inch, Class 3A	ARG, AUS, CAN, GBR, IND, JPN, NOR, PAK, SWE and others
Unified **N**ational **E**xtra **F**ine Thread		UNEF	$^1/_4$–32 UNEF – 3A	ISO-UNEF thread with $^1/_4$ inch nominal diameter, 32 threads/inch, Class 3A	ARG, AUS, CAN, IND, NOR, PAK, SWE and others
Unified **N**ational **S**pecial Thread, special diameter/lead combinations		UNS	$^1/_4$–27 UNS	UNS threads with $^1/_4$ inch nominal diameter, 27 threads/inch	ARG, AUS, CAN, NZL, USA
National Standard **S**traight **P**ipe Threads for **M**echanical Joints	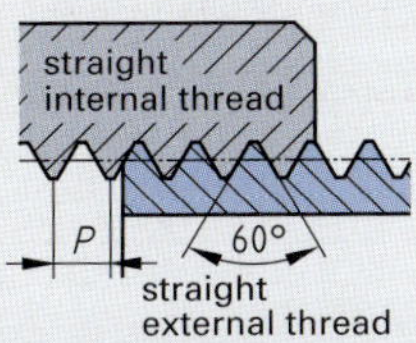	NPSM	$^1/_2$–14 NPSM	NPSM threads with $^1/_2$ inch nominal diameter, 14 threads/inch	USA, CAN
American **N**ational Standard **T**aper **P**ipe Thread	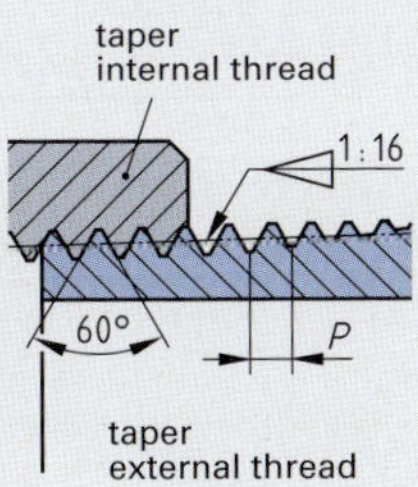	NPT	$^3/_8$–18 NPT	NPT thread with $^3/_8$ inch nominal diameter, 18 threads/inch	BRA, CAN, FRA, USA and others
American **N**ational Standard **T**aper **P**ipe Thread, **F**uel		NPTF	$^1/_2$–14 NPTF (dryseal)	NPTF threads with $^1/_2$ inch nominal diameter, 14 threads/inch, (dry sealing)	BRA, CAN, USA
American trapezoidal threads $h = 0.5 \cdot P$	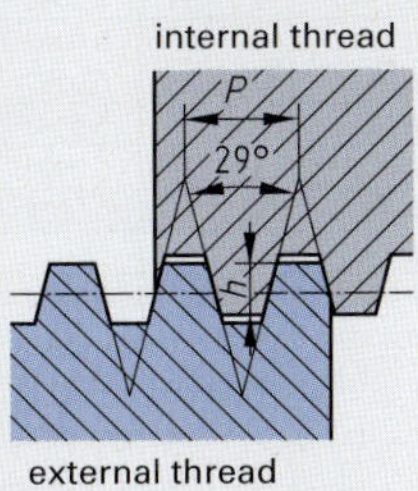	Acme	$1^3/_4$–4 Acme – 2G	Acme threads with $1^3/_4$ inch nominal diameter 4 threads/inch, Class 2G	AUS, CAN, GBR, NZL, USA
American truncated trapezoidal threads $h = 0.3 \cdot P$		Stub Acme	$^1/_2$–20 Stub Acme	Stub Acme threads with $^1/_2$ inch nominal diameter, 20 threads/inch	CAN, USA

[1] cf. Kaufmann, Manfred: "Wegweiser zu den Gewindenormen verschiedener Länder" DIN, Beuth-Verlag (2009-09)

[2] Three-letter codes for countries, cf. DIN EN ISO 3166-1 (2008-06)

ME

Imperial Threads

Imperial Threads for general purposes

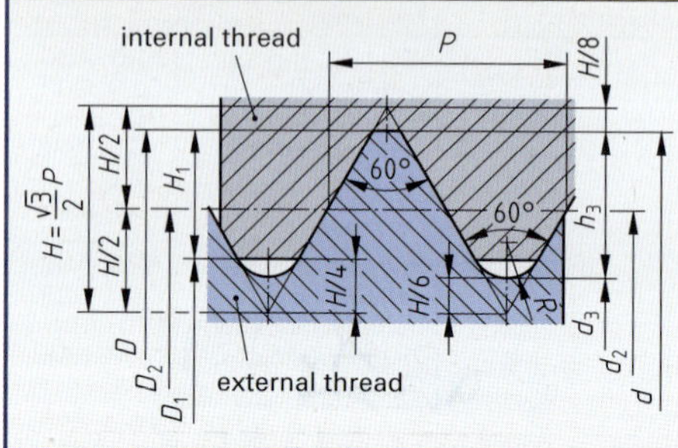

Major diameter	$d = D$
Pitch	P
Depth of external thread	$h_3 = 0.6134 \cdot P$
Depth of internal thread	$H_1 = 0.5413 \cdot P$
Radius at root	$R = 0.1443 \cdot P$
Basic pitch Ø	$d_2 = D_2 = d - 0.6495 \cdot P$
Minor Ø of external thread	$d_3 = d - 1.1904 \cdot P$
Minor Ø of internal thread	$D_1 = d - 1.0825 \cdot P$
Tap hole drill Ø	$= d - P$
Thread angle	60°
Stress area	$S = \frac{\pi}{4} \cdot \left(\frac{d_2 + d_3}{2}\right)^2$

$H = \frac{\sqrt{3}}{2} P$

Basic sizes for Unified National Coarse Threads (UNC)

ANSI/ASME B1.1 (1989)

No. size or inches	Threads per inch	Major diameter D inches	Pitch P inches	Pitch diameter $d_2 = D_2$ inches	Minor: External threads d_3 inches	Minor: Internal threads D_1 inches	Thread depth: External threads h_3 inches	Thread depth: Internal threads H_1 inches	Radius R inches	Stress area S inch²	Drill bit for tap hole: Drill size	Drill bit for tap hole: Decimal equiv.
6	32	0.1380	0.0313	0.1177	0.1008	0.1042	0.01920	0.01691	0.0045	0.0093	#36	0.1065
8	32	0.1640	0.0313	0.1437	0.1268	0.1302	0.01920	0.01691	0.0045	0.0142	#29	0.1360
10	24	0.1900	0.0417	0.1629	0.1404	0.1449	0.02558	0.02255	0.0060	0.0179	#25	0.1495
12	24	0.2160	0.0417	0.1889	0.1664	0.1709	0.02558	0.02255	0.0060	0.0246	#16	0.1770
1/4	20	0.2500	0.0500	0.2175	0.1905	0.1959	0.03067	0.02706	0.0072	0.0324	#7	0.2010
5/16	18	0.3125	0.0556	0.2764	0.2464	0.2524	0.03411	0.03007	0.0080	0.0532	F	0.2579
3/8	16	0.3750	0.0625	0.3344	0.3006	0.3073	0.03834	0.03383	0.0090	0.0786	5/16	0.3125
7/16	14	0.4375	0.0714	0.3911	0.3525	0.3602	0.04380	0.03866	0.0103	0.1078	U	0.3680
1/2	13	0.5000	0.0769	0.4500	0.4084	0.4167	0.04717	0.04164	0.0111	0.1438	27/64	0.4219
9/16	12	0.5625	0.0833	0.5084	0.4633	0.4723	0.05110	0.04511	0.0120	0.1842	31/64	0.4843
5/8	11	0.6250	0.0909	0.5660	0.5168	0.5266	0.05576	0.04921	0.0131	0.2288	17/32	0.5313
3/4	10	0.7500	0.1000	0.6851	0.6310	0.6418	0.06134	0.05413	0.0144	0.3382	21/32	0.6562
7/8	9	0.8750	0.1111	0.8028	0.7427	0.7547	0.06815	0.06014	0.0160	0.4666	49/64	0.7656
1	8	1.0000	0.1250	0.9188	0.8512	0.8647	0.07668	0.06766	0.0180	0.6120	7/8	0.8750
1 1/8	7	1.1250	0.1429	1.0322	0.9549	0.9704	0.08765	0.07732	0.0206	0.7713	63/64	0.9844
1 1/4	7	1.2500	0.1429	1.1572	1.0799	1.0954	0.08765	0.07732	0.0206	0.9781	1 7/64	1.1093
1 3/8	6	1.3750	0.1667	1.2668	1.1766	1.1946	0.10225	0.09021	0.0241	1.1664	1 7/32	1.2187
1 1/2	6	1.5000	0.1667	1.3918	1.3016	1.3196	0.10225	0.09021	0.0241	1.4179	1 11/32	1.3437
1 3/4	5	1.7500	0.2000	1.6201	1.5119	1.5335	0.12268	0.10825	0.0289	1.9171	1 9/16	1.5625
2	4.5	2.0000	0.2222	1.8557	1.7355	1.7594	0.13630	0.12028	0.0321	2.5207	1 25/32	1.7812

Basic sizes for Unified National Fine Threads (UNF)

ANSI/ASME B1.1 (1989)

No. size or inches	Threads per inch	Major diameter D inches	Pitch P inches	Pitch diameter $d_2 = D_2$ inches	Minor: External threads d_3 inches	Minor: Internal threads D_1 inches	Thread depth: External threads h_3 inches	Thread depth: Internal threads H_1 inches	Radius R inches	Stress area S inch²	Drill bit for tap hole: Drill size	Drill bit for tap hole: Decimal equiv.
6	40	0.1380	0.0250	0.1218	0.1082	0.1109	0.0153	0.01353	0.0036	0.0103	#33	0.1130
8	36	0.1640	0.0278	0.1460	0.1309	0.1339	0.0170	0.01504	0.0040	0.0149	#29	0.1360
10	32	0.1900	0.0313	0.1697	0.1528	0.1562	0.0192	0.01691	0.0045	0.0203	#21	0.1590
12	28	0.2160	0.0357	0.1928	0.1735	0.1773	0.0219	0.01933	0.0052	0.0262	#14	0.1820
1/4	28	0.2500	0.0357	0.2268	0.2075	0.2113	0.0219	0.01933	0.0052	0.0368	I	0.2720
5/16	24	0.3125	0.0417	0.2854	0.2629	0.2674	0.0256	0.02255	0.0060	0.0587	I	0.2720
3/8	24	0.3750	0.0417	0.3479	0.3254	0.3299	0.0256	0.02255	0.0060	0.0886	Q	0.3320
7/16	20	0.4375	0.0500	0.4050	0.3780	0.3834	0.0307	0.02706	0.0072	0.1198	25/64	0.3906
1/2	20	0.5000	0.0500	0.4675	0.4405	0.4459	0.0307	0.02706	0.0072	0.1612	29/64	0.4531
9/16	18	0.5625	0.0556	0.5264	0.4964	0.5024	0.0341	0.03007	0.0080	0.2046	33/64	0.5156
5/8	18	0.6250	0.0556	0.5889	0.5589	0.5649	0.0341	0.03007	0.0080	0.2578	37/64	0.5781
3/4	16	0.7500	0.0625	0.7094	0.6756	0.6823	0.0383	0.03383	0.0090	0.3754	11/16	0.6875
7/8	14	0.8750	0.0714	0.8286	0.7900	0.7977	0.0438	0.03866	0.0103	0.5127	13/16	0.8125
1	12	1.0000	0.0833	0.9459	0.9008	0.9098	0.0511	0.04511	0.0120	0.6674	59/64	0.9219
1 1/8	12	1.1250	0.0833	1.0709	1.0258	1.0348	0.0511	0.04511	0.0120	0.8607	1 3/64	1.0469
1 1/4	12	1.2500	0.0833	1.1959	1.1508	1.1598	0.0511	0.04511	0.0120	1.0785	1 11/64	1.1719
1 3/8	12	1.3750	0.0833	1.3209	1.2758	1.2848	0.0511	0.04511	0.0120	1.3208	1 19/64	1.2968
1 1/2	12	1.5000	0.0833	1.4459	1.4008	1.4098	0.0511	0.04511	0.0120	1.5877	1 27/64	1.4219

Imperial Threads

Basic sizes National Pipe Taper (NPT)

ANSI/ASME B1.20.1 – 1983 (R 1992)

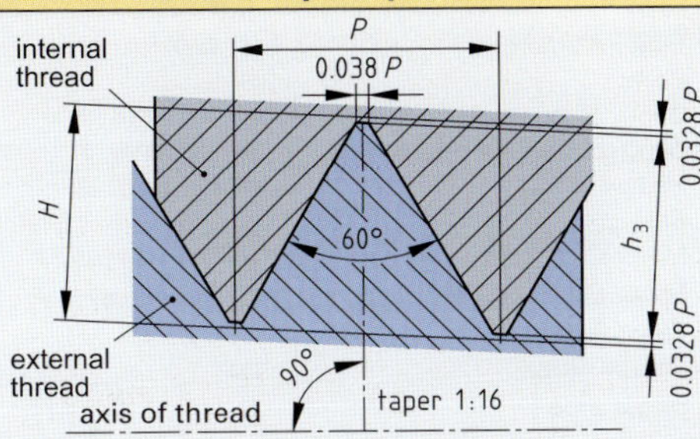

Thread depth $h_3 = 0.8 \cdot P$

Hight $H = 0.865 \cdot P$

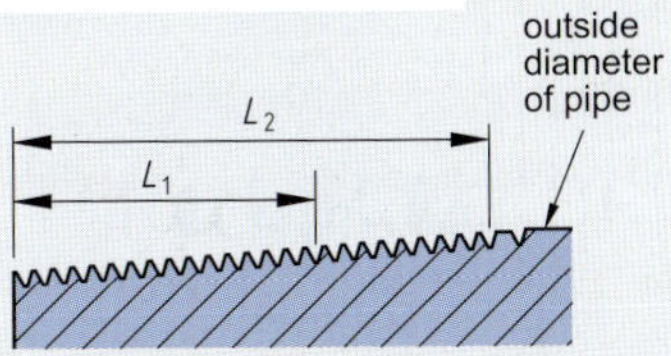

No. size	Threads	Outside diam. of pipe D	Pitch P	Pitch diameter $d_2 = D_2$	Gauge length L_1	Usuable length of ext. thread L_2	Depth of external thread $h_3 = 8P$	Drill bit for tap hole Drill size	Decimal equival.
	all dimensions in inches								
1/16	27	0.3125	0.03704	0.28120	0.1598	0.2611	0.02963	C	0.2420
1/ 8	27	0.4050	0.03704	0.37360	0.1613	0.2639	0.02963	Q	0.3320
1/ 4	18	0.5400	0.05556	0.49163	0.2275	0.4018	0.04444	7/16	0.4380
3/ 8	18	0.6750	0.05556	0.62701	0.2398	0.0478	0.04444	9/16	0.5620
1/ 2	14	0.0625	0.07143	0.77843	0.3199	0.5337	0.05714	45/64	0.7030
3/4	14	1.0500	0.07143	0.98887	0.3391	0.5457	0.05714	29/32	0.9060
1	11 1/2	1.3150	0.08696	1.23863	0.3997	0.6828	0.06957	1 9/64	1.1410
1 1/4	11 1/2	1.6600	0.08696	1.58338	0.4197	0.7068	0.06957	1 31/64	1.484
1 1/2	11 1/2	1.9000	0.08696	1.82234	0.4197	0.7235	0.06957	1 23/32	1.7190
2	11 1/2	2.3750	0.08696	2.29627	0.4354	0.7565	0.06957	2 3/16	2.1880
2 1/2	8	2.8750	0.12500	2.76215	0.6825	1.1375	0.10000	2 39/64	2.6090

Basic sizes American National Standard General Purp. Acme Screw Thread

ANSI/ASME B1.5 – 1988 (R 1994)

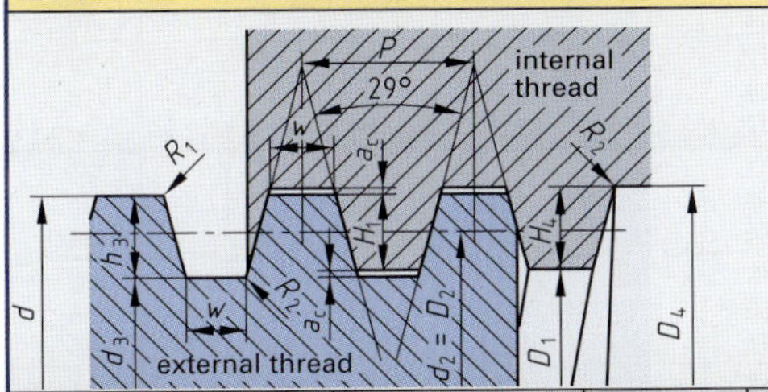

a_c	up to 10 tpi = 0.020
a_c	over 10 tpi = 0.010
R_1	$0.06 \cdot P$
R_2	$0.12 \cdot P$
Minor ∅ external threads	$d_3 = d - (P + 2 \cdot a_c)$
Major ∅ internal threads	$D_4 = d + 2 \cdot a_c$
Minor ∅ internal threads	$D_1 = d - P$
Pitch ∅	$d_2 = D_2 = d - 0.5 \cdot P$
Thread depth	$h_3 = H_4 = 0.5 \cdot P + a_c$
Width of flat	$w = 0.370 \cdot P - 0.259 \cdot a_c$

No. size	Threads per inch	Nominal diameter d	Pitch P	Pitch diameter $d_2 = D_2$	Minor diameter External thread d_3	Minor diameter Internal thread D_1	Thread depth $h_3 = H4$
	all dimensions in inches						
3/8	12	0.3750	0.0833	0.3333	0.2717	0.2917	0.0517
7/16	12	0.4375	0.0833	0.3958	0.3342	0.3542	0.0517
1/2	10	0.5000	0.1000	0.4500	0.3600	0.4000	0.0700
5/8	8	0.6250	0.1250	0.5625	0.4600	0.5000	0.0825
3/4	6	0.7500	0.1667	0.6667	0.5433	0.5833	0.1033
7/8	6	0.8750	0.1667	0.7917	0.6683	0.7083	0.1033
1	5	1.0000	0.2000	0.9000	0.7600	0.8000	0.1200
1 1/8	5	1.1250	0.2000	1.0250	0.8850	0.9250	0.1200
1 1/4	5	1.2500	0.2000	1.1500	1.0100	1.0500	0.1200
1 3/8	4	1.3750	0.2500	1.2500	1.0850	1.1250	0.1450
1 1/2	4	1.5000	0.2500	1.3750	1.2100	1.2500	0.1450
1 3/4	4	1.7500	0.2500	1.6250	1.4600	1.5000	0.1450
2	4	2.0000	0.2500	1.8750	1.7100	1.7500	0.1450
2 1/4	3	2.2500	0.3333	2.0833	1.8767	1.9167	0.1867
2 1/2	3	2.5000	0.3333	2.3333	2.1267	2.1667	0.1867
2 3/4	3	2.7500	0.3333	2.5833	2.3767	2.4167	0.1867
3	2	3.0000	0.5000	2.7500	2.4600	2.5000	0.2700
3 1/2	2	3.5000	0.5000	3.2500	2.9600	3.0000	0.2700
4	2	4.0000	0.5000	3.7500	3.4600	3.5000	0.2700
4 1/2	2	4.5000	0.5000	4.2500	3.9600	4.0000	0.2700
5	2	5.0000	0.5000	4.7500	4.4600	4.5000	0.2700

ME

Metric threads and fine threads

Metric ISO threads for general purpose application, basic profiles

cf. DIN 13-19 (1999-11)

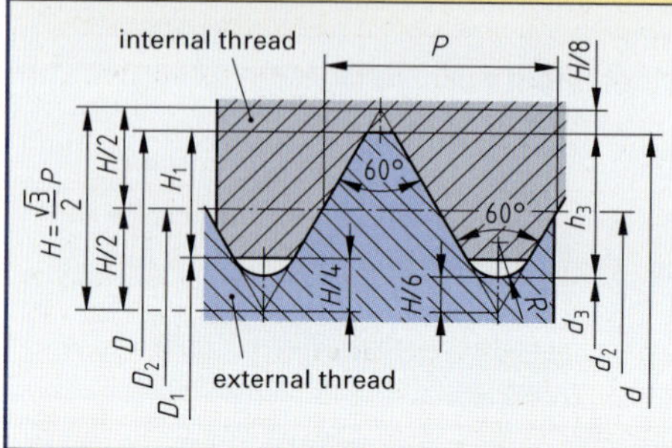

Major diameter	$d = D$
Pitch	P
Depth of external thread	$h_3 = 0.6134 \cdot P$
Depth of internal thread	$H_1 = 0.5413 \cdot P$
Radius at root	$R = 0.1443 \cdot P$
Basic pitch Ø	$d_2 = D_2 = d - 0.6495 \cdot P$
Minor Ø of external thread	$d_3 = d - 1.2269 \cdot P$
Minor Ø of internal thread	$D_1 = d - 1.0825 \cdot P$
Tap hole drill Ø	$= d - P$
Thread angle	60°
Stress area	$S = \frac{\pi}{4} \cdot \left(\frac{d_2 + d_3}{2}\right)^2$

Basic sizes for coarse threads Series 1[1] (dimensions in mm)

cf. DIN 13-1 (1999-11)

Thread-designation $d = D$	Pitch P	Pitch Ø $d_2 = D_2$	Minor Ø external threads d_3	Minor Ø internal threads D_1	Thread depth external threads h_3	Thread depth internal threads H_1	Rounded root R	Stress area S mm²	Drill bit Ø for tap hole[2]	Hexagonal width across flats[3]
M 1	0.25	0.84	0.69	0.73	0.15	0.14	0.04	0.46	0.75	–
M 1.2	0.25	1.04	0.89	0.93	0.15	0.14	0.04	0.73	0.95	–
M 1.6	0.35	1.38	1.17	1.22	0.22	0.19	0.05	1.27	1.25	3.2
M 2	0.4	1.74	1.51	1.57	0.25	0.22	0.06	2.07	1.6	4
M 2.5	0.45	2.21	1.95	2.01	0.28	0.24	0.07	3.39	2.05	5
M 3	0.5	2.68	2.39	2.46	0.31	0.27	0.07	5.03	2.5	5.5
M 4	0.7	3.55	3.14	3.24	0.43	0.38	0.10	8.78	3.3	7
M 5	0.8	4.48	4.02	4.13	0.49	0.43	0.12	14.2	4.2	8
M 6	1	5.35	4.77	4.92	0.61	0.54	0.14	20.1	5.0	10
M 8	1.25	7.19	6.47	6.65	0.77	0.68	0.18	36.6	6.8	13
M 10	1.5	9.03	8.16	8.38	0.92	0.81	0.22	58.0	8.5	16
M 12	1.75	10.86	9.85	10.11	1.07	0.95	0.25	84.3	10.2	18
M 16	2	14.70	13.55	13.84	1.23	1.08	0.29	157	14	24
M 20	2.5	18.38	16.93	17.29	1.53	1.35	0.36	245	17.5	30
M 24	3	22.05	20.32	20.75	1.84	1.62	0.43	353	21	36
M 30	3.5	27.73	25.71	26.21	2.15	1.89	0.51	561	26.5	46
M 36	4	33.40	31.09	31.67	2.45	2.17	0.58	817	32	55
M 42	4.5	39.08	36.48	37.13	2.76	2.44	0.65	1121	37.5	65
M 48	5	44.75	41.87	42.59	3.07	2.71	0.72	1473	43	75
M 56	5.5	52.43	49.25	50.05	3.37	2.98	0.79	2030	50.5	85
M 64	6	60.10	56.64	57.51	3.68	3.25	0.87	2676	58	95

Basic sizes for fine threads (dimensions in mm)

cf. DIN 13-2 – 10 (1999-11)

Thread designation $d \times P$	Pitch Ø $d_2 = D_2$	Minor Ø ext. th. d_3	Minor Ø int. th. D_1	Thread designation $d \times P$	Pitch Ø $d_2 = D_2$	Minor Ø ext. th. d_3	Minor Ø int. th. D_1	Thread designation $d \times P$	Pitch Ø $d_2 = D_2$	Minor Ø ext. th. d_3	Minor Ø int. th. D_1
M 2 × 0.25	1.84	1.69	1.73	M 10 × 0.25	9.84	9.69	9.73	M 24 × 2	22.70	21.55	21.84
M 3 × 0.25	2.84	2.69	2.73	M 10 × 0.5	9.68	9.39	9.46	M 30 × 1.5	29.03	28.16	28.38
M 4 × 0.2	3.87	3.76	3.78	M 10 × 1	9.35	8.77	8.92	M 30 × 2	28.70	27.55	27.84
M 4 × 0.35	3.77	3.57	3.62	M 12 × 0.35	11.77	11.57	11.62	M 36 × 1.5	35.03	34.16	34.38
M 5 × 0.25	4.84	4.69	4.73	M 12 × 0.5	11.68	11.39	11.46	M 36 × 2	34.70	33.55	33.84
M 5 × 0.5	4.68	4.39	4.46	M 12 × 1	11.35	10.77	10.92	M 42 × 1.5	41.03	40.16	40.38
M 6 × 0.25	5.84	5.69	5.73	M 16 × 0.5	15.68	15.39	15.46	M 42 × 2	40.70	39.55	39.84
M 6 × 0.5	5.68	5.39	5.46	M 16 × 1	15.35	14.77	14.92	M 48 × 1.5	47.03	46.16	46.38
M 6 × 0.75	5.51	5.08	5.19	M 16 × 1.5	15.03	14.16	14.38	M 48 × 2	46.70	45.55	45.84
M 8 × 0.25	7.84	7.69	7.73	M 20 × 1	19.35	18.77	18.92	M 56 × 1.5	55.03	54.16	54.38
M 8 × 0.5	7.68	7.39	7.46	M 20 × 1.5	19.03	18.16	18.38	M 56 × 2	54.70	53.55	53.84
M 8 × 1	7.35	6.77	6.92	M 24 × 1.5	23.03	22.16	22.38	M 64 × 2	62.70	61.55	61.84

[1] Series 2 and Series 3 also have intermediate sizes (e.g. M7, M9, M14).
[2] cf. DIN 336 (2003-07) [3] cf. DIN ISO 272 (1979-10)

Tapered external thread, Trapezoidal thread

Metric tapered thread with corresponding cylindrical internal thread[1]

cf. DIN 158-1 (1997-06)

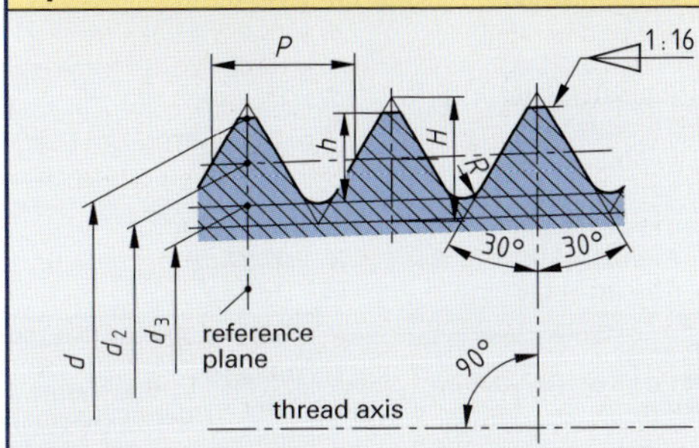

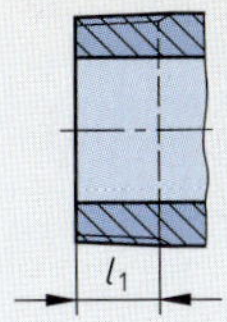

Thread dimensions of tapered external threads

Outer diameter	d
Pitch	P
Height	$H = 0.866 \cdot P$
Thread depth	$h = 0.613 \cdot P$
Pitch diameter	$d_2 = d - 0.650 \cdot P$
Minor diameter	$d_3 = d - 1.23 \cdot P$
Root radius	$R = 0.144 \cdot P$

Thread designation $d \times P$	Thread length l_1	Thread depth h	Pitch diameter d_2	Thread designation $d \times P$	Thread length l_1	Thread depth h	Pitch diameter d_2
M 5 x 0.8 keg	5	0.52	4.48	M 24 x 1.5 keg	8.5	0.98	23.03
M 6 x 1 keg	5.5	0.66	5.35	M 30 x 1.5 keg	10.5	1.01	29.03
M 8 x 1 keg			7.35	M 36 x 1.5 keg			35.03
M 10 x 1 keg			9.35	M 42 x 1.5 keg			41.03
M 12 x 1 keg			11.35	M 48 x 1.5 keg			47.03
M 16 x 1.5 keg	8.5	0.98	15.03	M 56 x 2 keg	13	1.34	54.70
M 20 x 1.5 keg			19.03	M 60 x 2 keg			58.70

[1] For self-sealing joints (e. g. drain plugs, grease nipples). For larger nominal diameters, a sealing compound acting inside the thread is recommended.

Metric ISO trapezoidal screw thread

cf. DIN 103-1 (1977-04)

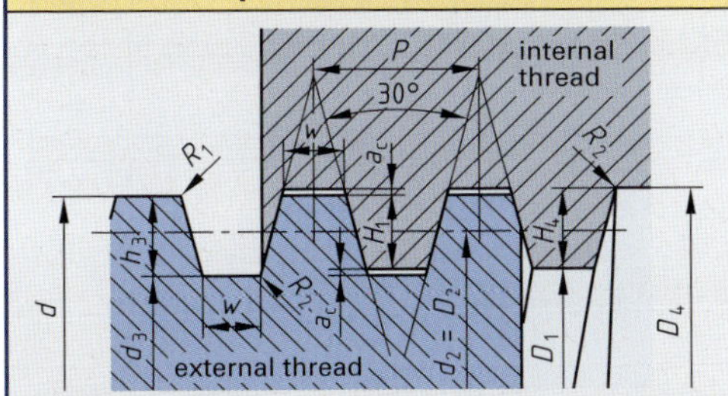

Nominal diameter	d
Single start pitch and multiple start lead	P
Multiple start pitch	P_h
No. of threads	$n = P_h : P$
Minor ∅ external threads	$d_3 = d - (P + 2 \cdot a_c)$
Major ∅ internal threads	$D_4 = d + 2 \cdot a_c$
Minor ∅ internal threads	$D_1 = d - P$
Pitch diameter	$d_2 = D_2 = d - 0.5 \cdot P$
Thread depth	$h_3 = H_4 = 0.5 \cdot P + a_c$
Thread overlap	$H_1 = 0.5 \cdot P$
Crest clearance	a_c
Radii	R_1 and R_2
Width of flat	$w = 0.366 \cdot P - 0.54 \cdot a_c$
Thread angle	30°

Dimension	For pitch P in mm: 1.5	2–5	6–12	14–44
a_c	0.15	0.25	0.5	1
R_1	0.075	0.125	0.25	0.5
R_2	0.15	0.25	0.5	1

Thread designation $d \times P$	Thread dimensions in mm: Pitch ∅ $d_2 = D_2$	Minor ∅ ext. th. d_3	Minor ∅ int. th. D_1	Major ∅ D_4	Thread depth $h_3 = H_4$	Width of flat w	Thread designation $d \times P$	Thread dimensions in mm: Pitch ∅ $d_2 = D_2$	Minor ∅ ext. th. d_3	Minor ∅ int. th. D_1	Major ∅ D_4	Thread depth $h_3 = H_4$	Width of flat w
Tr 10 × 2	9	7.5	8	10.5	1.25	0.60	Tr 40 × 7	36.5	32	33	41	4	2.29
Tr 12 × 3	10.5	8.5	9	12.5	1.75	0.96	Tr 44 × 7	40.5	36	37	45	4	2.29
Tr 16 × 4	14	11.5	12	16.5	2.25	1.33	Tr 48 × 8	44	39	40	49	4.5	2.66
Tr 20 × 4	18	15.5	16	20.5	2.25	1.33	Tr 52 × 8	48	43	44	53	4.5	2.66
Tr 24 × 5	21.5	18.5	19	24.5	2.75	1.70	Tr 60 × 9	55.5	50	51	61	5	3.02
Tr 28 × 5	25.5	22.5	23	28.5	2.75	1.70	Tr 70 × 10	65	59	60	71	5.5	3.39
Tr 32 × 6	29	25	26	33	3.5	1.93	Tr 80 × 10	75	69	70	81	5.5	3.39
Tr 36 × 6	33	29	30	37	3.5	1.93	Tr 100 × 12	94	87	88	101	6.5	4.12

Whitworth threads, Pipe threads

Whitworth threads

(not standardized)

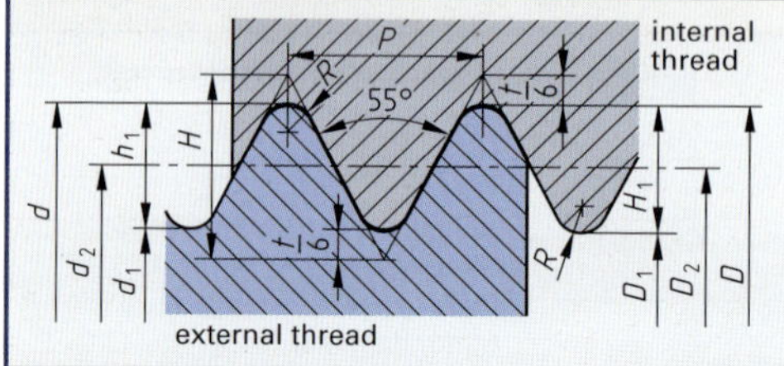

Major diameter	$d = D$
Minor diameter	$d_1 = D_1 = d - 1.28 \cdot P$
	$= d - 2 \cdot t_1$
Pitch diameter	$d_2 = D_2 = d - 0.640 \cdot P$
Threads/inch	N
Pitch	$P = \dfrac{25.4\ \text{mm}}{N}$
Thread depth	$h_1 = H_1 = 0.640 \cdot P$
Radius	$R = 0.137 \cdot P$
Thread angle	55°

Thread designation d	Dimensions in mm for external and internal threads: Major Ø $d = D$	Minor Ø $d_1 = D_1$	Pitch Ø $d_2 = D_2$	Threads per inch N	Thread depth $h_1 = H_1$	Core cross section mm^2
1/4"	6.35	4.72	5.54	20	0.81	17.5
5/16"	7.94	6.13	7.03	18	0.90	29.5
3/8"	9.53	7.49	8.51	16	1.02	44.1
1/2"	12.70	9.99	11.35	12	1.36	78.4
5/8"	15.88	12.92	14.40	11	1.48	131
3/4"	19.05	15.80	17.42	10	1.63	196
7/8"	22.23	18.61	20.42	9	1.81	272
1"	25.40	21.34	23.37	8	2.03	358
1 1/4"	31.75	27.10	29.43	7	2.32	577
1 1/2"	38.10	32.68	35.39	6	2.71	839
1 3/4"	44.45	37.95	41.20	5	3.25	1 131
2"	50.80	43.57	47.19	4.5	3.61	1 491
2 1/4"	57.15	49.02	53.09	4	4.07	1 886
2 1/2"	63.50	55.37	59.44	4	4.07	2 408
3"	76.20	66.91	72.56	3.5	4.65	3 516
3 1/2"	88.90	78.89	83.89	3.25	5.00	4 888

Pipe threads

cf. DIN ISO 228-1 (2003-05), DIN EN 10226-1 (2004-10)

Pipe threads DIN ISO 228-1

for joints not sealed by threads; straight internal and external threads

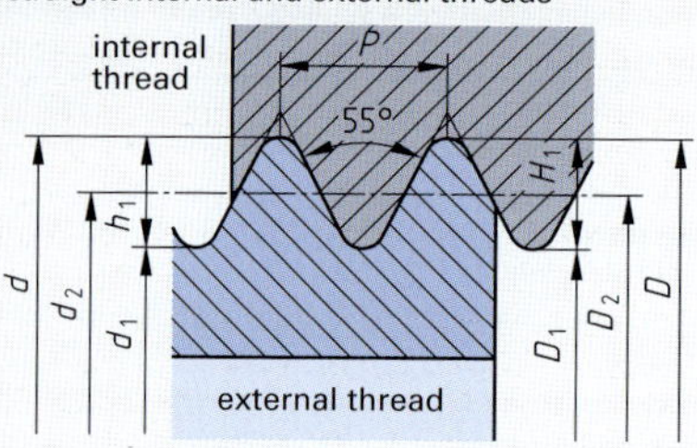

Pipe threads DIN EN 10226-1

sealed by threads; straight internal threads, taper external threads

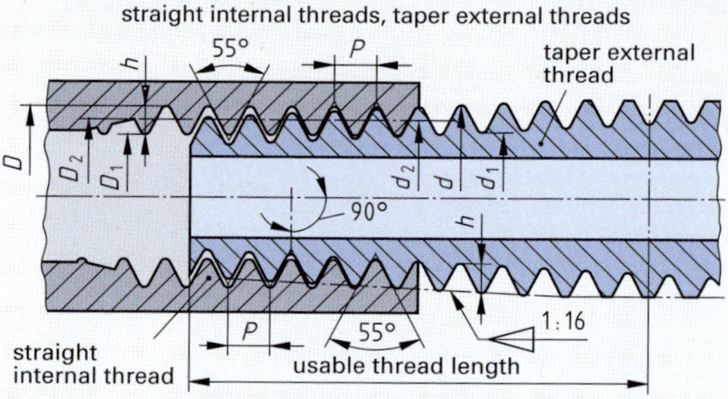

cf. American Taper Standard-Pipe Threads NPT: page 203

Thread designation DIN ISO 228-1: External and internal threads	DIN EN10226-1: External threads	DIN EN10226-1: Internal threads	Major diameter $d = D$	Pitch diameter $d_2 = D_2$	Minor diameter $d_1 = D_1$	Pitch P	Threads per inch N	Profile height $h = h_1 = H_1$	Usable length of external threads ≥
G1/16	R1/16	Rp1/16	7.723	7.142	6.561	0.907	28	0.581	6.5
G1/8	R1/8	Rp1/8	9.728	9.147	8.566	0.907	28	0.581	6.5
G1/4	R1/4	Rp1/4	13.157	12.301	11.445	1.337	19	0.856	9.7
G3/8	R3/8	Rp3/8	16.662	15.806	14.950	1.337	19	0.856	10.1
G1/2	R1/2	Rp1/2	20.955	19.793	18.631	1.814	14	1.162	13.2
G3/4	R3/4	Rp3/4	26.441	25.279	24.117	1.814	14	1.162	14.5
G1	R1	Rp1	33.249	31.770	30.291	2.309	11	1.479	16.8
G1 1/4	R1 1/4	Rp1 1/4	41.910	40.431	38.952	2.309	11	1.479	19.1
G1 1/2	R1 1/2	Rp1 1/2	47.803	46.324	44.845	2.309	11	1.479	19.1
G2	R2	Rp2	59.614	58.135	56.656	2.309	11	1.479	23.4
G2 1/2	R2 1/2	Rp2 1/2	75.184	73.705	72.226	2.309	11	1.479	26.7
G3	R3	Rp3	87.884	86.405	84.926	2.309	11	1.479	29.8
G4	R4	Rp4	113.030	111.551	110.072	2.309	11	1.479	35.8
G5	R5	Rp5	138.430	136.951	135.472	2.309	11	1.479	40.1
G6	R6	Rp6	163.830	162.351	160.872	2.309	11	1.479	40.1

Thread tolerances

Tolerance classes for metric ISO threads

cf. DIN ISO 965-1 (1999-11)

Screw thread tolerances are to ensure the function and interchangeability of internal and external threads. They are dependent on the diameter tolerances set in this standard and on the precision of the pitch and the thread angle.

The tolerance class (fine, medium and coarse) is also dependent on the **surface finish** of the threads. Thick electroplated protective coatings require more clearance (e.g. Tolerance Class 6G) than bright or phosphatized surfaces (Tolerance Class 5H).

Thread tolerance	Internal threads	External threads
Applies to	pitch and minor diameters	pitch and major diameters
Labeled by	upper case letters	lower case letters
Tolerance class (example)	**5H**	**6g**
Tolerance grade (size of tolerance)	5	6
Tolerance zone (position of zero line)	H	g

Designation examples	Explanations
M12 x 1 – 5g 6g	External fine threads, nominal ∅ 12 mm, pitch 1 mm; 5g → Tolerance class for pitch ∅; 6g → Tolerance class for major ∅
M12 – 6g	External coarse threads, nominal ∅ 12 mm; 6g → Tolerance class for pitch and major ∅
M24 – 6G/6e	Thread fit for coarse threads, nominal ∅ 24 mm, 6G → Tolerance class of the internal threads, 6e → Tolerance class of the external threads
M16	Tolerance class medium 6H/6g applies to threads without tolerance indication

Tolerance Class 6H/6g is assigned to the "medium" (general purpose) tolerance class and "normal" engagement length in DIN ISO 965-1 (see table below).

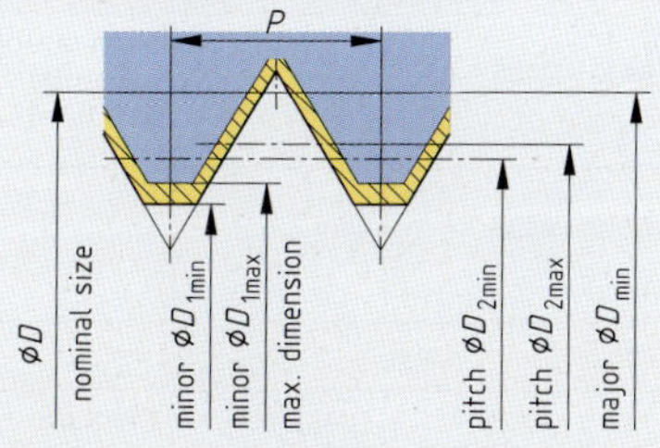

Internal threads, tolerance zone location H

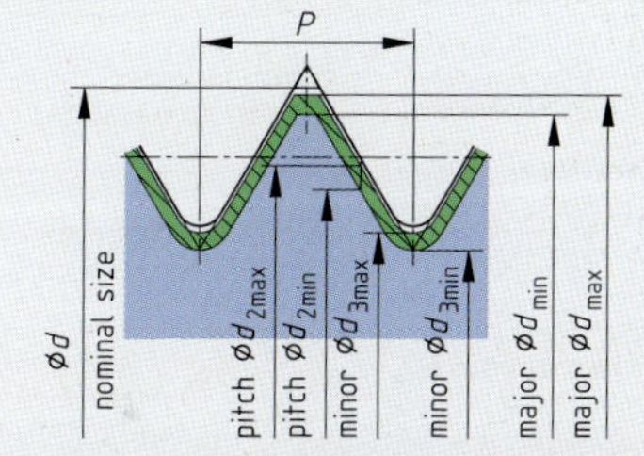

External threads, tolerance zone location g

Limits for external and internal threads (selection)

cf. DIN ISO 965-2 (1999-11)

Threads	Internal threads – Tolerance class 6H					External threads – Tolerance class 6g					
	Major Ø D	Pitch Ø D_2		Minor Ø D_1		Major Ø d		Pitch Ø d_2		Minor Ø[1] d_3	
	min.	min.	max.	min.	max.	max.	min.	max.	min.	max.	min.
M3	3.0	2.675	2.775	2.459	2.599	2.980	2.874	2.655	2.580	2.367	2.273
M4	4.0	3.545	3.663	3.242	3.422	3.978	3.838	3.523	3.433	3.119	3.002
M5	5.0	4.480	4.605	4.134	4.334	4.976	4.826	4.456	4.361	3.995	3.869
M6	6.0	5.350	5.500	4.917	5.135	5.974	5.794	5.324	5.212	4.747	4.596
M8	8.0	7.188	7.348	6.647	6.912	7.972	7.760	7.160	7.042	6.438	6.272
M8 × 1	8.0	7.350	7.500	6.917	7.153	7.974	7.794	7.324	7.212	6.747	6.596
M10	10.0	9.026	9.206	8.376	8.676	9.968	9.732	8.994	8.862	8.128	7.938
M10 ×1	10.0	9.350	9.500	8.917	9.153	9.974	9.794	9.324	9.212	8.747	8.596
M12	12.0	10.863	11.063	10.106	10.441	11.966	11.701	10.829	10.679	9.819	9.602
M12 × 1.5	12.0	11.026	11.216	10.376	10.676	11.968	11.732	10.994	10.854	10.128	9.930
M16	16.0	14.701	14.913	13.385	14.210	15.962	15.682	14.663	14.503	13.508	13.271
M16 × 1.5	16.0	15.026	15.216	14.376	14.676	15.968	15.732	14.994	14.854	14.128	13.930
M20	20.0	18.376	18.600	17.294	17.744	19.958	19.623	18.334	18.164	16.891	16.625
M20 × 1.5	20.0	19.026	19.216	18.376	18.676	19.968	19.732	18.994	18.854	18.128	18.930
M24	24.0	22.051	22.316	20.752	21.252	23.952	23.577	22.003	21.803	20.271	19.955
M24 × 2	24.0	22.701	22.925	21.835	22.210	23.962	23.682	22.663	22.493	21.508	21.261
M30	30.0	27.727	28.007	26.211	26.771	29.947	29.522	27.674	27.462	25.653	25.306
M30 × 2	30.0	28.701	28.925	27.835	28.210	29.962	29.682	28.663	28.493	27.508	27.261
M36	36.0	33.402	33.702	31.670	32.270	35.940	35.465	33.342	33.118	31.033	30.655
M36 × 3	36.0	34.051	34.316	32.752	33.252	35.952	35.577	34.003	33.803	32.271	31.955

[1] cf. DIN 13-20 (2000-08) and DIN 13-21 (2005-08)

Bolts and screws – Overview

Illustration	Design	Standard range from–to	Standard	Application, properties
Hexagon head bolts and screws				pages 211–213
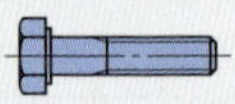	Partly threaded with coarse threads	M1.6 – M64	DIN EN ISO 4014	The most commonly used bolts/screws in machine, equipment and automotive industry **Fully threaded type:** higher fatigue strength
	Fully threaded with coarse threads	M1.6 – M64	DIN EN ISO 4017	
	Partly threaded with fine threads	M8x1 – M64x4	DIN EN ISO 8765	**Compared to coarse threads:** smaller thread depth, smaller pitch, higher load capacity, larger minimum engagement depth l_e
	Fully threaded with fine threads	M8x1 – M64x4	DIN EN ISO 8676	
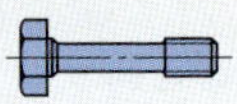	With reduced shank	M3 – M20	DIN EN ISO 24015	Expansion bolts, for dynamic loads, no bolt retaining device required when properly installed
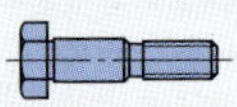	Fit bolt	M8 – M48	DIN 609	Fixing position of parts against movement, fit shank transmits transverse loads
Hexagon bolts and screws for steel structures				page 213
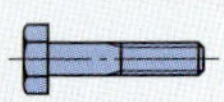	With larger width across flats	M12 – M36	DIN EN 14399-4	High-strength structural bolting assemblies (HV), with nuts as per DIN EN 14399-4 (page 229)
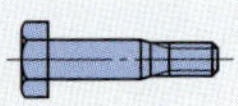	Fit bolt with large widths across flats	M12 – M30	DIN 7999	Friction grip (FG) joints, shear/bearing stress connection
Cap screws				pages 214, 215
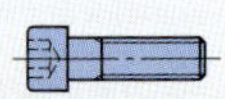	With hexagon socket, coarse thread	M1.6 – M36	DIN EN ISO 4762	Machine, equipment and automotive industry; low space requirements, head sinkable **With low head:** smaller height, lower resistance **Internal serration:** good torque transmission, low fitting space requirements **Slotted screws/bolts:** small screws, low resistance
	With hexagon socket and low head	M6 – M16	DIN 7984	
	With internal serration, coarse, fine thread	M6 – M16	DIN 34821	
	With slotted head, coarse thread	M1.6 – M10	DIN EN ISO 1207	
Countersunk head screws				pages 215, 216
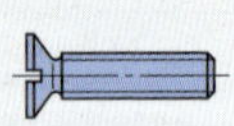	Slotted	M1.6 – M10	DIN EN ISO 2009	Variety of applications in machine, equipment and automotive industry **For screws with hexagon socket:** greater load capacity **For screws with cross recess:** Secure tightening and loosening compared to slotted screws
	With hexagon socket	M3 – M20	DIN EN ISO 10642	
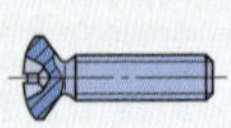	Slotted raised head countersunk	M1.6 – M10	DIN EN ISO 2010	
	Recessed raised head countersunk cross	M1.6 – M10	DIN EN ISO 7047	
Sheet metal screws with tapping threads				pages 216, 217
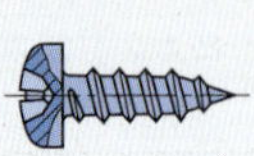	Round head screw	ST2.2 – ST9.5	DIN ISO 7049	Vehicle body and sheet metal manufacturing. The sheets to be joined have tap holes. The threads are formed by the screw. Locking fasteners are only needed for thin sheets.
	Countersunk head screw	ST2.2 – ST6.3	DIN ISO 7050	
	Round head countersunk screws	ST2.2 – ST9.9	DIN ISO 7051	

Bolts and screws – Overview, Designation of bolts and screws

Illustration	Design	Standard range from–to	Standard	Application, properties
Drilling screws with tapping threads				
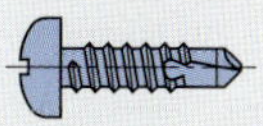	Flat head with cross recess	ST2.2–ST6.3	DIN EN ISO 15481	Vehicle body and sheet metal manufacturing drilling screws bore the tap hole while being screwed in and form the threads.
	Round head counter-sunk with cross/recess	ST2.2–ST6.3	DIN EN ISO 15483	
Studs				page 218
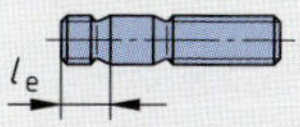	$l_e \approx 2 \cdot d$	M4–M24	DIN 835	For aluminum alloys
	$l_e \approx 1.25 \cdot d$	M4–M48	DIN 939	For cast iron materials
	$l_e \approx 1 \cdot d$	M4–M48	DIN 938	For steel
Set screws				page 219
	With dog point and slotted	M1.6–M12	DIN EN 27435	Compression loadable screws for securing position of parts, e.g. levers, bearing bushings, hubs Set screws are not suitable for power transmission of torques, e.g. for joining shafts to hubs.
	With dog point and hex socket	M1.6–M24	DIN EN ISO 4028	
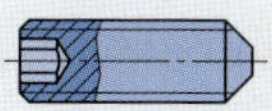	With cone point and slotted	M1.6–M12	DIN EN 27434	
	With cone point and hex socket	M1.6–M24	DIN EN ISO 4027	
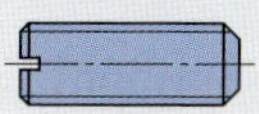	With flat point and slotted	M1.6–M12	DIN EN 24766	
	With flat point and hex socket	M1.6–M24	DIN EN ISO 4026	
Drain plugs				page 218
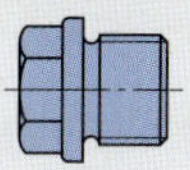	Heavy type with hexagon socket or hexagon head	M10x1–M52x1.5	DIN 908 DIN 910	Gearbox manufacturing; Fill, overflow and drain screws for gear oil; machining of seating surface necessary
Thread forming screws				page 217
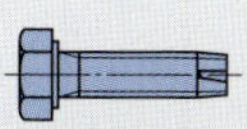	Various head forms e.g. hexagon, cheese head	M2–M10	DIN 7500-1	For low loading in malleable materials, e.g. S235, DC01–DC04, non-ferrous metals; use without locking fastener
Eye bolts				page 218
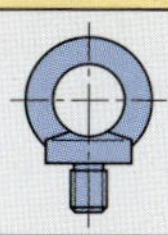	With coarse threads	M8–M100x6	DIN 580	Transport eyes on machines and equipment; stress depends on the angle of the applied load, milling of seating surface necessary

Designation of bolts and screws

cf. DIN 962 (2001-11)

Examples:

Hex screw	**ISO 4017**	**– M12 x 80**	**– A2-70**
Drain plug	**DIN 910**	**– M24 x 1.5**	**– St**
Cap screws	**ISO 4762**	**– M10 x 55**	**– 8.8**

Type	Reference standard, e.g. ISO, DIN, EN; Sheet number of the standard[1]	Nominal data, e.g. M → metric screw thread; 12 → nominal diameter d; 80 → shank length l	Property class, e.g. 8.8, 10.9, A2-70, A4-70; Material, e.g. St steel, CuZn copper-zinc-alloy

[1] Bolts and screws standardized according to ISO or DIN EN ISO have the abbreviation **ISO** in their designation. Bolts and screws standardized according to DIN have the abbreviation **DIN** in their designation.

ME

Property classes, Product grades, Clearance holes, Minimum engagement depths

Property classes of screws and bolts

cf. DIN EN ISO 898-1 (2009-08), DIN EN ISO 3506-1 (2010-03)

Example:

Unalloyed and alloyed steels DIN EN ISO 898-1: **9 . 8**

Tensile strength R_m	Yield strength R_e
R_m = **9** · 100 N/mm² = 900 N/mm²	R_e = **9** · **8** · 10 N/mm² = 720 N/mm²

Stainless steels DIN EN ISO 3506-1: **A 2 – 70**

Tensile strength R_m
R_m = **70** · 10 N/mm² = 700 N/mm²

Steel type	Note
A → austenitic steel A2 → rustproof screws A4 → rustproof and acid-resistant screws	Observe relevant codes of practice

Property classes and material properties

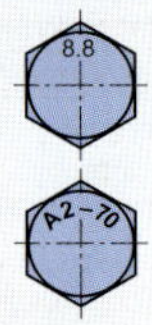

Material property	Property classes for bolts and screws made of unalloyed and alloyed steels						stainless steels[1]		
	5.8	6.8	8.8	9.8	10.9	12.9	A2-50	A4-50	A2-70
Tens. strength R_m in N/mm²	500	600	800	900	1000	1200	500	500	700
Yield strength R_e in N/mm²	400	480	640	720	900	1080	210	210	450
Elong. at fracture *EL* in %	–	–	12	10	9	8	20	20	13

[1] Material properties apply to threads ≤ M20.

Product grades for bolts and nuts

cf. DIN EN ISO 4759-1 (2001-04)

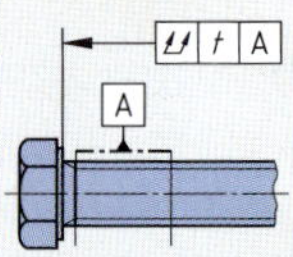

Product grade	Tole-rances	Explanation, application
A	fine	Dimensional, form and positional tolerances for bolts and nuts with ISO threads are specified in tolerance grades A, B, C.
B	medium	
C	coarse	

Clearance holes for bolts

cf. DIN EN 20273 (1992-02)

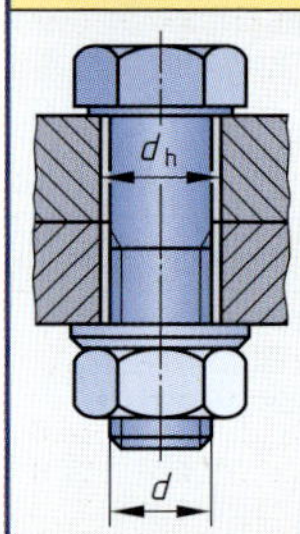

Thread d	Clearance hole d_h[1] Series fine	med.	coarse	Thread d	Clearance hole d_h[1] Series fine	med.	coarse	Thread d	Clearance hole d_h[1] Series fine	med.	coarse
M1	1.1	1.2	1.3	M5	5.3	5.5	5.8	M24	25	26	28
M1.2	1.3	1.4	1.5	M6	6.4	6.6	7	M30	31	33	35
M1.6	1.7	1.8	2	M8	8.4	9	10	M36	37	39	42
M2	2.2	2.4	2.6	M10	10.5	11	12	M42	43	45	48
M2.5	2.7	2.9	3.1	M12	13	13.5	14.5	M48	50	52	56
M3	3.2	3.4	3.6	M16	17	17.5	18.5	M56	58	62	66
M4	4.3	4.5	4.8	M20	21	22	24	M64	66	70	74

[1] Tolerance grades for d_h; fine series: H12, medium series: H13, coarse series: H14

ME

Minimum engagement depth in blind hole

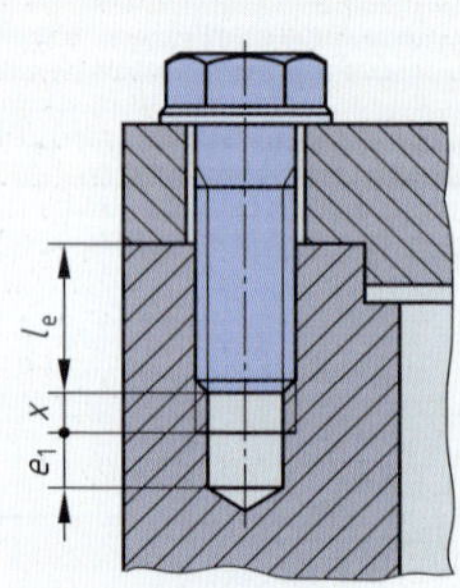

x ≈ 3 · *P* (thread pitch)
e_1 according to DIN 76, see page 84

Area of application		Minimum engagement depth l_e[1] for coarse threads and property class 3.6, 4.6	4.8–6.8	8.8	10.9
Struc. steel	$R_m \leq 400$ N/mm²	0.8 · *d*	1.2 · *d*	–	–
	R_m = 400–600 N/mm²	0.8 · *d*	1.2 · *d*	1.2 · *d*	–
	R_m > 600–800 N/mm²	0.8 · *d*	1.2 · *d*	1.2 · *d*	1.2 · *d*
	R_m > 800 N/mm²	0.8 · *d*	1.2 · *d*	1.0 · *d*	1.0 · *d*
Cast iron materials		1.3 · *d*	1.5 · *d*	1.5 · *d*	–
Copper alloys		1.3 · *d*	1.3 · *d*	–	–
Aluminum casting alloys		1.6 · *d*	2.2 · *d*	–	–
Al alloys, age-hardened		0.8 · *d*	1.2 · *d*	1.6 · *d*	–
Al alloys, not age-hardened		1.2 · *d*	1.6 · *d*	–	–
Plastics		2.5 · *d*	–	–	

[1] Engagement depth for fine threads l_e = 1.25 · Engagement depth for coarse threads

Hexagon head bolts

Hexagon head bolt with shank and coarse threads

cf. DIN EN ISO 4014 (2001-03)

Valid standard DIN EN ISO	Replaces DIN EN	DIN
4014	24014	931

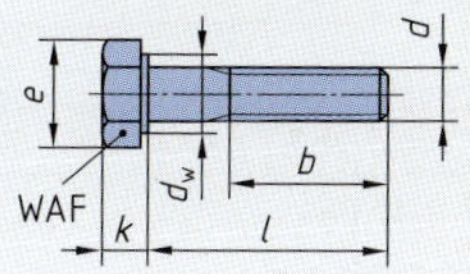

1) for l < 125 mm
2) for l = 125–200 mm
3) for l > 200 mm

Product grades (page 210)

Threads d	l in mm	Grade
≤ M12	all	A
M16–M24	l ≤ 150	A
	l ≥ 160	B
≥ M30	all	B

Thread d	M1.6	M2	M2.5	M3	M4	M5	M6	M8	M10
WAF	3.2	4	5	5.5	7	8	10	13	16
k	1.1	1.4	1.7	2	2.8	3.5	4	5.3	6.4
d_w	2.3	3.1	4.1	4.6	5.9	6.9	8.9	11.6	14.6
e	3.4	4.3	5.5	6	7.7	8.8	11.1	14.4	17.8
b	9	10	11	12	14	16	18	22	26
l from	12	16	16	20	25	25	30	40	45
l to	16	20	25	30	40	50	60	80	100
Property classes	5.6, 8.8, 9.8, 10.9, A2-70, A4-70								

Thread d	M12	M16	M20	M24	M30	M36	M42	M48	M56
WAF	18	24	30	36	46	55	65	75	85
k	7.5	10	12.5	15	18.7	22.5	26	30	35
d_w	16.6	22	27.7	33.3	42.8	51.1	60	69.5	78.7
e	20	26.2	33	39.6	50.9	60.8	71.3	82.6	93.6
b 1)	30	38	46	54	66	–	–	–	–
b 2)	–	44	52	60	72	84	96	108	–
b 3)	–	–	–	73	85	97	109	121	137
l from	50	65	80	90	110	140	160	180	220
l to	120	160	200	240	300	360	440	500	500
Property classes	5.6, 8.8, 9.8, 10.9						as per agreement		
	A2-70, A4-70				A2-50, A4-50				
Nominal lengths l	12, 16, 20, 25, 30, 35–60, 65, 70, 80, 90–140, 150, 160, 180, 200–460, 480, 500 mm								
⇒	**Hexagon head bolt ISO 4014 – M10 x 60 – 8.8:** d = M10, l = 60 mm, property class 8.8								

Hexagon head bolts with coarse threads, fully threaded

cf. DIN EN ISO 4017 (2001-03)

Valid standard DIN EN ISO	Replaces DIN EN	DIN
4017	24017	933

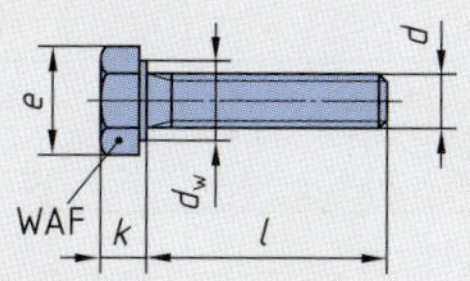

Product grades (page 210)

Threads d	l in mm	Grade
≤ M12	all	A
M16–M24	l ≤ 150	A
	l ≥ 160	B
≥ M30	all	B

Thread d	M1.6	M2	M2.5	M3	M4	M5	M6	M8	M10
WAF	3.2	4	5	5.5	7	8	10	13	16
k	1.1	1.4	1.7	2	2.8	3.5	4	5.3	6.4
d_w	2.3	3.1	4.1	4.6	5.9	6.9	8.9	11.6	14.6
e	3.4	4.3	5.5	6	7.7	8.8	11.1	14.4	17.8
l from	2	4	5	6	8	10	12	16	20
l to	16	20	25	30	40	50	60	80	100
Property classes	5.6, 8.8, 9.8, 10.9, A2-70, A4-70								

Thread d	M12	M16	M20	M24	M30	M36	M42	M48	M56
WAF	18	24	30	36	46	55	65	75	85
k	7.5	10	12.5	15	18.7	22.5	26	30	35
d_w	16.6	22	27.7	33.3	42.8	51.1	60	69.5	78.7
e	20	26.2	33	39.6	50.9	60.8	71.3	82.6	93.6
l from	25	30	40	50	60	70	80	100	110
l to	120	200	200	200	200	200	200	200	200
Property classes	5.6, 8.8, 9.8, 10.9						as per agreement		
	A2-70, A4-70				A2-50, A4-50				
Nominal lengths l	2, 3, 4, 5, 6, 8, 10, 12, 16, 20, 25, 30, 35–60, 65, 70, 80, 90–140, 150, 160, 180, 200 mm								
⇒	**Hexagon head bolt ISO 4017 – M8 x 40 – A4-50:** d = M8, l = 40 mm, property class A4-50								

Hexagon head bolts

Hexagon head bolt with shank and fine threads

cf. DIN EN ISO 8765 (2001-03)

Valid standard DIN EN ISO	Replaces DIN EN	Replaces DIN
8765	28765	960

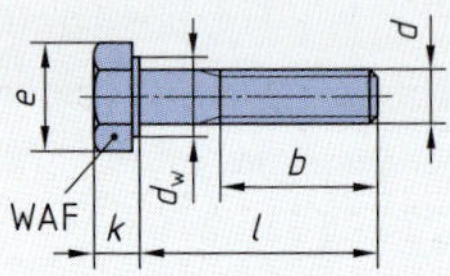

Product grades (page 210)

Threads *d*	*l* in mm	Grade
≤ M12x1.5	all	A
M16x1.5–M24x2	≤ 150	A
	> 150	B
≥ M30x2	all	B

Thread *d*	M8 x1	M10 x1	M12 x1.5	M16 x1.5	M20 x1.5	M24 x2	M30 x2	M36 x3	M42 x3	M48 x3	M56 x4
WAF	13	16	18	24	30	36	46	55	65	75	85
k	5.3	6.4	7.5	10	12.5	15	18.7	22.5	26	30	35
d_w	11.6	14.6	16.6	22.5	28.2	33.6	42.8	51.1	60	69.5	78.7
e	14.4	17.8	20	26.2	33	39.6	50.9	60.8	71.3	82.6	93.6
b[1]	22	26	30	38	46	54	66	–	–	–	–
b[2]	–	–	–	44	52	60	72	84	96	108	–
b[3]	–	–	–	–	–	73	85	97	109	121	137
l from	40	45	50	65	80	100	120	140	160	200	220
l to	80	100	120	160	200	240	300	360	440	480	500

Nominal lengths *l*	40, 45, 50, 55, 60, 65, 70, 80, 90–140, 150, 160, 180, 200, 220–460, 480, 500 mm	
Property classes	*d* ≤ M24x2: 5.6, 8.8, 10.9, A2-70, A4-70 *d* = M30x2–M36x3: 5.6, 8.8, 10.9, A2-50, A4-50	d ≥ M42x3: as per agreement
Explanations	[1] for *l* < 125 mm [2] for *l* = 125–200 mm [3] for *l* > 200 mm	
⇒	**Hexagon head bolt ISO 8765–M20 x 1.5 x 120 – 5.6:** *d* = M20 x 1.5, *l* = 120 mm, property class 5.6	

Hexagon head bolts with fine threads, fully threaded

cf. DIN EN ISO 8676 (2001-03)

Valid standard DIN EN ISO	Replaces DIN EN	Replaces DIN
8676	28676	961

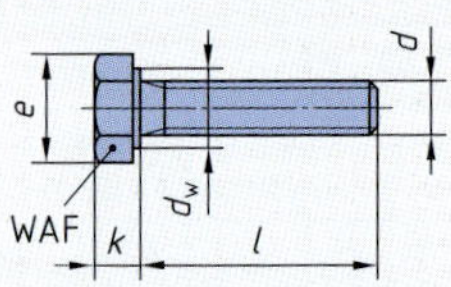

Product grades according to DIN EN ISO 8765

Thread *d*	M8 x1	M10 x1	M12 x1.5	M16 x1.5	M20 x1.5	M24 x2	M30 x2	M36 x3	M42 x3	M48 x3	M56 x4
WAF	13	16	18	24	30	36	46	55	65	75	85
k	5.3	6.4	7.5	10	12.5	15	18.7	22.5	26	30	35
d_w	11.6	14.6	16.6	22.5	28.2	33.6	42.8	51.1	60	69.5	78.7
e	14.4	17.8	20	26.2	33	39.6	50.9	60.8	71.3	82.6	93.6
l from	16	20	25	35	40	40	40	40	90	100	120
l to	80	100	120	160	200	200	200	200	420	480	500

Nominal lengths *l*	16, 20, 25, 30, 35–60, 65, 70, 80, 90–140, 150, 160, 180, 200, 220–460, 480, 500 mm	
Property classes	d ≤ M24x2: 5.6, 8.8, 10.9, A2-70, A4-70 *d* = M30x2–M36x3: 5.6, 8.8, 10.9, A2-50, A4-50	d ≥ M42x3: as per agreement
⇒	**Hexagon head bolt ISO 8676 – M8 x 1 x 55 – 8.8:** *d* = M8 x 1, *l* = 55 mm, property class 8.8	

ME

Hex head bolt with reduced shank

cf. DIN EN 24015 (1991-12)

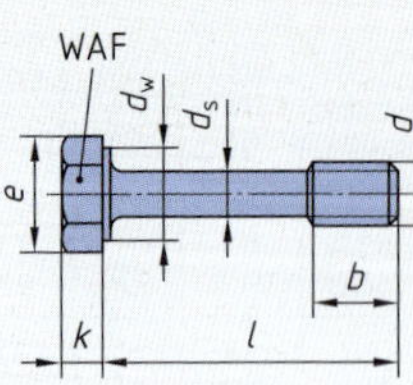

Product grades (page 210)

Threads *d*	*l* in mm	Grade
≤ M20	all	B

Thread *d*	M3	M4	M5	M6	M8	M10	M12	M16	M20
WAF	5.5	7	8	10	13	16	18	24	30
k	2	2.8	3.5	4	5.3	6.4	7.5	10	12.5
d_w	4.4	5.7	6.7	8.7	11.4	14.4	16.4	22	27.7
d_s	2.6	3.5	4.4	5.3	7.1	8.9	10.7	14.5	18.2
e	6	7.5	8.7	10.9	14.2	17.6	19.9	26.2	33
b[1]	12	14	16	18	22	26	30	38	46
b[2]	–	–	–	–	28	32	36	44	52
l from	20	20	25	25	30	40	45	55	65
l to	30	40	50	60	80	100	120	150	150

Nominal lengths *l*	20, 25, 30–65, 70, 75, 80, 90, 100–130, 140, 150 mm
Property classes	5.8, 6.8, 8.8, A2-70
Explanations	[1] for *l* ≤ 120 mm [2] for *l* > 125 mm
⇒	**Hexagon head bolt ISO 4015 – M8 x 45 – 8.8:** *d* = M8, *l* = 45 mm, property class 8.8

Hexagon head bolts

Hexagon head fit bolts with long thread

cf. DIN 609 (1995-02)

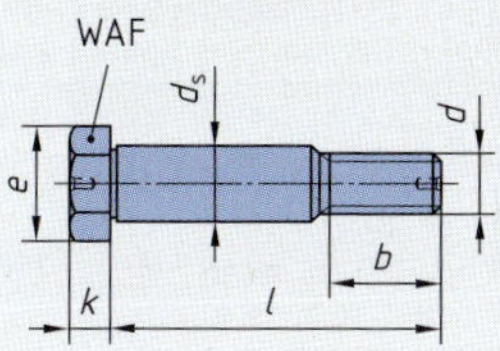

Thread d	M8 M8 x1	M10 M10 x1	M12 M12 x1.5	M16 M16 x1.5	M20 M20 x1.5	M24 M24 x2	M30 M30 x2	M36 M36 x3	M42 M42 x3	M48 M48 x3
WAF	13	16	18	24	30	36	46	55	65	75
k	5.3	6.4	7.5	10	12.5	15	19	22	26	30
d_s k6	9	11	13	17	21	25	32	38	44	50
e	14.4	17.8	19.9	26.2	33	39.6	50.9	60.8	71.3	82.6
b 1)	14.5	17.5	20.5	25	28.5	–	–	–	–	–
b 2)	16.5	19.5	22.5	27	30.5	36.5	43	49	56	63
b 3)	–	–	–	32	35.5	41.5	48	54	61	68
l from	25	30	32	38	45	55	65	70	80	85
l to	80	100	120	150	150	150	200	200	200	200
Nominal lengths l	25, 28, 30, 32, 35, 38, 40, 42, 45, 48, 50, 55, 60–150, 160–200 mm									
Property classes	8.8								as per agreement	
	A2-70					A2-50				
Explanations	1) for $l \leq 50$ mm 2) for $l = 50–150$ mm 3) for $l > 150$ mm									
⇒	**Fit bolt DIN 609 – M16 x 1.5 x 125 – A2-70:** d = M16 x 1.5, l = 125 mm, property class A2-70									

Product grades (page 210)

d in mm	l in mm	Grade
≤ 10	all	A
≥ 12	all	B

Hexagon head bolts with large width across flats
for high-strength structural bolting assemblies (HV)

cf. DIN EN 14399-4 (2006-06), replaces DIN 6914

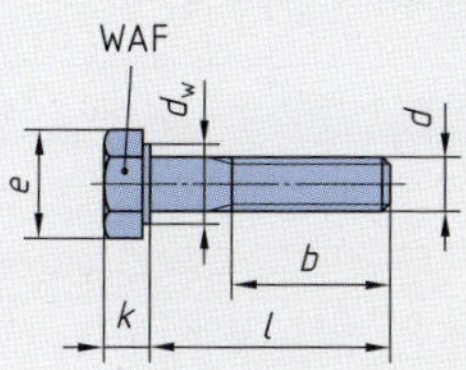

Product grade C (page 210)

Thread d	M12	M16	M20	M22	M24	M27	M30	M36
WAF	22	27	32	36	41	46	50	60
k	8	10	13	14	15	17	19	23
d_w	20.1	24.9	29.5	33.3	38	42.8	46.6	55.9
e	23.9	29.6	35	39.6	45.2	50.9	55.4	66.4
b_{min}	23	28	33	34	39	41	44	52
l from	35	40	45	50	60	70	75	85
l to	95	130	155	165	195	200	200	200
Nominal lengths l	35, 40, 45, 50, 55, 60, 65, 70–175, 180, 185, 190, 195, 200 mm							
Property class, surface	10.9 normal → with thin oil film, hot-galvanized → code: tZn							
⇒	**Hexagon head bolt EN 14399-4 – M12 x 65 – 10.9 – HV – tZn:** d = M12, l = 65 mm, property class 10.9, for high-strength bolting assemblies, with hot-galvanized surface							

Hexagon fit bolts with large width across flats
for high-strength structural bolt assemblies (HV)

cf. DIN EN 14399-8 (2008-03) replaces DIN 7999

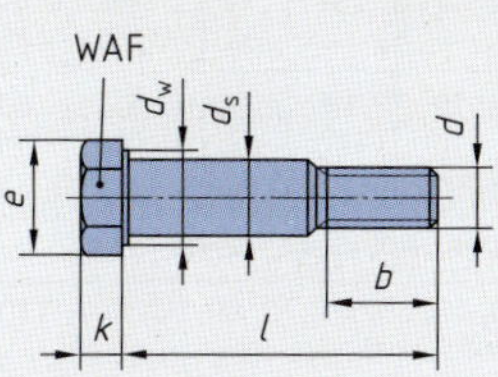

Product grade C (page 210)

Thread d	M12	M16	M20	M22	M24	M27	M30	M36
WAF	22	27	32	36	41	46	50	60
k	8	10	13	14	15	17	19	23
d_w	20	25	29.5	33.3	38	42.8	46.6	56
d_s b11	13	17	21	23	25	28	31	37
e	23.9	29.6	35	39.6	45.2	50.9	55.4	66.4
b	23	28	33	34	39	41	44	52
l from	50	65	75	80	90	95	105	125
l to	95	125	155	155	185	200	200	200
Property class	10.9							
⇒	**Hexagon fit bolt EN 14399-8 – M24 x 120 – 10.9 – HVP:** d = M24, l = 120 mm, property class 10.9, fit bolt for high-strength bolting assemblies							

ME

Hexagon socket head cap screws

Hexagon socket head cap screws with coarse threads

cf. DIN EN ISO 4762 (2004-06)

Valid standard DIN EN ISO	Replaces DIN
4762	912

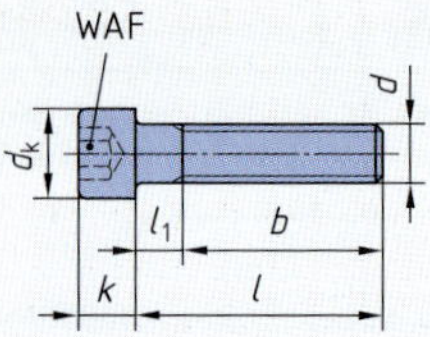

Thread *d*	M1.6	M2	M2.5	M3	M4	M5	M6	M8	M10
WAF	1.5	1.5	2	2.5	3	4	5	6	8
k	1.6	2	2.5	3	4	5	6	8	10
d_k	3	3.8	4.5	5.5	7	8.5	10	13	16
b	–	16	17	18	20	22	24	28	32
for *l*	–	20	25	≥ 25	≥ 30	≥ 30	≥ 35	≥ 40	≥ 45
l_1	1.1	1.2	1.4	1.5	2.1	2.4	3	3.8	4.5
for *l*	≤ 16	≤ 16	≤ 20	≤ 20	≤ 25	≤ 25	≤ 30	≤ 35	≤ 40
l from	2.5	3	4	5	6	8	10	12	16
l to	16	20	25	30	40	50	60	80	100
Property classes	by agreement			8.8, 10.9, 12.9					
	Stainless steels A2-70, A4-70								

Thread *d*	M12	M16	M20	M24	M30	M36	M42	M48	M56
WAF	10	14	17	19	22	27	32	36	41
k	12	16	20	24	30	36	42	48	56
d_k	18	24	30	36	45	54	63	72	84
b	36	44	52	60	72	84	96	108	124
for *l*	≥ 55	≥ 65	≥ 80	≥ 90	≥ 110	≥ 120	≥ 140	≥ 160	≥ 180
l_1	5.3	6	7.5	9	10.5	12	13.5	15	16.5
for *l*	≤ 50	≤ 60	≤ 70	≤ 80	≤ 100	≤ 110	≤ 130	≤ 150	≤ 160
l from	20	25	30	40	45	45	60	70	80
l to	120	160	200	200	200	200	300	300	300
Property classes	8.8, 10.9, 12.9						as per agreement		
	A2-70, A4-70				A2-50, A4-50				
Nominal lengths *l*	2.5, 3, 4, 5, 6, 8, 10, 12, 16, 20, 25, 30–65, 70, 80–150, 160, 180, 200, 220, 240, 260, 280, 300 mm								
⇒	**Cap screw ISO 4762 – M10 x 55 – 10.9:** *d* = M10, *l* = 55 mm, property class 10.9								

Product grades (page 210)

Thread *d*	Grade
M1.6 – M56	A

ME

Hexagon socket cap screws, low head
Hexagon socket head cap screws, low head and pilot recess

cf. DIN 7984 (2009-06)
cf. DIN 6912 (2009-06)

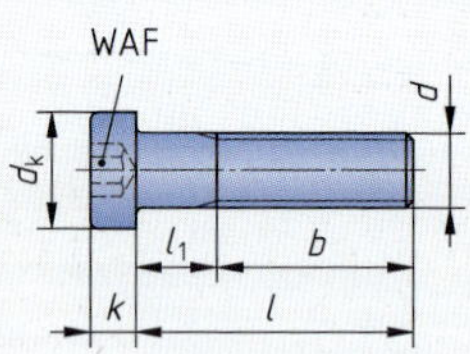

DIN 6912 – pilot recess

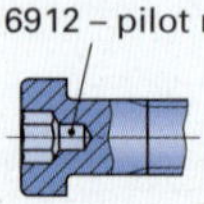

	Thread d	M3	M4	M5	M6	M8	M10	M12	M16	M20	M24
	WAF	2	2.5	3	4	5	7	8	12	14	17
	k	2	2.8	3.5	4	5	6	7	9	11	13
	d_k	5,5	7	8.5	10	13	16	18	24	30	36
DIN 7984	*b*	12	14	16	18	22	26	30	38	46	54
	for *l*	≥ 20	≥ 25	≥ 30	≥ 30	≥ 35	≥ 40	≥ 50	≥ 60	≥ 70	≥ 90
	l_1	1.5	2.1	2.4	3	3.8	4.5	5.3	6	7.5	9
	for *l*	≤ 16	≤ 20	≤ 25	≤ 25	≤ 30	≤ 35	≤ 45	≤ 50	≤ 60	≤ 80
	l from	5	6	8	10	12	16	20	30	30	40
	l to	20	25	30	40	80	100	80	80	100	100
DIN 6912	$b^{1)}$	–	14	16	18	22	26	30	38	46	50
	$b^{2)}$	–	–	–	–	–	–	–	40	52	60
	l from	–	10	10	10	12	16	16	20	30	60
	l to	–	50	60	70	80	90	100	140	180	200
Property classes		08.8 → approx. 25% less resistance as with class 8.8. A2 → approx. 25% less resistance as with class A2-70									
Nominal lengths *l*		5, 6, 8, 10, 12, 16, 20, 25, 30, 35, 40, 45, 50, 60, 70, 80, 90, 100, 120, 140, 160, 180, 200 mm									
Explanation		1) *b* for *l* ≤ 125 mm, 2) *b* for 125 < *l* ≤ 200 mm									
⇒		**Cap screw DIN 7984 – M12 x 50 – 08.8:** *d* = M12, *l* = 50 mm, Property class 08.8 (08 → lower resistance)									

Product grades (page 210)

Thread *d*	Grade
M3 – M24	A

Cap screws, Flat head head screws

Cap screws with internal serration

cf. DIN 34821 (2005-11)

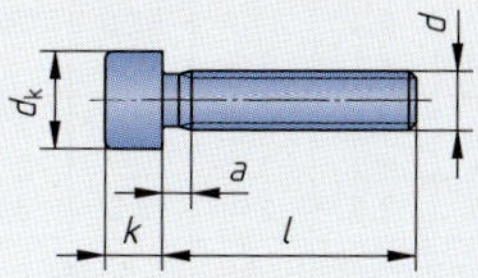

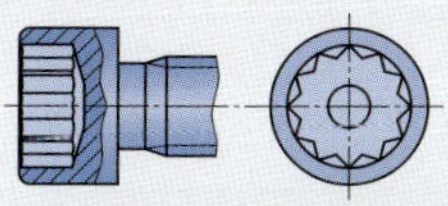

Internal serration

Thread d	M6 –	M8 –	M10 –	M12 M12x1.25	M14 M14x1.5	M16 M16x1.5
Rated size[1]	N8	N10	N12	N14	N16	N18
k d_k	6 10	8 13	10 16	12 18	14 21	16 24
a_{max} for l	2.0 ≤ 16	2.5 ≤ 20	3.0 ≤ 25	3.5 ≤ 25	4.0 ≤ 30	4.0 ≤ 30
a_{max} for l	3.0 > 16	3.8 > 20	4.5 > 25	5.3 > 25	6.0 > 30	6.0 > 30
l from to	12 70	16 80	20 90	20 90	20 100	20 100
Property class	8.8, 10.9, A2-70					
Nominal lengths l	12, 16, 20, 25, 30, 35, 40, 45, 50, 55, 60, 65, 70, 80, 90, 100 mm					
Explanation	[1] Rated size of internal serration (nominal tool size))					
⇒	**Cap screw DN 34821 – M10 x 35 – 8.8:** d = M10, l = 35 mm, property class 8.8					

Product grade A (page 210)

Slotted cheese head screws

cf. DIN EN ISO 1207 (1994-10)

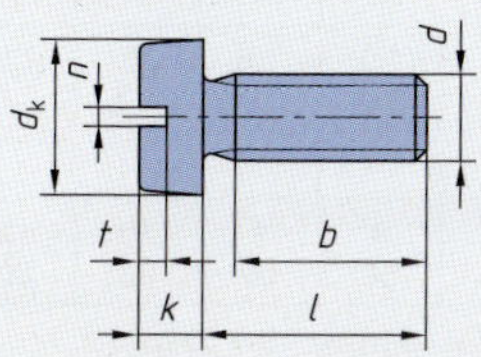

Thread d	M1.6	M2	M2.5	M3	M4	M5	M6	M8	M10
d_k k	3 1.1	3.8 1.4	4.5 1.8	5.5 2	7 2.6	8.5 3.3	10 3.9	13 5	16 6
n t	0.4 0.5	0.5 0.6	0.6 0.7	0.8 0.9	1.2 1.1	1.2 1.3	1.6 1.6	2 2	2.5 2.4
l from to	2 16	3 20	3 25	4 30	5 40	6 50	8 60	10 80	12 80
b	for l < 45 mm → threads near to head for l ≥ 45 mm → b = 38 mm								
Nominal lengths l	2, 3, 4, 5, 6, 8, 10, 12, 16, 20, 25–45, 50, 60, 70, 80 mm								
Property classes	4.8, 5.8, A2-50, A4-50								
⇒	**Cheese head screw ISO 1207 – M6 x 25 – 5.8:** d = M6, l = 25 mm, property class 5.8								

Product grade A (page 210)

Hexagon socket flat head screws

cf. DIN EN ISO 10642 (2004-06), replaces DIN 7991

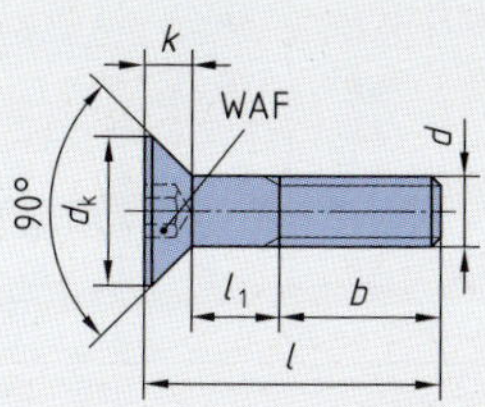

Thread d	M3	M4	M5	M6	M8	M10	M12	M16	M20
WAF d_k k	2 5.5 1.9	2.5 7.5 2.5	3 9.4 3.1	4 11.3 3.7	5 15.2 5	6 19.2 6.2	8 23.1 7.4	10 29 8.8	12 36 10.2
b for l	18 ≥ 30	20 ≥ 30	22 ≥ 35	24 ≥ 40	28 ≥ 50	32 ≥ 55	36 ≥ 65	44 ≥ 80	52 100
l_1 for l	1.5 ≤ 25	2.1 ≤ 25	2.4 ≤ 30	3 ≤ 35	3.8 ≤ 45	4.5 ≤ 50	5.3 ≤ 60	6 ≤ 70	7.5 ≤ 90
l from to	8 30	8 40	8 50	8 60	10 80	12 100	20 100	30 100	35 100
Property classes	8.8, 10.9, 12.9								
Nominal lengths l	8, 10, 12, 16, 20, 25, 30, 35, 40, 45, 50, 55, 60, 65, 70, 80, 90, 100 mm								
⇒	**Flat head screw ISO 10642 – M5 x 30 – 8.8:** d = M5, l = 30 mm, property class 8.8								

Product grade A (page 210)

ME

Flat head screws, Raised head countersunk screws, Tapping screws

Slotted raised head countersunk screws — cf. DIN EN ISO 2010 (1994-10)
Raised head countersunk screws with cross recess — cf. DIN EN ISO 7047 (1994-10)

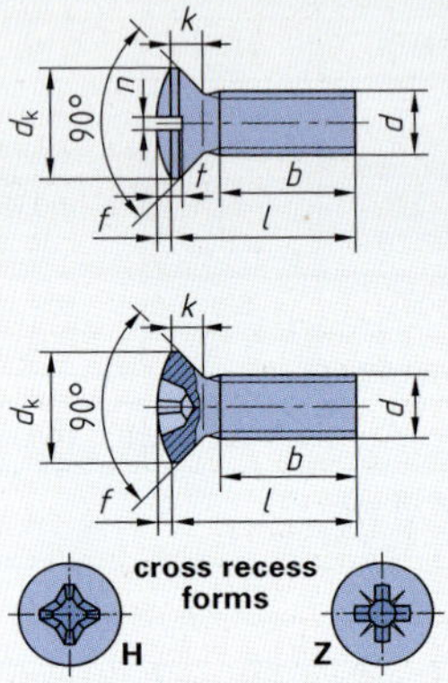

Product grade A (page 210)

Thread d	M1.6	M2	M2.5	M3	M4	M5	M6	M8	M10
d_k	3	3.8	4.7	5.5	8.4	9.3	11.3	15.8	18.3
k	1	1.2	1.5	1.7	2.7	2.7	3.3	4.7	5
n	0.4	0.5	0.6	0.8	1.2	1.2	1.6	2	2.5
f	0.4	0.5	0.6	0.7	1.0	1.2	1.4	2	2.3
t	0.6	0.8	1.0	1.2	1.6	2.0	2.4	3.2	3.8
C[1]	0		1		2		3	4	
l from	2.5	3	4	5	6	8	8	10	12
l to	16	20	25	30	40	50	60	80	80
b	for l < 45 mm → $b \approx l$; for $l \geq$ 45 mm → b = 38 mm								
Property classes	DIN EN ISO 2010: 4.8, 5.8, A2-50, A2-70 DIN EN ISO 7047: 4.8, A2-50, A2-70								
Nominal lengths l	2.5, 3, 4, 5, 6, 8, 10, 12, 16, 20, 25–45, 50, 60, 70, 80 mm								
Explanation	[1] C cross recess size, forms H and Z								
⇒	**Countersunk head screw ISO 7047 – M3 x 20 – 4.8 – H:** d = M3, l = 20 mm, property class 4.8, cross recess form H								

Slotted flat head screws — cf. DIN EN ISO 2009 (1994-10)
Flat head screws with cross recess — cf. DIN EN ISO 7046-1 (1994-10)

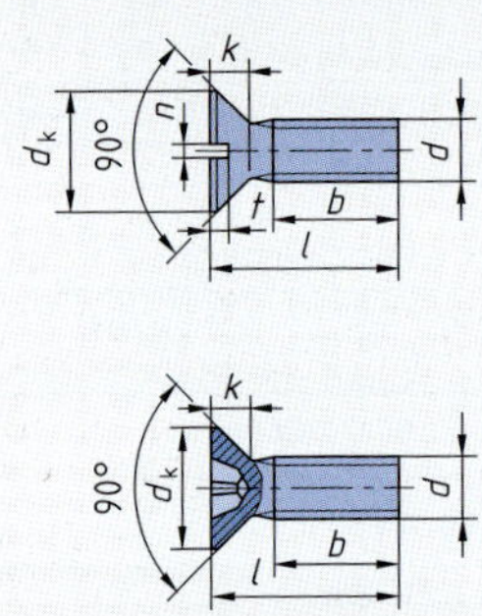

Product grade A (page 210)

Thread d	M1.6	M2	M2.5	M3	M4	M5	M6	M8	M10
d_k	3	3.8	4.7	5.5	8.4	9.3	11.3	15.8	18.3
k	1	1.2	1.5	1.7	2.7	2.7	3.3	4.7	5
n	0.4	0.5	0.6	0.8	1.2	1.2	1.6	2	2.5
t	0.5	0.6	0.8	0.9	1.3	1.4	1.6	2.3	2.6
C[1]	0		1		2		3	4	
l from	2.5	3	4	5	6	8	8	10	12
l to	16	20	25	30	40	50	60	80	80
b	for l < 45 mm → $b \approx l$; for $l \geq$ 45 mm → b = 38 mm								
Property classes	DIN EN ISO 2009: 4.8, 5.8, A2-50, A2-70 DIN EN ISO 7046-1: 4.8, A2-50, A2-70								
Nominal lengths l	2.5, 3, 4, 5, 6, 8, 10, 12, 16, 20, 25–45, 50, 60, 70, 80 mm								
Explanation	[1] C cross recess size, forms H and Z (see DIN EN 2010)								
⇒	**Flat head screw ISO 7046-1 – M5 x 40 – 4.8 – H:** d = M5, l = 40 mm, property class 4.8, cross recess form H								

ME

Flat head tapping screws — cf. DIN EN ISO 7050 (1990-08)
Raised head countersunk tapping screws — cf. DIN EN ISO 7051 (1990-08)

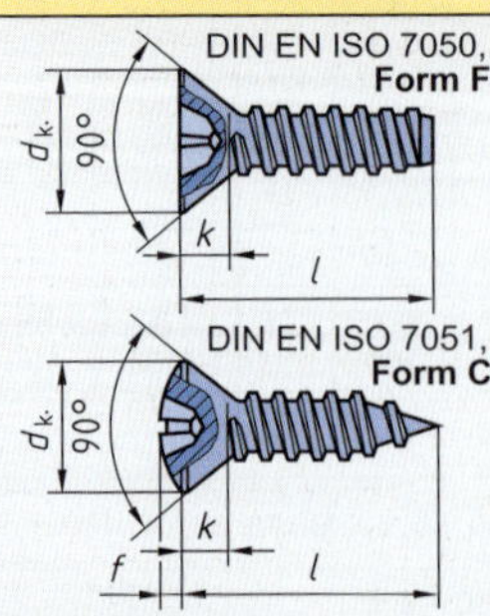

Product grade A (page 210)

Thread d	ST2.2	ST2.9	ST3.5	ST4.2	ST4.8	ST5.5	ST6.3
d_k	3.8	5.5	7.3	8.4	9.3	10.3	11.3
k	1.1	1.7	2.4	2.6	2.8	3	3.2
f	0.5	0.7	0.8	1.0	1.2	1.3	1.4
l from	4.5	6.5	9.5	9.5	9.5	13	13
l to	16	19	25	32	32	38	38
C[1]	0	1	2			3	
Nominal lengths l	4.5, 6.5, 9.5, 13, 16, 19, 22, 25, 32, 38 mm						
Forms	Form C with cone point, form F with dog point						
Explanation	[1] C cross recess size, forms H and Z (see DIN EN 2010)						
⇒	**Tapping screw ISO 7050 – ST4.8 x 32 – F – Z:** d = ST4.8, l = 32 mm, form F, cross recess form Z						

Tapping screws, Thread forming screws

Pan head tapping screws

cf. DIN EN ISO 7049 (1990-08)

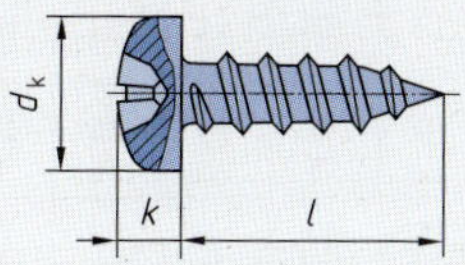

Product grade A (page 210)

Thread d	ST2.2	ST2.9	ST3.5	ST4.2	ST4.8	ST5.5	ST6.3
d_k	4	5.6	7	8	9.5	11	13
k	1.6	2.4	2.6	3.1	3.7	4	4.6
l from	4.5	6.5	9.5	9.5	9.5	13	13
l to	16	19	25	32	32	38	38
C[1)]	0	1	2			3	
Nominal lengths l	4.5, 6.5, 9.5, 13, 16, 19, 22, 25, 32, 38 mm						
Forms	Form C with cone point, form F with dog point						
Explanation	1) C cross recess size, forms H and Z (see DIN EN 2010)						
⇒	**Tapping screw ISO 7049 – ST2.9 x 13 – C – H:** d = ST2.9, l = 13 mm, form C, cross recess form H						

Tap hole diameter for tapping screws (selection)

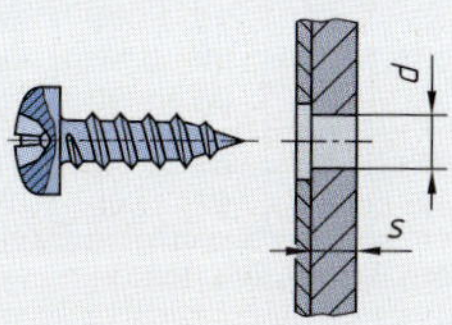

1) Holes bored or punched in steel or copper alloy sheet

Sheet metal thickness s in mm from–to	Tap hole diameter d for tapping screw threads[1]						
	ST2.2	ST2.9	ST3.5	ST4.2	ST4.8	ST5.5	ST6.3
0–0.5	1.6	2.2	2.6	–	–	–	–
0.6–0.8	1.7	2.3	2.7	3.2	3.7	–	–
0.9–1.1	1.8	2.4	2.8	3.2	3.7	4.2	4.9
1.2–1.4	1.8	2.4	2.8	3.3	3.9	4.3	4.9
1.5–1.7	–	2.5	2.9	3.5	3.9	4.5	5.0
1.8–2.0	–	2.6	3.0	3.5	4.0	4.6	5.2
2.0–2.5	–	–	3.0	3.5	4.0	4.6	5.3
2.6–3.0	–	–	3.0	3.8	4.1	4.7	5.3
3.1–3.5	–	–	–	3.9	4.3	5.0	5.8

Thread forming screws

cf. DIN 7500-1 (2009-06)

Form DE: hexagon head bolt

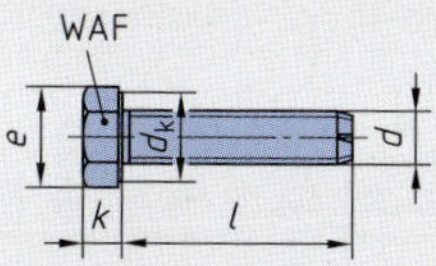

Form EE: hexagon socket head cap bolt

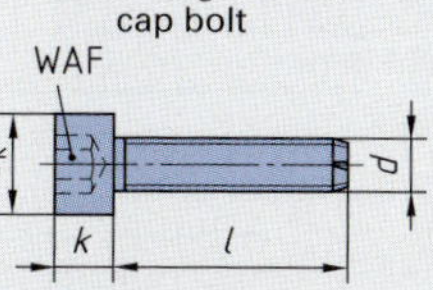

Form NE: raised countersunk head bolt with cross recess

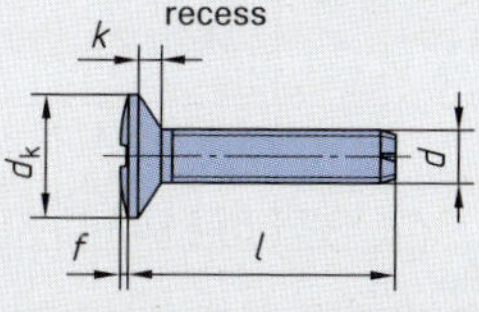

Product grade A (page 210)

Form	Thread d	M2	M2.5	M3	M4	M5	M6	M8	M10
DE	WAF	4	5	5,5	7	8	10	13	16
	k	1.4	1.7	2	2.8	3.5	4	5.3	6.4
	d_k	2.3	3.1	4.1	4.6	6	6.9	11.6	14.6
	e	3.4	4.3	5.5	6	7.7	11.1	14.4	17.8
	l from	3	4	4	6	8	8	10	12
	l to	16	20	25	30	40	50	60	80
EE	WAF	1.5	2	2.5	3	4	5	6	8
	k	2	2.5	3	4	5	6	8	10
	d_k	3.8	4.5	5.5	7	8.5	10	13	16
	l from	3	4	4	6	8	8	10	12
	l to	16	20	25	30	40	50	60	80
NE	d_k	3.8	4.7	5.5	8.4	9.3	11.3	15.8	18.3
	k	1.2	1.5	1.7	2.7	2.7	3.3	4.7	5
	f	0.4	0.5	1	1.2	1.4	1.4	2	2.3
	l from	4	5	6	8	10	10	12	20
	l to	16	20	25	30	40	50	60	80
	C[1)]	0	1		2		3	4	
Nominal lengths l		3, 4, 5, 6, 8, 10, 12, 16, 20, 25, 30–50, 55, 60, 70, 80 mm							
Explanation		1) C cross recess size, forms H and Z (DIN EN 2010)							
⇒		**Screw DIN 7500 – DE – M8 x 25 – St:** DE Hex head, d = M8, l = 25 mm (material: case-hardened steel)							

ME

Studs, Eye bolts, Drain plugs

Studs

cf. DIN 835 (2010-07), DIN 938, 939 (1995-02)

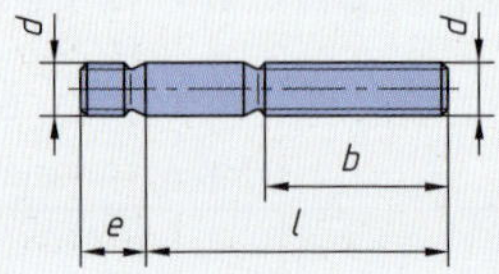

Product grade A (page 210)

Application	
DIN	For fastening into
835	Aluminum alloys
938	Steel
939	Cast iron

Thread d	M3	M4	M5	M6	M8 M8 x1	M10 M10 x1.25	M12 M12 x1.25	M16 M16 x1.5	M20 M20 x1.5	M24 M24 x2
b for l < 125	12	14	16	18	22	26	30	38	46	54
b for l > 125	18	20	22	24	28	32	36	44	52	60
e DIN 835	–	8	10	12	16	20	24	32	40	48
e DIN 938	3	4	5	6	8	10	12	16	20	24
e DIN 939	–	5	6.5	7.5	10	12	15	20	25	30
l from	20	20	25	25	30	35	40	50	60	70
l to	30	40	50	60	80	100	120	160	200	200
Property classes	5.6, 8.8, 10.9									
Nominal lengths l	20, 25, 30–75, 80, 90–180, 190, 200 mm									
⇒	**Stud ISO 939 – M10 x 65 – 8.8:** d = M10, l = 65 mm, property class 8.8									

Eye bolts

cf. DIN 580 (2010-09)

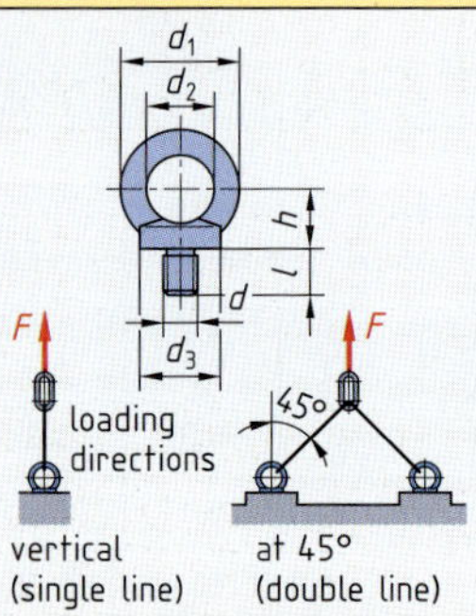

Thread d	M8	M10	M12	M16	M20	M24	M30	M36	M42	M48	M56
h	18	22.5	26	30.5	35	45	55	65	75	85	95
d_1	36	45	54	63	72	90	108	126	144	166	184
d_2	20	25	30	35	40	50	60	70	80	90	100
d_3	20	25	30	35	40	50	65	75	85	100	110
l	13	17	20.5	27	30	36	45	54	63	68	78
Materials	Case hardened steel C15E, A2, A3, A4, A5										
	Lifting capacity in t for loading direction										
Vertical	0.14	0.23	0.34	0.70	1.20	1.80	3.20	4.60	6.30	8.60	11.5
At 45°	0.10	0.17	0.24	0.50	0.86	1.29	2.30	3.30	4.50	6.10	8.20
⇒	**Eye bolt DIN 580 – M20 – C15E:** d = M20, material C15E										

ME

Hexagon head Drain plugs

cf. DIN 910 (1992-01)

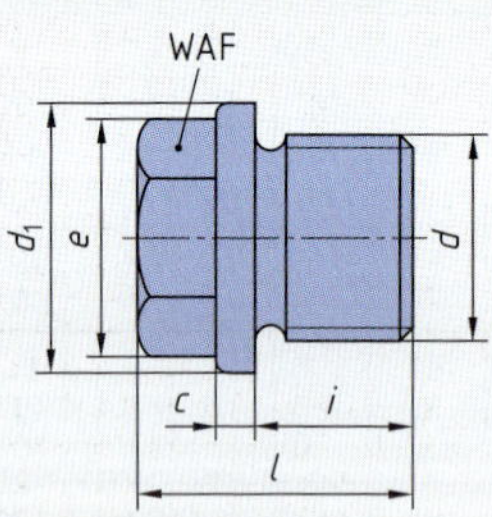

Thread d	M10 x1	M12 x1.5	M16 x1.5	M20 x1.5	M24 x1.5	M30 x1.5	M36 x1.5	M42 x1.5	M48 x1.5	M52 x1.5
d_1	14	17	21	25	29	36	42	49	55	60
l	17	21	21	26	27	30	32	33	33	33
i	8	12	12	14	14	16	16	16	16	16
c	3	3	3	4	4	4	4	5	5	5
WAF	10	13	17	19	22	24	27	30	30	30
e	10.9	14.2	18.7	20.9	23.9	26.1	29.6	33	33	33
Materials	St steel, Al Al-alloy, CuZn copper-zinc-alloy									
⇒	**Screw plug DIN 910 – M24 x 1.5 – St:** d = M24 x 1.5, material: steel									

Hexagon socket Drain plugs

cf. DIN 908 (1992-01)

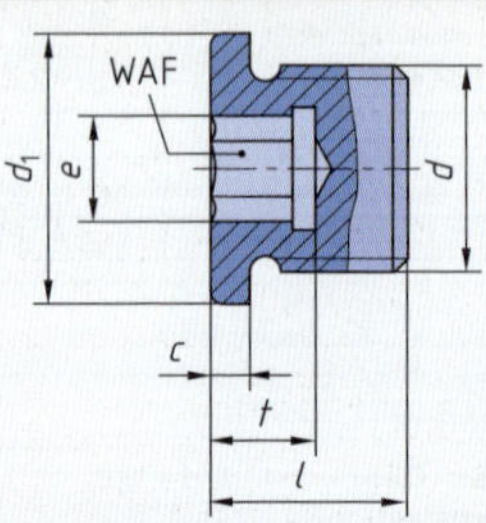

Thread d	M10 x1	M12 x1.5	M16 x1.5	M20 x1.5	M24 x1.5	M30 x1.5	M36 x1.5	M42 x1.5	M48 x1.5	M52 x1.5
d_1	14	17	21	25	29	36	42	49	55	60
l	11	15	15	18	18	20	21	21	21	21
c	3	3	3	4	4	4	5	5	5	5
WAF	5	6	8	10	12	17	19	22	24	24
t	5	7	7.5	7.5	7.5	9	10.5	10.5	10.5	10.5
e	5.7	6.9	9.2	11.4	13.7	19.4	21.7	25.2	27.4	27.4
Materials	St steel, Al Al-alloy, CuZn copper-zinc-alloy									
⇒	**Screw plug DIN 908 – M20 x 1.5 – CuZn:** d = M20 x 1.5, material: copper-zinc-alloy									

Set screws

Slotted set screws

cf. DIN EN 27434, 27435, 24766 (all 1992-10)

with cone point

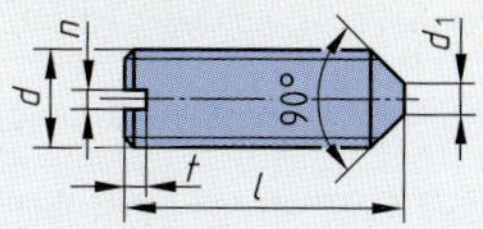

with dog point

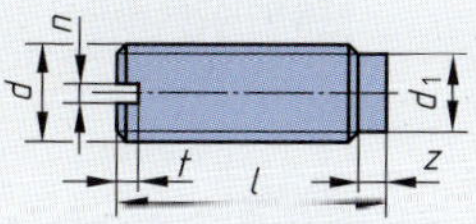

with flat point

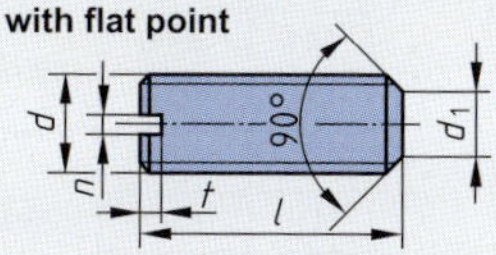

Product grade A (page 210)

Valid standard	Replaces
DIN EN 27434	DIN 553
DIN EN 27435	DIN 417
DIN EN 24766	DIN 551

	Thread d	M1.2	M1.6	M2	M2.5	M3	M4	M5	M6	M8	M10	M12
DIN EN 27434	d_1	0.1	0.2	0.2	0.3	0.3	0.4	0.5	1.5	2	2.5	3.6
	n	0.2	0.3	0.3	0.4	0.4	0.6	0.8	1	1.2	1.6	2
	t	0.5	0.7	0.8	1	1.1	1.4	1.6	2	2.5	3	3
	l from	2	2	3	3	4	6	8	5	10	12	16
	l to	6	8	10	12	16	25	30	35	40	55	60
DIN EN 27435	d_1	–	0.8	1	1.5	2	2.5	3.5	4.3	5.5	7	8.5
	z	–	1.1	1.3	1.5	1.8	2.3	2.8	3.3	4.3	5.3	6.3
	n	–	0.3	0.3	0.4	0.4	0.6	0.8	1	1.2	1.6	2
	t	–	0.7	0.8	1	1.1	1.4	1.6	2	2.5	3	3
	l from	–	2.5	3	4	5	6	8	8	10	12	16
	l to	–	8	10	12	16	20	25	30	40	50	60
DIN EN 24766	d_1	0.6	0.8	1	1.5	2	2.5	3.5	4	5.5	7	8.5
	n	0.2	0.3	0.3	0.4	0.4	0.6	0.8	1	1.2	1.6	2
	t	0.5	0.7	0.8	1	1.1	1.4	1.6	2	2.5	3	3.6
	l from	2	2	2	2.5	3	4	5	6	8	10	12
	l to	6	8	10	12	16	20	25	30	40	50	60
Property classes		14H, 22H, A1-50										
Nominal lengths l		2, 2.5, 3, 4, 5, 6, 8, 10, 12, 16, 20, 25, 30–50, 55, 60 mm										
⇒		**Set screw ISO 7434 – M6 x 25 – 14H:** d = M6, l = 25 mm, property class 14H										

Set screws with hexagon socket

cf. DIN EN ISO 4026, 4027, 4028 (2004-05)

with cone point

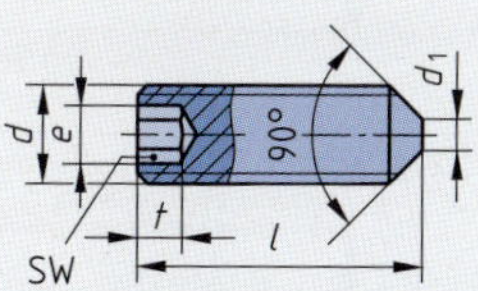

with dog point

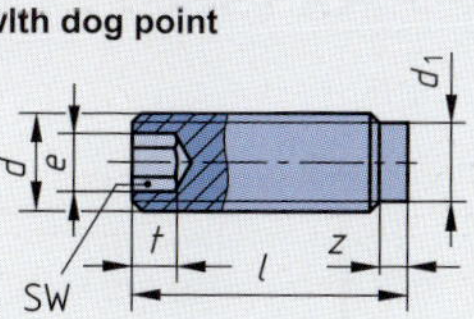

with flat point

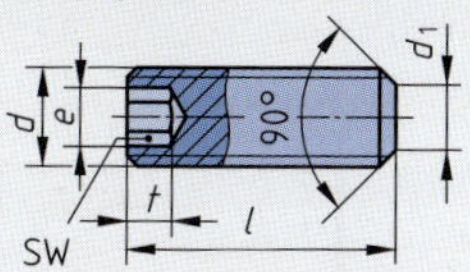

Product grade A (page 210)

Valid standard	Replaces
DIN EN ISO 4026	DIN 913
DIN EN ISO 4027	DIN 914
DIN EN ISO 4028	DIN 915

	Thread d	M2	M2.5	M3	M4	M5	M6	M8	M10	M12	M16	M20
DIN EN ISO 4027	d_1	0.5	0.7	0.8	1	1.3	1.5	2	2.5	3	4	5
	WAF	0.9	1.3	1.5	2	2.5	3	4	5	6	8	10
	e	1	1.5	1.7	2.3	2.9	3.4	4.6	5.7	6.9	9.1	11.4
	t	0.8	1.2	1.2	1.5	2	2	3	4	4.8	6.4	8
	l from	2	2.5	3	4	5	6	8	10	12	16	20
	l to	10	12	16	20	25	30	40	50	60	60	60
DIN EN ISO 4028	d_1	1	1.5	2	2.5	3.5	4	5.5	7	8.5	12	15
	z	1.3	1.5	1.8	2.3	2.8	3.3	4.3	5.3	6.3	8.4	10.4
	WAF	0.9	1.3	1.5	2	2.5	3	4	5	6	8	10
	e	1	1.5	1.7	2.3	2.9	3.4	4.6	5.7	6.9	9.1	11.4
	t	0.8	1.2	1.2	1.5	2	2	3	4	4.8	6.4	8
	l from	2.5	3	4	5	6	8	8	20	12	16	20
	l to	10	12	16	20	25	30	40	50	60	60	60
DIN EN ISO 4026	d_1	1	1.5	2	2.5	3.5	4	5.5	7	8.5	12	15
	WAF	0.9	1.3	1.5	2	2.5	3	4	5	6	8	10
	e	1	1.5	1.7	2.3	2.9	3.4	4.6	5.7	6.9	9.2	11.4
	t	0.8	1.2	1.2	1.5	2	2	3	4	4.8	6.4	8
	l from	2	2.5	3	4	5	6	8	10	12	16	20
	l to	10	12	16	20	25	30	40	50	60	60	60
Property classes		45H, A1-12H, A2-21H, A3-21H, A4-21H, A5-21H										
Nominal lengths l		2, 2.5, 3, 4, 5, 6, 8, 10, 12, 16, 20, 25, 30–50, 60 mm										
⇒		**Set screw ISO 4026 – M6 x 25 – A5 – 45H:** d = M6, l = 25 mm, A5 stainless steel, property class 45H										

ME

Simplified calculation of screw joints

Most screws (simple screw joints) are fitted without verifying the tightening torque, however a reliable screw joint can still be ensured by observing some values based on experience.

If screw joints are fitted without verifying the tightening torque, the preload force F_p, the preloading σ_p and the contact pressure P_v cannot be determined. Experience values are available for the preloading σ_p, however, (cf. table below).

It is recommended to use preferably screws of strength class 8.8 with smaller diameters (cf. example) and apply a safety factor of 1.5 to determine the minimum yield strength.

The calculation of the screw joint is exclusively based on the axially applied operating force. A high safety factor (e.g. $\nu = 2.5$) results in the rather inaccurate calculation of the total screw joint force.

Operating force in axial direction

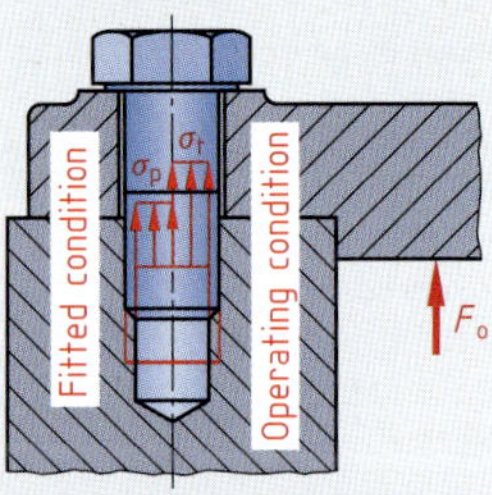

Applied force perpendicular to the axial direction

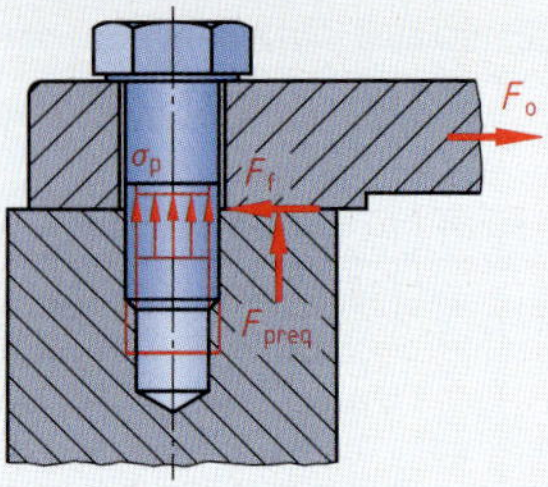

Example: disk clutch

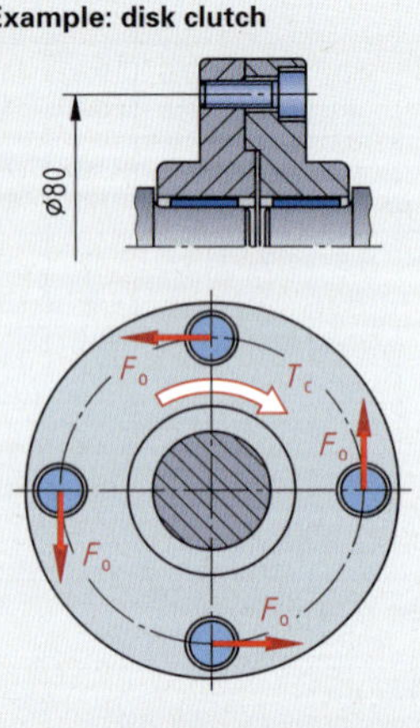

F_o	operating force in N
F_p	preload force in N
R_e	yield strength in N/mm²
σ_p	preloading in N/mm²
σ_t	tensile stress in N/mm²
σ_{all}	allowed stress in N/mm²
S	tensile stress area in mm²
d	thread e.g. M10
F_f	frictional force in N
F_{preq}	required preload force in kN
μ	friction coefficient
φ	slip resistance
ν	safety factor
T_c	coupling torque

Example for operating force applied in axial direction

$F_o = 1875$ N per bolt, $\nu = 2.5$,
hexagon bolt ISO 4014 – 8.8;
thread diameter = ?

$R_e = 640$ N/mm² for 8.8

$$\sigma_{all} = \frac{R_e}{\nu} = \frac{640\ \text{N/mm}^2}{2.5} = 256\ \frac{\text{N}}{\text{mm}^2}$$

$$S = \frac{F_o}{\sigma_{all}} = \frac{1875\ \text{N}}{256\ \text{N/mm}^2} = 7.32\ \text{mm}^2$$

M4 is selected (cf. page 204)
With $\sigma_v = 350$ N/mm² (cf. table above)
$R_{e\,req} \geq 1.5 \cdot \sigma_p = 1.5 \cdot 350$ N/mm²
$= \mathbf{525\ N/mm^2} < R_e$

$$R_e = 640\ \frac{\text{N}}{\text{mm}^2} > R_{e\,req}$$

Example for operating force applied perpendicular to the axial direction (disk clutch):

Disk clutch made of S235JR with four cylinder bolts ISO 4762 – 8.8,
$T_c = 256$ N · m, $\mu = 0.2$, $\varphi = 2$;
thread diameter = ?

$$F_o = \frac{T_c}{n \cdot \frac{d}{2}} = \frac{256\ \text{N} \cdot \text{m}}{4 \cdot \frac{0.08\ \text{m}}{2}} = 1600\ \text{N}$$

$$F_{preq} = \frac{\varphi \cdot F_o}{\mu} = \frac{2 \cdot 1600\ \text{N}}{0.2} = 16\,000\ \text{N}$$

M10 is selected (cf. table above)

Experience values for the preload force F_p and the preloading σ_p

Thread d	Preload force F_p N	Pre-loading σ_p N/mm²
M4	3000	350
M6	7000	
M8	10 000	280
M10	16 000	
M12	23 000	
M16	28 000	180
M20	44 000	

Minimum yield strength

$$R_{e\,req} \geq 1.5 \cdot \sigma_p$$

Tensile stress area

$$S = \frac{F_o}{\sigma_{all}}$$

Allowable stress

$$\sigma_{all} = \frac{R_e}{\nu}$$

Required preload force

$$F_{preq} = \frac{\varphi \cdot F_o}{\mu}$$

ME

Locking fasteners

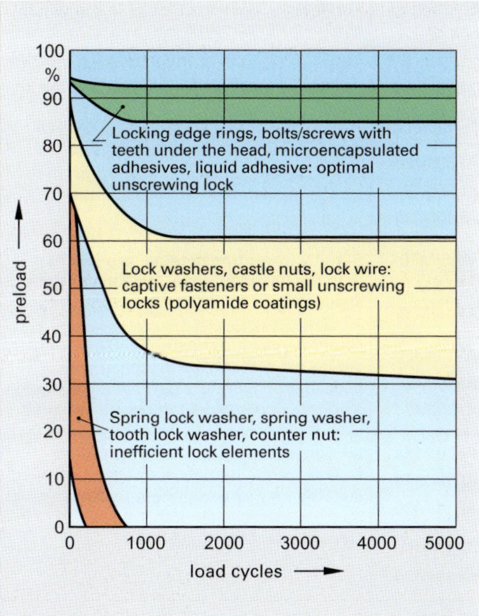

Vibration test DIN 65151 performed on various locking elements

The locking behavior of screw joints under transverse loading on the bolt is tested
ISO 4014-M10.

A locking fastener is generally not necessary for screw joints which are sufficiently dimensioned and securely mounted. The clamping forces prevent the slipping of the screwed parts or loosening of the bolts and nuts. In practice a loss of clamping force can still occur due to the following causes:

- **Loosening of the screw joint** caused by high surface contact pressures which initiate plastic deformation (so-called settling) and reduce the preload of the screw joint.

  Remedy: As little seperation as possible, minimal surface roughness, use of high-strength bolts (large preload).

- **Unscrewing of the screw joint**: For joints dynamically loaded transverse to the bolt axis a fully self-actuated unscrewing can occur.

  This is remedied with locking elements. These are divided into three groups based on their effectiveness.

  Ineffective locking elements (e.g. spring lock washers and tooth lock washers).

  Captive fasteners, which allow a partial unscrewing, but prevent the screw joint from coming completely apart.

  Threadlocking (e.g. glue or corrugated head screws). The preload remains approximately constant. The nut or bolt cannot loosen by itself (best method of locking).

Overview of locking fasteners

Joint	Locking element	Standard	Type, property
Loaded together, spring loaded	spring lock washer spring washer tooth lock washer serrated lock washer	withdrawn withdrawn withdrawn withdrawn	ineffective ineffective ineffective ineffective
Interlocking	lock washer castle nut with cotter pin lock wire	withdrawn DIN 935-1 (2000-10) –	captive fastener captive fastener captive fastener
Force-fit (gripping)	jam nut	–	ineffective, loosening possible
	bolts and nuts with gripping polyamide coating	DIN 267-28 (2009-09) ISO 2320 (2009-03)	captive fastener or slight anti-rotation lock
Blocking (force-fit and interlocking)	bolts with teeth under the head	–	anti-rotation lock, not suitable for hardened parts
	detent edged ring detent washer self-locking pair of washers	– –	anti-rotation lock, not suitable for hardened parts anti-rotation lock
Bonded	microencapsulated adhesives in threads	DIN 267-27 (2009-09)	anti-rotation lock, sealing joint; temperature range –50 °C to 150 °C
	liquid adhesive	–	anti-rotation lock

ME

Screw drive systems

Illustration, designation	Nominal sizes Tool sizes/threads *d*		Properties, application examples
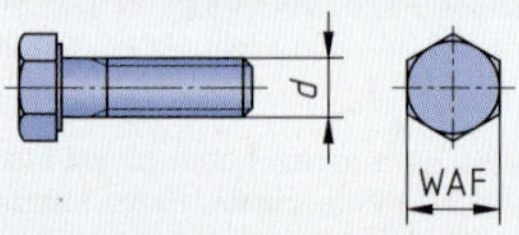	WAF5/M2.5 WAF5.5/M3 WAF7/M4	WAF16/M10 WAF18/M12 WAF24/M16	High transferable torques, no axial force required, same tool for the bolt and the nut
Hexagon head	WAF8/M5 WAF10/M6 WAF13/M8	WAF30/M20 WAF36/M24 WAF46/M30	General machine construction, automotive, vehicle manufacturing
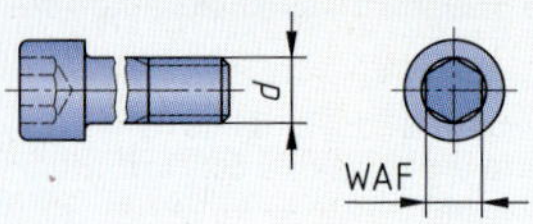	WAF2/2.5 WAF2.5/M3 WAF3/M4	WAF8/M10 WAF10/M12 WAF14/M16	Slightly lower transferable torques than with the hexagon drive, suitable for tight spaces
Hexagon socket	WAF4/M5 WAF5/M6 WAF6/M8	WAF17/M20 WAF19/M24 WAF22/M30	General machine construction, hex with centre pin can only be loosened with special tool, therefore particularly suitable as protection against tampering and theft
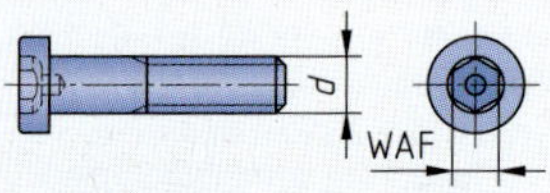	WAF3/M4 WAF4/M5 WAF5/M6	WAF10/M12 WAF12/M14 WAF14/M16	Like internal hexagon drive, the transferable torque is slightly smaller, however; suitable for low component thicknesses
Hexagon socket with low head and pilot recess	WAF6/M8 WAF8/M10	WAF17/M20 WAF19/M24	General machine construction
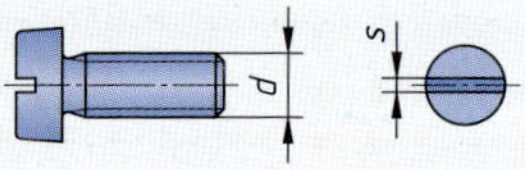	S0.5/M2 S0.6/M2.5	S1.2/M5 S1.6/M6	Drive difficult to center, low transferable torque, high contact pressure at the force application areas
Slotted	S0.8/M3 S1.2/M4	S2/M8 S2.5/M10	Electric machine building industry
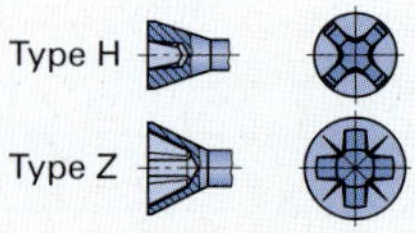	PH0/M2 PH1/M2.5–3 PH2/M3.5–5	PZ0/M2 PZ1/M2.5–3 PZ2/M3.5–5	Higher torque and better centering of the tool as with simple slots, lower contact pressure
Cross slotted	PH3/M5.5–8 PH4/M8–10	PZ3/M5.5–8 PZ4/M8–10	Electric machine building industry, apparatus construction
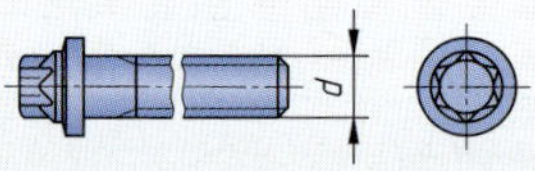	E5/M4 E6/M5 E8/M6	E14/M12 E18/M14 E20/M16	Favorable for transmission, easy positioning and good engagement of the driving tool
External hexalobular (Torx)	E10/M8 E12/M10	E24/M18 E32/M20	Machine construction, automotive, vehicle manufacturing
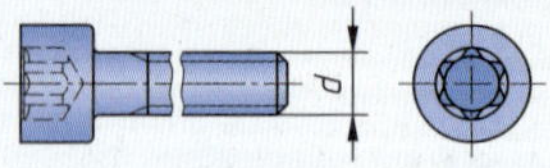	T6/M2 T8/M2.5 T10/M3	T30/M6 T40/M8 T45/M8	Good torque transfer, low space requirements for the tool; torx with center pin can only be loosened with special tool, therefore particularly suitable as protection against tampering and theft.
Internal hexalobular (Torx)	T15/M3,5 T20/M4 T25/M5	T50/M10 T55/M12 T60/M16	General machine construction, apparatus construction, electric machine building industry
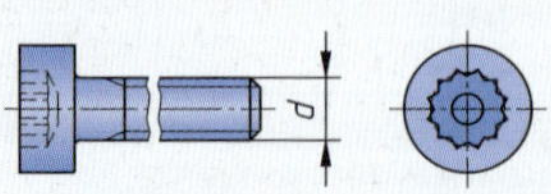	N4/M4 N5/M5 N6/M6	N12/M10–12 N14/M12–14 N16/M14–16	Safety profile with twelve small tips, favorable force distribution due to a broad application of the driving force, transfer of medium-sized torques
Internal serration	N8/M6–8 N10/M8–10	N18/M16	Automotive and vehicle manufacturing

ME

Countersinks for countersunk head screws

Countersinks for countersunk screws with head forms as per ISO 7721

cf. DIN EN ISO 15065 (2005-05)

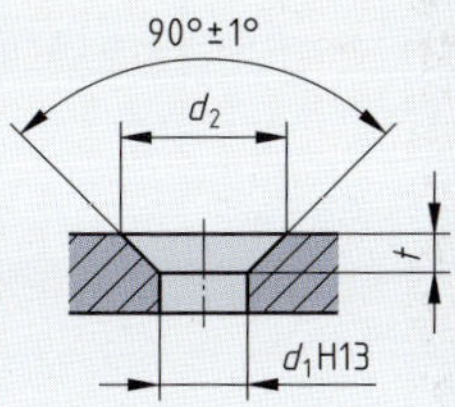

Graphical representation, see page 78;

Nominal sizes	1.6	2	2.5	3	3.5	4
Metric screws	M1.6	M2	M2.5	M3	M3.5	M4
Tapping screws	–	ST2.2	–	ST2.9	ST3.5	ST4.2
d_1 H13	1.8	2.4	2.9	3.4	3.9	4.5
d_2 min.	3.6	4.4	5.5	6.3	8.2	9.4
d_2 max.	3.7	4.5	5.6	6.5	8.4	9.6
$t_1 \approx$	1.0	1.1	1.4	1.6	2.3	2.6
Nominal sizes	**5**	**5.5**	**6**	**8**	**10**	**–**
Metric screws	M5	–	M6	M8	M10	–
Tapping screws	ST4.8	ST5.5	ST6.3	ST8	ST9.5	–
d_1 H13	5.5	6	6.6	9	11	–
d_2 min.	10.4	11.5	12.6	17.3	20	–
d_2 max.	10.7	11.8	12.9	17.6	20.3	–
$t_1 \approx$	2.6	2.9	3.1	4.3	4.7	–

⇒ **Countersink ISO 15065-8:** Nominal size 8 (metric threads M8 or tapping screw threads ST8)

Application for:

Slotted flat head countersunk screws	DIN EN ISO 2009
Cross recessed flat head countersunk screws	DIN EN ISO 7046-1
Slotted raised head countersunk screws	DIN EN ISO 2010
Cross rec. raised head countersunk screws	DIN EN ISO 7047
Slotted flat head countersunk tapping screws	DIN ISO 1482
Cross rec. flat head counters. tapping screws	DIN ISO 7050
Slotted raised head countersunk tapping screws	DIN ISO 1483
Cross rec. raised head counters. tapping screws	DIN ISO 7051
Cross recessed flat head countersunk tapping screws	ISO 15482
Cross recessed raised head countersunk tapping screws	ISO 15483

Countersinks for countersunk head screws

cf. DIN 74 (2003-04)

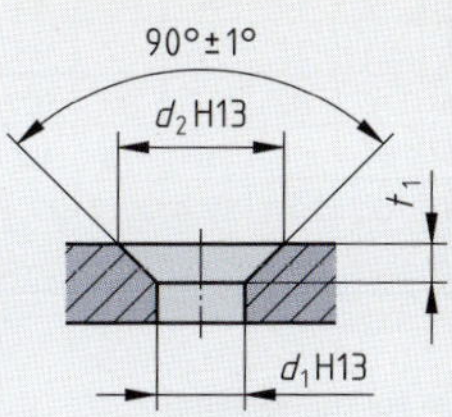

Form A and Form F

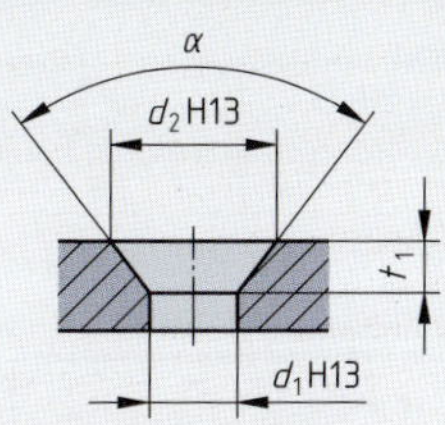

Form E

Graphical representation, see page 78;

Forms B, C and D are no longer standardized

	Thread Ø	1.6	2	2.5	3	4	4.5	5	6	7	8
Form A	d_1 H13[1)]	1.8	2.4	2.9	3.4	4.5	5	5.5	6.6	7.6	9
Form A	d_2 H13	3.7	4.6	5.7	6.5	8.6	9.5	10.4	12.4	14.4	16.4
Form A	$t_1 \approx$	0.9	1.1	1.4	1.6	2.1	2.3	2.5	2.9	3.3	3.7

⇒ **Countersink DIN 74 – A4:** Form A, thread diameter 4 mm

Application of Form A for:

Countersunk flat head wood screws	DIN 97 and DIN 7997
Raised head countersunk wood screws	DIN 95 and DIN 7995

	Thread Ø	10	12	16	20	22	24
Form E	d_1 H13[1)]	10.5	13	17	21	23	25
Form E	d_2 H13	19	24	31	34	37	40
Form E	$t_1 \approx$	5.5	7	9	11.5	12	13
Form E	α	75° ± 1°			60° ± 1°		

⇒ **Countersink DIN 74 – E12:** Form E, thread diameter 12 mm

Application of Form E for: Countersunk head bolts for steel structures DIN 7969

	Thread Ø	3	4	5	6	8	10	12	14	16	20
Shape F	d_1 H13[1)]	3.4	4.5	5.5	6.6	9	11	13.5	15.5	17.5	22
Shape F	d_2 H13	6.9	9.2	11.5	13.7	18.3	22.7	27.2	31.2	34.0	40.7
Shape F	$t_1 \approx$	1.8	2.3	3.0	3.6	4.6	5.9	6.9	7.8	8.2	9.4

⇒ **Countersink DIN 74 – F12:** Form F, thread diameter 12 mm

Application of Form F for: Hexagon socket head countersunk screws DIN EN ISO 10642 (replaces DIN 7991)

1) Medium size clearance hole according to DIN EN 20273, page 210

ME

Counterbores for cap screws and Hexagon head bolts

Counterbores for cap screws

cf. DIN 974-1 (2008-02)

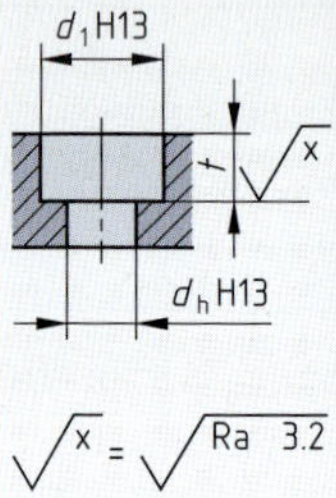

	d	3	4	5	6	8	10	12	16	20	24	27	30	36
	d_h H13[1]	3.4	4.5	5.5	6.6	9	11	13.5	17.5	22	26	30	33	39
d_1 H13	Series 1	6.5	8	10	11	15	18	20	26	33	40	46	50	58
	Series 2	7	9	11	13	18	24	–	–	–	–	–	–	–
	Series 3	6.5	8	10	11	15	18	20	26	33	40	46	50	58
	Series 4	7	9	11	13	16	20	24	30	36	43	46	54	63
	Series 5	9	10	13	15	18	24	26	33	40	48	54	61	69
	Series 6	8	10	13	15	20	24	33	43	48	58	63	73	–
t[2]	ISO 1207	2.4	3.0	3.7	4.3	5.6	6.6	–	–	–	–	–	–	–
	ISO 4762	3.4	4.4	5.4	6.4	8.6	10.6	12.6	16.6	20.6	24.8	–	31.0	37.0
	DIN 7984	2.4	3.2	3.9	4.4	5.6	6.6	7.6	9.6	11.6	13.8	–	–	–

⇒ DIN 974 provides no code designations for counterbores.

Series	Cap screws without washer components	
1	Screws (bolts) ISO 1207, ISO 4762, DIN 6912, DIN 7984, DIN 34821, ISO 4579, ISO 4580	
2	Screws (bolts) ISO 1580, DIN EN ISO 7045, DIN EN ISO 14583	
	Cap screws and the following washer components:	
3	Screws (bolts) ISO 1207, ISO 4762, DIN 7984 with spring lock washers DIN 7980[3]	
4	Washers DIN EN ISO 7092 Spring washers DIN 137 Form A[3] Spring lock washers DIN 128 + DIN 6905[3]	Tooth lock washers DIN 6797[3] Serrated lock washers DIN 6798[3] Serrated lock washers DIN 6907[3]
5	Washers DIN EN ISO 7089 + 7090 Washers DIN 6902 Form A[3]	Spring washers DIN 137 Form B[3] Spring washers DIN 6904[3]
6	Conical spring washers DIN 6796, DIN 6908	

Graphical representation, see page 78;

1) Clearance hole according to DIN EN 20273, series medium, page 210
2) For screws/bolts without washer components 3) Standards withdrawn

Counterbore for hexagon bolts/screws and hexagon nuts

cf. DIN 974-2 (1991-05)

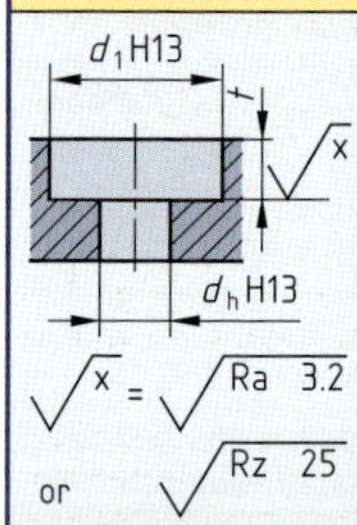

	d	4	5	6	8	10	12	14	16	20	24	27	30	33	36	42
	Width across flats	7	8	10	13	16	18	21	24	30	36	41	46	50	55	65
	d_h H13	4.5	5.5	6.6	9	11	13.5	15.5	17.5	22	26	30	33	36	39	45
d_1 H13	Series 1	13	15	18	24	28	33	36	40	46	58	61	73	76	82	98
	Series 2	15	18	20	26	33	36	43	46	54	73	76	82	89	93	107
	Series 3	10	11	13	18	22	26	30	33	40	48	54	61	69	73	82
t[1]	Hex bolt	3.3	4.1	4.6	6.1	7.2	8.3	9.6	10.8	13.3	16.0	18.2	20.1	22.4	23.9	27.4

⇒ DIN 974 provides no code designations for counterbores.

Graphical representation, see page 78;

Series 1: For socket wrench DIN 659, DIN 896, DIN 3112 or socket DIN 3124
Series 2: For box wrench DIN 838, DIN 897 or socket DIN 3129
Series 3: For recesses in tight space conditions (not suitable for conical spring washers)
1) For hexagon bolts/screws ISO 4014, ISO 4017, ISO 8765, ISO 8676 without washer components

Calculation of counterbore depth for flush mounting (for DIN 974-1 and DIN 974-2)

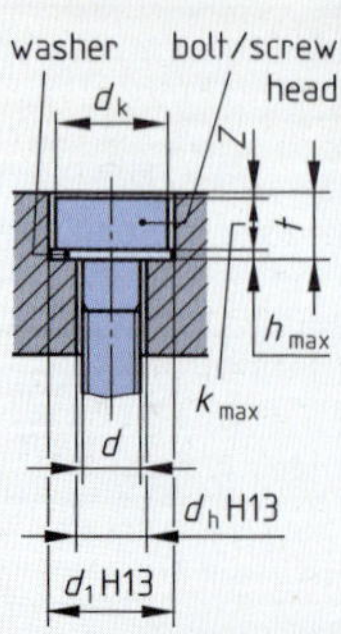

Determining the allowance *Z*

Thread nominal ∅ *d*	over 1 to 1.4	over 1.4 to 6	over 6 to 20	over 20 to 27	over 27 to 100
Allowance *Z*	0.2	0.4	0.6	0.8	1.0

t counterbore depth
k_{max} maximum height of the screw/bolt head
h_{max} maximum height of the washer component
Z allowance based on thread nominal diameter (see table)

Counterbore depth[1]

$$t = k_{max} + h_{max} + Z$$

1) If values k_{max} and h_{max} are unavailable, values k and h can be used as approximations.

Nuts – Overview

Illustration	Design	Standard range from–to	Standard	Applications, properties
Hexagon nuts, type 1				pages 227, 228
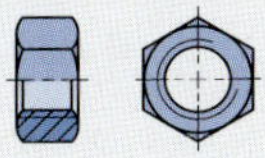	with coarse threads	M1.6–M64	DIN EN ISO 4032	Most commonly used nuts, used with bolts up to equal property class **Fine threads:** greater transmitted force than for coarse threads
	with fine threads	M8x1–M64x4	DIN EN ISO 8673	
Hexagon nuts, type 2				page 228
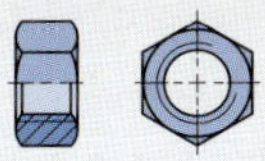	with coarse threads	M5–M36	DIN EN ISO 4033	Nut height *m* is approx. 10% higher than nuts of type 1, used with bolts up to equal property class **Fine threads:** greater transmitted force than for coarse threads
	with fine threads	M8x1–M36x3	DIN EN ISO 8674	
Low hexagon nuts				pages 228, 229
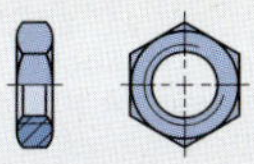	with coarse threads	M1.6–M64	DIN EN ISO 4035	Use with low installation heights and low stresses **Fine threads:** higher transmission of force than coarse threads
	with fine threads	M8x1–M64x4	DIN EN ISO 8675	
Prevailing torque hexagon nuts with locking insert				page 229
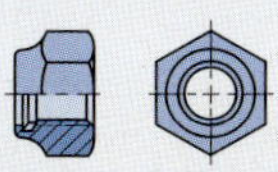	with coarse threads	M3–M36	DIN EN ISO 7040	Self-locking nuts with full loading capacity and non-metallic insert, up to operating temperatures of 120 °C **Fine threads:** greater transmitted force than for coarse threads
	with fine threads	M8x1–M36x3	DIN EN ISO 10512	
	with coarse threads	M5–M36	DIN EN ISO 7719	Self-locking all-metal nuts with full loading capacity **Fine threads:** greater transmitted force than for coarse threads
	with fine threads	M8x1–M36x3	DIN EN ISO 10513	
Hexagon nuts, other forms				pages 229, 231
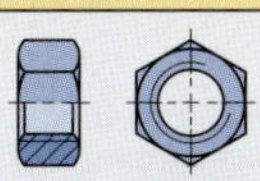	with large width across flats, coarse threads	M12–M36	DIN EN 14399-4	Metal construction: high-strength custom preloaded joints (HV), with hexagon head bolts DIN EN 14999-4 (page 213)
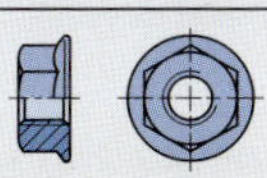	with flange, coarse threads	M5–M20	DIN EN 1661	Might be used with large clearance holes or to reduce contact pressure
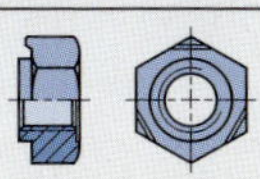	weld nuts, coarse threads	M3–M16 M8x1–M16x1.5	DIN 929	Used in sheet metal structures; nuts are usually joined to metal sheets by projection welding
Castle nuts, cotter pins				page 231
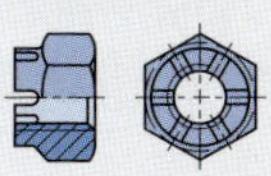	high form, coarse or fine threads	M4–M100 M8x1–M100x4	DIN 935	Might be used for axial fixing of bearings, hubs in safety joints (steering area of vehicles) Locking with cotter pin and transverse hole in the bolt. At full load of the bolt, the cotter pin is sheared off above property class 8.8.
	low form, coarse or fine threads	M6–M48 M8x1–M48x3	DIN 979	
	cotter pins	0.6x12–20x280	DIN EN ISO 1234	

Nuts – Overview, Designation of nuts

Illustration	Design	Standard range from–to	Standard	Application, properties
Acorn nuts				page 230
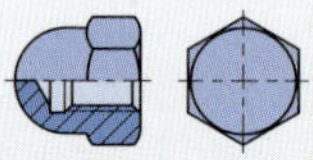	high form, coarse or fine threads	M4–M36 M8x1–M24x2	DIN 1587	Decorative and sealing external joint closures, protection for threads, protection from injuries
	low form, coarse or fine threads	M4–M48 M8x1–M48x3	DIN 917	
Eye nuts, eye bolts				page 230
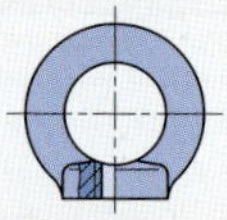	eye nuts, coarse or fine threads	M8–M100x6 M20x2–M100x4	DIN 582	Transport eyes on machines and equipment; stress depends on the angle of the applied load, milling of seating surface necessary
Lock nuts, lock washers				page 230
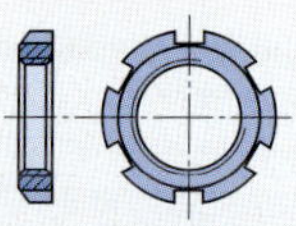	lock nuts with fine threads	M10x1–M200x1.5	DIN 70852	For axial positioning, e.g. of hubs, with small mounting heights and low stresses, locking with lock washers
	lock washers	10–200	DIN 70952	
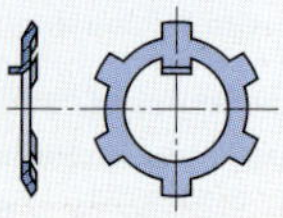	lock nuts with fine threads	M10x0.75–M115x2 (KM0–KM23)	DIN 981	For axial positioning of roller bearings, for adjustment of the bearing clearance, e.g. with tapered roller bearings that are locked with lock washers
	lock washers	10–115 (MB0–MB23)	DIN 5406	
Knurled nuts				page 231
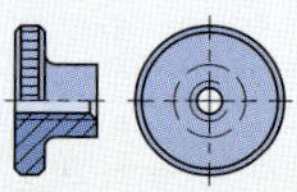	high form, coarse threads	M1–M10	DIN 466	Used in joints that are opened frequently, e.g. in manufacturing of jigs and fixtures, in control cabinets
	low form, coarse threads	M1–M10	DIN 467	
Hexagon turnbuckle nuts				
	coarse threads	M6–M30	DIN 1479	For joining and adjusting, e.g. of threaded and connecting bars, with left-hand and right-hand threads; locked by jam nuts

Designation of nuts

cf. DIN 962 (2001-11)

Examples:

Hexagon nut	**ISO 4032**	**– M12**	**– 8**
Castle nut	**DIN 929**	**– M8 x 1**	**– St**
Hexagon nut	**EN 1661**	**– M12**	**– 10**

- Type
- Reference standard, e.g. ISO, DIN, EN; sheet number of the standard[1)]
- Nominal data, e.g.
  M → metric threads
  8 → nominal diameter d
  1 → thread pitch P for fine threads
- Property class, e.g. 05, 8, 10
  Material, e.g.: St steel
  GT malleable cast iron

[1)] Nuts standardized according to ISO or DIN EN ISO, have the code **ISO** in their designation.
Nuts standardized according to DIN, have the code **DIN** in their designation.
Nuts standardized according to DIN EN, have the code **EN** in their designation.

Property classes, hexagon nuts with coarse threads

Property classes of nuts

cf. DIN EN 20898-2 (1994-02), DIN EN ISO 3506-2 (2010-04)

Examples:

Unalloyed and alloy steels
DIN EN 29898-2

nut height $m \geq 0.8 \cdot d$: **8**
nut height $m < 0.8 \cdot d$: **04**

Stainless steels
DIN EN ISO 3506-2

nut height $m \geq 0.8 \cdot d$: **A 2 – 70**
nut height $m < 0.8 \cdot d$: **A 4 – 035**

Code
8 property class
04 low nuts, test load = $4 \cdot 100$ N/mm²

Steel type
A → austenitic steel
A2 → rustproof nuts
A4 → rustproof and acid-resistant nuts

Code
70 proof stress = $70 \cdot 10$ N/mm²
035 low nut, proof stress = $35 \cdot 10$ N/mm²

Allowable combinations of nuts and bolts

cf. DIN EN 20898-2 (1994-02)

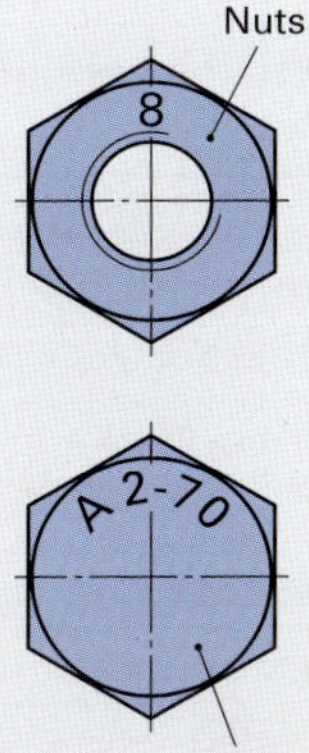

Property class of the nut	Usable bolts up to property class: Unalloyed and alloy steels 4.8	5.8	6.8	8.8	9.8	10.9	12.9	Stainless steels A2-50	A2-70	A4-50	A4-70
4	■										
5	■	■									
6	■	■	■								
8	■	■	■	■							
9	■	■	■	■	■						
10	■	■	■	■	■	■					
12	■	■	■	■	■	■	■				
A2-50								■			
A2-70								■	■		
A4-50										■	
A4-70										■	■
04, 05, A2-025, A4-025	Property classes for low nuts. The nuts are designed for smaller load capacity. Bolts and nuts of the same material group, e.g. stainless steel, can be combined with each other.										

■ allowable combinations of property classes for nuts and bolts

Hexagon nuts with coarse threads, Type [1]

cf. DIN EN ISO 4032 (2001-03)

Valid standard DIN EN ISO	Replaces DIN EN	Replaces DIN
4032	24032	934

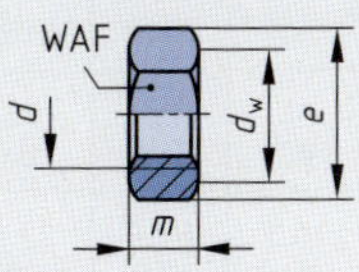

Thread d	M1.6	M2	M2.5	M3	M4	M5	M6	M8	M10
WAF	3.2	4	5	5.5	7	8	10	13	16
d_w	2.4	3.1	4.1	4.6	5.9	6.9	8.9	11.6	14.6
e	3.4	4.3	5.5	6	7.7	8.8	11.1	14.4	17.8
m	1.3	1.6	2	2.4	3.2	4.7	5.2	6.8	8.4
Property classes	as per agreement			6, 8, 10					
	A2-70, A4-70								
Thread d	M12	M16	M20	M24	M30	M36	M42	M48	M56
WAF	18	24	30	36	46	55	65	75	85
d_w	16.6	22.5	27.7	33.3	42.8	51.1	60	69.5	78.7
e	20	26.8	33	39.6	50.9	60.8	71.3	82.6	93.6
m	10.8	14.8	18	21.5	25.6	31	34	38	45
Property classes	6, 8, 10						as per agreement		
	A2-70, A4-70				A2-50, A4-50		–		
Explanation	[1] Type 1: Nut height $m \geq 0.8 \cdot d$								
⇒	**Hexagon nut ISO 4032 – M10 – 10:** d = M10, property class 10								

Product grades (page 210)	
Thread d	Grade
M1.6–M16	A
M20–M64	B

ME

Hexagon nuts

Hexagon nuts with coarse threads, type 2[1)]

cf. DIN EN ISO 4033 (2001-03), replaces DIN EN 24033

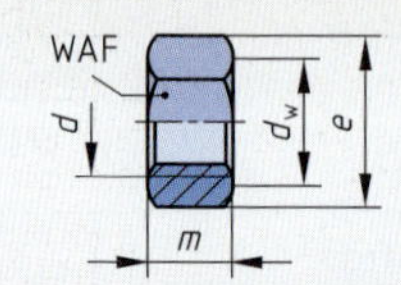

Thread d	M5	M6	M8	M10	M12	M16	M20	M24	M30	M36
WAF	8	10	13	16	18	24	30	36	46	55
d_w	6.9	8.9	11.6	14.8	14.6	22.5	27.7	33.2	42.7	51.1
e	8.8	11.1	14.4	17.8	20	26.8	33	39.6	50.9	60.8
m	5.1	5.7	7.5	9.3	12	16.4	20.3	23.9	28.6	34.7
Property classes	9, 12									

Product grades (page 210)

Thread d	Grade
M1.6–M16	A
M20–M64	B

Explanation: 1) Hexagon nuts of type 2 are approx. 10% higher than nuts of type 1.

⇒ **Hexagon nut ISO 4033 – M24 – 9:** d = M24, property class 9

Hexagon nuts with fine threads, type 1 and type 2[1)]

cf. DIN EN ISO 8673 and 8674 (2001-03)

Valid standard DIN EN ISO	Replaces DIN EN	Replaces DIN
8673	28673	934
8674	28674	971

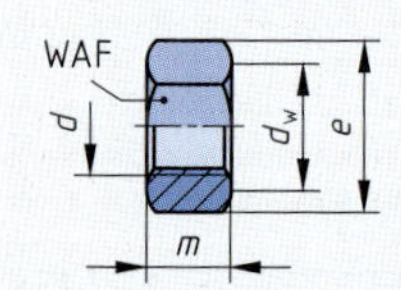

Thread d	M8x1	M10x1	M12x1.5	M16x1.5	M20x1.5	M24x2	M30x2	M36x3	M42x3	M48x3	M56x4
WAF	13	16	18	24	30	36	46	55	65	75	85
d_w	11.6	14.6	16.6	22.5	27.7	33.3	42.8	51.1	60	69.5	78.6
e	14.4	17.8	20	26.8	33	39.6	50.9	60.8	71.3	82.6	93.6
m_1[1)]	6.8	8.4	10.8	14.8	18	21.5	25.6	31	34	38	45
m_2[1)]	7.5	9.3	12	16.4	20.3	23.9	28.6	34.7	–	–	–
Property classes, Type 1	6, 8								as per agreement		
	A2-70, A4-70						A2-50, A4-50				
Property classes, Type 2	8, 10, 12				10				–		

Product grades (page 210)

Thread d	Grade
M8x1–M16x1.5	A
M20x1.5–M64x3	B

Explanation: 1) Hexagon nut type 1: DIN EN ISO 8673, nut height $m_1 \geq 0.8 \cdot d$
Hexagon nut type 2: DIN EN ISO 8674, nut height m_2 is approx. 10% larger than nuts of type 1.

⇒ **Hexagon nut ISO 8673 – M8x1 – 6:** d = M8x1, property class 6

Low hexagon nuts with coarse threads[1)]

cf. DIN EN ISO 4035 (2001-03)

Valid standard DIN EN ISO	Replaces DIN EN
4035	24035

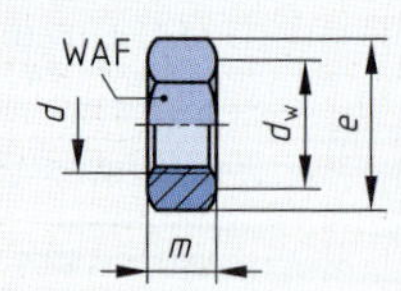

Thread d	M1.6	M2	M2.5	M3	M4	M5	M6	M8	M10
WAF	3.2	4	5	5.5	7	8	10	13	16
d_w	2.4	3.1	4.1	4.6	5.9	6.9	8.9	11.6	14.6
e	3.4	4.3	5.5	6	7.7	8.8	11.1	14.4	17.8
m	1	1.2	1.6	1.8	2.2	2.7	3.2	4	5
Property classes	as per agreement			04, 05					
	A2-035, A4-035								

Thread d	M12	M16	M20	M24	M30	M36	M42	M48	M56
WAF	18	24	30	36	46	55	65	75	85
d_w	16.6	22.5	27.7	33.2	42.8	51.1	60	69.5	78.7
e	20	26.8	33	39.6	50.9	60.8	71.3	82.6	93.6
m	6	8	10	12	15	18	21	24	28
Property classes	04, 05						as per agreement		
	A2-035, A4-035				A2-025, A4-025		–		

Product grades (page 210)

Thread d	Grade
M1.6–M16	A
M20–M36	B

Explanation: 1) Low hexagon nuts (nut height $m < 0.8 \cdot d$) have a smaller load capacity as type 1 nuts.

⇒ **Hexagon nut ISO 4035 – M16 – A2-035:**
d = M16, property class A2-035

Hexagon nuts

Low hexagon nuts with fine threads[1)]

cf. DIN EN ISO 8675 (2001-03)

Valid standard DIN EN ISO	Replaces DIN EN
8675	28675

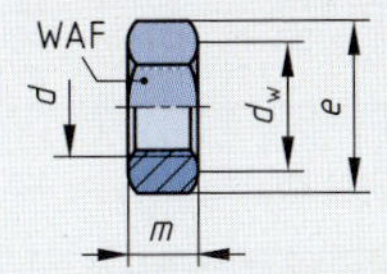

Product grades (page 210)

Thread *d*	Grade
M8x1–M16x1.5	A
M20x1.5–M64x3	B

Thread *d*	M8 x1	M10 x1	M12 x1.5	M16 x1.5	M20 x1.5	M24 x2	M30 x2	M36 x3	M42 x3	M48 x4	M56 x4
WAF	13	16	18	24	30	36	46	55	65	75	85
d_w	11.6	14.6	16.6	22.5	27.7	33.3	42.8	51.1	60	69.5	76.7
e	14.4	17.8	20	26.8	33	39.6	50.9	60.8	71.3	82.6	93.6
m	4	5	6	8	10	12	15	18	21	24	28
Property classes	04, 05								as per agreement		
	A2-035, A4-035						2)				

Explanations
1) Low hexagon nuts (nut height m < 0.8 · *d*) have a smaller load capacity of type 1 nuts (page 228).
2) Property classes for stainless steels: A2-025, A4-025

⇒ **Hexagon nut ISO 8675 – M20x1.5 – A2-035:** *d* = M20x1.5, property class A2-035

Hexagon nuts with insert, type 1[1)]

cf. DIN EN ISO 7040 and 10512 (2001-03)

Valid standard DIN EN ISO	Replaces DIN EN	Replaces DIN
7040	27040	982
10512		

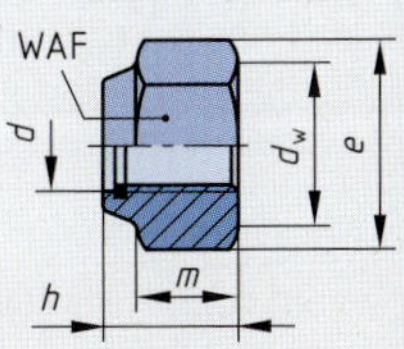

Product grades see DIN EN ISO 4032

Thread *d*	M4 –	M5 –	M6 –	M8 M8 x1	M10 M10 x1	M12 M12 x1.5	M16 M16 x1.5	M20 M20 x1.5	M24 M24 x2	M30 M30 x2	M36 M36 x3
WAF	7	8	10	13	16	18	24	30	36	46	55
d_w	5.9	8.9	8.9	11.6	14.6	16.6	22.5	27.7	33.3	42.8	51.1
e	7.7	8.8	11.1	14.4	17.8	20	26.8	33	39.6	50.9	60.8
h	6	6.8	8	9.5	11.9	14.9	19.1	22.8	27.1	32.6	38.9
m	2.9	4.4	4.9	6.4	8	10.4	14.1	16.9	20.2	24.3	29.4
Property cl.	for DIN EN ISO 7040: 5, 8, 10					for DIN EN ISO 10512: 6, 8, 10					

Explanation
1) Hexagon nuts type 1 (nut height $m \geq 0.8 \cdot d$)
DIN EN ISO 7040: Nuts with coarse threads
DIN EN ISO 10512: Nuts with fine threads

⇒ **Hexagon nut ISO 7040 – M16-10:** *d* = M10, property class 10

Hexagon nuts with large width across flats[1)]

cf. DIN EN 14399-4 (2006-06)

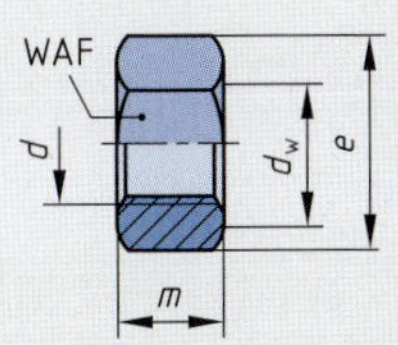

Product grade B

Thread *d*	M12	M16	M20	M22	M24	M27	M30	M36
WAF	22	27	32	36	41	46	50	60
d_w	20.1	24.9	29.5	33.3	38	42.8	46.6	55.9
e	23.9	29.6	35	39.6	45.2	50.9	55.4	66.4
m	10	13	16	18	20	22	24	29
Property cl., surface	10 normal –> lightly oiled, hot-galvanized –> code: tZn							

Explanation
1) for high-strength structural bolting assemblies (HV) in metal construction. Used in combination with hexagon head bolts as per DIN EN 14399-4 (page 213).

⇒ **Hexagon nut DIN EN 14399-4 – M16 – 10 – HV:** *d* = M24, property class 10, high-strength preloaded

Hexagon nuts with flange

cf. DIN EN 1661 (1998-02)

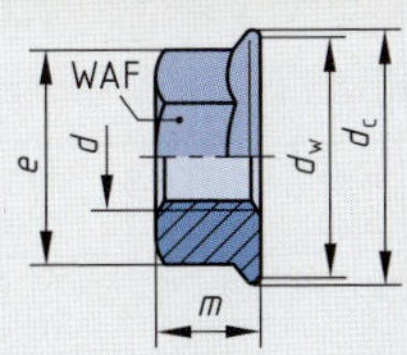

Product grades see DIN EN ISO 4032

Thread *d*	M5	M6	M8	M10	M12	M16	M20
WAF	8	10	13	16	18	24	30
d_w	9.8	12.2	15.8	19.6	23.8	31.9	39.9
d_c	11.8	14.2	17.9	21.8	26	34.5	42.8
e	8.8	11.1	14.4	17.8	20	26.8	33
m	5	6	8	10	12	16	20
Property classes	8, 10, A2-70						

⇒ **Hexagon nut EN 1661 – M16-8:** *d* = M16, property class 8

Hexagon acorn nuts, Lock nuts, Eye nuts

Hexagon acorn nuts, high form

cf. DIN 1587 (2000-10)

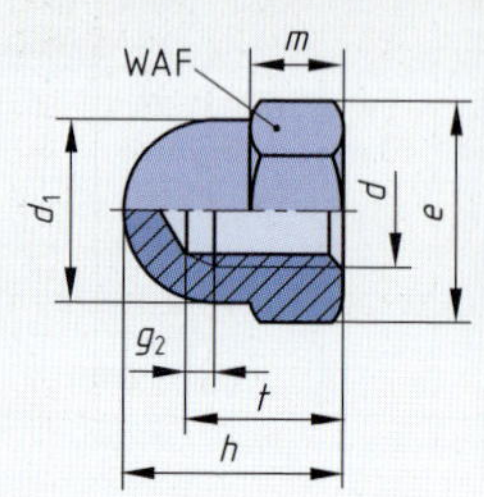

Product grade A or B by choice of manufacturer

Thread d	M4 –	M5 –	M6 –	M8 M8 x1	M10 M10 x1	M12 M12 x1.5	M16 M16 x1.5	M20 M20 x2	M24 M24 x2
WAF	7	8	10	13	16	18	24	30	36
d_1	6.5	7.5	9.5	12.5	15	17	23	28	34
m	3.2	4	5	6.5	8	10	13	16	19
e	7.7	8.8	11.1	14.4	17.8	20	26.8	33.5	40
h	8	10	12	15	18	22	28	34	42
t	5.3	7.2	7.8	10.7	13.3	16.3	20.6	25.6	30.5
g_2	$g \approx 2 \cdot P$ (P thread pitch)					Thread undercut DIN 76-D			
Property classes	6, A1-50								
⇒	**Acorn nut DIN 1587 – M20 – 6:** d = M20, property class 6								

Lock nuts

cf. DIN 70852 (1989-06)

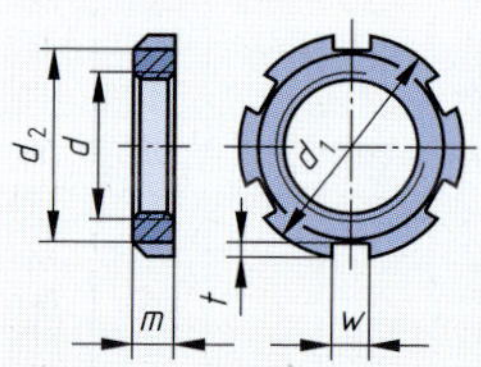

Thread d	M12 x1.5	M16 x1.5	M20 x1.5	M24 x1.5	M30 x1.5	M35 x1.5	M40 x1.5	M48 x1.5	M55 x1.5	M60 x1.5	M65 x1.5
d_1	22	28	32	38	44	50	56	65	75	80	85
d_2	18	23	27	32	38	43	49	57	67	71	76
m	6	6	6	7	7	8	8	8	8	9	9
w	4.5	5.5	5.5	6.5	6.5	7	7	8	8	11	11
t	1.8	2.3	2.3	2.8	2.8	3.3	3.3	3.8	3.8	4.3	4.3
Material	St (steel)										
⇒	**Lock nut DIN 70852 – M16x1.5 – St:** d = M16x1.5, material steel										

Lock washers

cf. DIN 70952 (1976-05)

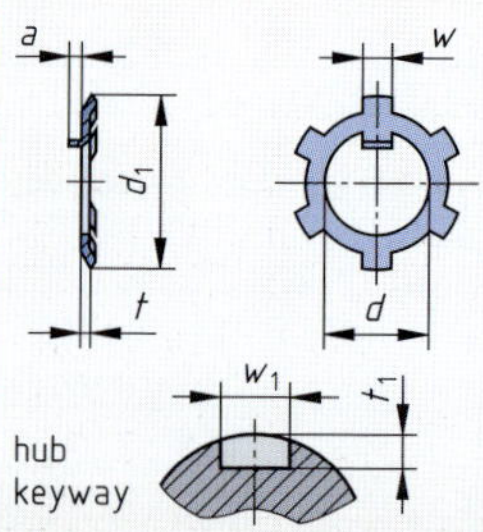

d	12	16	20	24	30	35	40	48	55	60	65
d_1	24	29	35	40	48	53	59	67	79	83	88
t	0.75	1	1	1	1.2	1.2	1.2	1.2	1.2	1.5	1.5
a	3	3	4	4	5	5	5	5	6	6	6
w	4	5	5	6	7	7	8	8	10	10	10
w_1 C11	4	5	5	6	7	7	8	8	10	10	10
t_1	1.2	1.2	1.2	1.2	1.5	1.5	1.5	1.5	1.5	2	2
Material	St (steel sheet)										
⇒	**Lock washer DIN 70952-16 – St:** d = 16 mm, material steel										

Eye nuts

cf. DIN 582 (2010-09)

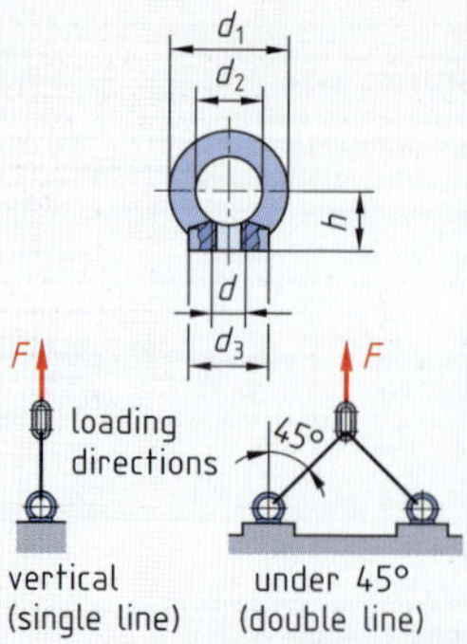

Thread d	M8	M10	M12	M16	M20	M24	M30	M36	M42	M48	M56
h	18	22.5	26	30.5	35	45	55	65	75	85	95
d_1	36	45	54	63	72	90	108	126	144	166	184
d_2	20	25	30	35	40	50	60	70	80	90	100
d_3	20	25	30	35	40	50	65	75	85	100	110
	Load capacity[1] in t for direction of load application										
Vertical	0.14	0.23	0.34	0.70	1.20	1.80	3.20	4.60	6.30	8.60	11.5
under 45°	0.10	0.17	0.24	0.50	0.86	1.29	2.30	3.30	4.50	6.10	8.20
Materials	Case hardened steel C15, A2, A3, A4, A5										
Explanation	[1] The values include a safety factor ν = 6, based on the ultimate load.										
⇒	**Eye nut DIN 582 – M36 – C15E:** d = M36x3, material C15E										

Castle nuts, Cotter pins, Weld nuts, Knurled nuts

Castle nuts, high form

cf. DIN 935-1 (2000-10)

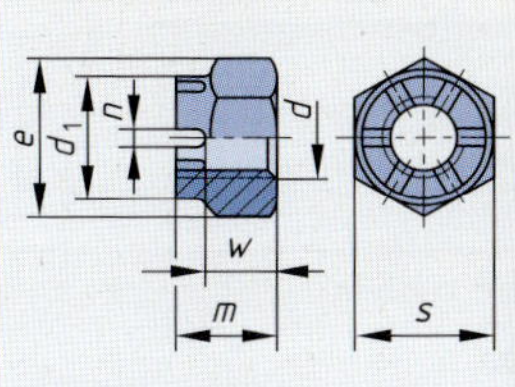

Product grades (page 210)

Thread *d*	Grade
M1.6–M16	A
M20–M100	B

Thread *d*	M4 –	M5 –	M6 –	M8 M8 x1	M10 M10 x1	M12 M12 x1.5	M16 M16 x1.5	M20 M20 x2	M24 M24 x2	M30 M30 x2
s	7	8	10	13	16	18	24	30	36	46
e	7.7	8.8	11.1	14.4	17.8	20	26.8	33	39.6	50.9
m	5	6	7.5	9.5	12	15	19	22	27	33
d_1	no cylindrical shoulder					15.6	21.5	27.7	33.2	42.7
n	1.2	1.4	2	2.5	2.8	3.5	4.5	4.5	5.5	7
w	3.2	4	5	6.5	8	10	13	16	19	24
Property classes	6, 8, 10									
	A2-70								A2-50	
⇒	**Castle nut DIN 935 – M20 – 8:** *d* = M20, property class 8									

Cotter pins

cf. DIN EN ISO 1234 (1998-02)

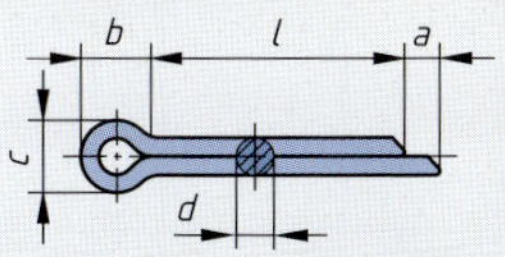

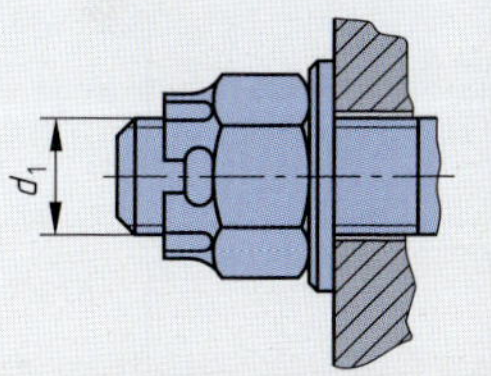

d [1]	1	1.2	1.6	2	2.5	3.2	4	5	6.3	8
b	3	3	3.2	4	5	6.4	8	10	12.6	16
c	1.6	2	2.8	3.6	4.6	5.8	7.4	9.2	11.8	15
a	1.6	2.5	2.5	2.5	2.5	3.2	4	4	4	4
l from	6	8	8	10	12	14	18	22	28	36
l to	20	25	32	40	50	63	80	100	125	160
d_1 [2] over	3.5	4.5	5.5	7	9	11	14	20	27	39
d_1 [2] to	4.5	5.5	7	9	11	14	20	27	39	56
Nominal lengths	6, 8, 10, 12, 14, 16, 18, 20, 22, 25, 28, 32, 36, 40, 45, 50, 56, 63, 71, 80, 90, 100, 112, 125, 140, 160 mm									
Explanations	[1] *d* Nominal sizes = cotter pin hole diameter [2] d_1 applicable bolt diameter									
⇒	**Cotter pin ISO 1234 – 2.5x32 – St:** *d* = 2.5 mm, *l* = 32 mm, material steel									

Hexagon weld nuts

cf. DIN 929 (2000-01)

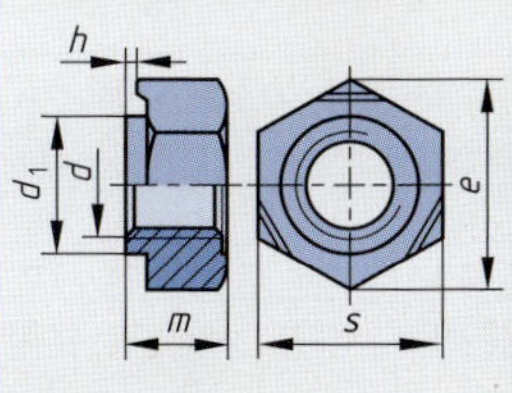

Product grade A

Thread *d*	M3	M4	M5	M6	M8	M10	M12	M16
s	7.5	9	10	11	14	17	19	24
d_1	4.5	6	7	8	10.5	12.5	14.8	18.8
e	8.2	9.8	11	12	15.4	18.7	20.9	26.5
m	3	3.5	4	5	6.5	8	10	13
h	0.3	0.3	0.3	0.4	0.4	0.5	0.6	0.8
Material	St – steel with a maximum carbon content of 0.25%							
⇒	**Weld nut DIN 929 – M16 – St:** *d* = M16, material steel							

Knurled nuts

cf. DIN 466 and 467 (2006-08)

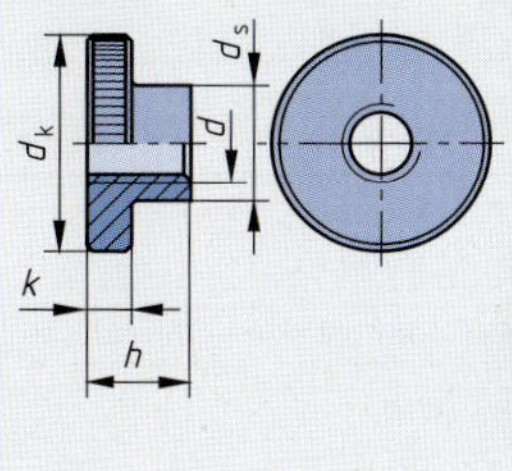

Thread *d*	M1.2	M1.6	M2	M2.5	M3	M4	M5	M6	M8	M10
d_k	6	7.5	9	11	12	16	20	24	30	36
d_s	3	3.8	4.5	5	6	8	10	12	16	20
k	1.5	2	2	2.5	2.5	3.5	4	5	6	8
h [1]	4	5	5.3	6.5	7.5	9.5	11.5	15	18	23
h [2]	2	2.5	2.5	3	3	4	5	6	8	10
Property classes	St (steel), A1-50									
Explanations	[1] Nut height for DIN 466 high form [2] Nut height for DIN 467 low form									
⇒	**Knurled nut DIN 467 – M6 – A1-50:** *d* = M6, property class A1-50									

Flat washers, Overview

Designation example: **Washer ISO 7090 – 8 – 300 HV – A2**[1)]

Washer	ISO 7090	8	300 HV	A2
Name	Standard	Nominal size (Thread nominal ∅)	Hardness grade	Material

[1)] Stainless steel, steel group A2

Overview

Illustration	Design Standard range from–to	M[1)]	Standard	Illustration	Design Standard range from–to	M[1)]	Standard
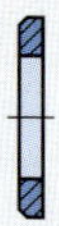	Flat washers with chamfer Product grade A[2)] M5–M64 table below	Steel, stainless steel	DIN EN ISO 7090	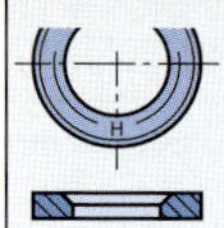	Flat washers with chamfer, for HV bolts M12–M36 page 234	Steel	DIN EN 14399-6
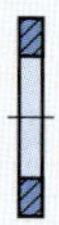	Flat washers small series Product grade A[2)] M1.6–M36 page 233	Steel, stainless steel	DIN EN ISO 7092	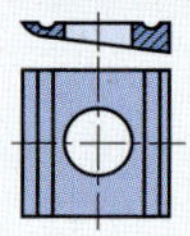	Washers, square, for channels and *I* beams M8–M27 page 234	Steel	DIN 434 DIN 435
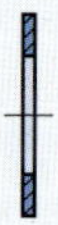	Flat washers normal series Product grade C[2)] M1.6–M64 page 233	Steel	DIN EN ISO 7091	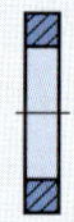	Plain washers for clevis pins Product grade A[2)] d = 3–100 mm page 234	Steel	DIN EN 28738
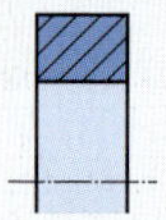	Washers for steel structures Product grade A[2)], C[2)] M10–M30 page 233	Steel	DIN 7989-1	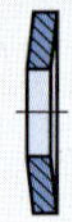	Conical spring washers for screw joints d = 2–30 mm page 234	Spring steel	DIN 6796

[1)] Material is steel with corresponding hardness grade (e.g. 200 HV; 300 HV); other materials as agreed upon.
[2)] Product grades are differentiated by tolerance and by manufacturing process.

Flat washers with chamfer, normal series

cf. DIN EN ISO 7090 (2000-11)

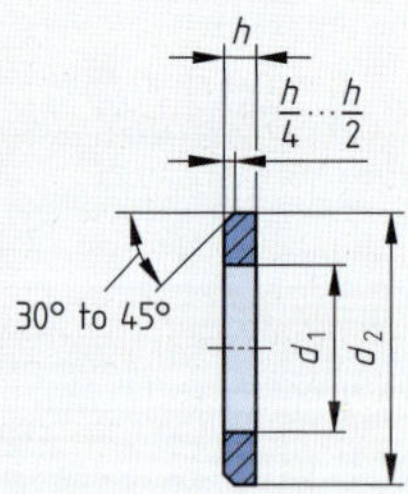

For threads	M5	M6	M8	M10	M12	M16	M20
Nominal size	5	6	8	10	12	16	20
d_1 min.[1)]	5.3	6.4	8.4	10.5	13.0	17.0	21.0
d_2 max.[1)]	10.0	12.0	16.0	20.0	24.0	30.0	37.0
h[1)]	1	1.6	1.6	2	2.5	3	3
For threads	**M24**	**M30**	**M36**	**M42**	**M48**	**M56**	**M64**
Nominal size	24	30	36	42	48	56	64
d_1 min.[1)]	25.0	31.0	37.0	45.0	52.0	62.0	70.0
d_2 max.[1)]	44.0	56.0	66.0	78.0	92.0	105.0	115.0
h[1)]	4	4	5	8	8	10	10

Material[2)]	Steel		Stainless steel
Type	–	–	A2, A4, F1, C1, C4 (ISO 3506)[3)]
Hardness grade	200 HV	300 HV (quenched and tempered)	200 HV
⇒	**Washer ISO 7090-20-200 HV:** Nominal size (= thread nominal ∅) = 20 mm, hardness grade 200 HV, steel		

[1)] These are all nominal dimensions
[2)] Non-ferrous metals and other materials as per agreement
[3)] Compare to page 210

Hardness grade 200 HV suitable for:
- Hexagon bolts and nuts of property classes ≤ 8.8 or ≤ 8 (nut)
- Hexagon bolts and nuts made of stainless steel

Hardness grade 300 HV suitable for:
- Hexagon bolts and nuts of property classes ≤ 10.9 or ≤ 10 (nut)

ME

Flat washers, Washers for steel structures

Flat washers, small series

cf. DIN EN ISO 7092 (2000-11)

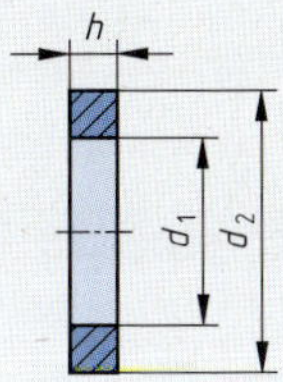

Hardness grade 200 HV suitable for:
- Cap screws with property classes ≤ 8.8 or of stainless steel
- Cap screws with hexagon socket and property classes ≤ 8.8 or of stainless steel

Hardness grade 300 HV suitable for:
- Cap screws with hexagon socket and property classes ≤ 10.9

For threads	**M1.6**	**M2**	**M2.5**	**M3**	**M4**	**M5**	**M6**	**M8**
Nominal size	1.6	2	2.5	3	4	5	6	8
d_1 min.[1]	1.7	2.2	2.7	3.2	4.3	5.3	6.4	8.4
d_2 max.[1]	3.5	4.5	5	6	8	9	11	15
h_{max}	0.35	0.35	0.55	0.55	0.55	1.1	1.8	1.8
For threads	**M10**	**M12**	**M14**[2]	**M16**	**M20**	**M24**	**M30**	**M36**
Nominal size	10	12	14	16	20	24	30	36
d_1 min.[1]	10.5	13.0	15.0	17.0	21.0	25.0	31.0	37.0
d_2 max.[1]	18.0	20.0	24.0	28.0	34.0	39.0	50.0	60.0
h_{max}	1.8	2.2	2.7	2.7	3.3	4.3	4.3	5.6
Material[3]	**Steel**				**Stainless steel**			
Type	–		–		A2, A4, F1, C1, C4 (ISO 3506)[4]			
Hardness grade	200 HV		300 HV (quenched and tempered)		200 HV			
⇒	**Washer ISO 7092-8-200 HV-A2:** Nominal size (= thread nominal ∅) = 8 mm, small series, hardness grade 200 HV, of stainless steel A2							

1) These are all nominal dimensions
2) Avoid this size if at all possible
3) Non-ferrous metals and other materials as per agreement
4) Compare to page 210

Flat washers, normal series

cf. DIN EN ISO 7091 (2000-11)

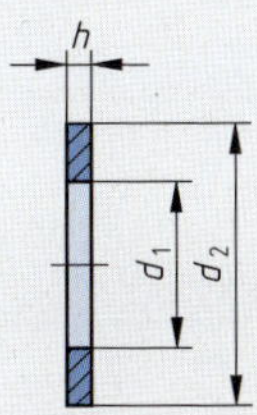

Hardness grade 100 HV suitable for:
- Hexagon bolts/screws, product grade C, with property classes ≤ 6.8
- Hexagon nuts, product grade C, with property classes ≤ 6

For threads	**M2**	**M3**	**M4**	**M5**	**M6**	**M8**	**M10**	**M12**
Nominal size	2	3	4	5	6	8	10	12
d_1 min.[1]	2.4	3.4	4.5	5.5	6.6	9.0	11.0	13.5
d_2 max.[1]	5.0	7.0	9.0	10.0	12.0	16.0	20.0	24.0
h[1]	0.3	0.5	0.8	1.0	1.6	1.6	2	2.5
For threads	**M16**	**M20**	**M24**	**M30**	**M36**	**M42**	**M48**	**M64**
Nominal size	16	20	24	30	36	42	48	64
d_1 min.[1]	17.5	22.0	26.0	33.0	39.0	45.0	52.0	70.0
d_2 max.[1]	30.0	37.0	44.0	56.0	66.0	78.0	92.0	115.0
h[1]	3	3	4	4	5	8	8	10
⇒	**Washer ISO 7091-12-100 HV:** Nominal size (= thread nominal ∅), d = 12 mm, hardness grade 100 HV							

1) These are all nominal dimensions

Washers for steel structures

cf. DIN 7989-1 and DIN 7989-2 (2001-04)

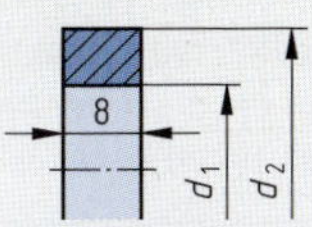

Suitable for bolts according to DIN 7968, DIN 7969, DIN 7990 joined with nuts according to ISO 4032 and ISO 4034.

For threads[1]	**M10**	**M12**	**M16**	**M20**	**M24**	**M27**	**M30**
d_1 min.	11.0	13.5	17.5	22.0	26.0	30.0	33.0
d_2 max.	20.0	24.0	30.0	37.0	44.0	50.0	56.0
⇒	**Washer DIN 7989-16-C-100 HV:** Thread nominal ∅ d = 16 mm, product grade C, hardness grade 100						

Versions: Product grade C (stamped version) thickness h = (8 ± 1.2) mm
Product grade A (turned version) thickness h = (8 ± 1) mm

1) Nominal dimensions

Washers for HV bolts, Channels and I beams, Clevis pins, Conical spring washers

Flat washers with chamfer for HV screw joints

cf. DIN EN 14399-6 (2006-06)

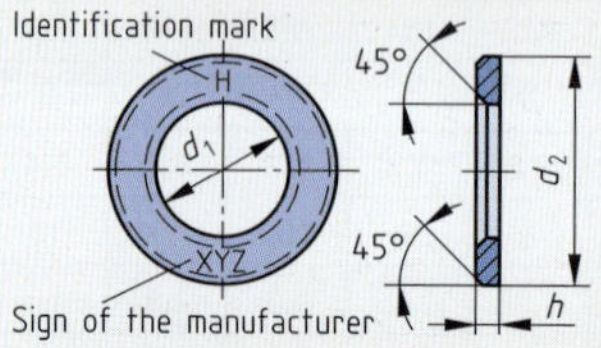

For threads	M12	M16	M20	M22	M24	M27	M30
d_1 min.	13	17	21	23	25	28	31
d_2 max.	24	30	37	39	44	50	56
h	3	4	4	4	4	5	5
⇒	**Washer DIN EN 14399-6 – 20:** Nominal size d = 20 mm (the nominal size d corresponds to thread diameter)						

Material: steel, quenched and tempered to 300 HV–370 HV.

Square, tapered washers for channels and I beams

cf. DIN 434 (2000-04), DIN 435 (2000-01)

channel washer DIN 434 | **I-beam washer** DIN 435

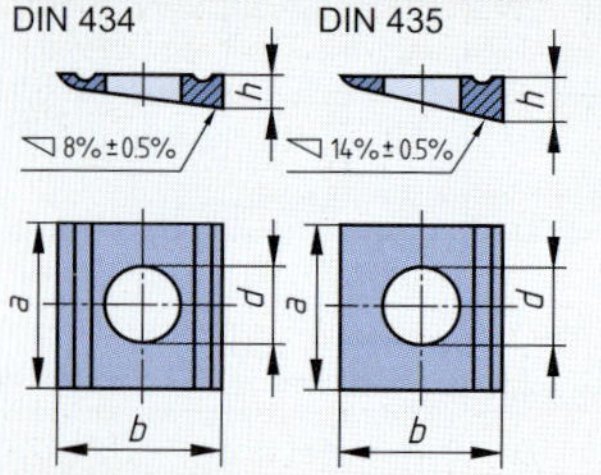

For threads	M8	M10	M12	M16	M20	M22	M24
d_1 min.[1]	9	11	13.5	17.5	22	24	26
a	22	22	26	32	40	44	56
b	22	22	30	36	44	50	56
h DIN 434	3.8	3.8	4.9	5.8	7	8	8.5
h DIN 435	4.6	4.6	6.2	7.5	9.2	10	10.8
⇒	**I-Washer DIN 435-13.5:** Nominal sizes d_1 = 13.5 mm						

Material: Steel, hardness 100 HV 10 to 250 HV 10

[1] Nominal diameter

Washers for clevis pins, product grade A[1]

cf. DIN EN 28738 (1992-10)

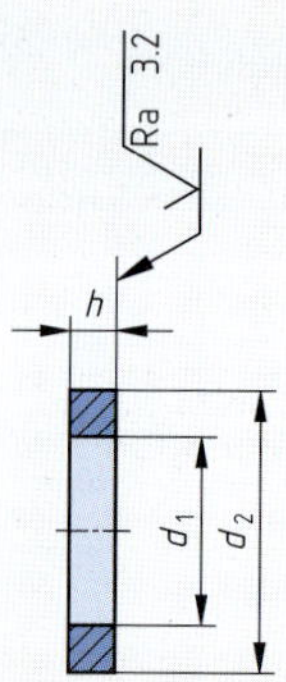

d_1 min.[2]	**3**	**4**	**5**	**6**	**8**	**10**	**12**
d_2 max.	6	8	10	12	15	18	20
h	0.8		1	1.6	2	2.5	3
d_1 min.[2]	**14**	**16**	**18**	**20**	**22**	**24**	**27**
d_2 max.	22	24	28	30	34	37	39
h	3		4				5
d_1 min.[2]	**30**	**36**	**40**	**50**	**60**	**80**	**100**
d_2 max.	44	50	56	66	78	98	120
h	5	6		8	10	12	
⇒	**Washer ISO 8738-14-160 HV:** d_1 min. = 14 mm, hardness grade 160 HV						

Material: Steel, hardness 160 to 250 HV
Application: For clevis pins according to ISO 2340 and ISO 2341 (page 237), used only on the cotter pin end.

[1] Product grades are differentiated by tolerance and manufacturing process
[2] nominal dimensions

Conical spring washers for screw joints

cf. DIN 6796 (2009-08)

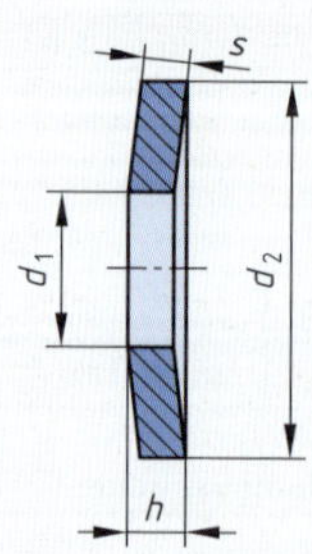

For threads	M2	M3	M4	M5	M6	M8	M10
d_1 H14	2.2	3.2	4.3	5.3	6.4	8.4	10.5
d_2 h14	5	7	9	11	14	18	23
h max.	0.6	0.85	1.3	1.55	2	2.6	3.2
s	0.4	0.6	1	1.2	1.5	2	2.5
For threads	**M12**	**M16**	**M20**	**M22**	**M24**	**M27**	**M30**
d_1 H14	13	17	21	23	25	28	31
d_2 h14	29	39	45	49	56	60	70
h max.	3.95	5.25	6.4	7.05	7.75	8.35	9.2
s	3	4	5	5.5	6	6.5	7
⇒	**Conical spring washer DIN 6796-10-FSt:** for threads M10, of spring steel						

Material: Spring steel (FSt) according to DIN 267-26 or stainless steel
Application: Conical spring washers should prevent loosening of the screw joints. This does not apply to alternating transverse loads. Its application is therefore limited to predominantly axially loaded, short bolts/screws of property classes 8.8 to 10.9.

Pins and clevis pins, Overview

Designation example: Taper pin ISO 2339 – A – 10x40 – St

- Taper pin → Name
- ISO 2339 → Standard
- A → Form or Type[1]
- 10x40 → Nominal ∅ x nominal length
- St → Material

e.g. St = steel
Stainless steels:
A1 = austenitic
C1 = martensitic

Pins with DIN-EN main numbers are designated with ISO numbers.
ISO number = DIN-EN number – 20000; example: DIN EN 22338 = ISO 2338

[1] if available

Illustration	Designation, Standard range from–to	Standard	Illustration	Designation, Standard range from–to	Standard
Pins					
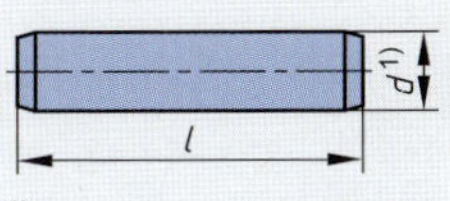 [1] tolerance m6 or h8	Dowel pin, not hardened d = 1–50 mm	DIN EN ISO 2338	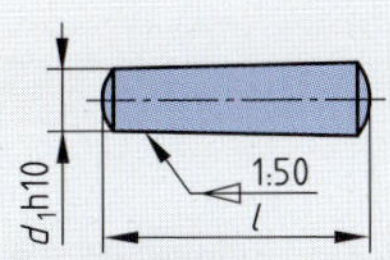	Taper pin d_1 = 0.6–50 mm	DIN EN 22339
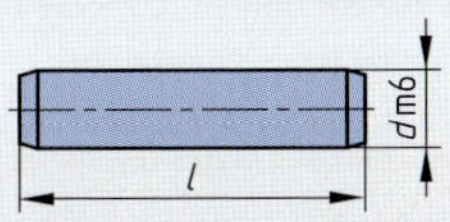	Dowel pin, hardened d = 0.8–20 mm	DIN EN ISO 8734	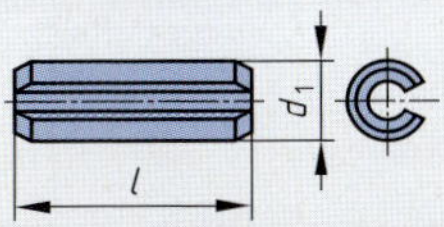	Spring pin (clamping sleeves), slotted d_1 = 1–50 mm	DIN EN ISO 8752 DIN EN ISO 13337
Grooved pins, grooved drive studs					
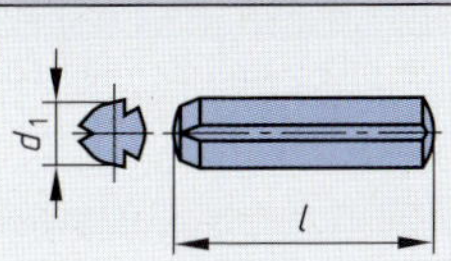	Straight grooved pin with chamfer d_1 = 1.5–25 mm	DIN EN ISO 8740	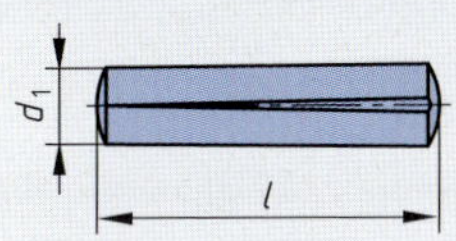	Tapered grooved pin d_1 = 1.5–25 mm	DIN EN ISO 8744
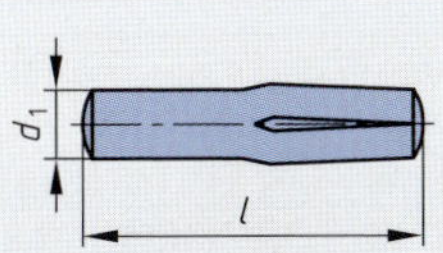	Half length reversed taper grooved pin d_1 = 1.5–25 mm	DIN EN ISO 8741	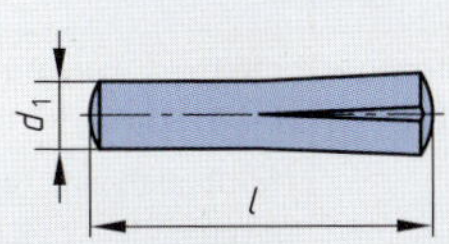	Half length taper grooved pin d_1 = 1.2–25 mm	DIN EN ISO 8745
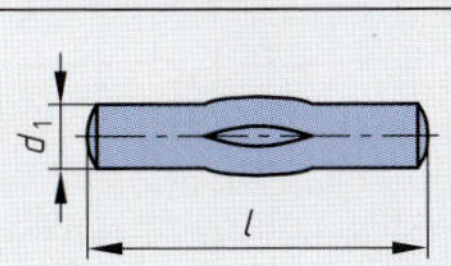	Center grooved pin, grooved 1/3 the length d_1 = 1.2–25 mm	DIN EN ISO 8742	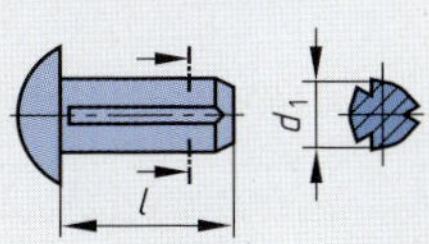	Round head grooved pin d_1 = 1.4–20 mm	DIN EN ISO 8746
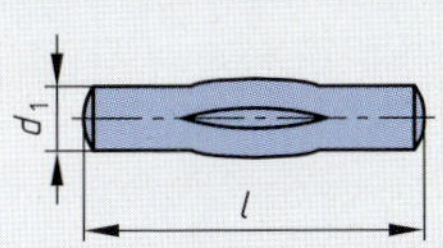	Center grooved pin, with long grooves d_1 = 1.2–25 mm	DIN EN ISO 8743	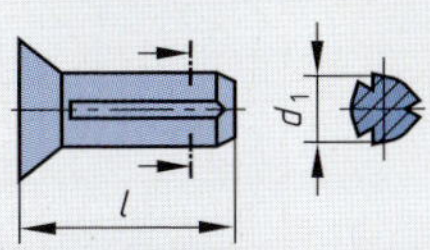	Grooved pin with countersunk head d_1 = 1.4–20 mm	DIN EN ISO 8747
Clevis pins					
Form A 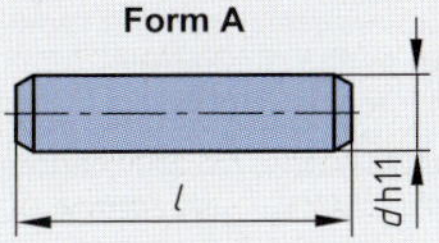	Clevis pins without head, form A without cotter pin hole, form B with d = 3–100 mm	DIN EN 22340	**Form A** 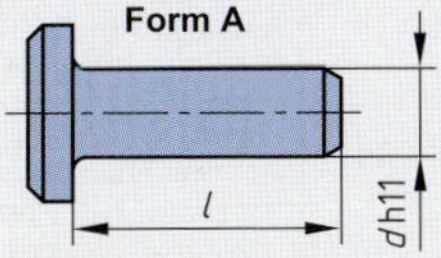	Clevis pins with head, form A without cotter pin hole, form B with d = 3–100 mm	DIN EN 22341

Dowel, Taper and spring pins

Dowel pins of unhardened steel and austenitic stainless steel

cf. DIN EN ISO 2338 (1998-02)

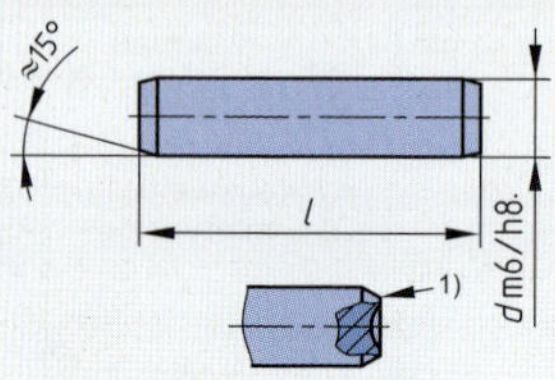

[1] Radius and hollow allowed at end of pin

d m6/h8[2]	0.6	0.8	1	1.2	1.5	2	2.5	3	4	5
l from	2	2	4	4	4	6	6	8	8	10
l to	6	8	10	12	16	20	24	30	40	50
d m6/h8[2]	**6**	**8**	**10**	**12**	**16**	**20**	**25**	**30**	**40**	**50**
l from	12	14	18	22	26	35	50	60	80	95
l to	60	80	95	140	180	200	200	200	200	200
Nominal lengths *l*	2, 3, 4, 5, 6, 8, 10, 12, 14, 16, 18, 20, 22, 24, 26, 28, 30, 32, 35, 40–95, 100, 120, 140, 160, 180, 200 mm.									
⇒	**Dowel pin ISO 2338 – 6 m6 x 30 – St**: *d* = 6 mm, tolerance class m6, *l* = 30 mm, of steel									

[2] Available in tolerance classes m6 and h8

Dowel pins, hardened

cf. DIN EN ISO 8734 (1998-03)

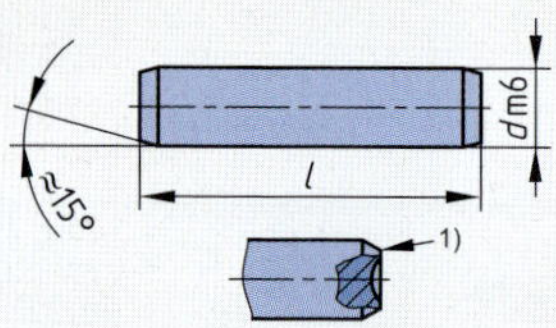

[1] Radius and hollow allowed on end of pin

d m6	1	1.5	2	2.5	3	4	5	6	8	10	12	16	20
l from	3	4	5	6	8	10	12	14	18	22	26	40	50
l to	10	16	20	24	30	40	50	60	80	100			
Nominal lengths *l*	3, 4, 5, 6, 8, 10, 12, 14, 16, 18, 20, 22, 24, 26, 28, 30, 32, 35, 40, 45, 50, 55, 60, 65, 70, 75, 80, 85, 90, 95, 100 mm												
Materials	• Steel: Type A pin fully hardened, type B case hardened • Stainless steel type C1												
⇒	**Dowel pin ISO 8734 – 6 x 30 – C1**: *d* = 6 mm, *l* = 30 mm, of stainless steel of type C1												

Taper pin, unhardened

cf. DIN EN 22339 (1992-10)

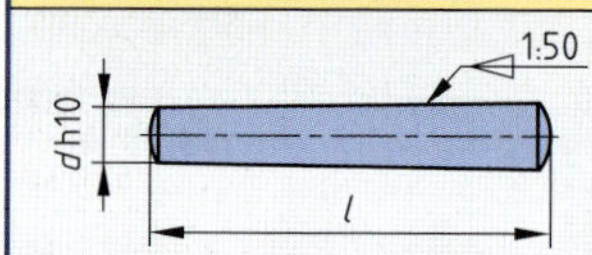

Type A ground, Ra = 0.8 µm;
Type B turned, Ra = 3.2 µm

d h10	1	2	3	4	5	6	8	10	12	16	20	25	30
l from	6	10	12	14	18	22	22	26	32	40	45	50	55
l to	10	35	45	55	60	90	120	160	180	200			
Nominal lengths *l*	2, 3, 4, 5, 6, 8, 10, 12, 14, 16, 18, 20, 22, 24, 26, 28, 30, 32, 35, 40, 45–95, 100, 120–180, 200 mm												
⇒	**Taper pin ISO 2339 – A – 10 x 40 – St**: Type A, *d* = 10 mm, *l* = 40 mm, of steel												

Spring pins (clamping sleeves), slotted, heavy duty
Spring pins (clamping sleeves), slotted, light duty

cf. DIN EN ISO 8752 (2009-10)
cf. DIN EN ISO 13337 (2009-10)

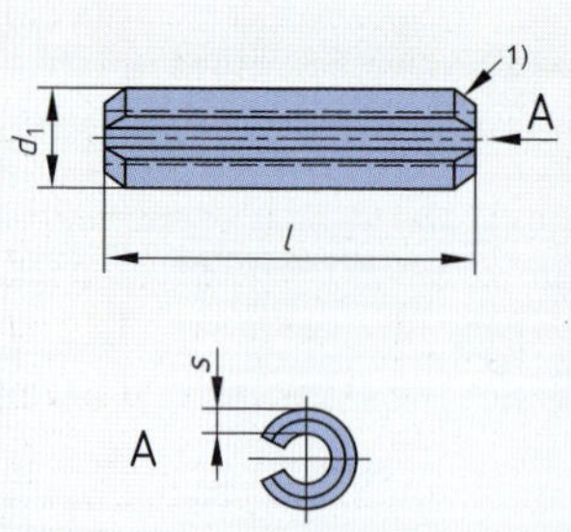

[1] Only one chamfer is allowed for spring pins with nominal diameter d_1 ≥ 10 mm.

Nominal Ø d_1	2	2.5	3	4	5	6	8	10	12
d_1 max.	2.4	2.9	3.5	4.6	5.6	6.7	8.8	10.8	12.8
s ISO 8752	0.4	0.5	0.6	0.8	1	1.2	1.5	2	2.5
s ISO 13337	0.2	0.25	0.3	0.5	0.5	0.75	0.75	1	1
l from	4	4	4	4	5	10	10	10	10
l to	20	30	40	50	80	100	120	160	180
Nominal Ø d_1	**14**	**16**	**20**	**25**	**30**	**35**	**40**	**45**	**50**
d_1 max.	14.8	16.8	20.9	25.9	30.9	35.9	40.9	45.9	50.9
s ISO 8752	3	3	4	5	6	7	7.5	8.5	9.5
s ISO 13337	1.5	1.5	2	2	2.5	3.5	4	4	5
l from	10			14		20			
l to	200			200		200			
Nominal lengths *l*	4, 5, 6, 8, 10, 12, 14, 16, 18, 20, 22, 24, 26, 28, 30, 32, 35, 40, 45–95, 100, 120, 140, 160, 180, 200 mm								
Materials	• Steel: Hardened and tempered 420 HV – 520 HV • Stainless steel: Type A or type C								
Application	The diameter of the location hole (tolerance class H12) must have the same nominal diameter d_1 as the mating pin. After installing the pin in the smallest receiving hole, the slot should not be completely closed.								
⇒	**Spring pin ISO 8752 – 6 x 30 – St**: d_1 = 6 mm, *l* = 30 mm, of steel								

Grooved pins, Grooved drive studs, Clevis pins

Grooved pins, grooved drive studs

cf. DIN EN ISO 8740–8747 (1998-03)

Type	d_1	1.5	2	2.5	3	4	5	6	8	10	12	16	20	25
Full length straight grooved pin with chamfer ISO 8740 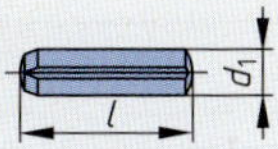	l from	8	8	10	10	10	14	14	14	14	18	22	26	26
	l to	20	30	30	40	60	60	80	100	100	100	100	100	100
1/2 length reverse-taper grooved pin ISO 8741 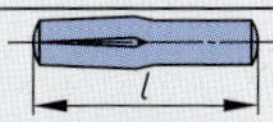	l from	8	8	8	8	10	10	12	14	18	26	26	26	26
	l to	20	30	30	40	60	60	80	100	160	200	200	200	200
1/3 length center grooved pins ISO 8742+8743 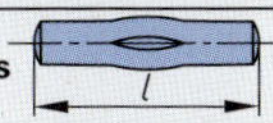	l from	8	12	12	12	18	18	22	26	32	40	45	45	45
	l to	20	30	30	40	60	60	80	100	160	200	200	200	200
Tapered groove pin ISO 8744 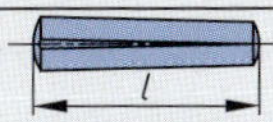	l from	8	8	8	8	8	8	10	12	14	14	24	26	26
	l to	20	30	30	40	60	60	80	100	120	120	120	120	120
Full length taper grooved pins ISO 8745 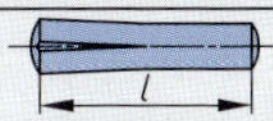	l from	8	8	8	8	10	10	10	14	14	18	26	26	26
	l to	20	30	30	40	60	60	80	100	200	200	200	200	200

Type	d_1	1.4	1.6	2	2.5	3	4	5	6	8	10	12	16	20
Grooved pins with round head ISO 8746 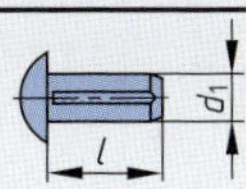	l from	3	3	3	3	4	5	6	8	10	12	16	20	25
	l to	6	8	10	12	16	20	25	30	40	40	40	40	40
Grooved pins with countersunk head ISO 8747 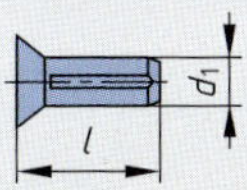	l from	3	3	4	4	5	6	8	8	10	12	16	20	25
	l to	6	8	10	12	16	20	25	30	40	40	40	40	40
	Nominal lengths l	Pins: 8, 10–30, 32, 35, 40–100, 120, 140–180, 200 mm Studs: 3, 4, 5, 6, 8, 10, 12, 16, 20, 25, 30, 35, 40 mm												
	⇒	**Grooved pin ISO 8740 – 6 x 50 – St:** d_1 = 6 mm, l = 50 mm, of steel												

Clevis pins with and without head

cf. DIN EN 22340, 22341 (1992-10)

Clevis pins without head ISO 2340

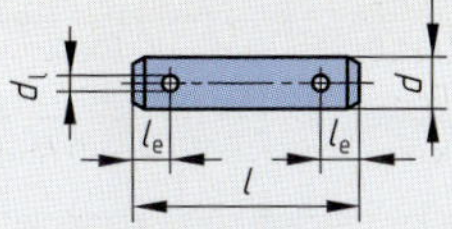

Clevis pins with head ISO 2341

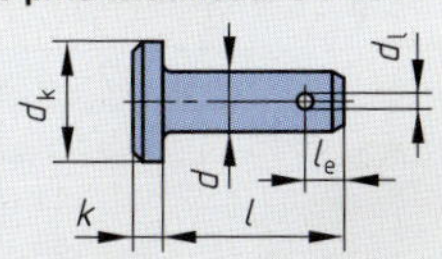

Form A without cotter pin hole
Form B with cotter pin hole

d h11	3	4	5	6	8	10	12	14	16	18	20	22	24
d_l H13	0.8	1	1.2	1.6	2	3.2	3.2	4	4	5	5	5	6.3
d_k h14	5	6	8	10	14	18	20	22	25	28	30	33	36
k js14	1	1	1.6	2	3	4	4	4	4.5	5	5	5.5	6
l_e	1.6	2.2	2.9	3.2	3.5	4.5	5.5	6	6	7	8	8	9
l from	6	8	10	12	16	20	24	28	30	35	40	45	50
l to	30	40	50	60	80	100	120	140	160	180	200	200	200
Nominal lengths l	6, 8, 10–30, 32, 35, 40–95, 100, 120, 140–180, 200 mm												
⇒	**Clevis pin ISO 2340 – B – 20 x 100 – St**: Form B, d = 20 mm, l = 100 mm, of free-cutting steel												

Clevis pins with head and threaded stud end

cf. DIN 1445 (1977-02)

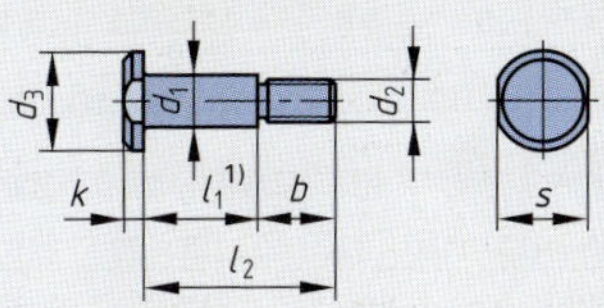

d_1 h11	8	10	12	14	16	18	20	24	30	40	50
b min	11	14	17	20	20	20	25	29	36	42	49
d_2	M6	M8	M10	M12	M12	M12	M16	M20	M24	M30	M36
d_3 h14	14	18	20	22	25	28	30	36	44	55	66
k js14	3	4	4	4	4.5	5	5	6	8	8	9
s	11	13	17	19	22	24	27	32	36	50	60
Nominal lengths l_2	16, 20, 25, 30, 35–125, 130, 140, 150–190, 200 mm										
⇒	**Clevis pin DIN 1445 – 12h11 x 30 x 50 – St:** d_1 = 12 mm, tolerance class h11, l_1 = 30 mm, l_2 = 50 mm, of 9SMnPb28 (St)										

[1] gripping length

Keys, Gib-head keys

Designation example: Feather key DIN 6885 – A – 12x8x56 – E295

- Feather key: Name
- DIN 6885: Standard
- A: Form or Type
- 12x8x56: Width x height x length
- E295: Material, e.g. steel

Illustration	Designation, Standard range from–to	Standard	Illustration	Designation, Standard range from–to	Standard
Overview of tapered keys					table below
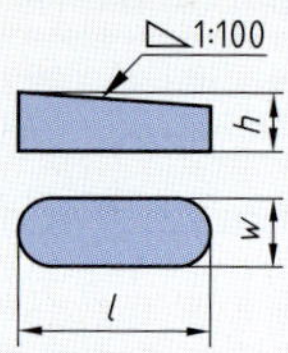	Tapered key *w* x *h* = 2 x 2–100 x 50	DIN 6886 Form A: sunk key Form B: driving key	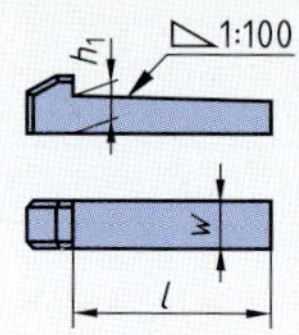	Gib-head tapered key *w* x *h* = 4 x 4–100 x 50	DIN 6887
Overview of feather keys					page 239
Form A 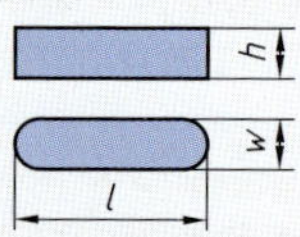	Feather key *w* x *h* = 2 x 2–100 x 50	DIN 6885 Form A–J	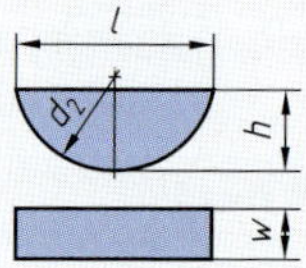	Woodruff keys *w* x *h* = 2.5x3.7–10x16	DIN 6888

Tapered keys, Gib-head tapered keys

cf. DIN 6886 (1967-12) or DIN 6887 (1968-04)

Form A (sunk key) **Form B (driving key)** **Gib head tapered key**

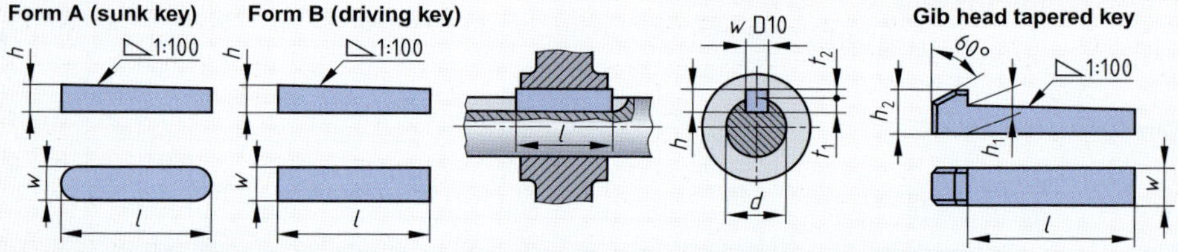

For shaft diameter *d* over / to	10 12	12 17	17 22	22 30	30 38	38 44	44 50	50 58	58 65	65 75	75 85	85 95	95 110
Tapered keys w D10 h	4 4	5 5	6 6	8 7	10 8	12 8	14 9	16 10	18 11	20 12	22 14	25 14	28 16
Gib-head tapered keys h_1 h_2	4.1 7	5.1 8	6.1 10	7.2 11	8.2 12	8.2 12	9.2 14	10.2 16	11.2 18	12.2 20	14.2 22	14.2 22	16.2 25
Shaft keyway depth t_1 Hub keyway depth t_2	2.5 1.2	3 1.7	3.5 2.2	4 2.4	5 2.4	5 2.4	5.5 2.9	6 3.4	7 3.4	7.5 3.9	9 4.4	9 4.4	10 5.4
Allow. deviation t_1, t_2	+0.1			+0.2									
Key length l from / to	10[1)] 45	12[1)] 56	16 70	20 90	25 110	32 140	40 160	45 180	50 200	56 220	63 250	70 280	80 320
Nominal lengths l	6, 8–20, 22, 25, 28, 32, 40, 45, 50, 56, 63, 70, 80–100, 110, 125, 140, 160–200, 220, 250, 280, 320, 360, 400 mm												
Length tolerances	Key length l, from–to	6–28		32–80			90–400						
Tolerances for	Key length	–0.2		–0.3			–0.5						
	Keyway length (sunk key)	+0.2		+0.3			+0.5						

1) Gib-head key lengths from 14 mm

Feather keys, Splined shaft joints

Feather keys (high form) — cf. DIN 6885-1 (1968-08)

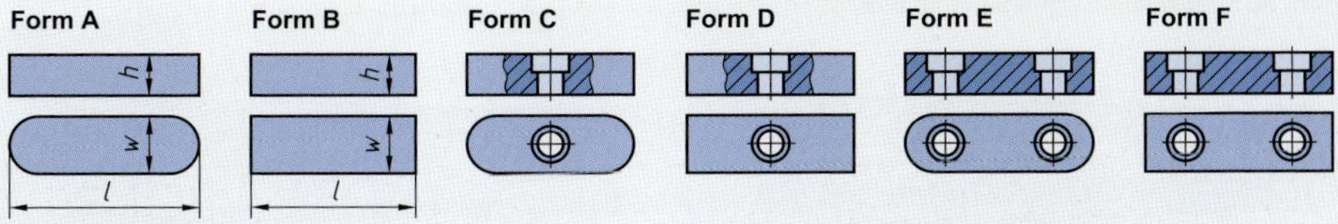

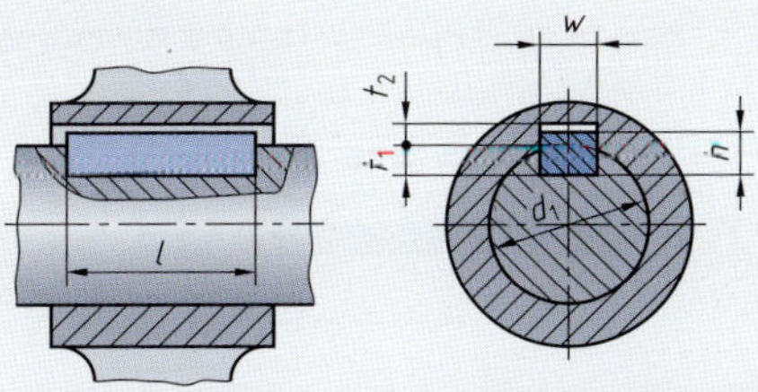

Tolerances for feather keyways			
Shaft keyway width w	tight fit normal fit		P 9 N 9
Hub keyway width w	tight fit normal fit		P 9 JS 9
Allow. deviation for d_1	≤ 22	≤ 130	> 130
Shaft keyway depth t_1 Hub keyway depth t_2	+0.1 +0.1	+0.2 +0.2	+0.3 +0.3
Alllow. deviation for length l	6–28	32–80	90–400
Length tolerances for key	−0.2	−0.3	−0.5
Length tolerances for keyway	+0.2	+0.3	+0.5

d_1 over	6	8	10	12	17	22	30	38	44	50	58	65	75	85	95	110
d_1 to	8	10	12	17	22	30	38	44	50	58	65	75	85	95	110	130
w	2	3	4	5	6	8	10	12	14	16	18	20	22	25	28	32
h	2	3	4	5	6	7	8	8	9	10	11	12	14	14	16	18
t_1	1.2	1.8	2.5	3	3.5	4	5	5	5.5	6	7	7.5	9	9	10	11
t_2	1	1.4	1.8	2.3	2.8	3.3	3.3	3.3	3.8	4.3	4.4	4.9	5.4	5.4	6.4	7.4
l from	6	6	8	10	14	18	20	28	36	45	50	56	63	70	80	90
l to	20	36	45	56	70	90	110	140	160	180	200	220	250	280	320	360
Nominal lengths l	6, 8, 10, 12, 14, 16, 18, 20, 22, 25, 28, 32, 36, 40, 45, 50, 56, 63, 70, 80, 90, 100, 110, 125, 140, 160, 180, 200, 220, 250, 280, 320 mm															
⇒	**Feather key DIN 6885 – A – 12 x 8 x 56**: Form A, w = 12 mm, h = 8 mm, l = 56 mm															

Splined shaft joints with straight flanks and internal centering — cf. DIN ISO 14 (1986-12)

Hub

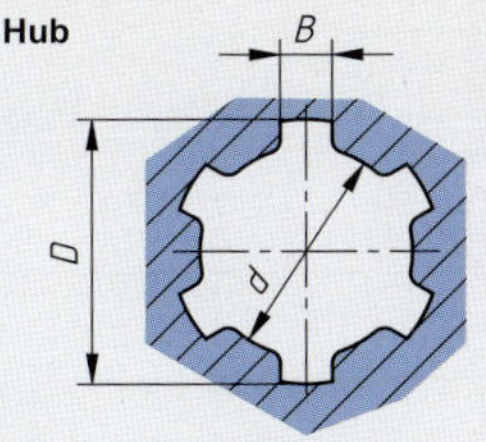

Shaft

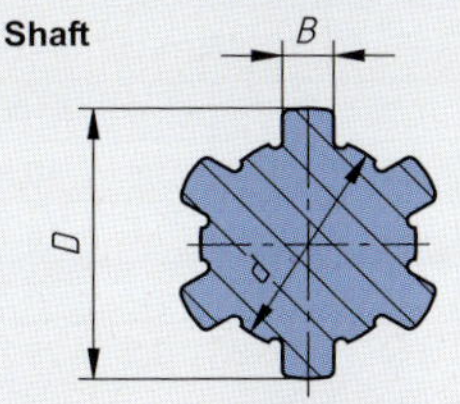

Internal centering

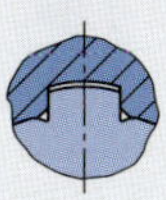

d	Light series N[1)]	Light series D	Light series B	Medium series N[1)]	Medium series D	Medium series B	d	Light series N[1)]	Light series D	Light series B	Medium series N[1)]	Medium series D	Medium series B
11	–	–	–	6	14	3	42	8	46	8	8	48	8
13	–	–	–	6	16	3.5	46	8	50	9	8	54	9
16	–	–	–	6	20	4	52	8	58	10	8	60	10
18	–	–	–	6	22	5	56	8	62	10	8	65	10
21	–	–	–	6	25	5	62	8	68	12	8	72	12
23	6	26	6	6	28	6	72	10	78	12	10	82	12
26	6	30	6	6	32	6	82	10	88	12	10	92	12
28	6	32	7	6	34	7	92	10	98	14	10	102	14
32	8	36	6	8	38	6	102	10	108	16	10	112	16
36	8	40	7	8	42	7	112	10	120	18	10	125	18

Tolerance classes for the hub

Not heat treated dimensions B	D	d	Heat treated dimensions B	D	d
H9	H10	H7	H11	H10	H7

Tolerance classes for the shaft

Dimen.	Type of fit: Sliding fit	Transition fit	Press fit
B	d10	f9	h10
D	a11	a11	a11
d	f7	g7	h7

⇒ **Shaft (or hub) DIN ISO 14 – 6 x 23 x 26**: N = 6, d = 23 mm, D = 26 mm

1) N number of splines

ME

Metric tapers, Morse tapers, Steep tapers

Morse tapers and metric tapers

cf. DIN 228-1 (1987-05); DIN 228-2 (1987-03)

Form A: Taper shank with tightening thread

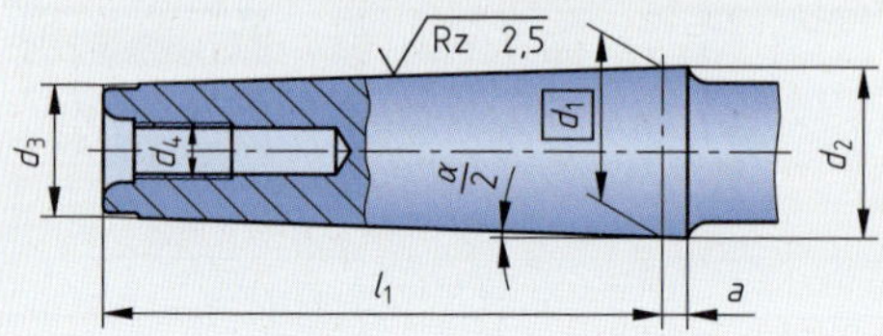

Form B: Taper shank with tang

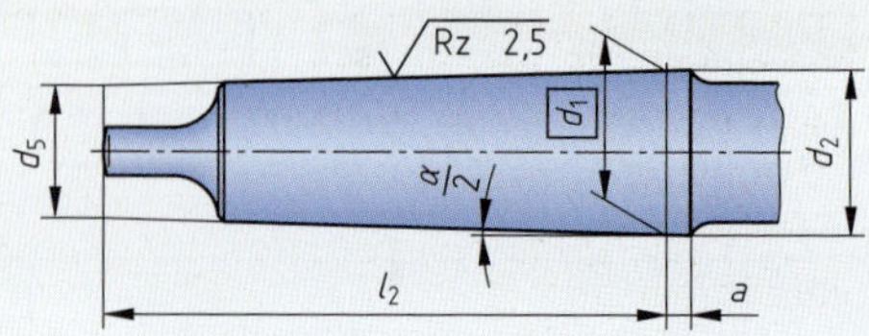

Form C: Taper sleeve for taper shanks with draw-in threads

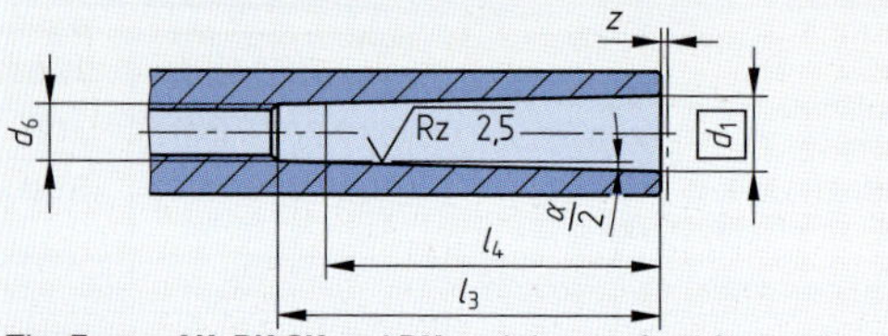

Form D: Taper sleeve for taper shanks with tang

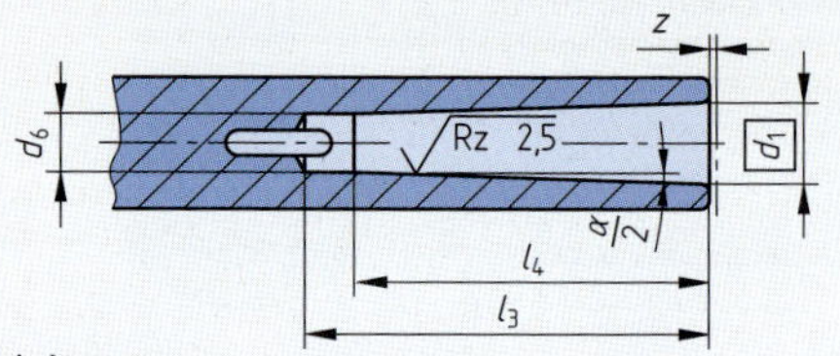

The **Forms AK, BK CK** and **DK** each have a feed for cooling lubricants.

Type of taper	Size	Taper shank								Taper shank				Taper	
		d_1	d_2	d_3	d_4	d_5	l_1	a	l_2	d_6 H11	l_3	l_4	z 1)	Taper ratio	$\frac{\alpha}{2}$
Metric taper (ME)	4	4	4.1	2.9	–	–	23	2	–	3	25	20	0.5	1 : 20	1.432°
	6	6	6.2	4.4	–	–	32	3	–	4.6	34	28	0.5		
Morse taper (MT)	0	9.045	9.2	6.4	–	6.1	50	3	56.5	6.7	52	45	1	1 : 19.212	1.491°
	1	12.065	12.2	9.4	M6	9	53.5	3.5	62	9.7	56	47	1	1 : 20.047	1.429°
	2	17.780	18.0	14.6	M10	14	64	5	75	14.9	67	58	1	1 : 20.020	1.431°
	3	23.825	24.1	19.8	M12	19.1	81	5	94	20.2	84	72	1	1 : 19.922	1.438°
	4	31.267	31.6	25.9	M16	25.2	102.5	6.5	117.5	26.5	107	92	1	1 : 19.254	1.488°
	5	44.399	44.7	37.6	M20	36.5	129.5	6.5	149.5	38.2	135	118	1	1 : 19.002	1.507°
	6	63.348	63.8	53.9	M24	52.4	182	8	210	54.8	188	164	1	1 : 19.180	1.493°
Metric taper (ME)	80	80	80.4	70.2	M30	69	196	8	220	71.5	202	170	1.5	1 : 20	1.432°
	100	100	100.5	88.4	M36	87	232	10	260	90	240	200	1.5		
	120	120	120.6	106.6	M36	105	268	12	300	108.5	276	230	1.5		
	160	160	160.8	143	M48	141	340	16	380	145.5	350	290	2		
	200	200	201.0	179.4	M48	177	412	20	460	182.5	424	350	2		

⇒ **Taper shank DIN 228 – ME – B 80 AT6:** Metric taper shank, Form B, Size 80, Taper angle tolerance quality AT6

1) Control dimension d_1 may lie a maximum distance z in front of the taper sleeve.

Steep taper shanks for tools and chucks form A

cf. DIN 2080-1 (1978-12)

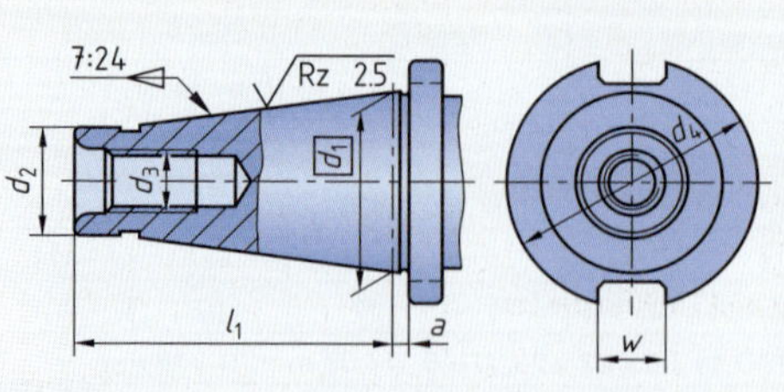

No.	d_1	d_2 a10	d_3	d_4 – 0.4	l_1	a ± 0.2	b H12
30	31.75	17.4	M12	50	68.4	1.6	16.1
40	44.45	25.3	M16	63	93.4	1.6	16.1
50	69.85	39.6	M24	97.5	126.8	3.2	25.7
60	107.95	60.2	M30	156	206.8	3.2	25.7
70	165.1	92	M36	230	296	4	32.4
80	254	140	M48	350	469	6	40.5

⇒ **Steep taper shank DIN 2080 – A 40 AT4:** Form A, No. 40, Taper angle tolerance quality AT4

Tool holding fixtures

Tool holding fixtures join the tool with the spindle of the machine tool. They transmit the torque and are responsible for precise concentric running.

Type of design	Function, advantages (+) and disadvantages (–)	Application, sizes
Metric taper (ME) and Morse taper (MT)		cf. DIN 228-1 (1987-05) and -2 (1987-03)
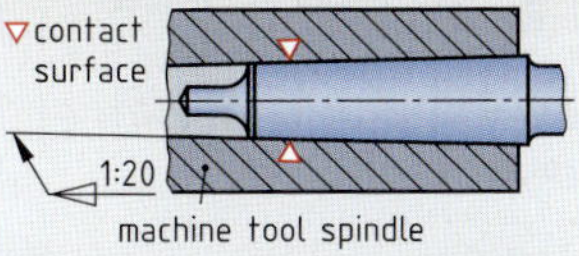 Metric taper 1: 20; Morse taper 1: 19.002 to 1: 20.047	Torque transmission: • force-fit over the taper surface + reduction sleeves fit different taper diameters – not suitable for automatic tool change	Clamping device for conventional drilling and milling. Taper shank numbers: • ME 4; 6 • MT 0; 1; 2; 3; 4; 5; 6 • ME 80; 100; 120; (140); 160; (180); 200
Steep taper shank (SK)		cf. DIN 2080-1 (1978-12) and -2 (1979-09) and DIN 69871-1 (1995-10)
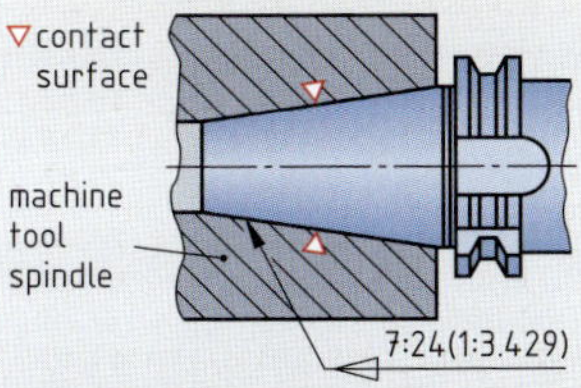 Fastening in the machine spindle: Form A: with draw-in bar Form B: by front fastener Taper 7: 24 (1: 3.429) according to DIN 254	Torque transmission: • grooves on taper edge produce interlock. The steep taper is not meant for transmission of forces, it only centers the tool. Axial locking is achieved by the thread or the ring groove. + DIN 69871-1 suitable for automatic tool change – high weight, therefore less suited for quick tool change with high axial repeating clamping accuracy and for high revolution speeds	Use with CNC machine tools, especially machining centers; less suited for high-speed cutting (HSC) Steep taper numbers: • DIN 2080-1 (form A): 30; 40; 45; 50; 55; 60; 65; 70; 75; 80 • DIN 69871-1: 30; 40; 45; 50; 60
Hollow taper shanks (designation HSK)		cf. DIN 69893-1 (2011-02)
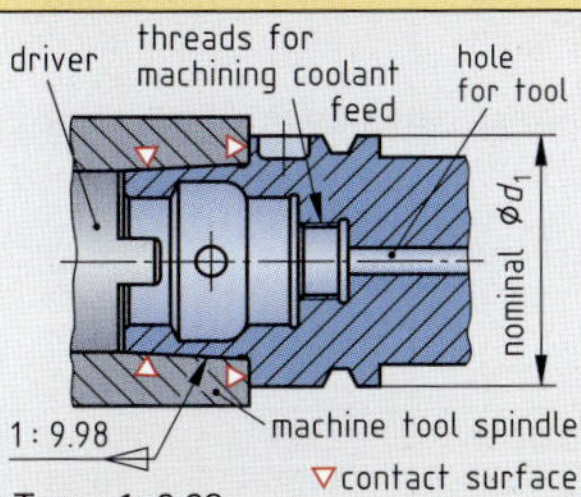 Taper 1: 9.98	Torque transmission: • force-fit using the taper and contact surfaces • drive slots on shaft end produce interlock. + low weight, therefore + high static and dynamic rigidity + high repeated clamping accuracy (3 µm) + high rotational speeds – more expensive than steep taper	Safer use with high-speed cutting Nominal sizes: d_1 = 25; 32; 40; 50; 63; 80; 100; 125; 160 mm **Form A:** with shoulder and clamping keyway for automatic and manual tool change **Form C:** only manual change is possible
Shrinkage chucks		
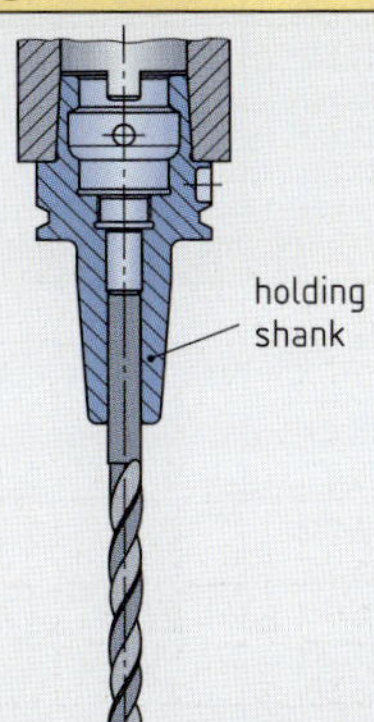 available with HSK or steep taper	Torque transmission like HSK. Clamping the tool by quick, inductive heating (approx. 340 °C) of the holding shank in the shrinkage chuck. A shrinkage joint is formed by the oversize of the tool (approx. 3–7 µm) after the joining and cooling. + transmission of high torques + high radial rigidity + higher cutting values possible + shorter machining times + good runout + greater running smoothness + better surface quality + reliable tool changes – relatively expensive – additional induction and cooling devices required	Universally applicable in machine tools with steep taper or hollow shank tool holders; suitable for tools with cylindrical shank of HSS or carbide. Shank diameters: 6; 8; 10; 12; 14; 16; 18; 20; 25 mm

Cylindrical helical tension springs

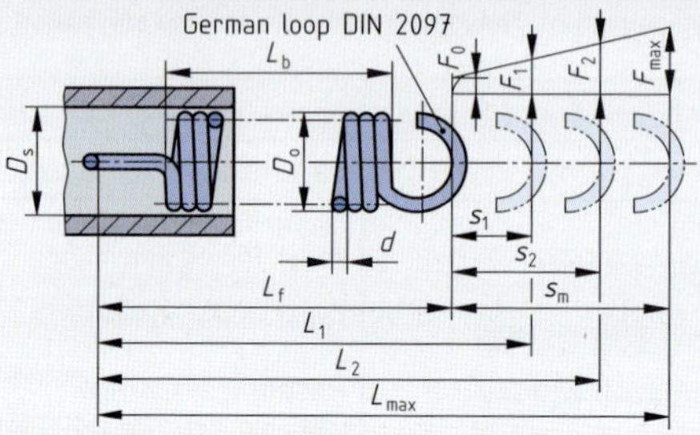

d wire diameter in mm
D_o outside coil diameter
D_s minimum sleeve diameter in mm
L_f free length, with no load on spring in mm
L_b length of spring body with no load in mm
L_{max} maximum spring length
F_0 internal prestress in N
F_{max} maximum allowable spring force in N
R spring rate in N/mm
s_m maximum allowable spring displacement for F_{max} in mm

d	D_o	D_s	L_f	L_b	F_0	F_{max}	R	s_m
Tension springs of patented drawn unalloyed spring steel wire[1)]						cf. DIN EN 10270-1 (2001-12)		
0.20	3.00	3.50	8.6	4.35	0.06	1.26	0.036	33.37
0.25	5.00	5.70	10.0	2.63	0.03	1.46	0.039	36.51
0.32	5.50	6.30	10.0	2.08	0.08	2.71	0.140	18.85
0.36	6.00	6.90	11.0	2.34	0.16	3.50	0.173	19.23
0.40	7.00	8.00	12.7	2.60	0.16	4.06	0.165	23.67
0.45	7.50	8.60	13.7	3.04	0.25	5.31	0.207	24.41
0.50	10.00	11.10	20.0	5.25	0.02	5.40	0.078	68.79
0.55	6.00	7.10	13.9	5.78	0.88	11.66	0.606	17.78
0.63	8.60	9.90	19.9	7.88	0.79	12.13	0.276	41.15
0.70	10.00	11.40	23.6	9.63	0.83	14.13	0.239	55.78
0.80	10.80	12.30	25.1	10.20	1.22	19.10	0.355	50.36
0.90	10.00	11.70	23.0	9.45	1.99	28.59	0.934	28.49
1.00	13.50	15.40	31.4	12.50	1.77	28.63	0.454	59.22
1.10	12.00	14.00	27.8	11.83	2.99	41.95	1.181	32.98
1.25	17.20	19.50	39.8	15.63	2.77	42.35	0.533	74.25
1.30	11.30	13.50	134.0	118.95	5.771	70.59	0.322	201.60
1.40	15.00	17.50	34.9	15.05	5.44	66.08	1.596	38.00
1.50	20.00	22.70	48.9	21.75	3.99	60.54	0.603	93.72
1.60	21.60	24.50	50.2	20.00	3.99	67.40	0.726	87.38
1.80	20.00	23.20	46.0	19.35	6.88	100.90	1.819	51.70
2.00	27.00	30.50	62.8	25.00	6.88	101.20	0.907	104.00
2.20	24.00	27.80	55.6	23.10	9.81	148.00	2.425	57.02
2.50	34.50	38.90	79.7	31.25	9.88	148.50	1.056	131.33
2.80	30.00	34.70	69.8	29.40	17.77	233.40	3.257	65.85
3.00	40.00	45.10	140.0	86.25	11.50	214.20	0.587	345.31
3.20	43.20	46.60	100.0	40.00	11.88	238.40	1.451	156.13
3.60	40.00	46.00	92.1	37.80	19.60	357.10	3.735	90.38
4.00	44.00	50.60	117.0	58.00	24.50	436.30	3.019	136.43
4.50	50.00	57.60	194.0	128.25	28.00	532.30	1.613	312.74
5.00	50.00	58.30	207.0	142.50	47.00	707.90	2.541	260.12
5.50	60.00	69.30	236.0	156.75	38.00	774.50	2.094	351.72
6.30	70.00	80.00	272.0	179.55	45.00	968.50	2.258	429.00
7.00	80.00	92.00	306.0	199.50	70.00	1132.00	2.286	464.83
8.00	80.00	94.00	330.0	228.00	120.00	1627.00	4.065	370.91
Tension springs of stainless steel spring steel wire[1)]						cf. DIN EN 10270-3 (2001-08)		
0.20	3.00	3.50	8.60	4.35	0.05	0.99	0.031	30.54
0.40	7.00	8.00	12.70	2.60	0.121	3.251	0.142	22.11
0.63	8.60	9.90	19.90	7.88	0.631	9.861	0.237	38.97
0.80	10.80	12.30	25.1	10.20	0.971	15.67	0.305	48.19
1.00	13.50	15.40	31.4	12.50	1.411	23.77	0.390	57.40
1.25	17.20	19.50	39.8	15.63	2.211	35.50	0.458	72.73
1.40	15.00	17.50	34.9	15.05	4.351	55.72	1.371	37.48
1.60	21.60	24.50	50.2	20.00	3.211	56.93	0.623	86.19
2.00	27.00	30.50	62.8	25.00	5.501	84.86	0.779	101.86
4.00	44.00	50.60	117.0	58.00	19.600	366.50	2.593	133.83

[1)] In addition to the springs listed, other springs with different outside diameters and lengths are commercially available for each wire diameter.

ME

Cylindrical helical compression springs

cf. DIN 2098-1 (1968-10), -2 (1970-08)

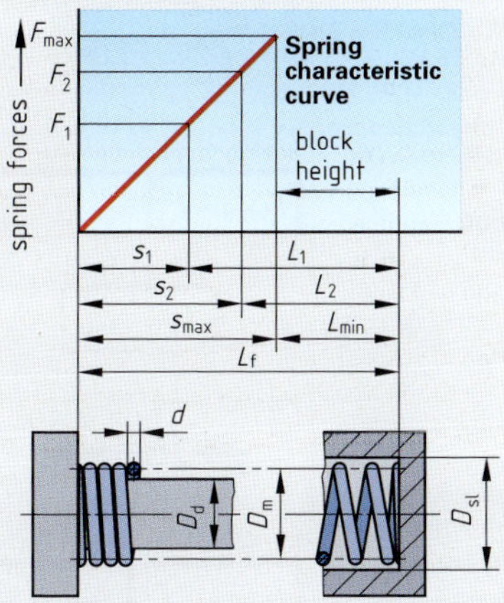

d	wire diameter
D_m	mean coil diameter
D_d	mandrel diameter
D_{sl}	sleeve diameter
L_f	free length, unloaded spring
L_1, L_2	length of loaded spring at F_1, F_2
L_{min}	minimum allowable test length of the spring
F_1, F_2	spring force at L_1, L_2
F_{max}	maximum allowable spring force at s_{max}
s_1, s_2	spring displacement at F_1, F_2
s_{max}	maximum allowable spring displacement at F_{max}
i_s	number of spring coils
i_t	total number of coils (ends ground)
R	spring rate in N/mm

Total number of coils

$$i_t = i_s + 2$$

⇒ **Compression spring DIN 2098 – 2 x 20 x 94:**
d = 2 mm, D_m = 20 mm and L_f = 94 mm

d	D_m	D_d max.	D_{sl} min.	F_{max} in N	i_s = 3.5 L_f	s_{max}	R	i_s = 5.5 L_f	s_{max}	R	i_s = 8.5 L_f	s_{max}	R	i_s = 12.5 L_f	s_{max}	R
0.2	2.5	2.0	3.1	1.00	5.4	3.8	0.26	8.2	6.0	0.17	12.4	9.3	0.11	17.9	13.7	0.07
	2	1.5	2.6	1.24	4.0	2.4	0.51	5.9	3.8	0.33	8.7	5.9	0.21	12.6	8.6	0.15
	1.6	1.1	2.1	1.50	3.0	1.5	1.0	4.4	2.4	0.65	6.4	3.6	0.42	9.2	5.4	0.28
0.5	6.3	5.3	7.5	6.6	13.5	9.2	0.73	20.0	14.0	0.46	30.0	21.3	0.30	44.0	31.8	0.21
	4	3.1	5.0	9.3	7.0	3.3	2.84	10.0	4.9	1.81	15.0	7.9	1.17	21.5	11.7	0.79
	2.5	1.7	3.4	10.4	4.4	0.9	11.6	6.1	1.4	7.43	8.7	2.2	4.80	12.0	3.0	3.27
1	12.5	10.8	14.4	22	24.0	14.6	1.49	36.5	23.1	0.95	55.5	36.1	0.61	80.5	53.1	0.41
	8	6.5	9.6	33.2	13.0	5.7	5.68	19.0	8.9	3.61	28.5	14.2	2.33	40.5	20.6	1.59
	5	3.6	6.5	43.8	8.5	1.9	23.2	12.0	3.0	14.8	17.0	4.4	9.57	24.0	6.6	6.51
1.6	20	17.5	22.6	84.9	48.0	35.6	2.38	73.5	55.9	1.52	110	84.5	0.99	165	129	0.67
	12.5	10.3	14.7	135	24.0	14.0	9.76	36.0	21.9	6.23	53.5	33.4	4.0	78.0	50.0	2.73
	8	5.9	10.1	212	14.5	5.5	37.3	21.5	8.9	23.7	31.5	13.6	15.4	45.0	20.2	10.4
2	25	22.0	28.0	128	58.0	43.0	2.98	88.5	67.1	1.90	135	104	1.23	195	151	0.83
	16	13.4	18.6	198	30.0	17.5	11.4	45.0	27.3	7.24	68.0	42.5	4.69	98	62.1	3.19
	10	7.5	12.5	318	18.0	6.8	46.6	26.5	10.9	29.7	38.5	16.5	19.2	55	24.4	13.0
2.5	32	28.3	36.0	182	71.5	52.2	3.48	110	82.1	2.22	170	129	1.43	245	187	0.97
	25	21.6	28.4	233	49.0	32.2	7.29	74.5	50.5	4.64	115	80.2	3.0	165	116	2.04
	20	16.8	23.2	292	36.0	20.5	14.2	54.0	32.1	9.05	81.5	50.0	5.86	120	75.7	3.98
	16	12.9	19.1	365	27.5	12.9	27.8	41.0	20.5	17.7	61.0	31.7	11.5	88.0	49.9	7.78
3.2	40	35.6	44.6	288	82.0	60.8	4.76	125	95.3	3.03	190	148	1.96	275	216	1.33
	32	27.6	36.5	361	58.5	38.7	9.3	88.5	61.1	5.92	135	96.2	3.82	190	136	2.61
	25	21.1	28.9	461	42.5	23.4	19.4	63.5	37.2	12.4	94.5	57.4	8.0	135	83.4	5.45
	20	16.1	23.9	577	33.5	15.0	38.2	49.5	23.6	24.2	74.0	36.9	15.7	105	53.4	10.7
4	50	44.0	56.0	427	99.0	71.6	5.95	150	111	3.79	230	175	2.45	335	257	1.65
	40	34.8	45.2	533	71.0	45.8	11.7	105	69.9	7.41	160	110	4.79	235	165	3.26
	32	27.0	37.0	666	53.5	29.5	22.8	79.5	46.2	14.4	120	72.8	9.35	170	104	6.36
	25	20.3	29.7	852	41.0	18.1	47.7	60.5	28.3	30.3	89.5	43.5	19.6	130	65.5	13.3
5	63	56.0	70.0	623	120	87.7	7.27	180	135	4.63	275	210	2.99	395	304	2.03
	50	43.0	57.0	785	85.0	54.1	14.5	130	86.8	9.25	195	133	5.98	280	194	4.07
	40	34.0	46.0	981	64.0	34.4	28.4	95.5	54.5	18.1	140	81.6	11.7	205	124	7.95
	32	26.0	38.0	1226	51.0	22.3	55.4	75.0	34.8	35.3	110	52.5	22.9	160	79.5	15.5
6.3	80	71.0	89.0	932	145	103	8.96	220	160	5.70	335	250	3.69	490	370	2.51
	63	55.0	71.5	1177	105	65.0	18.3	155	99.0	11.7	235	155	7.55	340	277	5.13
	50	42.0	58.0	1481	80.0	42.0	36.7	115	62.0	23.3	175	100	15.1	250	145	10.3
	40	32.6	47.5	1854	60.0	24.0	71.7	90.0	39.7	45.6	135	63.2	29.5	195	95.0	20.1
8	100	89.0	111	1413	170	118	11.9	260	187	7.58	390	286	4.9	570	423	3.34
	80	69.0	91.0	1766	125	76.0	23.2	180	111	14.8	285	186	9.58	410	271	6.51
	63	53.0	73.0	2237	95.0	48.0	47.0	140	74.0	30.3	205	112	19.6	300	169	13.3
	50	40.5	60.0	2825	75.0	30.0	95.4	110	46.8	60.8	160	70.0	39.2	230	103	26.7

Disc springs

cf. DIN 2093 (2006-03)

Single spring

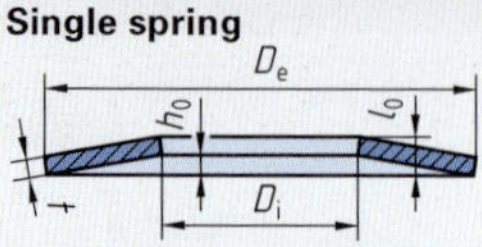

$h_0 \approx l_0 - t$

without contact surface:
Groups 1 & 2

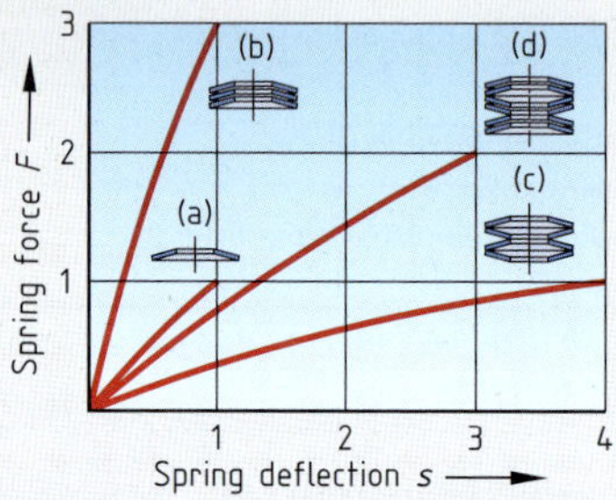

Spring force graph for various disc spring combinations: (a) single spring;
(b) parallel stack of 3 single springs: 3 times force;
(c) series stack of 4 single springs: 4-fold deflection;
(d) series stack of 3 parallel stacks with 2 single springs each: 3-fold deflection, 2-fold force

D_e	outside diameter
D_i	inside diameter
t	thickness of the single disc spring
h_0	spring height (theoretic spring displacement to flat position)
l_0	overall height of the unloaded single spring
s	spring deflection of a single spring
s_{total}	spring deflection of stack of disc springs
F	load generated by a single disc spring
F_{total}	total load generated by stack of disc springs
L_0	length of unloaded spring stack
n	number of disc springs in parallel stack
i	number of disc springs in series stack

Series stack

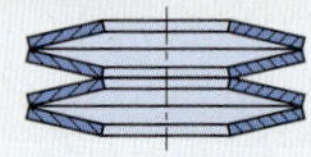

Spring force

$F_{total} = F$

Spring deflection

$s_{total} = i \cdot s$

Spring length

$L_0 = i \cdot l_0$

Parallel stack

Spring force

$F_{total} = n \cdot F$

Spring deflection

$s_{total} = s$

Spring length

$L_0 = l_0 + (n - 1) \cdot t$

ME

Group[3]	D_e h12	D_i H12	Series A: hard springs $D_e/t \approx 18$; $h_0/t \approx 0.4$				Series B: medium hard springs $D_e/t \approx 28$; $h_0/t \approx 0.75$				Series C: soft springs $D_e/t \approx 40$; $h_0/t \approx 1.3$			
			t	l_0	F in kN[1]	s[2]	t	l_0	F in kN[1]	s[2]	t	l_0	F in kN[1]	s[2]
Group. 1: $t < 1.25$ mm without contact surface	8	4.2	0.4	0.6	0.21	0.15	0.3	0.55	0.12	0.19	0.2	0.45	0.04	0.19
	10	5.2	0.5	0.75	0.33	0.19	0.4	0.7	0.21	0.23	0.25	0.55	0.06	0.23
	14	7.2	0.8	1.1	0.81	0.23	0.5	0.9	0.28	0.30	0.35	0.8	0.12	0.34
	16	8.2	0.9	1.25	1.00	0.26	0.6	1.05	0.41	0.34	0.4	0.9	0.15	0.38
	20	10.2	1.1	1.55	1.53	0.34	0.8	1.35	0.75	0.41	0.5	1.15	0.25	0.49
	25	12.2	–	–	–	–	0.9	1.6	0.87	0.53	0.7	1.6	0.60	0.68
	28	14.2	–	–	–	–	1.0	1.8	1.11	0.60	0.8	1.8	0.80	0.75
	40	20.4	–	–	–	–	–	–	–	–	1	2.3	1.02	0.98
Group 2: $t = 1.25 – 6$ mm without contact surface	25	12.2	1.5	2.05	2.93	0.41	–	–	–	–	–	–	–	–
	28	14.2	1.5	2.15	2.84	0.49	–	–	–	–	–	–	–	–
	40	20.4	2.2	3.15	6.50	0.68	1.5	2.6	2.62	0.86	–	–	–	–
	45	22.4	2.5	3.5	7.72	0.75	1.7	3.0	3.66	0.98	1.25	2.85	1.89	1.20
	50	25.4	3	4.1	12.0	0.83	2	3.4	4.76	1.05	1.25	2.85	1.55	1.20
	56	28.5	3	4.3	11.4	0.98	2	3.6	4.44	1.20	1.5	3.45	2.62	1.46
	63	31	3.5	4.9	15.0	1.05	2.5	4.2	7.19	1.31	1.8	4.15	4.24	1.76
	71	36	4	5.6	20.5	1.20	2.5	4.5	6.73	1.50	2	4.6	5.14	1.95
	80	41	5	6.7	33.6	1.28	3	5.3	10.5	1.73	2.25	5.2	6.61	2.21
	90	46	5	7.0	31.4	1.50	3.5	6	14.2	1.88	2.5	5.7	7.68	2.40
	100	51	6	8.2	48.0	1.65	3.5	6.3	13.1	2.10	2.7	6.2	8.61	2.63
	125	64	–	–	–	–	5	8.5	29.9	2.63	3.5	8	15.4	3.38
	140	72	–	–	–	–	5	9	27.9	3.00	3.8	8.7	17.2	3.68
	160	82	–	–	–	–	6	10.5	41.0	3.38	4.3	9.9	21.8	4.20
	180	92	–	–	–	–	6	11.1	37.5	3.83	4.8	11	26.4	4.65

⇒ **Disc spring DIN 2093 – A 16:** Series A, outside diameter D_e = 16 mm

[1] Spring force F of a single disc with spring deflection $s \approx 0.75 \cdot h_0$
[2] $s \approx 0{,}75 \cdot h_0$
[3] Group 3: $t > 6$–14 mm, with contact surface, D_e = 125, 140, 160, 180, 200, 225, 250 mm

Grub screws, Thrust pads, Ball knobs

Grub screws with thrust point

cf. DIN 6332 (2003-04)

Form S (M6 to M20)

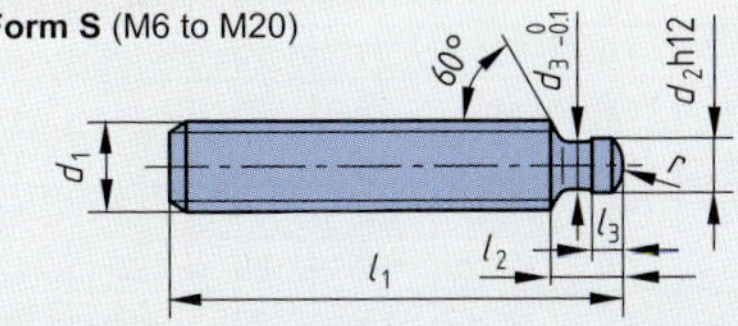

Application examples as clamping screws

with star knob[1)] DIN 6335 M6 to M20

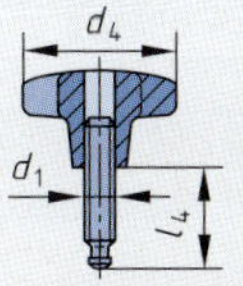

with knurled nut DIN 6303 M6 to M10

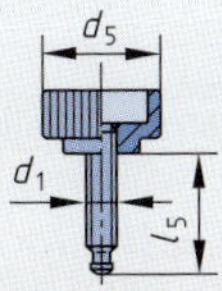

with wing nut DIN 315 M6 to M10

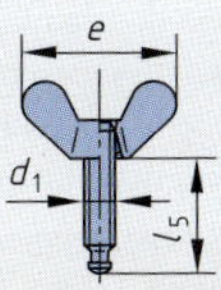

d_1	M6		M8		M10		M12			M16		
d_2	4.8		6		8		8			12		
d_3	4		5.4		7.2		7.2			11		
r	3		5		6		6			9		
l_2	6		7.5		9		10			12		
l_3	2.5		3		4.5		4.5			5		
d_4	32		40		50		63			80		
d_5	24		30		36		–			–		
e	33		39		51		65			73		
l_1	30	50	40	60	60	80	60	80	100	80	100	125
l_4	20	40	27	47	44	64	40	60	80	–	–	–
l_5	22	42	30	50	48	68	–	–	–	–	–	–

⇒ **Grub screw DIN 6332 – S M 12 x 60:** Form S with threads d_1 = M12, l_1 = 60 mm

[1)] or scallop knob DIN 6336 M6 to M16

Thrust pads

cf. DIN 6311 (2002-06)

Form S with snap ring

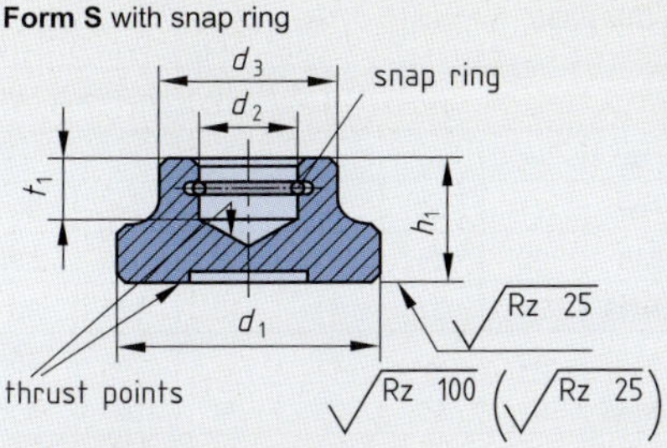

EHT (450 HV 1) 0.3 + 0.2 mm,
surface hardness 550 + 100 HV 10

d_1	d_2 H12	d_3	h_1	t_1	Snap ring DIN 7993	Grub screw DIN 6332
12	4.6	10	7	4	–	M6
16	6.1	12	9	5	–	M8
20	8.1	15	11	6	8	M10
25	8.1	18	13	7	8	M12
32	12.1	22	15	7.5	12	M16
40	15.6	28	16	8	16	M20

⇒ **Thrust pad DIN 6311 – S 40:** Form S, d_1 = 40 mm, with inserted snap ring

Ball knobs

cf. DIN 319 (2002-04)

Form C
with threads

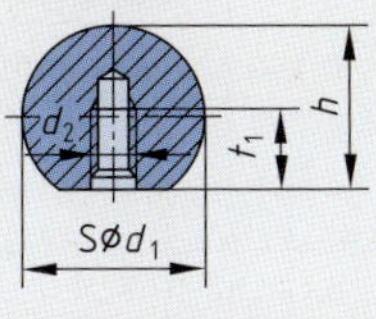

Form L
with clamping sleeve

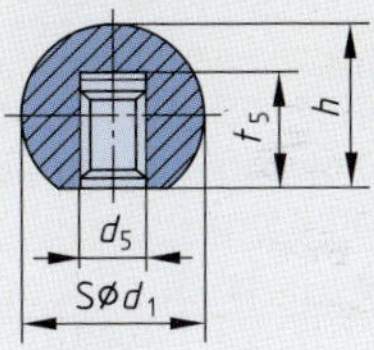

Form M
with tapered hole

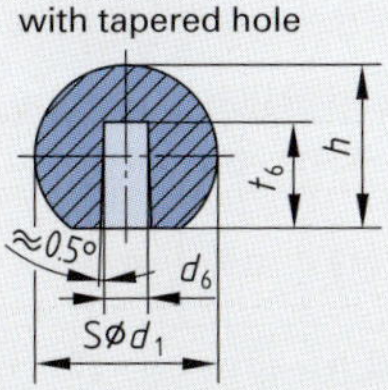

Form E
with threaded bushing

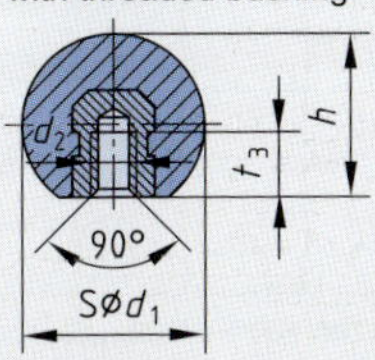

Other forms no longer standardized.

d_1	16	20	25			32			40			50		
d_2	M4	M5	M6			M8			M10			M12		
t_1	7	9	11			14.5			18			21		
t_3	6	7.5	9			12			15			18		
d_5	4	5	6	8	10	8	10	12	10	12	16	12	16	20
t_5	11	13	16	15	15	15	20	20	20	23	23	20	23	28
d_6	4	5	6	8	–	8	10	–	10	12	–	12	16	–
t_6	9	12	15	15	–	15	15	–	20	20	–	22	22	–
h	15	18	22.5			29			37			46		

⇒ **Ball knob DIN 319 – E 25 PF:** Form E, d_1 = 25 mm, of phenolic molding compound PF (thermoset plastic)

Material: Ball knob of phenolic molding compound PF (thermoset plastic); threaded bushing of steel (St) by choice of manufacturer; other materials by agreement.

Color: black

ME

Knobs, Locating and seating pins

Star knobs

cf. DIN 6335 (2008-05)

Form A

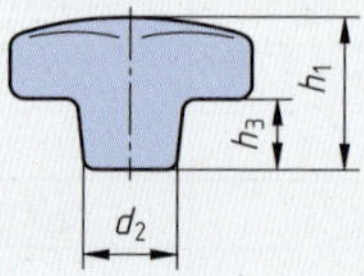

Form B

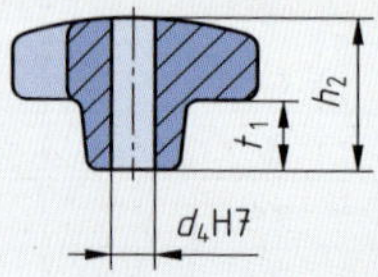

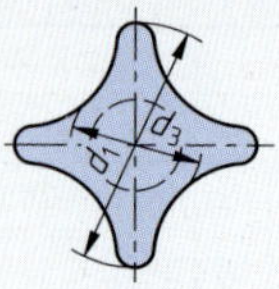

Form E

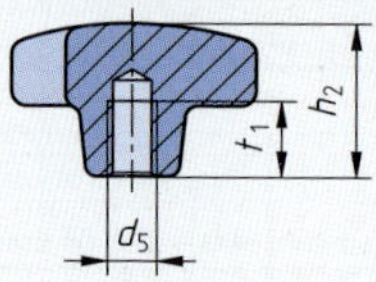

Form C

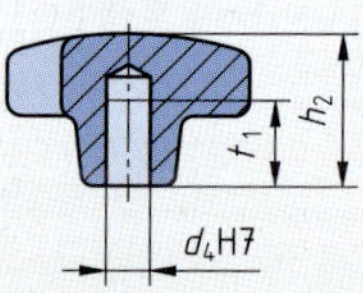

Form K

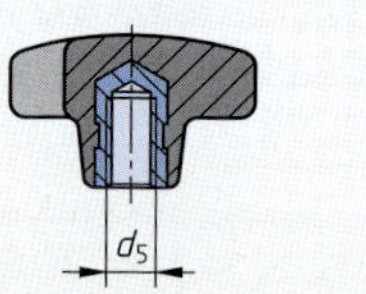

d_1	d_2	d_3	d_4	d_5	h_1	h_2	h_3	t_1
32	12	18	6	M6	21	20	10	12
40	14	21	8	M8	26	25	14	15
50	18	25	10	M10	34	32	20	18
63	20	32	12	M12	42	40	25	22
80	25	40	16	M16	52	50	30	28
100[1]	32	48	20	M20	65	60	38	36

Form	Description
A to E	Metal knobs
A	rough part of metal
B	with through bore d_4
C	with blind bore d_4
D	with through threaded bore d_5
E	with blind threaded bore d_5
K[2]	of molding mat. (plastic) with threaded bushing d_5 (of metal)
L[2]	of molding material (plastic) with threaded pin d_5 (of metal)
⇒	**Star knob DIN 6335 – A 50 AL:** Form A, d_1 = 50 mm, of aluminum

[1] This size is not available in molding material.
[2] Sometimes with insignificant other dimensions; material like fluted knobs DIN 6336

Fluted knobs

cf. DIN 6336 (2008-05)

Form A

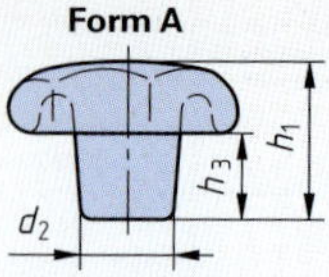

Form E

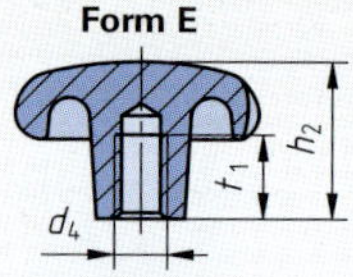

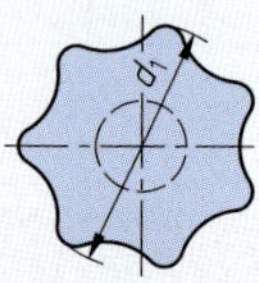

Form L

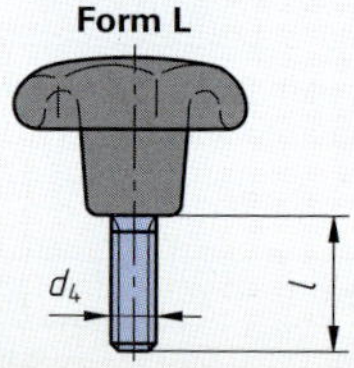

ME

d_1	d_2	d_4	h_1	h_2	h_3	t_1	l	
32	12	M6	21	20	10	12	20	30
40	14	M8	26	25	13	15	20	30
50	18	M10	34	32	17	18	25	30
63	20	M12	42	40	21	22	30	40
80	25	M16	52	50	25	28	30	40

⇒ **Fluted knob DIN 6336 – L 40 x 30:** Form L (molding material) d_1 = 40 mm, l = 30 mm

Forms A to E (metal knobs) as well as K and L (knobs of molding material) correspond to star knobs DIN 6335.

Materials: Cast iron, aluminum, phenolic molding compound (PF) or polyamide (PA)

Locating and seating pins

cf. DIN 6321 (2002-10)

Form A Seating pin

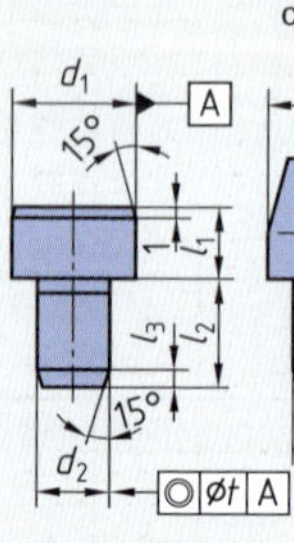

hardened 53 + 6 HRC

Form B Locating pin cylindrical

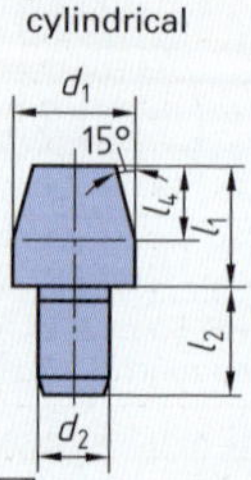

Form C Locating pin truncated

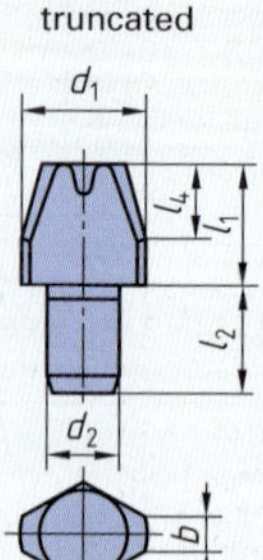

d_1 g6	l_1 Form A h9	l_1 Form B and C short	l_1 Form B and C long	b	d_2[1] n6	l_2	l_3	l_4	t
6	5	7	12	1	4	6	1.2	4	0.02
8	–	10	16	1.6	6	9	1.6	6	
10	6		18	2.5					
12	–								
16	8	13	22	3.5	8	12	2	8	0.04
20	–	15	25	5	12	18	2.5	9	
25	10								

⇒ **Clevis pins DIN 6321 – C 20 x 25:** Form C, d_1 = 20 mm, l_1 = 25 mm

[1] Appropriate bore tolerance: H7

T-slots and accessories, Spherical washers, Conical seats

T-slots and nuts for T-slots

cf. DIN 650 (1989-10) and 508 (2002-06)

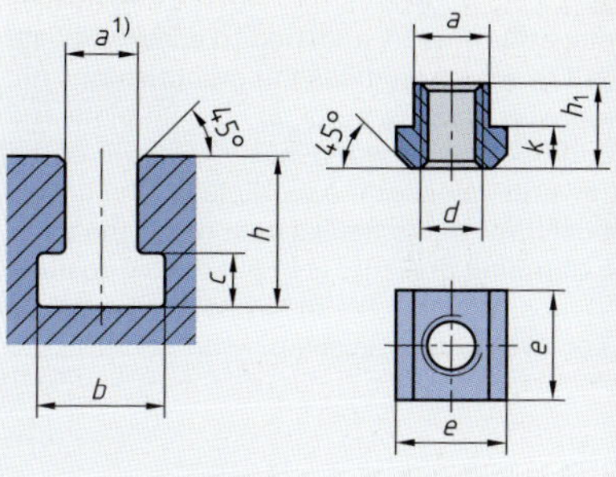

1) Tolerance class H8 for pilot T-slots and clamping slots; H12 for clamping slots

Width a	8	10	12	14	18	22	28	36	42
Deviation from a	−0.3/−0.5		−0.3/−0.6					−0.4/−0.7	
b	14.5	16	19	23	30	37	46	56	68
Deviation from b	1.5/0	+2/0				+3/0	+4/0		
c	7	7	8	9	12	16	20	25	32
Deviation from c	+1/0			+2/0				+3/0	
h max.	18	21	25	28	36	45	56	71	85
h min.	15	17	20	23	30	38	48	61	74
Thread d	M6	M8	M10	M12	M16	M20	M24	M30	M36
e	13	15	18	22	28	35	44	54	65
h_1	10	12	14	16	20	28	36	44	52
k	6	6	7	8	10	14	18	22	26
Deviation from k	0/−0.5					0/−1			
⇒	**Nut DIN 508 – M10 x 12:** d = M10, a = 12 mm								

Bolts for T-slots

cf. DIN 787 (2005-02)

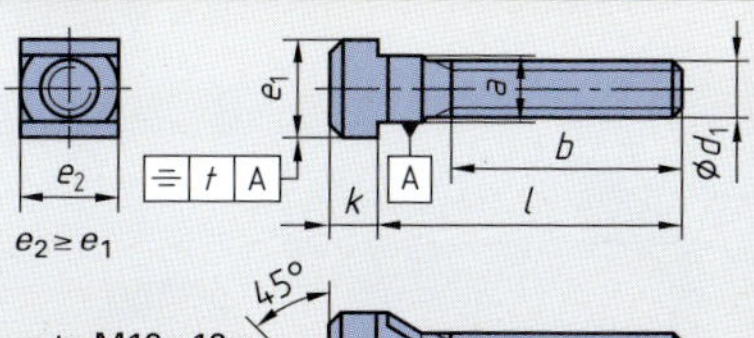

$e_2 \geq e_1$

up to M12 x 12: $a \leq d_1$

M12 x 14 and up: $a > d_1$

d_1	M8	M10	M12		M16	M20	M24	M30
a	8	10	12	14	18	22	28	36
b from	22	30	35		45	55	70	80
b to	50	60	120		150	190	240	300
e_1	13	15	18	22	28	35	44	54
h_1	12	14	16	20	24	32	41	50
k	6	6	7	8	10	14	18	22
Nominal lengths l	25, 32, 40, 50, 63, 80, 100, 125, 160, 200, 250, 315, 400, 500 mm							
⇒	**Bolt DIN 787 – M10 x 10 x 100 – 8.8:** d_1 = M10, a = 10 mm, l = 100 mm, property class 8.8							

Loose slot tenons

vgl. DIN 6323 (2003-08)

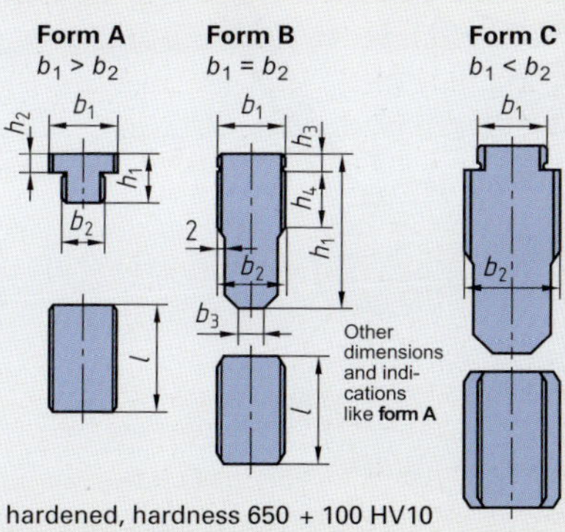

hardened, hardness 650 + 100 HV10

b_1 h6	b_2 h6	Form	b_3	h_1	h_2	h_3	h_4	l
12	6	A	–	12	3.6	–	–	20
	8							
	10							
	12	B	5	28.6	–	5.5	9	20
20	12	A	–	14	5.5	–	–	32
	14							
	18							
	22	C	9	50.5	–	7	18	40
	28		12	61.5			24	
	36		16	76.5			30	50
	42		19	90.5			36	
⇒	**Slot tenon DIN 6323 – C 20 x 28:** Form C, b_1 = 20 mm, b_2 = 28 mm							

Spherical washers and conical seats

cf. DIN 6319 (2001-10)

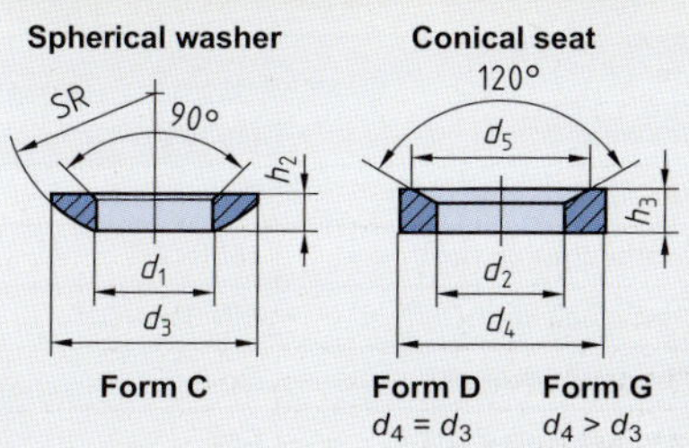

d_1 H13	d_2 H13	d_3	d_4 Form D	d_4 Form G	d_5	h_2	h_3 Form D	h_3 Form G	R Sphere
6.4	7.1	12	12	17	11	2.3	2.8	4	9
8.4	9.6	17	17	24	14.5	3.2	3.5	5	12
10.5	12	21	21	30	18.5	4	4.2	5	15
13	14.2	24	24	36	20	4.6	5	6	17
17	19	30	30	44	26	5.3	6.2	7	22
21	23.2	36	36	50	31	6.3	7.5	8	27
⇒	**Spherical washer DIN 6319 – C 17:** Form C, d_1 = 17 mm								

ME

Quick-release drilling fixture

cf. DIN 6348 (1971-01)

Quick-release drilling fixture

The fixture is available in nine standard sizes and allows the user to clamp even small workpieces quickly and precisely for drilling operations. The jig plate (5) and base plate (4) must be adjusted to accommodate the workpiece. The jig plate, which is fixed to the two guide posts (7), accommodates the drill bushings (6). Commonly, the workpiece is fixed in the accurate drilling position with locating bolts (3). The jig plate and base plate can be exchanged quickly to prepare the fixture for another type of workpiece. Clamping of the workpiece is achieved by lowering the lever. Turning the lever upwards releases the workpiece. The helical pinion shaft (10) is fitted with opposite tapers on each end. The axial force of the helical gear drags the taper of the pinion shaft into the internal taper of the housing when the clamping pressure is applied. This ensures safe and reliable clamping, even under vibrations. The opposite taper ensures the safe position of the jig plate when the workpiece is released. Clamping could also be achieved hydraulically or with compressed air.

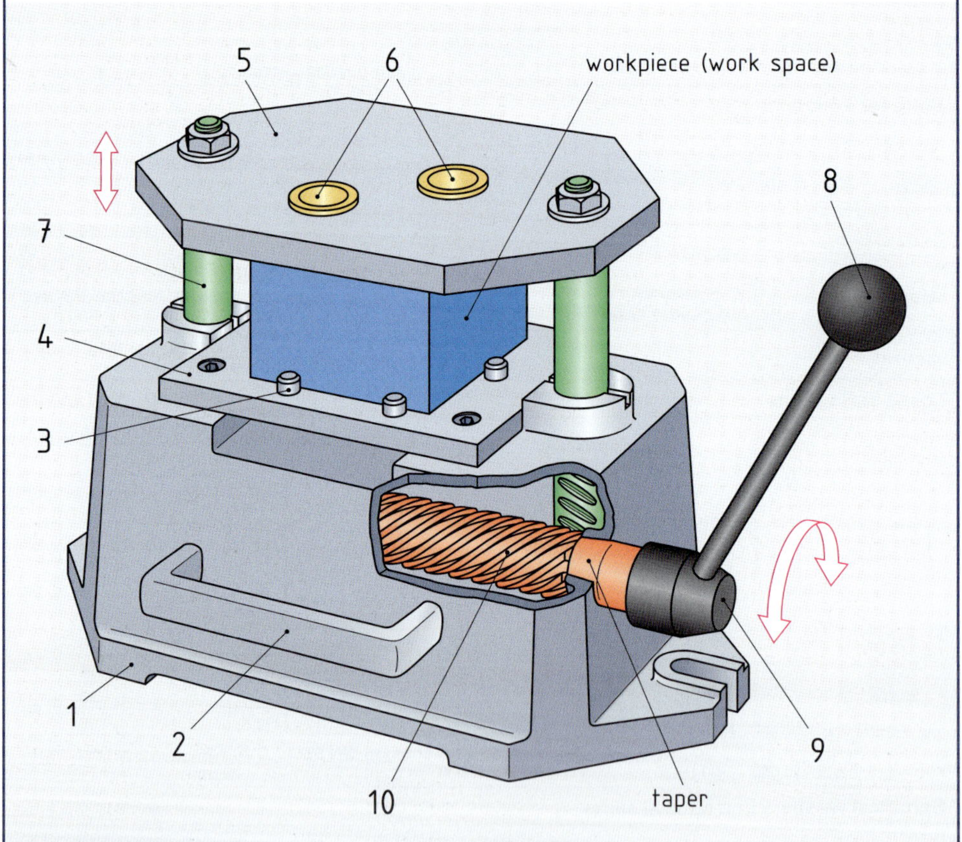

ME

Part list (excerpt)

Item	Designation	Standard/material	Item	Designation	Standard/material
1	Body	EN-GJL 250	6	Headed drill bushing	DIN 172 - A
2	Handle	AlMg3	7	Toothed guide post	16MnCr5
3	Locating bolt	DIN 6321 - A	8	Plastic ball knob	DIN 319 - C
4	Base plate	DIN 6348 - A	9	Clamping lever	E295
5	Jig plate	DIN 6348 - B	10	Toothed pinion shaft	C45

Standards parts for fixtures

<table>
<tr><th>Illustration</th><th>Dimensions from – to in mm</th><th>Material, standard</th><th>Function, properties</th></tr>
<tr><td colspan="4">Drill bushings</td></tr>
<tr><td>Form A
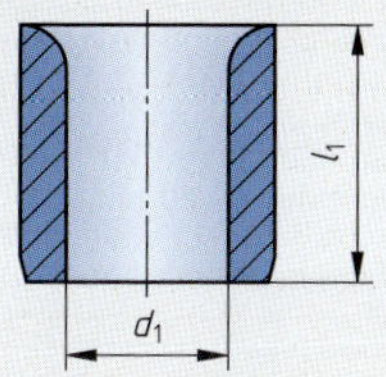</td><td>d_1 = 0.4 to 48.0 mm
graduation: 0.1 mm

d_1 > 15 mm
graduation:
0.5 mm in addition

l_1 is available in three lengths matched to d_1: short, medium, long</td><td>Tools steel
hardness:
740 + 80
HV 10

DIN 179
(withdrawn)</td><td>• Drill bushings
to guide twist drills, countersinks, step drills
• Head drill bushings
also used as tool stop
• Slip-type drill bushings
for workpieces, where counter-boring or countersinking is required after drilling</td></tr>
<tr><td colspan="4">Jig plates</td></tr>
<tr><td>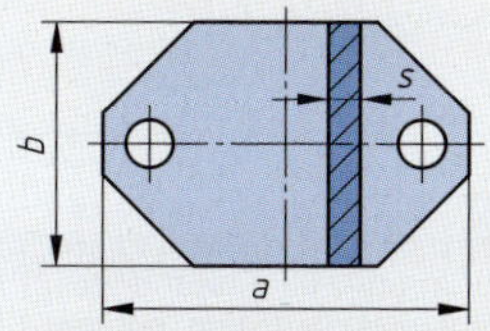</td><td><table><tr><th colspan="3">Sizes</th></tr><tr><th>a</th><th>b</th><th>s</th></tr><tr><td>60</td><td>32</td><td>8</td></tr><tr><td>80</td><td>50</td><td>11</td></tr><tr><td>100</td><td>60</td><td>14</td></tr><tr><td>100</td><td>125</td><td>16</td></tr><tr><td>200</td><td>160</td><td>16</td></tr><tr><td>300</td><td>190</td><td>20</td></tr><tr><td>400</td><td>215</td><td>27</td></tr></table></td><td>Structural steel, black-finished

DIN 6348</td><td>The jig plate, which is fixed to the two guide posts, accommodates the drill bushings. A lever mechanism allows the clamping and releasing of the workpiece.</td></tr>
<tr><td colspan="4">Fixture bases</td></tr>
<tr><td>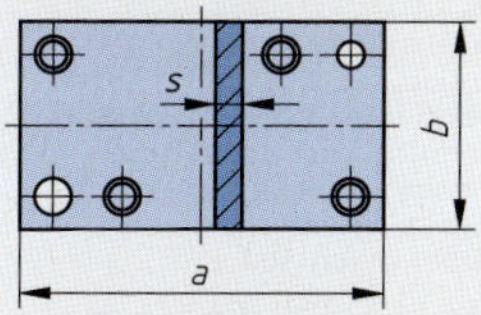</td><td><table><tr><th colspan="3">Sizes</th></tr><tr><th>a</th><th>b</th><th>s</th></tr><tr><td>60</td><td>32</td><td>6</td></tr><tr><td>80</td><td>50</td><td>10</td></tr><tr><td>100</td><td>60</td><td>10</td></tr><tr><td>100</td><td>125</td><td>10</td></tr><tr><td>200</td><td>160</td><td>15</td></tr><tr><td>300</td><td>190</td><td>15</td></tr><tr><td>400</td><td>215</td><td>18</td></tr></table></td><td>Structural steel, black-finished

DIN 6348</td><td>Supports and holds the workpiece in its position by means of stops, locating bolts or positioning pins.</td></tr>
<tr><td colspan="4">Toggle locators with external thread</td></tr>
<tr><td>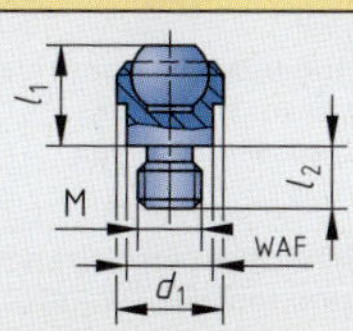</td><td>Thread:
M8 to M20
d_1 = 13 to 50 mm
l_1 = 13 to 35 mm
l_2 = 8 to 20 mm</td><td>Body:
quenched & tempered steel

Ball:
ball-bearing steel (hardened)</td><td>• Stops
• Supports
• Plungers

Particularly suited for slanted or unmachined workpiece surfaces</td></tr>
<tr><td colspan="4">Spring plungers</td></tr>
<tr><td>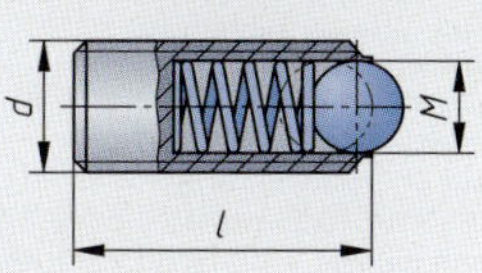</td><td>Thread:
M3 to M24
l = 9 to 48 mm
d = 1.5 to 15 mm</td><td>Sleeve:
steel strength class 5.8

Ball:
hardened spring steel

Spring:
spring steel</td><td>• Indexing
• Locking
• Positioning
• Thrust pins, draw-off pins</td></tr>
<tr><td colspan="4">Feet</td></tr>
<tr><td>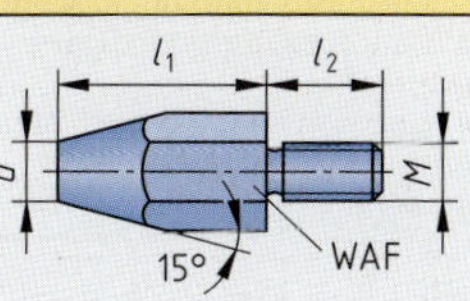</td><td>Thread:
M6 to M12
l_1 = 10 to 50 mm
l_2 = 11 to 20 mm
width across flats (WAF) = 10 to 19 mm</td><td>Quenched & tempered steel

DIN 6320</td><td>Feet are primarily used for fixtures. They can be of great advantage when higher distances (up to 50 mm) are required.</td></tr>
</table>

ME

V-belts, Positive drive belts

Design types

Designation Standard for the belts	Range of dimensions h[1] in mm	L[2] in mm	Speed range v_{max} in m/s	Power range P'_{max} in kW[3]	Properties, application
	Standard for pulleys				
Classic V-belts 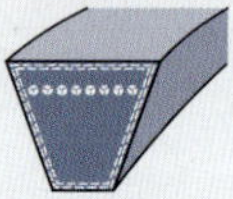	4–25	185–19000	30	65	For higher maximum tensile strengths, reliable tractive power; construction equipment, variable drives for the mining industry, agricultural machinery, conveyors, general machine construction
DIN 2215, ISO 4184	DIN 2217, ISO 4183				
Narrow V-belts 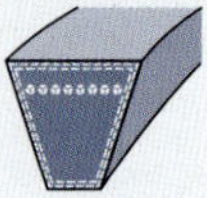	8–18	630–12500	40	70	Good power transmission, twice the power with the same width as classic V-belts; gearbox manufacturing, machine tools, HVAC
DIN 7753, ISO 4184	DIN 2211, ISO 4183				
Cogged V-belts 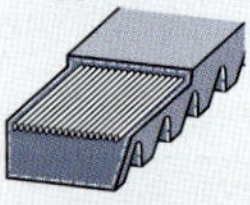	4–25	800–3150	50	70	Low elongation, small pulley diameter, high temperature resistance from −30 °C to +80 °C; automotive alternator drives, transmission design, pumps, HVAC
DIN 2215, DIN 7753	DIN 2211, DIN 2217				
Joined V-belts (Power Band) 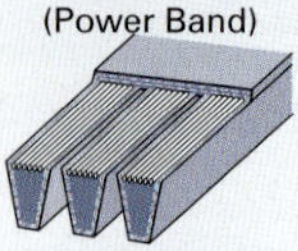	10–26	1250–15000	30	65	Insensitive to vibration or impact, no twisting of single belts in the pulleys, absolutely uniform force distribution, high tensile strength, for long distances between axles; paper machines
	DIN 2211, DIN 2217				
V-ribbed belts (ribbed belts) 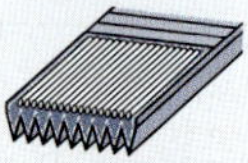	3–17	600–15000	60	20	Large transmission ratios possible, low vibration running behavior; automotive alternator drives, compressor drives in HVAC, small machines
DIN 7867	DIN 7867				
Wide V-belts 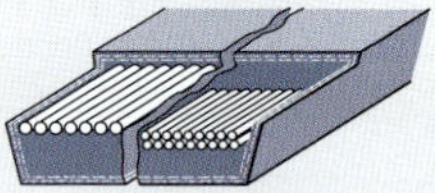	6–18	468–2500	30	85	Excellent transverse strength, very high tensile strength, flexible; speed control gears, machine tools, textile machines, printing machines, agricultural machinery
DIN 7719	DIN 7719				
Double V-belts (Hexagonal belts) 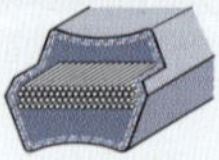	10–25	2000–6900	30	20	Good power transmission for drives with several pulleys and alternating direction of rotation, 10% less efficiency than classic V-belts; agricultural machinery, textile machines, general machine building
DIN 7722, ISO 5289	DIN 2217				
Positive drive belts 	0.7–5.0	100–3620	40–80	0.5–900	Efficiency $\eta_{max} \geq 0.98$, synchronous running, low prestress forces, therefore lower bearing load; precision machine drives, office machine drives, automotive industry, CNC spindle drives
DIN 7721, DIN ISO 5296	DIN ISO 5294				

[1] Belt height (pages 251, 252) [2] Belt length [3] Transmittable power per belt

ME

Narrow V-belts

Narrow V-belts DIN 7753-1 (1988-01)	Narrow V-belt pulley DIN 2211-1 (1984-03)
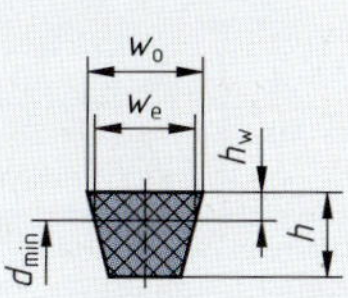	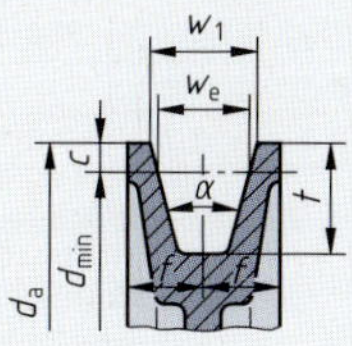

Effective diameter $d_{min} = d_a - 2 \cdot c$

→ **Narrow V-belt DIN 7753 – XPZ 710:**
Narrow V-belt, cogged profile, reference length 710 mm

Designations		Narrow V-belts, V-belt pulleys			
	Belt profile (ISO designation codes)	SPZ	SPA	SPB	SPC
w_u w_e	upper belt width effective width	9.7 8.5	12.7 11	16.3 14	22 19
h h_w	belt height distance	8 2	10 2.8	13 3.5	18 4.8
d_{min} w_1	minimum allowable effective ∅ upper groove width	63 9.7	90 12.7	140 16.3	224 22
c t	distance from effective ∅ to outer ∅ minimum allowable groove depth	2 11	2.8 13.8	3.5 17.5	4.8 23.8
e f	groove spacing for multi-groove pulleys groove spacing from outer edge	12 8	15 10	19 12.5	25.5 17
α	34° for effective ∅ up to	80	118	190	315
	38° for effective ∅ over	80	118	190	315

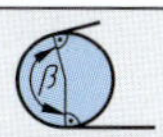

Angle factor c_1	1	1.02	1.05	1.08	1.12	1.16	1.22	1.28	1.37	1.47
Wrap angle β	180°	170°	160°	150°	140°	130°	120°	110°	100°	90°

Service factor c_2

Daily operating time in hours: up to 10	from 10 to 16	over 16	Driven machines (examples)
1.0	1.1	1.2	Centrifugal pumps, fans, conveyor belts for light material
1.1	1.2	1.3	Machine tools, presses, sheet metal shearers, printing machines
1.2	1.3	1.4	Grinding gears, piston pumps, textile and paper machines
1.3	1.4	1.5	Stone crushers, mixers, winches, cranes, excavators

Efficiency values for narrow V-belts

cf. DIN 7753-2 (1976-04)

Belt profile	SPZ			SPA			SPB			SPC		
smaller pulley d_{min}	63	100	180	90	160	250	140	250	400	224	400	630
smaller pulley n_s	Power rating P_{rated} in kW per belt											
400	0.35	0.79	1.71	0.75	2.04	3.62	1.92	4.86	8.64	5.19	12.56	21.42
700	0.54	1.28	2.81	1.17	3.30	5.88	3.02	7.84	13.82	8.13	19.79	32.37
950	0.68	1.66	3.65	1.48	4.27	7.60	3.83	10.04	17.39	10.19	24.52	37.37
1450	0.93	2.36	5.19	2.02	6.01	10.53	5.19	13.66	22.02	13.22	29.46	31.74
2000	1.17	3.05	6.63	2.49	7.60	12.85	6.31	16.19	22.07	14.58	25.81	–
2800	1.45	3.90	8.20	3.00	9.24	14.13	7.15	16.44	9.37	11.89	–	–

Profile selection for narrow V-belts

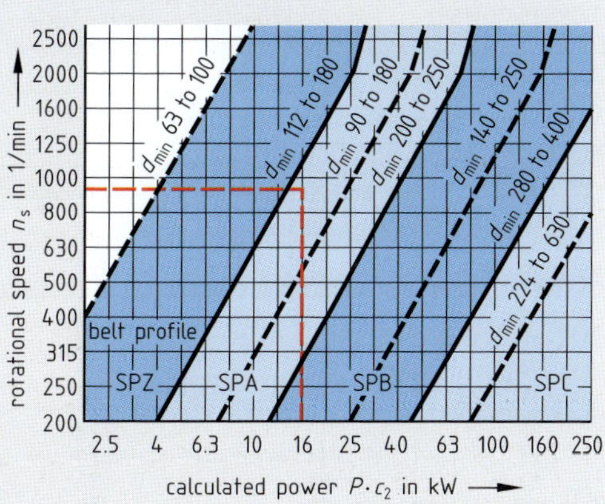

P power to be transmitted
P_{rated} power rating per belt
N number of belts
c_1 angle factor
c_2 service factor

Number of belts

$$N = \frac{P \cdot c_1 \cdot c_2}{P_{rated}}$$

Example:

Transmission parameters P = 12 kW with c_1 = 1.12; c_2 = 1.4; d_{min} = 160 mm, n_s = 950 1/min; β_s = ?, N = ?

1. $P \cdot c_2$ = 12 kW · 1.4 = 16.8 kW
2. From the diagram n_s = 950 1/min and $P \cdot c_2$ = 16.8 kW → profile **SPA**
3. P_{rated} = 4.27 kW from the table
4. $N = \frac{P \cdot c_1 \cdot c_2}{P_{rated}} = \frac{12\ \text{kW} \cdot 1.12 \cdot 1.4}{4.27\ \text{kW}} = 4.4$
5. Selected: N = **5 belts**

ME

Positive drive belts

Positive drive belts (timing belts)

cf. DIN 7721-1 (1989-06)

Single-sided

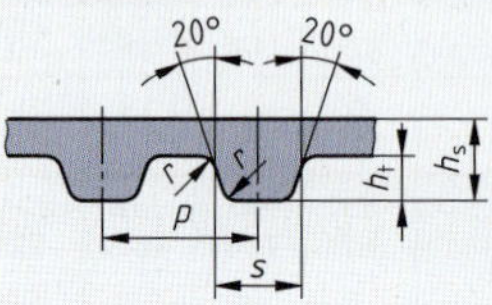

Double-sided

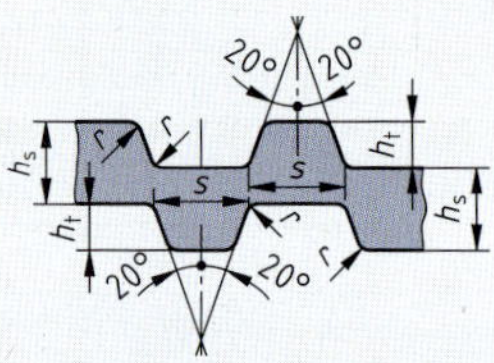

Non-standardized tooth forms

HT profile

LAHN profile

Tooth spacing: Code	Tooth spacing: p	Tooth size: s	Tooth size: h_t	Tooth size: r	Nominal thickness h_s	Positive drive belt width w			
T2.5	2.5	1.5	0.7	0.2	1.3	–	4	6	10
T5	5	2.7	1.2	0.4	2.2	6	10	16	25
T10	10	5.3	2.5	0.6	4.5	16	25	32	50

Effective length 1)	No. of teeth for T2.5	No. of teeth for T5	Effective length 1)	No. of teeth for T5	No. of teeth for T10	Effective length 1)	No. of teeth for T10
120	48	–	530	–	53	1010	101
150	–	30	560	112	56	1080	108
160	64	–	610	122	61	1150	115
200	80	40	630	126	63	1210	121
245	98	49	660	–	66	1250	125
270	–	54	700	–	70	1320	132
285	114	–	720	144	72	1390	139
305	–	61	780	156	78	1460	146
330	132	66	840	168	84	1560	156
390	–	78	880	–	88	1610	161
420	168	84	900	180	–	1780	178
455	–	91	920	184	92	1880	188
480	192	96	960	–	96	1960	196
500	200	100	990	198	–	2250	225

⇒ **Belt DIN 7721 – 6 T2.5 x 480:** w = 6 mm, spacing p = 2.5 mm, effective length = 480 mm, single-sided

The code letter D is added for double-sided positive drive belts.

1) Effective lengths from 100–3620 mm, in custom-made products up to 25 000 mm

Timing belt pulleys

cf. DIN 7721-2 (1989-06)

Pulley groove dimensions

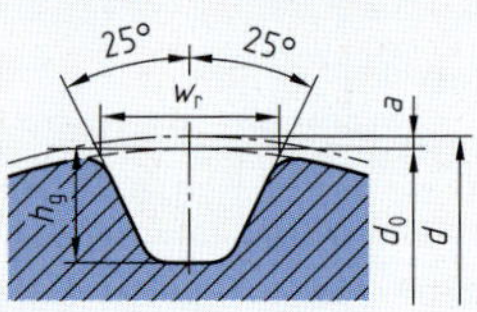

Effective diameter

$$d = d_0 + 2 \cdot a$$

1) Form SE for ≤ 20 grooves
2) Form N for > 20 grooves

Pully dimensions

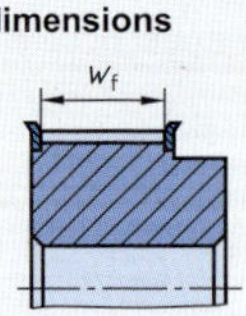

with pulley flange

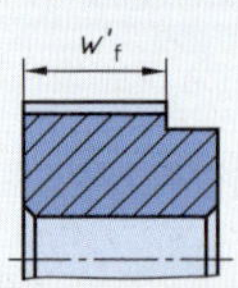

without pulley flange

Pulley groove	Pulley outer Ø d_0 for T2.5	T5	T10	Pulley groove	Pulley outer Ø d_0 for T2.5	T5	T10	Pulley groove	Pulley outer Ø d_0 for T2.5	T5	T10
10	7.4	15.0	–	17	13.0	26.2	52.2	32	24.9	50.1	100.0
11	8.2	16.6	–	18	13.8	27.8	55.4	36	28.1	56.4	112.7
12	9.0	18.2	36.3	19	14.6	29.4	58.6	40	31.3	62.8	125.4
13	9.8	19.8	39.5	20	15.4	31.0	61.8	48	37.7	75.5	150.9
14	10.6	21.4	42.7	22	17.0	34.1	68.2	60	47.2	94.6	189.1
15	11.4	23.0	45.9	25	19.3	38.9	77.7	72	56.8	113.7	227.3
16	12.2	24.6	49.1	28	21.7	43.7	87.2	84	66.3	132.9	265.5

Code	Pulley groove dimensions: Groove width w_r Form SE 1)	Groove width w_r Form N 2)	Groove height h_g Form SE 1)	Groove height h_g Form N 2)	2 a
T2.5	1.75	1.83	0.75	1	0.6
T5	2.96	3.32	1.25	1.95	1
T10	6.02	6.57	2.6	3.4	2

Letter symbols	Belt width w	Pulley width with flange w_f	Pulley width without flange w'_f
T2.5	4	5.5	8
	6	7.5	10
	10	11.5	14
T5	6	7.5	10
	10	11.5	14
	16	17.5	20
	25	26.5	29
T10	16	18	21
	25	27	30
	32	34	37
	50	52	55

ME

Straight-toothed spur gears

Unmodified spur gears with straight teeth

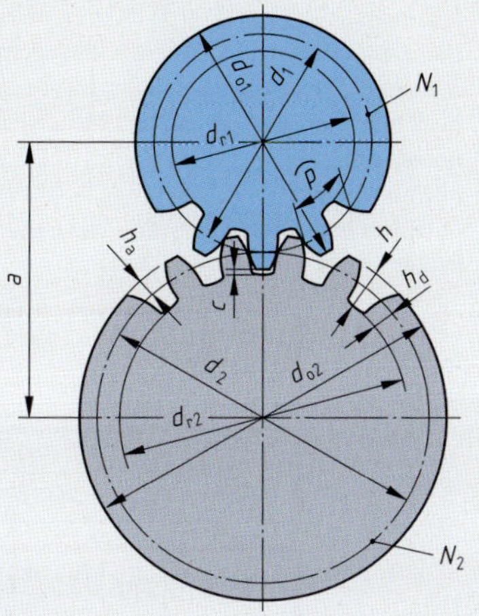

m	module	N, N_1, N_2	no. of teeth
p	pitch	d, d_1, d_2	pitch diameter
c	clearance	d_o, d_{o1}, d_{o2}	outside diameter
h	whole depth	d_r, d_{r1}, d_{r2}	root diameter
h_a	addendum		
h_d	dedendum		
a	center distance		

Example:

External spur gear,
m = 2 mm; N = 32; c = 0.167 · m; d = ?; d_o = ?; h = ?

d = $m \cdot N$ = 2 mm · 32 = **64 mm**

d_o = $d + 2 \cdot m$ = 64 mm + 2 · 2 mm = **68 mm**

h = $2 \cdot m + c$ = 2 · 2 mm + 0.167 · 2 mm = **4.33 mm**

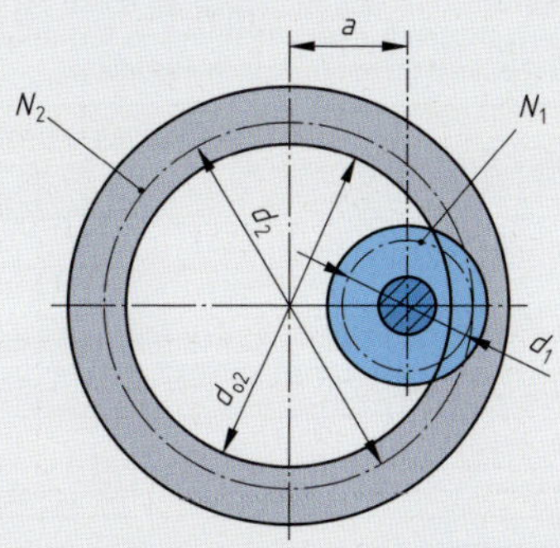

External teeth

Number of teeth

$$N = \frac{d}{m} = \frac{d_o - 2 \cdot m}{m}$$

Outside diameter

$$d_o = d + 2 \cdot m = m \cdot (N + 2)$$

Root diameter

$$d_r = d - 2 \cdot (m + c)$$

Center distance

$$a = \frac{d_1 + d_2}{2} = \frac{m \cdot (N_1 + N_2)}{2}$$

External and internal teeth

Module

$$m = \frac{p}{\pi} = \frac{d}{N}$$

Pitch

$$p = \pi \cdot m$$

Pitch diameter

$$d = m \cdot N$$

Clearance

$$c = 0.1 \cdot m \text{ to } 0.3 \cdot m$$
often $c = 0.167 \cdot m$

Addendum

$$h_a = m$$

Dedendum

$$h_d = m + c$$

Whole depth

$$h = 2 \cdot m + c$$

Internal teeth

Number of teeth

$$N = \frac{d}{m} = \frac{d_o + 2 \cdot m}{m}$$

Outside diameter

$$d_o = d + 2 \cdot m = m \cdot (N + 2)$$

Root diameter

$$d_r = d - 2 \cdot (m + c)$$

Center distance

$$a = \frac{d_2 - d_1}{2} = \frac{m \cdot (N_2 - N_1)}{2}$$

Example:

Internal spur gear, m = 1.5 mm; N = 80;
c = 0.167 · m; d = ?; d_o = ?; h = ?

d = $m \cdot N$ = 1.5 mm · 80 = **120 mm**

d_o = $d - 2 \cdot m$ = 120 mm − 2 · 1.5 mm = **117 mm**

h = $2 \cdot m + c$ = 2 · 1.5 mm + 0.167 · 1.5 mm = **3.25 mm**

Helical gears, Module series for spur gears

Unmodified helical gears

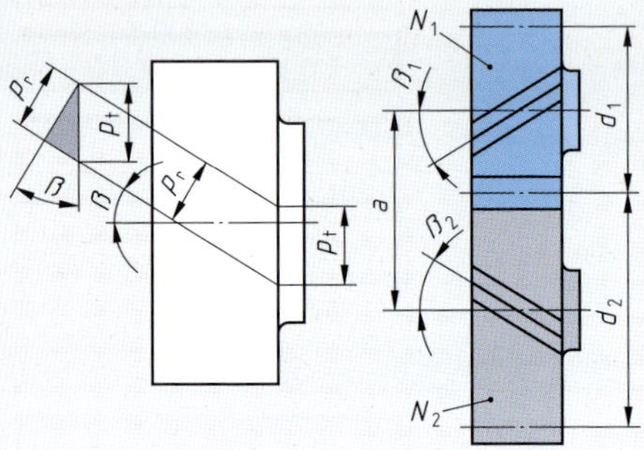

In helical gears the teeth run in a screw-like pattern on the cylindrical wheel body. The tools for manufacturing spur gears and helical gears conform to the real pitch module.

In the case of parallel shafts the two gears have the same helix angle, but opposite direction of rotation, i.e., one gear has a right-hand helix and the other a left-hand helix ($\beta_1 = -\beta_2$).

Example:

Helical gear, $N = 32$; $m_r = 1.5$ mm;
$\beta = 19.5°$; $c = 0.167 \cdot m$; $m_t = ?$; $d_o = ?$; $d = ?$; $h = ?$

$$\boldsymbol{m_t} = \frac{m_r}{\cos\beta} = \frac{1.5\ \text{mm}}{\cos 19.5°} = \mathbf{1.591\ mm}$$

$\boldsymbol{d_o} = d + 2 \cdot m_r = 50.9\ \text{mm} + 2 \cdot 1.5\ \text{mm} = \mathbf{53.9\ mm}$

$\boldsymbol{d} = m_t \cdot N = 1.591\ \text{mm} \cdot 32 = \mathbf{50.9\ mm}$

$\boldsymbol{h} = 2 \cdot m_r + c = 2 \cdot 1.5\ \text{mm} + 0.167 \cdot 1.5\ \text{mm} = \mathbf{3.25\ mm}$

m_t	transverse module
m_r	real pitch module
p_t	transverse pitch
p_r	real pitch
β	helix angle (normally $\beta = 8°$ to $25°$)
N, N_1, N_2	no. of teeth
d, d_1, d_2	pitch diameter
d_o	outside diameter
a	center distance

Transverse module

$$m_t = \frac{m_r}{\cos\beta} = \frac{p_t}{\pi}$$

Transverse pitch

$$p_t = \frac{p_r}{\cos\beta} = \frac{\pi \cdot m_r}{\cos\beta}$$

Pitch diameter

$$d = m_t \cdot N = \frac{N \cdot m_r}{\cos\beta}$$

Number of teeth

$$N = \frac{d}{m_t} = \frac{\pi \cdot d}{p_t}$$

Real pitch module

$$m_r = \frac{p_r}{\pi} = m_t \cdot \cos\beta$$

Real pitch

$$p_r = \pi \cdot m_r = p_t \cdot \cos\beta$$

Outside diameter

$$d_o = d + 2 \cdot m_r$$

Center distance

$$a = \frac{d_1 + d_2}{2}$$

Calculations of whole depth, addendum, dedendum, clearance and root diameter are the same as those for spur gears with straight teeth (page 253). In the formulae the module m is replaced by the real pitch module m_r.

Module series for spur gears (Series I)

cf. DIN 780-1 (1977-05)

Module	0.2	0.25	0.3	0.4	0.5	0.6	0.7	0.8	0.9	1.0	1.25
Pitch	0.628	0.785	0.943	1.257	1.571	1.885	2.199	2.513	2.827	3.142	3.927
Module	1.5	2.0	2.5	3.0	4.0	5.0	6.0	8.0	10.0	12.0	16.0
Pitch	4.712	6.283	7.854	9.425	12.566	15.708	18.850	25.132	31.416	37.699	50.265

Classification of a tool set of 8 module side milling cutters (up to $m = 9$ mm)[1)]

Cutter no.	1	2	3	4	5	6	7	8
No. of teeth	12–13	14–16	17–20	21–25	26–34	35–54	55–134	135 to toothed rack

1) The manufacture of gears with side milling cutters is not an involute process. Only an approximate involute form of the tooth flank is produced. Therefore this manufacturing process is only suitable for secondary gears. For gears with $m > 9$ mm a tool set with 15 module side milling cutters is used.

Bevel gears, Worm drive

Unmodified bevel gears with straight teeth

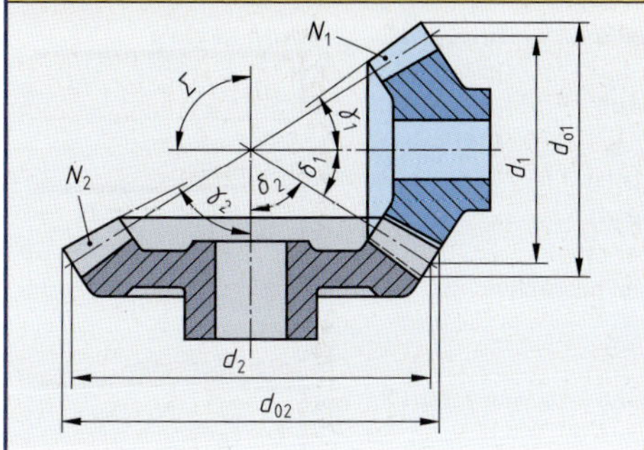

m	module	N, N_1, N_2	no. of teeth
d, d_1, d_2	pitch diameter	δ, δ_1, δ_2	pitch angle
d_o, d_{o1}, d_{o2}	outside diameter	γ_1, γ_2	tip angle
Σ	shaft angle (normally 90°)		

Pitch and whole depth narrow to the cone point, so that at every point of the tooth width a bevel gear has another module, outside diameter, etc. The outermost module corresponds to the standard module.

In addition to the dimensions given on the outside edges, the dimensions in the centers and inner edges of gear teeth are also important for manufacturing.

Example:

Bevel gear drive, m = 2 mm; N_1 = 30; N_2 = 120; Σ = 90°. Calculate the dimensions for turning the driving bevel gear.

$\tan\delta_1 = \frac{N_1}{N_2} = \frac{30}{120} = \mathbf{0.2500}$; $\delta_1 = \mathbf{14.04°}$

$d_1 = m \cdot N_1 = 2\text{ mm} \cdot 30 = \mathbf{60\text{ mm}}$

$d_{o1} = d_1 + 2 \cdot m \cdot \cos\delta_1$
$= 60\text{ mm} + 2 \cdot 2\text{ mm} \cdot \cos 14.04° = \mathbf{63.88\text{ mm}}$

$\tan\gamma_1 = \frac{N_1 + 2 \cdot \cos\delta_1}{N_2 - 2 \cdot \sin\delta_1} = \frac{30 + 2 \cdot \cos 14.04°}{120 - 2 \cdot \sin 14.04°} = \mathbf{0.267}$

$\gamma_1 = \mathbf{14.95°}$

Pitch diameter	$d = m \cdot N$
Outside diameter	$d_o = d + 2 \cdot m \cdot \cos\delta$
Tip angle gear 1	$\tan\gamma_1 = \frac{N_1 + 2 \cdot \cos\delta_1}{N_2 - 2 \cdot \sin\delta_1}$
Tip angle gear 2	$\tan\gamma_2 = \frac{N_2 + 2 \cdot \cos\delta_2}{N_1 - 2 \cdot \sin\delta_2}$
Pitch angle gear 1	$\tan\delta_1 = \frac{d_1}{d_2} = \frac{N_1}{N_2} = \frac{1}{i}$
Pitch angle gear 2	$\tan\delta_2 = \frac{d_2}{d_1} = \frac{N_2}{N_1} = i$
Shaft angle	$\Sigma = \delta_1 + \delta_2$

Whole depth, addendum, clearance, etc. are calculated like spur gears with straight teeth (page 253).

Worm drive

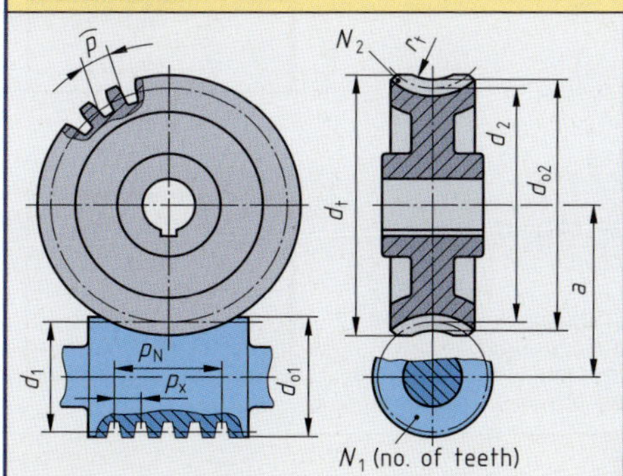

m	module	N_1, N_2	no. of teeth
d, d_1, d_2	pitch diameter	p_n	lead
d_o, d_{o1}, d_{o2}	outside diameter	p_x, p	(axial) pitch
r_t	throat radius	d_t	tip ∅

Example:

Worm drive m = 2.5 mm; N_1 = 2; d_1 = 40 mm; N_2 = 40; d_{o1} = ?; d_2 = ?; d_t = ?; r_t = ?; a = ?

$d_{o1} = d_1 + 2 \cdot m = 40\text{ mm} + 2 \cdot 2.5\text{ mm} = \mathbf{45\text{ mm}}$

$d_2 = m \cdot N_2 = 2.5\text{ mm} \cdot 40 = \mathbf{100\text{ mm}}$

$d_{o2} = d_2 + 2 \cdot m = 100\text{ mm} + 2 \cdot 2.5\text{ mm} = \mathbf{105\text{ mm}}$

$d_t \approx d_{o2} + m = 105\text{ mm} + 2.5\text{ mm} = \mathbf{107.5\text{ mm}}$

$r_t = \frac{d_1}{2} - m = \frac{40\text{ mm}}{2} - 2.5\text{ mm} = \mathbf{17.5\text{ mm}}$

$a = \frac{d_1 + d_2}{2} = \frac{40\text{ mm} + 100\text{ mm}}{2} = \mathbf{70\text{ mm}}$

Worm

Pitch diameter	d_1 = nominal size
Axial pitch – worm	$p_x = \pi \cdot m$
Outside diameter	$d_{o1} = d_1 + 2 \cdot m$
Lead	$p_n = p_x \cdot N_1 = \pi \cdot m \cdot N_1$

Worm gear

Pitch diameter	$d_2 = m \cdot N_2$
Pitch	$p = \pi \cdot m$
Outside diameter	$d_{o2} = d_2 + 2 \cdot m$
Tip diameter	$d_t \approx d_{o2} + m$
Throat radius	$r_t = \frac{d_1}{2} - m$

Clearance, whole depth, addendum, dedendum and center distance like spur gears (page 253).

ME

Transmission ratios

Gear drives

Single gear ratio

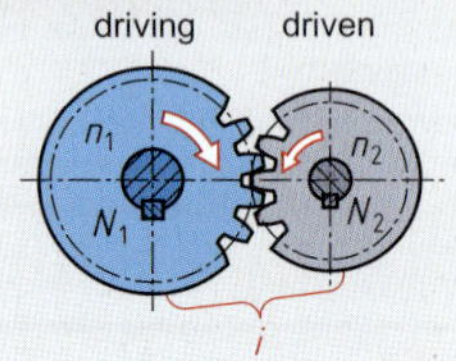

Multiple gear ratio

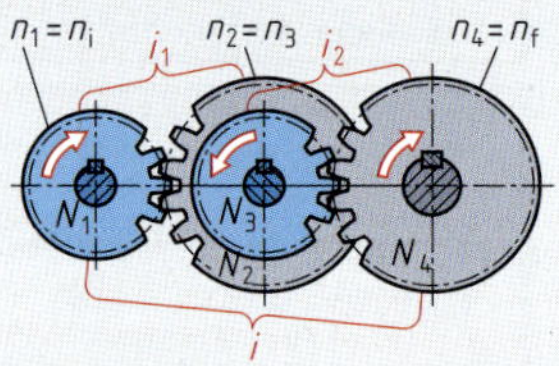

$N_1, N_3, N_5 \dots$	no. of teeth	driving gears
$n_1, n_3, n_5 \dots$	speeds	
$N_2, N_4, N_6 \dots$	no. of teeth	driven gears
$n_2, n_4, n_6 \dots$	speeds	
n_i	initial speed	
n_f	final speed	
i	total gear ratio	
$i_1, i_2, i_3 \dots$	individual gear ratios	

Example:

$i = 0.4$; $n_1 = 180/\text{min}$; $N_2 = 24$; $n_2 = ?$; $N_1 = ?$

$$\boldsymbol{n_2} = \frac{n_1}{i} = \frac{180/\text{min}}{0.4} = \mathbf{450/min}$$

$$\boldsymbol{N_1} = \frac{n_2 \cdot N_2}{n_1} = \frac{450/\text{min} \cdot 24}{180/\text{min}} = \mathbf{60}$$

Torque for gears, page 34

Drive formula

$$n_1 \cdot N_1 = n_2 \cdot N_2$$

Gear ratio

$$i = \frac{N_2}{N_1} = \frac{n_1}{n_2} = \frac{n_i}{n_f}$$

Total gear ratio

$$i = \frac{N_2 \cdot N_4 \cdot N_6 \dots}{N_1 \cdot N_3 \cdot N_5 \dots}$$

$$i = i_1 \cdot i_2 \cdot i_3, \dots$$

Belt drives

Single gear ratio

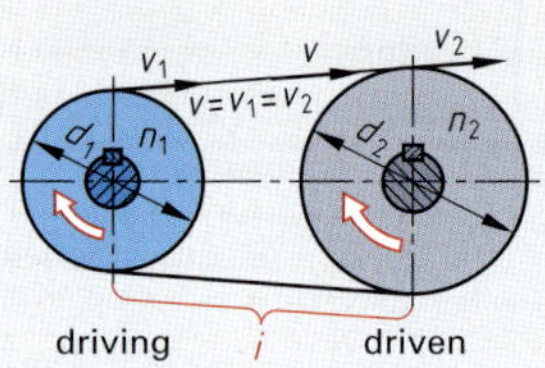

Multiple gear ratio

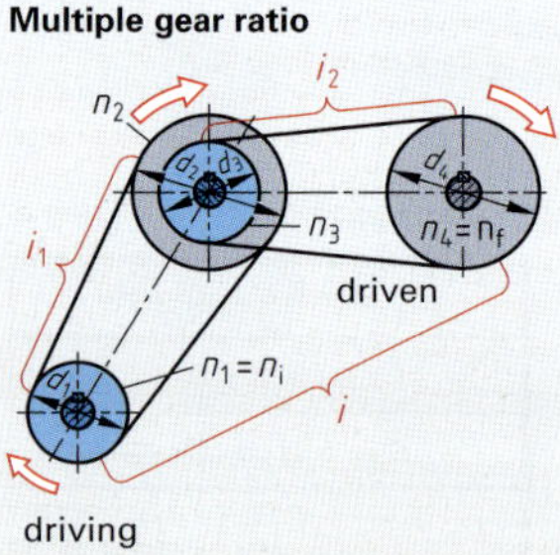

$d_1, d_3, d_5 \dots$	diameters[1)]	driving pulleys
$n_1, n_3, n_5 \dots$	speeds	
$d_2, d_4, d_6 \dots$	diameters[1)]	driven pulleys
$n_2, n_4, n_6 \dots$	speeds	
n_i	initial speed	
n_f	final speed	
i	total gear ratio	
$i_1, i_2, i_3 \dots$	individual gear ratios	
v, v_1, v_2	circumferential velocity	

Example:

$n_1 = 600/\text{min}$; $n_2 = 400/\text{min}$;
$d_1 = 240$ mm; $i = ?$; $d_2 = ?$

$$\boldsymbol{i} = \frac{n_1}{n_2} = \frac{600/\text{min}}{400/\text{min}} = \frac{1{,}5}{1} = \mathbf{1{,}5}$$

$$\boldsymbol{d_2} = \frac{n_1 \cdot d_1}{n_2} = \frac{600/\text{min} \cdot 240\ \text{mm}}{400/\text{min}} = \mathbf{360\ mm}$$

1) For V-belts (page 251) calculate with the effective diameter d_e; for positive drive belts (page 252) calculate with the number of teeth on the pulley.

Velocity

$$v = v_1 = v_2$$

Drive formula

$$n_1 \cdot d_1 = n_2 \cdot d_2$$

Gear ratio

$$i = \frac{d_2}{d_1} = \frac{n_1}{n_2} = \frac{n_i}{n_f}$$

Total gear ratio

$$i = \frac{d_2 \cdot d_4 \cdot d_6 \dots}{d_1 \cdot d_3 \cdot d_5 \dots}$$

$$i = i_1 \cdot i_2 \cdot i_3 \dots$$

ME

Worm drives

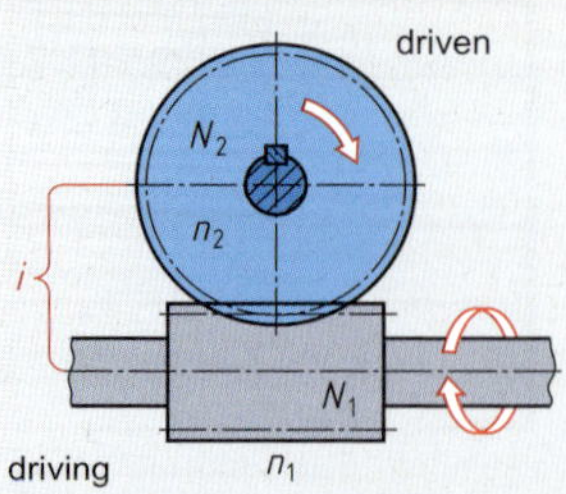

N_1	no. of teeth (no. of threads) of the worm
n_1	speed of the worm
N_2	no. of teeth of the worm gear
n_2	speed of the worm gear
i	gear ratio

Example:

$i = 25$; $n_1 = 1500/\text{min}$; $N_1 = 3$; $n_2 = ?$

$$\boldsymbol{n_2} = \frac{n_1}{i} = \frac{1500/\text{min}}{25} = \mathbf{60/min}$$

Drive formula

$$n_1 \cdot N_1 = n_2 \cdot N_2$$

Gear ratio

$$i = \frac{n_1}{n_2} = \frac{N_2}{N_1}$$

Plain bearings, Overview

Plain bearings[1] (Selection by type of lubrication)

Hydrodynamic plain bearings	Hydrostatic plain bearings	Dry-running plain bearings
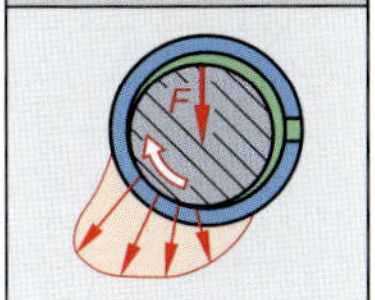	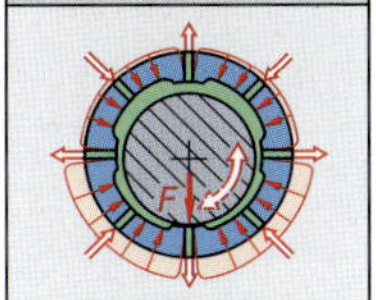	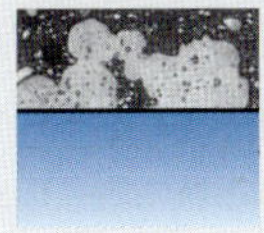
Suitable for	**Suitable for**	**Suitable for**
– low-wear continuous operation – high speeds – high impact loads	– wear-free continuous operation – low friction losses – low speeds possible	– maintenance free or low maintenance operation – with or without lubrication
Areas of application	**Areas of application**	**Areas of application**
– main and big end bearings – gearboxes – electric motors – turbines, compressors – lifting equipm., agricul. machinery	– precision bearings – space telescopes and antennae – machine tools – axial bearings for high forces	– construction equipment – armatures and devices – packaging machines – jet engines – household appliances

[1] Other plain bearings: air or gas and water lubricated plain bearings, magnetic bearings

Properties of plain bearing materials

Designation, Material number	Elongation limit $R_{p\,0.2}$ N/mm²	Specific bearing load p_L[1] N/mm²	Shaft min. hardness	Sliding properties	Sliding speed	Emergency-running behavior	Properties, application
Lead and tin casting alloys							cf. DIN ISO 4381 (2001-02)
G-PbSb15Sn10[2] 2.3391	43	7	160 HB	◔	◐	◕	Medium loading; all purpose plain bearing
G-SnSb12Cu6Pb 2.3790	61	10	160 HB	●	●	◕	Good impact loading; turbines, compressors, electric machines
Cast copper alloys and copper wrought alloys							cf. DIN ISO 4382-1 and -2 (1992-11)
CuSn8Pb2-C 2.1810	130	21	280 HB	◕	◕	◐	Low to moderate loading, sufficient lubrication
CuZn31Si1 2.1831	250	58	55 HRC				High loading, high vertical and horizontal impact loading
CuPb10Sn10-C[2] 2.1816	80	18	250 HB	◕	◕	◐	High surface pressures; vehicle bearings, bearings in hot-rolling mills
CuPb20Sn5-C 2.1818	60	11	150 HB	●	●	●	Suitable for water lubrication, resistant to sulfuric acid
Thermoplastics							cf. DIN ISO 6691 (2001-05)
PA 6 (Polyamide)	–	12	50 HRC	●	○	●	Impact and wear resistant; bearings in farm machinery
POM (Polyoxymethylene	–	18	50 HRC				Harder and capable of higher compressive loads than PA; bearings in precision mechanics, suitable for dry-running

[1] Bearing force based on the projected bearing surface
[2] Composite material according to DIN ISO 4383 for thin-walled plain bearings

● very good ◕ good ◐ normal ◔ limited ○ poor

Plain bearing bushings

Bushings made of copper alloys

cf. DIN ISO 4379 (1995-10)

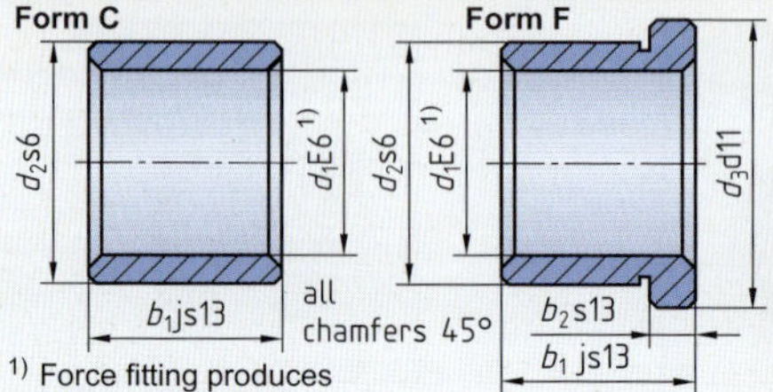

1) Force fitting produces tolerance class H8

Recommended tolerance classes for mounting dimensions	
Location hole	H7
Shaft	e7 or g7 (depending on application)

d_1	Form C d_2			Form F Series 1 d_2	Form F Series 1 d_3	Form F Series 1 b_2	Form F Series 2 d_2	Form F Series 2 d_3	Form F Series 2 b_2	Lengths b_1		
10	12	14	16	12	14	1	16	20	3	–	10	–
12	14	16	18	14	16	1	18	22	3	10	15	20
15	17	19	21	17	19	1	21	27	3	10	15	20
18	20	22	24	20	22	1	24	30	3	12	20	30
20	23	24	26	23	26	1.5	26	32	3	15	20	30
22	25	26	28	25	28	1.5	28	34	3	15	20	30
25	28	30	32	28	31	1.5	32	38	4	20	30	40
30	34	36	38	34	38	2	38	44	4	20	30	40
35	39	41	45	39	43	2	45	50	5	30	40	50
40	44	48	50	44	48	2	50	58	5	30	40	60

Diameter range d_1: 6–200

⇒ **Bushing ISO 4379 – F22 x 25 x 30 – CuSn8P:** Form F, d_1 = 22 mm, d_2 = 25 mm, b_1 = 30 mm, of CuSn8P

Bushings made of sintered metal

cf. DIN 1850-3 (1998-07)

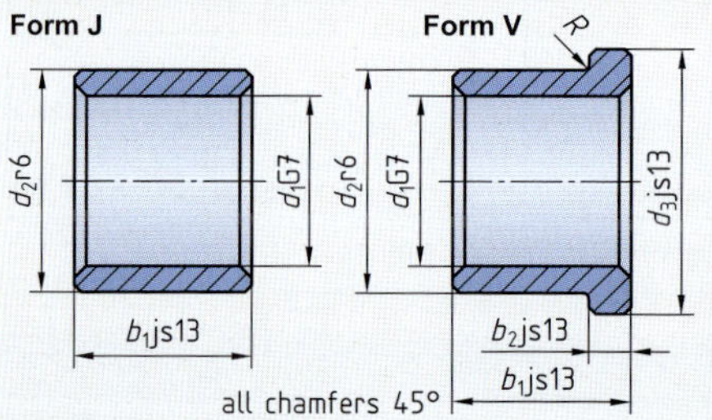

Recommended tolerance classes for mounting dimensions	
Location hole	H7
Shaft	–

d_1	Form J d_2		Form V d_2	Form V d_3	Form V b_2	Form V R_{max}	Lengths b_1		
10	16	14	16	22	2	0.6	8	10	16
12	18	16	18	24	3	0.6	8	12	20
15	21	19	21	27	3	0.6	10	15	25
18	24	22	24	30	3	0.6	12	18	30
20	26	25	26	32	3	0.6	15	20	25
22	28	27	28	34	3	0.6	15	20	25
25	32	30	32	39	3.5	0.8	20	25	30
30	38	35	38	46	4	0.8	20	25	30
35	45	41	45	55	5	0.8	25	35	40
40	50	46	50	60	5	0.8	30	40	50

Diameter range d_1: 1–60

⇒ **Bushing DIN 1850 – V18 x 24 x 18 – Sint-B50:** d_1 = 18 mm, d_2 = 24 mm, b_1 = 18 mm, sintered bronze Sint-B50

ME

Bushings made of thermosets and thermoplastics

cf. DIN 1850-5 and -6 (1998-07)

Thermoset plastics

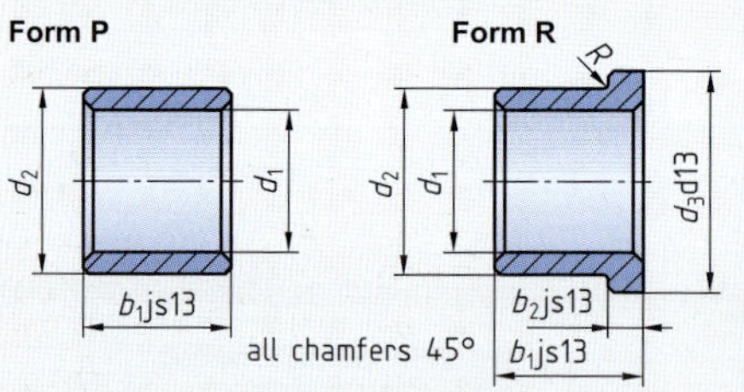

d_1	d_2	d_3	b_2	R_{max}	Lengths b_1		
10	16	20	3	0.3	6	10	–
12	18	22	3	0.5	10	15	20
15	21	27	3	0.5	10	15	20
18	24	30	3	0.5	12	20	30
20	26	32	3	0.5	15	20	30
22	28	34	3	0.5	15	20	30
25	32	38	4	0.5	20	30	40
30	38	44	4	0.5	20	30	40
35	45	50	5	0.8	30	40	50

Diameter range d_1 for thermosets: 3–250, for thermoplastics: 6–200

Thermoplastics

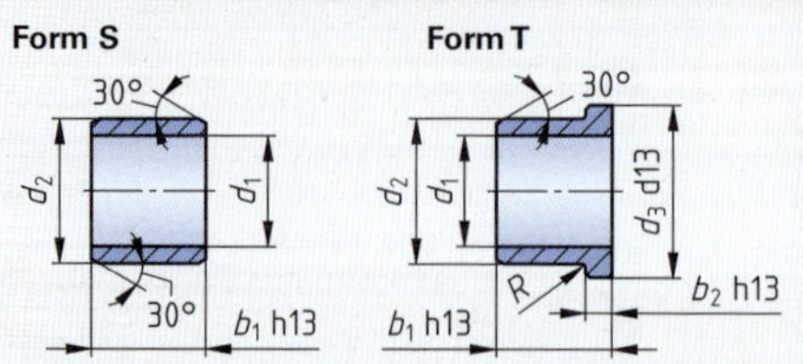

Recommended tolerance classes for mounting dimensions		
	Thermosets	Thermoplastics
Location hole	H7	H7
Shaft	h7	h9

Limit deviations d_2 and d_1 of tolerance classes A and B for bushings made of thermoplastics

	d_2						Fabrication method	Tolerance class resulting after force fitting d_1
from to	10 14	15 18	20 25	28 32	35 40	42 55		
A	+0.21 +0.07	+0.2 0	+0.4 +0.1	+0.6 +0.2	+0.69 +0.23	+0.90 +0.30	injection molded	D12
B	Tolerance class zb11						machined	C11

Additional codes for bushings made of thermoset plastics

W	Circular grooves on outer diameter d_2	Y	Assembly bevel 15° (inst. of 45°)
		Z	Undercut instead of radius R

⇒ **Bushing DIN 1850 – S20 A20 – PA 6:** Form S; d_1 = 20 mm, tolerance cl. A, b_1 = 20 mm, polyamide 6

Other stand. designs: Wrapped bushings DIN 1494, internal tension bushings DIN 1498, external tension bushings DIN 1499

Antifriction bearings, Overview

Roller bearings (selection)

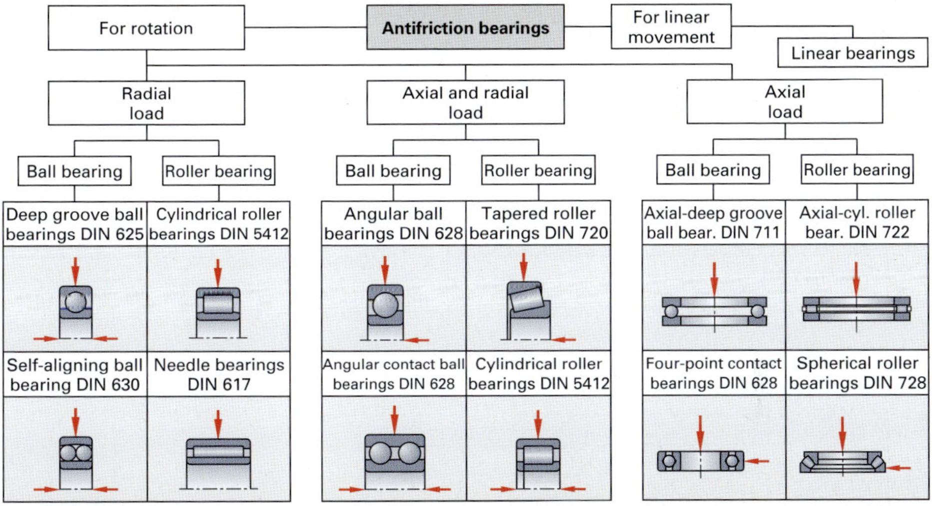

Properties of roller bearings

Bearing design[1]	Inside ∅ *d*	Radial loading	Axial loading	High speed	High loads	Quiet running	Application
Ball bearings							
Deep groove ball bearings	1.5–600	◕	◐	●	◐	●	Universal bearings in machine and automotive manufacturing
Self-aligning ball bearings	5–120	◕	◔	◕	◔	◔	Compensation with misalignment
Angular contact ball bearings single-row	10–170	◕	◕	●[2]	◕[3]	◕	Only used in pairs, large forces, automotive manufacturing
Angular contact ball bearings double-row	10–110	◕	◕	◐	◕	◔	Large forces, automotive manufacturing, with limited space requirements
Axial deep groove ball bearings	8–360	○	◕	◐	◐	◔	Acceptance of very high axial forces, drill spindles, tail stock centers
Four-point contact bearings	20–240	◔	◕	◔	◐	◔	Very tight spaces, spindle bearing layouts, gear and roller bearing assemblies
Roller bearings							
Cylindrical roller bearings (form N)	17–240	●	○	●	◕	◐	Acceptance of very large radial forces, roller bearing assemblies, transmissions
Cylindrical roller bearings (form NUP)	15–240	●	◐	◕	◕	◔	Like Form N, with flanged wheel additional acceptance of axial forces
Needle bearings	90–360	●	○	◔	●	◐	High carrying capacity with tight mounting space
Tapered roller bearings	15–360	●	●	◐[2]	●[3]	◔	Usually mounted in pairs, wheel bearings in automobiles, spindle bearings
Axial cylindrical roller bearings	15–600	○	●	◔	◕	○	Stiff bearing requiring minimal axial space, high friction
Spherical roller bearings	60–1060	◔	●	◔	◕	○	Angular displacement thrust bearings, thrust bearings in cranes

[1] For all radial bearings the prefix "radial" is omitted.
[2] Reduced suitability with paired mounting
[3] Mounted in pairs

Suitability levels:
● very good ◕ good ◐ normal
◔ limited ○ not suitable

ME

ME

Antifriction bearings, Designation

Designation of antifriction bearings

cf. DIN 623-1 (1993-05)

Example: **Tapered roller bearings DIN 720 – S 30208 P2**

- Tapered roller bearings → Name
- DIN 720 → Standard
- S → Prefix symbol
- 30208 → Basic numbers
- P2 → Suffix symbol

Prefix symbols	
K	cage with roller elements
L	free ring
R	ring with roller set
S	stainless steel

Suffix symbols (selection)	
K	bearing with tapered bore
Z	bearing with shield on one side
2Z	bearing with shield on both sides
E	reinforced design
RS	bearing with seal on one side
2RS	bearing with seal on both sides
P2	highest precision: dimensional, form and running

Example of basic numbers: **3 0 2 08**

- 3 → Bearing type 3
- 3 0 2 → Bearing series 302
  - 0 → Width series 0
  - 2 → Diameter series 2
  - 0 2 → Dimension series 02
- 08 → Bore code 08

Bearing type	Design
0	Angular contact ball bear., double row
1	Self-aligning ball bearing
2	Barrel and spherical roller bearings
3	Tapered roller bearings
4	Deep groove ball bear., double row
5	Axial deep groove ball bearings
6	Deep groove ball bear., single row
7	Angular contact ball bear., single row
8	Axial cylindrical roller bearings
NA	Needle bearings
QJ	Four-point contact bearing
N, NJ, NJP, NN, NNU, NU, NUP	Cylindrical roller bearings

Bore-code	Bore ∅ *d*	Bore code	Bore ∅ *d*
00	10	12	60
01	12	13	65
02	15	14	70
03	17	15	75
04	20	16	80
05	25	17	85
06	30	18	90
07	35	19	95
08	40	20	100
09	45	21	105
10	50	22	110
11	55	23	115

Dimension series (selection)

cf. DIN 616 (2000-06)

Explanation

The dimension plans in DIN 616 contain diameter series in which each nominal diameter of a bearing bore *d* (= shaft diameter) is assigned a number of:

- outside diameters and
- width series (for radial bearings) or
- height series (for axial bearings).

Structure of the dimension series

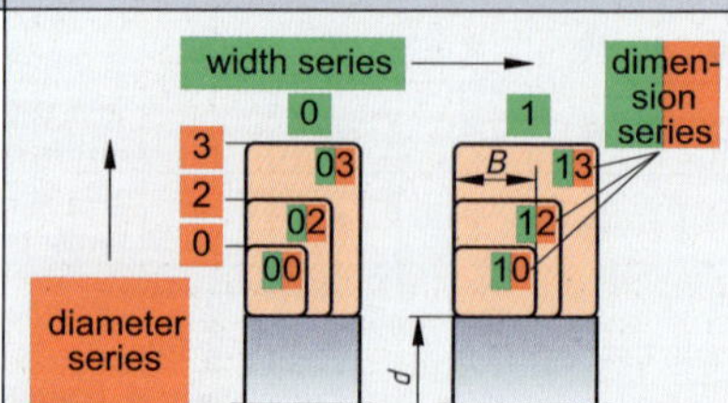

Example: Tapered roller bearings[1]

Dimension series 02			
Bore code	Bore ∅ *d*	*D*	*B*
07	35	72	17
08	40	80	18
09	45	85	19
10	50	90	20

[1] other dimensions, see page 263

Ball bearings

Deep groove ball bearings (selection)

cf. DIN 625-1 (1989-04)

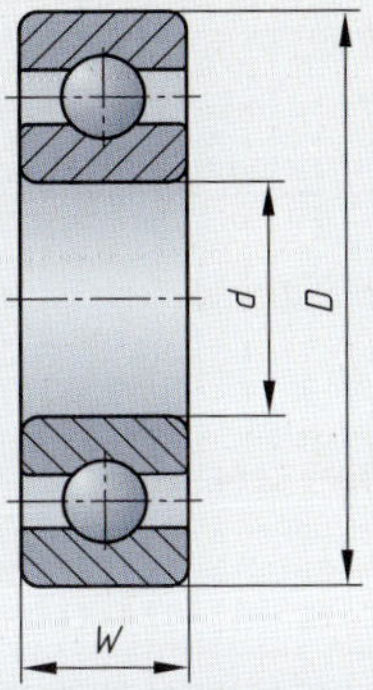

d from 1.5 to 600 mm

Mounting dimensions according to DIN 5418:

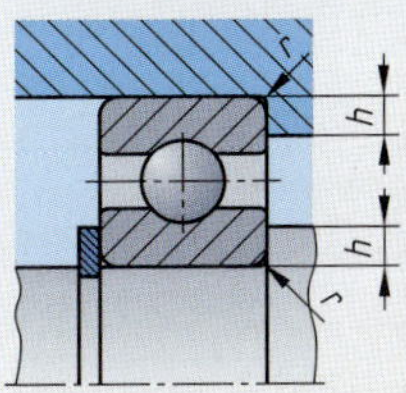

	Bearing series 60					Bearing series 62					Bearing series 63				
d	*D*	*W*	*r* max	*h* min	Basic number	*D*	*W*	*r* max	*h* min	Basic number	*D*	*W*	*r* max	*h* min	Basic number
10	26	8	0.3	1	6000	30	9	0.6	2.1	6200	35	11	0.6	2.1	6300
12	28	8	0.3	1	6001	32	10	0.6	2.1	6201	37	12	1	2.8	6301
15	32	9	0.3	1	6002	35	11	0.6	2.1	6202	42	13	1	2.8	6302
17	35	10	0.3	1	6003	40	12	0.6	2.1	6203	47	14	1	2.8	6303
20	42	12	0.6	1.6	6004	47	14	1	2	6204	52	15	1	3.5	6304
25	47	12	0.6	1.6	6005	52	15	1	2	6205	62	17	1	3.5	6305
30	55	13	1	2.3	6006	62	16	1	2	6206	72	19	1	3.5	6306
35	62	14	1	2.3	6007	72	17	1	2	6207	80	21	1.5	4.5	6307
40	68	15	1	2.3	6008	80	18	1	3.5	6208	90	23	1.5	4.5	6308
45	75	16	1	2.3	6009	85	19	1	3.5	6209	100	25	1.5	4.5	6309
50	80	16	1	2.3	6010	90	20	1	3.5	6210	110	27	2	5.5	6310
55	90	18	1	3	6011	100	21	1.5	4.5	6211	120	29	2	5.5	6311
60	95	18	1	3	6012	110	22	1.5	4.5	6212	130	31	2.1	6	6312
65	100	18	1	3	6013	120	23	1.5	4.5	6213	140	33	2.1	6	6313
70	110	20	1	3	6014	125	24	1.5	4.5	6214	150	35	2.1	6	6314
75	115	20	1	3	6015	130	25	2	5.5	6215	160	37	2.1	6	6315
80	125	22	1	3	6016	140	26	2	5.5	6216	170	39	2.5	7	6316
85	130	22	1.5	3.5	6017	150	28	2.1	6	6217	180	41	2.5	7	6317
90	140	24	1.5	3.5	6018	160	30	2.1	6	6218	190	43	2.5	7	6318
95	145	24	1.5	3.5	6019	170	32	2.1	6	6219	200	45	2.5	7	6319
100	150	24	1.5	3.5	6020	180	34	2.1	6	6220	215	47	2.5	7	6320

⇒ **Deep groove ball bearing DIN 625 – 6208 – 2Z – P2:** Deep groove ball bearing (bearing type 6), width series 0[1)], diameter series 2, bore code 08 ($d = 8 \cdot 5$ mm = 40 mm), design with 2 shields, bearing with highest precision (tolerance class 2)

Angular contact ball bearings (selection)

cf. DIN 628-1 (2008-01)

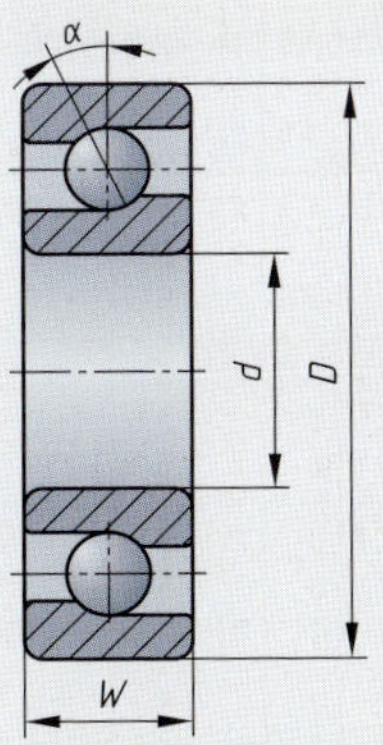

d from 10 to 170 mm

Mounting dimensions according to DIN 5418:

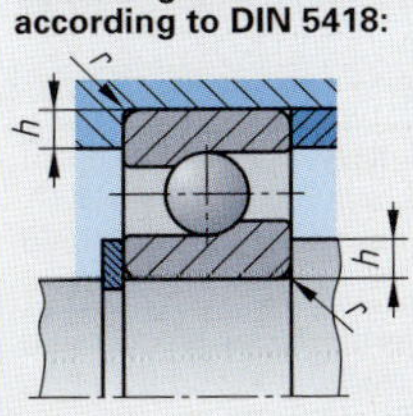

	Bearing series 72					Bearing series 73					Bearing ser. 33 (double row)				
d	*D*	*W*	*r* max	*h* min	Basic number[2)]	*D*	*W*	*r* max	*h* min	Basic number[2)]	*D*	*W*	*r* max	*h* min	Basic number[3)]
15	35	11	0.6	2.1	7202B	42	13	1	2.8	7302B	42	19	1	2.8	3302
17	40	12	0.6	2.1	7203B	47	14	1	2.8	7303B	47	22.2	1	2.8	3303
20	47	14	1	2.8	7204B	52	15	1	3.5	7304B	52	22.2	1	3.5	3304
25	52	15	1	2.8	7205B	62	17	1	3.5	7305B	62	25.4	1	3.5	3305
30	62	16	1	2.8	7206B	72	19	1	3.5	7306B	72	30.2	1	3.5	3306
35	72	17	1	3.5	7207B	80	21	1.5	4.5	7307B	80	34.9	1.5	4.5	3307
40	80	18	1	3.5	7208B	90	23	1.5	4.5	7308B	90	36.5	1.5	4.5	3308
45	85	19	1	3.5	7209B	100	25	1.5	4.5	7309B	100	39.7	1.5	4.5	3309
50	90	20	1	3.5	7210B	110	27	2	5.5	7310B	110	44.4	2	5.5	3310
55	100	21	1.5	4.5	7211B	120	29	2	5.5	7311B	120	49.2	2	5.5	3311
60	110	22	1.5	4.5	7212B	130	31	2.1	6	7312B	130	54	2.1	6	3312
65	120	23	1.5	4.5	7213B	140	33	2.1	6	7313B	140	58.7	2.1	6	3313
70	125	24	1.5	4.5	7214B	150	35	2.1	6	7314B	150	63.5	2.1	6	3314
75	130	25	1.5	4.5	7215B	160	37	2.1	6	7315B	160	68.3	2.1	6	3315
80	140	26	2	5.5	7216B	170	39	2.1	6	7316B	170	68.3	2.1	6	3316
85	150	28	2	5.5	7217B	180	41	2.5	7	7317B	180	73	2.5	7	3317
90	160	30	2	5.5	7218B	190	43	2.5	7	7318B	190	73	2.5	7	3318
95	170	32	2.1	6	7219B	200	45	2.5	7	7319B	200	77.8	2.5	7	3319
100	180	34	2.1	6	7220B	215	47	2.5	7	7320B	215	82.6	2.5	7	3320

⇒ **Angular contact ball bearing DIN 628 – 7309B:** Angular contact ball bearing (Bearing type 7), width series 0[1)], diameter series 3, bore code 09 (bore diameter $d = 9 \cdot 5$ mm = 45 mm), contact angle $\alpha = 40°$ (B)

[1)] In the designations for deep groove and angular contact ball bearings the 0 for the width series is sometimes omitted according to DIN 623-1.

[2)] Contact angle $\alpha = 40°$ [3)] Contact angle not standardized

Ball bearings, Roller bearings

Axial deep groove ball bearings (selection)

cf. DIN 711 (2010-05)

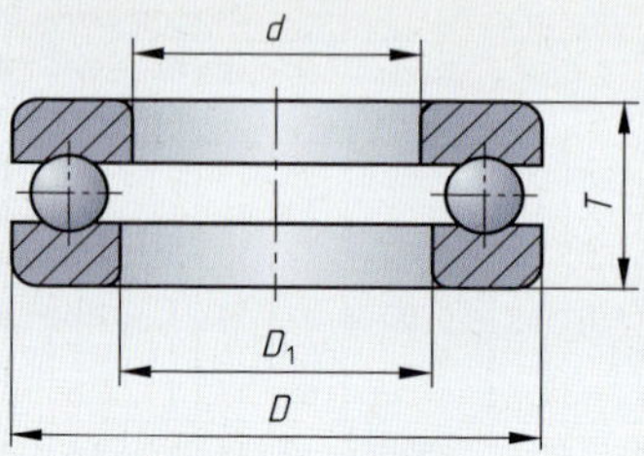

d from 8 to 360 mm

Mounting dimensions according to DIN 5418:

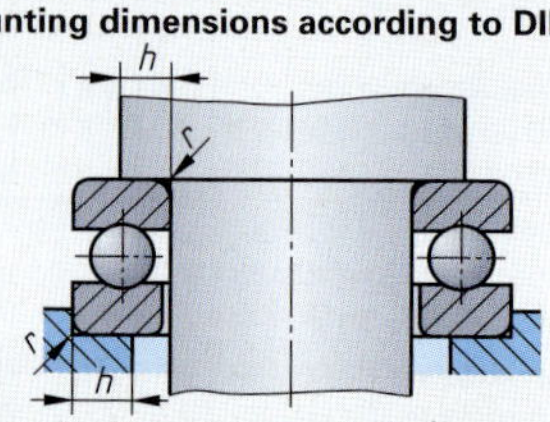

d	D_1	Bearing series 512 *D*	*T*	*r* max	*h* min	Basic number	Bearing series 513 *D*	*T*	*r* max	*h* min	Basic number
25	27	47	15	0.6	6	51205	52	18	1	7	51305
30	32	52	16	0.6	6	51206	60	21	1	8	51306
35	37	62	18	1	7	51207	68	24	1	9	51307
40	42	68	19	1	7	51208	78	26	1	10	51308
45	47	73	20	1	7	51209	85	28	1	10	51309
50	52	78	22	1	7	51210	95	31	1	12	51310
55	57	90	25	1	9	51211	105	35	1	13	51311
60	62	95	26	1	9	51212	110	35	1	13	51312
65	67	100	27	1	9	51213	115	36	1	13	51313
70	72	105	27	1	9	51214	125	40	1	14	51314
75	77	110	27	1	9	51215	135	44	1.5	15	51315
80	82	115	28	1	9	51216	140	44	1.5	15	51316

⇒ **Axial deep groove ball bearing DIN 711 – 51210:** Axial-deep groove ball bearing of the bearing series 512 with bearing type 5, width series 1, diameter series 2 and bore code 10

Cylindrical roller bearings (selection)

cf. DIN 5412-1 (2005-08)

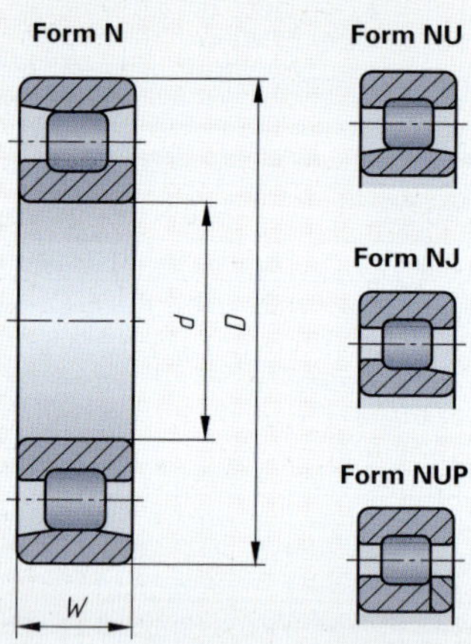

d from 15 to 500 mm

Mounting dimensions according to DIN 5418:

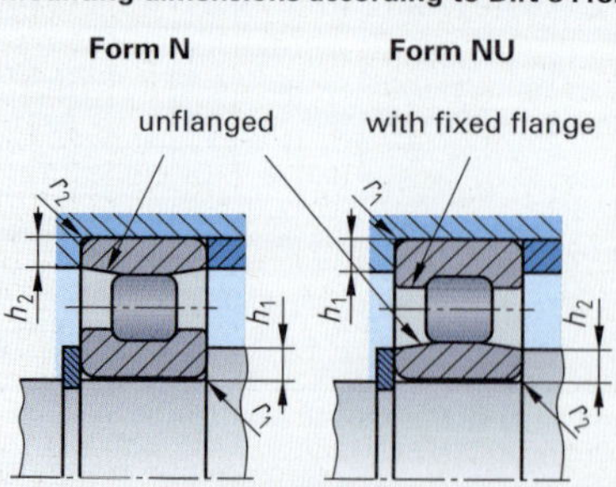

d	Bearing series N2, NU2, NJ2, NUP2 *D*	*W*	r_1 max	h_1 min	r_2 max	h_2 min	Bearing series N3, NU3, NJ3, NUP3 *D*	*W*	r_1 max	h_1 min	r_2 max	h_2 min	Bore code
17	40	12	0.6	2.1	0.3	1.2	47	14	1	2.8	1	2.8	03
20	47	14	1	2.8	0.6	2.1	52	15	1.1	3.5	1	2.8	04
25	52	15	1	2.8	0.6	2.1	62	17	1.1	3.5	1	2.8	05
30	62	16	1	2.8	0.6	2.1	72	19	1.1	3.5	1	2.8	06
35	72	17	1	3.5	0.6	2.1	80	21	1.5	4.5	1	2.8	07
40	80	18	1	3.5	1	3.5	90	23	1.5	4.5	2	5.5	08
45	85	19	1	3.5	1	3.5	100	25	1.5	4.5	2	5.5	09
50	90	20	1	3.5	1	3.5	110	27	2	5.5	2	5.5	10
55	100	21	1.5	4.5	1	3.5	120	29	2	5.5	2	5.5	11
60	110	22	1.5	4.5	1.5	4.5	130	31	2.1	6	2	5.5	12
65	120	23	1.5	4.5	1.5	4.5	140	33	2.1	6	2	5.5	13
70	125	24	1.5	4.5	1.5	4.5	150	35	2.1	6	2	5.5	14
75	130	25	1.5	4.5	1.5	4.5	160	37	2.1	6	2	5.5	15
80	140	26	2	5.5	2	5.5	170	39	2.1	6	2	5.5	16
85	150	28	2	5.5	2	5.5	180	41	3	7	3	7	17
90	160	30	2	5.5	2	5.5	190	43	3	7	3	7	18
95	170	32	2.1	6	2.1	6	200	45	3	7	3	7	19
100	180	34	2.1	6	2.1	6	215	47	3	7	3	7	20
105	–	–	–	–	–	–	225	49	3	7	3	7	21
110	200	38	2.1	6	2.1	6	240	50	3	7	3	7	22
120	215	40	2.1	6	2.1	6	260	55	3	7	3	7	24

⇒ **Cylindrical roller bearing DIN 5412 – NUP 312 E:** Cylindrical roller bearing of bearing series NUP3 with bearing type NUP, width series 0, diameter series 3 and bore code 12, reinforced design

The normal design of the dimension series 02, 22, 03 and 23 were deleted from the standard with no replacement and then replaced with the reinforced design (suffix symbol E).

Roller bearings

Tapered roller bearings (selection) cf. DIN 720 (2008-08) and DIN 5418 (1993-02)

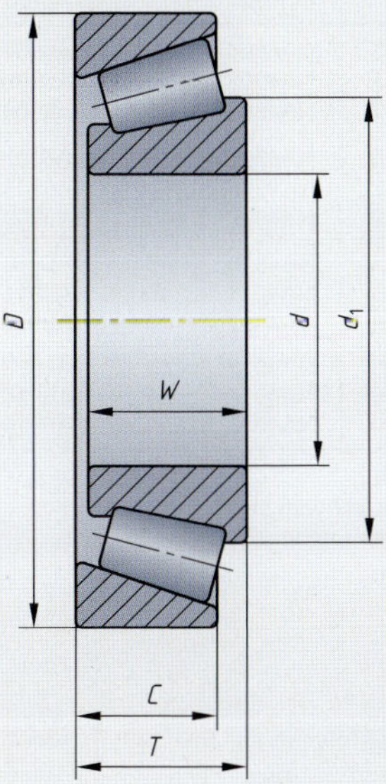

Mounting dimensions according to DIN 5418:

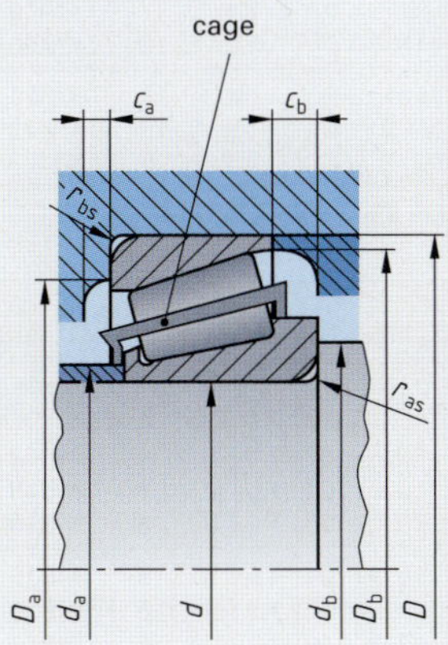

In the case of tapered roller bearings the cage projects beyond the lateral face of the outer ring.

The mounting dimensions of DIN 5418 must be maintained so that the cage does not rub against other parts.

Bearing series 302

Dimensions						Mounting dimension									
d	D	W	C	T	d_1	d_a max	d_b min	D_a min	D_a max	D_b min	c_a min	c_b min	r_{as} max	r_{bs} max	Basic no.
20	47	14	12	15.25	33.2	27	26	40	41	43	2	3	1	1	30204
25	52	15	13	16.25	37.4	31	31	44	46	48	2	2	1	1	30205
30	62	16	14	17.25	44.6	37	36	53	56	57	2	3	1	1	30206
35	72	17	15	18.15	51.8	44	42	62	65	67	3	3	1.5	1.5	30207
40	80	18	16	19.75	57.5	49	47	69	73	74	3	3.5	1.5	1.5	30208
45	85	19	16	20.75	63	54	52	74	78	80	3	4.5	1.5	1.5	30209
50	90	20	17	21.75	67.9	58	57	79	83	85	3	4.5	1.5	1.5	30210
55	100	21	18	22.75	74.6	64	64	88	91	94	4	4.5	2	1.5	30211
60	110	22	19	23.75	81.5	70	69	96	101	103	4	4.5	2	1.5	30212
65	120	23	20	24.75	89	77	74	106	111	113	4	4.5	2	1.5	30213
70	125	24	21	26.25	93.9	81	79	110	116	118	4	5	2	1.5	30214
75	130	25	22	27.25	99.2	86	84	115	121	124	4	5	2	1.5	30215
80	140	26	22	28.25	105	91	90	124	130	132	4	6	2.5	2	30216
85	150	28	24	30.5	112	97	95	132	140	141	5	6.5	2.5	2	30217
90	160	30	26	32.5	118	103	100	140	150	150	5	6.5	2.5	2	30218
95	170	32	27	34.5	126	110	107	149	158	159	5	7.5	3	2.5	30219
100	180	34	29	37	133	116	112	157	168	168	5	8	3	2.5	30220
105	190	36	30	39	141	122	117	165	178	177	6	9	3	2.5	30221
110	200	38	32	41	148	129	122	174	188	187	6	9	3	2.5	30222
120	215	40	34	43.5	161	140	132	187	203	201	6	9.5	3	2.5	30224

Bearing series 303

Dimensions						Mounting dimension									
d	D	W	C	T	d_1	d_a max	d_b min	D_a min	D_a max	D_b min	c_a min	c_b min	r_{as} max	r_{bs} max	Basic no.
20	52	15	13	16.25	34.3	28	27	44	45	47	2	3	1.5	1.5	30304
25	62	17	15	18.25	41.5	34	32	54	55	57	2	3	1.5	1.5	30305
30	72	19	16	20.75	44.8	40	37	62	65	66	3	4.5	1.5	1.5	30306
35	80	21	18	22.75	54.5	45	44	70	71	74	3	4.5	2	1.5	30307
40	90	23	20	25.25	62.5	52	49	77	81	82	3	5	2	1.5	30308
45	100	25	22	27.25	70.1	59	54	86	91	92	3	5	2	1.5	30309
50	110	27	23	29.25	77.2	65	60	95	100	102	4	6	2.5	2	30310
55	120	29	25	31.5	84	71	65	104	110	111	4	6.5	2.5	2	30311
60	130	31	26	33.5	91.9	77	72	112	118	120	5	7.5	3	2.5	30312
65	140	33	28	36	98.6	83	77	122	128	130	5	8	3	2.5	30313
70	150	35	30	38	105	89	82	120	138	140	5	8	3	2.5	30314
75	160	37	31	40	112	95	87	139	148	149	5	9	3	2.5	30315
80	170	39	33	42.5	120	102	92	148	158	159	5	9.5	3	2.5	30316
85	180	41	34	44.5	126	107	99	156	166	167	6	10.5	4	3	30317
90	190	43	36	46.5	132	113	104	165	176	176	6	10.5	4	3	30318
95	200	45	38	49.5	139	118	109	172	186	184	6	11.5	4	3	30319
100	215	47	39	51.5	148	127	114	184	201	197	6	12.5	4	3	30320
105	225	49	41	53.5	155	132	119	193	211	206	7	12.5	4	3	30321
110	240	50	42	54.5	165	141	124	206	226	220	8	12.5	4	3	30322
120	260	55	46	59.5	178	152	134	221	246	237	8	13.5	4	3	30324

⇒ **Tapered roller bearing DIN 720 – 30212:** Tapered roller bearing of bearing series 302 with bearing type 3, width series 0, diameter series 2, bore code 12

Needle bearings, Lock nuts, Lock washers

Needle bearings (selection)

cf. DIN 617 (2008-10)

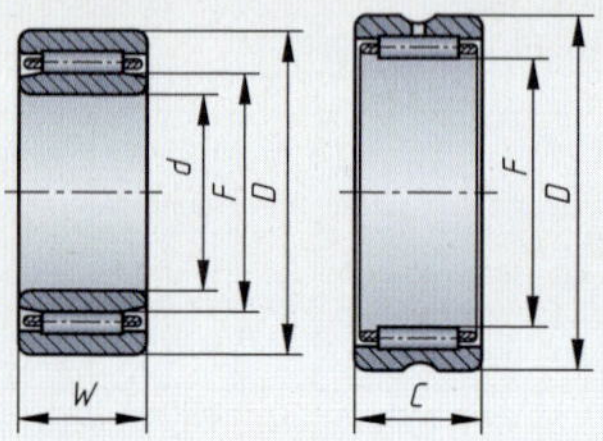

Mounting dimensions according to DIN 5418:

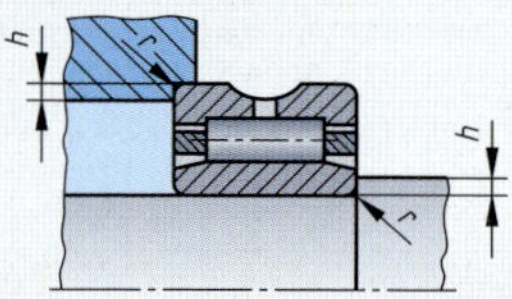

d	D	F	r max	h min	Bearing series NA49 W	Bearing series NA49 Basic number	Bearing series NA69 W	Bearing series NA69 Basic number
20	37	25	0.3	1	17	NA4904	30	NA6904
25	42	28	0.3	1	17	NA4905	30	NA6905
30	47	30	0.3	1	17	NA4906	30	NA6906
35	55	42	0.6	1.6	20	NA4907	36	NA6907
40	62	48	0.6	1.6	22	NA4908	40	NA6908
45	68	52	0.6	1.6	22	NA4909	40	NA6909
50	72	58	0.6	1.6	22	NA4910	40	NA6910
55	80	63	1	2.3	25	NA4911	45	NA6911
60	85	68	1	2.3	25	NA4912	45	NA6912
65	90	72	1	2.3	25	NA4913	45	NA6913
70	100	80	1	2.3	30	NA4914	54	NA6914
75	105	85	1	2.3	30	NA4915	54	NA6915

⇒ **Needle bearing DIN 617 – NA4909:** Needle bearing of bearing series NA49 with bearing type NA, width series 4, diameter series 9, bore code 09

NA6907 and up: double row

Lock nuts for antifriction bearings (selection)

cf. DIN 981 (2009-06)

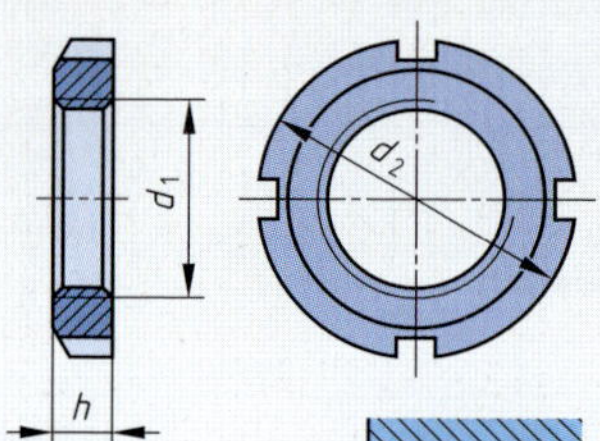

Mounting example:

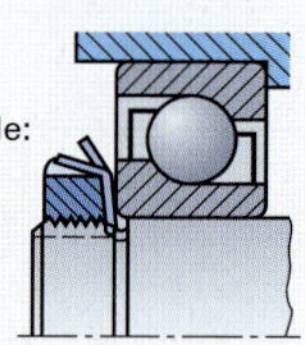

d_1 from M10 to M200

d_1	d_2	h	Code	d_1	d_2	h	Code
M10 × 0.75	18	4	KM0	M60 × 2	80	11	KM12
M12 × 1	22	4	KM1	M65 × 2	85	12	KM13
M15 × 1	25	5	KM2	M70 × 2	92	12	KM14
M17 × 1	28	5	KM3	M75 × 2	98	13	KM15
M20 × 1	32	6	KM4	M80 × 2	105	15	KM16
M25 × 1.5	38	7	KM5	M85 × 2	110	16	KM17
M30 × 1.5	45	7	KM6	M90 × 2	120	16	KM18
M35 × 1.5	52	8	KM7	M95 × 2	125	17	KM19
M40 × 1.5	58	9	KM8	M100 × 2	130	18	KM20
M45 × 1.5	65	10	KM9	M105 × 2	140	18	KM21
M50 × 1.5	70	11	KM10	M110 × 2	145	19	KM22
M55 × 2	75	11	KM11	M115 × 2	150	19	KM23

⇒ **Lock nut DIN 981 – KM6:** Lock nut of d_1 = M30 x 1.5

Lock washers (selection)

cf. DIN 5406 (2009-05)

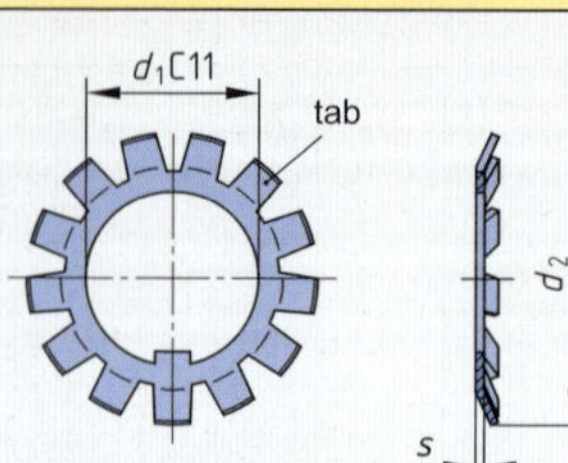

Mounting dimensions

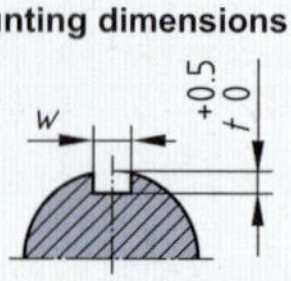

d_1 from 10 to 200 mm

d_1	d_2	s	w_1	t	Code	d_1	d_2	s	w_2	t	Code
10	21	1	4	2	MB0	60	86	1.5	9	4	MB12
12	25	1	4	2	MB1	65	92	1.5	9	4	MB13
15	28	1	5	2	MB2	70	98	1.5	9	5	MB14
17	32	1	5	2	MB3	75	104	1.5	9	5	MB15
20	36	1	5	2	MB4	80	112	1.7	11	5	MB16
25	42	1.2	6	3	MB5	85	119	1.7	11	5	MB17
30	49	1.2	6	4	MB6	90	126	1.7	11	5	MB18
35	57	1.2	7	4	MB7	95	133	1.7	11	5	MB19
40	62	1.2	7	4	MB8	100	142	1.7	14	6	MB20
45	69	1.2	7	4	MB9	105	145	1.7	14	6	MB21
50	74	1.2	7	4	MB10	110	154	1.7	14	6	MB22
55	81	1.5	9	4	MB11	115	159	2	14	6	MB23

⇒ **Lock washer DIN 5406 – MB6:** Lock washer of d_1 = 30 mm

Internal and external retaining rings, Circlips

Retaining rings in standard design[1] (selection)

For shafts (external) cf. DIN 471 (1981-09)

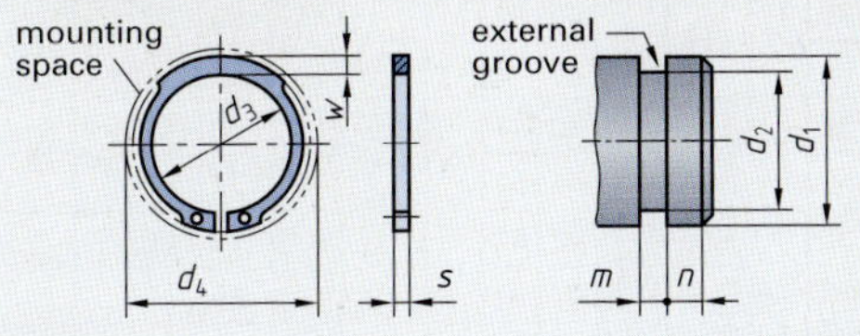

For bores (internal) cf. DIN 472 (1981-09)

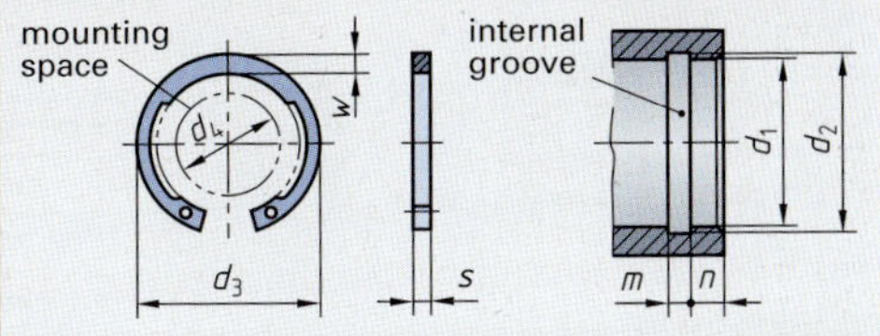

Nominal size d_1 mm	Ring s	Ring d_3	Ring d_4	Ring w ≈	Slot d_2	Slot m H13	Slot n min	Nominal size d_1 mm	Ring s	Ring d_3	Ring d_4	Ring w ≈	Slot d_2	Slot m H13	Slot n min
10	1	9.3	17	1.8	9.6	1.1	0.6	10	1	10.8	3.3	1.4	10.4	1.1	0.6
12	1	11	19	1.8	11.5	1.1	0.8	12	1	13	4.9	1.7	12.5	1.1	0.8
15	1	13.8	22.6	2.2	14.3	1.1	1.1	15	1	16.2	7.2	2	15.7	1.1	1.1
18	1.2	16.5	26.2	2.4	17	1.3	1.5	18	1	19.5	9.4	2.2	19	1.1	1.5
20	1.2	18.5	28.4	2.6	19	1.3	1.5	20	1	21.5	11.2	2.3	21	1.1	1.5
22	1.2	20.5	30.8	2.8	21	1.3	1.5	22	1	23.5	13.2	2.5	23	1.1	1.5
25	1.2	23.2	34.2	3	23.9	1.3	1.7	25	1.2	26.9	15.5	2.7	26.2	1.3	1.8
28	1.5	25.9	37.9	3.2	26.6	1.6	2.1	28	1.2	30.1	17.9	2.9	29.4	1.3	2.1
30	1.5	27.9	40.5	3.5	28.6	1.6	2.1	30	1.2	32.1	19.9	3	31.4	1.3	2.1
32	1.5	29.6	43	3.6	30.3	1.6	2.6	32	1.2	34.4	20.6	3.2	33.7	1.3	2.6
35	1.5	32.2	46.8	3.9	33	1.6	3	35	1.5	37.8	23.6	3.4	37	1.6	3
38	1.75	35.2	50.2	4.2	36	1.85	3	38	1.5	40.8	26.4	3.7	40	1.6	3
40	1.75	36.5	52.6	4.4	37.5	1.85	3.8	40	1.75	43.5	27.8	3.9	42.5	1.85	3.8
42	1.75	38.5	55.7	4.5	39.5	1.85	3.8	42	1.75	45.5	29.6	4.1	44.5	1.85	3.8
45	1.75	41.5	59.1	4.7	42.5	1.85	3.8	45	1.75	48.5	32	4.3	47.5	1.85	3.8
48	1.75	44.5	62.5	5	45.5	1.85	3.8	48	1.75	51.5	34.5	4.5	50.5	1.85	3.8
50	2.0	45.8	64.5	5.1	47.0	2.15	4.5	50	2.0	54.2	36.3	4.6	53.0	2.15	4.5
60	2.0	55.8	75.6	5.8	57.0	2.15	4.5	60	2.0	64.2	44.7	5.4	63.0	2.15	4.5
65	2.5	60.8	81.4	6.3	62.0	2.65	4.5	65	2.5	69.2	49.0	5.8	68.0	2.65	4.5
70	2.5	65.5	87	6.6	67.0	2.65	4.5	72	2.5	76.5	55.6	6.4	75.0	2.65	4.5
75	2.5	70.5	92.7	7.0	72.0	2.65	4.5	75	2.5	79.5	58.6	6.6	78.0	2.65	4.5
80	2.5	74.5	98.1	7.4	76.5	2.65	5.3	80	2.5	85.5	62.1	7.0	83.5	2.65	5.3
90	3.0	84.5	108.5	8.2	86.5	3.15	5.3	90	3.0	95.5	71.9	7.6	93.5	3.15	5.3
100	3.0	94.5	120.2	9	96.5	3.15	5.3	100	3.0	105.5	80.6	8.4	103.5	3.15	5.3

⇒ **Retaining ring DIN 471 – 40 x 1.75:** d_1 = 40 mm, s = 1.75 mm

⇒ **Retaining ring DIN 472 – 80 x 2.5:** d_1 = 80 mm, s = 2.5 mm

Tolerance classes for d_2 (shafts)

d_1 in mm	3–10	12–22	24–100
d_2	h10	h11	h12

Tolerance classes for d_2 (bores)

d_1 in mm	8–22	24–100	100–300
d_2	H11	H12	H13

[1] Standard design: d_1 from 3–300 mm; heavy duty design: d_1 from 15–100 mm

Circlips (selection)

cf. DIN 6799 (1981-09)

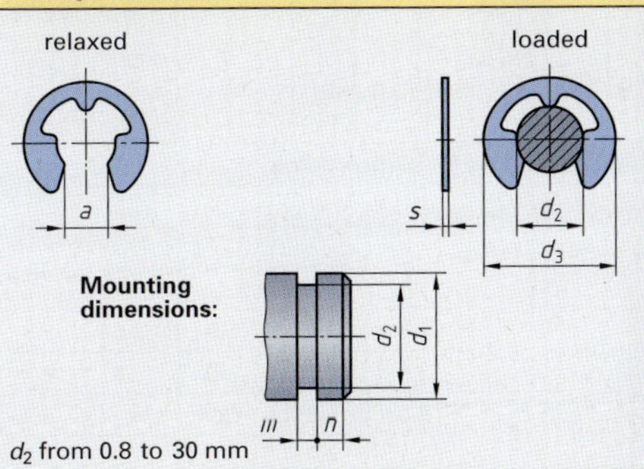

Circlips d_2 h11	Circlips d_3 loaded	Circlips a	Circlips s	Shaft d_1 from–to	Shaft m		Shaft n min
6	12.3	5.26	0.7	7– 9	0.74	+ 0.05 / 0	1.2
7	14.3	5.84	0.9	8–11	0.94		1.5
8	16.3	6.52	1	9–12	1.05	+ 0.08 / 0	1.8
9	18.8	7.63	1.1	10–14	1.15		2
10	20.4	8.32	1.2	11–15	1.25		2
12	23.4	10.45	1.3	13–18	1.35		2.5
15	29.4	12.61	1.5	16–24	1.55		3
19	37.6	15.92	1.75	20–31	1.80		3.5
24	44.6	21.88	2	25–38	2.05		4

⇒ **Circlip DIN 6799 – 15:** d_2 = 15 mm

ME

Sealing elements

Radial seals (selection)

cf. DIN 3760 (1996-09)

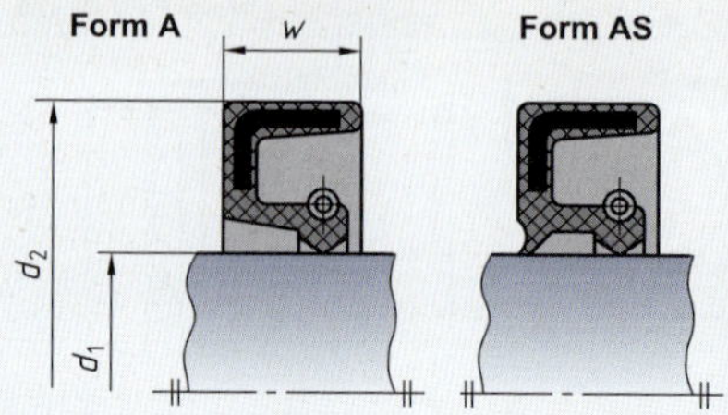

Mounting dimensions:

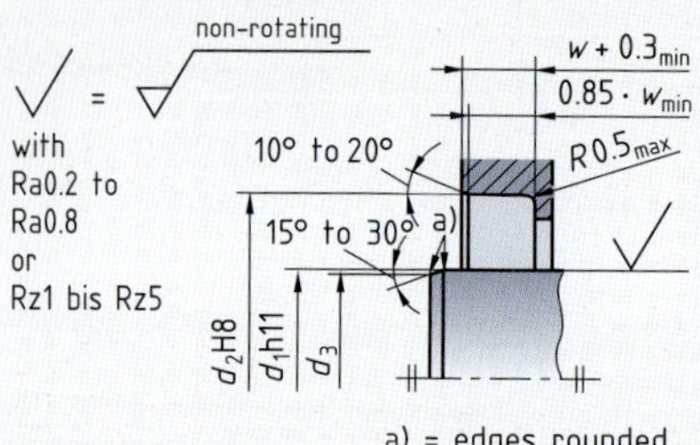

d_1 from 6 to 500 mm

d_1	d_2		w	d_3	d_1	d_2		w	d_3	d_1	d_2		w	d_3
10	22	26	7	8.5	28	40	52	7	25.5	50	65	72	8	46.5
	25	–				47	–				68	–		
12	22	30	7	10	30	40	47	8	27.5	55	70	80	8	51
	25	–				42	52				72	–		
14	24	30	7	12	32	45	52	8	29	60	75	85	8	56
15	26	35	7	13		47	–				80	–		
	30	–			35	47	52	8	32	65	85	90	10	61
16	30	35	7	14		50	55	8	35	70	90	95	10	66
18	30	35	7	16	38	55	62			75	95	100	10	70.5
20	30	40	7	18	40	52	62	8	37	80	100	110	10	75.5
	35	–				55	–	8	38.5	85	110	120	12	80.5
22	35	47	7	19.5	42	55	62			90	110	120	12	85.5
	40	–			45	60	65	8	41.5	95	120	125	12	90.5
25	35	47	7	22.5		62	–	8	44.5	100	120	130	12	94.5
	40	52			48	62	–				125	–		

⇒ **RWDR DIN 3760 – A25 x 40 x 7 – NBR:** Radial seal (RWDR) of form A with d_1 = 25 mm, d_2 = 40 mm and w = 7 mm, elastomer part of Nitrile-Butadiene rubber (NBR)

Felt rings (selection)

cf. DIN 5419 (2010-05)

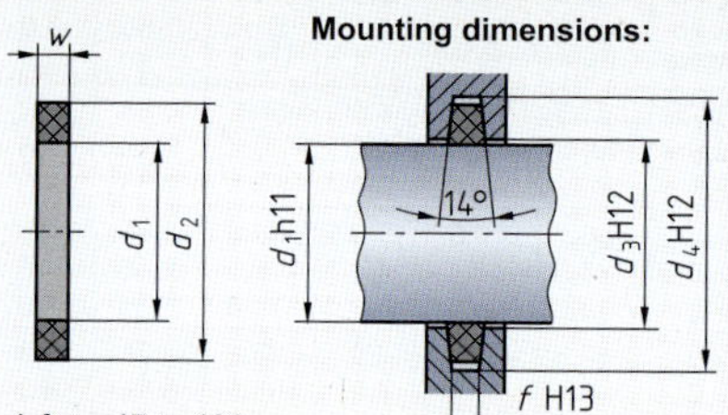

d_1 from 17 to 180 mm

Dimensions			Mounting dim.			Dimensions			Mounting dim.		
d_1	d_2	w	d_3	d_4	f	d_1	d_2	w	d_3	d_4	f
20	30	4	21	31	3	60	76	6.5	61.5	77	5
25	37	5	26	38	4	65	81	6.5	66.5	82	5
30	42	5	31	43	4	70	88	7.5	71.5	89	6
35	47	5	36	48	4	75	93	7.5	76.5	94	6
40	52	5	41	53	4	80	99	7.5	81.5	99	6
45	57	5	46	58	4	85	103	7.5	86.5	104	6
50	66	6.5	51	67	5	90	110	9.5	92	111	7
55	71	6.5	56	72	5	100	124	10	102	125	8

⇒ **Felt ring DIN 5419 M5-40:** Felt ring of d_1 = 40 mm, felt hardn. M5

O-rings

DIN 3771-1 (1984-12)

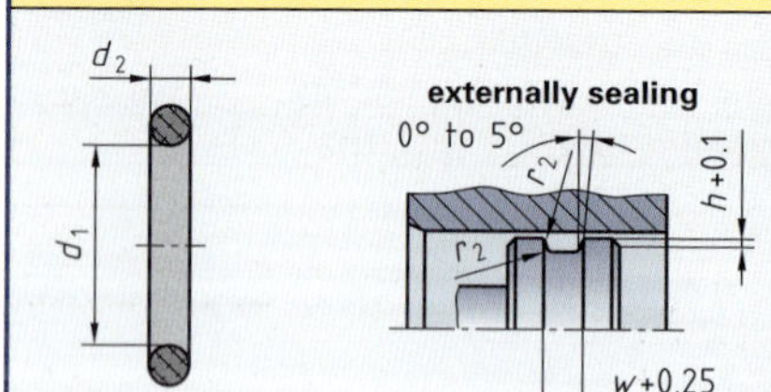

d_1 from 1.8 to 670 mm,
d_2 from 1.8 to 7 mm

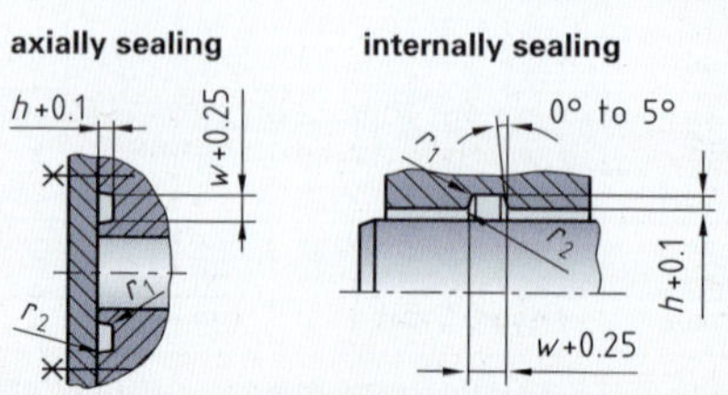

d_1	d_2		d_1	d_2		d_1	d_2		d_1	d_2	
5			18			56			85		
6			20			58			90		
8	1.8		25	2.65	3.55	60			95		
9			28			63			100		
10			30			67	3.55	5.3	103	3.55	5.3
14			40			69			106		
15			45			71			109		
16	1.8	2.65	50	3.55	5.3	75			112		
17			53			80			115		

Mounting dimensions for static loading

d_2	r_1	r_2	internally & extern. sealing			axially sealing	
			w	internal h	external h	w	h
1.8	0.3	0.2	2.4	1.4	1.3	2.6	1.3
2.65			3.6	2.1	1.95	3.8	2
3.55	0.6	0.2	4.8	2.85	2.65	5	2.75
5.3			7.1	4.3	4.15	7.3	4.25

ME

Lubricants

cf. DIN 51502 (1990-08)

Designation of lubricating oils

Designation using code letters

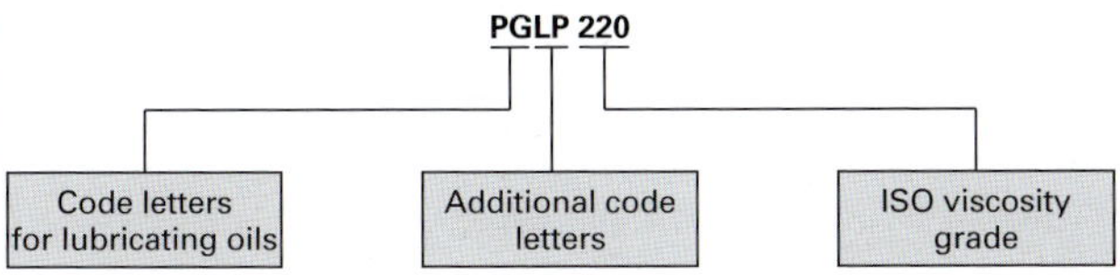

Designation using symbols

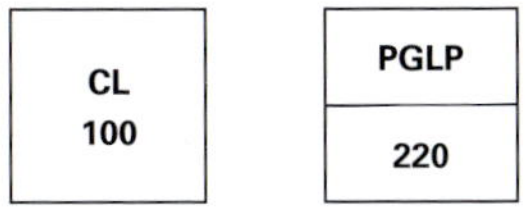

Mineral oil based lubricating oil — Synthetic oil based lubricating oil

⇒ **Lubricating oil DIN 51517 – CL 100:** Circulating mineral oil based lubricating oil (C), increased corrosion and aging resistance (L), ISO viscosity grade VG 100 (100)

⇒ **Lubricating oil DIN 51517 – PGLP 220:** Polyglycol oil (PG), increased corrosion and aging resistance (L), increased wear protection (P), ISO viscosity grade VG 220 (220)

Types of lubrication oils

cf. DIN 51502 (1990-08)

Code letters	Type of lubricant and properties	Standard	Application
Mineral oils			
AN	Normal lubricating oils without additives	DIN 51501	Continuous and circulating lubrication at oil temperatures up to 50 °C
B	Bitumen containing lubricating oils with high adhesion	DIN 51513	Manual, continuous and immersion lubrications, mainly for open lubrication points
C	Circulating lubricating oil, without additives	DIN 51517	Plain bearings, antifriction bearings, gears
CG	Sliding track oil with active ingredients for reducing wear	DIN 8659 T2	In mixed friction operations for slideways and guideways, and for worm gears
Synthetic liquids			
E	Ester oils with especially low change in viscosity	–	Bearings with widely varying temperatures
PG	Polyglycol oils with high aging resistance	–	Bearings with frequent mixed friction conditions
SI	Silicon oils with high aging resistance	–	Bearings with very high and low temperatures, very water repellant

Additional code letters

cf. DIN 51502 (1990-08)

Additional code letters	Application and explanation
E	For lubricants that are mixed with water, e.g. cooling lubricant SE
F	For lubricants with solid lubricant additive, e.g. graphite, molybdenum sulfide
L	For lubricants with active ingredients to improve corrosion protection and/or aging resistance
P	For lubricants with active ingredients for reducing friction and wear in mixed friction areas and/or to increase the load capacity

Liquid industrial lubricants – ISO viscosity grades

cf. DIN ISO 3448 (2010-02)

The lubricant classification of DIN ISO 3448 ist based on a kinematic viscosity of 40° C (104° F).

ISO viscosity grade	Kinematic viscosity at 40° C in mm²/s	ISO viscosity grade	Kinematic viscosity at 40° C in mm²/s	ISO viscosity grade	Kinematic viscosity at 40° C in mm²/s
ISO VG 2	1.98 to 2.42	ISO VG 22	19.8 to 24.2	ISO VG 220	198 to 242
ISO VG 3	2.88 to 3.52	ISO VG 32	28.8 to 35.2	ISO VG 320	288 to 352
ISO VG 5	4.14 to 5.06	ISO VG 46	41.4 to 50.6	ISO VG 460	414 to 506
ISO VG 7	6.12 to 7.48	ISO VG 68	61.2 to 74.8	ISO VG 680	612 to 748
ISO VG 10	9.00 to 11.0	ISO VG 100	90.0 to 110	ISO VG 1000	900 to 1100
ISO VG 15	13.5 to 16.5	ISO VG 150	135 to 165	ISO VG 1500	1350 to 1650

Lubricants

cf. DIN 51502 (1990-08)

Designation of lubricating greases and solid lubricants

Designation by code letters | **Designation by symbols**

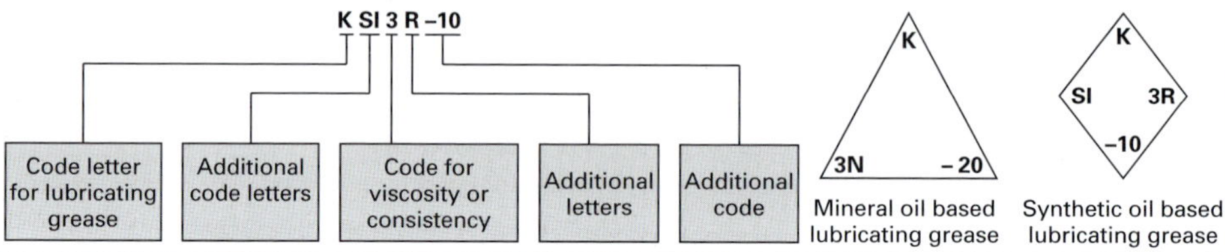

⇒ **Lubricating grease DIN 51825 – K3N –20:** Lubricating grease for antifriction and plain bearings (K) based on mineral oil (NLGI grade 3) (3), upper working temperature +140 °C (N), lower working temperature –20 °C (–20)

⇒ **Lubricating grease DIN 51825 – KSI3R –10:** Silicon based lubricating grease for antifriction and plain bearings (K) (SI), NLGI-grade 3 (3), upper working temperature +180 °C (R), lower working temperature –10 °C (–10)

Lubricating greases

Code letters	Application/additives	Code letters	Application
K	General: antifriction bearings, plain bearing, sliding surfaces	G	Closed gears
KP	Like K, but with additives for reducing friction	OG	Open gears (adhesive lubricant without bitumen)
KF	Like K, but with solid lubricant additives	M	For plain bearings and seals (low requirements)

Consistency[1] classification for lubricating greases

NLGI-grade[3]	Worked penetration[2]	NLGI-grade[3]	Worked penetration[2]	NLGI-grade[3]	Worked penetration[2]
000	445–475 (very soft)	1	310–340	4	175–205
00	400–430	2	265–295	5	130–160
0	355–385	3	220–250	6	85–115 (very firm)

[1] Code for the viscoelasticity
[2] Measure of the penetration depth of a standardized test ball in the kneaded (worked) grease
[3] National Lubrication Grease Institute (NLGI)

Additional letters for lubricating greases

Addit. letter[1])	Upper working temperature °C	Grade [2]	Addit. letter[1])	Upper working temperature °C	Grade [2]	Addit. letter[1])	Upper working temperature °C	Grade [2]
C	+60	0 or 1	G	+100	0 or 1	N	+140	as per agreement
D	+60	2 or 3	H	+100	2 or 3	P	+160	
						R	+180	
						S	+200	
E	+80	0 or 1	K	+120	0 or 1	T	+220	
F	+80	2 or 3	M	+120	2 or 3	U	+220	

[1] The number value for the lower working temperature can be appended to the additional code letters; e. g. –20 for –20 °C
[2] Grades for behavior when subjected to water, cf. DIN 51807-1:
0: no change; 1: small change; 2: moderate change; 3: large change

Solid lubricants

Lubricant	Code	Working temperature	Application
Graphite	C	– 18 to + 450 °C	As powder or paste and as an additive to lubricating oils and lubricating greases, not in oxygen, nitrogen and vacuums
Molybdenum	MoS_2	– 180 to + 400 °C	As mineral oil-free paste, sliding lacquer or additive to lubricating oils sulfide and lubricating greases, suitable for very high surface pressures
Polytetra-fluorethylene	PTFE	– 250 to + 260 °C	As powder in sliding lacquer and synthetic lubricating greases and as bearing material, very low coefficient of sliding friction μ = 0.04 to 0.09

6 Production Engineering

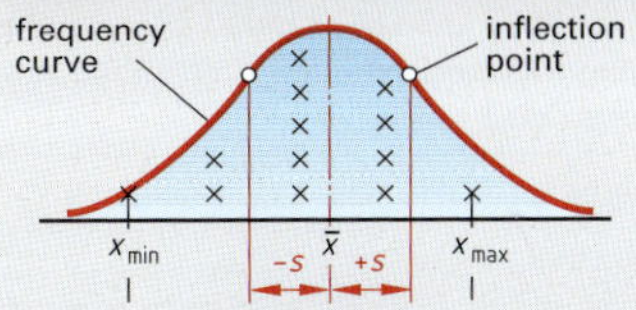

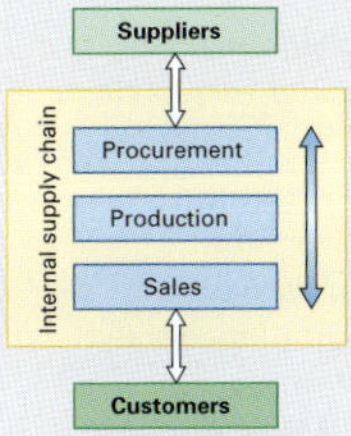

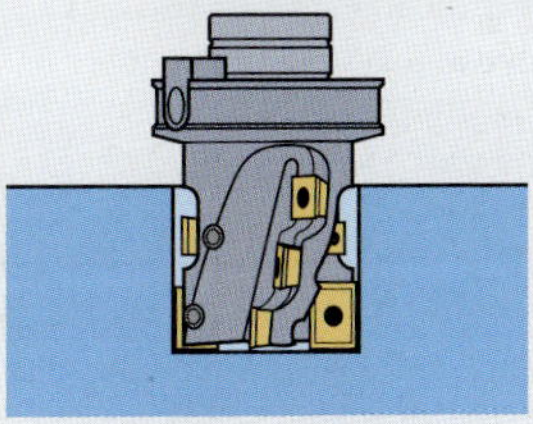

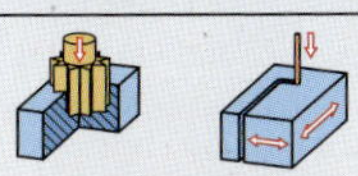

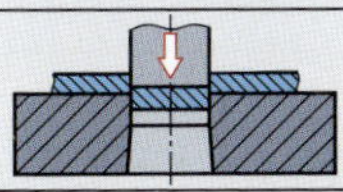

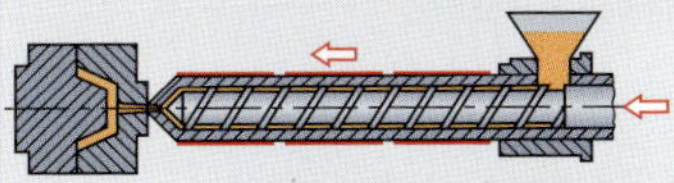

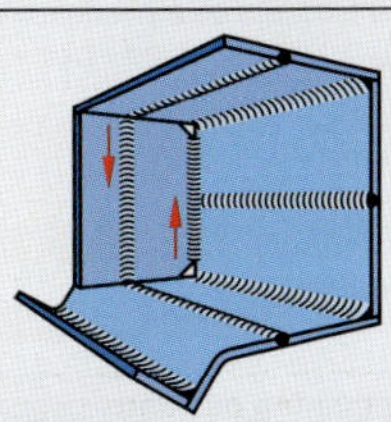

Standards ISO 9000, 9001, 9004

Standards of the ISO-9000 family should help organizations of all types and sizes to implement quality management systems, to work with existing quality management systems, and to facilitate mutual understanding in national and international trade.

Quality management standards

cf. DIN EN ISO 9000 (2005-12), 9001 (2008-12), 9004 (2009-12)

Standard	Explanation, contents
DIN EN ISO 9000	**Fundamentals of quality management systems**
	Principle of quality management • customer focus • leadership • involvement of people • process approach • system approach to management • continuous improvement • factual approach to decision making • mutually beneficial supplier relationships **Fundamentals of quality management systems (QM systems)** • reasons for QM systems • requirements of QM systems and products • progressive implementation of QM systems • process oriented evaluation • quality policies and goals • role of top management in the QM system • documentation; advantages and types • evaluation of QM systems • continuous improvement • role of statistical methods • QM systems as part of the total management system • requirements of QM systems and comparative evaluation of organizations based on criteria of excellence models
	Terminology for quality management systems
	For a selection of definitions and explanations of terms, see page 271.
DIN EN ISO 9001[1])	**Requirements of a quality management system**
	This international standard applies to organizations in any industry or business sector regardless of products offered. It establishes requirements for a QM system, based on fundamentals outlined in ISO 9000, if an organization: • must demonstrate capability to offer products which fulfill both customer and legal and official requirements, • strives to improve customer satisfaction, including the process of continuous improvement of the system. Specified requirements can be used for: • internal applications by organizations • certification purposes • contract purposes The standard is based on **a process oriented evaluation**, i.e. every activity or sequence of activities which uses resources to convert input into results is regarded as a process. **Requirements** The organization must: • establish all necessary processes for the QM system and their use in the organization, • establish the flows and interdependencies of these processes, • establish criteria and methods for ensuring implementation and control of these processes, • ensure availability of resources and information for these processes, • monitor and, if applicable, measure and analyze these processes, • take necessary actions for continuous improvement of these processes, • fulfill documentation requirements for the QM system, and • observe regulations for document control.
	[1]) This standard also replaces previous standards 9002 and 9003.
DIN EN ISO 9004	**Managing and controlling for the long-lasting success of an organization – an approach to quality management**
	The standard includes guidelines for organizations on how to pursue long-term goals in a demanding, changing and uncertain environment. The standard is not intended for certification purposes.

PE

Terminology

Terms (selection)	**Definitions & explanations** cf. DIN EN ISO 9000 (2005-12)
Quality-related terms	
Quality	Extent to which the characteristics of a product fulfill the requirements for that product.
Requirement	Specified or mandatory demands for characteristics of a unit, e.g. nominal values, tolerances, functional capability, reliability or safety.
Customer satisfaction	Customer's perception of degree to which its requirements have been fulfilled.
Capability	Suitability of an organization, system or process to provide a product that fulfills that product's quality requirements.
Characteristic and conformity related terms	
Quality characteristic	Identifying attribute of a product or process, which is utilized in assessing quality based on the specified quality requirements. • Quantitative (variable) characteristics: discrete characteristics (whole numbers), i.e. number of holes, piece count continuous characteristics (counter values), e.g. length, position, mass • Qualitative characteristics: ordinal characteristics (with ranking), e.g. light blue – blue – dark blue nominal characteristics (without ranking), e.g. good – bad, blue – yellow Identifying attribute of a product, a process or system relating to a requirement.
Conformity	Fulfilling a specified requirement, e.g. a dimensional tolerance.
Defect	Not fulfilling a specified requirement, e.g. not conforming to a required dimensional tolerance or surface quality.
Rework	Action taken on a defective product so that it fulfills requirements.
Process and product related terms	
Process	Mutually interactive resources and activities which convert inputs into results. Some examples of resources are personnel, finances, facilities and manufacturing methods.
Method	Defined manner in which an activity or process is performed. In written form also referred to as process instructions.
Product	Result of a process, e.g. part, assembly, service, processed item, knowledge, concept, document, contract, pollutant.
Terms related to organization	
Organization	Group of persons and facilities with a matrix of responsibilities, authorities and relationships.
Customer	Organization or person which receives a product from a supplier.
Supplier	Organization or person which provides a product to a customer.
Terms relating to management	
Quality management system	Organization and organizational structures, methods and processes of an operation required to put a quality management into practice.
Quality management	All coordinated activities for managing and controlling the quality-related aspects of an organization by: • establishing a quality policy • quality control • setting quality goals • quality assurance • quality planning • quality improvement
Quality planning	Activities directed toward establishing quality goals and required implementation processes, as well as associated resources for attaining quality goals.
Quality control	Work activities and techniques to continually fulfill requirements despite unavoidable variations in quality. Consists primarily of process monitoring and elimination of weak points.
Quality assurance	Performing and generating required documentation for all activities relating to the QM system, with the goal of creating an atmosphere of trust, both in-house and with the customer, that quality requirements will be fulfilled.
Quality improvement	Actions taken throughout the organization to increase product quality.
Quality manual	Document describing the quality policy, quality goals and quality management system of an organization.

Quality planning, Quality control, Quality testing

Quality planning

Rule-of-ten (for costs)

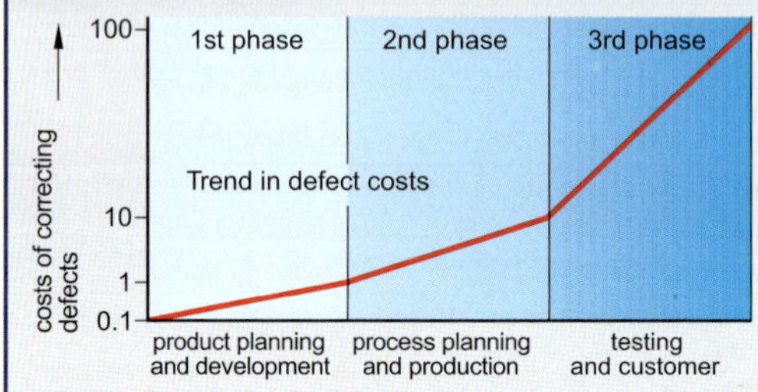

Costs required to eliminate defects or costs resulting from defects increase by about a factor of 10 from phase to phase in the product life cycle.

Example: A tolerance error on a single part can be corrected during the design phase with negligible increase of costs. If the defect is first noticed in production, much larger costs result. If the defect leads to problems in assembly or has an adverse impact on the functionality of the finished product or even leads to a recall, enormous costs are incurred.

Quality control

Quality control circle

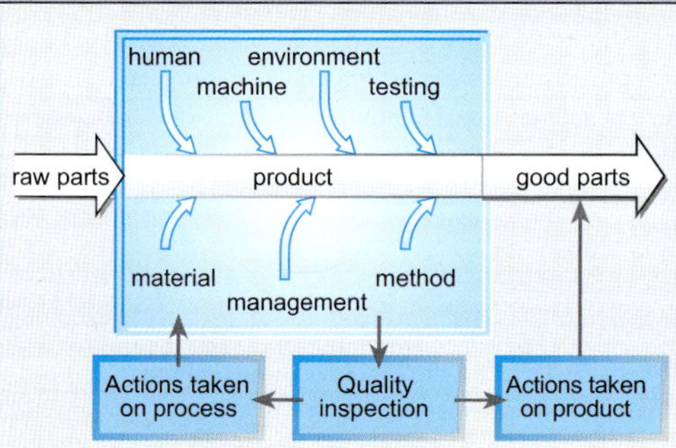

Factors causing variance in quality

Factor	Examples
Human	qualification, motivation, degree of utilization
Machine	machine rigidity, positioning accuracy, wear condition
Material	deviations, material properties, material variations
Method	work steps, production process, test conditions
Surroundings (environment)	temperature, vibrations, light, noise, dust
Management	poor quality goals or policies
Measurability	measurement inaccuracy

Quality testing

cf. DIN 55350-17 (1988-08), DIN 55350-14 and -31 (1985-12)

Concepts	Explanations
Quality testing	Determine to what extent a unit meets specified quality requirements.
Test plan **Test instructions**	Define and describe the type and scope of testing, e.g. measuring and monitoring devices, frequency of testing, test personnel, testing location.
Complete testing	Testing of a unit for all specified quality characteristics, e.g. complete inspection of a single workpiece regarding all requirements.
100% testing	Testing of all units within a test lot, e.g. visual inspection of all delivered parts.
Statistical testing (sampling test)	Quality testing with the aid of statistical methods, e.g. evaluation of a large quantity of parts by analyzing a number of sampled parts.
Test lot (sampling test)	The entirety of the considered units, e.g. a production of 5000 identical workpieces.
Sample	One or more units which are taken from the population or a subset of the population, e.g. 50 parts from a daily production of 400 parts.

PE

Probability (Probability of defect)

Probability of a defective part within a defined total number of parts.

P probability in %
n number of defective parts
m total number of parts

Example:

In a crate there are $m = 400$ parts, where $n = 10$ parts have a dimensional defect. What is the probability P of obtaining a defective part when taking one part out of the crate?

Probability $P = \frac{n}{m} \cdot 100\,\% = \frac{10}{400} \cdot 100\,\% = \mathbf{2.5\%}$

Probability

$$P = \frac{n}{m} \cdot 100\,\%$$

Statistical analysis

Statistical analysis of continuous characteristics

cf. DIN 53 804-1 (2002-04)

Presentation of test data	Example

Raw data list

Raw data is the documentation of all observed values from the test lot or sample in the sequence in which they occur.

Sample size: 40 parts

Test characteristic: part diameter $d = 8 \pm 0.05$ mm

Measured part diameter d in mm										
Parts 1–10	7.98	7.96	7.99	8.01	8.02	7.96	8.03	7.99	7.99	8.01
Parts 11–20	7.96	7.99	8.00	8.02	8.02	7.99	8.02	8.00	8.01	8.01
Parts 21–30	7.99	8.05	8.03	8.00	8.03	7.99	7.98	7.99	8.01	8.02
Parts 31–40	8.02	8.01	8.05	7.94	7.98	8.00	8.01	8.01	8.02	8.00

Tally sheet

The tally sheet provides a clear presentation of the observed values and assignment into classes (ranges) of a specific class interval size.

- n number of individual values
- k number of classes
- i class interval
- R range (page 274)
- n_j absolute frequency
- h_j relative frequency in %

Class no.	Measured value ≥	Measured value <	Tally sheet	n_j	h_j in %
1	7.94	7.96	I	1	2.5
2	7.96	7.98	III	3	7.5
3	7.98	8.00	𝍸 𝍸 I	11	27.5
4	8.00	8.02	𝍸 𝍸 III	13	32.5
5	8.02	8.04	𝍸 𝍸	10	25
6	8.04	8.06	II	2	5
			Σ =	40	100

$c = \sqrt{n} = \sqrt{40} = 6.3 \approx 6$

$i = \frac{R}{c} = \frac{0.11\ \text{mm}}{6} = 0.018\ \text{mm} \approx 0.02\ \text{mm}$

Number of classes

$$k \approx \sqrt{n}$$

Class interval size

$$i \approx \frac{R}{k}$$

Relative frequency

$$h_j = \frac{n_j}{n} \cdot 100\%$$

Histogram

A histogram is a bar graph for visualizing the distribution of individual test data.

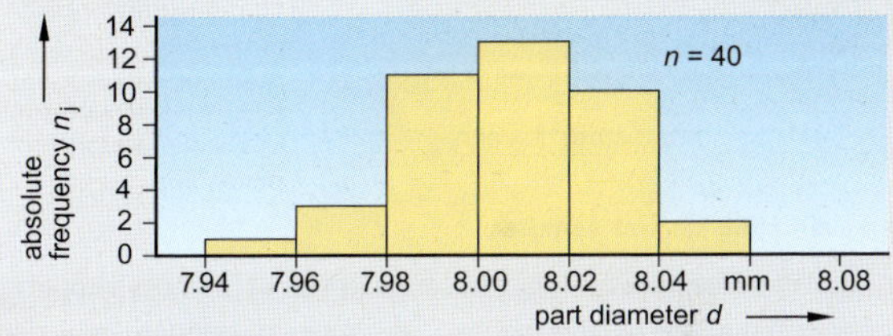

Cumulative frequency curve in probability system

The cumulative frequency curve in the probability system is a simple and clear graphical method used to check for the existence of a normal distribution (page 274).

If the cumulative relative frequency in the probability system approximates a straight line, then a normal distribution of the individual values can be assumed, i.e. a further evaluation can be conducted per DIN 53 804-1 (page 274).

In this case specific values can additionally be determined from the samples.

Example of problem solving using the graph:

Arithmetic mean $\bar{x}$ (for F_j = 50%) and standard deviation s (as difference 68.26% ÷ 2 between F_j = 50% and 84.13%):

$\bar{x} \approx 8.003$ mm; $s \approx 0.02$ mm

The probability model of the example shows that in the entire lot approximately 0.6% of parts can be expected to be too thin and 3% too thick.

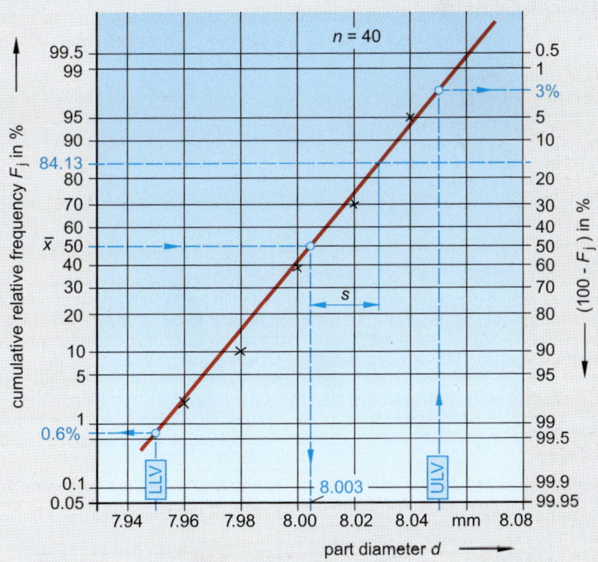

LLV lower limit value; ULV upper limit value

PE

Normal distribution

Gaussian distribution

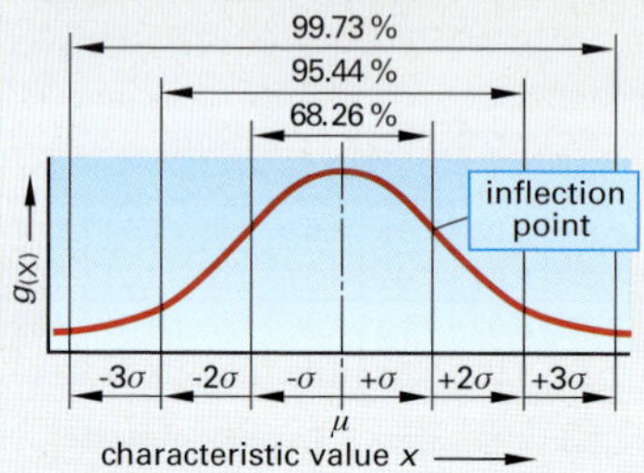

Continuous data values often exhibit a characteristic in their distribution which is approximated mathematically by the **Gaussian normal distribution** model. For an infinite number of individual values the probability density of a normal distribution yields the typical **bell curve.** This symmetrical and continuous distribution curve is clearly described by the following parameters:

The **mean** μ lies on the curve maximum and identifies the position of the distribution.

The **standard deviation** σ is a measure of the variations, i.e. how values deviate from the mean.

[1] Carl Friedrich Gauß (1777–1855), German mathematician

Normal distribution in sampling

cf. DIN 53804-1 (2002-04) or DGQ 16-31 (1990)

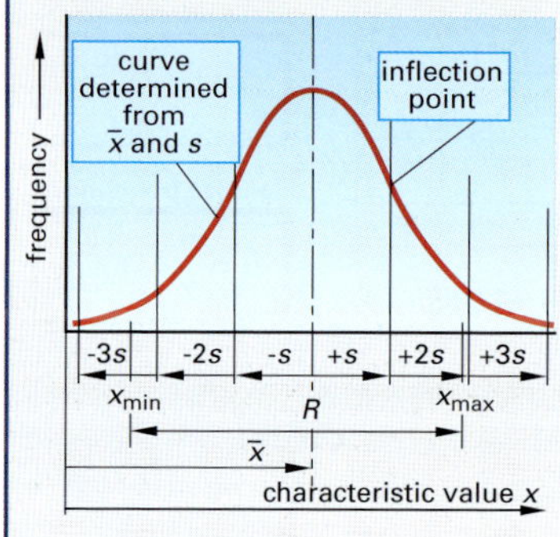

n	number of individual values (sample size)
x_i	value of measurable properties, e.g. individual value
x_{max}	largest measurement value
x_{min}	smallest measurement value
$\bar{x}$	arithmetic mean
$\tilde{x}$	median value[1], middle value of measured values arranged in order of magnitude
s	standard deviation
R	range
D	mode (measurement value occurring most frequently in a test series)
$g_{(x)}$	probability density

Arithmetic mean[2)]

$$\bar{x} = \frac{x_1 + x_2 + \ldots + x_n}{n}$$

Standard deviation[2)]

$$s = \sqrt{\frac{\sum(x_i - \bar{x})^2}{n-1}}$$

Range

$$R = x_{max} - x_{min}$$

When evaluating several samples:

m	number of samples	$\bar{R}$	mean of multiple sample ranges
$\bar{\bar{x}}$	mean of multiple sample means	$\bar{s}$	mean of standard deviations

Example: Evaluation of sample values from page 273:

$\bar{x}$ = **8.00225 mm** R = **0.11 mm** $\tilde{x}$ = **8.005 mm** s = **0.02348 mm** D = **7.99 mm**

Mean of sample ranges

$$\bar{R} = \frac{R_1 + R_2 + \ldots + R_m}{m}$$

Mean of sample means

$$\bar{\bar{x}} = \frac{\bar{x}_1 + \bar{x}_2 + \ldots + \bar{x}_m}{m}$$

Mean of standard deviations

$$\bar{s} = \frac{s_1 + s_2 + \ldots + s_m}{m}$$

[1] Median value for
odd number of individual values: e.g. x_1; x_2; x_3; x_4; x_5: $\tilde{x} = x_3$
even number of individual values: e.g. x_1; x_2; x_3; x_4; x_5; x_6: $\tilde{x} = (x_3 + x_4)/2$

[2] Many pocket calculators have special functions for calculating the mean and standard deviation.
Repeated occurrences of identical measurement values can be represented by a suitable factor.

Normal distribution in an inspection lot

Parameters of the population are estimated using a sampling method based on characteristic values from the sample (confirmatory statistics). To differentiate sampling characteristics clearly from parameters of the population, other designations are used. These estimated values are distinguished from the calculated process values for a 100 % inspection (descriptive statistics) by adding a ^ mark.

Characteristic values and designations in quality testing

Sampling test (confirmatory statistics)		100 % inspection (descriptive statistics)
Sample	**Population**	
Number of measured values n	Number of measured values $m \cdot n$	Number of measured values N
Arithmetic mean $\bar{x}$	Estimated process mean $\hat{\mu}$	Process mean μ
Standard deviation s	Estimated process standard deviation $\hat{\sigma}$ (calculator σ_{n-1})	Process standard deviation σ (calculator σ_n)

Statistical process control

Quality control charts

Process control charts	Acceptance control charts
Process control charts are used for monitoring a process for changes compared to a target value or a previous process value. The intervention and warning limits are determined by the process estimated value of a population or a preliminary run.	Acceptance control charts are used to monitor a process in reference to set specification limits (limit values). Control limits are calculated as tolerance limits for the location of the process mean and a tolerance range for process variance.

Process control charts for quantitative characteristics (Shewhart-control charts)[1)]

Raw data chart

The raw data chart is a documentation of all measurement values by entering directly on the chart. It assumes an approximate normal distribution process and is relatively complex because of the number of entries.

Control limits

$\bar{x}$	characteristic mean (mean of the characteristic, target value, ideal value)
UWL	upper warning limit
LWL	lower warning limit
UCL	upper control limit
LCL	lower control limit
USL	upper specification limit
LSL	lower specification limit

Example: 5 individual values for each sample

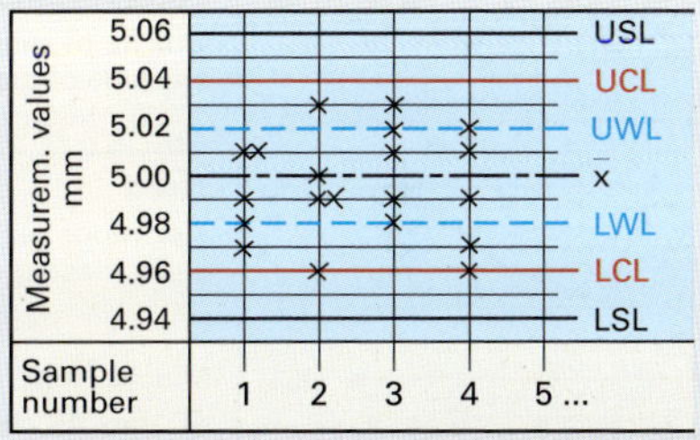

Median value range chart ($\tilde{x}$-R-chart)

These charts are used to clearly represent production dispersion without requiring much calculation. They are suitable for manual control chart management.

Example:

Inspect. characteristic: diameter		Control dimension: 5±0.05			
Sample size $n = 5$		Control interval 60 min			
Measurem. values mm	x_1	4.98	4.96	5.03	4.97
	x_2	4.97	4.99	5.01	4.96
	x_3	4.99	5.03	5.02	5.01
	x_4	5.01	4.99	4.99	4.99
	x_5	5.01	5.00	4.98	5.02
	Σx	24.96	24.97	25.03	24.95
	$\tilde{x}$	4.99	4.99	5.01	4.99
	R	0.04	0.07	0.05	0.06

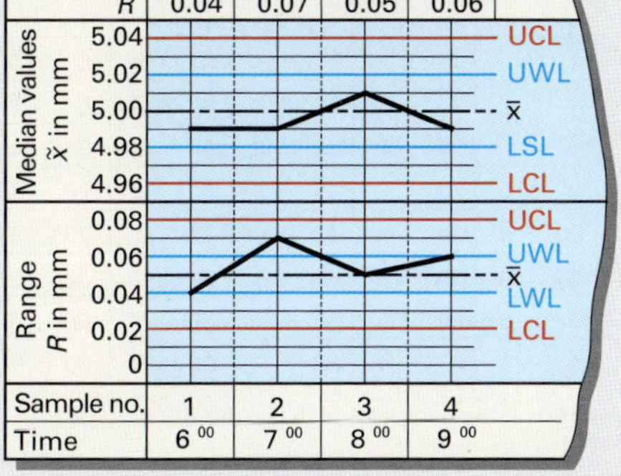

Mean standard deviation chart ($\bar{x}$-s-chart)

These charts are used to show the trend of the mean and exhibit greater sensitivity than $\tilde{x}$-R-charts. They require computer-aided control chart management.

Example:

Inspect. characteristic: diameter		Control dimension: 5±0.05			
Sample size: $n = 5$		Control intervall: 60 min			
Measurem. values mm	x_1	4.98	4.96	5.03	4.97
	x_2	4.97	4.99	5.01	4.96
	x_3	4.99	5.03	5.02	5.01
	x_4	5.01	4.99	4.99	4.99
	x_5	5.01	5.00	4.98	5.02
	$\bar{x}$	4.992	4.994	5.006	4.990
	s	0.018	0.025	0.021	0.025

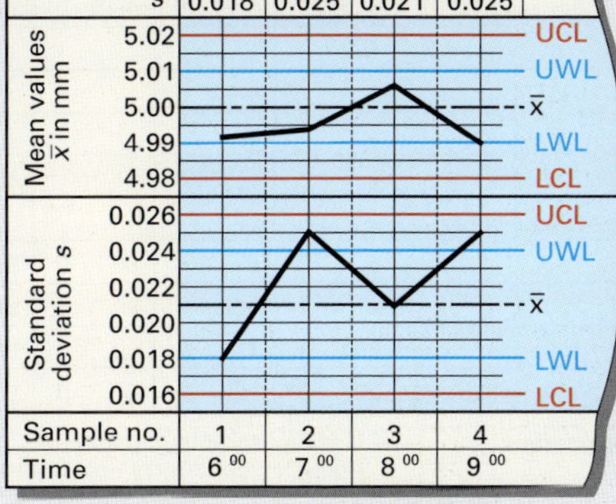

[1)] Walter Andrew Shewhart (1891–1967), American scientist

PE

Process trend, Acceptance sampling and plan

Process trends

Process trend (e.g. from an $\bar{x}$ trace)	Designation/observations	Possible causes → Actions
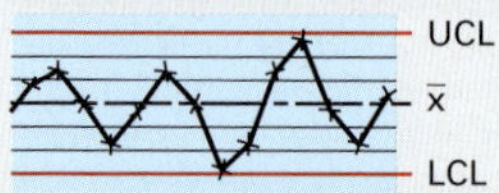	**Natural run** 2/3 of all values lie in the range ± standard deviation *s* and all values lie within the control limits.	The process is under control and can continue without interruption.
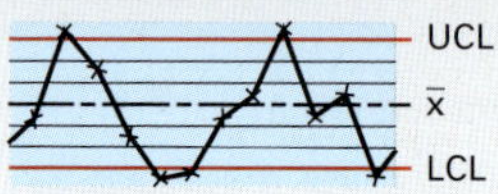	**Exceeding the control limits** The values are outside of the control limits.	Over-adjusted machine, different material, damaged or worn equipment → Stop process and 100% inspect parts since the last sampling
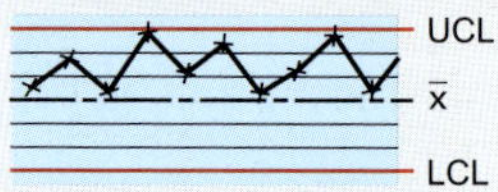	**RUN (sequential)** 7 or more sequential values lie on one side of the mean line.	Tool wear, different material batch, new tool, new personnel → Tightened observation of the process
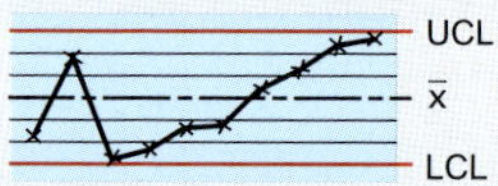	**Trend** 7 or more sequential values show an increasing or decreasing trend.	Wear on tool, equipment or measuring devices, operator fatigue → Stop process to determine reasons for adjustment
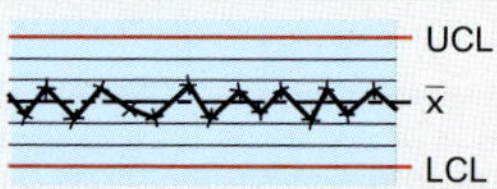	**Middle Third** At least 15 consecutive values lie within ± standard deviation *s*.	Improved production, better supervision, corrected test results → Determine how the process was improved or check the test results
	**Cyclical** The values cross the mean line periodically .	Different measuring devices, systematic spread of the data → Examine manufacturing process for influences

Acceptance sampling (attribute sampling)

cf. DIN ISO 2859-1 (2004-01)

An attribute inspection is an acceptance sampling inspection in which the acceptability of the inspection lot is determined based on defective units or defects in individual sampling.

The **percentage of nonconforming units or the number of defects per hundred units of the lot** identifies the **quality level**. The acceptable quality level is the quality level defined for continuously presented lots; it is a quality level that is specified by the customer in most cases. The associated sampling instructions are summarized in control tables.

Acceptance sampling plan for single sampling inspection as the normal inspection (excerpt from a control table)

PE

Lot size	Acceptable quality level AQL (preferred values)																			
	0.04		0.065		0.10		0.15		0.25		0.40		0.65		1.0		1.5		2.5	
2– 8	↓		↓		↓		↓		↓		↓		↓		↓		↓		↓	
9– 15	↓		↓		↓		↓		↓		↓		↓		↓		8	0	5	0
16– 25	↓		↓		↓		↓		↓		↓		↓		13	0	8	0	5	0
26– 50	↓		↓		↓		↓		↓		↓		20	0	13	0	8	0	5	0
51– 90	↓		↓		↓		↓		50	0	32	0	20	0	13	0	8	0	20	1
91– 150	↓		↓		↓		80	0	50	0	32	0	20	0	13	0	32	1	20	1
151– 280	↓		↓		125	0	80	0	50	0	32	0	20	0	50	1	32	1	32	2
281– 500	↓		200	0	125	0	80	0	50	0	32	0	80	1	50	1	50	2	50	3
501–1200	315	0	200	0	125	0	80	0	50	0	125	1	80	1	80	2	80	3	80	5

Explanation:

↓ — Use first sampling instruction of this column. If the sample size is greater than or equal to the batch size: Carry out a 100% inspection.

50 2

Second number: Acceptance number = number of the accepted delivered defective units

First number: Sample size = number of units to be tested

Process and machine capability, Quality control charts

Capability, Quality control charts

During an evaluation of the quality-related capability of a process through **capability characteristics** (capability indices), differentiation must be made between **short-term capability (machine capability)** and **long-term capability (process capability)**.

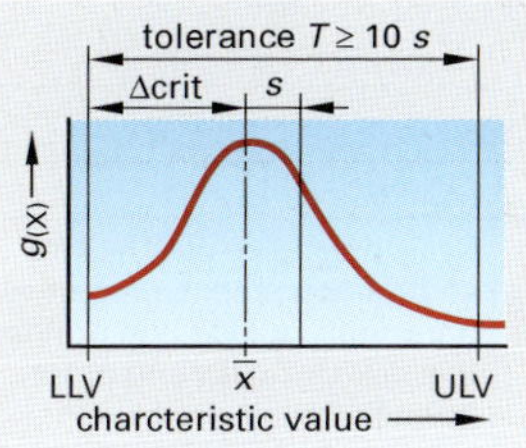

Machine capability is an evaluation of the machine, i.e. whether there is sufficient probability that it can produce within specified limits given its normal fluctuations.

If $C_m \geq 1.67$ and $C_{mk} \geq 1.67$, this means that 99.9999 4% (range ± 5 *s*) of the quality characteristics lie within the limits and the mean $\bar{x}$ lies at least an amount of 5 *s* away from the tolerance limits.

LLV lower limit value
ULV upper limit value
$\bar{x}$ arithmetic mean
s standard deviation
Δcrit smallest interval between mean and a tolerance limit
C_m, C_{mk} machine capability index

Process capability is an assessment of the manufacturing process, i.e. whether there is sufficient probability that it can fulfill specified requirements given its normal fluctuations.

$\hat{\sigma}$ estimated standard deviation
C_p, C_{pk} process capability index

Example:

Examination of machine capability for production dimension 80 ± 0.05;

Values from preliminary run: $s = 0.009$ mm; $\bar{x} = 79.997$ mm

$$C_m = \frac{T}{6 \cdot s} = \frac{0{,}1\,\text{mm}}{6 \cdot 0.009\ \text{mm}} = \mathbf{1.852};\quad C_{mk} = \frac{\Delta\text{crit}}{3 \cdot s} = \frac{0.047\ \text{mm}}{3 \cdot 0.009\ \text{mm}} = \mathbf{1.74}$$

The machine capability for this process is confirmed.

Machine capability index

$$C_m = \frac{T}{6 \cdot s}$$

$$C_{mk} = \frac{\Delta\text{crit}}{3 \cdot s}$$

Requirement[1] e.g.

$C_m \geq$ **1.67 and** $C_{mk} \geq$ **1.67.**

Process capability index

$$C_p = \frac{T}{6 \cdot \hat{\sigma}}$$

$$C_{pk} = \frac{\Delta\text{crit}}{3 \cdot \hat{\sigma}}$$

Requirement[1] e.g.

$C_p \geq$ **1.33 and** $C_{pk} \geq$ **1.33**

[1] Customer or contract specific requirements; in large scale production, e.g. automotive industry, tendency to higher requirements, e.g. $C_m \geq 2.0$.

Quality control charts for qualitative characteristics

cf. DGQ 16-33 (1990); DGQ 11-19 (1994)

Defect chart

Defect charts record the defective units, the defect types and their frequency in samples.

n scope of sampling
m number of samples

Reading example for F3:

$m \cdot n = 9 \cdot 50 = 450$

$$\text{defects in \%} = \frac{\Sigma i_j}{n} \cdot 100\,\% = \frac{3}{450} \cdot 100\,\% = \mathbf{0.66\,\%}$$

Example:

Part: **Cover**		Sample size $n = 50$										Test interval: 60 min		
Defect type		Frequency of defect i_j										Σi_j	%	Perc. of total
Paint damage	F1		1					1				2	0.44	
Dents	F2	1	2		2	1	2	2	2	2		14	3.11	
Corrosion	F3		1			1			1			3	0.66	
Burr	F4	1										1	0.22	
Crackings	F5		1									1	0.22	
Angle error	F6	2		3	1		3	1		2		12	2.66	
Bent	F7					1						1	0.22	
Threads missing	F8		1									1	0.22	
Defects per sample		4	6	3	3	3	5	4	3	4		35		
Sample no.		1	2	3	4	5	6	7	8	9				

Pareto[1] diagram

The Pareto diagram classifies criteria (e.g. defects) according to type and frequency and is therefore an important aid in analyzing criteria and establishing priorities.

Example for F2:

Percentage of total defects

$$= \frac{14}{35} \cdot 100\,\% = 40\,\%$$

[1] Pareto – Italian sociologist

Example:

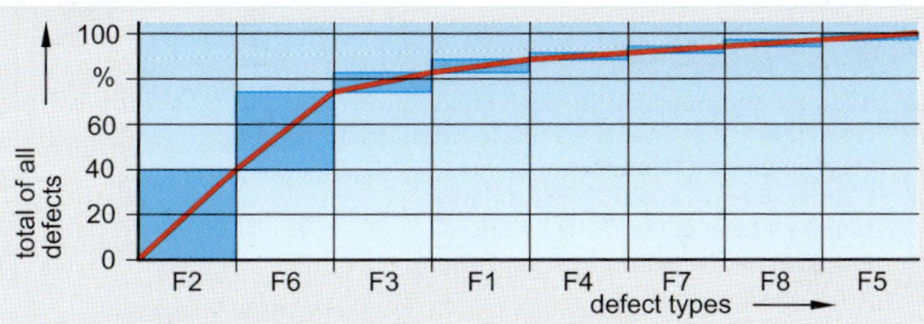

Example of graphic representation: Dents (F2) and angle error (F6) together make up for approx. 74% of the total errors.

PE

EC Machinery Directive

Structure and contents

DIRECTIVE 42/2006/EC

The purpose of the EC Machinery Directive is to reduce accidents in connection with the operation of machines. This aim should be reached by observing safety aspects in the engineering and construction of machines as well as by proper installation and maintenance.

Only machines that comply with the requirements of the EC Machinery Directive may be traded freely in the European market.

Overview

Articles	Content
Article 1, 2 and 3 Area of application, definition of terms	Enumeration and definition of the products to which the Directive does or does not apply
Article 4 to 11 Market surveillance, marketing and commissioning	Description of the measures to be undertaken before the marketing and commissioning of a machine in the free economy
Article 12 to 15 Conformity evaluation procedures	Information concerning the type, scope and execution of evaluation procedures
Article 16 and 17 CE marking	Definition of the CE marking
Article 18 to 29 Observance of secrecy, sanctions, implementation	General notes concerning the implementation, sanctions and the coming into force of the directive

Annexes	Content
Annex I Safety and health protection requirements	Basic principles of safety for control equipment, protective measures, guards and protective devices, risks and other hazards, repair and maintenance, information, **warning notes** and signs, operating instructions
Annex II to V EC Declaration of Conformity, CE marking, list of machines and safety components	Minimum specifications in the declaration of conformity, representation of the CE marking, particularly dangerous machines
Annex VI and VII Technical documents for machinery	Content and scope of the technical documentation for complete and partly completed machines
Annex VIII to XI EC type examination, comprehensive quality assurance	Description of the type examination, basic principles for the evaluation of the quality assurance system

Products to which the EC Machinery Directive does and does not apply

As per article 1 (1) and article 2, the EC Machinery Directive applies to the following products:

a) Machines
b) Safety components
c) Lifting accessories
d) Chains, ropes and webbing
e) Removable mechanical transmission devices
f) Partly completed machines that are intended for the installation in a machine in the sense of the EC Machinery Directive

The EC Machinery Directive **does not apply** to the following products (excerpt):

a) Safety components intended as spare parts
b) Specific equipment for fairgrounds and amusement parks
c) Machinery designed for nuclear purposes
d) Weapons
e) Means of transportation and sea-going vessels
f) Machinery specially designed for research purposes
g) Electrical and electronic products, such as household appliances, IT equipment, office machinery, low-voltage control equipment, electrical motors

PE

Approach to compliance with the EC Machinery Directive

1. Check applicable standards and regulations, particularly the safety and health requirements specified in Annex I (cf. page 279).
2. Evaluate the machine in view of compliance with the conformity assessment described in the EC Machinery Directive
   - in the factory "First Party",
   - or by the customer "Second Party",
   - or by a trust center "Third Party"
3. Prepare the declaration of conformity
4. Apply the CE mark
5. Prepare the operating instructions
6. Prepare additional technical documents, such as assembly instructions and the declaration of incorporation for partly completed machines, if applicable.

CE Marking

Declaration of conformity

The manufacturer must provide evidence that he/she has respected the regulations in the required EC directives (Declaration of conformity).

Conformity assessment

- Machinery directive and its safety and health regulations specified in Annex I → **Declaration of conformity / CE marking**
- Regulations of other directives, e. g. Low Voltage Directive 2006/95/EC → **Declaration of conformity / CE marking**

CE marking

CE mark

Identification plate

"CE" = Communauté européenne[1)]

By applying the CE mark, the manufacturer confirms to the customer compliance of the product with the relevant EC directives and the requirements contained therein. The identification plate must contain the following specifications:

- Name and address of the manufacturer
- CE marking
- Type and serial number of the machine
- Year of manufacture

1) European Community

Safety and health protection regulations

The regulations refer to hazards as a potential source of injury and damage to health and to impairments by noise and vibration as well as to ergonomic (human-oriented) principles.

Measures to avoid hazards

Assess and evaluate risks
↓
Avoid or minimize hazards
↓
Undertake protective measures
↓
Inform operators and users about hazards

Safety standards (excerpt from Annex I)

- Operation, set-up and maintenance must not involve any hazard to people
- The materials used must not imply any hazards
- All components must be sufficiently stable and the connections among them must be able to withstand the loads and stresses acting on them
- Ergonomic principles must be taken into consideration
- Commissioning of machines only by deliberate activation
- EMERGENCY-OFF devices must be in existence
- Any risk of contact with moving parts must be avoided by suitable guards
- Hazards by noise emissions must be reduced to the lowest achievable level
- Hazards produced by gases and dusts must be avoided
- Transport of machines must be safe. Corresponding handles or lifting equipment must be available
- Lighting must not produce disturbing shadows or glare
- Changes in the power supply (mains failure) must not produce hazardous situations

PE

Technical documents for machines (excerpt from Annex VII)

- General description
- Overview drawing, wiring diagrams
- Detail drawings and calculations
- Risk assessment
- List of applicable standards
- Operating instructions for the machine
- Assembly instructions and incorporation declaration, if applicable
- EC declaration of conformity

Production organization - Overview

Production process (supply chain management)

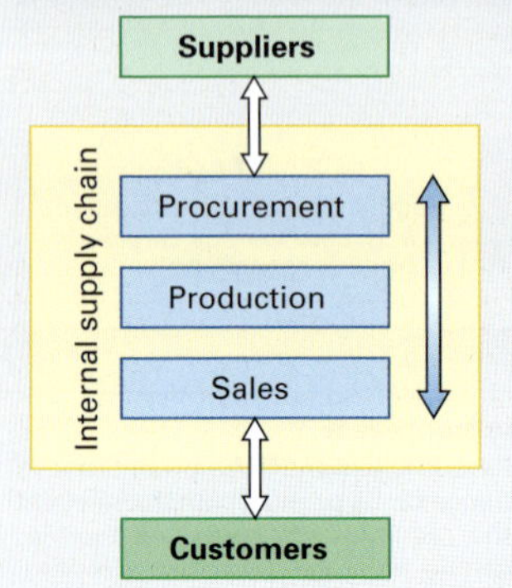

Product life cycle (life cycle management)

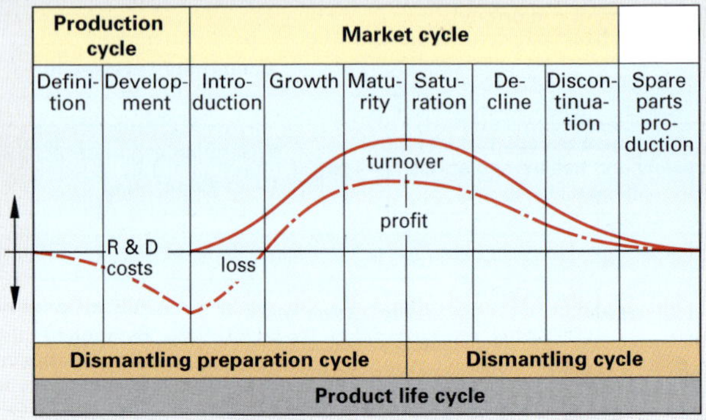

Product management systems

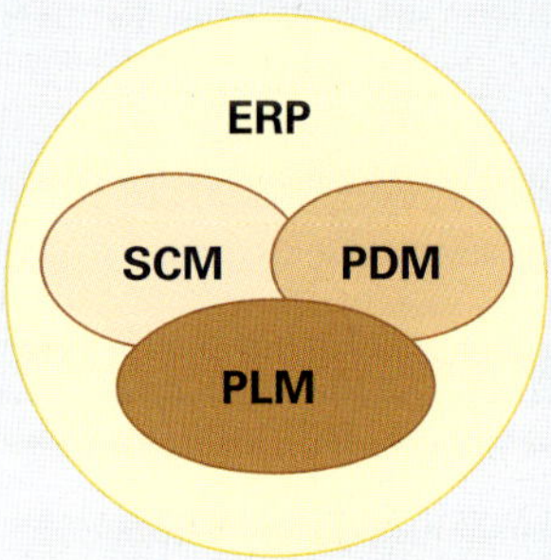

ERP	**Enterprise Recource Planning** Information system for the planning and organization of personnel and manufacturing capacities and the ability to deliver
PLM	**Life Cycle Management** Strategy concept for the management of a product over its entire life cycle, i. e. from production to recycling
PDM	**Product Data Management** System for the unified storage and management of all data that accrue during the development, production, storage and distribution of a product
SCM	**Supply Chain Management** Planning and organization of all tasks arising in connection with procurement, production and logistics

Structuring of production planning

The product data include all order-neutral production data. They are supplemented with order-related data.

Production planning

Documentation
- Product breakdown structure
- Bills of materials
- Work-processing sheets
- Testing schedules
- Deadline schedules

Production resources
- Personnel
- Operating resources
- Material
- Information

Production process
- Determination and description of the production process
- Assessment of the target time and the cycle time

Order-neutral
- Product breakdown structure
- Drawings
- Bills of materials
- Work-processing sheets
- Testing schedules
- Planning costs

Order-specific
- Production schedules
- Calculation
- Procurement

→

- Utilization plan
- Time schedule

PE

Product breakdown structure, Bills of materials

In most cases, products are composed of several parts that could be summarized into groups. In order to obtain a better overview, structure plans are drawn up according to function, processing, assembly or procurement. The product breakdown structure is also the basis for the preparation of the bills of materials.

Product breakdown structure according to functional levels

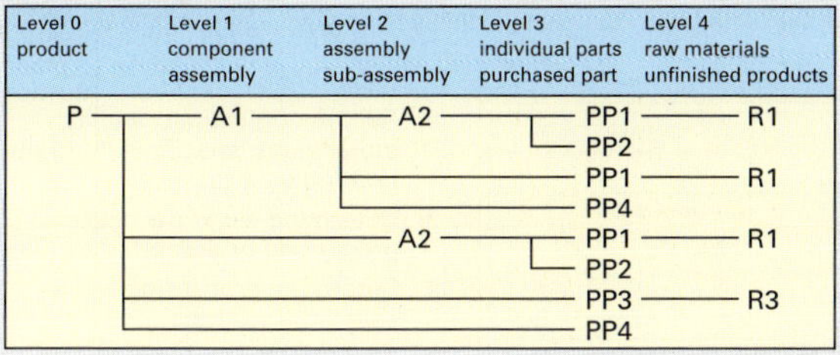

P = Product, C = Component;
A = Assembly/sub-assembly
PP = Individual part/purchased part;
R = Raw part/unfinished product

Each assembly, subassembly, individual or purchased part or raw material is on the same functional level and level of consideration. The structure shows the detailed composition of the entire product. This type of breakdown structure is mainly used in the design phase.

Product breakdown structure according to processing stages

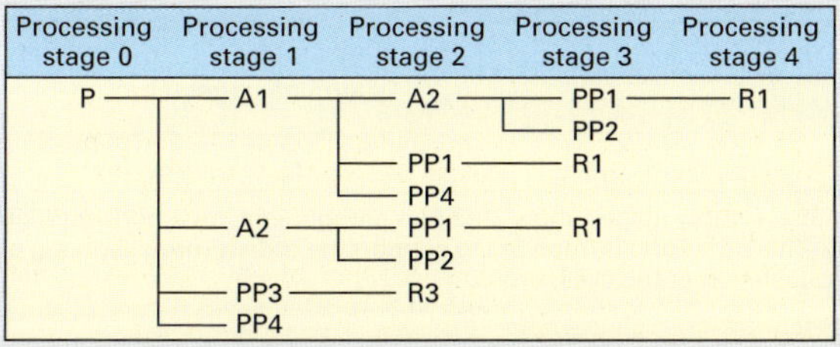

The manufacturing process determines the mapping in this structure. Each component appears on the level where it is needed in the manufacturing or assembly process. The structure forms the basis of the production planning. The multi-level BOM and the assembly drawing are derived from this breakdown structure.

Bills of materials (order-neutral)

The bill of materials is the basis for the assessment of the demand for parts and raw materials. It is used for the preparation of the work-processing sheet. The structure of the BOM depends on the intended use and is not standardized. Types: design and manufacturing BOM, single-level and multi-level BOM, quantified BOM.

Design and manufacturing BOM

The manufacturing parts list includes the specifications of the design parts list and additional details concerning manufacturing. It often replaces the work-processing sheet.

Bill of materials — Sheet 1 of 1
Art. no. 12.000 — Designation: P - Product — Date: 11/03/12

Item	Qty	Designation	Material	Unfinished/DIN
10	3	Individual part PP1	S235JR	Rd 30x18
20	2	Purchased part PP2		DIN EN
30	1	Individual part PP3	S235JR	Rd 125x65
40	2	Purchased part PP4		DIN EN

Single-level BOM (simplified)

Single-level BOMs only contain items of the same processing structure levels. There are always several BOMs required for a single product.

Single-level BOM P - Product — Sheet 1 of 3

Designation	Quantity
A1	1
A2	1
PP3	1
PP4	1

Single-level BOM A1 - Assembly 1 — Sheet 2 of 3

Designation	Quantity
A2	1
PP1	1
PP4	1

Single-level BOM A2 - Assembly 2 — Sheet 3 of 3

Designation	Quantity
PP1	1
PP2	1

Multi-level BOM (simplified))

Each assembly is broken down to the lowest product level (is composed of ...).

Multi-level BOM P - Product

Stage 1	Stage 2	Stage 3	Stage 4	Designation
1				A1
	2			A2
		3		PP1
			4	R1
		3		PP2
	2			PP1
		3		R1
	2			PP4
1				A2
	2			PP1
		3		R1
	2			PP2
1				PP3
	2			R3
1				PP4

PE

Work processing sheet, Order schedule, Processing control

Work processing or assembly sheet (order-neutral)

The work-processing sheet is used for information and instruction in manufacturing processes. It describes the workflow for the manufacturing or processing of a part, an assembly or a product. The material to be used, the workplace for each processing step, the resources and the target times are specified as a minimum requirement. The structure of the work-processing sheet is not standardized.

Order-related work processing sheets are supplemented with the order number, the batch size and the deadlines.

	Work-Processing Sheet				Prepared by Go	
					Date 10/01/02	
1	Art. No. 12.001	Article PP1 - individual part 1			Drawing 12.001-1	
2	Hz. no.	Original part Rd EN 10060-30 x 18 – S235JR				
	Process. step no.	**Cost group**	**Description**	**Aids/NC software**	**Set-up time [min]**	**Time per unit [min]**
3	10	Turning	Complete turning work	NC_12_001	15	5.25
	20	Drilling	Drill cross hole	Prism	3	4
	4		5	6		7

1. Record initial data.
2. Assess original part and its dimensions; specify BOM of the assembly parts.
3. Determine workflow of the processing or assembly.
4. Determine processing or assembly systems.
5. Prepare workflow description of the processing or assembly.
6. Determine auxiliary equipment and material for the processing or assembly.
7. Determine target time.

Order schedule

The product structure based on processing and assembly stages shows the part components on a horizontal time axis from the start point to the date of delivery. This order grid is intended to support the order time scheduling and facilitates the order processing. It allows the calculation of the cycle times.

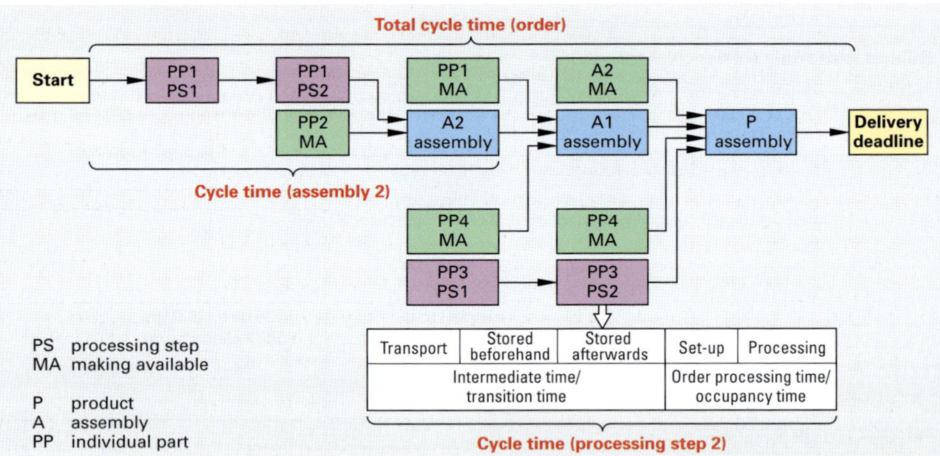

Centralized processing control

Push principle: Orders are launched by the process control and pushed forward.

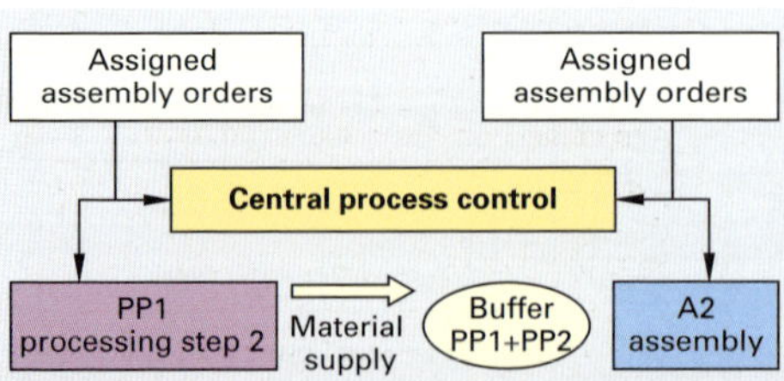

Decentralized processing control

Pull or get principle: According to the Kanban principle, processing is launched on demand (Kanban card) of the subsequent step (Kanban = signboard, billboard).

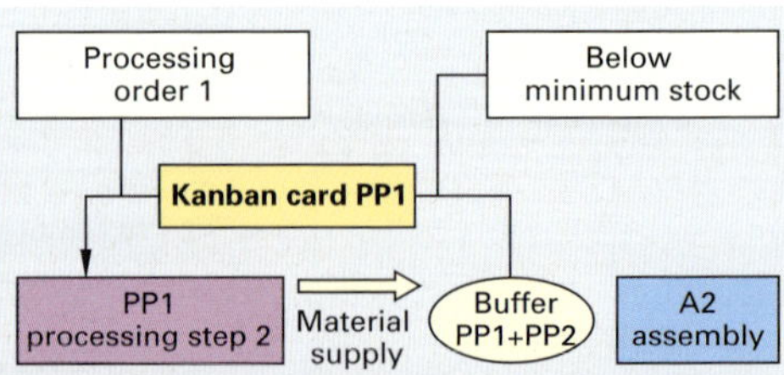

Cycle time[1)]

Structuring of time types in a work system (S)

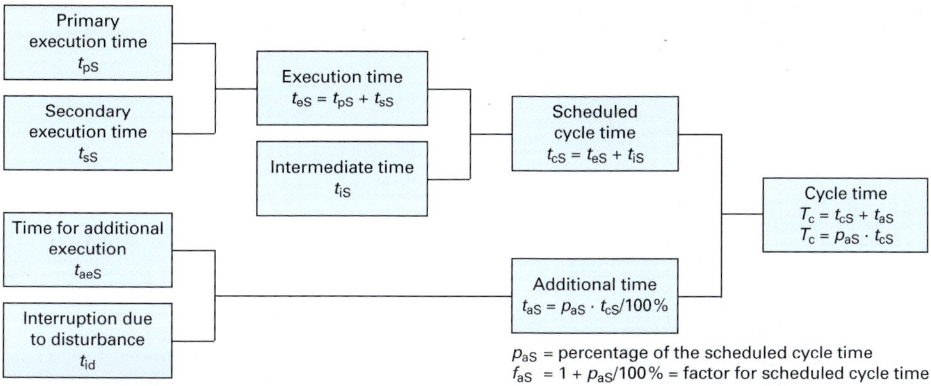

p_{aS} = percentage of the scheduled cycle time
f_{aS} = 1 + $p_{aS}/100\,\%$ = factor for scheduled cycle time

Symbol	Designation	Description with examples
T_c	Cycle time	Target time (allowed time) for the completion of a task in one or more work systems
t_{cS}	Scheduled cycle time	Sum of the target times for the processing of a batch quantity in a work system S in accordance with the scheduled cycle times
t_{eS}	Execution time	Target time for the execution of a batch quantity in a work system S • Job time T referenced to the executing person (cf. page 284) • Utilization time t_{ur} referenced to the operating resource (cf. page 285)
t_{pS}	Primary execution time	Time during which the task is carried out in a work system S, it often corresponds to the time types: • Activity time $t_{aC} = t_{tf} + t_{tv}$ (cf. page 284) • Main production time $t_{mp} = L \cdot i / n \cdot f$ (cf. page 285)
t_{sS}	Secondary execution time	Time during which • the main execution is prepared, set up, fed and emptied in a work system S • operators check their work or repose
t_{iS}	Intermediate time	Target time during which the execution of the job is interrupted in accordance with the schedule: • Wait time (t_w) after processing in work system S1 • Transport (t_{tr}) from S1 to S2 • Wait time (t_w) before processing in work system S2
t_{aS}	Additional time	Unscheduled times are added to the scheduled cycle times in form of a safety addition determined by experience values or a multiplication factor. Additional times are mainly caused by: • Additional executions t_{aeS} • Interruptions due to disturbances t_{id}

Calculation methods for the cycle time

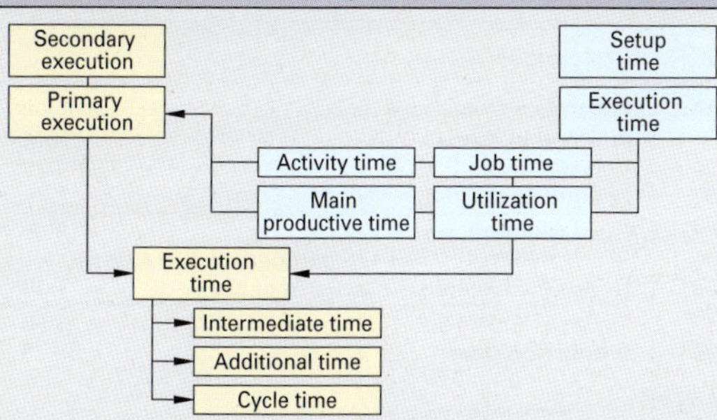

Example:

A CNC machine is utilized for 6.5 hr. The wait and transport times amount to 3 hr each. Calculate the cycle time with a safety addition of 20%.

Execution time	$t_e = t_{ur}$	=	6.5 hr
Intermediate time	$t_i = 2 \cdot t_w + t_{tr}$	=	9.0 hr
Scheduled cycle time	$t_c = t_e + t_i$	=	15.5 hr
Additional time	$t_a = p_a \cdot t_c/100\,\%$	=	3.1 hr
Cycle time	$\mathbf{T_c = t_c \cdot t_a}$	=	**18.6 hr**

Cycle time in days:
18.6 hrs/6 hrs per work day = **3.1 days**

1) According to REFA International Association for Work Design, Industrial Organization and Company Development

PE

Job time[1)]

Structure of types of time for workers

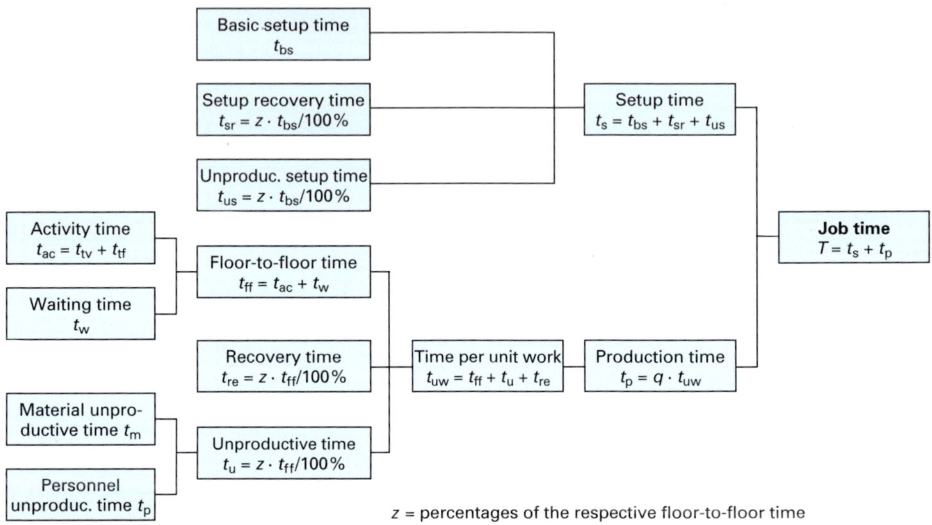

z = percentages of the respective floor-to-floor time

Symbol	Designation	Explanation with examples
T	Job time	Time allowed for manufacturing a lot size
t_s	Setup time	Setup for an entire job • basic setup time t_{bs} → turn on machine • setup recovery time t_{sr} → recovery time after strenuous changeover • setup unproductive time t_{us} → repair of brief machine malfunction
t_p	Production time	Time allowed for production of a lot size (without setup)
t_{re}	Recovery time	Personnel break time to reduce work-related fatigue
t_u	Unproductive time	• job-related interruption time t_m → unforeseen tool sharpening • personnel interruption time t_p → checking work times, taking care of needs
t_{ac}	Activity time	Times in which the actual job is processed • variable times t_{tv} → assembly or deburring work • fixed times t_{tf} → cycle of a CNC program
t_w	Waiting time	Waiting for the next workpiece in the continuous flow production
q	Job volume	Number of units to be produced for a job (lot size)

PE

Example: Turning three shafts on a lathe

Setup times:		min
Setup job		= 4.50
Setup of machine		= 10.00
Setup of tool		= 12.50
Basic setup time	t_{bs}	= 27.00
Setup recovery time	t_{sr} = 4% of t_{bs}	= 1.08
Unproduc. setup time	t_{us} = 14% of t_{bs}	= 3.78
Setup time	$\mathbf{t_s = t_{bs} + t_{sr} + t_{us}}$	**= 31.86**

Production times:		min
Activity time	t_{ac}	= 14.70
Waiting time	t_w	= 3.75
Floor-to-floor time	$t_{ff} = t_{ac} + t_w$	= 18.45
Recovery time	t_{re} compens. for in t_w	–
Unproductive time	t_u = 8% of t_{ff}	= 1.48
Time per unit work	$t_{uw} = t_{ff} + t_{re} + t_u$	= 19.93
Production time	$\mathbf{t_p = q \cdot t_{uw}}$	**= 59.79**

Job time $\boldsymbol{T = t_s + t_p}$ ≈ 32 min + 60 min = **92 min** (= 1.53 hr)

[1)] According to REFA International Association for Work Design, Industrial Organization and Corporate Development

Utilization time[1)]

Structure of the types of times for production resources (PR)

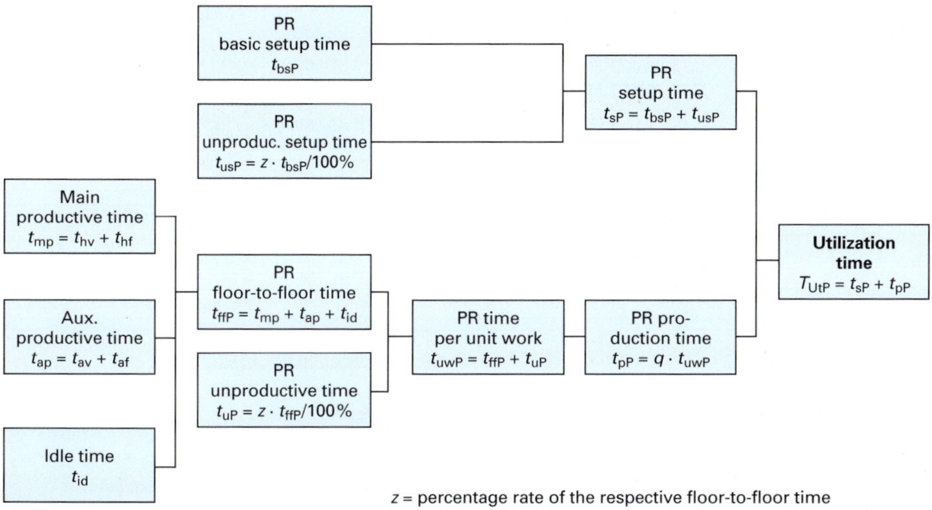

z = percentage rate of the respective floor-to-floor time

Symbol	Designation	Explanation with examples
T_{UtP}	Utilization time	Time allowed for utilization of a production resource for manufacturing a lot size
t_{sP}	Production resource setup time	Setup of production resource for completing an entire job • PR basic setup time t_{bsP} → clamping equipment on a machine • unproductive setup time t_{usP} → optimization of CNC program
t_{pP}	Production resource production time	Time allowed for the production time of a lot size (without setup)
t_{uP}	Production resource interruption time	Time in which the production resource is not utilized or additionally utilized; power outage, un-planned repair work, etc.
t_{mp}	Main productive time	Times in which the work object is processed according to plan • variable times t_{hv} → manual drilling • fixed times t_{hf} → cycle of CNC program
t_{ap}	Auxiliary productive time	Production resources are prep., loaded or emptied for the main productive time • variable times t_{av} → manual clamping • fixed times t_{af} → automatic workpiece change
t_{id}	Idle time	Process or recovery related down time, e.g. filling of a magazine
q	Job volume	Number of units to be produced for a job (lot size)

Example: Milling a contact surface on 20 base plates using a vertical milling machine

Setup times:	min
Read the job order and drawing	= 4.54
Set up and store the surface cutter	= 3.65
Clamp and unclamp the cutter	= 3.10
Set up the machine	= 2.84
Production resources basic setup time t_{bsP}	= 14.13
Prod. res. unproductive s. time t_{usP} = 10% of t_{bsP}	= 1.41
Production resources setup time $t_{sP} = t_{bsP} + t_{usP}$	**= 15.54**

Production times:	min
Milling ≙ main productive time t_{mp}	= 3.52
Clamp workpiece ≙ aux. productive time t_{ap}	= 4.00
Transport workpiece ≙ idle time t_{id}	= 1.20
Prod. res. floor-to-floor time $t_{ffP} = t_{mp} + t_{ap} + t_{id}$	= 8.72
Prod. res. unproductive time t_{uP} = 10% of t_{ffP}	= 0.87
Prod. resource time per unit $t_{uwP} = t_{ffP} + t_{uP}$	= 9.59
Production resource prod. time $t_{pP} = q \cdot t_{uwP}$	**= 191.80**

Utilization time $T_{UtP} = t_{sP} + t_{pP}$ ≈ 16 min + 192 min = **208 min** (= 3.47 hr)

1) According to REFA International Association for Work Design, Industrial Organization and Corporate Development

PE

Cost accounting

Simple calculation (numerical example)

	Direct costs[1)] *directly* attributable to a specific product	Overhead[1)] *Not directly* attributable to a specific product	Surcharge in percent of wage costs
Types of costs[1)]	Material costs $ 80 000.00 Labor costs $ 120 000.00	Depreciation $ 50 000.00 Salaries (incl. management salaries) $ 80 000.00 Interest $ 40 000.00 Other costs $ 50 000.00 ∑ Overhead $ 220 000.00	$\frac{\$ 220\,000.00 \cdot 100\%}{\$ 120\,000.00} = 183.33\%$ A surcharge rounded off to 185% is applied to each wage hour to cover overhead costs.
Cost calcula-tion	Wage hours = 10 000 hrs	Labor costs/hr = $/hr 12.00	Material costs of order $ 124.75
	Rate per hour = $/hr 12.00 + 185% = $/hr 34.20 (for independent contractor invoices; management salaries = profit)		Working time 5 hr x $/hr 34.20 $ 171.00
[1)] Costs must be determined periodically for every operation.			Price without VAT $ 295.75

Expanded calculation (schematic)

Material direct costs
Procurement costs
\+
Material overhead
Percent of material direct costs, e.g. purchasing costs, storage costs, etc.
↓
Material costs → (to Material costs)

Design costs
Salaries etc.
\+
Equipment costs
Drilling equipment molds etc.
\+
Special tools
Special drills etc.
\+
Out-of-house processing
Heat treatment etc.
↓
Special direct costs of production → (to Special direct costs of production)

Material costs
\+
Direct production costs
Production wages attributable to one product
\+
Production overhead[1)]
Machine costs
Depreciation, interest, occupancy, energy and maintenance costs
Remaining overhead
Percent of production wages, e.g. fringe benefits, occupancy, operating materials, etc.
↓
Production costs
\+
Special direct costs of production
↓
Manufacturing costs
\+
Management and sales overhead
Percent of manufacturing costs
↓
Prime cost
\+
Profit
Percent of prime cost
↓
Raw price
\+
Commissions, discounts, rebates
Percent of sales price
↓
Sales price without VAT

[1)] *If **no** machine hourly rates are calculated, these are included in the production overhead and increase the surcharge rate. The overhead surcharge rates are taken from the operational accounting sheet.*

Example:

Material direct costs	$ 1225.00
Material overhead 5%	$ 61.25
Production wages 10 hr x $/hr 15.–	$ 150.00
Machine costs 8 hr x $/hr 30.–	$ 240.00
Residual overhead 200% of production wages	$ 300.00
Special tools	$ 125.00
Manufacturing costs	**$ 2101.25**
Management and sales overhead 12% of manufacturing costs	$ 252.15
Prime cost	**$ 2353.40**
Profit addition 10% of the prime cost	$ 235.34
Raw price	**$ 2588.74**
Commissions 5% of sales price	$ 136.25
Sales price before VAT	**$ 2724.99**

PE

Machine hourly rate calculation

Machine hourly rate calculation

Average production overhead does not take into consideration various machine costs attributable to a specific product. This type of cost accounting would be misleading.
If machine costs are taken out of production overhead and converted to hours the machine was utilized, this yields the **machine hourly rate**.

Compilation of machine costs

Machine costs are:

- **Calculated depreciation**
  Linear loss of value over the service life of the machine relative to replacement cost
- **Calculated interest**
  Average interest for capital invested for the machine
- **Occupancy costs**
  Costs incurred by floor and traffic space of the machine
- **Energy costs**
  Costs incurred by electricity, natural gas, steam or gasoline consumption
- **Maintenance costs**
  Costs for repairs and regular service
- **Other types of costs**
  Costs for tool wear, insurance premiums, disposal of coolants and lubricants etc.

Machine running time, Machine hourly rates

according to VDI Directive 3258

T_{RT} machine running time in hours/period
T_T total theoretical machine time in hours/period
T_{ST} down times, e.g. work free days, work interruptions etc., usually in % of T_T
T_{SM} times for service and maintenance, usually in % of T_T

C_M sum of machine costs per period (usually per year)
C_{Mhr} machine costs per hour; machine hourly rate
C_f machine fixed costs per year; e.g. depreciation
C_v/hr machine variable costs per hour; e.g. electrical consumption

Machine running time

$$T_{RT} = T_T - T_{ST} - T_{SM}$$

Machine hourly rates

$$C_{Mhr} = \frac{C_f}{T_{RT}} + C_v/hr$$

Calculation of machine hourly rate (example)

Tool machine:

Procurement value \$ 160 000.00
Power consumption 8 kW
Occupancy costs \$/m² 10.00 × month
Additional maintenance \$/hr 5.00

Service life 10 years
Cost per kWh \$ 0.15
Space req. 15 m²
Normal utilization
T_{RT} = 1200 hr/year (100%)

Assumed interest rate 8%
Base charge \$/month 20.00
Maintenance \$/year 8 000.00
Actual utilization 80%

What would be the machine hourly rate for normal utilization and 80% utilization?

Type of cost	Calculation	Fixed costs \$/year	Variable costs \$/hr
Calculated depreciation	$\frac{\text{procurement value}}{\text{service life in years}} = \frac{\$\ 160\,000.00}{10\ \text{years}}$	\$ 16 000.00	
Calculated interest	$\frac{{}^1/_2\ \text{procurement value in \$} \times \text{interest}}{100\%} = \frac{\$\ 80\,000.- \times 8\%}{100\ \%}$	\$ 6 400.00	
Maintenance costs	maintenance factor × depreciation – e.g. 0.5 × \$ 16 000.00 maintenance is dependent upon utilization.	\$ 8 000.00	\$ 5.00
Energy costs	base charge for power supply \$/month 20.00 x 12 mon. power consumption × energy costs 8 kW × \$/kWh 0.15	\$ 240.00	\$ 1.20
Proportional occupancy costs	space cost rate x space requirement = \$/m² 10.00 × month × 15 m² × 12 months	\$ 1 800.00	
	Total machine costs (C_M)	**\$ 32 440.00**	**\$ 6.20**

Machine hourly rate (C_{Mhr}) at 100% utilization = $\frac{C_f}{T_{RT}} + C_v/hr = \frac{\$\ 32\,440.00}{1200\ hr}$ + \$/hr 6.20 = **\$/hr 33.23**

Machine hourly rate (C_{Mhr}) at 80% utilization = $\frac{C_f}{0.8 \cdot T_{RT}} + C_v/hr = \frac{\$\ 32\,440.00}{0.8 \cdot 1\,200\ hr}$ + \$/hr 6.20 = **\$/hr 40.00**

The machine hourly rate does not include costs for operator.

PE

Direct costing[1)]

Marginal costing (with numerical example)

Marginal costing takes the market price of a product into consideration. The market price must at least cover variable costs (lower price limit). The remainder is the contribution margin. Contribution margins of all products carry the costs of operational readiness.

R/piece	market price; revenue per piece	C_f	fixed costs
R	revenue (sales) of product	C_v	variable costs
CM	contribution margin of product	*P*	profit or gain
CM/piece	contribution margin per piece	*Bp*	breakeven point

Contribution margin

$$\frac{CM}{\text{piece}} = \frac{R}{\text{piece}} - \frac{C_v}{\text{piece}}$$

$$CM = \frac{CM}{\text{piece}} \cdot \text{volume}$$

Profit

$$P = CM - C_f$$

	Variable costs (C_v)[2)] depends on production volume		Fixed costs (C_f) independent of production volume		Contribution margin (CM) $CM = R/\text{piece} - C_v/\text{piece}$
Types of costs	Material costs	\$/piece 30.00	Depreciation	\$ 50 000.00	Revenue of \$/piece 110.00 must cover all variable costs first. The remainder is used to cover total fixed costs and includes profit.
	Labor costs	\$/piece 20.00	Wages	\$ 80 000.00	
	Energy costs	\$/piece 10.00	Interest	\$ 40 000.00	
			Others C_f	\$ 30 000.00	
	Σ Variable costs	\$/piece 60.00	Σ Fixed costs	\$ 200 000.00	

Cost calculation

No. of pieces produced 5 000 pieces Contribution margin \$ 110.00 – \$ 60.00 = \$/piece 50.00

Total contribution margin 5 000 pieces · \$/piece 50.00 = \$ 250 000.00

Σ Fixed costs \$ 200 000.00

Profit \$ 50 000.00

$$\text{Breakeven point } Bp = \frac{C_f}{CM/\text{piece}} = \frac{\$\ 200\ 000.00}{\$/\text{piece } 50.00} = 4\ 000 \text{ pieces}$$

Breakeven point

$$Bp = \frac{C_f}{CM/\text{piece}}$$

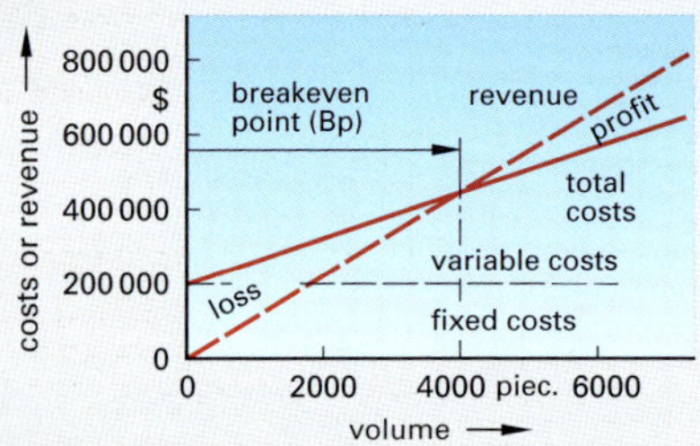

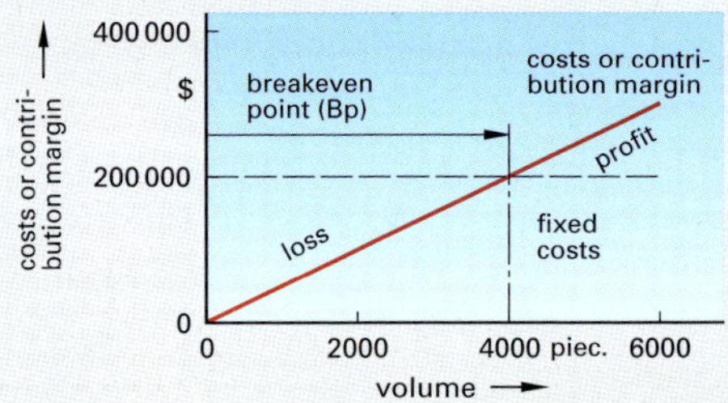

Cost comparison method

In the cost comparison method, the machine or facility that incurs the lowest costs for a given production volume should be selected.

Example for 5 000 pieces

Machine 1: C_{f1} = \$/year 100 000.–; C_{v1} = \$/piece 75.00
\$/year 100 000.– + \$/piece 75 × 5 000 pieces = \$ 475 000
Machine 2: C_{f2} = \$/year 200 000.00; C_{v2} = \$/piece 50.00
\$/year 200 000.– + \$/piece 50.00 × 5000 pieces = \$ 450 000
Machine 1 costs > machine 2 costs

$$\text{Piece count limit } M_{lim} = \frac{C_{f2} - C_{f1}}{C_{v1}/\text{piece} - C_{v2}/\text{piece}}$$

$$M_{lim} = \frac{\$\ 200\ 000.00 - \$\ 100\ 000.00}{\$/\text{piece } 75.00 - \$/\text{piece } 50.00} = 4000 \text{ pieces}$$

Machine 2 is more economical at volumes above 4000 pieces.

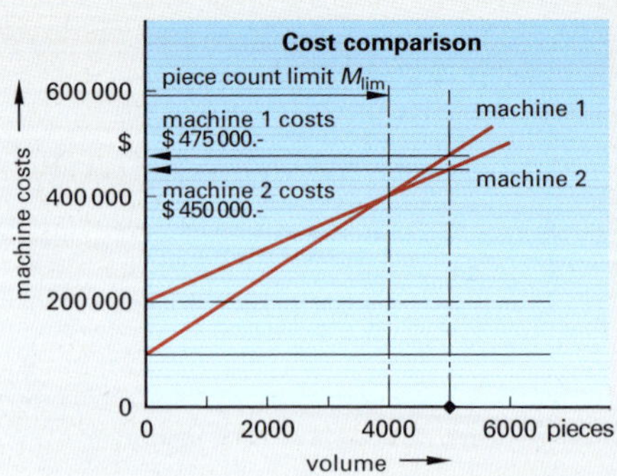

PE

1) Direct costing separates costs into fixed costs (costs of operating readiness) and variable costs (direct costs).
2) Variable costs are calculated for each job and compared to revenue.

Maintenance, Inspection, Repair, Improvement

MRO and wear

cf. DIN 31051 (2003-06)

In accordance with DIN 31051, maintenance, repair and overhaul (MRO) comprise all actions which have the objective of retaining or restoring an item of a technical system in or to the desired state as well as to evaluate the actual condition of it. MRO measures: **maintenance, inspection, repair and improvement**

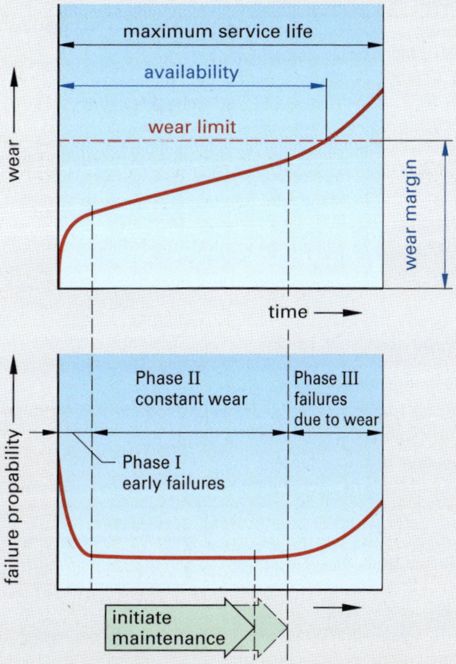

The **wear limit** is set to a state where the work output is principally not affected and the quality criteria are still complied with.

The **wear margin** is determined by the wear limit. The availability of a machine and the service life of a tool are determined by the wear margin.

Maintenance, repair and overhaul measures

Type	Measure
Maintenance delays the exhaustion of the wear margin	• Cleaning • Lubrication/greasing • Refilling • Adjusting
Inspection helps to identify and evaluate the actual condition and detect the reasons of wear	• Testing and measuring • Diagnosis and evaluation • Planning of MRO measures
Repair restores to the desired condition	• Repair by restoring and adjusting • Replacement of spare parts or renewal of tools
Improvement increases functional safety or enhances the wear margin	• Evaluation of errors • Analysis of weak points • Selection of better materials or tools

Optimization of MRO

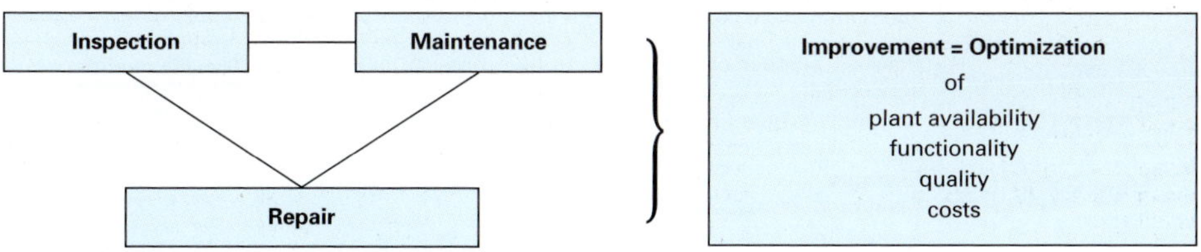

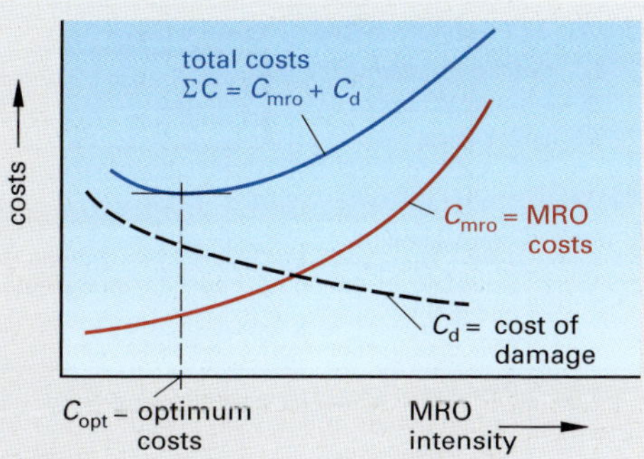

Frequent maintenance, inspection, repair and improvement increases the MRO costs and decreases the cost of damage (costs due to failure etc.). The costs are at the optimum level if the total of the MRO and damage costs has attained a minimum.

Optimization of MRO should pursue

- economic
- environmental (ecologic) and
- human objectives

MRO concepts

Interval-driven MRO

Preventive MRO measures scheduled at maintenance and inspection intervals of 8, 40, 160 and 2000 operating hours. The equipment operator carries out the regular MRO at the end of a shift or workday.

MRO intervals

Period	Maintenance work, examples	Period	Maintenance work, examples
Day 6 to 8 operating hours	• Cleaning of the workspace, i. e. removal of chips and cooling lubricant residues • Verification of the oil levels • Performance of vibration and sound checks	**Month** 140 to 160 operating hours	• Daily and weekly maintenance operations • Greasing of slideways • Replacement of cooling lubricants • Testing of hose connections
Week 35 to 40 operating hours	• Daily maintenance operations • Thorough cleaning of the machine • Inspection of the cooling lubricant and cleaning if required • Replacement of filters (e.g. blower fans)	**Year** 1400 to 2000 operating hours	• Daily, weekly and monthly maintenance operations • Inspection for wear and readjustment of slideways • Oil change (central lubrication, hydraulic system)

Condition-driven MRO

Process run trend

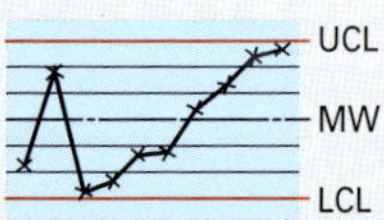

When the wear margin of mechanical parts, such as plain or roller bearings or slideways reaches its limit, the processing quality of the machine deteriorates. The work process should be observed in regard to eventual indications.

With machines:
- Changes to smooth running, rattling or whistling sounds are generated
- The process run exhibits a rising or falling trend

With tools:
- Wear marks become distinctly larger
- The workpiece surface is in poorer condition

Service life

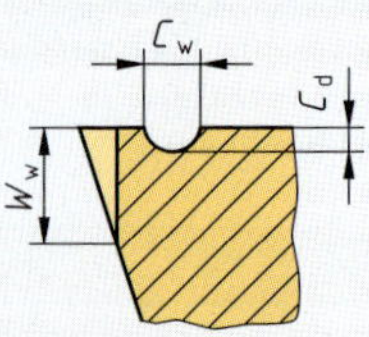

The wear mark width (WW), the crater width (C_w) and the crater depth (C_d) are considered as criteria for the service life. When the tool reaches the preset limit, its wear margin is exhausted.

W_W wear mark width in mm
C_w crater width in mm
C_d crater depth in mm
T service life in minutes
v_0 cutting velocity in m/min

Example:

$T_{v200\,VB0.2}$ = 15 min

Specified service life of 15 minutes at a cutting velocity v_0 = 200 m/min and a wear mark width W_W = 0.2 mm

Service life

$$T_{v200\,VB0.2} = 15\ min$$

Tool life quantity

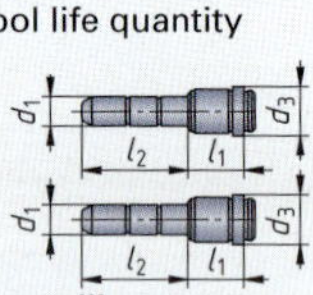

Possible number of workpieces to be processed at a given service life

N tool life quantity in piece numbers
t_{mp} main productive time in minutes

Example:

Main productive time t_{mp} = 1.3 minutes,
service life T = 15 minutes

$N = T/t_{mp}$ = 15 min/1.3 min = 11.5; $\boldsymbol{N}$ = **11 pieces**

Tool life quantity

$$N = T/t_{mp}$$

Tool life travel

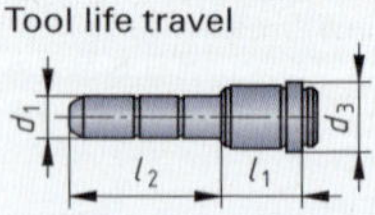

Possible feed distance that a tool travels during its service life

L_f tool live travel in m
v_f feed velocity in mm/min
f_{th} feed per tooth
th number of teeth
n speed 1/min

Tool life travel

$$L_f = T \cdot v_f$$
$$L_f = T \cdot n \cdot f_{th} \cdot th$$

Terms referring to time

Term	Explanation	Application
Life cycle	Period during which plant or machinery can be used without interruption	Specified in operating hours for machinery and plant Specified in kilometers or miles for cars
MTTF Mean Time To Failure	Mean operating time until failure as statistical average value	Parameter for the evaluation of the machine or component safety (roller bearing) in conformity examinations as per EN ISO 13849-1

PE

Workplace health and safety, Documentation

Corrective MRO

Repair after failure	Failure despite inspection and maintenance, due to (examples) • incorrect operation • overload • undetected wear Lack of inspection and maintenance, possible failure is accepted and repaired on occurrence • with non-critical parts, e. g. defect lights • to avoid downtimes due to repair • if maintenance and inspection are not feasible or not economical

Risk-based MRO

RBM risk-based maintenance	Risk-based MRO strives for the reduction of MRO operations and costs while ensuring compliance with the pre-defined safety standard in order to avoid **plant failure.** • Evaluation of failure risks • Assessment of failure frequencies • Definition of effective maintenance measures • Definition of priorities for failure risks and corresponding maintenance measures

Reliability-centered MRO

RCM reliability-centered maintenance	Reliability-centered MRO makes best possible use of various MRO strategies to prevent **functional disturbance.** The applied strategy depends on the situation and the type of plant or machinery. • Description of the machine or plant and its interaction with coupled plant parts • Weak point analysis on each machine • Definition of a MRO strategy

Workplace safety and health protection

Accident prevention regulations	**Instruction** An operator who is not familiar with his/her workplace bears a higher risk of accident, even if maintenance and repair measures are carried out regularly and properly. Personnel must be made familiar with the workplace and the working processes. This also includes instructions about safety-oriented behavior and health protection. Written confirmation of the instruction must be drawn up for documentation purposes. **The following should be observed in particular:** • Always use personal protective equipment, e. g. safety shoes, protective gloves or ear protection • Workplace organization and tidiness • Behavior in compliance with safety regulations, e. g. never touch running machine parts, never put protective equipment and guards out of operation • Proper handling of hazardous materials in accordance with the operating instructions • Correct transport of loads, use of lifting equipment, if required

Technical documentation

Systematic approach to documentation DIN 6789 (1990-09)

Documentation of machines, products and production

Machine or plant		Product and production
• General description of the machine • Drawings • Installation/assembly drawing	• Declaration of Incorporation for the machine • Operating instructions • EC declaration of conformity	• Drawings and product creation • BOMs • Work processing sheets • Deadline schedules

MRO documentation

All malfunctions, maintenance operations, inspections and repairs must be documented as evidence for warranty claims and for quality management. Specifications in the MRO document:

General	Actual condition	Maintenance works	Commissioning	Confirmation
Details of the machine, maintenance personnel	Visual inspection, noise, …	Cleaning, replacement of parts, …	Functional testing, acceptance report, …	Date, signature

PE

Documentation system

Computer-aided documentation system (simplified)

Technical documentation is supported by electronic data processing. Details of all machines are stored in a database together with specifications concerning MRO measures and their execution. The records could be used to provide quality evidence or for other evaluations.

List of machines

Document	Designation
MW	**Machine tools**

No.	Machine	Year of construction	Manu-facturer	Cost group	Location
MW01	Drill press	2005	Flott	WB	WB-105
MW02	CNC machining center	2004	Chiron	WB	WB-105

Machine card

Document	Designation	Year of Con-struction	Manufacturer	Cost group	Location
MW01	**Drill press**	**2005**	**Flott**	**WB**	**WB-105**

Item no.	Documents	Notes
01	Tool list	with specification of the storage location
02	Operating instructions	contain assembly and other drawings
03	Maintenance schedule	includes forms for documentation
04	Disturbances/repairs	only repairs after failure
05	Notes	
06		

Maintenance schedule

Document	Designation	Year of Con-struction	Manufacturer	Cost group	Location
MW 01-03	**Drill press**	**2005**	**Flott**	**WB**	**WB-105**

Interval	Part	Maintenance work	Notes
Daily	Machine	Cleaning, testing of operational ability and operational safety	
Weekly	Column	Lubrication	Oil: standard
Weekly	Table adjustment	Lubrication	Grease: standard
Weekly	Feed	Lubrication	Oil: standard
Monthly	Spindle sleeve	Lubrication	Grease: standard
Monthly	Sleeve	Lubrication	Grease: HD
Yearly	–	–	–
Others	–	–	–

Maintenance report

Maintenance after 6 to 8 operating hours (daily)

Machine	Designation
MW01	**Drill press**

Cleaning
Testing of operational ability and safety

Date	Name	Signature

Repair report

Machine	Designation
MW01	**Drill press**

Cause:
Name of maintenance personnel:

Type	Measure
Assessment of actual condition	
Visual inspection	
Noise generation	
Maintenance works	
Cleaning	
Oil level	
Replacement of parts	
Commissioning	
Operational testing	
Serviceability	

Confirmed by
Date: Signature:

PE

Selection of tool and machining data (overview), Chip removal rate

Selection of tool and machining data
Example: external cylindrical turning with indexable cutter insert

Tool selection

Influencing parameters

Workpiece		
Werkstoff	**Bearbeitung**	**Cutting**
– Material group, pages 298 to 302	**Pre-machining** – roughing **Finish-machining** – finishing – fine finishing	– continuous – discontinuous due to grooves, bores

Machine
- tool clamping system, tool holder
- stability
- clamping of workpiece
- driving power[1)]

Indexable cutter insert
- cutter material • geometry • coating • cooling lubrication

→ Selection on pages 308 to 310 or in company catalogs

Selection of machining data

Parameters influencing the guide values
- workpiece • machine • cutter material

Machining data
- cutting velocity v_c
- cutting depth a_p
- feed f

Standard values from the tables on the pages 298 et seq. or manufacturer data

have an influence on
- chip formation
- service life
- surface quality
- chip removal rate
- cutting force
- driving power

[1)] **Driving power**
The driving power of modern machine tools allows the operation of most of the tools designed for the clamping system. In the planning of the machining process, the provision of a driving power is only necessary if pre-machining processes such as roughing are planned.

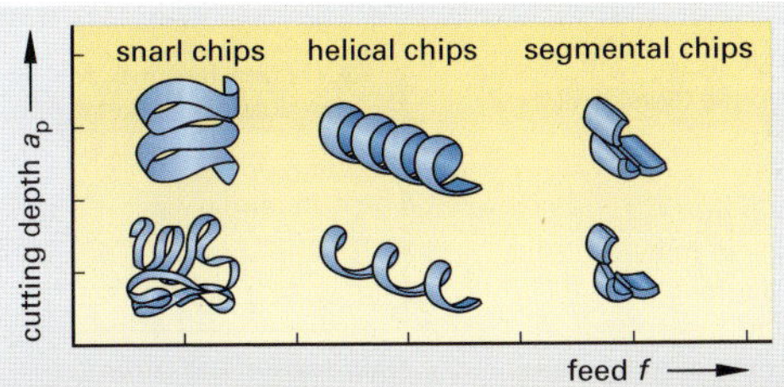

Shapes of chips, feed

Regardless whether pre-machining or fine finishing is concerned, the control of the chip discharge is of particular importance in all chip removal processes. The illustration on the left shows the influence of the feed f and the cutting depth a_p on the shape of the chips.

An increased feed produces also the following effects:
- reduction of the cutting forces (page 295)
- reduction of the load on the cutting edges
- reduction of the driving power

Chip removal rate

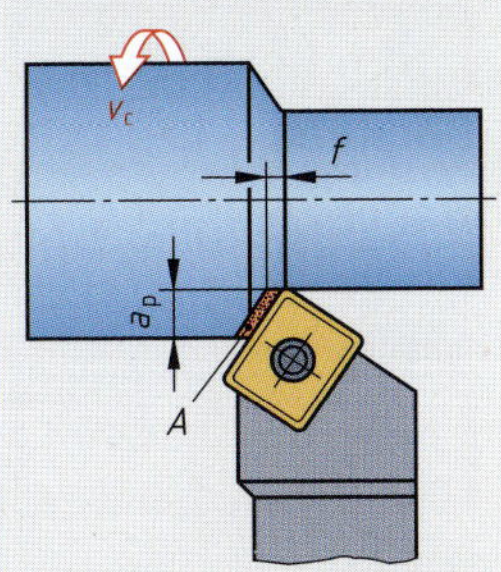

In the pre-machining of workpieces (roughing), the volume of removed metal is an important comparative parameter. The chip removal rate Q refers to the volume of removed metal in cm^3/min.

A	machining cross section in mm^2
a_p	cutting depth in mm
f	feed in mm
Q	chip removal rate in cm^3/min
v_c	cutting velocity in m/min

Machining cross section

$$A = a_p \cdot f$$

Chip removal rate

$$Q = A \cdot v_c$$

Example:

Cutting velocity $v_c = 125$ m/min, cutting depth $a_p = 5$ mm, feed $f = 0.8$ mm

Find: chip removal rate Q

Solution:

$$Q = A \cdot v_c = a_p \cdot f \cdot v_c$$

$$= 0.5\ \text{cm} \cdot 0.08\ \text{cm} \cdot 12500\ \frac{\text{cm}}{\text{min}} = \mathbf{500\ \frac{cm^3}{min}}$$

PE

Forces and power in turning and drilling

Turning

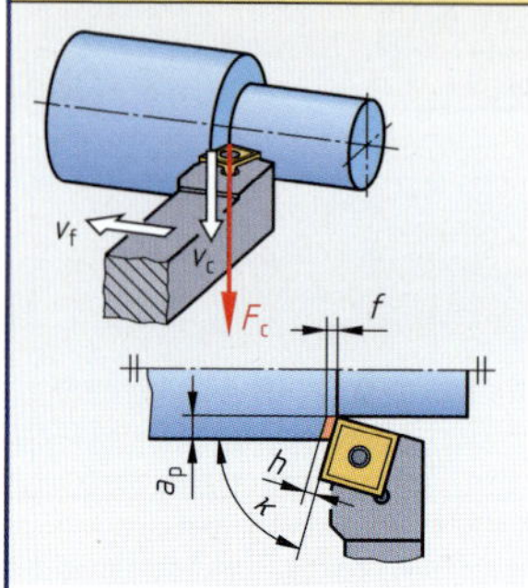

- F_c cutting force in N
- A chip section in mm²
- a_p cutting depth in mm
- f feed per revolution in mm
- h chip thickness in mm
- $\varkappa$ cutting edge angle in degrees (°)
- C correction factor for the cutting speed
- v_c cutting velocity in m/min
- k_c specific cutting force in N/mm² (page 295)
- P_c cutting power in kW
- P_1 drive power of the machine tool in kW
- η efficiency of the machine tool

Correction factor C for the cutting velocity

Cutting velocity v_c in m/min	C
10–30	1.3
31–80	1.1
81–400	1.0

Chip section

$$A = a_p \cdot f$$

Cutting force

$$F_c = A \cdot k_c \cdot C$$

Chip thickness

$$h = f \cdot \sin \varkappa$$

Cutting power

$$P_c = F_c \cdot v_c$$

Drive power

$$P_1 = \frac{P_c}{\eta}$$

Example:

A shaft of 16MnCr5, $a_p = 5$ mm, $f = 0.32$ mm, $v_c = 110$ m/min, $\varkappa = 75°$

Sought after: h; k_c; C; A; F_c; P_1 with $\eta = 0.75$

Solution: $\boldsymbol{h} = f \cdot \sin\varkappa = 0.32\text{ mm} \cdot \sin 75° = \mathbf{0.31\text{ mm}}$

$\boldsymbol{k_c} = \mathbf{3735\text{ N/mm}^2}$ (see table on page 295),

$\boldsymbol{C} = \mathbf{1.0}$ (see correction factor table)

$\boldsymbol{A} = a_p \cdot f = 5\text{ mm} \cdot 0.32\text{ mm} = \mathbf{1.6\text{ mm}^2}$

$$\boldsymbol{F_c} = A \cdot k_c \cdot C = 1.6\text{ mm}^2 \cdot 3735\,\frac{\text{N}}{\text{mm}^2} \cdot 1.0 = \mathbf{5976\text{ N}}$$

$$\boldsymbol{P_1} = \frac{P_c}{\eta} = \frac{F_c \cdot v_c}{\eta} = \frac{5976\text{ N} \cdot 110\text{ m}}{0.75 \cdot 60\text{ s}} = 14608\text{ W} = \mathbf{14.6\text{ kW}}$$

Drilling

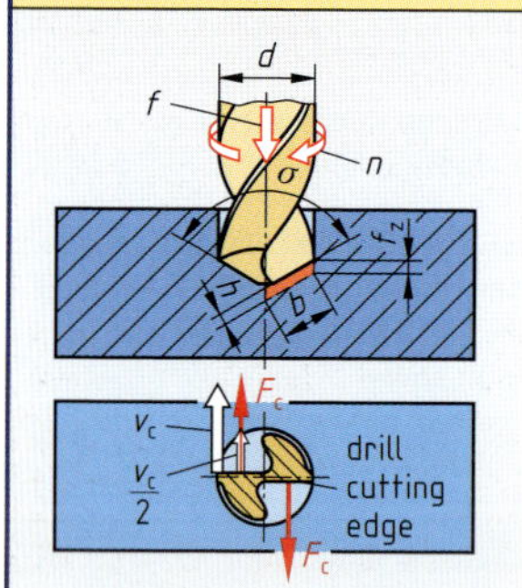

- F_c cutting force per edge in N
- z number of cutting edges (twist drill $z = 2$)
- A chip section in mm²
- d drill diameter in mm
- f feed per revolution in mm
- f_z feed per cutting edge in mm
- σ drill point angle in degrees (°)
- h chip thickness in mm
- C correction factor for the cutting speed
- v_c cutting velocity in m/min
- k_c specific cutting force in N/mm² (page 295)
- P_c cutting power in kW
- P_1 drive power of the machine tool in kW
- η efficiency of the machine tool

Correction factor C for the cutting velocity

Cutting velocity v_c in m/min	C
10–30	1.3
31–80	1.1

Chip section per cutting edge

$$A = \frac{d \cdot f}{4}$$

Cutting force per cutting edge[1]

$$F_c = 1.2 \cdot A \cdot k_c \cdot C$$

Chip thickness

$$h = \frac{f}{2} \cdot \sin\frac{\sigma}{2}$$

Cutting power

$$P_c = \frac{z \cdot F_c \cdot v_c}{2}$$

Drive power

$$P_1 = \frac{P_c}{\eta}$$

Example:

Material 42CrMo4, $d = 16$ mm, $v_c = 28$ m/min, $f = 0.18$ mm, $\sigma = 118°$

Sought after: h; k_c; C; A; F_c; P_c

Solution: $$\boldsymbol{h} = \frac{f}{2} \cdot \sin\frac{\sigma}{2} = \frac{0.18\text{ mm}}{2} \cdot \sin 59° = \mathbf{0.08\text{ mm}}$$

$\boldsymbol{k_c} = \mathbf{6265\text{ N/mm}^2}$ (see table on page 295)

$$\boldsymbol{A} = \frac{d \cdot f}{4} \cdot \frac{16\text{ mm} \cdot 0.18\text{ mm}}{4} = \mathbf{0.72\text{ mm}^2}$$

$\boldsymbol{C} = \mathbf{1.3}$ (see correction factor table)

$$\boldsymbol{F_c} = 1.2 \cdot A \cdot k_c \cdot C = 1.2 \cdot 0.72\text{ mm}^2 \cdot 6265\,\frac{\text{N}}{\text{mm}^2} \cdot 1.3 = \mathbf{7037\text{ N}}$$

$$\boldsymbol{P_c} = \frac{z \cdot F_c \cdot v_c}{2} \cdot \frac{2 \cdot 7037\text{ N} \cdot 28\text{ m}}{60\text{ s} \cdot 2} = 3284\,\frac{\text{N} \cdot \text{m}}{\text{s}} = 3284\text{W} = \mathbf{3.3\text{ kW}}$$

[1] The specific cutting force values k_e are assessed in turning tests. The conversion to drilling is realized via the factor 1.2 in the formula.

PE

Specific cutting force

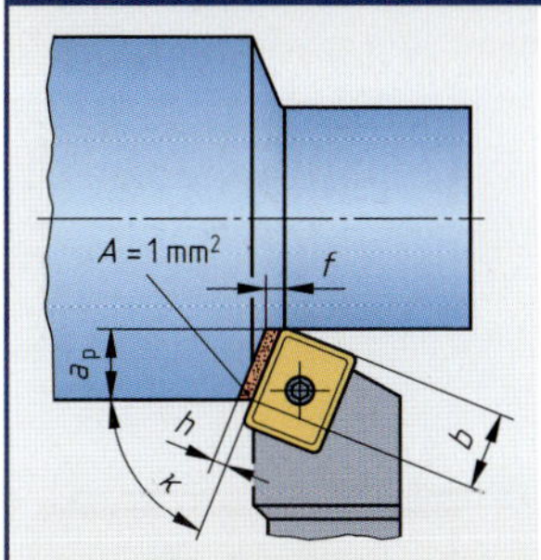

The specific cutting force k_c is the the force that is required to separate a chip with a cross section of $A = 1\ mm^2$ from a workpiece. The values are assessed in turning tests and form the basis of the calculation of the cutting forces and the drive power in chip-removing machining processes.

k_c specific cutting force N/mm^2
h chip thickness in mm
f feed in mm
a_p cutting depth in mm
$\varkappa$ angle of incidence in degrees (°)

The chip thickness h depends on the applied machining process. Calculation of chip thicknesses: pages 294 and 296.

Standard values for the specific cutting force[1]

Material	Specific cutting force k_c in N/mm^2 for the chip thickness h in mm												
	0.05	0.08	0.10	0.15	0.20	0.25	0.30	0.40	0.50	0.80	1.00	1.50	2.00
S235	3850	3555	3425	3195	3040	2930	2840	2705	2605	2405	2315	2160	2055
E295	5635	4990	4705	4235	3930	3710	3535	3285	3100	2740	2585	2330	2160
E355	4565	4215	4055	3785	3605	3470	3365	3205	3085	2850	2745	2560	2340
C15	4575	4125	3925	3590	3370	3210	3085	2895	2755	2485	2365	2165	2030
C35	4425	3895	3670	3290	3045	2865	2725	2525	2375	2095	1970	1765	1635
C45	4760	4210	3975	3575	3320	3130	2985	2770	2615	2315	2185	1965	1825
C60, C60E	4750	4365	4190	3895	3700	3555	3440	3265	3135	2880	2770	2575	2445
11SMnPb30	2675	2460	2360	2195	2085	2000	1935	1840	1765	1625	1560	1450	1375
16MnCr5	5950	5265	4965	4470	4150	3915	3735	3465	3270	2895	2730	2455	2280
20MnCr5	5775	5135	4855	4385	4085	3860	3690	3435	3245	2885	2730	2475	2295
18CrMo4	4955	4575	4405	4110	3915	3770	3655	3480	3350	3095	2975	2780	2645
34CrAlMo5	4930	4360	4115	3705	3435	3245	3095	2870	2710	2395	2260	2035	1890
42CrMo4	7080	6265	5915	5320	4940	4660	4445	4125	3890	3445	3250	2925	2715
50CrV4	6290	5565	5250	4725	4385	4140	3945	3660	3455	3060	2885	2595	2410
102Cr6	5895	4910	4500	3840	3435	3145	2930	2620	2400	2000	1835	1565	1400
90MnCrV8	5610	5080	4850	4455	4195	4000	3850	3625	3460	3135	2990	2745	2585
X210CrW12	5155	4565	4305	3875	3595	3395	3235	3005	2835	2510	2365	2130	1975
X5CrNi18-10	5730	5190	4955	4550	4285	4085	3935	3705	3535	3200	3055	2805	2640
X30Cr13	5155	4565	4305	3875	3595	3395	3235	3005	2835	2510	2365	2130	1975
TiAl6V4	3340	3025	2890	2655	2495	2385	2295	2160	2060	1985	1780	1635	1540
GJL-150	2315	2100	2005	1840	1730	1650	1590	1500	1430	1295	1235	1135	1065
GJL-200	2805	2495	2360	2130	1985	1875	1790	1670	1575	1405	1325	1200	1115
GJL-400	4165	3685	3480	3130	2905	2740	2615	2425	2290	2025	1910	1720	1595
GJS-400	2765	2455	2325	2100	1955	1845	1765	1645	1555	1380	1305	1180	1100
GJS-600	3200	2955	2845	2655	2530	2435	2360	2250	2165	2000	1925	1795	1710
GJS-800	5500	4470	4055	3390	2985	2710	2500	2200	1995	1625	1470	1230	1085
AlCuMg1	2150	1930	1835	1670	1565	1485	1425	1335	1265	1135	1080	985	920
AlMg3	2020	1810	1725	1570	1470	1395	1340	1250	1190	1065	1015	925	865
AC-AlSi12	2150	1930	1835	1670	1565	1485	1425	1335	1265	1135	1080	985	920
MgAl8Zn	895	820	785	725	690	660	635	605	580	530	505	470	445
CuZn40Pb2	1740	1600	1535	1425	1355	1300	1260	1195	1150	1055	1015	945	895
CuSn7ZnPb	1760	1565	1480	1335	1245	1175	1125	1045	990	880	830	750	700

[1] The standard values apply to carbide tooling. Variations in tensile strength, even if in compliance with standards, the degree of purity and the delivery condition (e.g. hot-rolled, cold-rolled, quenched & tempered ...) have an influence on the specific cutting force. Tool wear increases the specific cutting force by approximately 30 %. The values specified in the table above include this addition.

Forces and power in milling

Face milling

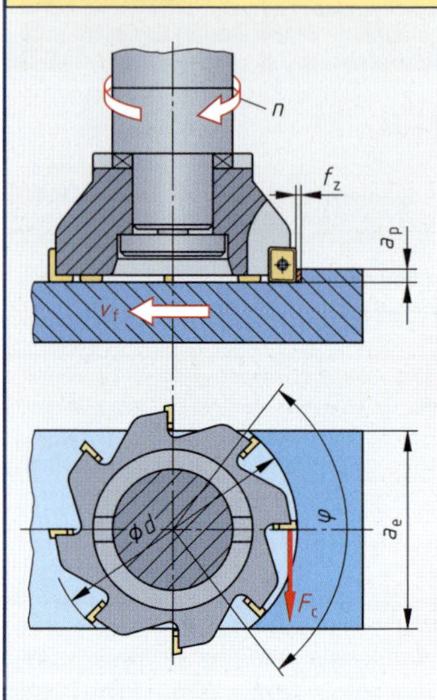

F_c	cutting force per tooth in N
A	chip section per tooth in mm²
a_p	cutting depth in mm
a_e	engagement (milling width) in mm
h	chip thickness in mm
f	feed per revolution in mm
f_z	feed per tooth in mm
d	cutter diameter in mm
v_c	cutting velocity in m/min
v_f	feed rate in mm/min
N	number of teeth
N_e	number of teeth engaged
φ	angle of engagement in degrees (°)
k_c	specific cutting force in N/mm² (page 295)
C	correction factor for the cutting speed
P_c	cutting power in kW
P_1	drive power in kW
η	effective power of the machine tool

Feed rate

$$v_f = N \cdot f_z \cdot n$$

Chip cross section per tooth

$$A = a_p \cdot f_z$$

Cutting force per tooth[1)]

$$F_c = 1.2 \cdot A \cdot k_c \cdot C$$

Chip thickness for d = (1.2–1.6)· a_e[2)]

$$h \approx f_z$$

Number of teeth engaged

$$N_e = N \cdot \frac{\varphi}{360°}$$

Cutting power

$$P_c = N_e \cdot F_c \cdot v_c$$

Drive power

$$P_1 = \frac{P_c}{\eta}$$

Example:

Material 16MnCr5; d = 180 mm; N = 12; a_e = 120 mm; a_p = 6 mm; f_z = 0.10 mm; v_c = 85 m/min; η = 0.8.

Sought after: A; h; k_c; F_c; φ; N_e; P_c; P_1

Solution:

$A = a_p \cdot f_z$ = 6 mm · 0.1 mm = **0.6 mm²**

$h \approx f_z$ = **0.1 mm**

k_c = **4965 $\frac{\textbf{N}}{\textbf{mm}^2}$** (table on page 295)

$F_c = 1.2 \cdot A \cdot k_c \cdot C$; C = 1.0 (table of correction factors C)

$F_c = 1.2 \cdot 0.6\ \text{mm}^2 \cdot 4965\ \frac{\text{N}}{\text{mm}^2} \cdot 1.0$ mm = **3575 N**

$\frac{d}{a_e} = \frac{180\ \text{mm}}{120\ \text{mm}} = 1.5$; φ = **83°** (angle of engagement φ table)

$N_e = N \cdot \frac{\varphi}{360°} = 12 \cdot \frac{83°}{360°}$ = **2.8**

$P_c = N_e \cdot F_c \cdot v_c = 2.8 \cdot 3575\ \text{N} \cdot \frac{85\ \text{m}}{60\ \text{s}} = 14181\ \frac{\text{N}\cdot\text{m}}{\text{s}} \approx$ **14.2 kW**

$P_1 = \frac{P_c}{\eta} = \frac{14.2\ \text{kW}}{0.8}$ = **17.8 kW**

Angle of engagement φ					
d/a_e	φ in °	d/a_e	φ in °	d/a_e	φ in °
1.20	113	1.35	96	1.50	83
1.25	106	1.40	91	1.55	80
1.30	100	1.45	87	1.60	77
$\frac{d}{a_e} = \frac{\text{cutter diameter}}{\text{engagement}}$					

Correction factor C for the cutting velocity	
Cutting velocity v_c in m/min	C
30–80	1.1
81–400	1.0

1) The values of the specific cutting force k_c (page 295) are assessed in turning tests. The conversion to milling is achieved via the factor 1.2 in the formula.

2) In order to ensure favorable cutting conditions, the cutter diameter should be selected in the range $d = (1.2–1.6) \cdot a_e$.

Speed graph

The speed n of a machine tool from the workpiece or tool diameter d and the selected cutting velocity v_c can be determined

- on a computer/calculator using the formula, or
- graphically using the speed graph.

Speed graphs have the speeds under load which can be set on the machine. These are stepped geometrically. For infinitely variable drives the calculated speed can be set precisely.

Speed

$$n = \frac{v_c}{\pi \cdot d}$$

Speed graph with logarithmically scaled coordinates

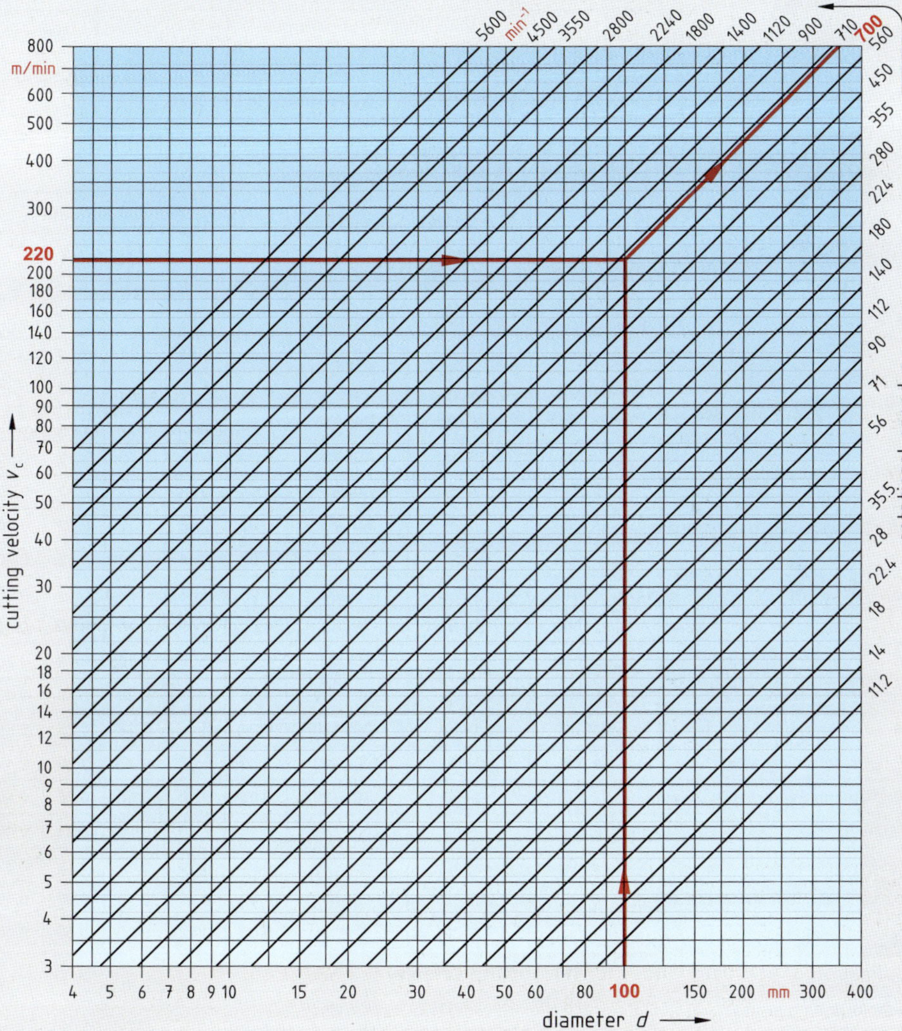

Example: $d = 100$ mm; $v_c = 220 \frac{\text{m}}{\text{min}}$; $n = ?$

Calculation: $\boldsymbol{n} = \frac{v_c}{\pi \cdot d} = \frac{220 \frac{\text{m}}{\text{min}}}{\pi \cdot 0.1\ \text{m}} = \mathbf{700.3} \frac{\mathbf{1}}{\mathbf{min}}$; read from the speed graph above: $\boldsymbol{n} \approx \mathbf{700} \frac{\mathbf{1}}{\mathbf{min}}$

Drilling

Twist drills made of high-speed steel (HSS)

Helix angle

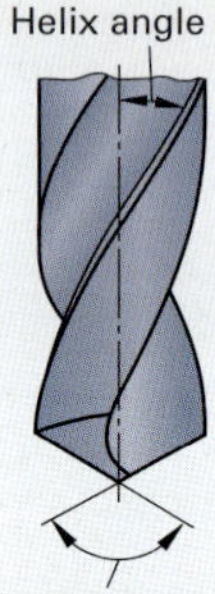

Point angle

Type[1]	Application	Helix angle[2]	Point angle[3]
N	Universal application for materials up to $R_m \approx$ 1000 N/mm², e.g. structural, case-hardened, quenched and tempered steels	30° – 40°	118°
H	Drilling of brittle, short-chipping non-ferrous metals and plastics, e.g. CuZn alloys and PMMA (Plexiglas)	13° – 19°	118°
W	Drilling of soft, long-chipping non-ferrous metals and plastics, e.g. Al and Mg alloys, PA (polyamide) and PVC	40° – 47°	130°

[1] Tool application groups for HSS tools according to DIN 1836
[2] Depends on drill diameter and pitch
[3] Standard version

Standard values for drilling with HSS twist drills and carbide drills[1]

Workpiece material: Material group	Tensile strength R_m in N/mm² or Hardness HB	Cutting velocity v_c in m/min: HSS drill	Carbide drill	Drill diameter d in mm: 3	6	10	16	25
				Feed f in mm/revolutions HSS and carbide drill				
Structural steel	$R_m \le 500$	25–40	60–80	0.05	0.14	0.25	0.32	0.43
	$R_m > 500$	20–30	50–70	0.05	0.14	0.25	0.30	0.40
Free-cutting steel	$R_m < 900$	20–30	60–80	0.05	0.14	0.25	0.32	0.40
Case-hardened steel	$R_m \le 700$	15–20	40–70	0.04	0.10	0.15	0.25	0.30
	$R_m > 700$	10–15	30–60	0.04	0.08	0.15	0.22	0.30
Quenched & tempered steel	$R_m \le 1000$	20–25	50–70	0.03	0.06	0.12	0.22	0.30
	$R_m > 1000$	15–20	40–60	0.03	0.06	0.12	0.15	0.20
Tools steel	$R_m < 1000$	8–12	25–40	0.02	0.05	0.10	0.16	0.20
Corrosion-proof steel, austenitic	$R_m < 1000$	10–15	25–40	0.02	0.05	0.10	0.15	0.20
martensitic	$R_m < 1000$	6–8	20–40	0.02	0.05	0.10	0.15	0.20
Malleable cast iron	< 250 HB	20–25	60–100	0.05	0.13	0.20	0.30	0.40
Spheroidal graphite iron	< 300 HB	20–25	60–100	0.04	0.12	0.18	0.25	0.35
Wrought aluminum alloy	$R_m < 300$	40–60	100–200	0.05	0.16	0.20	0.35	0.45
Cast aluminum alloy (> 10% Si)	$R_m < 500$	0–50	80–120	0.03	0.10	0.16	0.30	0.40
Cu alloy, brass	$R_m < 600$	40–60	100–160	0.06	0.10	0.20	0.35	0.45
Cu alloy, bronze	$R_m < 700$	25–40	80–120	0.05	0.10	0.15	0.25	0.32
Thermoplastics, thermoset plastics	–	20–40	25–60	0.04	0.12	0.20	0.30	0.40
Fiber-reinforced plastics	–	15–20	–	0.04	0.12	0.20	0.30	0.40

PE

Application of standard values for the cutting velocity

Example: drilling in structural steel Rm > 500 N/mm², using HSS twist drills

Upper value	Application	Lower value	Application
v_c = 30 m/min	Steels of low strength, e. g. S235; low drilling depths	v_c = 20 m/min	Steels of high strengths, e. g. S960; deep holes

Machining often starts with the medium value (start value) as a test. In the example: $v_c \approx$ 25 m/min.

[1] Cooling lubrication on pages 311 and 314.

Reaming, tapping and thread forming

Standard values for reaming with HSS reamers and reamers with carbide blades[1)]

Workpiece material: Material group	Tensile strength R_m in N/mm² or Hardness HB	Cutting velocity v_c in m/min: HSS	Cutting velocity v_c in m/min: Carbide	Feed f in mm/revolution, HSS and carbide reamers, tool diameter d = 3 mm	6	10	16	25	Reaming allowance for d in mm, HSS & carbide: ≤ 20	> 20
Structural steel	$R_m \le 500$	12–18	25–40	0.10	0.12	0.15	0.20	0.25	0.2	0.3
	$R_m > 500$	10–16	25–35	0.10	0.12	0.15	0.20	0.25		
Free-cutting steel	$R_m < 900$	10–15	20–30	0.10	0.12	0.15	0.20	0.25		
Case-hardened steel	$R_m \le 700$	8–12	15–20	0.08	0.10	0.12	0.15	0.20		
	$R_m > 700$	6–10	10–15	0.10	0.13	0.15	0.20	0.25		
Quenched & tempered steel	$R_m \le 1000$	8–12	20–30	0.08	0.10	0.12	0.15	0.25		
	$R_m > 1000$	6–10	15–25	0.10	0.12	0.15	0.20	0.25		
Tool steel	$R_m < 1000$	–	15–25	0.08	0.10	0.12	0.15	0.25		
Corrosion-proof steel, austenitic	$R_m < 1000$	–	20–30	0.10	0.12	0.15	0.20	0.25		
martensitic	$R_m < 1000$	–	15–40	0.08	0.10	0.12	0.15	0.25		
Malleable cast iron	< 250 HB	8–14	50–80	0.18	0.20	0.23	0.25	0.30		
Spheroidal graphite iron	< 300 HB	8–14	30–60	0.12	0.15	0.18	0.20	0.25		
Wrought aluminum alloy	$R_m < 300$	14–20	30–60	0.15	0.18	0.20	0.25	0.30	0.3	0.4
Cast aluminum alloy (> 10% Si)	$R_m < 500$	14–20	30–60	0.15	0.18	0.20	0.25	0.30		
Cu alloy, brass	$R_m < 600$	14–20	20–50	0.20	0.25	0.30	0.35	0.40		
Cu alloy, bronze	$R_m < 700$	–	20–40	0.20	0.25	0.30	0.35	0.40		
Thermoplastics, thermoset plastics	–	8–14	20–30	–	–	–	–	–	–	–
Fibre-reinforced plastics	–	–	15–25	–	–	–	–	–	–	–

Standard values for tapping and thread forming[1)]

Workpiece material: Material group	Tensile strength R_m in N/mm² or Hardness HB	**Material group** Tapping velocity v_c in m/min: HSS	Carbide	**Thread forming** Tapping velocity v_c in m/min: HSS	Carbide
Steels, low strength	$R_m \le 600$	16–20	–	16–20	24–40
Steel, high strength	$R_m > 600$	10–15	–	10–15	15–25
Stainless steels	$R_m \ge 800$	4–6	–	10–15	15–25
Cast iron (malleable cast iron)	≤ 250 HB	10–16	20–30	–	–
Al alloys	$R_m \le 350$	16–30	40–60	16–20	30–50
Cu alloys	$R_m \le 500$	16–25	25–35	16–20	30–45
Thermoplastics	–	15–20	30–50	–	–
Thermoset plastics, reinforced	–	10–15	25–40	–	–

Application of standard values for the cutting velocity

Example: reaming of aluminum alloy workpieces with HSS tools

Upper value	**Application**	**Lower value**	**Application**
v_c = 20 m/min	Al alloys with low strength, e.g. AlMg1	v_c = 14 m/min	Al alloys with high strength or high Si content, e.g. AC-AlSi12

Machining often starts with the medium value (start value) as a test. In the example: $v_c \approx 17$ m/min.

[1)] Cooling lubrication on pages 311 and 314.

PE

Drilling

Main productive time for drilling, reaming and counterboring

t_p	main productive time	L	travel
d	tool diameter	f	feed per revolution
l	bore depth	n	speed
l_{si}	starting idle	v_c	cutting velocity
l_{oi}	overrun idle travel	i	number of cuts
l_c	lead	σ	point angle

Lead l_c

σ	l_c
80°	$0.6 \cdot d$
118°	$0.3 \cdot d$
130°	$0.23 \cdot d$
140°	$0.18 \cdot d$

Productive time

$$t_p = \frac{L \cdot i}{n \cdot f}$$

Speed

$$n = \frac{v_c}{\pi \cdot d}$$

Calculating travel L

for drilling and reaming

Through hole

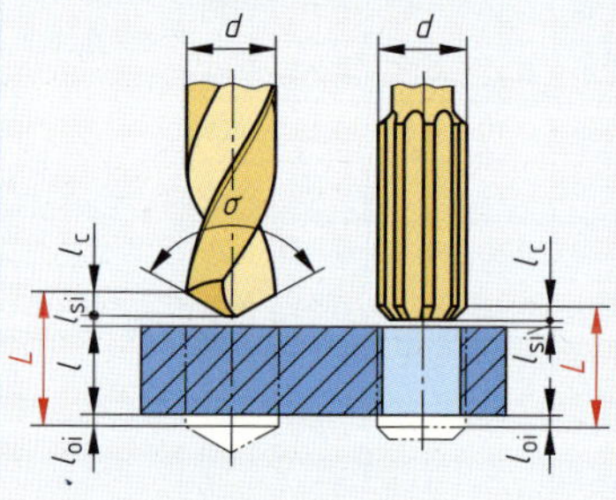

$$L = l + l_c + l_{si} + l_{oi}$$

Blind hole

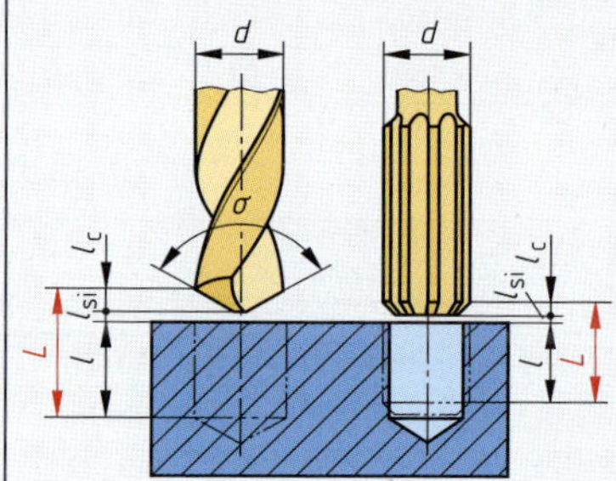

$$L = l + l_c + l_{si}$$

for counterboring

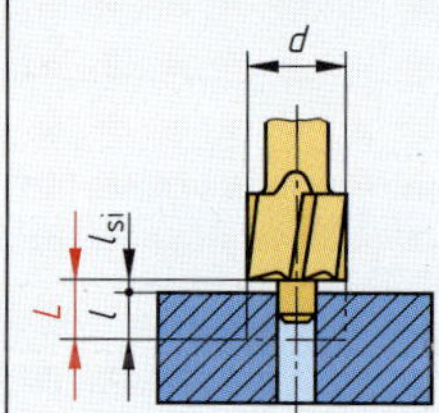

$$L = l + l_{si}$$

Example:

Blind hole of d = 30 mm;
l = 90 mm; f = 0.15 mm;
n = 450/min; i = 15; l_{si} = 1 mm;
σ = 130°; L = ?; t_p = ?

$\mathbf{L} = l + l_c + l_{si} = 90\text{ mm} + 0.23 \cdot 30\text{ mm} + 1\text{ mm} = \mathbf{98\text{ mm}}$

$$\mathbf{t_p} = \frac{L \cdot i}{n \cdot f} = \frac{98\text{ mm} \cdot 15}{450\,\frac{1}{\text{min}} \cdot 0.15\text{ mm}} = \mathbf{21.78\text{ min}}$$

Troubleshooting and corrective measures for drilling problems

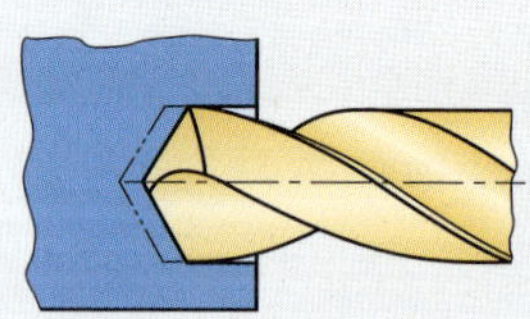

Drill point damaged	Wear on outside diameter	Hole too large	Chips stuck in flutes	Chipping of cutting edge	Noncircular hole	Short tool life	Vibrations	Possible corrective measures
•	•	•		•				Check cutting geometry
			•			•		Increase supply of lubricant
		•	•		•		•	Decrease feed f
			•	•				Increase cutting velocity v_c
•	•		•			•	•	Decrease projection length
•	•	•	•			•	•	Check cutting parameters
•	•			•		•		Check type of carbide

PE

Turning

Roughness depth depending on tool nose radius and feed

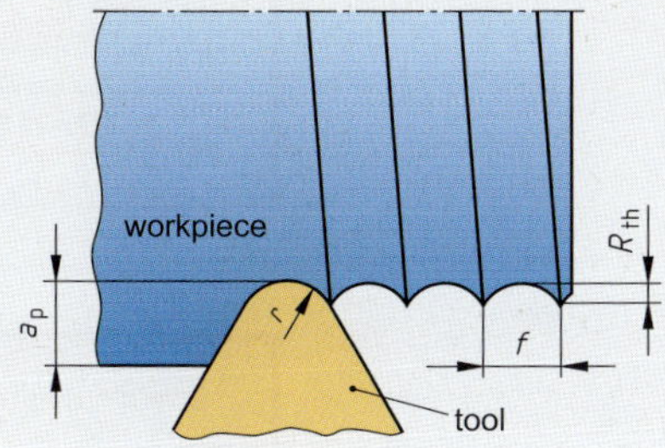

R_{th} theoretical roughness depth
r tool nose radius
f feed
a_p cutting depth

Example:

R_{th} = 25 µm; r = 1.2 mm; f = ?

$f \approx \sqrt{8 \cdot r \cdot R_{th}}$
$= \sqrt{8 \cdot 1.2\ \text{mm} \cdot 0.025\ \text{mm}} \approx \mathbf{0.5\ mm}$

Theor. roughness depth

$$R_{th} \approx \frac{f^2}{8 \cdot r}$$

$R_{th} \approx R_z$

Roughn. depth R_{th} in µm	Tool nose radius r in mm: 0.4	0.8	1.2	1.6
	Feed f in mm			
1.6	0.07	0.10	0.12	0.14
4	0.11	0.16	0.20	0.23
10	0.18	0.25	0.31	0.36
16	0.23	0.32	0.39	0.45
25	0.28	0.40	0.49	0.57

Standard values for turning with HSS and carbide tools

Workpiece material: Material group	Tensile strength R_m in N/mm² or Hardness HB	Cutting velocity v_c in m/min: HSS	Carbide	Feed f in mm (HSS and Carbide)	Cutting depth a_p in mm (HSS and Carbide)
Structural steel	$R_m \leq 500$	40–80	150–200	0.1–0.5	0.5–4.0
	$R_m > 500$	30–60	120–180		
Free-cutting steel	$R_m < 900$	40–80	120–180		
Case-hardened steel	$R_m \leq 700$	40–80	150–220		
	$R_m > 700$	30–60	100–160		
Quenched & tempered steel	$R_m \leq 1000$	40–80	120–180		
	$R_m > 1000$	30–60	120–180		
Tool steel	$R_m < 1000$	30–60	100–180		
Corrosion-proof steel, austenitic	$R_m < 1000$	30–60	100–180		
martensitic	$R_m < 1000$	20–40	80–150		
Malleable cast iron	< 250 HB	20–40	100–180		
Spheroidal graphite iron	< 300 HB	15–30	80–160		
Wrought aluminum alloy	$R_m < 300$	120–180	400–800		
Cast aluminum alloy (> 10% Si)	$R_m < 500$	80–120	300–600		
Cu-Cu alloy, brass	$R_m < 600$	80–120	300–600		
Cu alloy, bronze	$R_m < 700$	60–100	200–400		
Thermoplastics, thermoset plastics	–	200–300	500–1500		
Fiber-reinforced plastics	–	150–250	400–1000		

Application of standard values for the cutting velocity, feed and cutting depth

Example: turning of structural steel with low strength using carbide tools

Upper values	Application	Lower values	Application
v_c = 200 m/min	• Finish-machining (finishing) • Stable tool and workpiece	v_c = 150 m/min	• Pre-machining (roughing) • Difficult clamping conditions
f = 0.5 mm, a_p = 4.0 mm	• Pre-machining (roughing) • Stable tool and workpiece	f = 0.1 mm, a_p = 0.5 mm	• Finish-machining (finishing) • Unstable tool or workpiece

Machining often starts with the medium values. In the example: $v_c \approx$ 170 m/min, $f \approx$ 0.3 mm, $a_p \approx$ 2 mm.

[1)] Cooling lubrication on pages 311 and 314.

PE

Turning and taper turning

Taper turning by setting the compound rest

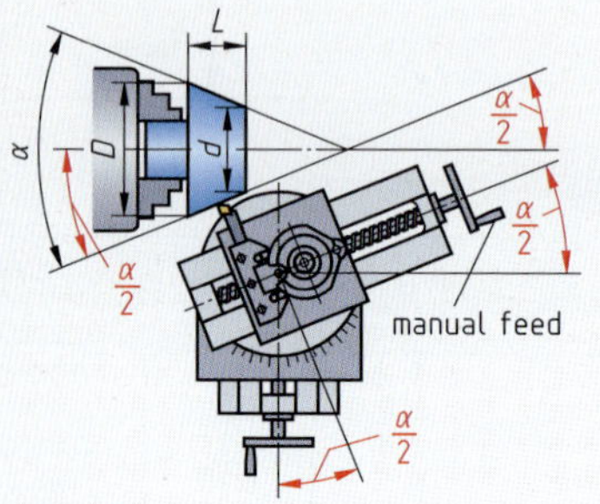

Taper turning on NC machines on page 412

D large taper diameter
d small taper diameter
L taper length
α taper angle
C taper ratio
$\frac{\alpha}{2}$ setting angle

Example:

$D = 225$ mm, $d = 150$ mm, $L = 100$ mm;

$\frac{\alpha}{2} = ?$; $C = ?$

$$\tan\frac{\alpha}{2} = \frac{D-d}{2 \cdot L}$$

$$= \frac{(225-150)\text{ mm}}{2 \cdot 100\text{ mm}} = 0.375$$

$$\frac{\alpha}{2} = \mathbf{20.556°} = \mathbf{20°\,33'\,22''}$$

$$C = \frac{D-d}{L} = \frac{(225-150)\text{ mm}}{100\text{ mm}} = 0.75 = \mathbf{1:1.33}$$

Setting angle

$$\tan\frac{\alpha}{2} = \frac{C}{2}$$

$$\tan\frac{\alpha}{2} = \frac{D-d}{2 \cdot L}$$

Taper angle

$$\alpha = 2 \cdot \frac{\alpha}{2}$$

Taper ratio

$$C = \frac{D-d}{L}$$

Tapered profiles on machine elements (examples)

Taper	Taper ratio	Taper angle	Setting angle	Application	Page
–	–	–	vd.	Bevel gears, circumferential surface of gearing	255
–	–	–	vd.	Cylindrical roller bearings, inclined feeds	262
–	–	–	vd.	Tapered roller bearings, running and sliding surfaces	263
40°	1:1.374	40°	20°	Draw-in collets	–
60°	1:0.866	60°	30°	Centering points on turning and grinding machines	–
136°	1:0.202	136°	68°	Indenters for Martens hardness test	195
7:24	1:3.429	16°36′	8°17′50″	Steep tapers for milling spindle heads	240
1:16	1:16	3°35′	1°47′24″	Tapered threads	205
MK 2	1:20.020	2.862°	1.431°	Morse tapers, size 2 for tools and jigs	240
1:20	1:20	2°52′	1°25′56″	Metric tapers for tools and jigs	240
1:50	1:50	1°9′	34′23″	Pins, taper gages	236

Troubleshooting for turning

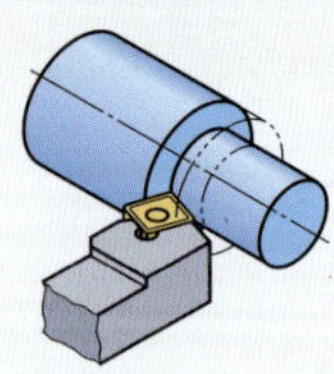

Problems: High wear at the flank and cutting faces	Deformation of the cutting edge	Galling of the cutting edges	Cracks perpendicular to the cutting edge	Chipping of cutting edge	Fracture of indexable insert	Long spiral chips	Vibrations	Possible corrective measures
⇓	⇓	⇑		⇑			⇓	Change cutting velocity v_c
⇓	⇓				⇓	⇑	⇑	Change feed f
					•		•	Decrease cutting depth
•	•							Choose a more wear-resistant carbide type
			•	•	•			Choose a tougher carbide type
•		•		•			•	Choose a positive cutting geometry

• problem to be solved ⇑ increase value of cutting parameter ⇓ decrease value of cutting parameter

PE

Turning, Thread cutting

Straight cylindrical turning and facing at constant rotational speed

t_p	productive time	l_{oi}	overrun idle travel
d	outside diameter	L	travel
d_1	inside diameter	f	feed per revolution
d_m	mean diameter[1)]	n	rotational speed
l	workpiece length	i	number of cuts
l_{si}	starting idle	v_c	cutting velocity

Productive time

$$t_p = \frac{L \cdot i}{n \cdot f}$$

Calculating travel L, mean diameter d_m and rotational speed n

Straight cylindrical turning		Facinwg		
without shoulder	with shoulder	Solid cylinder without shoulder	Solid cylinder with shoulder	Hollow cylinder
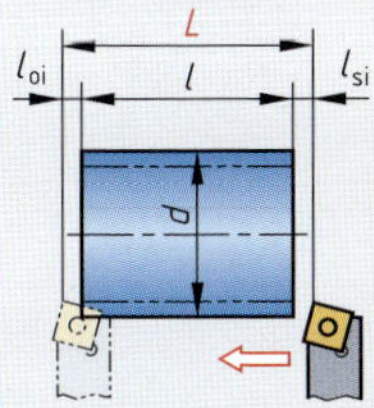	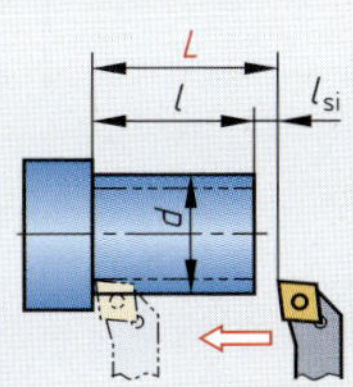	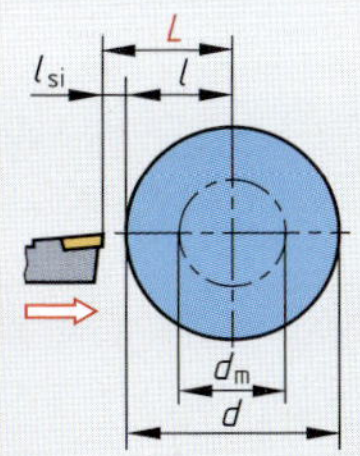	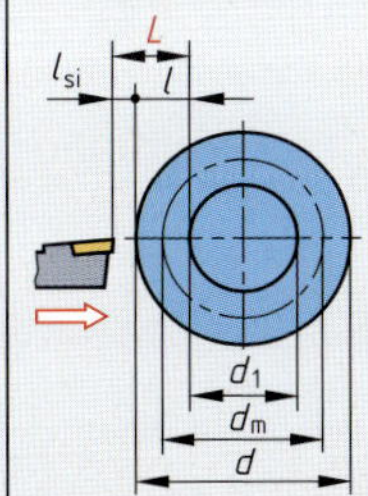	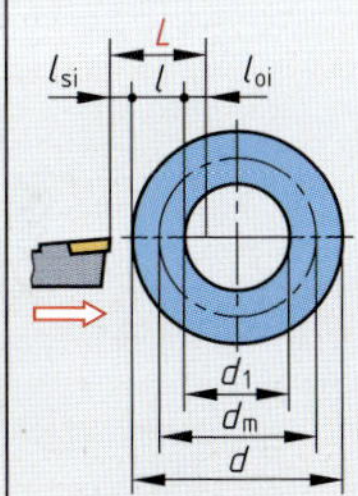
$L = l + l_{si} + l_{oi}$	$L = l + l_{si}$	$L = \frac{d}{2} + l_{si}$	$L = \frac{d - d_1}{2} + l_{si}$	$L = \frac{d - d_1}{2} + l_{si} + l_{oi}$
$n = \frac{v_c}{\pi \cdot d}$		$d_m = \frac{d}{2}$; $n = \frac{v_c}{\pi \cdot d_m}$	$d_m = \frac{d + d_1}{2}$; $n = \frac{v_c}{\pi \cdot d_m}$	

[1)] Use of mean diameter d_m leads to higher cutting velocities. This ensures acceptable cutting conditions for small diameters (inside area).

Example:

Straight cylindrical turning without shoulder, l = 1240 mm; $l_{si} = l_{oi}$ = 2 mm; f = 0.6 mm; v_c = 120 m/min; i = 2; d = 160 mm; L = ?; n = ? (for infinitely variable speed adjustment) t_p = ?

$$\boldsymbol{L} = l + l_{si} + l_{oi} = 1240\text{ mm} + 2\text{ mm} + 2\text{ mm} = \mathbf{1244\ mm}$$

$$\boldsymbol{n} = \frac{v_c}{\pi \cdot d} = \frac{120\,\frac{\text{m}}{\text{min}}}{\pi \cdot 0.16\text{ m}} \approx \mathbf{239\,\frac{1}{min}}$$

$$\boldsymbol{t_p} = \frac{L \cdot i}{n \cdot f} = \frac{1244\text{ mm} \cdot 2}{239\,\frac{1}{\text{min}} \cdot 0.6\text{ mm}} \approx \mathbf{17.4\ min}$$

Thread cutting

t_p	productive time	P	thread pitch
L	total travel of thread cutting tool	n	rotational speed
l	thread length	s	no. of starts
l_{si}	starting idle	h	thread depth
l_{oi}	overrun idle travel	a_p	cutting depth
i	number of cuts	v_c	cutting velocity

Productive time

$$t_p = \frac{L \cdot i \cdot s}{P \cdot n}$$

Number of cuts

$$i = \frac{h}{a_p}$$

Example:

Threads M 24; l = 76 mm; $l_{si} = l_{oi}$ = 2 mm; f = 0.6 mm; v_c = 6 m/min; i = 2; a_p = 0.15 mm; h = 1.84 mm; P = 3 mm; s = 1; L = ?; n = ?; i = ?; t_p = ?

$$i = \frac{h}{a_p} = \frac{1.84\text{ mm}}{0.15\text{ mm}} = 12.2 \approx \mathbf{13}$$

$$\boldsymbol{L} = l + l_{si} + l_{oi} = 76\text{ mm} + 2\text{ mm} + 2\text{ mm} = \mathbf{80\ mm}$$

$$\boldsymbol{n} = \frac{v_c}{\pi \cdot d} = \frac{6\,\frac{\text{m}}{\text{min}}}{\pi \cdot 0.024\text{ m}} \approx \mathbf{80\,\frac{1}{min}}$$

$$\boldsymbol{t_p} = \frac{L \cdot i \cdot s}{P \cdot n} = \frac{80\text{ mm} \cdot 13 \cdot 1}{3\text{ mm} \cdot 80\,\frac{1}{\text{min}}} = \mathbf{4.3\ min}$$

PE

Turning

Straight cylindrical turning and facing at constant cutting speed

If the rotational speed must be limited for safety reasons to a rotational speed limit n_{lim}, a turning diameter of d < transition diameter d_t is turned at constant rotational speed (page 303).

d_t	transition diameter	i	number of cuts
v_c	cutting velocity	d	outside diameter
n_{lim}	rotational speed limit	d_1	inside diameter
t_p	productive time	a_p	cutting depth
d_e	effective diameter	l_{si}	starting idle
L	travel	l_{oi}	overrun idle travel
f	feed		

Transition diameter

$$d_t = \frac{v_c}{\pi \cdot n_{lim}}$$

Productive time

$$t_p = \frac{\pi \cdot d_e \cdot L \cdot i}{v_c \cdot f}$$

Number of cuts for straight cylindrical turning

$$i = \frac{d - d_1}{2 \cdot a_p}$$

Calculating travel L and effective diameter d_e

Straight cylindrical turning

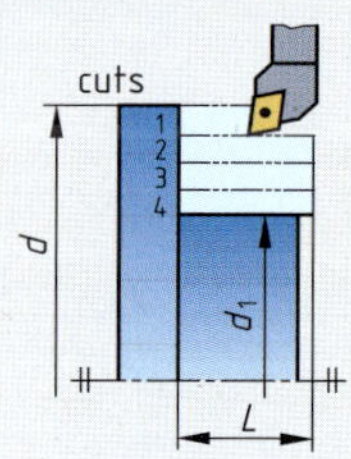

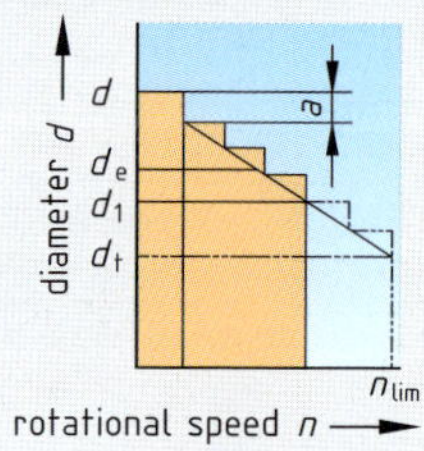

Facing

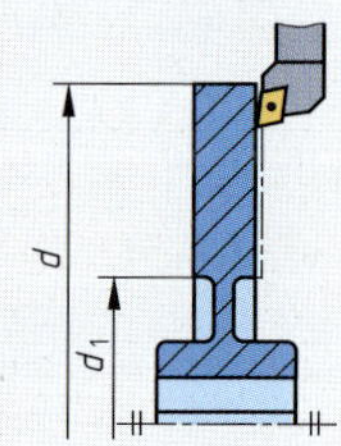

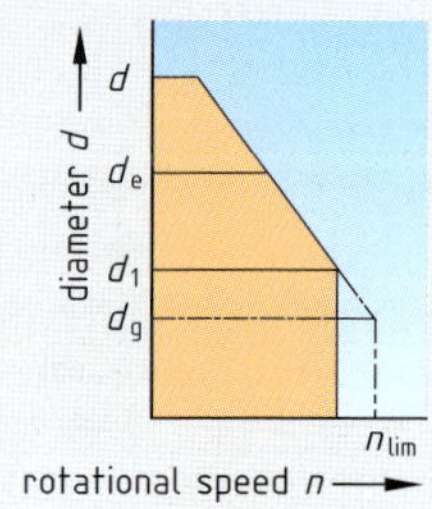

without shoulder

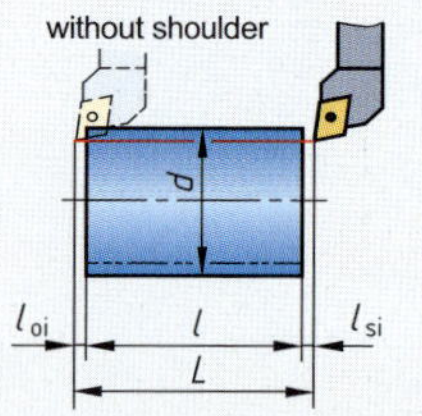

with shoulder

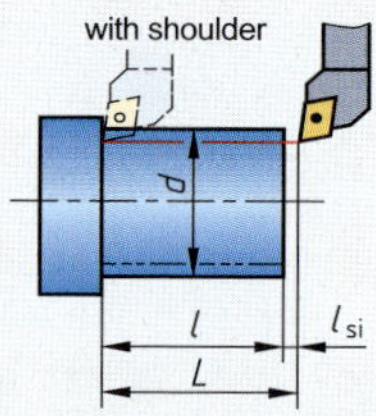

Solid cylinder with shoulder

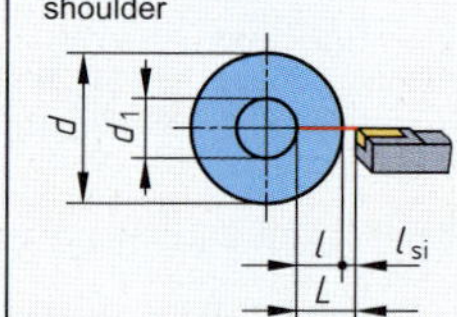

Hollow cylinder

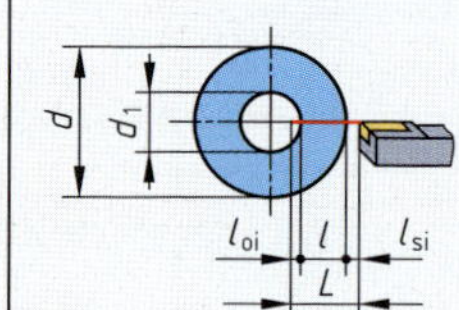

without shoulder	with shoulder	Solid cylinder with shoulder	Hollow cylinder
$L = l + l_{si} + l_{oi}$	$L = l + l_{si}$	$L = \frac{d - d_1}{2} + l_{si}$	$L = \frac{d - d_1}{2} + l_{si} + l_{oi}$
$d_e = d - a_p \cdot (i + 1)$		$d_e = \frac{d + d_1}{2} + l_{si}$	$d_e = \frac{d + d_1}{2} + l_{si} - l_{oi}$

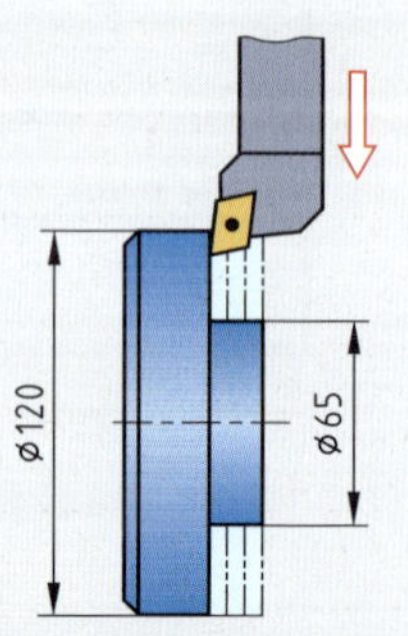

Example:

Facing; l_{si} = 1.5 mm; v_c = 220 m/min; f = 0.2 mm; i = 2; n_{lim} = 3000/min; d_t = ?; L = ?; d_e = ?; t_p = ?

$$\mathbf{d_t} = \frac{v_c}{\pi \cdot n_{lim}} = \frac{220\,000\ \frac{\text{mm}}{\text{min}}}{\pi \cdot 3000\ \frac{1}{\text{min}}} = \mathbf{23.3\ mm}\ (d_1 > d_t)$$

$$\mathbf{L} = \frac{d - d_1}{2} + l_{si} = \frac{120\ \text{mm} - 65\ \text{mm}}{2} + 1.5\ \text{mm} = \mathbf{29\ mm}$$

$$\mathbf{d_e} = \frac{d + d_1}{2} + l_{si} = \frac{120\ \text{mm} + 65\ \text{mm}}{2} + 1.5\ \text{mm} = \mathbf{94\ mm}$$

$$\mathbf{t_p} = \frac{\pi \cdot d_e \cdot L \cdot i}{v_c \cdot f} = \frac{\pi \cdot 94\ \text{mm} \cdot 29\ \text{mm} \cdot 2}{220\,000\ \frac{\text{mm}}{\text{min}} \cdot 0.2\ \text{mm}} = \mathbf{0.39\ min}$$

PE

Milling

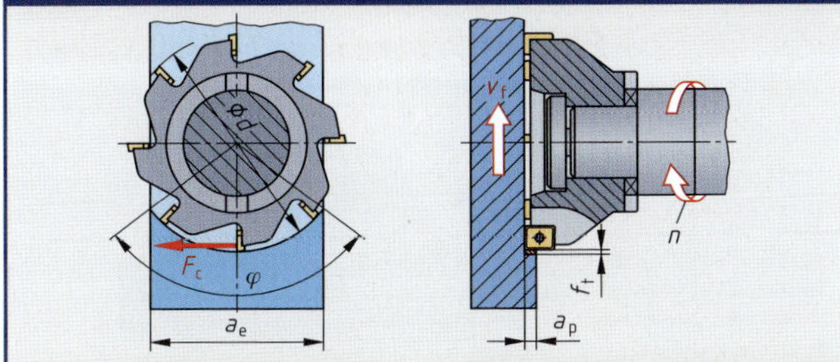

v_c cutting velocity in m/min
f_t feed per tooth in mm
a_p cutting depth in mm
a_e engagement (cutting width) in mm
d cutting diameter in mm

Standard values for milling with HSS and carbide milling cutters

Workpiece material: Material group	Tensile strength R_m in N/mm² or Hardness HB	Cutting velocity v_c in m/min: HSS	Carbide	Feed per tooth f_t in mm: for milling cutters (except end mill)	for end mills dia. 6–20 mm	Cutting depth a_p in mm
				HSS and carbide cutters		
Structural steel	$R_m \leq 500$	50–60	250–300	0.05–0.15	0.06–0.10	–5.0
	$R_m > 500$	40–50	200–250			
Free-cutting steel	$R_m < 900$	50–60	250–300			
Case-hardened steel	$R_m \leq 700$	50–60	120–150			
	$R_m > 700$	40–50	100–130			
Quenched & tempered steel	$R_m \leq 1000$	40–50	100–120			
	$R_m > 1000$	30–40	80–100			
Tools steel	$R_m < 1000$	30–40	100–120			
Corrosion-proof steel, austenitic	$R_m < 1000$	30–40	150–200			
martensitic	$R_m < 1000$	20–30	120–150			
Malleable cast iron	< 250 HB	50–60	180–220			
Spheroidal graphite iron	< 300 HB	50–60	180–240			
Wrought aluminum alloy	$R_m < 300$	200–300	600–1000			
Cast aluminum alloy (> 10% Si)	$R_m < 500$	80–100	400–600			
Cu alloy, brass	$R_m < 600$	70–90	400–600			
Cu alloy, bronze	$R_m < 700$	50–70	250–300			
Thermoplastics, thermoset plastics	–	100–200	500–800	0.10–0.20	0.10–0.20	
Fiber-reinforced plastics	–	40–60	300–600			

Increasing the recommended feed per cutting edge f_t for slotting with side milling cutters

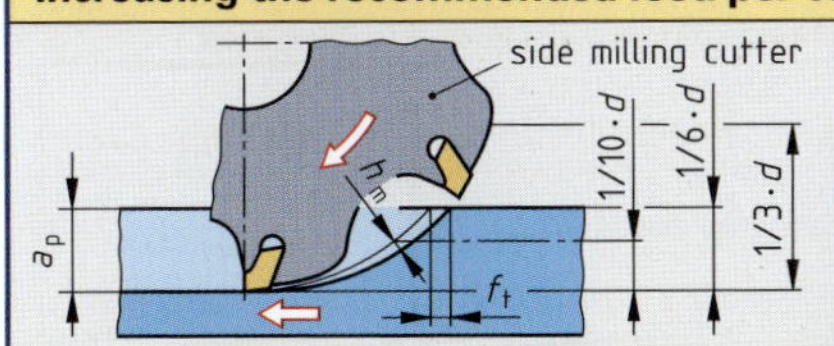

Feed per tooth	Cutting depth a_p referenced to the diameter d of the milling cutter: $1/3 \cdot d$	$1/6 \cdot d$	$1/10 \cdot d$	$1/20 \cdot d$
increase	$1 \cdot f_t$	$1.15 \cdot f_t$	$1.45 \cdot f_t$	$2 \cdot f_t$
to be adjusted	0.25 mm	0.29 mm	0.36 mm	0.50 mm

PE

Application of standard values for the cutting velocity, engagement and feed per tooth

Example: milling of structural steel with low strength using HSS cutters

Upper values	Application	Lower values	Application
v_c = 60 m/min	• finish-machining (finishing) • stable tool and workpiece • low cutting depth a_p	v_c = 50 m/min	• pre-machining (roughing) • unstable tool and workpiece • high cutting depth a_p
f_t = 0.15 mm	• pre-machining (roughing) • stable tool and workpiece	f_t = 0.05 mm	• finish-machining (finishing) • unstable tool and workpiece

Machining often starts with the medium values (start values). In the example: $v_c \approx 55$ m/min, $f_t \approx 0.10$ mm, $a_p \approx 3$ mm.

[1] Cooling lubrication on pages 311 and 314

Milling, Indexing with a dividing head

Troubleshooting for milling

Problems: High wear at the flank and cutting faces	Deformation of the cutting edge	Galling of the cutting edges	Cracks perpendicular to the cutting edge	Chipping of cutting edge	Fracture of the indexable insert	Poor surface quality	Vibrations	Possible corrective measures
⇓	⇓	⇑	⇓	⇑				Change cutting velocity v_c
⇑		⇑		⇑	⇓	⇓	⇑	Change feed per tooth f_t
	•					•		Choose a more wear-resistant carbide type
			•	•	•			Choose a tougher carbide type
							•	Use milling cutter with wider spacing
						•	•	Change milling cutter position
		•	•	•				Dry milling

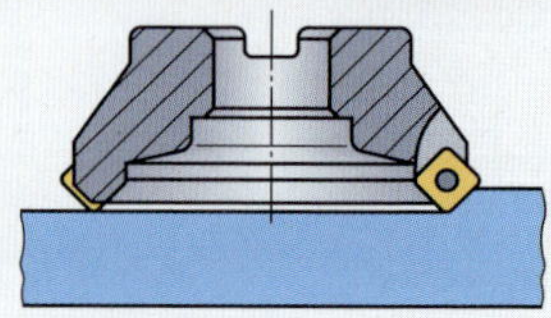

• problem to be solved ⇑ increase value of cutting parameter ⇓ decrease value of cutting parameter

Indexing with a dividing head

Direct indexing

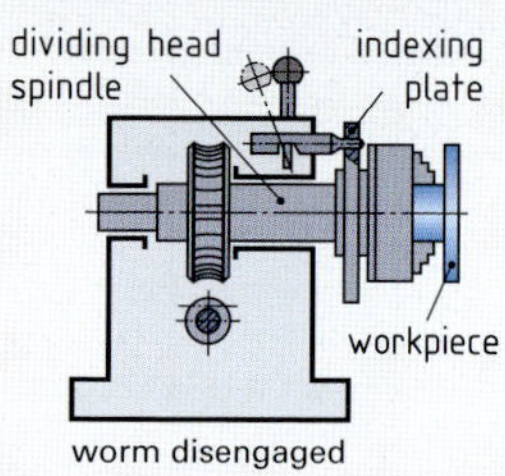

In direct indexing, the dividing head spindle, along with the indexing plate and workpiece, is turned to the desired indexing step. The worm is disengaged from the worm wheel.

D no. of divisions
α angular division
n_h no. of holes in the indexing plate
n_i indexing step; no. of hole spacings to be indexed

Indexing step

$$n_i = \frac{n_h}{D}$$

$$n_i = \frac{\alpha \cdot n_h}{360°}$$

Example:

$n_h = 24;\ D = 8;\ n_i = ?$ $\quad \boldsymbol{n_i} = \frac{n_h}{D} = \frac{24}{8} = \mathbf{3}$

Indirect indexing

PE

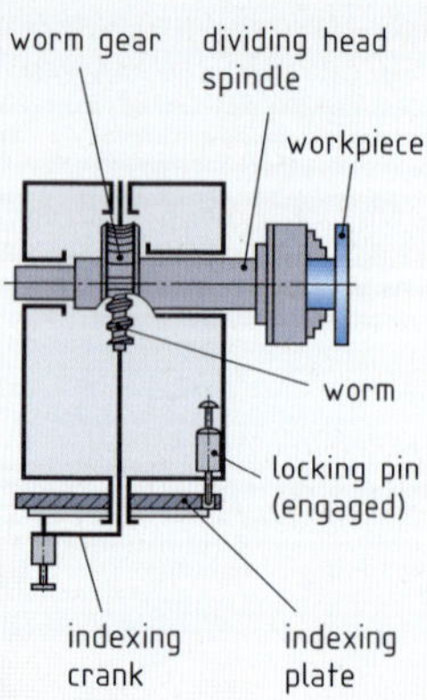

In indirect indexing the dividing head spindle is driven by the worm and worm wheel.

D no. of divisions
α angular division
i gear ratio of dividing head
n_c indexing step; no. of indexing crank revolutions for one division

Indexing step

$$n_c = \frac{i}{D}$$

$$n_c = \frac{i \cdot \alpha}{360°}$$

Example 1:

$D = 68;\ i = 40;\ n_c = ?$ $\quad \boldsymbol{n_c} = \frac{i}{D} = \frac{40}{68} = \mathbf{\frac{10}{17}}$

Example 2:

$\alpha = 37.2°;\ i = 40;\ n_c = ?$

$\boldsymbol{n_c} = \frac{i \cdot \alpha}{360°} = \frac{40 \cdot 37.2°}{360°} = \frac{37.2}{9} = \frac{186}{9 \cdot 5} = \mathbf{4\,\frac{2}{15}}$

Circles of holes on indexing plates					
15	16	17	18	19	20
21	23	27	29	31	33
37	39	41	43	47	49
or					
17	19	23	24	26	27
28	29	30	31	33	37
39	41	42	43	47	49
51	53	57	59	61	63

Milling

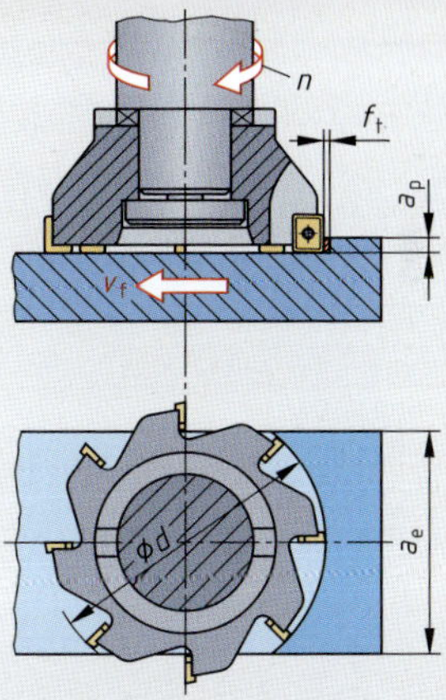

t_p productive time
l workpiece length
a_p cutting depth
a_e engagement (milling width)
l_a approach
l_{oi} overrun idle travel
l_{st} starting travel
L total travel
d cutter diameter
n rotational speed
f feed per revolution
f_t feed per tooth
N number of teeth
v_c cutting velocity
v_f feed rate
i number of cuts

Productive time

$$t_p = \frac{L \cdot i}{n \cdot f} \qquad t_p = \frac{L \cdot i}{v_f}$$

Feed per revolution of milling cutter

$$f = f_t \cdot N$$

Feed rate

$$v_f = n \cdot f \qquad v_f = n \cdot f_t \cdot N$$

Rotational speed

$$n = \frac{v_c}{\pi \cdot d}$$

Total travel L and starting travel l_{st} in relation to the milling process

<table>
<tr><th colspan="3">Face milling</th><th rowspan="3">Peripheral face milling</th></tr>
<tr><th rowspan="2">centric</th><th colspan="2">eccentric</th></tr>
<tr><th>$a_e > 0.5 \cdot d$</th><th>$a_e < 0.5 \cdot d$</th></tr>
<tr><td>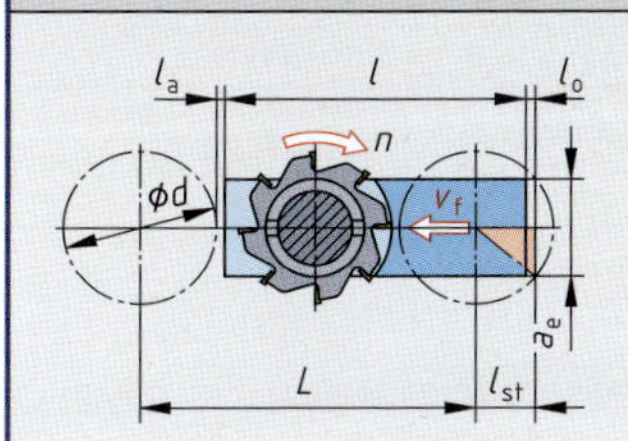</td><td>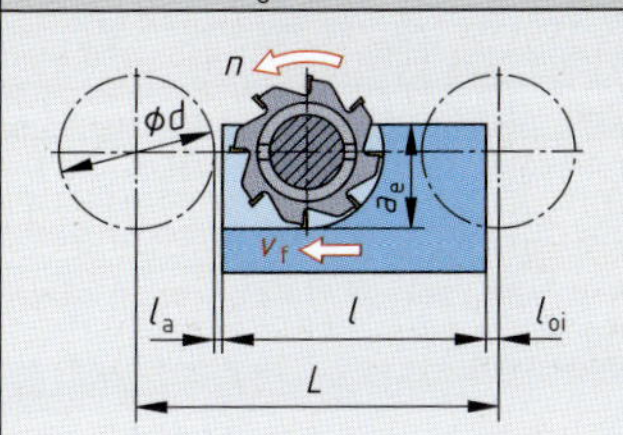</td><td colspan="2">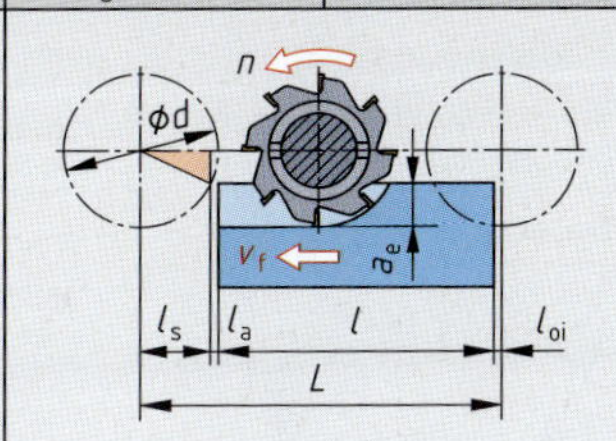</td></tr>
<tr><td>$L = l + 0.5 \cdot d + l_a + l_{oi} - l_{st}$</td><td rowspan="2">$L = l + 0.5 \cdot d + l_a + l_{oi}$</td><td colspan="2">$L = l + l_a + l_{oi} + l_{st}$</td></tr>
<tr><td>$l_{st} = 0.5 \cdot \sqrt{d^2 - a_e^2}$</td><td colspan="2">$l_{st} = \sqrt{a_e \cdot d - a_e^2}$</td></tr>
</table>

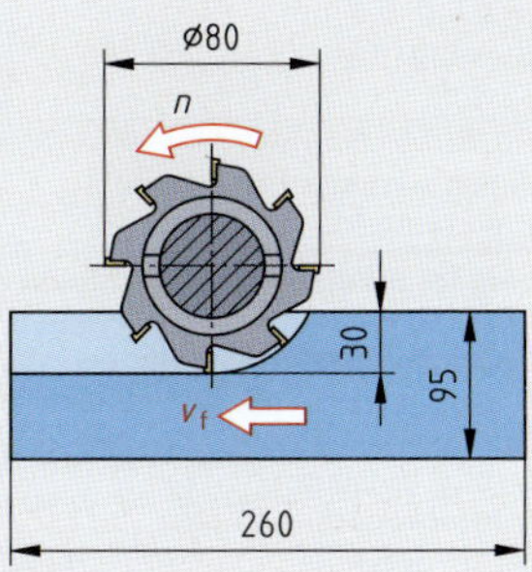

Example:

Face milling (see left illustration): $N = 10$, $f_t = 0.08$ mm, $v_c = 30$ m/min, $l_a = l_{oi} = 1.5$ mm, $i = 1$ cut

Sought after: n; v_f; L; t_p

Solution:

$$\boldsymbol{n} = \frac{v_c}{\pi \cdot d} = \frac{30\,\frac{\text{m}}{\text{min}}}{\pi \cdot 0.08\ \text{m}} = \mathbf{119}\,\frac{\mathbf{1}}{\mathbf{min}}$$

$$\boldsymbol{v_f} = n \cdot f_t \cdot N = 119\,\frac{1}{\text{min}} \cdot 0.08\ \text{mm} \cdot 10 = \mathbf{95.2}\,\frac{\mathbf{mm}}{\mathbf{min}}$$

$$\frac{\boldsymbol{a_e}}{\boldsymbol{d}} = \frac{30\ \text{mm}}{80\ \text{mm}} = 0.375\text{, it follows that } a_e < 0.5 \cdot d$$

$$\boldsymbol{L} = l + l_a + l_{oi} + l_{st}$$

$$\boldsymbol{l_{st}} = \sqrt{a_e \cdot d - a_e^2} = \sqrt{30\text{mm} \cdot 80\ \text{mm} - (30\ \text{mm})^2} = \mathbf{38.7\ mm}$$

$$\boldsymbol{L} = 260\ \text{mm} + 1.5\ \text{mm} + 1.5\ \text{mm} + 38.7\ \text{mm} = \mathbf{301.7 mm}$$

$$\boldsymbol{t_p} = \frac{L \cdot i}{v_f} = \frac{301.7\ \text{mm} \cdot 1}{95.2\,\frac{\text{mm}}{\text{min}}} = \mathbf{3.2\ min}$$

PE

Designation of indexable inserts for cutting tools

Designation of indexable inserts (selection)

cf. DIN ISO 1832 (2005-11)

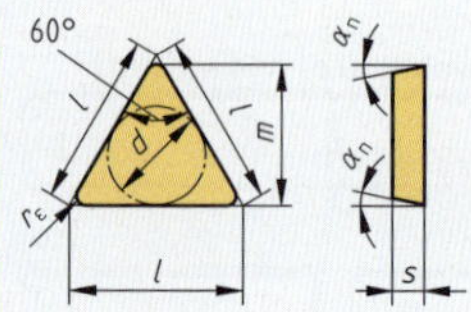

Designation examples:

Indexable carbide insert with rounded corners (DIN 4968) without mounting hole

Indexing insert DIN 4968 – T N G N 16 03 08 T – P20

Standard number (4968); T ①, N ②, G ③, N ④, 16 ⑤, 03 ⑥, 08 ⑦, T ⑧, – ⑨, P20 ⑩

①	basic shape	⑥	thickness of insert
②	clearance angle	⑦	design of the cutting edge corners
③	tolerance class	⑧	cutting edge
④	cutting face and clamping features	⑨	cutting direction
⑤	size of insert	⑩	cutting tool material

Basic shape

Equilateral, equiangular and round

Equilateral and non-equiangluar

Non-equilateral and
L equiangular
A, B, K non-equiangular

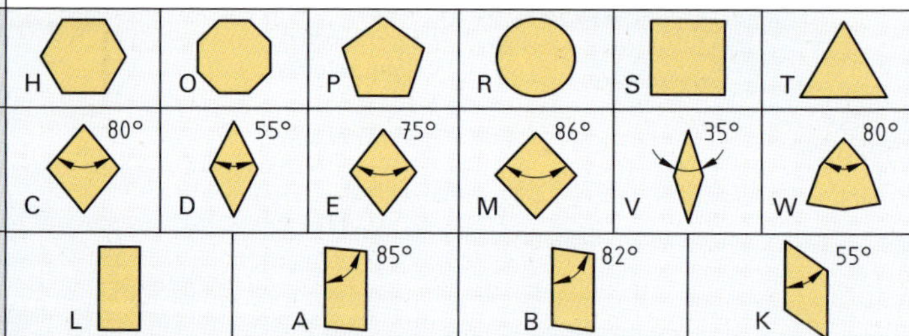

Many company-specific shapes are used in addition to the standard ones.

Standard clearance angle α_n to the insert	A	B	C	D	E	F	G	N	P	O
	3°	5°	7°	15°	20°	25°	30°	0°	11°	special specific.

Selection of indexing inserts according to the clearance angle and the cutting edge geometry

Basic shape	Examples of cutting edges (depending on manufacturer)	Application examples
		Rough turning for medium or heavy-duty machining
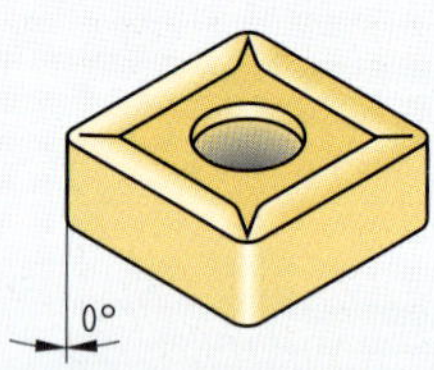 Indexable insert with a negative basic shape (clearance angle α_n = 0°)	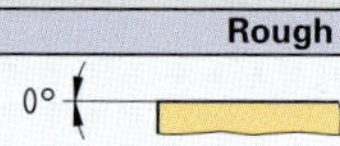	f = 0.1 to 1.2 mm, a_p = 0.3 to 12.0 mm longitudinal and face turning of cast iron materials
	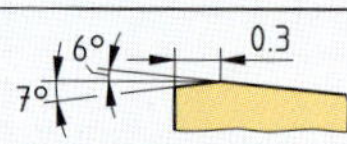	f = 0.3 to 1.3 mm, a_p = 0.8 to 6.5 mm longitudinal and face turning of steel, for forged and pre-machined workpieces with low allowance
	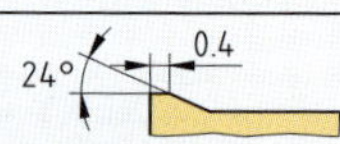	f = 0.2 to 1.4 mm, a_p = 0.7 to 15.0 mm longitudinal and face turning of steel, high cutting edge stability
		Finish-turning
	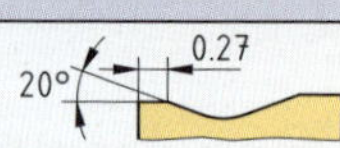	f = 0.15 to 0.9 mm, a_p = 0.6 to 5.0 mm longitudinal and face turning of steel and cast iron materials
	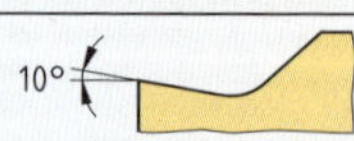	f = 0.2 to 1.4 mm, a_p = 0.7 to 15.0 mm longitudinal and face finish-turning of heat-resisting steel, easy-cutting geometry for low cutting forces
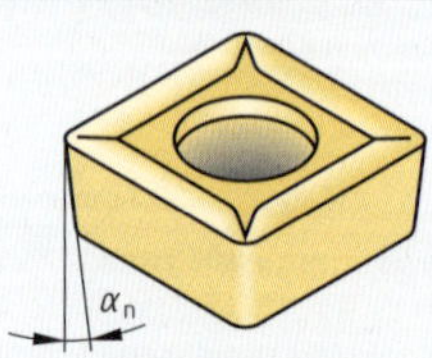 Indexing insert with a positive basic shape (e.g. clearance angle α_n = 1 to 11°, here α_n = 7°)		**Rough and finish-turning for medium-duty machining** f = 0.05 to 0.5 mm, a_p = 0.1 to 6.0 mm longitudinal and face turning of cast iron materials
	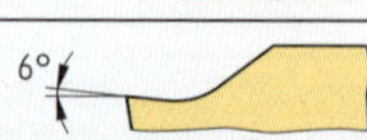	**Finish-turning** of steel, heat-resisting steels and cast iron materials f = 0.05 to 0.3 mm, a_p = 0.06 to 2.0 mm easy-cutting geometry for low cutting forces
	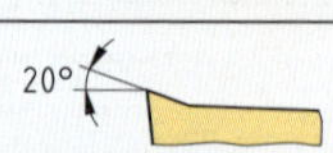	**Finish-turning** of aluminum and other non-ferrous metals f = 0.05 to 1.0 mm, a_p = 0.1 to 7.0 mm longitudinal and face turning at high cutting velocities
	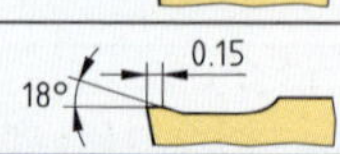	**Rough turning** with high chip removal rate f = 0.1 to 0.5 mm, a_p = 0.8 to 5.0 mm longitudinal and face turning with interrupted cutting

PE

Indexable inserts, Turning

Selecting the turning method in accordance with the shape of the indexable insert

Shapes of indexable inserts, designation cf. page 308	Roughing f = 0.5 to 1.5 mm, a_p = 5 to 15 mm	Light roughing/pre-finishing f = 0.3 to 0.5 mm, a_p = 1.5 to 5.0 mm	Finishing f = 0.1 to 0.3 mm, a_p = 0.5 to 2.0 mm	Longitudinal turning	Contour turning	Face turning	Versatile application options	Limited machine power	Tendency to vibration (workpiece)	Hard materials	Interrupted cutting	Large setting angle	Small setting angle
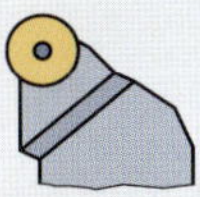 R	*d* in mm[1]												
	10 – 32	10 – 12	15 – 32	○	○	◐	◐	○	○	●	●	○	●
	●	○	○										
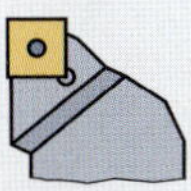 S	*l* in mm[1]												
	9 – 25	9 – 38	15 – 38	○	○	●	○	○	○	●	●	○	●
	●	◐	○										
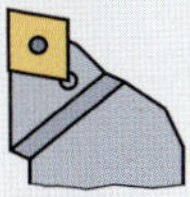 C	*l* in mm												
	6 – 12	6 – 25	16 – 25	●	◐	●	●	◐	○	○	◐	●	○
	●	●	○										
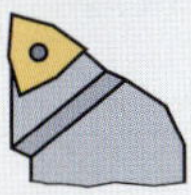 W	*l* in mm												
	6 – 8	6 – 8	–	◐	◐	●	◐	◐	◐	○	◐	●	●
	○	●	–										
 T	*l* in mm												
	11 – 22	11 – 33	22 – 33	◐	◐	◐	◐	●	●	○	◐	●	●
	○	●	●										
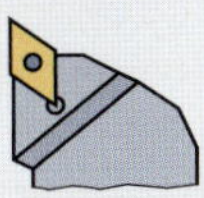 D	*l* in mm												
	6 – 15			●	●	◐	●	●	●	○	○	●	○
	○	●	●										
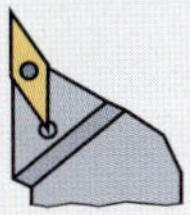 V	*l* in mm												
	11 – 22			●	●	○	◐	●	●	○	○	○	○
	○	○	●										

● very well suited ◐ suitable ○ not suitable

[1] Diameter and side length of indexable inserts, cf. page 308

The diamond-shaped indexable insert type C with a point angle of 80° is an efficient compromise of most various insert shapes and therefore suitable for many applications. It is used quite frequently.

PE

Milling tools with indexable inserts

Application examples for various milling cutters with indexable inserts

Milling cutter type Diameter in mm	Cutting depth a_p in mm	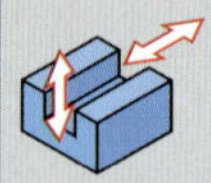 Slot milling, parting off	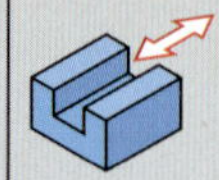 Bidirectional milling	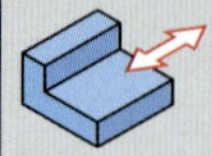 Corner milling	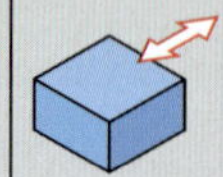 Face milling	 Profile milling	 Helical plunging
Face milling cutter 32–250	6–10	○	○	●	●	○	◐
Face milling cutter with close pitch (finishing cutter) 80–500	1–8	○	○	●	●	○	◐
Face milling cutter with coarse pitch (roughing cutter) 100–400	12	○	○	●	●	○	◐
Ball nose cutter (copy milling) 5–32	2–5	◐	○	○	○	●	●
Corner cutter 40–250	10	◐	○	○	●	○	◐
Side milling cutter 40–315	6–30	●	●	●	◐	○	○
Slotting cutter 80–315	2–6	●	○	○	○	○	○
End mill cutter 12–100	10–18	◐	○	●	◐	◐	●
Shell end mill cutter 20–100	5–100[1)]	○	○	●	◐	○	◐
Cutter with round indexable inserts 10–160	1–10	◐	○	◐	●	●	●

[1)] The upper value applies to corner milling of non-ferrous metals and finish corner milling.

● very well suited ◐ suitable ○ not suitable

PE

Cooling lubricants for cutting metals

Terminology and applications for cooling lubricants[1]

cf. DIN 51385 (1991-06)

Type of cooling lubricant	Effect	Group	Explanation	
			Composition	Applications
SESW cooling lubricants	↑ Increasing cooling effect (upwards); ↓ Increasing lubricating effect (downwards)	Solutions/ dispersions	Inorganic materials in water	Grinding
			Organic or synthetic materials in water	Machining at high cutting speed
SEMW cooling lubricants (oil in water)		Emulsions	2% – 20% emulsive (soluble) cooling lubricant in water	Good cooling effect, but low lubrication, e.g. machining (turning, milling, drilling) of easy-to-machine materials, at high cutting speed; for high working temperatures; susceptible to bacterial or fungal attack
SN cooling lubricants insoluble in water		Cutting oil	Mineral oils with polar additives (greases or synthetic esters) or EP additives[2] to increase lubricating performance	For lower cutting speed, higher surface quality, for difficult-to-machine materials; very good lubrication and corrosion protection

[1] Cooling lubricants may be hazardous to health (page 198) and are therefore only used in small quantities.
[2] EP = Extreme Pressure; additives to increase acceptance of high surface pressure between chip and tool

Guidelines for selecting cooling lubricants

Manufacturing process		Steel	Cast iron, malleable cast iron	Cu, Cu alloys	Al, Al alloys	Mg alloys
Turning	Roughing	emulsion, solution	dry	dry	emulsion, cutting oil	dry, cutting oil
	Finishing	emulsion, cutting oil	emulsion, cutting oil	dry, emulsion	dry, cutting oil	dry, cutting oil
Milling		emulsion, solution, cutting oil	dry, emulsion	dry, emulsion, cutting oil	cutting oil, emulsion	dry, cutting oil
Drilling		emulsion, cutting oil	dry, emulsion	dry, cutting oil, emulsion	cutting oil, emulsion	dry, cutting oil
Reaming		cutting oil, emulsion	dry, cutting oil	dry, cutting oil	cutting oil	cutting oil
Sawing		emulsion	dry, emulsion,	dry, cutting oil	cutting oil, emulsion	dry, cutting oil
Broaching		cutting oil, emulsion	emulsion	cutting oil	cutting oil	cutting oil
Hobbing, gear shaping		cutting oil	cutting oil, emulsion	–	–	–
Thread cutting		cutting oil	cutting oil, emulsion	cutting oil	cutting oil	cutting oil, dry
Grinding		emulsion, solution, cutting oil	solution, emulsion	emulsion, solution	emulsion	–
Honing, lapping		cutting oil	cutting oil	–	–	–

Waste types and disposal of cooling lubricants

Used up cooling lubricants, either mixed with water or not miscible, can be harmful to the ground water and affect the operation of sewage treatment plants if they are not disposed of properly. Therefore, a special treatment is required before their disposal. In Germany for instance, the Water Resources Act and the Waste Water Ordinance provide the legal basis. A possible solution to this problem is a minimum cooling lubrication system. Its low demand on cooling lubricant of 20 to 50 ml/hr makes sure that the machine tool, the workpiece and the chips remain dry and need no cleaning. Moreover, the amount of used cooling lubricant requiring treatment is very low.

Waste types and waste codes in accordance with the European Waste Catalog (EWC) and the Hazardous Waste List

Designation[1]	Example	Code letter as per DIN 51385	Waste code as per EWC
Cooling lubricants			
Drilling, cutting and grinding oils	Unserviceable and used up cooling lubricants, that are not water miscible, unserviceable cooling lubricants that are water-miscible, but without oil-water mixtures	SN SEM	120106 (containing halogen) 120107 (free of halogens)
Synthetic machining oils	Unserviceable or used up cooling lubricants on a synthetic basis without oil-water mixtures	SES	120110
Finishing oils	Unserviceable or used up honing, lapping and finishing oils	SN	120106 (containing halogen) 120107 (free of halogens)
Biogenic oils	Unserviceable vegetable oils	SN	130207
Drilling or grinding emulsions[2], emulsion mixtures or other oil-water mixtures	Unserviceable and used up cooling lubrication emulsion, unserviceable and used up cooling lubrication solutions	SEMW SESW	120108 (containing halogen) 120109 (free of halogens)
	Retentates[3] from diaphragm separation plants, evaporation residues from evaporation plants		130505
Other wastes			
Oil separator contents, sludge from oil separating plants	Sludge from oil and water separators		130502
Oils from oil or water separators	Oil from oil or water separators		130506
Sludge from honing and lapping and oil containing grinding sludge	Honing, lapping and grinding sludge from cooling lubricant maintenance facilities, such as filters, centrifuges or magnetic separators		120111 120202

[1] Terms that comply with the waste oil disposal regulations and are customary in practice.
[2] Emulsion: fine mixture of liquids that do not mix under normal conditions, such as oil and water.
[3] Retentate: fluid that is retained by the diaphragm in the separating processes.

PE

Treatment of cooling lubricants

Cooling lubricants mixed with water

1. Treatment with organic decomposing agents (demulsification) and splitting of the cooling lubricant into an oil and a water phase
   Duration of the separation process: approx. one day
2. Treatment by diaphragm filtration (cross-flow filtration) broken down according to their retaining capacity:
   - microfiltration
   - ultrafiltration

   Duration of the filtering process: approx. one week
3. Treatment by evaporation in a vacuum evaporator at a temperature of approximately 35 °C (95 °F)
   Duration of the evaporation process: a few hours
4. Post-treatment of non-evaporable residues by chemical treatment (combustion) and of the filtrate by nano filtration and reverse osmosis

Cooling lubricants not miscible with water

1. If extremely polluted with solid substances, removal of the metallic particles via suitable cleaning processes in the first place
2. If mixed with water, it must be checked whether the cooling lubricant is disposable without prior treatment or whether splitting into an oil and a water phase is required like with water-mixed cooling lubricants

Other (oil-containing) residues

1. Reduction of waste accrual (oil-polluted chips, grinding sludge) by removal of the oil or water in suitable centrifuges and presses
2. Reuse of the separated cooling lubricants
3. Collection and disposal of non-reusable, oil-containing wastes in accordance with local waste management laws (see page 197)

Designation of indexable and short indexable insert holders

cf. DIN 4983 (2004-07)

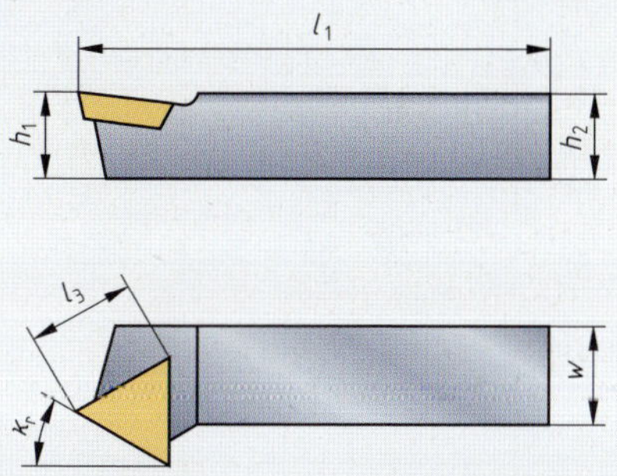

Designation example:

Holder DIN 4984 – C T W N R 32 25 M 16

- standard no. of holder – DIN 4984
- holding method – C
- insert shape[1)] – T
- design of holder – W
- normal clear. angle of insert[1)] α_n – N
- type of holder – R
- height of cutting edge $h_1 = h_2$ in mm – 32
- shank width w in mm – 25
- length of holder l_1 in mm – M
- indexable insert size[1)] – 16

1) For indexable inserts, see page 308

Designation		Configurations											
Insert holding	Letter symbol	C			M			P			S		
		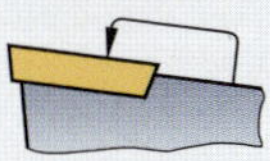			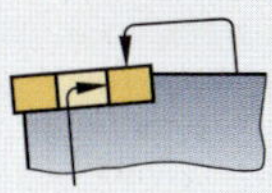			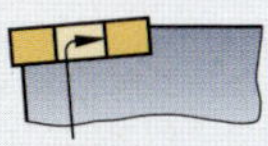					
	Holding of indexable insert	clamped from above			clamped from above and from hole			clamped from hole			countersink hole and screw		
Design of holder	Letter symbol	A	B	D	E	M	N	V	G	H	J	R	T
straight 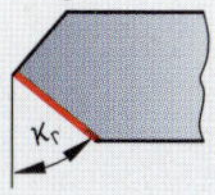	Side cutting edge angle κ_r	90°	75°	45°	60°	50°	63°	72.5°	90°	107.5°	93°	75°	60°
	Type of holder	straight							offset				
offset 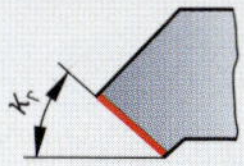	Letter symbol	C	F	K	S	U	W	Y	Forms D and S also available with round indexable inserts of basic form R				
	Side cutting edge angle κ_r	90°	90°	75°	45°	93°	60°	85°					
	Type of holder	straight	offset										
Type of holder	Letter symbol	R	right holder L				left holder N				neutral (both sides)		
Length of holder	Letter symbol	A	B	C	D	E	F	G	H	J	K	L	M
	l_1 in mm	32	40	50	60	70	80	90	100	110	125	140	150
	Letter symbol	N	P	Q	R	S	T	U	V	W	X	Y	
	l_1 in mm	160	170	180	200	250	300	350	400	450	Cust. lengths		500

⇒ **Holder DIN 4984 – CTWNR 3225 M 16:** holder with square shank, clamped above (C), triangular indexable insert (T), κ_r = 60° (W), α_n = 0° (N), right hand (R), $h_1 = h_2$ = 32 mm, w = 25 mm, l_1 = 150 mm (M), l_3 = 16.5 mm (16).

PE

Hard and dry machining, High-speed milling, MQCL

Hard turning with cubic boron nitride (CBN)

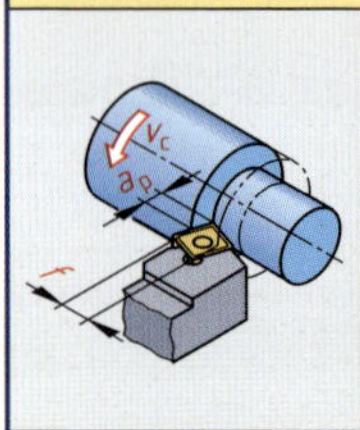

Turning process	Material hardened steel HRC	Cutting velocity v_c m/min	Feed f mm/revolution	Cutting depth a_p mm
External turning	45–58	60–220	0.05–0.3	0.05–0.5
Internal turning		60–180	0.05–0.2	0.05–0.2
External turning	> 58–65	50–190	0.05–0.25	0.05–0.4
Internal turning		50–150	0.05–0.2	0.05–0.2

Hard milling with coated solid carbide (VHM) tools

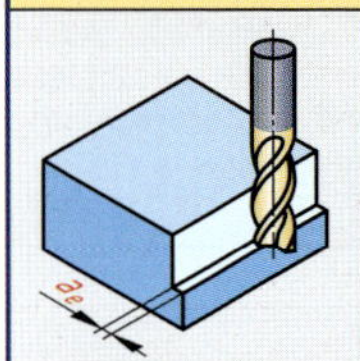

Material hardened steel HRC	Cutting velocity v_c m/min	working engagement $a_{e\,max}$ mm	Feed per tooth f_t in mm for lathe diameter d in mm: 2–8	> 8–12	> 12–20
to 35	80–90	0.05 · d	0.04	0.05	0.06
36–45	60–70	0.05 · d			
46–54	50–60	0.05 · d	0.03	0.04	0.05

High-speed cutting (HSC) with PCD

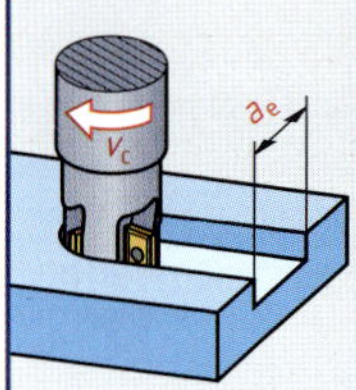

Material group	Cutting velocity v_c m/min	Cutter diameter d in mm: 10, a_e mm	10, f_t mm	20, a_e mm	20, f_t mm
Steel R_m 850–1100 > 1100–1400	280–360 210–270	0.25	0.09–0.13	0.40	0.13–0.18
Hardened steel 48–55 HRC > 55–67 HRC	90–240 75–120	0.25 0.20	0.09–0.13	0.40 0.35	0.13–0.18
EN-GJS > 180HB	300–360	0.25	0.09–0.13	0.40	0.13–0.18
Titanium alloy	90–270	0.20–0.25	0.09–0.13	0.35–0.40	0.13–0.18
Cu alloy	90–140	0.20	0.09–0.13	0.35	0.13–0.18

Dry machining

Process	Cutting tool material and cooling lubricant for: Iron materials: Quenched and tempered steels	Iron materials: High-alloy steels	Iron materials: Cast iron	Al materials: Cast alloy	Al materials: Wrought alloy
Drilling	TiN, dry	TiAlN[1], MQCL	TiN, dry	TiAlN, MQCL	TiAlN, MQCL
Reaming	PCD, MQCL	–[2]	PCD, MQCL	TiAlN, PCD, MQCL	TiAlN, MQCL
Milling	TiN, dry	TiAlN, MQCL	TiN, dry	TiAlN, dry	TiAlN, MQCL
Sawing	MQCL	MQCL	–[2]	TiAlN, MQCL	TiAlN, MQCL

PE

Minimum quantity cooling lubrication (MQCL or MQL)[3]

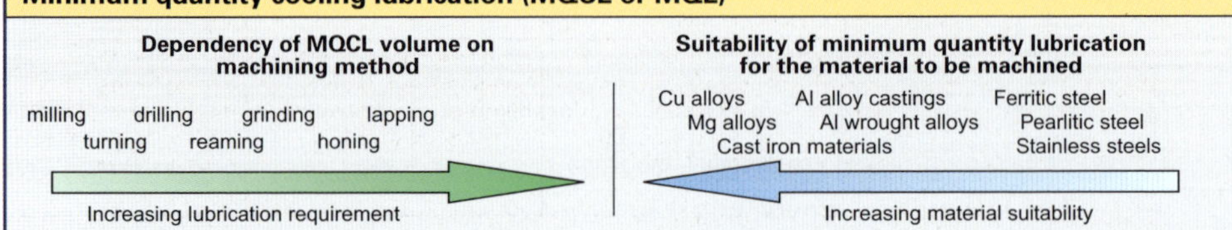

[1] Titanium aluminum nitride (super hard coating) [2] Application uncommon [3] Generally 0.01–3 l/hr

Cutting tool materials

Designation of hard cutting tool materials

cf. DIN ISO 513 (2005-11)

Example: Code letter (see the table below) — HC – K 20 — Application group

Cutting main group (below K): P (blue) | M (yellow) | K (red) | N (green) | S (brown) | H (gray)

Cutting tool material group	K[1]	Components	Properties	Applications
 Hard metals		Uncoated hard metal, main component is tungsten carbide (WC)	High hot hardness up to 1000 °C, high wear resistance, high compression strength, vibration damping	Indexable inserts for drilling, turning and milling tools, also for solid hard metal tools
	HW	Grain size > 1 µm		
	HF	Grain size < 1 µm		
	HT	Uncoated hard metal of titanium carbide (TiC), titanium nitride (TiN) or of both, also called cermet.	Like HW, but with high cutting edge stability, chemical resistance	Indexable inserts for lathe and milling tools for finishing at high cutting velocities
	HC	HW and HT, but coated with titanium carbonitride (TiCN)	Increase of wear resistance without reducing tough-ness	Increasingly replacing the uncoated hard metals
 Cutting ceramics	CA	Cutting ceramics, primarily of aluminum oxide (Al_2O_3)	High hardness and hot hardness up to 1200 °C sensitive to severe temperature changes	Cutting of cast iron, usually without cooling lubricant
	CM	Mixed ceramics with aluminum oxide (Al_2O_3) base, as well as other oxides	Tougher than pure ceramics, better resistance to temperature variations	Precision hard turning of hardened steel, cutting at high cutting velocity
	CN	Silicon nitride ceramics, primarily of silicon nitride (Si_3N_4)	High toughness, high cutting edge stability	Cutting of cast iron at high cutting velocity
	CR	Cutting ceramics with aluminum oxide (Al_2O_3), as a main component, reinforced	Tougher than pure ceramics due to reinforcement, improved resistance against temperature variations	Hard turning of har-dened steel, cutting at high cutting velocity
	CC	Cutting ceramics such as CA, CM and CN, but coated with titanium carbonitride (TiCN)	Increase of wear resistance without reducing tough-ness	Increasingly replacing the uncoated cutting ceramics
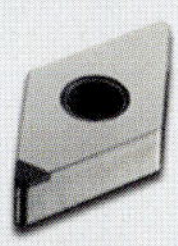 Boron nitride		Cubic crystalline boron nitride (BN), also designated CBN or PCB or "superhard cutting tool material"	Very high hardness and hot hardness up to 2000°C, high wear resistance, chemical resistance	Dressing of hard materials (HRC > 48) with high surface quality
	BL	With low boron nitride content		
	BH	With high boron nitride content		
	BC	BL and BH, but coated		
 Diamond		Cutting tool material of carbon (C), also designated CBN, PCB or „superhard cutting tool material"	High wear resistance, very brittle, temperature resistance up to 600 °C, reacts with alloying elements	Cutting of non-ferrous metals and Al alloys with high silicon content
	DP	Polycrystalline diamond (PCD)		
	DM	Monocrystalline diamond		
 Tool steel[2]	HS	High-performance high-speed steel with alloying elements tungsten (W), molybdenum (Mo), vanadium (V) and cobalt (Co), usually coated with titanium nitride (TiN)	High toughness, high bending strength, low hardness, temperature resistant up to 600 °C	For severe alternating cutting forces, machining of plastics, for the cutting of Al and Cu alloys

[1] Code letters according to DIN ISO 513

[2] Tool steels are not included in DIN ISO 513 but in ISO 4957

PE

Cutting tool materials

Classification and application of hard cutting tool materials

cf. DIN ISO 513 (2005-11)

Code letter color code	Application group		Workpiece – material	Cutting tool material properties[1)]		Possible cutting parameters[1)]	
				Wear resistance	Toughness	Cutting speed	Feed
Steel							
P blue	P01 P10 P20 P30 P40 P50	P05 P15 P25 P35 P45	All types of steels and cast steels, with the exception of stainless steel with austenitic structure	↑	↓	↑	↓
Stainless steel							
M yellow	M01 M10 M20 M30 M40	M05 M15 M25 M35	Austenitic and austenitic ferritic stainless steels and cast steels	↑	↓	↑	↓
Cast iron							
K red	K01 K10 K20 K30 K40	K05 K15 K25 K35	Cast iron with flake and spheroidal graphite malleable cast iron	↑	↓	↑	↓
Non-ferrous metals and other non-ferrous materials							
N green	N01 N10 N20 N30	N05 N15 N25	Aluminum and other non-ferrous metals (e.g. Cu, Mg), non-ferrous materials (e.g. GPR, CFRP)	↑	↓	↑	↓
Special alloys and titanium							
S brown	S01 S10 S20 S30	S05 S15 S25	High-temperature special alloy on the basis of iron, nickel and cobalt, titanium and titanium alloys	↑	↓	↑	↓
Hard materials							
H gray	H01 H10 H20 H30	H05 H15 H25	Hardened steel, hardened cast iron materials, cast iron for ingot casting	↑	↓	↑	↓

1) Increasing in direction of the arrow

Grinding

Surface grinding

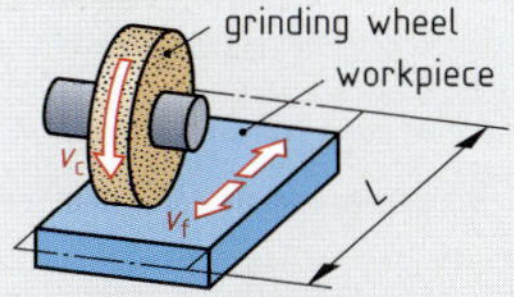

Cylindrical grinding

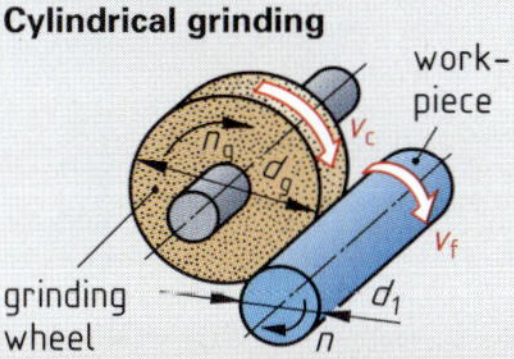

v_c cutting velocity
d_g diameter of grinding wheel
n_g rotational speed of grinding wheel
v_f feed rate
L travel
n_s no. of strokes
d_1 diameter of workpiece
n workpiece rotational speed
q speed ratio

Example:

v_c = 30 m/s; v_f = 20 m/min; q = ?

$$\boldsymbol{q} = \frac{v_c}{v_f} = \frac{30\ \text{m/s} \cdot 60\ \text{s/min}}{20\ \text{m/min}} = \frac{1800\ \text{m/min}}{20\ \text{m/min}} = \mathbf{90}$$

Cutting velocity

$$v_c = \pi \cdot d_g \cdot n_g$$

Feed rate

Surface grinding

$$v_f = L \cdot n_s$$

Cylindrical grinding

$$v_f = \pi \cdot d_1 \cdot n$$

Speed ratio

$$q = \frac{v_c}{v_f}$$

Standard values for cutting velocity v_c, feed rate v_f, speed ratio q

Material	Surface grinding						Cylindrical grinding					
	Peripheral grinding			Side wheeling			External cyl. grinding			Internal cyl. grinding		
	v_c m/s	v_f m/min	q	v_c m/s	v_f m/min	q	v_c m/s	v_f m/min	q	v_c m/s	v_f m/min	q
Steel	30	10–35	80	25	6–25	50	35	10	125	25	19–23	80
Cast iron	30	10–35	65	25	6–30	40	25	11	100	25	23	65
Carbide	10	4	115	8	4	115	8	4	100	8	8	60
Al alloys	18	15–40	30	18	24–45	20	18	24–30	50	16	30–40	30
Cu alloys	25	15–40	50	18	20–45	30	30	16	80	25	25	50

Grinding data for steel and cast iron with corundum or silicon carbide grinding wheels

Processes	Grain size	Grinding allowance	Depth of cut in mm	*Rz* in µm
Rough grind	30–46	0.5–0.2	0.02–0.1	3–10
Finishing	46–80	0.02–0.1	0.005–0.05	1–5
Precision grinding	80–120	0.005–0.02	0.002–0.008	1.6–3

Maximum speed of grinding wheels

cf. DIN EN 12413 (2007-09)

Shape of grinding wheel	Type of grinding machine	Guide[1]	Maximum speed v_c in m/s for bond type[2]							
			B	BF	E	M	R	RF	PL	V
Straight grinding wheel	stationary	pd or ho	50	63	40	25	50	–	50	40
	hand-held grinder	free-hand	50	80	–	–	50	80	50	–
Straight cutting wheel	stationary	pd or ho	80	100	63	–	63	80	–	–
	hand-held grinder	free-hand	–	80	–	–	–	–	–	–

[1] pd positively driven: feed by mechanical means; ho hand operated: feed by operator;
free-hand grinding: grinding machine is guided entirely by hand; [2] Type of bond, see page 318

Restrictions for use of grinding tools[3]*

cf. BGV D12[4] (2001-10)

VE	Meaning	VE	Meaning
VE1	Not allowed for free-hand or hand operated grinding	VE6	Not allowed for side wheeling
		VE7	Not allowed for free-hand grinding
VE2	Not allowed for free-hand abrasive cutting	VE8	Not allowed with backing pad
VE3	Not allowed for wet grinding	VE10	Not allowed for dry grinding
VE4	Not allowed in enclosed work area	VE11	Not allowed for free-hand or hand operated abrasive cutting
VE5	Not allowed without vacuum exhaust		

[3] If no restriction is given, the grinding tool is suitable for all applications.

Color stripes for maximum allowable peripheral speeds ≥ 50 m/s*

cf. BGV D12[4] (2001-10)

Color stripe	blue	yellow	red	green	blue & yellow	blue & red	blue & green
$v_{c\,max}$ in m/s	50	63	80	100	125	140	160
Color stripe	**yellow & red**	**yell. & green**	**red & green**	**blue & blue**	**yellow & yell.**	**red & red**	**green & green**
$v_{c\,max}$ in m/s	180	200	225	250	280	320	360

[4] BGV Berufsgenossenschaftliche Vorschrift (Employers' Liability Insurance Association Provisions)
*) According to European standards

Abrasives, Bonds

Abrasives

cf. DIN ISO 525 (2000-08)

Symbol	Abrasive	Chemical composition	Knoop-hardness	Areas of application
A	Norm. corundum	Al_2O_3 + additions	18000	Carb. steel, unhardened steel, cast steel, malleable cast iron
	white fused alumina	Al_2O_3 in crystalline form	21000	High and low alloyed steel, hardened steel, case hardened steel, tool steel, titanium
Z	zircon corundum	$Al_2O_3 + ZrO_2$	–	Stainless steels
C	silicon carbide	SiC + additions	24800	Hard materials: carbide, cast iron, HSS, ceramic, glass; soft materials: copper, aluminum, plastics
BK	boron carbide	B_4C in crystalline form	47000	Lapping, polishing of carbide and hardened steel
CBN	boron nitride	BN in crystalline form	60000	High-speed steels, cold and hot work steels
D	diamond	C in crystalline form	70000	Carbide, cast iron, glass, ceramic, stone, non-ferrous metals, not for steel; dressing of grinding wheels

Hardness grade

cf. DIN ISO 525 (2000-08)

Designation	Hardn. grade	Application	Designation	Hardn. grade	Application
extremely soft	A B C D	Deep and side wheeling of hard materials	hard	P Q R S	External cylindrical grinding; soft materials
very soft	E F G		very hard	T U V W	
soft	H I J K	Conventional metal grinding	extremely hard	X Y Z	
medium	L M N O				

Grain size

cf. DIN ISO 525 (2000-08)

Grain designation for bonded abrasives

Grain ranges	coarse	medium	fine	very fine
Grain designation	F4, F5, F6 to F24	F30, F36, F46 to F60	F70, F80, F90 to F220	F230 to F1200
Attainable *Rz* in µm	≈ 10–5	≈ 5–2.5	≈ 2.5–1.0	≈ 1.0–0.4

Structure

cf. DIN ISO 525 (2000-08)

Code	**0 1 2 3 4 5 6 7 8 9 10 11 12 13 14, etc. up to 30**
Structure	← **dense (nonporous)** **open (porous)** →

Bond

cf. DIN ISO 525 (2000-008) and VDI 3411 (2000-08)

Code	Type of bond	Properties	Areas of application
B BF	synthetic resin bond, fiber reinforced	Nonporous or porous, elastic, resistant to oil, cool grinding	Rough or cut-off grinding, form grinding with diam. and boron nitride, high pressure grinding
E	shellac bond	Sensitive to temperature, tough elastic, impact resistant	Saw tooth grinding, form grinding, control wheel for centerless grinding
G	galvanic bond	Tight grip due to protruding grains	Internal grinding of carbide, hand grinding
M	metal bond	Nonporous or porous, tough, insensitive to pressure and heat	Form and tool grinding using diamond or boron nitride, wet grinding
MG	magnesite bond	Soft, elastic, sensitive to water	Dry grinding, knife grinding
PL	plastic bond	Soft, elastic depending upon plastic and degree of hardening	Plastic abrasive material for finishing, precision finishing and polishing
R RF	rubber bond, fiber reinforced	Elastic, cold grinding, sensitive to oil and heat	Cut-off grinding
V	vitrified (ceramic) bond	Porous, brittle, insensitive to water, oil, heat	Rough and finish grinding of steels using corundum and silicon carbide

⇒ **Grinding wheel ISO 603-1 1 N-300 x 50 x 76.2 – A/F 36 L 5 V – 50:** Form 1 (straight grinding wheel), wheel face N, outside diameter 300 mm, width 50 mm, hole diameter 76.2 mm, abrasive A (normal corundum or white fused alumina), grain size F36 (medium), hardness grade L (medium), structure 5, vitrified (ceramic) bond (V), maximum peripheral speed 50 m/s.

Selecting grinding wheels

Standard values for selecting grinding wheels (excluding diamond and boron nitride)

Cylindrical grinding

Material	Abrasive	Roughing		Finishing with wheel diameter up to 500 mm		Finishing with wheel diameter over 500 mm		Fine finishing	
		Grain size	Hardness	Grain size	Hardness	Grain size	Hardness	Grain size	Hardness
Steel, unhardened	A	54	M–N	80	M–N	60	L–M	180	L–M
Steel, hard., unalloy. and alloy.	A	46	L–M	80	K–L	60	J–K	240–500	H–N
Steel, hardened, high alloyed	A, C	80	M–N	80	N–O	60	M–N	240–500	H–N
Carbide, ceramic	C	60	K	80	K	60	K	240–500	H–N
Cast iron	A, C	60	L	80	L	60	L	100	M
Non-ferr. met., e.g. Al, Cu, CuZn	C	46	K	60	K	60	K	–	–

Internal cylindrical grinding

Material	Abrasive	Grinding wheel diameter in mm: up to 20		from 20 to 40		from 40 to 80		over 80	
		Grain size	Hardness	Grain size	Hardness	Grain size	Hardness	Grain size	Hardness
Steel, unhardened	A	80	M	60	L–M	54	L–M	46	K
Steel, hard., unalloy. and alloy.	A	80	K–L	120	M–N	80	M–N	80	L
Steel, hardened, high alloyed	A, C	80	J–K	100	K	80	K	60	J
Carbide, ceramic	C	80	G	120	H	120	H	80	G
Cast iron	A	80	L–M	80	K–L	60	M	46	M
Non-ferr. met., e.g. Al, Cu, CuZn	C	80	I–J	120	K	60	J–K	54	J

Peripheral face grinding

Material	Abrasive	Cup wheel $D < 300$ mm		Straight grinding wheels $D \leq 300$ mm		Straight grinding wheels $D > 300$ mm		Abrasive segments	
		Grain size	Hardness	Grain size	Hardness	Grain size	Hardness	Grain size	Hardness
Steel, unhardened	A	46	J	46	J	36	J	24	J
Steel, hard., unalloy. and alloy.	A	46	J	60	J	46	J	36	J
Steel, hardened, high alloyed	A	46	H–J	60	I–J	46	I–J	36	I–J
Carbide, ceramic	C	46	J	60	J	60	J	46	J
Cast iron	A	46	J	46	J	46	J	24	J
Non-ferr. met., e.g. Al, Cu, CuZn	C	46	J	60	J	60	J	36	J

Tool grinding

Cutting tool material	Abrasive	Straight grinding wheels $D \leq 225$	Straight grinding wheels $D > 225$		Dish wheels $D \leq 100$	Dish wheels $D > 100$		Cup wheels	
		Grain size	Grain size	Hardness	Grain size	Grain size	Hardness	Grain size	Hardness
Tool steel	A	80	60	M	80	60	M	46	K
High-speed steel	A	60	46	K	60	46	K	46	H
Carbide	C	80	54	K	80	54	K	46	H

Cutting on stationary machines

Material	Abrasive	Straight cut-off wheels v_c up to 80 m/s, $D \leq 200$ mm		Straight cut-off wheels v_c up to 80 m/s, $D > 200$ mm		Straight cut-off wheels v_c up to 100 m/s, $D \leq 500$ mm		Straight cut-off wheels v_c up to 100 m/s, $D > 500$ mm	
		Grain size	Hardness	Grain size	Hardness	Grain size	Hardness	Grain size	Hardness
Steel, unhardened	A	80	Q–R	46	Q–R	24	U	20	Q–R
Cast iron	A	60	Q–R	46	Q–R	24	U–V	20	U–V
Non-ferr. met., e.g. Al, Cu, CuZn	A	60	Q–R	46	Q–R	30	S	24	S

Grinding and cutting with hand tools

Material	Abrasive	Cut-off wheels v_c up to 80 m/s		Rough grinding wheels v_c up to 45 m/s		Rough grinding wheels v_c up to 80 m/s		Mounted points	
		Grain size	Hardness	Grain size	Hardness	Grain size	Hardness	Grain size	Hardness
Steel, unhardened	A	30	T	24	M	24	R	36	Q–R
Steel, corrosion resistant	A	30	R	16	M	24	R	36	S
Cast iron	A, C	30	T	20	R	24	R	30	T
Non-ferr. met., e.g. Al, Cu, CuZn	A, C	30	R	20	R	–	–	–	–

Grinding with diamond and boron nitride

Grain designation

cf. DIN ISO 6106 (2006-03)

Areas of application		Rough grind	Finishing	Fine finishing	Lapping
Grain designation[1)]	diamond	D251–D151	D126–D76	D64, D54, D46	D20, D15, D7
	boron nitride	B251–B151	B126–B76	B64, B54, B46	B30, B6
Attainable *Ra* in µm		≈ 0.55–0.50	≈ 0.45–0.33	≈ 0.18–0.15	≈ 0.05–0.025

1) Mesh size of test sieve in µm; micro grits < D46, B46, not standardized by ISO 6106.

Standard values for cutting velocities

Process	Abrasive	Cutting velocity v_c in m/s by bond type[1)] B		M		G		V	
		dry	wet	dry	wet	dry	wet	dry	wet
Surface grinding	CBN	–	30–50	–	30–60	–	30–60	–	30–60
	D	–	22–50	–	22–27	20–30	22–50	–	25–50
External cylindrical grinding[2)]	CBN	–	30–50	–	30–60	–	30–60	–	30–60
	D	–	22–40	–	20–30	20–30	22–40	–	25–50
Internal cylindrical grinding	CBN	27–35	30–60	–	30–60	24–40	30–50	–	30–50
	D	12–18	15–30	8–15	18–27	12–20	18–40	–	25–50
Tool grinding	CBN	27–35	30–50	22–30	30–40	27–35	30–50	–	30–50
	D	15–22	22–50	15–22	15–27	15–30	22–35	–	–
Cut-off grinding	CBN	27–35	30–50	–	30–60	27–40	30–60	–	–
	D	12–18	22–35	–	22–27	18–30	22–40	–	–

1) Bond types, see page 318 2) Approx. four times the value for high speed grinding (HSG)

Standard values for depth of cut and feed of diamond grinding wheels

Process	Depth per stroke in mm for grain size D181	D126	D64	Feed m/min	Crossfeed relative to wheel width *w*
Face grinding[1)]	0.02–0.04	0.01–0.02	0.005–0.01	10–15	$^1/_4 - ^1/_2 \cdot w$
External cyl. grinding[1)]	0.01–0.03	0.0–0.02	0.005–0.01	0.3– 2.0	–
Internal cyl. grinding	0.002–0.007	0.002–0.005	0.001–0.003	0.5– 2.0	–
Tool grinding	0.01–0.03	0.005–0.015	0.002–0.005	0.3– 4.0	–
Groove grinding	–	1.0–5.0	0.5–3.0	0.01– 2.0	–

1) Approx. three times the value for high speed grinding (HSG)

Standard values for depth of cut and feed of CBN grinding wheels

Process	Depth per stroke in mm for grain size B252/B181	B151/B126	B91/B76	Feed m/min	Crossfeed relative to wheel width *w*
Surface grinding	0.03–0.05	0.02–0.04	0.01–0.015	20–30	$^1/_4 - ^1/_3 \cdot w$
External cyl. grinding	0.02–0.04	0.02–0.03	0.015–0.02	0.5–2.0	–
Internal cyl. grinding	0.005–0.015	0.005–0.01	0.002–0.005	0.5–2.0	–
Tool grinding	0.002–0.1	0.01–0.005	0.005–0.015	0.5–4.0	–
Groove grinding	1.0–10	1.0–5.0	0.5–3.0	0.01–2.0	–

High-performance grinding with CBN grinding wheels

cf. VDI 3411 (2000-08)

Grinding processes achieving extremely high material removal rates by utilization of special machines and tools with increased cutting velocities (> 80 m/s) and appropriate machine coolant. Predominantly used for side and external cylindrical grinding of metallic materials.

Grinding wheel preparation (conditioning)

Processing step	Dressing: Truing	Dressing: Sharpening	Cleaning
Action	Removal of grain and bond	Reduction of the bond	No effect on abrasive layer
Goal	Establishing concentricity and wheel profile	Creating the grinding wheel surface structure	Remove chips from pores

Maximum allowable peripheral speeds in high-performance grinding

Bond type[1)]	B	V	M	G
Highest allowable peripheral speed in m/s	140	200	180	280

1) Bond types, see page 318

Grinding, Productive time

Straight cylindrical grinding

t_p	productive time
L	travel
i	number of cuts
n	workpiece rotational speed
f	workpiece feed per revolution
v_f	feed rate
d_1	initial diameter of workpiece
d	final diameter of workpiece
a_p	cutting depth
l	workpiece length
w_g	grinding wheel width
l_{oi}	overrun idle travel
t	grinding allowance

Productive time

$$t_p = \frac{L \cdot i}{n \cdot f}$$

Workpiece rotational speed

$$n = \frac{v_f}{\pi \cdot d_1}$$

Number of cuts

for external straight grinding

$$i = \frac{d_1 - d}{2 \cdot a_p} + 2^{1)}$$

for internal straight grinding

$$i = \frac{d - d_1}{2 \cdot a_p} + 2^{1)}$$

1) 2 cuts to spark out, for lower tolerance grades additional cuts are necessary

Calculating travel L

Workpieces without shoulder

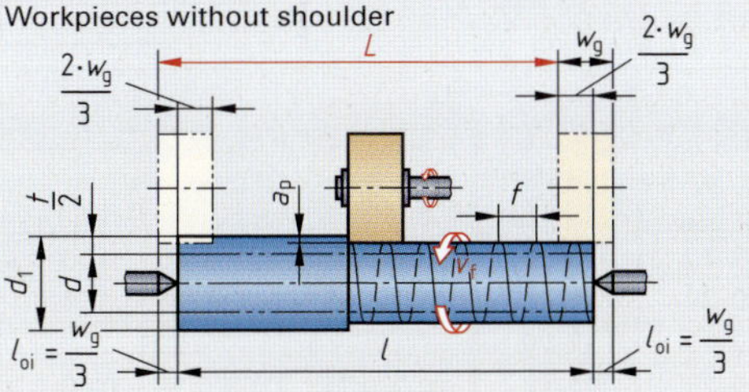

$$L = l - \frac{1}{3} \cdot w_g$$

Workpieces with shoulder

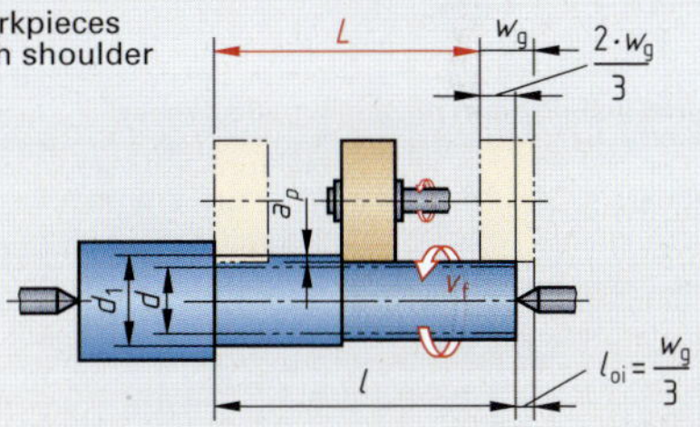

$$L = l - \frac{2}{3} \cdot w_g$$

Feed for roughing $f = 2/3 \cdot w_g$ to $3/4 \cdot w_g$; feed for finishing $f = 1/4 \cdot w_g$ to $1/2 \cdot w_g$

Surface grinding

t_p	productive time	f	transverse feed per stroke
l	workpiece length	n	no. of strokes per minute
l_i	start. idle, overrun idle travel	v_f	feed rate
L	travel	i	number of cuts
w	width of workpiece	t	grinding allowance
w_o	overrun width	w_g	grinding wheel width
W	grinding width	a_p	cutting depth

Number of cuts

$$i = \frac{t}{a_p} + 2^{1)}$$

1) 2 cuts to spark out

No. of strokes

$$n = \frac{v_f}{L}$$

Productive time

$$t_p = \frac{i}{n} \cdot \left(\frac{W}{f} + 1\right)$$

PE

Calculating travel L and grinding width W

Workpieces without shoulder

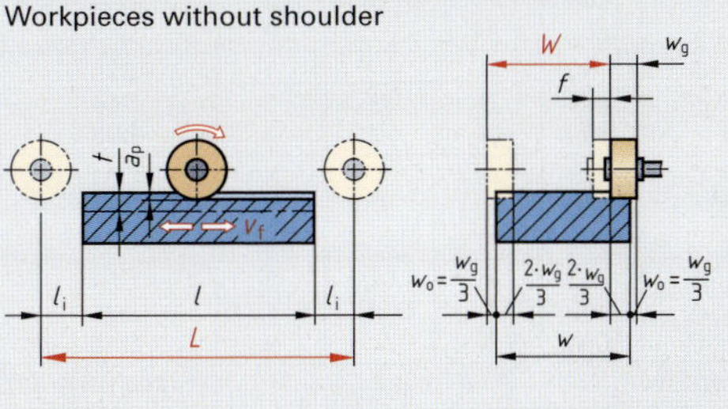

$L = l + 2 \cdot l_i$ $\quad l_i \approx 0.04 \cdot l$ $\quad W = w - \frac{1}{3} \cdot w_g$

Workpieces with shoulder

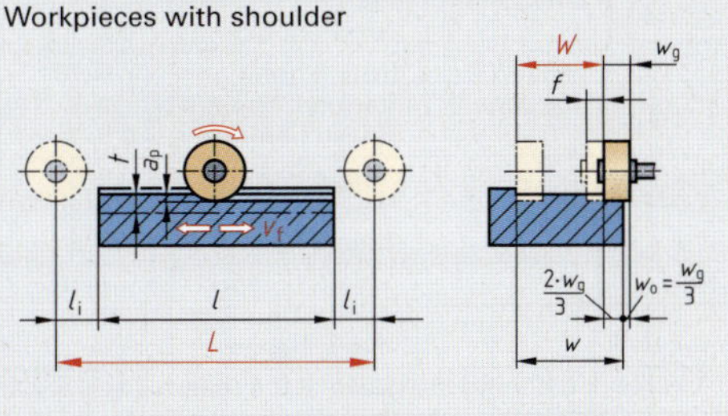

$L = l + 2 \cdot l_i$ $\quad l_i \approx 0.04 \cdot l$ $\quad W = w - \frac{2}{3} \cdot w_g$

Transverse feed for roughing $f = 2/3 \cdot w_g$ to $4/5 \cdot w_g$; feed for finishing $f = 1/2 \cdot w_g$ to $2/3 \cdot w_g$

Honing

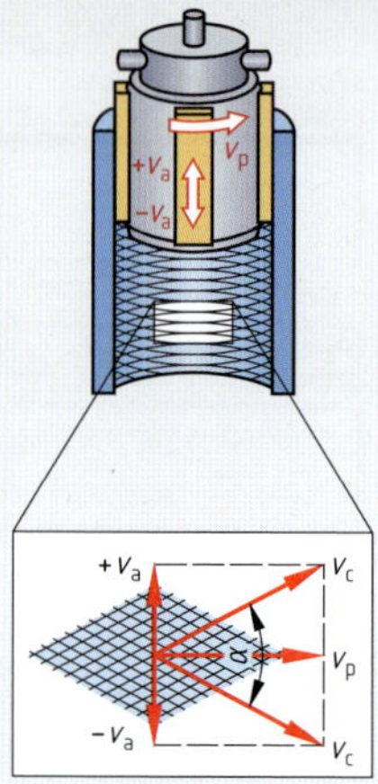

v_c cutting velocity
v_a axial speed
v_p peripheral speed
α angle of intersection betw. abrading tracks
p contact pressure
A contact area of honing stone
F_r radial infeed force
n number of honing stones
w width of honing stones
l length of honing stones

Example:

Hardened steel, finish honing, v_p = ?; v_a = ?; v_c = ?; α = ?
read from table: v_p = 25 m/min; v_a = 12 m/min

$$\boldsymbol{v_c} = \sqrt{v_a^2 + v_p^2} = \sqrt{\left(12\,\frac{\text{m}}{\text{min}}\right)^2 + \left(25\,\frac{\text{m}}{\text{min}}\right)^2} \approx \mathbf{28\,\frac{m}{min}}$$

$$\tan\frac{\alpha}{2} = \frac{v_a}{v_p} = \frac{12\text{ m/min}}{25\text{ m/min}} = 0.48;\quad \alpha = \mathbf{51.3°}$$

Cutting velocity

$$v_c = \sqrt{v_a^2 + v_p^2}$$

Angle of intersection

$$\tan\frac{\alpha}{2} = \frac{v_a}{v_p}$$

Contact pressure

$$p = \frac{F_r}{A}$$

$$p = \frac{F_r}{n \cdot w \cdot l}$$

Cutting velocity and machining allowances

Material	Peripheral velocity v_p in m/min		Axial velocity v_a in m/min		Machining allowances in mm for hole diameter in mm		
	Rough honing	Finish honing	Rough honing	Finish honing	2–15	15–100	100–500
Steel, unhardened	18–40	20–40	9–20	10–20	0.02–0.05	0.03–0.15	0.06–0.3
Steel, hardened	14–40	15–40	5–20	6–20	0.01–0.03	0.02–0.05	0.03–0.1
Alloy steels	23–40	25–40	10–20	11–20	0.02–0.05	0.03–0.15	0.06–0.3
Cast iron	23–40	25–40	10–20	11–20			
Aluminum alloys	22–40	24–40	9–20	10–20			

Honing with diamond grit v_p up to 40 m/min and v_a up to 60 m/min; α = 60°– 90°

Contact pressure of honing tools

Honing process	Contact pressure p in N/cm²			
	Ceramic honing stone	Plastic bonded honing stone	Diamond honing stick	Boron nitride honing stick
Rough honing	50–250	200–400	300–700	200–400
Finish honing	20–100	40–250	100–300	100–200

Selection of corundum, silicon carbide, CBN and diamond honing stones

Material	Tensile strength N/mm²	Process	Roughness depth Rz µm	Honing stone made of corundum and silicon carbide[2]					CBN or diamond
				Honing abrasive	Grain size	Hardness	Bond	Structure	Grain size
Steel	< 500 (unhardened)	rough honing intermed. honing finish honing	8–12 2–5 0.5–1.5	A	700 400 1200	R R M	B	1 5 2	D126 D54 D15
	500–700 (hardened)	rough honing intermed. honing finish honing	5–10 2–3 0.5–2	A	80 400 700	R O N	B	3 5 3	B76 B54 B30
Cast iron	–	rough honing finish honing plateau honing[1]	5–8 2–3 3–6	C	80 120 900	M K H	V	3 7 8	D91 D46 D25
Non-ferrous metals	–	rough honing intermed. honing finish honing	6–10 2–3 0.5–1	A A C	80 400 1000	O O N	V	3 1 5	D64 D35 D15

[1] In plateau honing the peaks of the material surface are removed.
[2] see page 318

PE

Selection of honing stone made of diamond and cubic boron nitride (CBN)

Abrasive	Natural diamond	Synthetic diamond	CBN
Material	Steel, carbide	Cast iron, nitrided steel, non-ferrous metals, glass, ceramic	Hardened steel

Productive time and Standard values for material removal

Electric discharge machining (wire EDM)

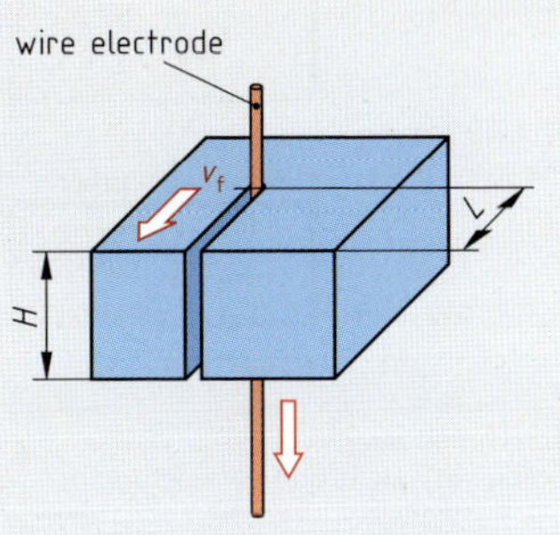

t_p productive time in min
v_f feed rate in mm/min
L travel, cutting length in mm
H cutting height in mm
T geometric tolerance in µm

Example:

Material: Steel, H = 30 mm; L = 320 mm;
T = 30 µm; v_f = ?; t_p = ?

$\mathbf{v_f}$ **= 1.8 mm/min** (from table)

$$\mathbf{t_p} = \frac{L}{v_f} = \frac{320\ \text{mm}}{1.8\ \text{mm/min}} = \mathbf{178\ min}$$

Productive time

$$t_p = \frac{L}{v_f}$$

Feed rate v_f (standard values)[1]

Cutting height H in mm	Feed rate v_f in mm/min										
	Steel eroding					Copper eroding			Carbide eroding		
	Desired geometric tolerance T in µm										
	60	40	30	20	10	40	20	10	80	20	10
10	9.0	8.5	4.0	3.9	2.1	7.5	3.5	2.0	4.5	0.7	0.6
20	5.1	5.5	2.5	2.5	1.5	4.7	2.4	1.5	3.1	0.3	0.3
30	3.7	4.0	1.8	1.8	1.1	4.0	1.9	1.1	2.3	0.2	0.2
50	2.5	2.5	1.2	1.2	0.8	2.6	1.4	0.7	1.4	0.2	0.2

[1] These standard values are average values from the main cut and all subsequent cuts required to reach geometric tolerance. With unfavorable flushing conditions the achievable feed rate drops considerably.

Characteristics and application of common wire electrodes

Wire material	El. conductivity in m/(Ω · mm²)	Tensile strength in N/mm²	Typical wire diameter in mm	Application
CuZn alloy	13.5	400–900	0.2–0.33	Universal
Molybdenum	18.5	1900	0.025–0.125	Cuts with very tight geometric tolerance
Tungsten	18.2	2500	0.025–0.125	Narrow slots, small corner radii

Electric discharge machining (sink EDM)

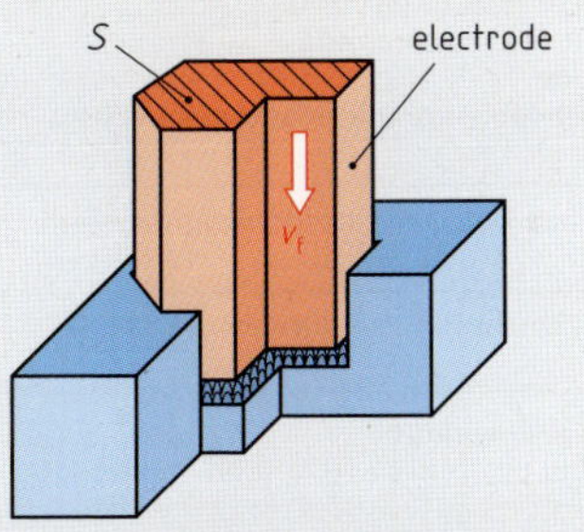

t_p productive time in min
S removal area of electrode in mm²
V removal volume in mm³
V_W removal rate in mm³/min

Example:

Roughing of steel; graphite electrode,
S = 150 mm²; V = 3060 mm³; V_W = ?; t_p = ?

$\mathbf{V_W}$ **= 31 mm³/min** (from table)

$$\mathbf{t_p} = \frac{V}{V_W} = \frac{3060\ \text{mm}^3}{31\ \text{mm}^3/\text{min}} = \mathbf{99\ min}$$

Productive time

$$t_p = \frac{V}{V_W}$$

Removal rate V_W (standard values)[1]

Workpiece material	Electrode	Removal rate V_w in mm³/min										
		Roughing removal area S in mm²						Finishing desired roughness depth Rz in µm				
		10 to 50	50 to 100	100 to 200	200 to 300	300 to 400	400 to 600	2 to 3	3 to 4	4 to 6	6 to 8	8 to 10
Steel	Graphite	7.0	18	31	62	81	105	–	–	–	2	5
	Copper	13.3	22	28	51	85	105	0.1	0.5	1.9	3.8	5
Carbide	Copper	6.0	15	18	28	30	33	–	0.1	0.5	2.2	5.2

[1] Actual values will vary widely due to the effects of different processing methods. Refer to page 324.

PE

Process parameters in EDM erosion

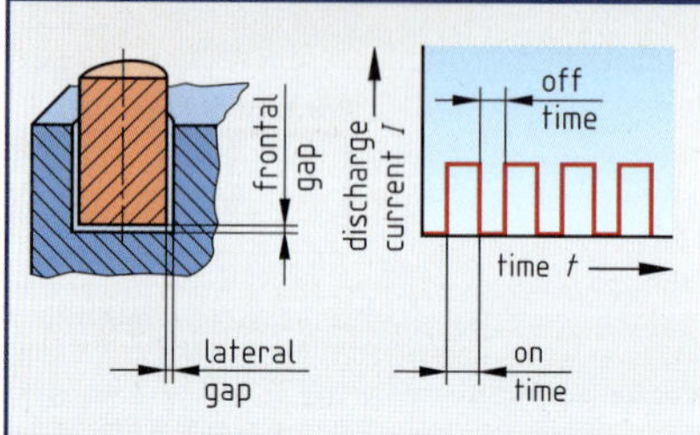

V_W removal rate in mm³/min
V removal volume in mm³
t removal time in min
V_E absolute tool wear in mm³
V_{rel} relative tool wear in %

Removal rate

$$V_W = \frac{V}{t}$$

Relative tool wear

$$V_{rel} = \frac{V_E}{V} \cdot 100\,\%$$

Parameter		Explanations, characteristics and applications
Electrode Material	**Electrolytic copper**	Universal application; low wear behavior; high removal rate; for finish and rough machining; difficult to manufacture electrode by machining; high thermal expansion; no cracked edges; tendency to warp
	Graphite in various grain sizes	Universal application; very low wear; greater current density than Cu; low electrode weight; easy to manufacture electrode by machining; non-warping; low thermal expansion; more detailed electrodes are made by selecting a finer graphite grain; unsuitable for carbide machining
	Tungsten-copper	Detailed electrodes; very low wear; very high material removal rate with relatively low discharge currents even with large current densities; only manufactured in limited sizes, high electrode weight
	Copper-graphite	Special applications involving small electrode dimensions with simultaneous high electrode strength; wear and material removal rate play a subordinate role in these special applications
Dielectric fluid	**Synthetic oils,** filtered and cooled; according to machine manufacturer	Requirements for dielectric fluids: • low and constant conductivity for stable sparking • low viscosity for filtrability and penetrating ability in narrow gaps • low evaporation to reduce hazardous vapors • high flash point to avoid fire hazard • high heat conductance value for good cooling • extremely low health hazard for operators
Flushing	**Replacement of dielectric fluid** at the erosion site **Remove eroded particles** from gap	Depending on requirements and available options, different flushing methods can be used to maintain stable erosion performance: • flooding (most commonly used method, simultaneous heat rejection) • pressure flushing through hollow electrodes or next to electrode • vacuum flushing through hollow electrode or next to electrode • interval flushing caused by retracting electrode • movement flushing by relative movement between workpiece and electrode, without interrupting erosion cycle
Polarity	**positive**	Electrode is positively polarized; for low electrode burn rate during roughing with long pulse duration and low frequency
	negative	Electrode is negatively polarized; for erosion with short pulse duration and high frequency
Gap	**face**	Kept constant during feed (controlled by discharge voltage). Control sensitivity set too high: Electrode continually pulses on and off, controlled discharge impossible. Control sensitivity set too low: Abnormal discharges increase or gap remains too large for discharge.
	side	Determined primarily by duration and size of discharge pulse, depends on material matching and no-load voltage
Discharge current	**low**	Low removal performance, low tool wear on copper electrodes, high wear on graphite electrodes
	high	High removal performance, high tool wear on copper electrodes, low wear on graphite electrodes
Pulse duration	**short**	Electrode wear with positive polarity is larger, lower removal rate
	long	Electrode wear with positive polarity is smaller, higher removal rate

PE

Cutting force, Operating conditions for presses

Cutting force, cutting work

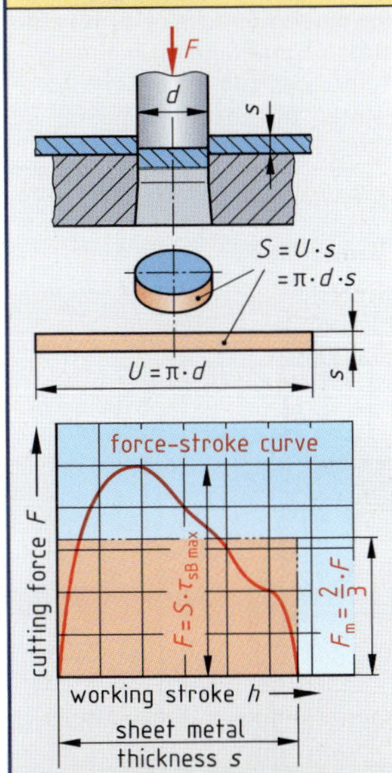

F	cutting force
F_m	calculated cutting force
S	shear area
$R_{m\,max}$	maximum tensile strength
$\tau_{sB\,max}$	maximum shear strength
W	cutting work
s	sheet metal thickness

Example:

$S = 236\ mm^2$; $s = 2.5\ mm$; $R_{m\,max} = 510\ N/mm^2$

Wanted: $\tau_{sB\,max}$; F; W

Solution: $\tau_{sB\,max} = 0.8 \cdot R_{m\,max}$
$= 0.8 \cdot 510\ N/mm^2 = \mathbf{408\ N/mm^2}$

$F = S \cdot \tau_{sB\,max} = 236\ mm^2 \cdot 408\ N/mm^2$
$= 96\,288\ N = \mathbf{96.288\ kN}$

$W = \frac{2}{3} \cdot F \cdot s = \frac{2}{3} \cdot 96.288\ kN \cdot 2.5\ mm$
$\approx 160\ kN \cdot mm = \mathbf{160\ N \cdot m}$

Cutting force

$$F = S \cdot \tau_{sB\,max}$$

Max. shear strength

$$\tau_{sB\,max} \approx 0.8 \cdot R_{m\,max}$$

Cutting work

$$W = \frac{2}{3} \cdot F \cdot s$$

Operating conditions for eccentric and crank presses

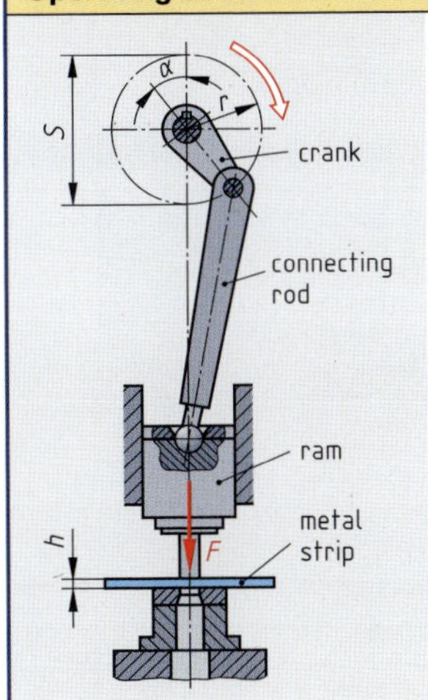

Press drives are usually designed such that the nominal pressing force is applied at crank angle $\alpha = 30°$.

Machines operate without interruption in continuous mode or can be stopped after each cycle in single-stroke mode. For presses with adjustable strokes, the allowable pressing force is less than the nominal pressing force.

F	cutting force, shaping force
F_n	nominal pressing force
F_{allow}	allow. pressing force for adjustable stroke
S	stroke, maximum stroke for adjustable stroke
S_a	adjusted stroke
h	working distance (≙ sheet metal thickness s)
α	crank angle
W	cutting work, shaping work
W_c	work capacity in continuous mode
W_s	work capacity in single-stroke mode

Work capacity in continuous mode

$$W_c = \frac{F_n \cdot S}{15}$$

Work capacity in single-stroke mode

$$W_s = 2 \cdot W_c$$

Operating conditions
Fixed stroke
$F \le F_n$ $W \le W_c$ or $W \le W_s$
Adjustable stroke
$F \le F_{allow}$ $F_{allow} = \frac{F_n \cdot S}{4 \cdot \sqrt{S_a \cdot h - h^2}}$ $W \le W_c$ or $W \le W_s$

Example:

Eccentric press with fixed stroke $F_n = 250\ kN$; $S = 30\ mm$;
$F = 207\ kN$; $s = 4\ mm$

Find: W; W_c. Can the press be put into continuous mode?

Solution: $W = \frac{2}{3} \cdot F \cdot s = \frac{2}{3} \cdot 207\ kN \cdot 4\ mm = 552\ kN \cdot mm = \mathbf{552\ N \cdot m}$

$W_c = \frac{F_n \cdot S}{15} = \frac{250\ kN \cdot 30\ mm}{15} = 500\ kN \cdot mm = \mathbf{500\ N \cdot m}$

If $F < F_n$, but $W > W_c$, the press cannot be used in continuous mode for this workpiece.

PE

Stamping tool

Stamping process: Lids made of sheet metal are produced on the post-guided progressive stamping tool. The metal strip to be cut is fed into the tool from the left. In progression A, the cutting punches (15, 16) cut the four round holes in the corners and the square hole in the middle in one stroke. Simultaneously, the cutting punch (20) notches the strip and sets the dimension for the progression. In progression B, the perimeter of the lid is blanked out with the help of the cutting punch (14). After each stroke, the sheet metal strip is advanced by the set progression feed dimension.

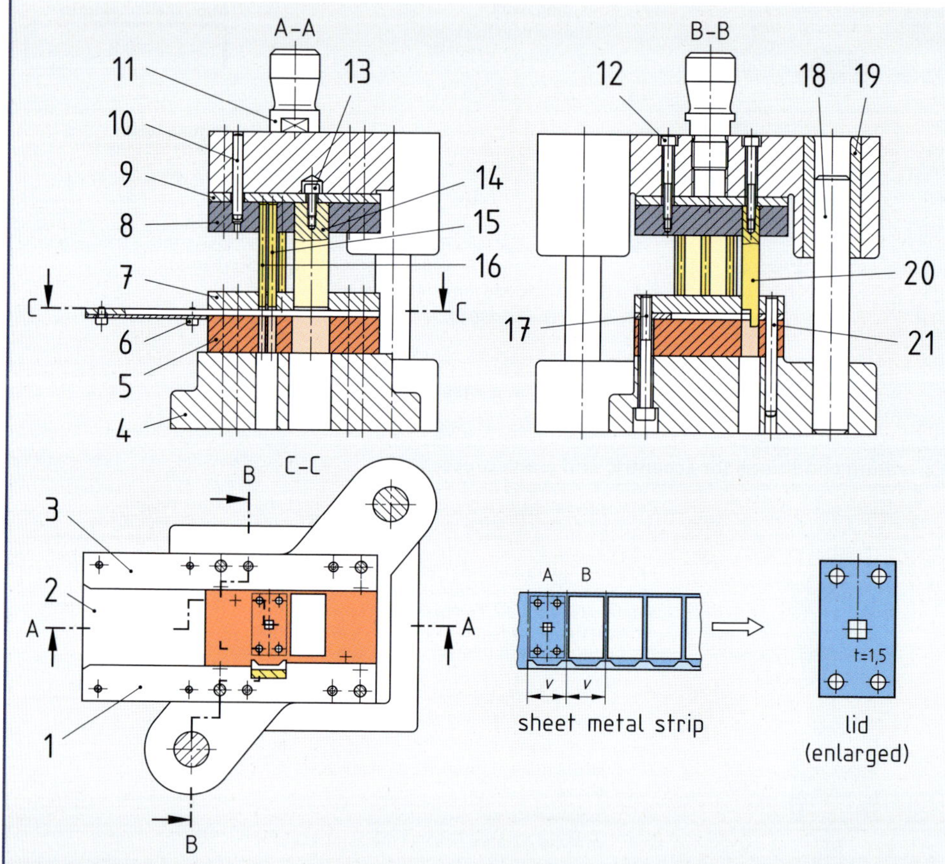

Bill of materials

Item	Designation	Standard/material	Item	Designation	Standard/material
1	Guide rail, front	S235JR	12	Socket head cap screw	ISO 4762
2	Stock shelf	DC01	13	Socket head cap screw	ISO 4762
3	Guide rail, rear	S235JR	14	Cutting punch, (hardened)	X210CrW12
4	Lower die shoe	DIN 9819	15	Cutting punch, (hardened)	S6-5-2
5	Die section, hardened	X210CrW12	16	Cutting punch, (hardened)	DIN 9861
6	Socket head cap screw	ISO 4762	17	Socket head cap screw	ISO 4762
7	Stripper	S235JR	18	Guide post, hardened	DIN 9825-2
8	Punch holder	S235JR	19	Guide bushing	DIN 9831-1
9	Backing plate, hardened	90MnCrV8	20	Notching punch, hardened	X210CrW12
10	Dowel pin, hardened	ISO 8734	21	Dowel pin, hardened	ISO 8734
11	Die shank	DIN 9859			

PE

Standard parts for punching tools

Illustration | Dimensions from – to in mm | Standard, material | Properties, function

Die sets

cf. DIN 9819 (1981-12)

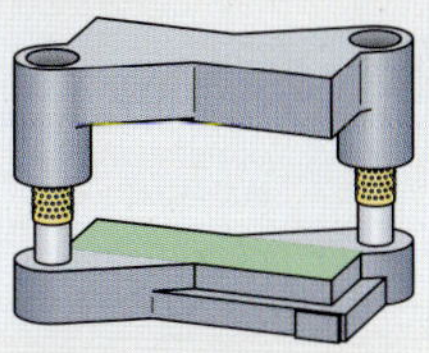

Dimensions from – to in mm	Standard, material	Properties, function
DIN 9819: Form C Working surfaces: *a* x *b* = 80 x 63 to 315 x 125	DIN 9819: diagonal posts DIN 9812: centrally positioned posts **Materials:** steel, cast iron, aluminum	Rectangular or round working surface with two or four guide posts, optionally available with plain or ball bearing guides.

Guide posts

cf. DIN 9825-2 (2004-10)

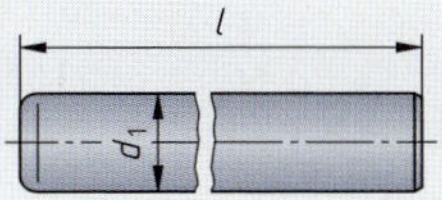

Dimensions from – to in mm	Standard, material	Properties, function
d_1 x l = 11 x 80 to 80 x 560	DIN 9825-2 **Materials:** e. g. C60E hardness 780+40 HV10 CHD = min. 0.8 mm	Guide diameter, tolerance zone h3, fine ground. The diameter of one post should be smaller than that of the other(s) in order to ensure that the die set always is assembled in the right position.

Guide bushings for die sets

cf. DIN 9831-1 (2004-10)

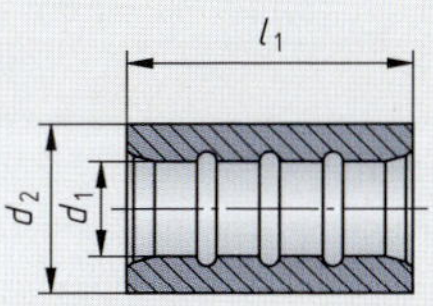

Dimensions from – to in mm	Standard, material	Properties, function
d_1 x d_2 x l_1 = 11 x 22 x 23 to 80 x 105 x 135	DIN 9831-1: with plain bearing guide DIN 9831-2: with ball bearing guide **Materials:** e. g. steel, bronze, cast resin, sintered material	Guide bushings are used for guide posts complying with DIN 9825-2. **Benefits:** • accurate guide • long service life

Round cutting punches with straight shaft

cf. DIN 9861-1 (1992-07)

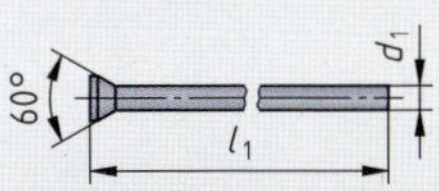

Dimensions from – to in mm	Standard, material	Properties, function
d_1 = 0.5 to 20 l_1 = 71, 80, 100	DIN 9861-1 ISO 6752 **Materials:** Tool steel, high-speed steel, optionally TiN-coated	Mostly used as piercing punch **Shaft hardness** with tool steel: HRC 62±2 high-speed steel: HRC 64±2 **Head hardness** with tool steel: HRC 50±5 high speed steel: HRC 50±5 ground head and shaft

Quick-change cutting punches

cf. DIN ISO 10071-1 (2010-11)

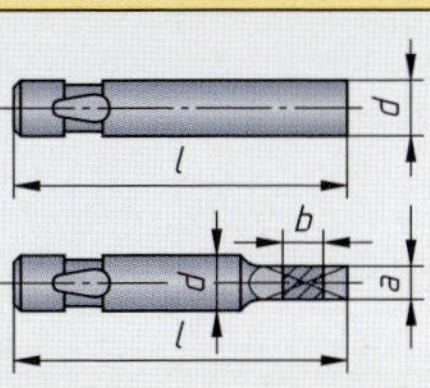

Dimensions from – to in mm	Standard, material	Properties, function
Form A: *d* x *l* = 6 x 63 to 32 x 100 **Form BS:** *a* = 1.6 to 22.5 **Form BR:** *a* = 1.6 to 12 *b* = 5.9 to 31.9	DIN ISO 10071-1 **Straight shaft:** Form A: cylindrical **Reduced shafts:** Form B: cylindrical Form BS: square Form BR: rectangular Form BO: oval	Short change-over times Workable sheet thickness: up to 3 mm Matching die buttons are available at specialized retailers for standard parts

Die set shank, form A

cf. DIN ISO 10242-1 (2000-03)

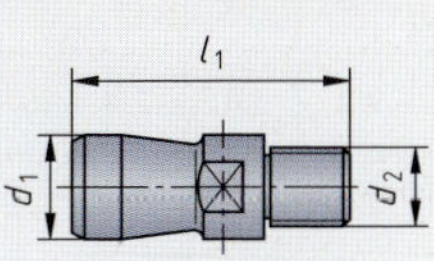

Dimensions from – to in mm	Standard, material	Properties, function
d_1 x d_2 x l_1 = 20 x M16 x 58 to 50 x M30 x 108	Form A: DIN ISO 10242-1 Form C: DIN ISO 10242-2 **Materials:** e.g. E295, C45	The punch holder shank is used to fix small and medium sized upper die shoes to the punch press ram. Punch press die sets are often fixed to the press ram with the help of a coupling shank and a receiving chuck.

PE

Tool and workpiece dimensions

Punch and cutting die dimensions

cf. VDI 3368 (1982-05)

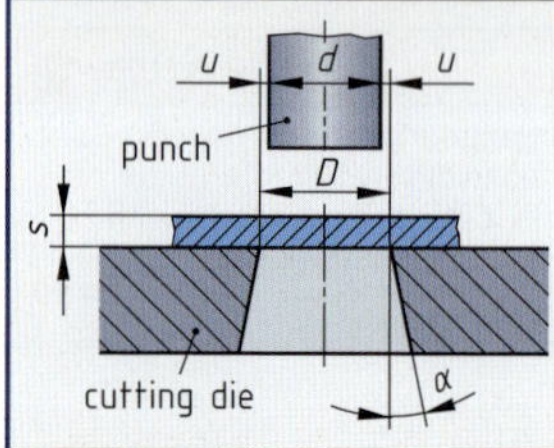

- *d* punch dimension
- *D* cutting die dimension
- *u* die clearance
- *s* sheet metal thickness
- *α* clearance angle

Process	Piercing	Blanking
Shape of workpiece	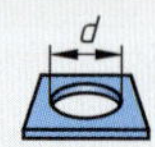	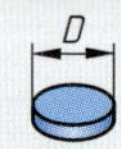
Governing specified size is:	dimension of punch *d*	dimension of cutting die *D*
Dimension of opposite tool	cutting die $D = d + 2 \cdot u$	punch $d = D - 2 \cdot u$

Die clearance *u* as a function of material and sheet metal thickness

sheet metal thickness *s* mm	Cutting die opening with clearance angle *α*; shear strength τ_{sB} in N/mm²; die clearance *u* in mm: up to 250	251–400	401–600	over 600	Cutting die opening without clearance angle *α*; shear strength τ_{sB} in N/mm²; die clearance *u* in mm: up to 250	251–400	401–600	over 600
0.4–0.6	0.01	0.015	0.02	0.025	0.015	0.02	0.025	0.03
0.7–0.8	0.015	0.02	0.03	0.04	0.025	0.03	0.04	0.05
0.9–1	0.02	0.03	0.04	0.05	0.03	0.04	0.05	0.05
1.5–2	0.03	0.05	0.06	0.08	0.05	0.07	0.09	0.11
2.5–3	0.04	0.07	0.10	0.12	0.08	0.11	0.14	0.17
3.5–4	0.06	0.09	0.12	0.16	0.11	0.15	0.19	0.23

Web width, edge width, trim stop waste for metallic materials

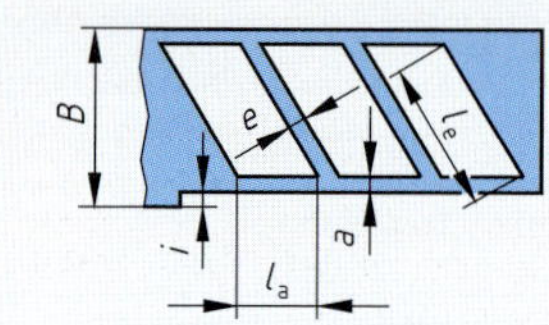

Polygonal workpieces

- *a* edge width
- *e* web width
- l_a edge length
- l_e web length
- *B* strip width
- *i* trim stop waste (french stop waste)

Polygonal workpieces:

The web or edge length, whichever is larger, is used to determine web and edge widths.

Round workpieces:

For all diameters values given for $l_e = l_a$ = 10 mm of polygonal workpieces apply to web and edge widths.

Strip width *B* mm	Web length l_e Edge length l_a mm	Web width *e* Edge width *a*	Sheet metal thickness *s* in mm: 0.1	0.3	0.5	0.75	1.0	1.25	1.5	1.75	2.0	2.5	3.0
up to 100 mm	up to 10	*e*	0.8	0.8	0.8	0.9	1.0	1.2	1.3	1.5	1.6	1.9	2.1
		a	1.0	0.9	0.9								
	11–50	*e*	1.6	1.2	0.9	1.0	1.1	1.4	1.4	1.6	1.7	2.0	2.3
		a	1.9	1.5	1.0								
	51–100	*e*	1.8	1.4	1.0	1.2	1.3	1.6	1.6	1.8	1.9	2.2	2.5
		a	2.2	1.7	1.2								
	over 100	*e*	2.0	1.6	1.2	1.4	1.5	1.8	1.8	2.0	2.1	2.4	2.7
		a	2.4	1.9	1.5								
	trim stop waste *i*		1.5					1.8	2.2	2.5	3.0	3.5	4.5
over 100 mm to 200 mm	up to 10	*e*	0.9	1.0	1.0	1.0	1.1	1.3	1.4	1.6	1.7	2.0	2.3
		a	1.2	1.1	1.1								
	11–50	*e*	1.8	1.4	1.0	1.2	1.3	1.6	1.6	1.8	1.9	2.2	2.5
		a	2.2	1.7	1.2								
	51–100	*e*	2.0	1.6	1.2	1.4	1.5	1.8	1.8	2.0	2.1	2.4	2.7
		a	2.4	1.9	1.5								
	101–200	*e*	2.2	1.8	1.4	1.6	1.7	2.0	2.0	2.2	2.3	2.6	2.9
		a	2.7	2.2	1.7								
	trim stop waste *i*		1.5				1.8	2.0	2.5	3.0	3.5	4.0	5.0

PE

Location of punch holder shank, Utilization of strip stock

Location of punch holder shank for punch geometry with known center of gravity

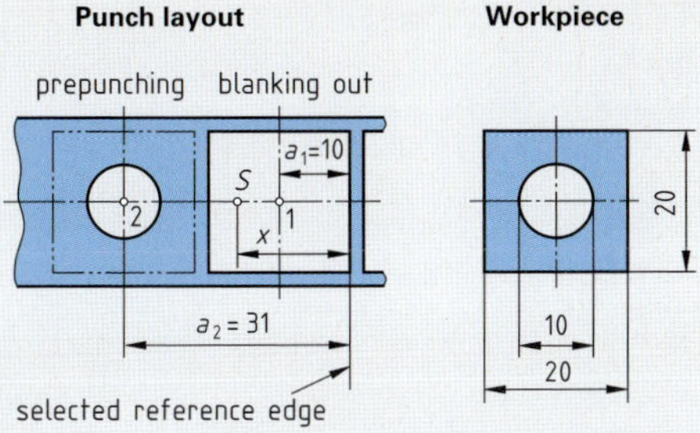

C_1, C_2, C_3 ... circumferences of individual punches

a_1, a_2, a_3 ... distances from punch centers of gravity to selected reference edge

x distance of center of forces S from chosen reference edge

Distance of the center of forces

$$x = \frac{C_1 \cdot a_1 + C_2 \cdot a_2 + C_3 \cdot a_3 + \ldots}{C_1 + C_2 + C_3 + \ldots}$$

Example:

Based on the figure at left, calculate the distance x of center of forces S.

Solution:

The outer perimeter of the cutting punch is chosen as reference edge.

Blanking punch: $C_1 = 4 \cdot 20$ mm = 80 mm; a_1 = 10 mm

Piercing punch: $C_2 = \pi \cdot 10$ mm = 31.4 mm; a_2 = 31 mm

$$x = \frac{C_1 \cdot a_1 + C_2 \cdot a_2}{C_1 + C_2}$$

$$\mathbf{x} = \frac{80\text{ mm} \cdot 10\text{ mm} + 31.4\text{ mm} \cdot 31\text{ mm}}{80\text{ mm} + 31.4\text{ mm}} \approx \mathbf{16\text{ mm}}$$

Location of punch holder shank for punch geometry with unknown center of gravity

Center of forces corresponds to centroid of the line[1)] of all cutting edges.

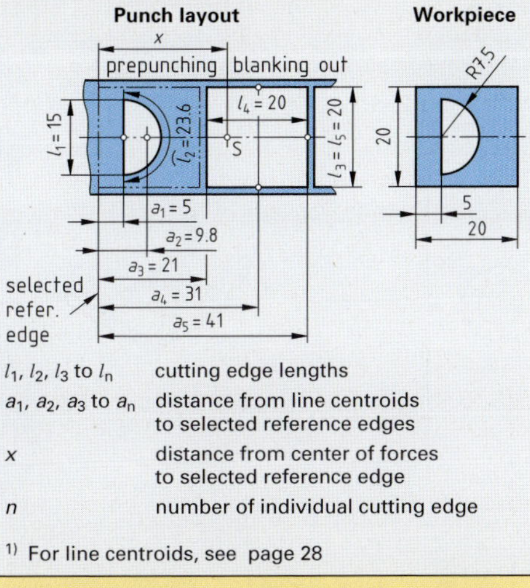

l_1, l_2, l_3 to l_n cutting edge lengths

a_1, a_2, a_3 to a_n distance from line centroids to selected reference edges

x distance from center of forces to selected reference edge

n number of individual cutting edge

[1)] For line centroids, see page 28

Distance of the center of forces

$$x = \frac{l_1 \cdot a_1 + l_2 \cdot a_2 + l_3 \cdot a_3 + \ldots}{l_1 + l_2 + l_3 + \ldots}$$

$$x = \frac{\sum l_n \cdot a_n}{\sum l_n}$$

Example:

Calculate the location of the punch holder shank on the progressive die for the workpiece shown in the figure at the left.

Solution:

n	l_n in mm	a_n in mm	$l_n \cdot a_n$ in mm²
1	15	5	75
2	23.6	9.8	231.28
3	20	21	420
4	2 · 20	31	1240
5	20	41	820
Σ	118.6	–	2786.28

$$\mathbf{x} = \frac{\sum l_n \cdot a_n}{\sum l_n} = \frac{2786.28\text{ mm}^2}{118.6\text{ mm}} = \mathbf{23.5\text{ mm}}$$

Utilization of strip stock for single row stamping

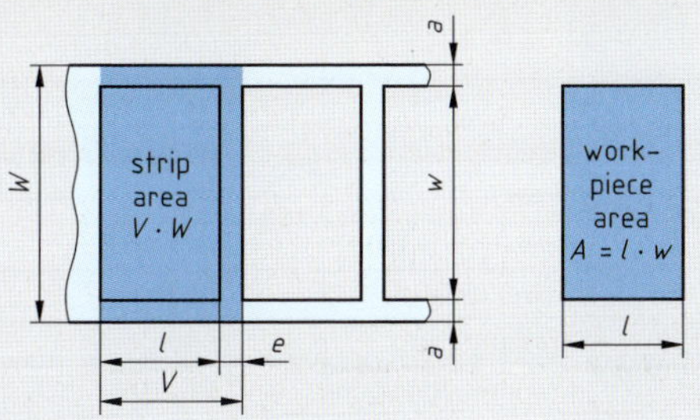

l workpiece length

w workpiece width

W strip width

a edge width

e web width

V strip feed

A area of workpiece (including holes)

R number of rows

η degree of utilization

Strip width

$$W = w + 2 \cdot a$$

Strip feed

$$V = l + e$$

Utilization factor

$$\eta = \frac{R \cdot A}{V \cdot W}$$

PE

Bending tool

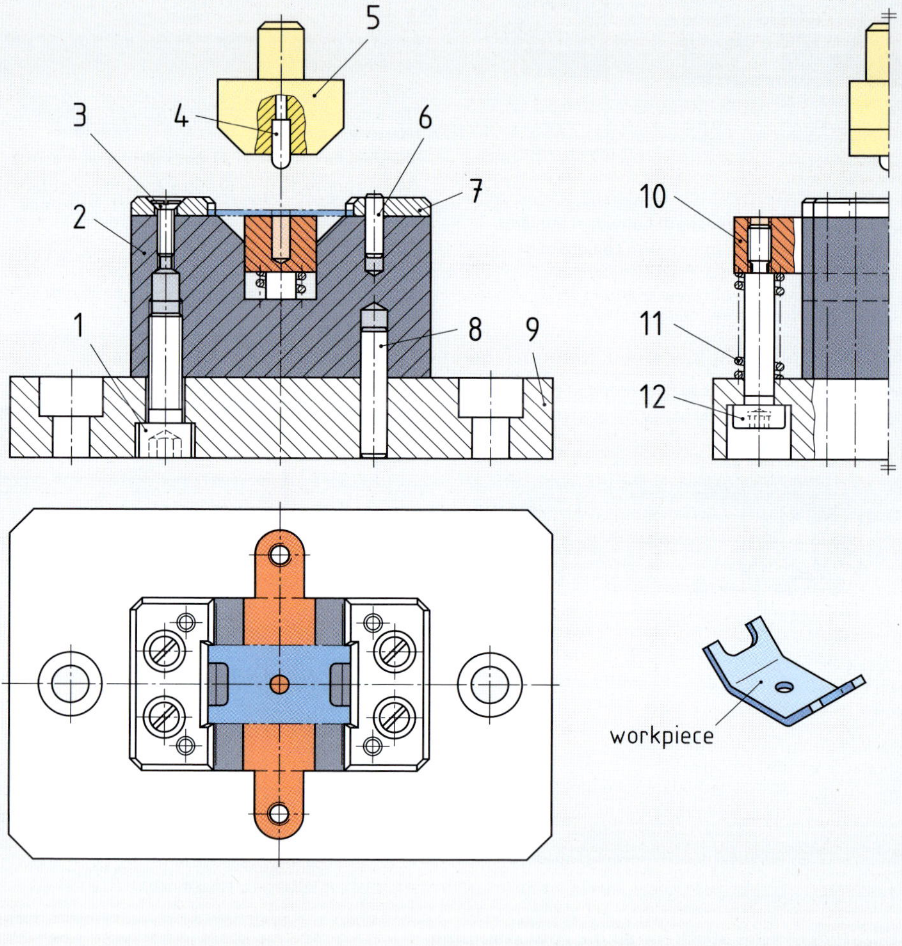

PE

Bill of materials					
Item	Designation	Standard/material	Item	Designation	Standard/material
1	Socket head cap screw	ISO 4762	7	Locator	C60
2	Bending section	90MnCrV8	8	Dowel pin (6 x 18)	ISO 2338
3	Flat head screw	ISO 10642	9	Lower die shoe	S235JR
4	Pilot pin	C45	10	Pressure pad	C45
5	Bending punch	C45	11	Compression spring	DIN 2098
6	Dowel pin (4 x 8)	ISO 2338	12	Shoulder bolt	C60

Bending processes

Illustration	Function	Application
Free bending		
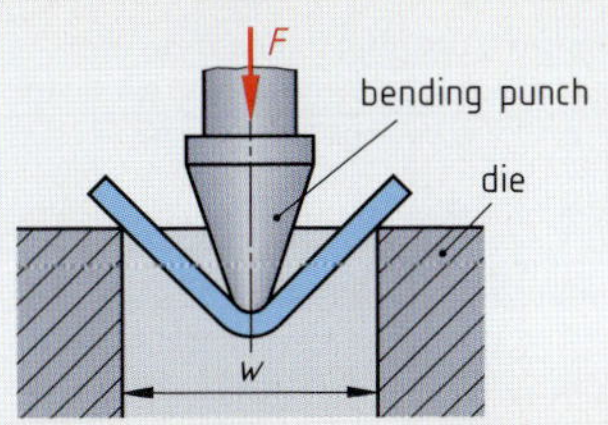	The die supports the sheet metal at two points. The punch forces the material downwards thereby producing a rounding, which shape primarily depends on the opening width of the die (w).	Various angles can be bent without having to change the tool. This method is also used for the straightening of workpieces.
Die bending		
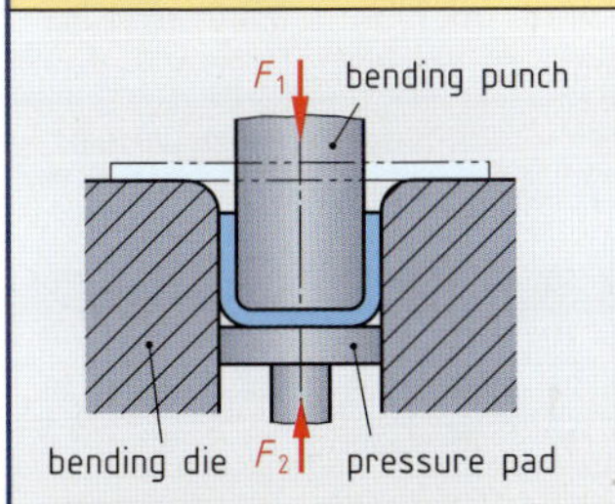	The punch forces the sheet metal downwards to match perfectly the shape of the die, causing a permanent deformation of the material. Depending on the shape of the die, a distinction is made between: • V-bending and • U-bending A pressure pad is often used in U-bending. The force F_2 prevents the blank from warping up at the bottom during the bending process.	This method is more accurate than free bending. More accurate inner dimensions are obtained through movable jaws, which are pressed inwards by cams. Accordingly, the cams press against the bending part from outside, if the outer dimensions need to be more accurate.
Swing folding		
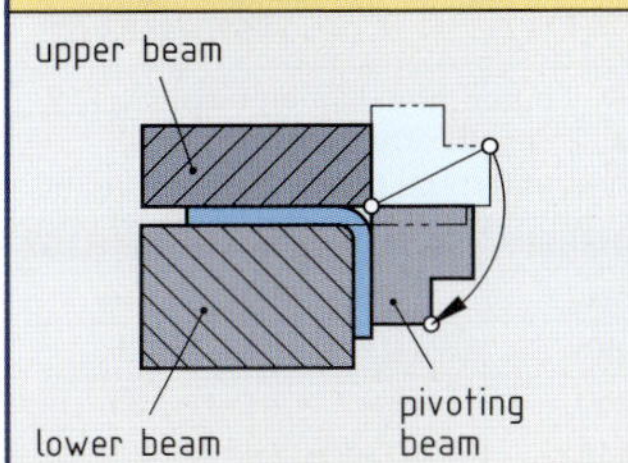	The sheet metal to be bent is clamped between the upper and the lower beam. The pivoting bending beam rotates with the sheet metal around the bend profile until the required bending angle is attained. The pivoting beam can be moved manually or motor driven. CNC-controls are widely used to produce sophisticated bent parts.	• Bending of short legs • Bending of delicate surfaces without scratches (e. g. copper and aluminum alloys, stainless steel and coated surfaces)
Curling		
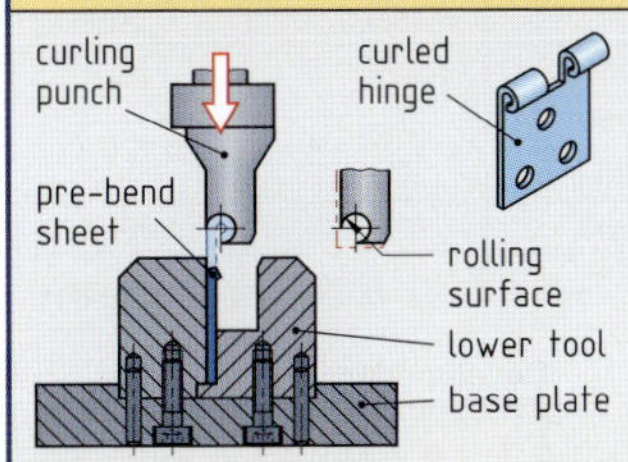	When the curling punch moves down it forces the material into its cylindrical cutout thereby producing the rolled shape. To facilitate rolling, the workpieces should be pre-bend beforehand. In progressive dies, workpieces are pre-bend in the first stage and rolled in the second.	Simple method to produce: • beads • hinges • joint hinges
Roll bending		
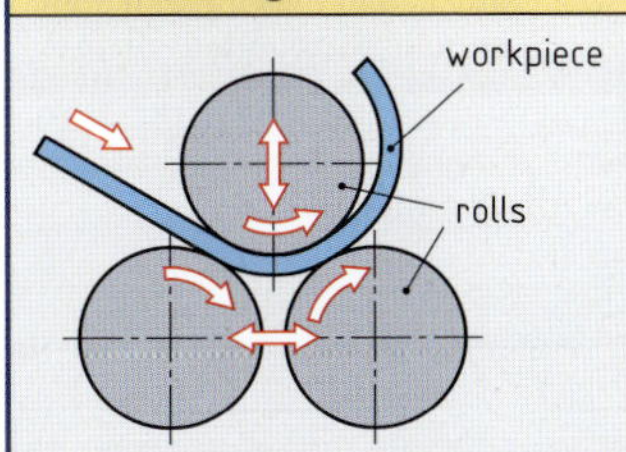	The metal sheet moves between three rolls. Various bending radii can be produced by adjusting the rolls accordingly.	Bending of metal sheets for boilers, vessels and tanks. A comparable method is **roll straightening.** Several roller sets of three straighten sheet metal, bars, wires or tubes.

PE

Bending radius, Bend allowances, Calculation of blank size

Smallest allowable bending radius for bent parts of non-ferrous metals

cf. DIN 5520 (2002-07)

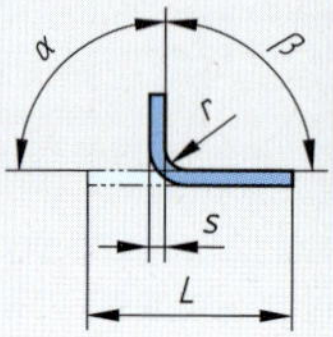

Material	Material condition	Thickness s in mm: 0.8	1	1.5	2	3	4	5	6
		Smallest allowable bending radius r[1] in mm							
AlMg3-01	spheroidized	0.6	1	2	3	4	6	8	10
AlMg3-H14	cold work hardened	1.6	2.5	4	6	10	14	18	–
AlMg3-H111	cold work hardened and annealed	1	1.5	3	4.5	6	8	10	–
AlMg4.5Mn-H112	spheroidized straightened	1	1.5	2.5	4	6	8	10	14
AlMg4.5Mn-H111	cold work hardened and annealed	1.6	2.5	4	6	10	16	20	25
AlMgSi1-T6	solution annealed and artificially aged	4	5	8	12	16	23	28	36
CuZn37-R600	hard	2.5	4	5	8	10	12	18	24

[1] For bending angle $\alpha = 90°$, regardless of rolling direction

Smallest allowable bending radius for cold bending steel

cf. DIN 6935 (2010-01)

Minimum tensile strength R_m in N/mm² over–to	Minimum bending radius[1] r for sheet metal thickness s in mm: 1	1.5	2.5	3	4	5	6	7	8	10	12	14	16	18	20
up to 390	1	1.6	2.5	3	5	6	8	10	12	16	20	25	28	36	40
390–490	1.2	2	3	4	5	8	10	12	16	20	25	28	32	40	45
490–640	1.6	2.5	4	5	6	8	10	12	16	20	25	32	36	45	50

[1] Values apply to bending angle $\alpha \le 120°$ and bending transverse to rolling direction. Value of the next larger sheet metal thickness should be selected for bending longitudinal to rolling direction and bending angle $\alpha > 120°$.

Bending radii for pipes

cf. DIN 25570 (2004-02)

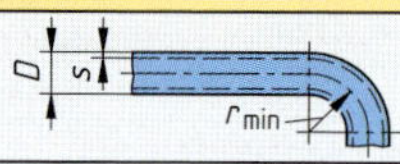

D outer pipe diameter in mm
s wall thickness in mm
r_{min} minimum bending radius in mm

<table>
<tr><th colspan="4">Steel pipes E235, X5CrNi18-10</th><th colspan="2">Aluminum pipes AlMgSi</th><th colspan="2">Copper pipes Cu-DHP-R250</th></tr>
<tr><th>D x s</th><th>r_min</th><th>D x s</th><th>r_min</th><th>D x s</th><th>r_min</th><th>D x s</th><th>r_min</th></tr>
<tr><td>6 x 1</td><td rowspan="3">20</td><td>22 x 1.5</td><td rowspan="2">50</td><td>16 x 1.5</td><td>80</td><td>6 x 1</td><td>25</td></tr>
<tr><td>8 x 1</td><td>22 x 2.5</td><td>20 x 1.5</td><td>100</td><td>8 x 1</td><td>35</td></tr>
<tr><td>10 x 1</td><td>25 x 1.5</td><td rowspan="2">55</td><td>25 x 3</td><td>110</td><td>10 x 1.5</td><td rowspan="2">40</td></tr>
<tr><td>12 x 1.5</td><td>25</td><td>25 x 2.5</td><td>28 x 2.5</td><td rowspan="2">125</td><td>12 x 1.5</td></tr>
<tr><td>15 x 2</td><td rowspan="2">35</td><td>30 x 1.5</td><td rowspan="2">80</td><td>30 x 1.5</td><td>16 x 1.5</td><td>60</td></tr>
<tr><td>16 x 2</td><td>30 x 2</td><td>40 x 3</td><td>180</td><td>18 x 1.5</td><td>80</td></tr>
<tr><td>18 x 2.5</td><td>40</td><td>40 x 2.5</td><td>100</td><td>50 x 3</td><td>250</td><td>28 x 1.5</td><td rowspan="2">125</td></tr>
<tr><td>20 x 1.5</td><td rowspan="2">45</td><td>45 x 2.5</td><td>125</td><td>60 x 2</td><td>300</td><td>30 x 2.5</td></tr>
<tr><td>20 x2</td><td>50 x 2.5</td><td>140</td><td>70 x 3</td><td>350</td><td>42 x 1.5</td><td>200</td></tr>
</table>

Calculation of blank size for 90° bent parts

cf. DIN 6935 (2010-01)

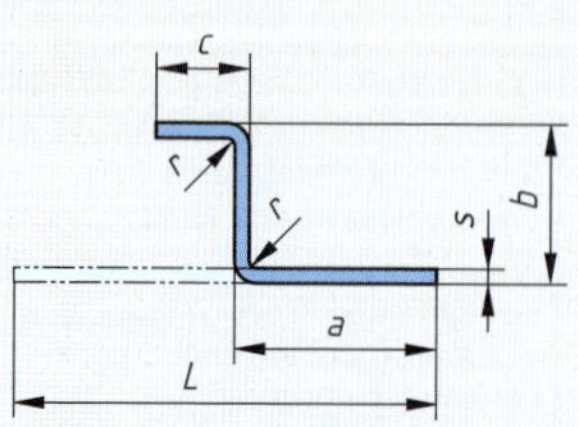

L developed length[1]
a, b, c length of leg
s thickness
r bending radius
n number of bends
v bend allowance (table page 333)

Developed length[2]

$$L = a + b + c + \ldots - n \cdot v$$

[2] Calculated developed length should be rounded off to a whole mm value.

Example (see illus.):

$a = 25$ mm; $b = 20$ mm; $c = 15$ mm; $n = 2$; $t = 2$ mm;
$r = 4$ mm; material S235JR; v = ?; L = ?
v = 4.5 mm
$L = a + b + c - n \cdot v = (25 + 20 + 15 - 2 \cdot 4.5)$ mm = **51 mm**

[1] If the ratio $r/s > 5$, the formula for developed length (page 20) can be used.

PE

Bend allowances, Calculation of a blank size, Springback

Bend allowances v for bending angle $\alpha = 90°$

cf. Supplement 2 to DIN 6935 (2010-01)

Bending radius r in mm	Bend allowance v per bend in mm for sheet metal thickness s in mm														
	0.4	0.6	0.8	1	1.5	2	2.5	3	3.5	4	4.5	5	6	8	10
1	1.0	1.3	1.6	1.9	–	–	–	–	–	–	–	–	–	–	–
1.6	1.2	1.5	1.8	2.1	2.9	–	–	–	–	–	–	–	–	–	–
2.5	1.5	1.8	2.1	2.4	3.2	4.0	4.8	–	–	–	–	–	–	–	–
4	–	2.4	2.7	3.0	3.7	4.5	5.2	6.0	6.9	–	–	–	–	–	–
6	–	–	3.5	3.8	4.5	5.2	5.9	6.7	7.5	8.3	9.1	9.9	–	–	–
10	–	–	–	5.5	6.1	6.7	7.4	8.1	8.9	9.6	10.4	11.2	12.7	–	–
16	–	–	–	8.1	8.7	9.3	9.9	10.5	11.2	11.9	12.6	13.3	14.8	17.8	21.0
20	–	–	–	9.8	10.4	11.0	11.6	12.2	12.8	13.4	14.1	14.9	16.3	19.3	22.3
25	–	–	–	11.9	12.6	13.2	13.8	14.4	15.0	15.6	16.2	16.8	18.2	21.1	24.1
32	–	–	–	15.0	15.6	16.2	16.8	17.4	18.0	18.6	19.2	19.8	21.0	23.8	26.7
40	–	–	–	18.4	19.0	19.6	20.2	20.8	21.4	22.0	22.6	23.2	24.5	26.9	29.7
50	–	–	–	22.7	23.3	23.9	24.5	25.1	25.7	26.3	26.9	27.5	28.8	31.2	33.6

Calculation of blank size for parts with any selected bending angle

cf. DIN 6935 (2010-01)

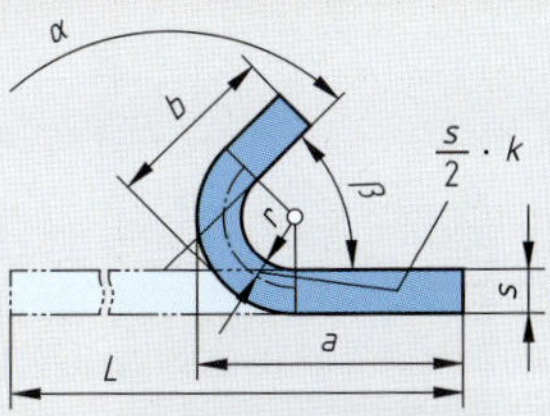

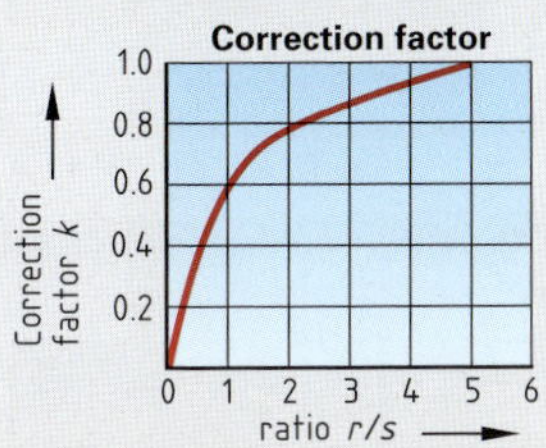

L developed length
a, b length of legs
v bend allowance
k correction factor
s sheet thickness
r bending radius
β aperture angle

Developed length[1)]

$$L = a + b - v$$

Bend allowance for $\beta = 0°$ to $90°$

$$v = 2 \cdot (r + s) - \pi \cdot \left(\frac{180° - \beta}{180°}\right) \cdot \left(r + \frac{s}{2} \cdot k\right)$$

Bend allowance for β over $90°$ to $165°$

$$v = 2 \cdot (r + s) \cdot \tan\frac{180° - \beta}{2} - \pi \cdot \left(\frac{180° - \beta}{180°}\right) \cdot \left(r + \frac{s}{2} \cdot k\right)$$

Bend allowance for β over $165°$ to $180°$ $v \approx 0$ (negligible)

[1)] For $r/s > 5$, the bend length (page 20) is sufficiently accurate for the calculation.

Correction factor

$$k = 0.65 + 0.5 \cdot \log\frac{r}{s}$$

Springback in bending

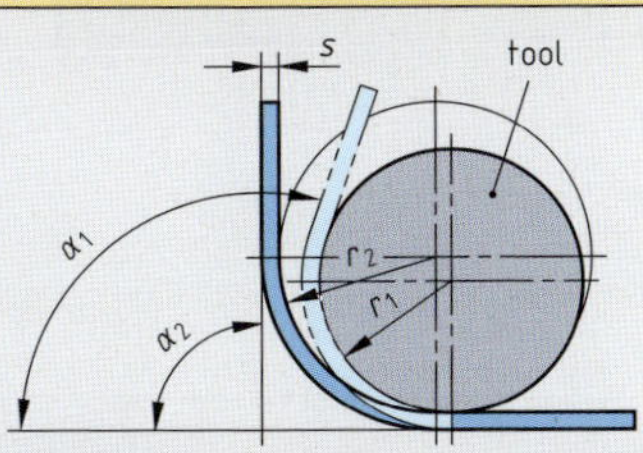

α_1 angle of bend before springback (on tool)
α_2 angle of bend after springback (on workpiece)
r_1 radius on tool
r_2 bending radius on workpiece
k_R springback factor
s sheet metal thickness

Radius on tool

$$r_1 = k_R \cdot (r_2 + 0.5 \cdot s) - 0.5 \cdot s$$

Angle of bend before springback

$$\alpha_1 = \frac{\alpha_2}{k_R}$$

Material of bent part	Springback factor k_R for the ratio r_2/s										
	1	1.6	2.5	4	6.3	10	16	25	40	63	100
DC04	0.99	0.99	0.99	0.98	0.97	0.97	0.96	0.94	0.91	0.87	0.83
DC01	0.99	0.99	0.99	0.97	0.96	0.96	0.93	0.90	0.85	0.77	0.66
X12CrNi18-8	0.99	0.98	0.97	0.95	0.93	0.89	0.84	0.76	0.63	–	–
E-Cu-R20	0.98	0.97	0.97	0.96	0.95	0.93	0.90	0.85	0.79	0.72	0.6
CuZn33-R29	0.97	0.97	0.96	0.95	0.94	0.93	0.89	0.86	0.83	0.77	0.73
CuNi18Zn20	–	–	–	0.97	0.96	0.95	0.92	0.87	0.82	0.72	–
Al99.00.99	0.99	0.99	0.99	0.98	0.98	0.97	0.97	0.96	0.95	0.93	
AlCuMg1	0.92	0.90	0.87	0.84	0.77	0.67	0.54	–	–	–	–
AlSiMgMn	0.98	0.98	0.97	0.96	0.95	0.93	0.90	0.86	0.82	0.76	0.72

PE

Deep drawing tool

Deep drawing process

The straight blank is cut to size (circular blank) and placed in the locator (14). The draw ring (13) keeps the blank on the blank holder (6) while retaining it at the outer edges. The draw ring moves down against the spring force and draws the material over the rounded drawing edge of the draw punch. A hollow round part is produced. The gap between the punch and the ring must be larger than the sheet thickness. A drawing gap that is too narrow might produce cracks in the material, whereas a gap that is too wide might produce wrinkles.

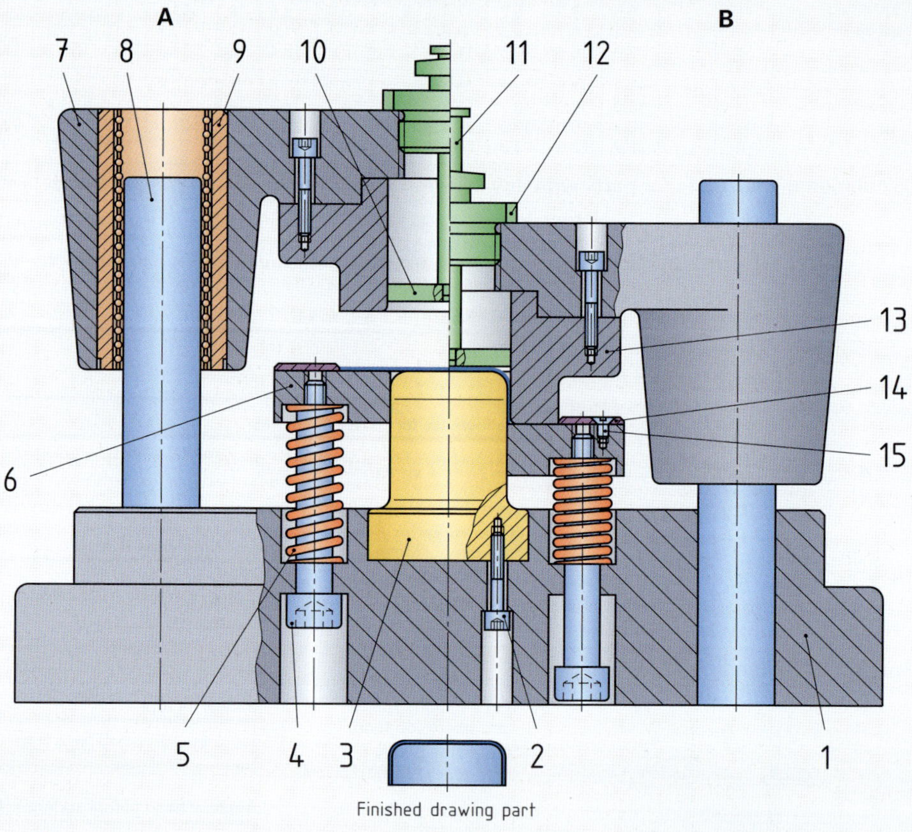

A Initial position (before deep drawing) **B** Final position (after deep drawing)

PE

Bill of materials

Item	Designation	Standard/material	Item	Designation	Standard/material
1	Lower die shoe	GJL – 250	9	Guiding bushing	CuSn8
2	Socket head cap screw	ISO 4762	10	Knock-out	S235JR
3	Draw punch	90Cr3	11	Knock-out rod	C60
4	Shoulder screw	ISO 4762	12	Coupling shank	E335
5	Compression spring	Spring steel	13	Draw ring	90Cr3
6	Blank holder	90Cr3	14	Locator	S235JR
7	Upper die shoe	GJL – 250	15	Flat head screw	ISO 2009
8	Guide post	16MnCr5			

Deep drawing processes

Illustration	Function	Properties, application
Mechanical deep drawing (first draw and redraw)		
First draw 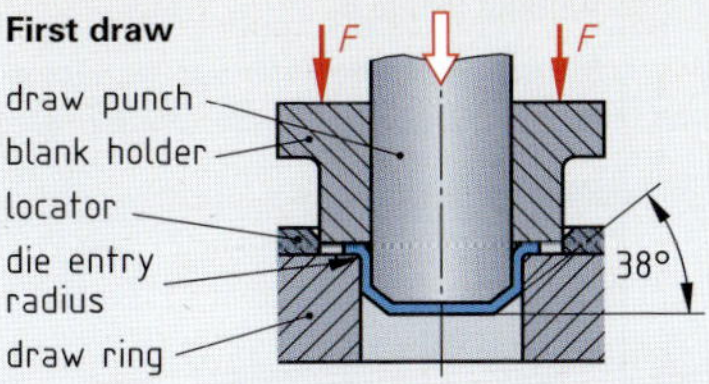 **Redraw** 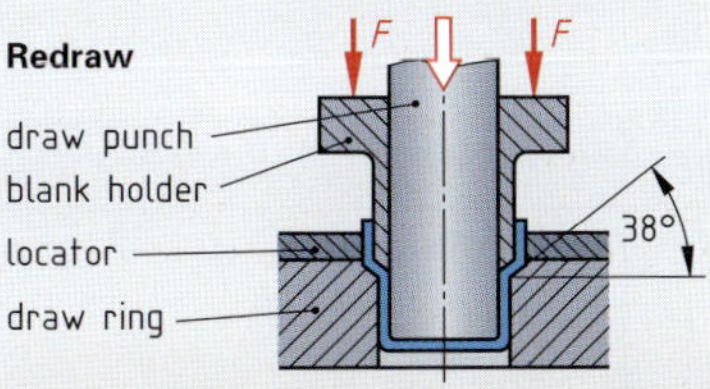	**Deep drawing workflow:** • Place the circular blank in the locator • The blank holder presses the circular blank down on the draw ring and locks it • The draw punch pulls the blank from the blank holder, forces it into the drawing gap to form a cup • If the drawing ratio *D/d* is too large, additional drawing stages are required (cf. page 337, first and redrawing) • Positioning of the pre-formed cup in the redrawing tool • The cup produced by redrawing has a smaller diameter and a greater height	Materials suitable for deep drawing must allow for major reshaping processes without being affected by splits. Examples: • DC01, DC03 • X15CrNiSi25-20 • CuZn38 • Cu95.5 • Al99.8 • AlMg1 Application for troughs, pots, cans, vehicle parts The material thickness is not affected by deep drawing.
Deep drawing with elastic cushion		
Initial shape of part to be drawn 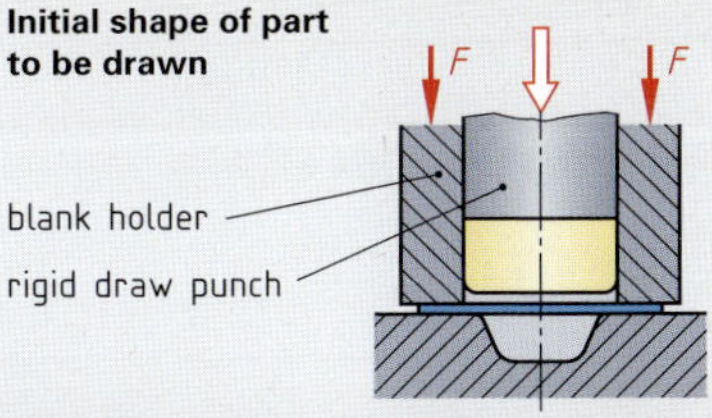 **Final shape of the drawn part** 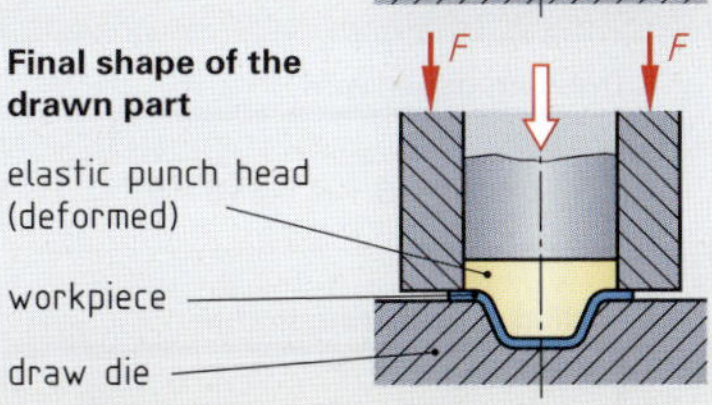	• A rigid draw die has the outline of the required workpiece shape • The head of the draw punch consists of an elastic rubber or elastomer cushion • When moving down the elastic cushion of the draw punch forces the blank into the cavity of the draw die.	• simple and inexpensive tool • low punch wear • no scratches on the workpiece surface • favorable drawing ratio • for the production of low quantities • for decorative parts
Hydromechanical deep drawing (hydromec drawing)		
Start of the first draw 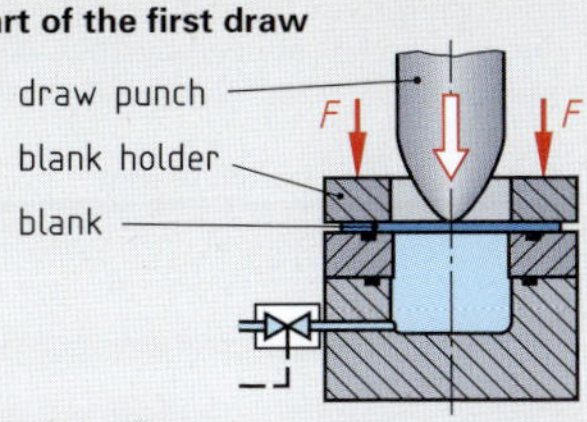 **End of the first draw** 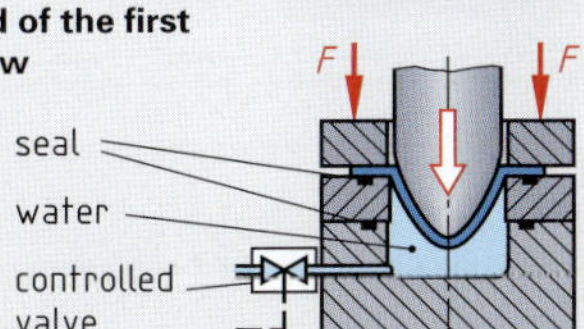	• The blank is retained by the blank holder • The draw punch forces the blank into the fluid, thereby generating a uniformly distributed pressure (200 to 700 bar) controlled by valves • The water pressure presses the blank against the draw punch • The blank adopts accurately the shape of the draw punch	• very favorable drawing ratio • low punch wear • low tool costs • requires fewer drawing stages • for sophisticated shapes (e. g. spheroidal or parabolic parts)

PE

Deep drawing

Calculation of blank diameter

Drawn part	Blank diameter D	Drawn part	Blank diameter D
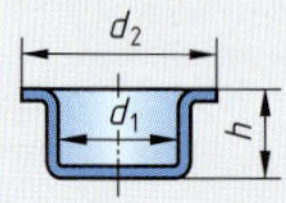	**without flange d_2** $D=\sqrt{d_1^2+4\cdot d_1\cdot h}$ **with flange d_2** $D=\sqrt{d_2^2+4\cdot d_1\cdot h}$	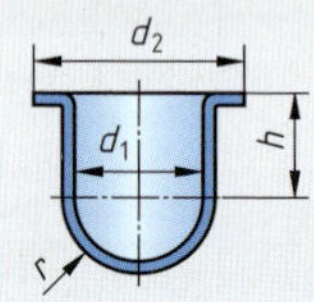	**without flange d_2** $D=\sqrt{2\cdot d_1^2+4\cdot d_1\cdot h}$ **with flange d_2** $D=\sqrt{2\cdot d_1^2+4\cdot d_1\cdot h+(d_2^2-d_1^2)}$
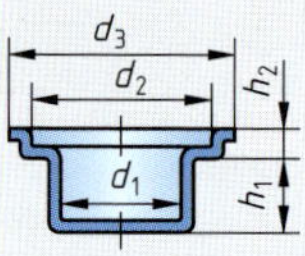	**without flange d_3** $D=\sqrt{d_2^2+4\cdot(d_1\cdot h_1+d_2\cdot h_2)}$ **with flange d_3** $D=\sqrt{d_3^2+4\cdot(d_1\cdot h_1+d_2\cdot h_2)}$	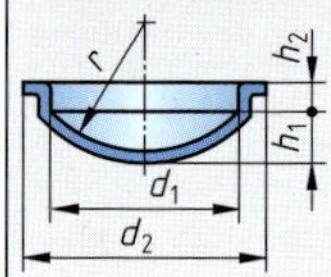	**without flange d_2** $D=\sqrt{d_1^2+4\cdot h_1^2+4\cdot d_1\cdot h_2}$ **with flange d_2** $D=\sqrt{d_1^2+4\cdot h_1^2+4\cdot d_1\cdot h_2+(d_2^2-d_1^2)}$
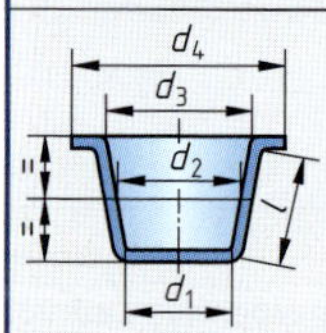	**without flange d_4** $D=\sqrt{d_1^2+4\cdot d_2\cdot l}$ **with flange d_4** $D=\sqrt{d_1^2+4\cdot d_2\cdot l+(d_4^2-d_3^2)}$	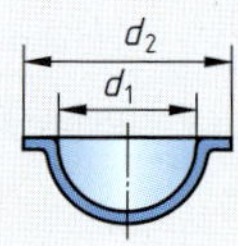	**without flange d_2** $D=\sqrt{2\cdot d_1^2}=1.414\cdot d$ **with flange d_2** $D=\sqrt{d_1^2+d_2^2}$

Example:

Cylindrical drawn part with flange d_2 (see figure, upper left) with d_1 = 50 mm, h = 30 mm; D = ?

$\boldsymbol{D}=\sqrt{d_1^2+4\cdot d_1\cdot h}=\sqrt{50^2\ \text{mm}^2+4\cdot 50\ \text{mm}\cdot 30\ \text{mm}}=\mathbf{92.2\ mm}$

Drawing gap and radii on draw ring and draw punch

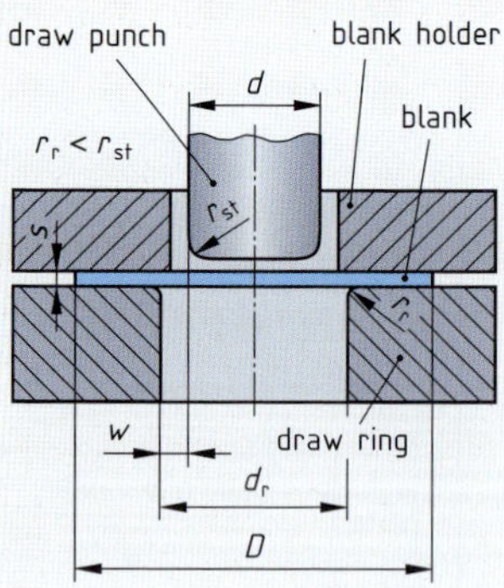

Material factor k	
Steel	0.07
Aluminum	0.02
Other non-ferrous metals	0.04

- w drawing gap
- s sheet metal thickness
- k material factor
- r_r radius on draw ring
- r_{st} radius of draw punch
- D blank diameter
- d punch diameter
- d_r draw ring diameter

Drawing gap in mm

$$w=s+k\cdot\sqrt{10\cdot s}$$

Radius of draw ring in mm

$$r_r=0.035\cdot[50+(D-d)]\cdot\sqrt{s}$$

For each redraw the radius of the draw ring should be reduced by 20 to 40 %.

Drawing gap

$$w=\frac{d_r-d}{2}$$

Radius of draw punch in mm

$$r_{st}=(4\text{ to }5)\cdot s$$

Example:

Steel sheet; D = 51 mm; d = 25 mm; s = 2 mm; w = ?; r_r = ?; r_{st} = ?

$\boldsymbol{k}$ = **0.07** (from table)

$\boldsymbol{w}=s+k\cdot\sqrt{10\cdot s}=2+0.07\cdot\sqrt{10\cdot 2}=\mathbf{2.3\ mm}$

$\boldsymbol{r_r}=0.035\cdot[50+(D-d)]\cdot\sqrt{s}=0.035\cdot[50+(51-25)]\cdot\sqrt{2}=\mathbf{3.8\ mm}$

$\boldsymbol{r_{st}}=4.5\cdot s=4.5\cdot 2\ \text{mm}=\mathbf{9\ mm}$

PE

Deep drawing

Drawing steps and drawing ratios

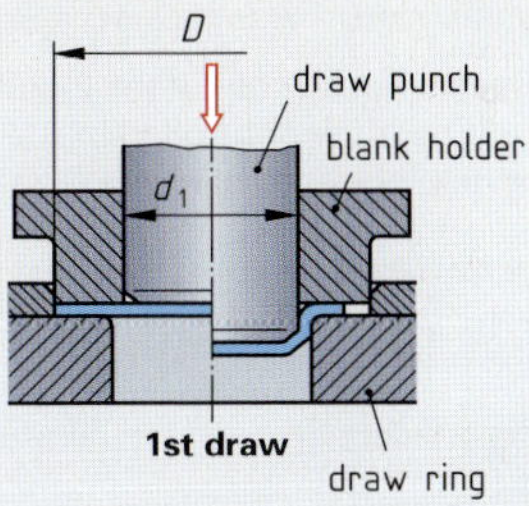

1st draw

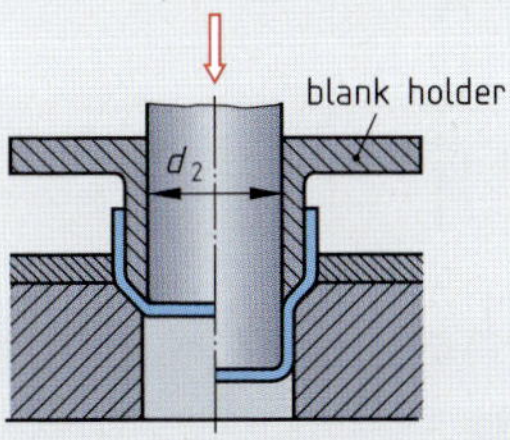

Redraw

D	blank diameter
d	inside diameter of finished drawn part
d_1	punch diameter for 1st draw
d_2	punch diameter for 2nd draw
d_n	punch diameter for nth draw
β_1	drawing ratio for 1st draw
β_2	drawing ratio for 2nd draw
β_{tot}	total drawing ratio
s	sheet metal thickness

Example:

Cup without flange made of DC04 (St 14) with d = 50 mm; h = 60 mm; D = ?; β_1 = ?; β_2 = ?; d_1 = ?; d_2 = ?

$$\mathbf{D} = \sqrt{d^2 + 4 \cdot d \cdot h}$$

$$= \sqrt{(50\ \text{mm})^2 + 4 \cdot 50\ \text{mm} \cdot 60\ \text{mm}} \approx \mathbf{120\ mm}$$

$\boldsymbol{\beta_1} = \mathbf{2.0}$; $\boldsymbol{\beta_2} = \mathbf{1.3}$ (according to table below)

$$\mathbf{d_1} = \frac{D}{\beta_1} = \frac{120\ \text{mm}}{2.0} = \mathbf{60\ mm}$$

$$\mathbf{d_2} = \frac{d_1}{\beta_2} = \frac{60\ \text{mm}}{1.3} = \mathbf{46\ mm}$$

Two draws sufficient since $d_2 < d$

Drawing ratio

1st draw

$$\beta_1 = \frac{D}{d_1}$$

2nd draw

$$\beta_2 = \frac{d_1}{d_2}$$

Total drawing ratio

$$\beta_{tot} = \beta_1 \cdot \beta_2 \cdot \ldots$$

$$\beta_{tot} = \frac{D}{d_n}$$

Material	Max. drawing ratios[1]		R_m[2]	Material	Max. drawing ratios[1]		R_m[2]	Material	Max. drawing ratios[1]		R_m[2]
	β_1	β_2	N/mm²		β_1	β_2	N/mm²		β_1	β_2	N/mm²
DCO1 (St12)	1.8	1.2	410	CuZn30-R270	2.1	1.3	270	Al99.5 H111	2.1	1.6	95
DCO3 (St13)	1.9	1.3	370	CuZn37-R300	2.1	1.4	300	AlMg1 H111	1.9	1.3	145
DCO4 (St14)	2.0	1.3	350	CuZn37-R410	1.9	1.2	410	AlCu4Mg1 T4	2.0	1.5	425
X10CrNi18-8	1.8	1.2	750	CuSn6-R350	1.5	1.2	350	AlSi1MgMn T6	2.1	1.4	310

[1] Values apply up to $d_1 : s$ = 300; they were determined for d_1 = 100 mm and s = 1 mm. Values change negligibly for other sheet metal thicknesses and punch diameters.
[2] maximum tensile strength

Tearing force, deep drawing force, blank holding force

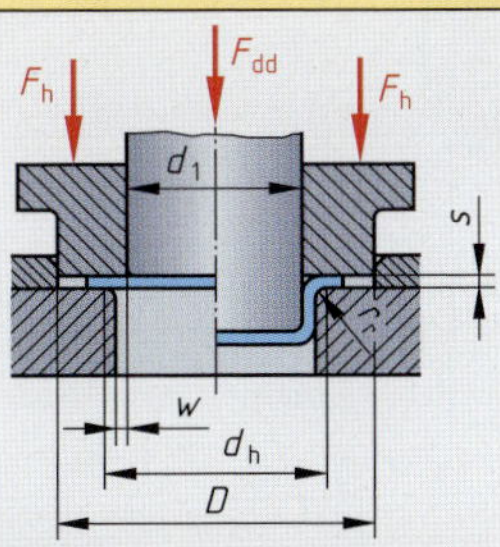

Blank holding pressure p in N/mm²	
Steel	2.5
Cu alloys	2.0–2.4
Al alloys	1.2–1.5

F_t	tearing force
F_{dd}	deep drawing force
d_1	punch diameter
s	sheet metal thickness
R_m	tensile strength
β	drawing ratio
β_{max}	max. possible drawing ratio
F_h	blank holding force
D	blank diameter
d_h	support diameter of blank holding force
p	blank holding pressure
r_r	radius on draw ring
w	drawing gap

Tearing force

$$F_t = \pi \cdot (d_1 + s) \cdot s \cdot R_m$$

Deep drawing force

$$F_{dd} = \pi \cdot (d_1 + s) \cdot s \cdot R_m \cdot 1.2 \cdot \frac{\beta - 1}{\beta_{max} - 1}$$

Blank holding force

$$F_h = \frac{\pi}{4} \cdot (D^2 - d_h^2) \cdot p$$

Support diameter of blank holding force

$$d_h = d_1 + 2 \cdot (r_r + w)$$

Example:

D = 210 mm; d_1 = 140 mm; s = 1 mm; R_m = 380 N/mm²; β = 1.5; β_{max} = 1.9; F_{dd} = ?

$$\mathbf{F_{dd}} = \pi \cdot (d_1 + s) \cdot s \cdot R_m \cdot 1.2 \cdot \frac{\beta - 1}{\beta_{max} - 1} = \pi \cdot (140\ \text{mm} + 1\ \text{mm}) \cdot 1\ \text{mm} \cdot 380 \frac{\text{N}}{\text{mm}^2} \cdot 1.2 \cdot \frac{1.5 - 1}{1.9 - 1} = \mathbf{112\,218\ N}$$

PE

Injection molding tool

Two-plate multi-cavity mold with tunnel gate

Injection molding process

Molten plastic compound is injected into the tool with high pressure via the sprue. The cavities of the tool fill with the molten material. The compound cools down under pressure in the temperature-controlled tool. As soon as it has sufficiently solidified to keep its shape, it is demolded with the help of ejector pins.

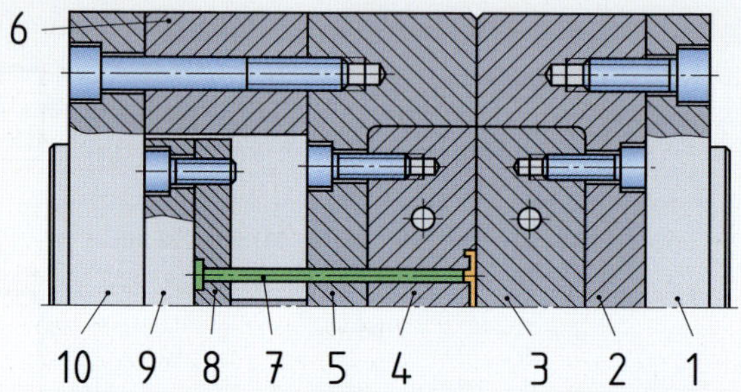

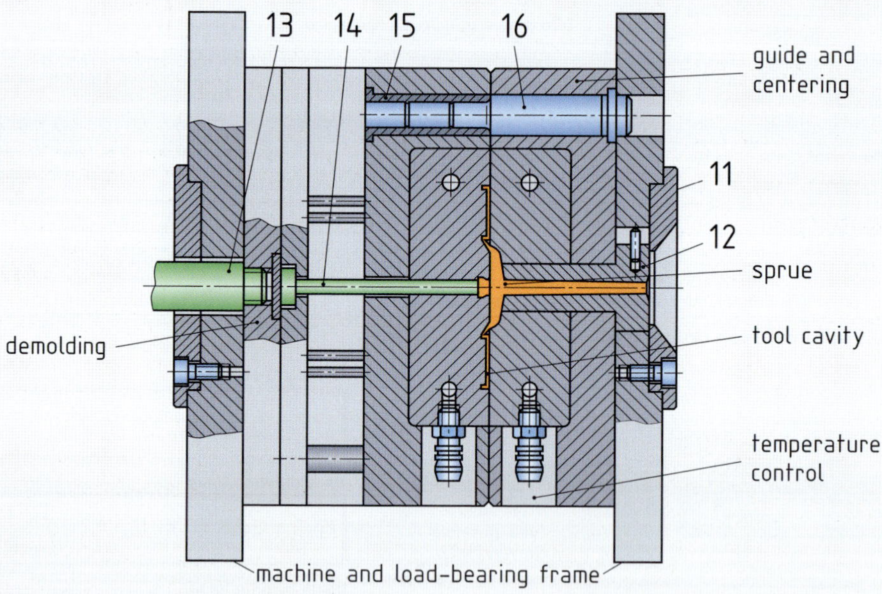

PE

Bill of materials

Item	Designation	Standard/material	Item	Designation	Standard/material
1	Clamping plate (stationary)	DIN 16760	9	Ejector base plate	DIN 16760
2	Mold plate (stationary)	DIN 16760	10	Clamping plate (movable)	DIN 16760
3	Mold insert	X19NiCrMo4	11	Locating ring	DIN ISO 10907
4	Mold insert	X19NiCrMo4	12	Sprue bushing	DIN ISO 10072
5	Mold plate (movable)	DIN 16760	13	Ejector bar	S235JR
6	Rail	DIN 16760	14	Sprue ejector pin	DIN 1530
7	Ejector pin	DIN 1530	15	Guide bushing	DIN 16761
8	Ejector retaining plate	DIN 16760	16	Guide post	DIN 16761

Standard parts for injection molding tools

Standard parts for injection molds (selection)

Illustration	Dimension in mm	Standard/material	Properties/function
Machined plates drilled or undrilled			
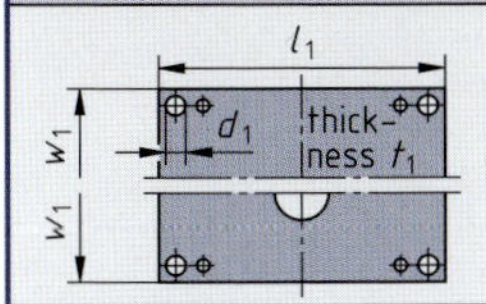	$w_1 \times l_1 \times t_1$ w_1 = 96–896 l_1 = 96–1116 t_1 = 12.5–200	DIN 16760 C45U 40CrMnMoS8-6	Plates to support the machine and bear the loads • clamping plates • mold plates • platens
Bars, rails			
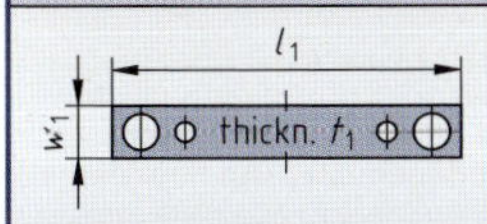	$w_1 \times l_1 \times t_1$ l_1 = 96–1116 w_1 = 26–74 t_1 = 25–160	DIN 16760 C45U	Rails for load bearing and spacing • in different shapes
Guide posts			
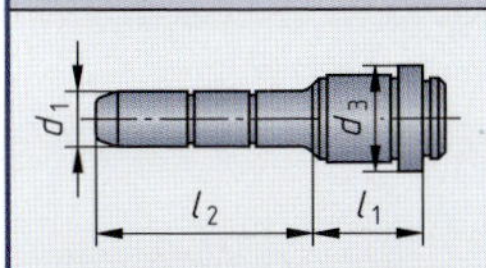	$d_1 \times l_1 \times l_2$ d_1 = 10–40 l_1 = 12.5–200 l_2 = 25–250	DIN 16761 Case-hardened steel (780+40) HV 10	Guidance and centering Guide posts with undercut shaft • form A with centering shoulder • form B without centering shoulder
Guide bushings			
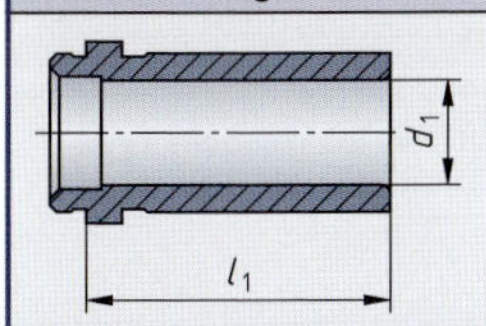	$d_1 \times l_1$ d_1 = 10–40 l_1 = 12.5–200	DIN 16761 Case-hardened steel (780+40) HV 10	Guidance and centering Guide posts with undercut shaft • form A with centering shoulder • form B without centering shoulder
Ejector pins			
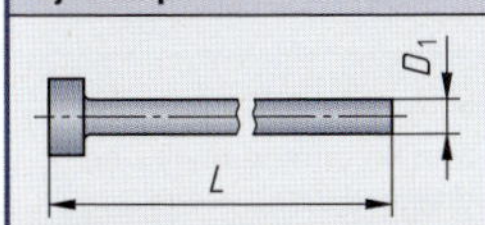	$D_1 \times L$ D_1 = 2–32 L = 100–1000	DIN 1530 Hot-worked steel 950 HV 0.3	Demolding system Ejector pins • with cylindrical head
Ejector sleeves			
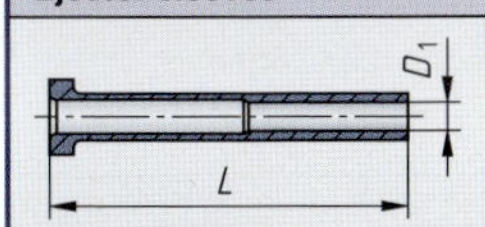	$D_1 \times L$ D_1 = 2–12 L = 75–300	DIN ISO 8405 Hot-worked steel 950 HV 0.3	Demolding system Ejector pins • with cylindrical head
Sprue bushings			
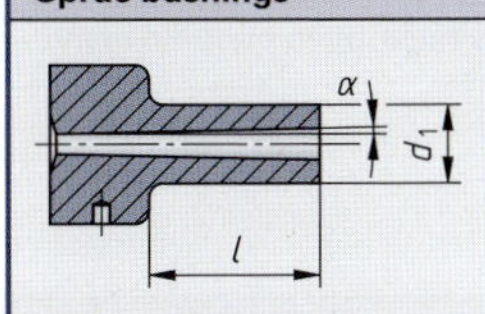	$d_1 \times l$ d_1 = 12–25 l = 20–100	DIN ISO 10072 Tool steel (50±5) HRC	Sprue system Sprue bushing • form A, with radius for machine nozzle • form B, straight for machine nozzle
Sprue retainer bushing			
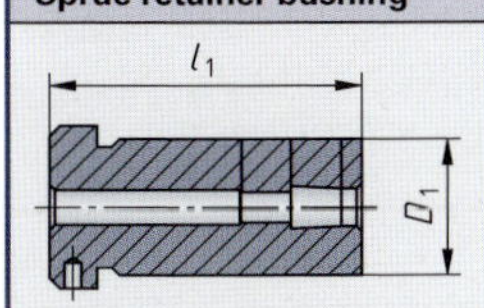	$D_1 \times l_1$ D_1 = 12–25 l_1 = 20–100	DIN ISO 16915 Tool steel 50 HRC	Sprue system Sprue retainer bushing

PE

Tool design

Hydraulic horizontal injection molding machine

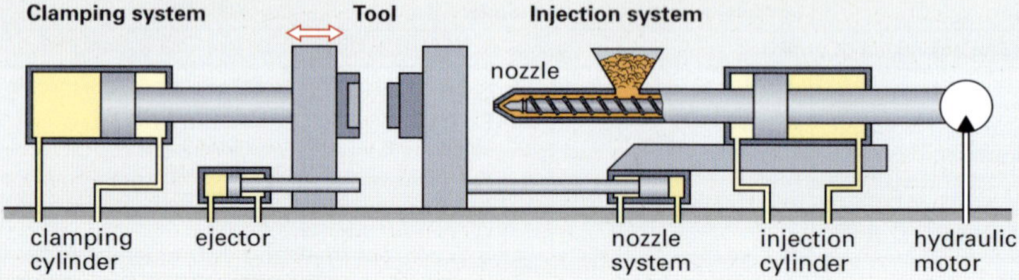

Types of tools

A distinction is made between single-cavity and multi-cavity tools, depending on the number of molding cavities. Multi-cavity tools have several cavities that are arranged in a symmetrical layout (molding surface) in most cases. The arrangement allows the molten material to fill all cavities simultaneously and homogeneously. The flow paths have the same length

Two-plate mold	Function and application
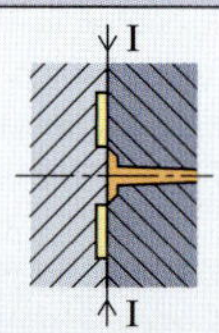	• Standard mold with a very simple design with two tool halves • A single parting surface (I) • Application: simple molded parts of all kinds • Special types: stripping plate mold, split mold and undercut mold
Three-plate mold	**Function und application**
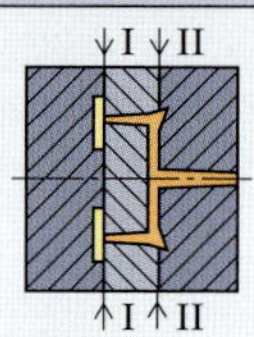	• Design similar to the standard mold with an additional intermediate plate which allows a separate demolding of the sprue, normally a pin gate. Tear-off mold. • Two parting surfaces (I, II) • Application: molded parts of all kinds; many cavities on a single manifold system, much waste
Stack mold	**Function and application**
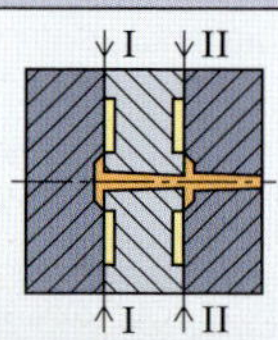	• The molded parts are arranged in several layers and lie immediately one behind the other. As the molded parts have identical projected areas, only the clamping force for one parting surface is required. • Two or more parting surfaces (I, II, etc.) • Application: flat molded parts of all kinds in large numbers, often as hot-runner versions
Insulated runner mold	**Function and application**
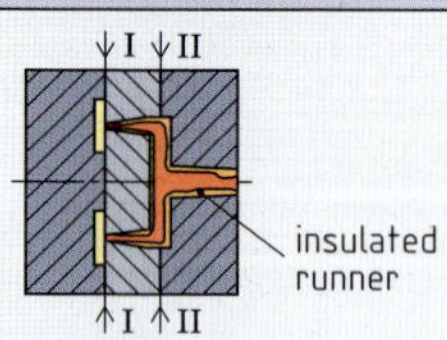	• The design corresponds to that of a three-plate mold. The intermediate plate is fitted with an insulated runner that keeps the molten material in the liquid state throughout the entire molding process • Two parting surfaces (I, II) • Application: compounds with a wide melting temperature range and a fast cycle sequence
Hot runner mold	**Function and application**
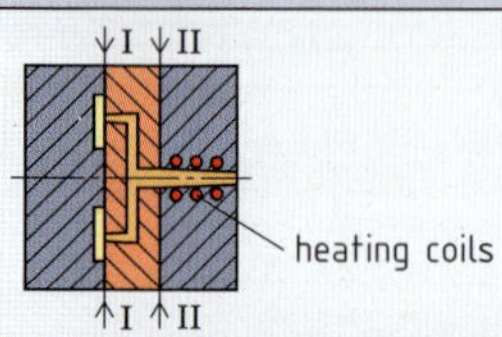	• Tools with heated nozzles and/or manifolds • One or two parting surfaces (I, II) • Application: technically sophisticated molded parts, also suitable for compounds that are difficult to process

PE

Shrinkage and Cooling

Shrinkage

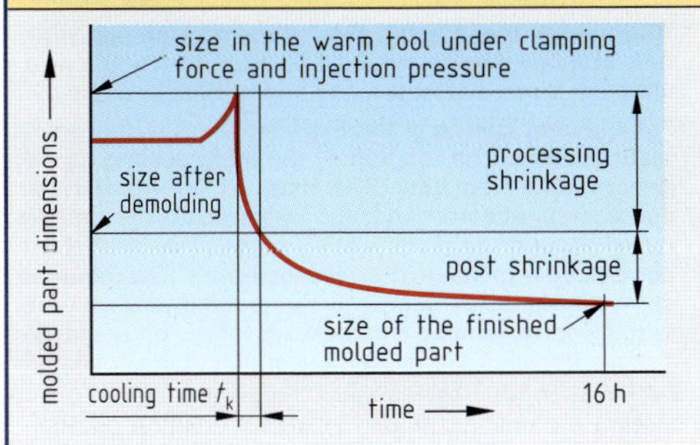

Shrinkage is the change in volume caused by cooling, i. e. the crystallization of the plastic material. A distinction is made between processing shrinkage and post shrinkage.

Shrinkage causes a difference in size between the tool cavity and the produced plastic part.

If thermoplastics are processed, the finished part must be measured 16 hours after its production. If thermoset plastics are used, the finished part must be checked between 24 and 168 hours after its production. Warm storage causes a change in structure in crystalline plastics, which also produces post shrinkage.

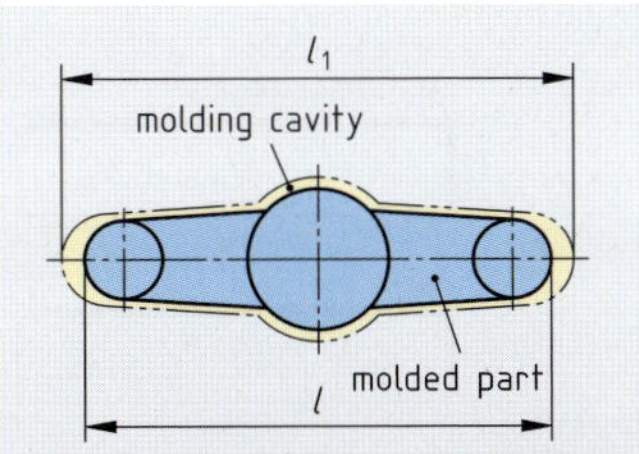

PS	processing shrinkage in %
AS	post shrinkage in %
S	total shrinkage in %
l	molded part dimension in mm
l_1	tool dimension in mm
l_{PS}	molded part dimension after processing in mm
l_{xh}	molded part dimension after x hours in mm

Processing shrinkage

$$PS = \frac{l_1 - l_{PS}}{l_1} \cdot 100\,\%$$

Size in tool

$$l_1 = \frac{l \cdot 100\,\%}{100\,\% - S}$$

Post shrinkage caused by warm demolding

$$AS = \frac{l_{PS} - l_{xh}}{l_1} \cdot 100\,\%$$

Examples for shrinkage due to cooling (cf. page 186)

Plastic material	Shrinkage in %	Plastic material	Shrinkage in %
Polyamide	1.3	Polypropylene	1.5
Polystyrene	0.45	PVC	0.6
Polyethylene	1.7	Polycarbonate	0.8

Cooling

Molding cycle

Close tool	Inject	Cooling			Open tool	Ejector forward/backward
		Packing	Batching	Holding		
1 s	2 s	7 s	2 s	12 s	0.8 s	1.4 s

Cycle time = 26.2 s

After the injection, the pressure is reduced to 30 to 70% of the injection pressure, depending on the plastic material, and applied until the gate solidifies. This happens after approximately 1/3 of the cooling time. The remaining cooling time is required to ensure a sufficient solidification of the material to keep its shape. During this solidification period, the batch volume for the next injection is prepared. The assessment of the required cooling time is based on diagrams. Also, an estimate calculation is sufficiently accurate.

PE

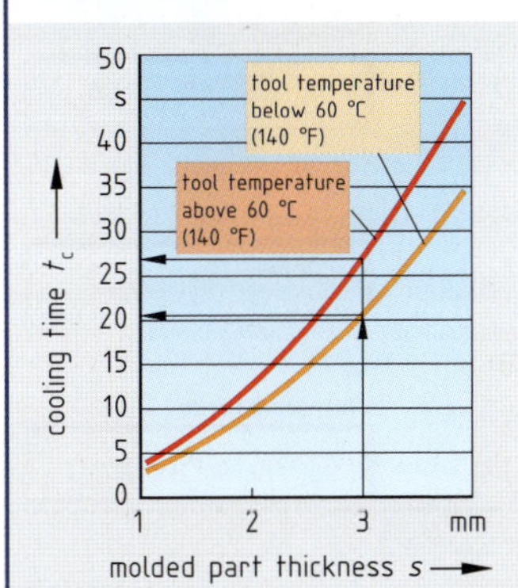

s	molded part thickness in mm
t_c	cooling time in sec.
t_p	post pressure time in sec.
t_{rc}	remaining cooling time in sec.
t_i	injection time in sec.

Example:

The cooling time of a thermoplastic part with a thickness of 3 mm and a tool temperature of 50 °C (122 °F) is to be calculated.

$t_c = s \cdot (1 + 2 \cdot s)$

$t_c = 3 \cdot (1 + 2 \cdot 3) =$ **21 sec.**

Total cooling time

$$t_c = t_p + t_{rc}$$

Estimate calculation of the cooling time:

- Tool temperature ≤ 60° (140 °F)

$$t_c = s \cdot (1 + 2 \cdot s)$$

- Tool temperature > 60 °C (140 °F)

$$t_c = 1.3 \cdot s\,(1 + 2 \cdot s)$$

Batching, Forces

Batching

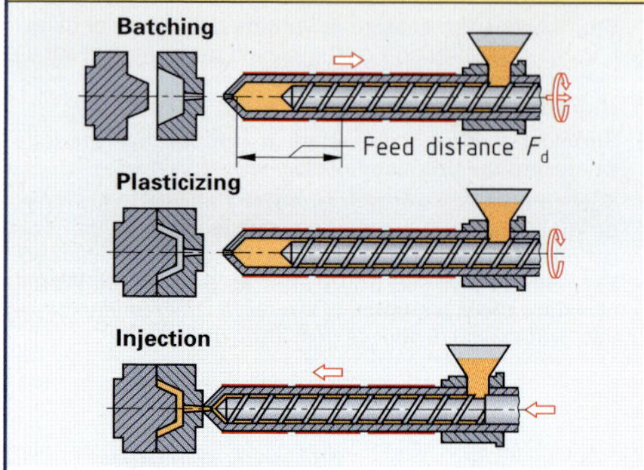

After injection and application of the packing pressure, new molding compound must be prepared for the next shot and made available in the appropriate amount.

For the **shot batching,** the extruder is reset and granulate filled in. The rotation of the reciprocating screw generates a **feed flow**. The material is warmed up, compacted, sheared and homogenized in the heated barrel around the reciprocating screw. The **plasticized compound** is driven to the injection nozzle. The required **shot volume** must comprise the **molded part volume** and the **sprue volume** and make up for the difference in density between the molten material and the solidified part (e.g. factor 1.25).

During the holding phase, a **packing cushion** (holding cushion) is required to compensate for shrinkage and prevent sink marks on the molded part.

- V_i injection volume in cm³
- V_{mp} molded part volume in cm³
- V_{sp} sprue volume in cm³
- V_b batch volume in cm³
- V_c packing cushion in cm³
- F_d feed distance in mm
- n number of cavities
- m_i injection mass in g
- m_b batch mass in g
- ϱ density in g/cm³
- Q_i injection flow rate in cm³/s
- Q_b batch flow rate in cm³/mm
- t_i injection time in sec.

Injection volume

$$V_i = n \cdot V_{mp} + V_{sp}$$

Batch volume

$$V_b = 1.25 \cdot V_i + V_c$$

Feed distance

$$F_d = V_b / Q_b$$

Injection time

$$t_i = V_i / Q_i$$

Injection mass

$$m_i = \varrho \cdot V_i$$

Examples of setting values and the maximum flow path

Code letters	Temperature in °C		Injection pressure in bar	Flow path length[1] in mm
	Mass	Tool		
PE	160–300	20–70	500	200–600
PP	170–300	20–100	1200	250–700
PVC	170–210	20–60	300[2], 1500[3]	250[3], 500[2]
PS	180–250	30–60	1000	400–500
PA	210–290	80–120	700–1200	200–500
ABS	200–240	40–85	800–1800	300

[1] maximum flow path length at a wall thickness of 2 mm [2] soft PVC [3] hard PVC

Forces

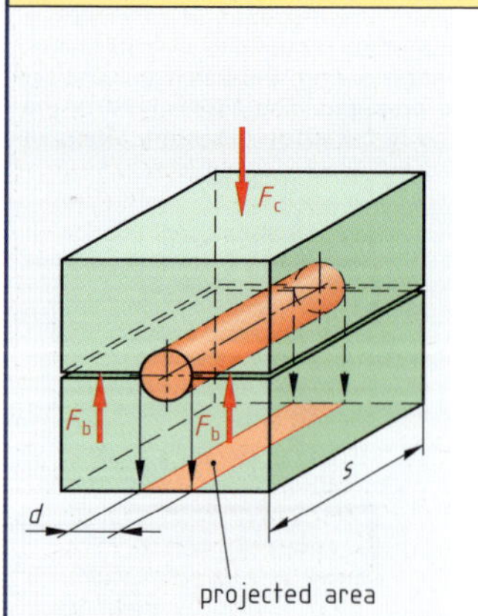

The injection pressure drives the compound into the tool. The pressure force is applied in the cavity and sprue perpendicular to the respective projected area (parting surface) and produces the buoyant force F_b. A clamping force F_c that is higher than the buoyant force must apply to prevent leakage of the compound. This force is applied mechanically, hydraulically or electromechanically.

PE

- n number of cavities
- A_p projected area in cm³
- A_{pc} projected area of a cavity in cm²
- A_{ps} projected area of the sprue in cm²
- p_t inner tool pressure in bar
- F_b buoyant force in kN
- F_c clamping force in kN
- φ safety factor

Projected area

$$A_p = n \cdot A_{pc} + A_{ps}$$

Buoyant force

$$F_b = p_t \cdot A_p$$

Clamping force

$$F_c \geq F_b \cdot \varphi$$

Example:

Polypropylene (PP), injection pressure p_i = 1200 bar, 1 bar = 10 N/cm²
Projected area A_p = 12.3 cm², φ = 1.3

$F_b = p_t\, A_p = 1200 \cdot 10\ \text{N/cm}^2 \cdot 12.3\ \text{cm}^2$

F_b = 147 600 N = **147.6 kN**

$F_c \geq F_b\, \varphi$ = 147.6 kN · 1.3 ≥ **191.9 kN**

Fusion welding, Welding processes (Overview)

Fusion welding is the permanent joining of material by means of locally limited fusion with the application of
- heat and/or
- metal fillers, such as electrodes, welding wire, for the filling of weld joints
- auxiliary materials, such as shielding gas in order to improve the welding conditions and the weld properties.

Welding processes (selection)

Illustration	Description	Process, application
Gas welding		(page 347)
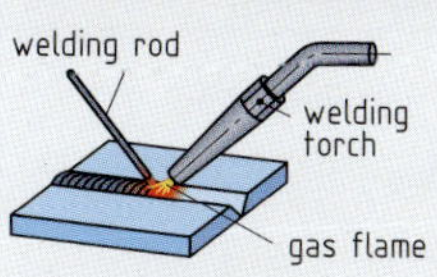	**Heat source** Gas flame of oxygen and acetylene Preferred flame adjustment: neutral flame with an equal portion of oxygen and acetylene **Metal fillers** Bare welding rods (page 347)	Manual welding; fusion welding of unalloyed and low-alloy steel tubes, repair welding on cast iron
Manual arc welding		(page 348)
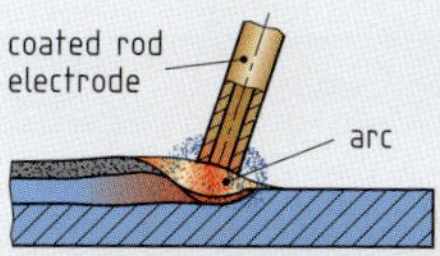	**Heat source** AC or DC arc between the consumable electrode and the base metal Temperatures in the fusion area: up to 4000 °C (7232 °F) **Metal fillers** Coated bar electrodes (page 348)	Manual welding; fusion welding of alloyed and unalloyed steels, fixed-position welding
Gas-shielded welding		
	Gas-shielded metal welding (MIG-, MAG-welding)	(page 346)
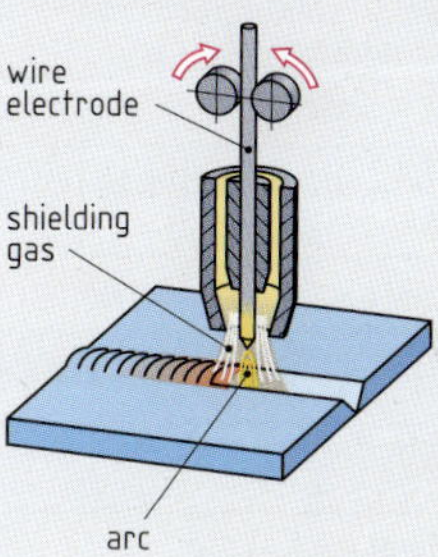	**Heat source** DC or AC arc between the consumable electrode and the workpiece **Metal fillers** Wire electrodes (page 346), tubular cored electrodes **Auxiliary material** Shielding gases are selected in accordance with the base metals to be welded and determine the welding method: Use of inert gases, such as argon or helium → **metal inert gas (MIG) welding,** no chemical reaction of the gas with the base metals Use of active shielding gases such as gases with a CO_2 or O_2 portion → **metal active gas (MAG) welding,** gases have an oxidizing effect on the base metals	Manual or automated welding, with welding robots for instance; high welding quality, high deposition efficiency **MIG-welding:** Aluminum and aluminum alloys, copper and copper alloys, nickel and nickel alloys **MAG-welding:** Unalloyed and alloyed steels, also stainless Cr-Ni steels
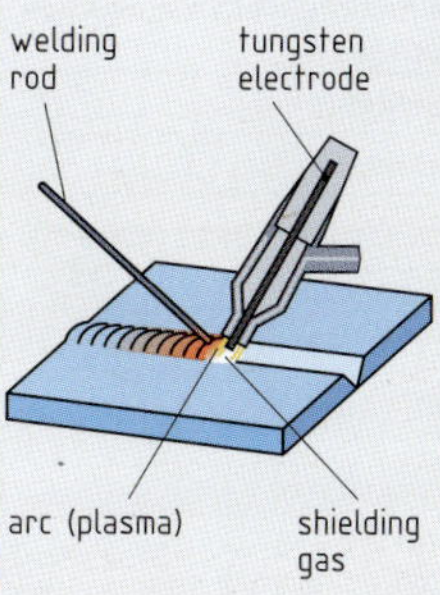	**Tungsten inert gas welding (TIG-welding)**	
	Heat source DC or AC arc between a non-consumable tungsten electrode and the base metal **Metal fillers** Welding rods are applied manually under normal conditions **Auxiliary material** Inert shielding gases, such as argon or helium → no chemical reaction of the gas with the base metals	Welding process for high-quality joints, particularly in combination with thin sheet metals; suitable for almost all materials, such as alloyed and unalloyed steels, also stainless Cr-Ni steels, non-ferrous metals, titanium, tantalum and zirconium
	Plasma arc welding (PAW)	
	Heat source → see TIG-welding, bundled arc (plasma) with high power **Metal fillers** → see TIG-welding **Auxiliary material** Two kinds of gas are needed for plasma welding: – plasma gas: mainly argon – shielding gas: argon with material-dependent components such as hydrogen, helium etc.	Welding process for high-quality joints with a thickness between 0.01 and 10 mm, build-up welding, e. g. on valve seats; metals as with TIG-welding

Welding processes, Positions, General tolerances

Welding, cutting, soldering and related processes

cf. DIN EN ISO 4063 (2000-04)

N[1]	Method, process	N[1]	Method, process	N[1]	Method, process
1	**Arc welding**	24 25	flash butt welding upset welding	**7**	**Other welding methods**
101 111	metal arc welding shielded metal arc welding	**3**	**Gas welding**	73 74	electrogas welding induction welding
11	metal arc welding without shielding gas	311	oxyacetylene welding	75 753	light beam welding infrared welding
12 13	submerged arc welding gas shielded metal arc welding	312	gas welding with oxygen/ propane flame	78 788	stud welding friction stud welding
131 135	metal inert gas welding (MIG) metal active gas welding (MAG)	**4**	**Pressure welding**	**8**	**Cutting**
136	flux cored arc welding with active gas shield	41 42	ultrasonic welding friction welding	81 82	oxygen cutting arc cutting
137	flux cored arc welding with inert gas shield	45 47	diffusion welding pressure gas welding	83 84	plasma cutting laser beam cutting
14 141	tungsten gas shield. arc welding tungsten intert gas welding (TIG)	**5**	**Beam welding**	**9**	**Brazing, soldering**
15 151	plasma arc welding plasma TIG welding	51 52	electron beam welding laser beam welding	91 912	brazing torch brazing
2	**Resistance welding**	512	electron beam welding, nonvacuum	914 924	metal bath brazing vacuum brazing
21 22	resistance spot welding seam welding	521	solid-state laser beam in atmosphere	94 944	soldering metal bath soldering
225 23	foil butt seam welding projection welding	522	gas laser beam welding	946 952	induction soldering iron soldering

⇒ **Process ISO 4063–111**: Specified welding process ⇒ manual arc welding (111)

[1] N Reference number for designating methods and processes in drawings, operating procedures and data processing

Welding positions

cf. DIN EN ISO 6947 (1997-05)

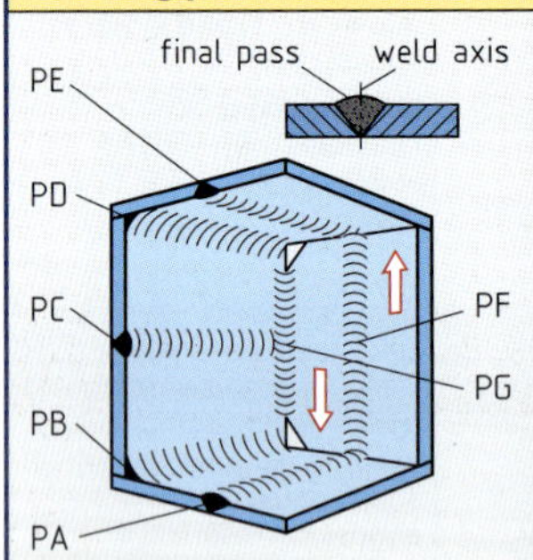

PE

Code	Name	Main position, description
PA	flat welding position	weld axis vertical, horizontal work, final pass at top
PB	horizontal position	horizontal work, final pass at top
PC	transverse position	weld axis horizontal, horizontal work direction
PD	horizontal overhead position	horizontal work direction, overhead, final pass at bottom
PE	overhead position	horizontal work direction, weld axis vertical, final pass at bottom
PF	vertical up position	upward work direction
PG	vertic. down position	downward work direction

General tolerances for weldments

cf. DIN EN ISO 13920 (1996-11)

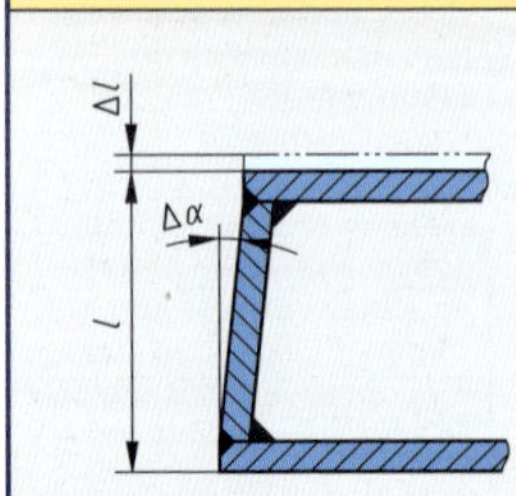

[1] l shorter leg

Degree of accuracy	Allowable deviations for length dimensions Δl in mm, nominal size range l[1]						for angle dimensions $\Delta\alpha$ in ° and ′, nominal size range l[1]		
	to 30	over 30 to 120	over 120 to 400	over 400 to 1000	over 1000 to 2000	over 2000 to 4000	to 400	over 400 to 1000	over 1000
A	±1	±1	±1	±2	± 3	± 4	±20′	±15′	±10′
B	±1	±2	±2	±3	± 4	± 6	±45′	±30′	±20′
C	±1	±3	±4	±6	± 8	±11	±1°	±45′	±30′
D	±1	±4	±7	±9	±12	±16	±1°30′	±1°15′	±1°

Weld preparation, Compressed gas cylinders

Weld for gas, gas-shielded and arc welding

cf. DIN EN ISO 9692-1 (2004-05)

Shape of the joint	Name, weld symbol	Thickness t mm	D[1]	Weld preparation: gap b mm	Weld preparation: web c mm	Weld preparation: angle α	Recommended welding method[2]	Remarks
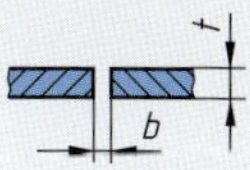	**Butt weld** ‖	0–4	s	≈ t	–	–	3, 111, 141	Little filler material, no weld preparation
		0–8	d	≈ $t/2$	–	–	111, 141	
		0–8	d	≈ $t/2$	–	–	13	
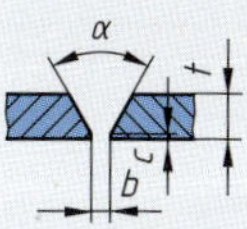	**V-groove weld** V	3–10	s	≤ 4	≤ 2	40°–60°	3	–
		3–40	d	≤ 3	≤ 2	≈ 60°	111, 141	With backing run
		3–40	d	≤ 3	≤ 2	40°–60°	13	With backing run
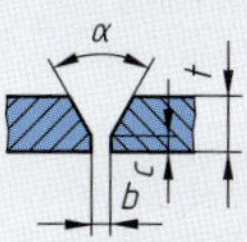	**Y-butt weld** Y	5–40	s	1–4	2–4	≈ 60°	111, 13, 141	–
		> 10	d	1–3	2–4	≈ 60°	111, 141	With root and backing run
		> 10	d	1–3	2–4	40°–60°	13	With root and backing run
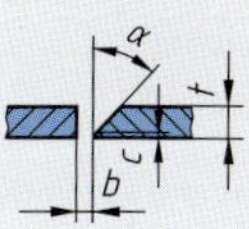	**Bevel groove weld** V	3–10	s	2–4	1–2	35°–60°	111, 13, 141	–
		3–30	d	1–4	≤ 2	35°–60°	111, 141	With backing run
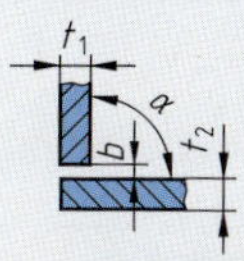	**Fillet weld** ◺	> 2	s	≤ 2	–	70°–100°	3, 111, 13, 141	T-joint

[1] D Design of the weld: s single-V weld; d double-V weld
[2] For welding methods, see page 344

Compressed gas cylinders

cf. DIN EN 1089-3 (2004-06)

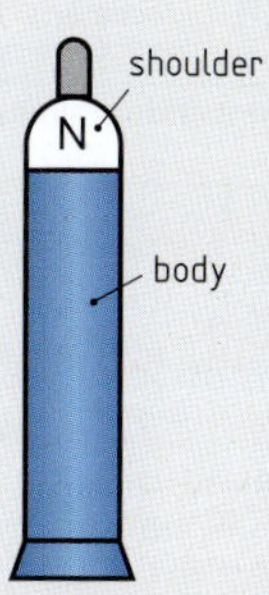

Type of gas	Color coding[1] as per DIN EN 1089-3: body	Color coding[1] as per DIN EN 1089-3: shoulder	Color coding[1]: previous	Connection threads	Volume V l	Filling pressure p_F bar	Filling quantity
Oxygen	blue	white	blue	R3/4	40 50	150 200	6 m³ 10 m³
Acetylene	chestnut-brown	chestnut-brown	yellow	Quick connect	40 50	19 19	8 kg 10 kg
Hydrogen	red	red	red	W21.80x1/14	10 50	200 200	2 m³ 10 m³
Argon	gray	dark-green	gray	W21.80x1/14	10 50	200 200	2 m³ 10 m³
Helium	gray	brown	gray	W21.80x1/14	10 50	200 200	2 m³ 10 m³
Argon-carbon dioxide mixture	gray	fluorescent green	gray	W21.80x1/14	20 50	200 200	4 m³ 10 m³
Carbon dioxide	gray	gray	gray	W21.80x1/14	10 50	58 58	7.5 kg 20 kg
Nitrogen	gray	black	dark-green	W24.32x1/14	40 50	150 200	6 m³ 10 m³

[1] Changeover to the new color coding should be completed by July 1, 2006. During the transition period the hazardous substance label (page 352) is the only legally valid designation.

PE

Gas-shielded welding

The quality of gas-shielded welds and the welding conditions are influenced by the selection of the wire electrode, the shielding gas and the arc adjustment.

Gas-shielded welding with consumable wire electrode

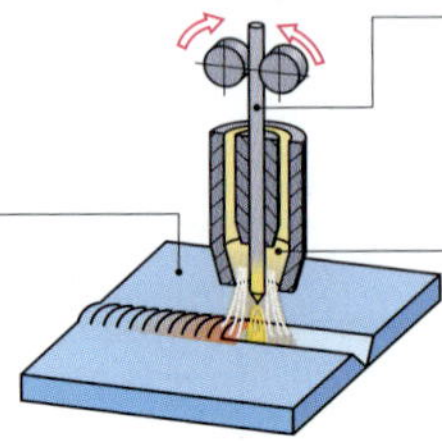

Base metal

Decisive for the welding method

steel → **MAG-welding**
aluminium → **MIG-welding**

Wire electrode

Effects, e. g.
- strength
- welding position
- current type
- toughness
- shielding gas type

Shielding gas

Determines, e. g.
- weld properties
- heat introduction
- arc adjustment
- weld penetration behavior
- pore formation

Arc types, arc adjustment[1)]

Arc type	Application, e. g.	Arc type	Application, e. g.
Short arc	Fixed-position and root welding, low welding power, thin sheet metal welding, MIG- and MAG-welding	Spray arc	High deposition efficiency, larger sheet metal thicknesses, higher welding speed, MIG- and MAG-welding
Transfer arc	Medium welding power, medium sheet metal thicknesses, MAG-welding	Pulsed arc	For all power ranges of MIG- and MAG-welding

1) The adjustment of the arc depends on
- the sheet metal thickness
- the shielding gas type
- the welding position
- the welding performance

MAG-welding of unalloyed and low-alloy steels

Designation of wire electrodes and weld metal for MAG-welding of unalloyed steels and fine-grain structural steel cf. DIN EN ISO 14341 (2008-08) replaces DIN EN 440

The strength and toughness of the weld metal are influenced by the electrode material and shielding gas. The designation of the weld metal therefore gives information about the strength, toughness, shielding gas and electrode material. Classification in two groups:
- Weld metal with a guaranteed **yield strength R_e** and a **notch impact energy of 47 J** (ISO 14341-A)
- Weld metal with a guaranteed **tensile strength R_m** and a **notch impact energy of 27 J** (ISO 14341-B)

Designation example (weld metal with guaranteed yield strength and notch impact energy of 47 J):

ISO 14341 – A – G 46 5 M G3Si

- Standard number: ISO 14341
- Code letter for the weld metal with guaranteed yield strength and notch impact energy of 47 J: A
- Code letter for gas-shielded metal-arc welding: G

Code digits for the minimum yield strength R_e	Code digit for the limit temperature of the notch impact energy	Code letter for the shielding gas[1)]	Wire electrode material G3Si
46 → R_e = 10 · 46 = 460 N/mm² 42 → R_e = 10 · 42 = 420 N/mm²	5 → notch impact energy of 47 J guaranteed down to −50 °C	M → shielding gas M21, M22 … A → shielding gas M13 C → shielding gas C1	Si → containing silicon

1) The code letters M, A and C refer to the shielding gases that ensure compliance with the temperature limits specified for the notch impact energy of 47 J.

Wire electrodes for unalloyed steels and fine-grain structural steels (selection) cf. DIN EN ISO 14341 (2008-08)

Weld metal/ wire electrode	Minimum yield strength R_e in N/mm²	t[1)] in °C	Shielding gas[2)]	Arc	Suitable for steels	Properties, application
G 42 3 C G3Si1	420	−30	C1	Short arc	S235–S355	Low-spatter material transfer with short and spray arcs, versatile range of application in manufacturing and repair
G 42 4 M G3Si1		−40	M21	Short arc, spray arc		
G 46 3 C G4Si1	460	−30	C1	Short arc	S235–S460	
G 46 4 M G4Si1		−40	M21	Short arc, spray arc		

1) Temperature limit for the notch impact resistance of 47 J
2) Shielding gases that ensure compliance with the required mechanical properties, types on page 347.

PE

Shielding gases for MAG-, MIG- and gas welding

Shielding gases for MAG-welding of unalloyed steels (selection)

cf. DIN EN ISO 14175 (2008-06)

Type[1], Code name	Component portions in % oxidizing (active) CO_2	O_2	inert Ar	Properties, application
M12 M13 M14	> 0.5–5 – > 0.5–5	– > 0.5–3 > 0.5–3	remain. remain. remain.	Low active portion (CO_2 and O_2), low slag and spatter formation, suitable for all arc types, sensitive to rust, cinder and contaminated sheets, preferred for bare, thin sheets.
M21 M22	> 15–25 –	– > 3–10	remain. remain.	Increased active part (CO_2 and O_2), higher slug and spatter formation, insensitive to rust, cinder and contaminated sheet surfaces, suitable for all arc types, higher sheet thicknesses **M22:** lower spatter, with pulsed arcs, CO_2-portion < 20%
M23 M24	> 0.5–5 > 5–15	> 3–10 > 0.5–3	remain. remain.	
C1 C2	100 remain.	– > 0.5–30	– –	High active portion (CO_2 and O_2), high slag and spatter formation, insensitive to rust, cinder and contamination, unsuitable for pulsed arcs, for higher sheet thicknesses with rust and scales.

[1] Gas suppliers often distribute a shielding gas type in several mixtures with company-specific names that are matched to the individual welding methods.

MIG-welding of aluminum and aluminum alloys

Wire electrodes for MIG-welding of aluminum and aluminum alloys

cf. DIN EN ISO 18273 (2004-05)

Designation example:

ISO 18273 – S Al 4047 (Al Si12)

- ISO 18273: Standard number
- S: Code letter for the wire electrode
- Al 4047: Numerical code
- (Al Si12): Chemical composition of the wire electrode: Al = main alloying element; Si12 = silicon portion, approx. 12%

Wire electrodes for MIG-welding of aluminum and Al-alloys (selection)

Code	Minimum yield strength R_e in N/mm²	Shielding gas	Arc (page 346)	Suitable for base metals (examples)
S Al 1450 (Al 99,5Ti)	20	I1	Short arc, spray arc, pulsed arc	Al 99, Al 99,5, Al 99,8, Al Mg0,5
S Al 4043 (Al Si5)	40	I1		Al MgSi0,5, Al MgSi0,7, Al MgSi1
S Al 4047 (Al Si12)	60	I1		G-Al Si11, G-Al Si10Mg(Cu), G-Al Si12(Cu)
S Al 5754 (Al Mg3)	80	I1		G-Al Mg3,5Si, Al Mg2,5, Al Mg3, Al Mg2Mn0,3
S Al 5356 (Al Mg5Cr(A))	110	I1		Al Mg3, Al Mg5, Al Zn4,5Mg1, Al Mg1SiCu

The strength properties are obtained when using shielding gas I1 (70% argon + 30% helium). The welding rods and wire electrodes could also be used in combination with shielding gas I3 (inert gas).

Gas welding rods for joint welding of steel

cf. DIN EN 12536 (2000-08)

Code	Minimum yield strength R_e in N/mm²	NI[1] N_i in J	Suitable for the steel types	Application area, welding behavior
O I	260	30	S235, S275	Metal sheets, tubes, pipes; low weld pool viscosity, many spatters, trend to pore formation
O II	300	47	S235, S275, P235GH, P265GH	Tanks, vessels, pipelines, higher weld pool viscosity as with O I, few spatters, trend to pore formation
O III	310	47	S235, S275, P235GH, P275GH	Tanks, vessels and pipelines, viscous weld pool, no splatters and no pores
O IV	260	47	S235, S275, S355, P235, P235GH, P265GH, P295GH, 16Mo3	Heat-resistant boilers and pipelines up to 450 °C; viscous weld pool, no splatters and no pores

[1] NI notch impact energy N_i at 20 °C, tested on an ISO-V specimen

Arc welding

Coated rod electrodes for unalloyed steels and fine grain steels
cf. DIN EN ISO 2560 (2010-03)

The electrode has a decisive influence on the mechanical properties of the welding joint and the welding behavior.

The **core wire** of the electrode has an influence on
- the chemical composition of the weld metal
- the strength and toughness

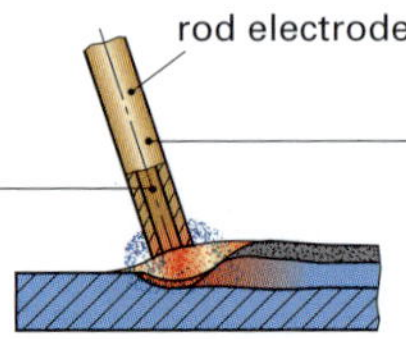

The **coating** has an influence on
- strength and toughness
- the ignition and welding behavior
- the look of the weld and the penetration behavior
- hot and cold cracking

Strength and toughness of the weld metal are the important parameters for the quality of the welding joint. The designation of the coated electrodes is based on these criteria and divided into two groups:
- Rod electrodes for weld metal with a guaranteed **yield strength** and a **notch impact energy of 47 J** (ISO 2560-A)
- Rod electrodes for weld metal with a guaranteed **tensile strength** and a **notch impact energy of 27 J** (ISO 2560-B)

Designation example: electrode for weld metal with a guaranteed yield strength and a notch impact energy of 47 J (**legally binding** specifications as per ISO 2560):

ISO 2560 – A – E 46 3 1Ni B

- ISO 2560: Standard number
- A: Code letters for weld metals with guaranteed yield strength and notch impact energy of 47 J
- E: Code letter for the coated rod electrode

Code digits for the minimum yield strength R_e	Code digit for the temperature limit of the notch impact energy	Code sings for the alloy:	Code letter for the rod coating type
46 → R_e = 10 · 46 = 460 N/mm² 42 → R_e = 10 · 42 = 420 N/mm² 38 → R_e = 10 · 38 = 380 N/mm²	3 → notch impact energy of 47 J guaranteed down to −30 °C	1Ni → approx. 0.9% nickel • if no alloy code is given, e. g. E 38 0 RC, the Mn portion is always 2%	

Coating types, properties and application (selection)

Coating type, code letters	Properties, application	Welding positions (page 344)
RA rutile acid coating	Mostly thick coated rod electrodes, flat and smooth welds, sensitive to solidification cracks	PA, PB, PC, PD, PE, PF
RB rutile basic coating	Mostly thick coated electrodes, good mechanical properties, good welding properties	
RC Rutile cellulose coating	Coarse drip transition, suitable for welding of thin sheets, also in vertical down positions	PA, PB, PC, PD, PE, PF, PG
RR thick rutile coating	Fine-scaled and even welds, good re-ignition of the arc, high deposition efficiency	PA, PB, PC, PD, PE, PF
B Basic coating	High toughness (notch impact energy) at low temperatures, well protected against cracking	PA, PB, PC, PD, PE, PF, PG

Coated rod electrodes for arc welding of unalloyed and low-alloy steels (selection)

Code name	Minimum yield strength R_e in N/mm²	t[1] in °C	Suitable for steels	Properties, application
E 38 0 RC	380	0	S235–S355	For fabrication and shop welding
E 42 0 RR	420	0	S235–S355	Excellent welding properties, proper welds
E 38 2 RB	380	−20	S235–S355	Suitable for root, filler and final passes
E 38 2 RA	380	−20	S235–S355	High deposition efficiency
E 46 6 1Ni	460	−60	S235–S460	Crack-free joint with high cold toughness

[1] Temperature limit for notch impact energy of 47 J

The manufacturers of electrodes offer various electrode types with the same code designation matched to special requirements

Weld design, Settings and performance values for welding

Weld design				Setting values				Performance values	
Weld seam type	Weld thickness *a* mm	Wire diameter mm	Number of passes	Voltage V	Current A	Wire feed rate[1] m/min	Schielding gas l/min	Filler material g/m	Main productive time min/m
Metal active gas (MAG) welding of unalloyed steels (standard values)									
Welding position: PA	Wire electrode: ISO 14341-A-G 46 4 M G3Si1						Shielding gas M21		
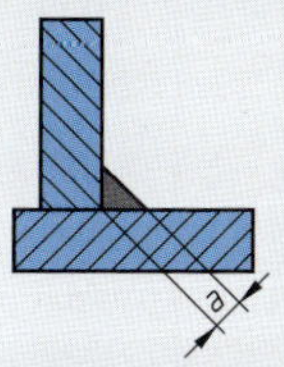	2 3 4	0.8 1.0 1.0	1	20 22 23	105 215 220	7 11 11	10	45 90 140	1.5 1.4 2.1
	5 6 7	1.0 1.0 1.2	1 1 3	30	300	10	15	215 300 390	2.6 3.5 4.6
	8 10	1.2	3 4	30	300	10	15	545 805	6.4 9.5
Metal inert gas (MIG) welding for aluminum alloys (standard values)									
Welding position: PA	Wire electrode: ISO 18273-A-S Al5754 (AlMg3)						Shielding gas: I1		
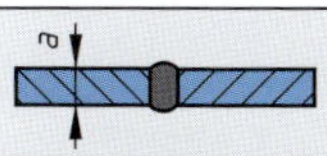	4 5 6	1.2 1.6 1.6	1	23 25 26	180 200 230	3 4 7	12 18 18	30 77 147	2.9 3.3 3.9
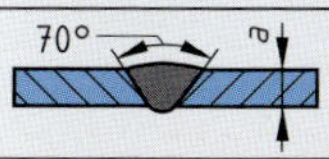	5 6 8	1.6	1 2 2	22 22 26	160 170 220	6 6 7	18	126 147 183	4.2 4.6 5.0

[1] For MIG-welding: welding travel speed

Arc welding, weld design of V welds

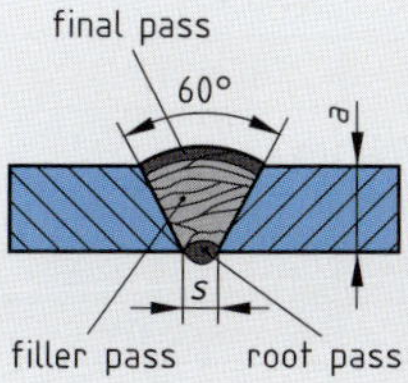

Weld thickness *a* mm	Gap *s* mm	Number and type of pass[1]	Electrode dimensions $d \times l$ mm	Spec. electrode consumption z_s piece/m	Weld weight per pass m_s g/m	Weld weight total *m* g/m
4	1	1 R 1 FP	3.2 x 450 4 x 450	3 2	75 80	155
5	1.5	1 R 1 FP	3.2 x 450 4 x 450	4 2.9	100 110	210
6	2	1 R 2 FP	3.2 x 450 4 x 450	4 4.7	100 185	285
8	2	1 R 1 F 1 FP	3.2 x 450 4 x 450 5 x 450	4 3.7 3.5	100 145 215	460
10	2	1 R 1 F 1 FP	3.2 x 450 4 x 450 5 x 450	4 4 6.2	100 195 380	675

Arc welding, weld design of fillet welds

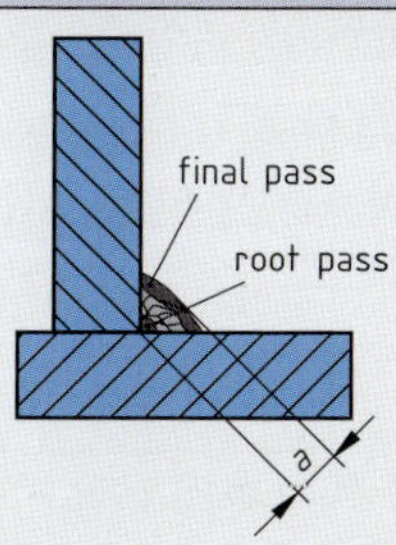

Weld thickness *a* mm	Gap *s* mm	Number and type of pass[1]	Electrode dimensions $d \times l$ mm	Spec. electrode consumption z_s piece/m	Weld weight per pass m_s g/m	Weld weight total *m* g/m
3 4	– –	1 1	3.2 x 450 4 x 450	3.2 3.6	80 140	80 140
5 6	– –	3 3	3.2 x 450 4 x 450	8.6 8	215 310	215 310
8	–	1 R 2 FP	4 x 450 5 x 450	3 7	120 430	550
10	–	1 R 4 FP	4 x 450 5 x 450	3 12.3	120 745	865
12	–	1 R 4 FP	4 x 450 5 x 450	3 18.5	120 1125	1245

[1] R root pass; F filler pass; FP final pass

PE

Standard values for beam cutting

Standard values for oxyacetylene cutting

Material: unalloyed structural steel; fuel gas: acetylene

Sheet met. thickn. s mm	Cutting nozzle mm	Width of cut mm	Oxygen pressure cutting bar	Oxygen pressure heating bar	Acetylene pressure bar	Total oxygen consumption m³/hr	Acetylene consumption m³/hr	Cutting rate quality cut m/min	Cutting rate standard cut m/min
5			2.0			1.67	0.27	0.69	0.84
8	3 – 10	1.5	2.5	2.0	0.2	1.92	0.32	0.64	0.78
10			3.0			2.14	0.34	0.60	0.74
10			2.5			2.46	0.36	0.62	0.75
15	10 – 25	1.8	3.0	2.5	0.2	2.67	0.37	0.52	0.69
20			3.5			2.98	0.38	0.45	0.64
25			4.0			3.20	0.40	0.41	0.60
30	25 – 40	2.0	4.3	2.5	0.2	3.42	0.42	0.38	0.57
35			4.5			3.54	0.44	0.36	0.55

Standard values for plasma cutting[1)]

	Material: high-alloyed structural steels Cutting method: argon-hydrogen							**Material: aluminum Cutting method: argon-hydrogen**					
Sheet met. thickn. s mm	Electrical current qual. cut A	Electrical current stand. cut A	Cutting rate quality cut m/min	Cutting rate stand. cut m/min	Consumption values argon m³/hr	Consumption values hydrogen m³/hr	Consumption values nitrogen m³/hr	Electrical current quality cut A	Electrical current stand. cut A	Cutting rate quality cut m/min	Cutting rate stand. cut m/min	Consumption values argon m³/hr	Consumption values hydrogen m³/hr
4			1.4	2.4	0.6	–	1.2			3.6	6.0		
5	70	120	1.1	2.0	0.6	–	1.2	70	120	1.9	5.0	1.2	0.5
10			0.65	0.95	1.2	0.24	–			1.1	1.6		
15			0.35	0.6	1.2	0.24	–			0.6	1.3		
20	70	120	0.25	0.45	1.2	0.24	–	70	120	0.35	0.75	1.2	0.5
25			0.35	0.35	1.5	0.48	–			0.2	0.5		

1) Values apply to an arc power of approx. 12 kW and 1.2 mm cutting noozle diameter.

Standard values for laser cutting[1)]

M[2)]	Sheet met. thickness s mm	Cutting velocity v m/min	Cutting gas	Cutting gas press. p bar	Cutting velocity v m/min	Cutting gas	Cutting gas press. p bar	Cutting velocity v m/min	Cutting gas	Cutting gas press. p bar
		Laser power 1 kW			Laser power 1.5 kW			Laser power 2 kW		
Unalloyed steel	1	5.0–8.0			7.0–10			7.0–10		
	1.5	4.0–7.0			5.5–7.5			5.6–7.4		
	2	4.0–6.0			4.8–6.2			4.8–6.1		
	2.5	3.5–5.0	O_2	1.5–3.5	4.2–5.0	O_2	1.5–3.5	4.2–5.0	O_2	1.5–3.5
	3	3.5–4.0			3.5–4.2			3.6–2.8		
	4	2.5–3.0			2.8–3.3			2.8–3.4		
	5	1.8–2.3			2.3–2.7			2.5–3.0		
	6	1.3–1.6			1.9–2.2			2.1–2.5		
Stainless steel	1	4.0–5.5		8	5.0–7.0		6	4.5–9.0		12
	1.5	2.8–3.6		10	3.5–5.2		10	3.8–6.6		13
	2	2.2–2.8		14	2.0–4.0		10	3.4–5.3		14
	2.5	1.6–2.0	N_2		1.9–3.2	N_2	14	2.7–3.8	N_2	
	3	1.3–1.4		15	1.8–.2.4		14	2.2–2.7		14
	4	–		–	1.0–1.1		15	1.4–1.8		16

1) The table values apply to a the focal length of f = 127 mm (5") and a cutting gap of w = 0.15 mm.
2) M material group

PE

Areas of application and standard values for beam cutting

Areas of application for cutting processes

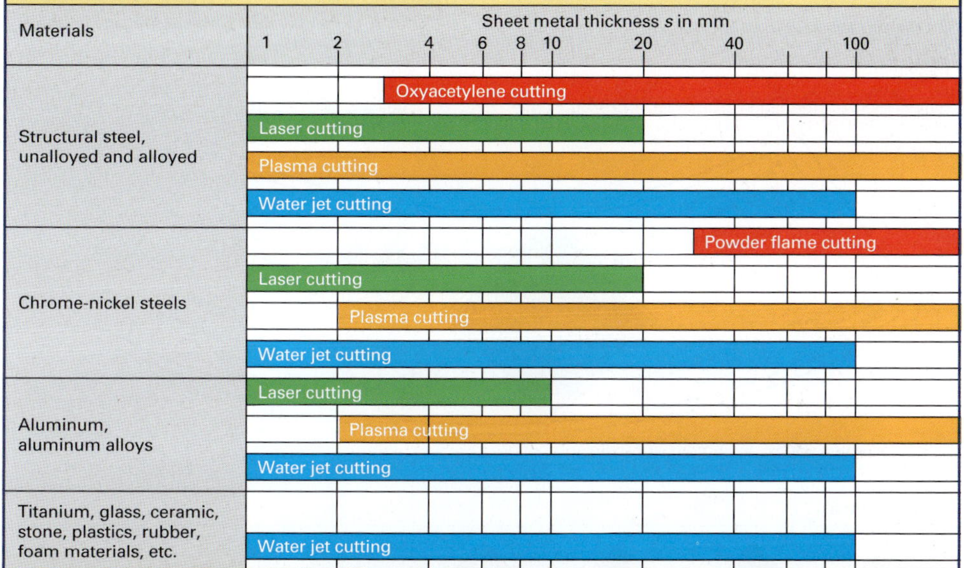

Cutting quality and dimensional tolerances for thermal cuts

cf. DIN EN ISO 9013 (2003-07)

The specifications apply to

- oxy-fuel gas cutting,
- plasma cutting,
- laser beam cutting.

The quality of the cut surfaces is determined by

- the perpendicularity tolerance u,
- the average surface roughness R_{z5}.

l nominal length
s workpiece thickness
u perpendicularity tolerance
R_{z5} average surface roughness
Δl limit deviations from the nominal length l

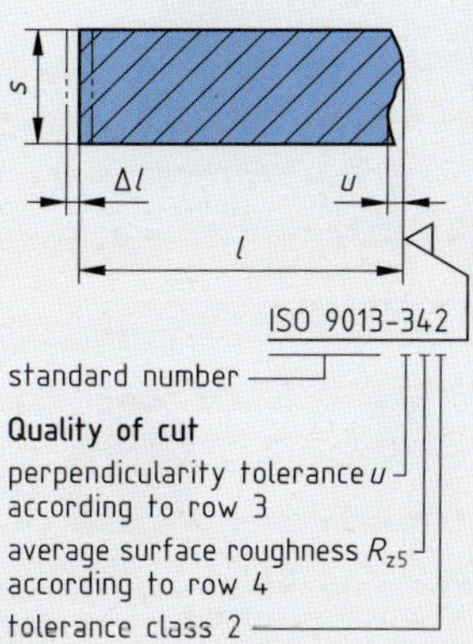

Quality of cut surfaces

Range	Perpendicularity tolerance u in mm	Average surface roughness R_{z5} in µm	Comments
1	$u < 0.05 + 0.03 \cdot s$	$R_{z5} < 10 + 0.6 \cdot s$	Put in workpiece thickness in mm
2	$u < 0.15 + 0.07 \cdot s$	$R_{z5} < 40 + 0.8 \cdot s$	
3	$u < 0.4 + 0.01 \cdot s$	$R_{z5} < 70 + 1.2 \cdot s$	
4	$u < 1.2 + 0.035 \cdot s$	$R_{z5} < 110 + 1.8 \cdot s$	

Limit deviations from the nominal length

Workpiece thickness s in mm	Limit deviations Δl from nominal lengths l in mm					
	Tolerance class 1			Tolerance class 2		
	> 35 ≤ 125	> 125 ≤ 315	> 315 ≤ 1000	> 35 ≤ 125	> 125 ≤ 315	> 315 ≤ 1000
> 1 ≤ 3.15	± 0.3	± 0.3	± 0.4	± 0.5	± 0.7	± 0.8
> 3.15 ≤ 6.3	± 0.4	± 0.4	± 0.5	± 0.8	± 0.9	± 1.1
> 6.3 ≤ 10	± 0.6	± 0.7	± 0.7	± 1.3	± 1.4	± 1.5
> 10 ≤ 50	± 0.7	± 0.7	± 0.8	± 1.8	± 1.9	± 2.3
> 50 ≤ 100	± 1.3	± 1.4	± 1.7	± 2.5	± 2.6	± 3.0
> 100 ≤ 150	± 1.9	± 2.0	± 2.1	± 3.3	± 3.4	± 3.7

Example: oxy-fuel gas cutting according to tolerance class 2, l = 450 mm, s = 12 mm, cutting quality according to range 4

Sought after: Δl; u; R_{z5}

Solution: $\Delta l = \pm$ **2.3 mm**

$u = 1.2 + 0.035 \cdot s = 1.2\ \text{mm} + 0.035 \cdot 12\ \text{mm} =$ **1.62 mm**

$R_{z5} = 110 + 1.8 \cdot s = 110\ \mu\text{m} + 1.8 \cdot 12\ \mu\text{m} =$ **131.6 µm**

PE

Gas cylinders – Identification*

Hazardous substance labels

cf. DIN EN ISO 7225 (2008-02)

A hazardous substance label must be applied to individual gas cylinders to identify their contents and any possible hazards from these contents. Up to three hazard labels warn of the main hazards.
Example:

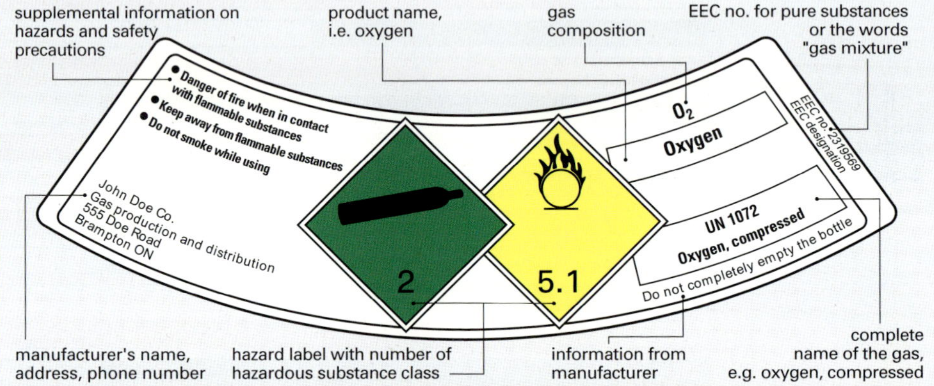

Hazard label

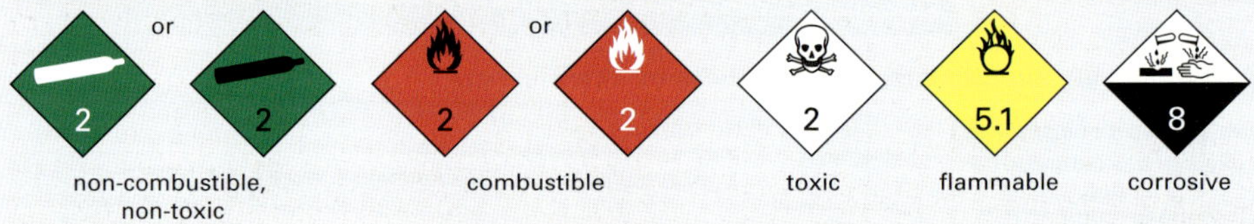

Color coding

cf. DIN EN 1089-3 (2004-06)

Color coding of the cylinder shoulder is used as additional information about the properties of the gases.
It is readily recognized when the hazardous substance label is illegible from a distance.
This color coding does not apply to liquid gases.

General color coding

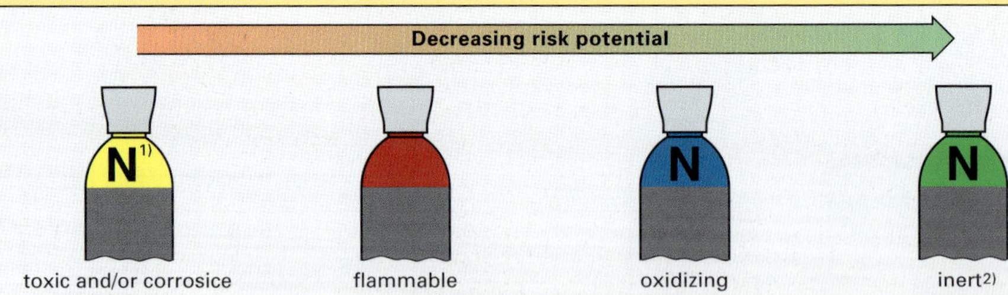

Color coding for special gases

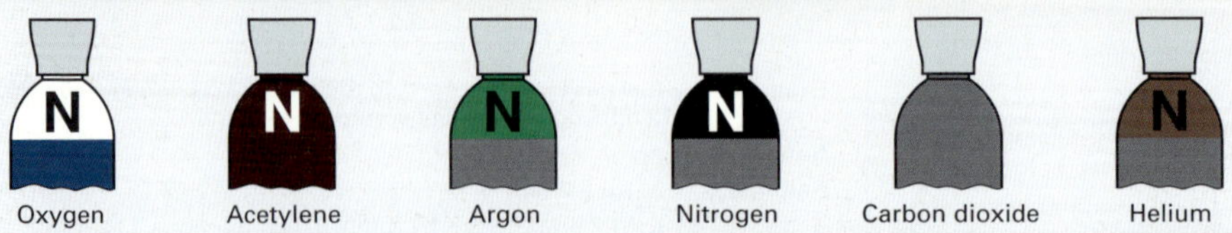

[1)] N = new [2)] Non-toxic, non-corrosive, non-flammable, non-oxidizing
*) According to European standards

PE

Gas cylinders – Identification*

Pure gases and gas mixtures for industrial use
Color coding (examples)

cf. Information sheet from Industrial Gases Association

Gas	Coding old (shoulder)	Coding old (body)	Coding new[1) 2)] (shoulder)	Coding new[1) 2)] (body)
Oxygen	blue	blue	white (N)	blue
Acetylene	yellow	yellow (black)	chestnut brown (N)	chestnut brown
Argon	gray	gray	dark green (N)	gray
Nitrogen	dark green	dark green	black (N)	gray
Carbon dioxide	gray	gray	gray	gray
Helium	gray	gray	brown (N)	gray
Xenon, Krypton, Neon	gray	gray (black)	flourescent green (N)	gray
Hydrogen	red	red	red	red
Forming gas (mixture of nitrogen/hydrogen)	red	red (dark green)	red (N)	gray
Mixture of argon/carbon dioxide	gray	gray	flourescent green (N)	gray
Compressed air	gray	gray	flourescent green (N)	gray

1) For gas cylinders color coded as per DIN EN 1089, the letter "N" (= new) must be put on the shoulder of the cylinder two times (opposite sides). The "N" is not required on cylinders whose color coding has not changed.

2) The cylinder body may be another color. However, this must not lead to confusion regarding the hazardous nature of the cylinder contents.

*) According to European standards

PE

Brazing

Brazing heavy non-ferrous metals

cf. DIN EN 1044 (1999-07)

Group	Brazing material: Designation[1]	Material number	Alloy designation as per ISO 3677[2]	Working temperature °C	Brazing joint[3]	Solder feed[4]	Information for use: Materials
Silver containing brazing materials							
AgCuCdZn	AG 301	2.5143	B-Ag50CdZnCu-620/640	640	G	f, l	precious metals, steels, copper alloys
	AG 302	2.5146	B-Ag45CdZnCu-605/620	620	G	f, l	
	AG 304	2.5141	B-Ag45ZnCdCu-595/630	610	G	f, l	steels, malleable cast iron, copper, copper alloys, nickel, nickel alloys
	AG 309	2.1215	B-Cu40ZnAgCd-605/765	750	G, V	f, l	
AgCuZn(Sn)	AG 104	2.5158	B-Ag45CuZnSn-640/680	670	G	f, l	steels, malleable cast iron, copper, copper alloys, nickel, nickel alloys
	AG 106	2.5157	B-Cu36AgZnSn-630/730	710	G	f, l	
	AG 203	2.5147	B-Ag44CuZn-675/735	730	G	f, l	
	AG 205	2.1216	B-Cu40ZnAg-700/790	780	G	f, l	
Silver content below 20%	AG 207	2.1207	B-Cu48ZnAg(Si)-800/830	830	G	f, l	steels, malleable cast iron, copper, copper alloys, nickel, nickel alloys
	AG 208	2.1205	B-Cu55ZnAg(Si)-820/870	860	G, V	f, l	
	CP 102	2.1210	B-Cu80AgP-645/800	710	G, V	f, l	copper and nickel-free copper alloys. **Unsuitable** for materials containing Fe or Ni
	CP 104	2.1466	B-Cu89PAg-645/815	710	G, V	f, l	
	CP 105	2.1467	B-Cu92PAg-645/825	710	G, V	f, l	
Special brazing	AG 351	2.5160	B-Ag50CdZnCuNi-635/655	660	G	f, l	Cu alloys
	AG 403	2.5162	B-Ag56CuInNi-600/710	730	G	f, l	chrome, chrome-nickel steels
	AG 502	2.5156	B-Ag49ZnCuMnNi-680/705	690	G	f, l	carbide onto steel, tungsten and molybdenum materials
Copper based brazing materials							
	CU 104	2.0091	B-Cu100(P)-1085	1100	G	l	steels
	CU 201	2.1021	B-Cu94Sn(P)-910/1040	1040	G	l	iron and nickel materials
	CU 202	2.1055	B-Cu88Sn(P)-825/990	990	G	l	
	CU 301	2.0367	B-Cu60Zn(Si)-875/895	900	G, V	f, l	steels, malleab. iron, Cu, Ni, Cu & Ni alloys
	CU 305	2.0711	B-Cu48ZnNi(Si)-890/920	910	G, V	f, l	steels, malleable iron, Ni, Ni alloys
					V	f	cast iron
	CP 202	2.1463	B-Cu93P-710/820	720	G	f, l	Cu, Fe-free and Ni-free Cu alloys
Nickel based brazing materials for high-temperature brazing							
	NI 101	2.4140	B-Ni73CrFeSiB(C)-960/1060	5)	5)	5)	nickel, cobalt, nickel and cobalt alloys, unalloyed and alloyed steels
	NI 103	2.4143	B-Ni92SiB-980/1040				
	NI 105	2.4148	B-Ni71CrSi-1080/1135				
	NI 107	2.4150	B-Ni76CrP-890				
Aluminum based brazing materials							
	AL 102	3.2280	B-Al92Si-575/615	610	G	f, l	aluminum and Al alloy types AlMn, AlMgMn, G-AlSi; especially for Al alloy types AlMg, AlMgSi up to 2% Mg content
	AL 103	3.2282	B-Al90Si-575/590	600	G	f, l	
	AL 104	3.2285	B-Al88Si-575/585	595	G	f, l	

PE

[1] The two letters indicate the alloy group, while the three digit numbers are purely numbers increasing sequentially.

[2] Numbers at the end indicate the melting range. Alloy components, see pages 112 and 113.

[3] G suitable for gap brazing; V suitable for V-joint brazing

[4] f filled brazing; l lapped brazing

[5] Refer to manufacturer's data.

Brazing joint

Gap brazing:
$w < 0.25$ mm

V-joint brazing:
$w > 0.3$ mm

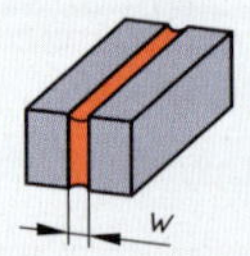

Solders and flux

Solders

cf. DIN EN ISO 9453 (2006-12)

Alloy group[1]	Alloy no.[2]	Alloy designation as per ISO 3677[3]	Previous designation DIN 1707	Working temperature °C	Application examples
tin-lead	101 102 103	S-Sn63Pb37 S-Sn63Pb37E S-Sn60Pb40	L-Sn63Pb L-Sn63Pb L-Sn60Pb	183 183 183–190	precision mechanics electronics, printed circuit boards printed circuit boards, high-grade steel
lead-tin	111 114 116 124	S-Pb50Sn50 S-Pb60Sn40 S-Pb70Sn30 S-Pb98Sn2	L-Sn50Pb L-PbSn40 – L-PbSn2	183–215 183–235 183–255 320–325	electronics industry, tin plating thin-sheet packaging, metal goods plumbing work, zinc, zinc alloys radiator manufacturing
tin-lead-antimony	131 132	S-Sn63Pb37Sb S-Sn60Pb40Sb	– L-Sn60Pb(Sb)	183 183–190	precision mechanics precision mechanics, electrical industry
	134 136	S-Pb58Sn40Sb2 S-Pb74Sn25Sb1	L-PbSn40Sb L-PbSn25Sb	185–231 185–263	radiator manufacturing, wiping solder wiping solder, lead solders
tin-lead-bismuth	141 142	S-Sn60Pb38Bi2 S-Pb49Sn48Bi3	– –	180–185 138	precision solders low-temperature solder, safety fuses
tin-lead-cadmium	151	S-Sn50Pb32Cd18	L-SnPbCd18	145	thermal fuses, cable joints
tin-lead-copper	161 162	S-Sn60Pb39Cu1 S-Sn50Pb49Cu1	L-SnPbCu3 L-Sn50PbCu	230–250 183–215	electronic devices, precision mechanics
tin-lead-silver	171	S-Sn60PbAg	L-Sn60PbAg	178–180	electrical devices, printed circuit boards
lead-tin-silver	182 191	S-Pb95Ag5 S-Pb93Sn5Ag2	L-PbAg5 –	304–365 296–301	for high operating temperatures electric motors, electrical equipment

[1] Filler metals for aluminium are no longer in EN ISO 9453.
[2] The alloy numbers replace the material numbers as per DIN 1707.
[3] With traces (<0.5%) of Sb, Bi, Cd, Au, In, Al, Fe, Ni, Zn: see pages 112 and 113.

Flux for soldering

cf. DIN EN 29454-1 (1994-02)

Designation by main constituents				Classification by effect		
Flux type	Flux basis	Flux activator	Flux form	Designations DIN EN	DIN 8511	Effect of residues
1 rosin	1 colophonium 2 without colophonium	1 without activator 2 activated by halogens 3 activated without halogens	A liquid B solid C paste	3.2.2... 3.1.1...	F-SW11 F-SW12	very corrosive
2 organic	1 water soluble 2 not water soluble			3.2.1... 3.1.1... 2.1.3... 2.1.2... 1.2.2...	F-SW13 F-SW21 F-SW23 F-SW25 F-SW28	somewhat corrosive
3 inorganic	1 salts	1 with ammonium chloride 2 without ammonium chloride				
	2 acids	1 phosphoric acid 2 other acids		1.1.1... 1.2.3...	F-SW31 F-SW33	non-corrosive
	3 alkaline	1 amine and/or ammonia				

⇒ **Flux ISO 9454 – 1.2.2.C:** Flux of type rosin (1), base without colophonium (2), activated by halogens (2), available in paste form (C)

Flux for brazing

cf. DIN EN 1045 (1997-08)

Flux	Activation temper.	Instructions for use
FH10 FH11 FH12	550–800 °C 550–800 °C 550–850 °C	Multi-purpose flux; residues rinsed off or chemically stripped. Cu-Al alloys; residues rinsed off or chemically stripped. Stainless and high-alloy steels, carbide; residues chemically stripped.
FH20 FH21 FH30 FH40	700–1000 °C 750–1100 °C over 1000 °C 650–1000 °C	Multi-purpose flux; residues rinsed off or chemically stripped. Multi-purpose flux; residues removed mechanically or chemically stripped. For copper and nickel solder; residues removed mechanically. Boron-free flux; residues rinsed off or chemically stripped.
FL10 FL20	400–700 °C 400–700 °C	Light alloys; residues are rinsed off or chemically stripped. Light alloys; residues are non-corrosive, but should be protected from moisture.

PE

Soldered and brazed joints

Classification of soldering and brazing processes

Differentiating characteristics	Soldering and brazing processes: Soldering	Brazing	High temperature brazing
Working temperature	< 450 °C	> 450 °C	> 900 °C
Energy source	soldering iron, soldering bath, electrical resistance	flame, furnace	flame, laser beam, electric induction
Base material	Cu, Ag, Al alloys, stainless steel, steel, Cu, Ni alloys	steel, carbide inserts	steel, carbide
Soldering or filler material	Sn, Pb alloys	Cu, Ag alloys	Ni-Cr alloys, Ag-Au-Pd alloys
Auxiliary materials	Flux	flux, vacuum	vacuum, shielding gas

Standard values for soldering gap widths

Base material	Soldering gap width in mm: for solders	for brazing materials primarily of: copper	brass	silver
unalloyed steel	0.05–0.2	0.05–0.15	0.1–0.3	0.05–0.2
Alloy steel	0.1–0.25	0.1–0.2	0.1–0.35	0.1–0.25
Cu, Cu alloys	0.05–0.2	–	–	0.05–0.25
Carbide	–	0.3–0.5	–	0.3–0.5

Design rules for soldered joints

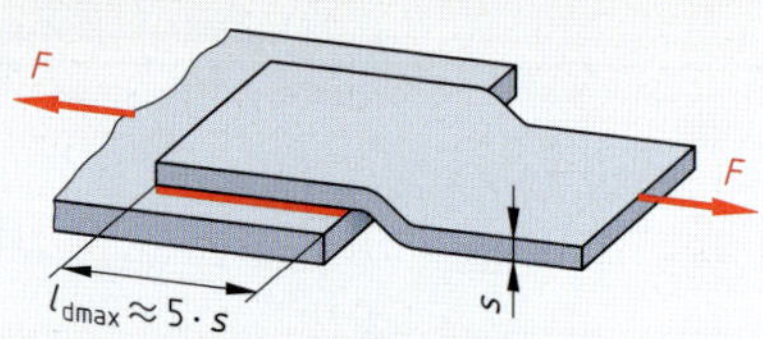

Soldered joint under shearing load

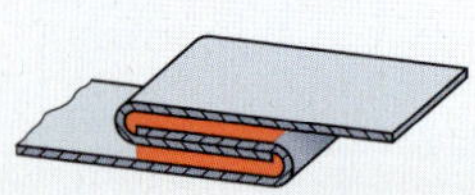

Load on solder joint reduced by folded seam

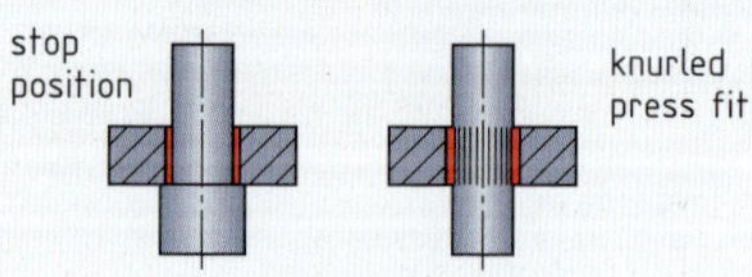

Production process simplification

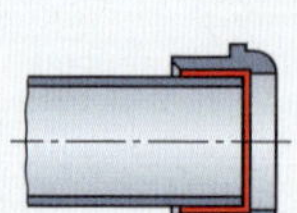

Soldered pipe fitting

Preconditions

- Soldering gap should be large enough so that flux and solder adequately fill the gap by capillary action (table above)
- The two surfaces to be soldered should be parallel.
- Surface roughness due to machining can remain for Cu soldering Rz = 10–16 µm, for Ag soldering at Rz = 25 µm.

Load transfer

- The load on the soldered joint should be in shear (transverse forces) if at all possible. In particular, solder seams should not be loaded with tensile or peeling stress.
- Soldering gap depths $l_d > 5 \cdot s$ do not fill with solder reliably. Therefore load capacity cannot be increased by a larger gap depth.
- Load capacity can be increased by design features such as folds

Production process simplification

- In soldering there should be a means for assuring proper positioning of the parts to be joined, e.g. by part shape or by knurled press fit.

Application examples

- pipes and fittings
- sheet metal parts
- tools with brazed carbide cutters

Adhesives, Preparation of joint surfaces

Properties and conditions of use for adhesives[1]

Adhesive	Trade name	Curing conditions Temperature °C	Time	max. operating temperature °C	Comb. tensile and shear strength τ_B N/mm²	Elasticity	Applications, special characteristics
Acrylic resins	Agomet M, Acronal, Stabilit-Express	20	24 hr	120	6–30	low	metals, thermosets, ceramics, glass
Epoxy resins (EP)	Araldit, Metallon, Uhu-Plus	20–200	1 hr to 12 hr	50–200	10–35	low	metals, thermosets, glass, ceramics, concrete, wood; long curing time
Phenolic resins (PF)	Porodur, Pertinax, Bakelite	120–200	60 s	140	20	low	metals, thermosets, glass, elastomers, wood, ceramics
Polyvinyl chloride (PVC)	Hostalit, Isodur, Macroplast	20	> 24 hr	60	60	low	metals, thermosets, glass, elastomers, wood, ceramics
Polyurethane (PUR)	Desmocoll, Delopur, Baydur	50	24 hr	40	50	present	metals, elastomers, glass, wood, some thermoplastics
Polyester resins (UP)	Fibron, Leguval, Verstopal	25	1 hr	170	60	low	metals, thermosets, ceramics, glass
Polychloroprene (CR)	Baypren, Contitec, Fastbond	50	1 hr	110	5	present	contact glue for metals and plastics
Cyanoacrylate	Permabond, Sicomet 77	20	40 s	85	20–25	low	fast-curing adhesive for metals, plastics, elastomers
Hot glue	Jet-Melt, Ecomelt, Vesta-Melt	20	> 30 s	50	2–5	present	all types of materials; adhesive action through cooling

[1] Due to varying chemical compositions of adhesives, the values given are only approximate values. For detailed information please refer to information from the manufacturer.

Preparation of parts for bonded joints

cf. VDI 2229 (1979-06)

Material	Treatment sequence[1] for load severity[2] low	medium	high	Material	Treatment sequence[1] for load severity[2] low	medium	high
Al alloys	1-2-3-4	1-6-5-3-4	1-2-7-8-3-4	Steel, bright	1-2-3-4	1-6-2-3-4	1-7-2-3-4
Mg alloys	1-2-3-4	1-6-2-3-4	1-7-2-9-3-4	Steel, galvanized	1-2-3-4	1-2-3-4	1-2-3-4
Ti alloys	1-2-3-4	1-6-2-3-4	1-2-10-3-4	Steel, phosphatized	1-2-3-4	1-2-3-4	1-6-2-3-4
Cu alloys	1-2-3-4	1-6-2-3-4	1-7-2-3-4	Other metals	1-2-3-4	1-6-2-3-4	1-7-2-3-4

[1] **Code numbers for type of treatment**

1 **Cleaning** of dirt, scale, rust
2 **Removing grease** with organic solvent or aqueous cleaning agent
3 **Rinsing** with clear water
4 **Drying** in hot air up to 65 °C
5 **Removing grease** with simultaneous etching
6 **Mechanical roughing** by grinding or brushing
7 **Mechanical roughing** by shot blasting
8 **Etching 30 min**, at 60 °C in 27.5 % sulfuric acid solution
9 **Etching 1 min**, at 20 °C in 20 % nitric acid solution
10 **Etching 3 min**, at 20 °C in 15 % hydrofluoric acid solution

[2] **Load severity for bonded joints**

Low: Tensile shear strength up to 5 N/mm²; dry environment; for precision mechanics, electrical equipment
Medium: Tensile shear strength up to 10 N/mm²; humid air; contact with oil; for machine and vehicule manufacturing
High: Tensile shear strength up to 10 N/mm²; direct contact with liquids; for aircraft, ship, and container manufacturing

PE

Design of adhesive bonded joints, Test methods

Design examples

Bonded joints should be loaded in compression or shearing if possible.
Tensile, peeling or bending loads should be avoided.

Butt joint/overlap joint

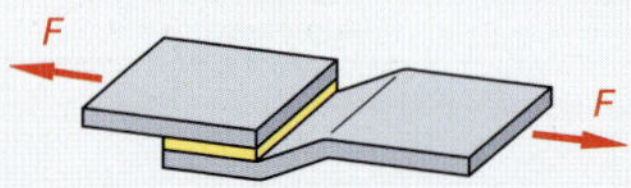

good, since the bonding surfaces only have a shear load

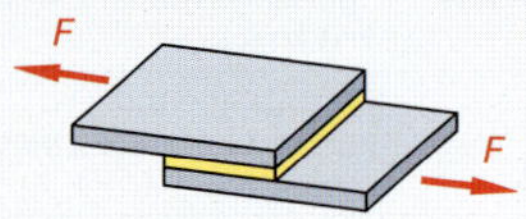

not as good, since peeling forces act due to off-center application of force

T-joint

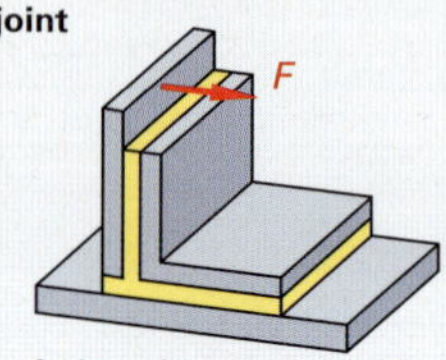

good, since the bonding surfaces only have a shear and compression load

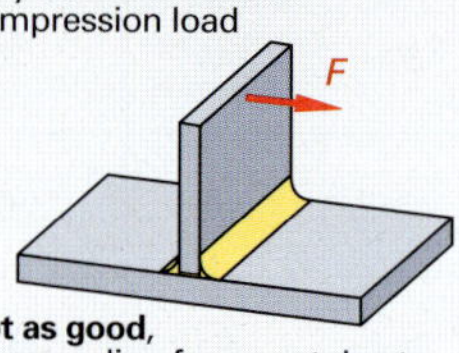

not as good, since peeling forces act due to bending load

Tube joint

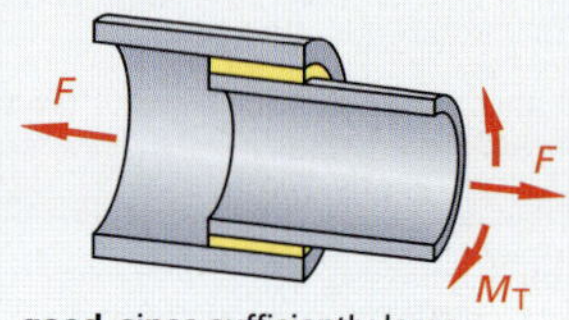

good, since sufficiently large bonding surfaces can withstand shear load

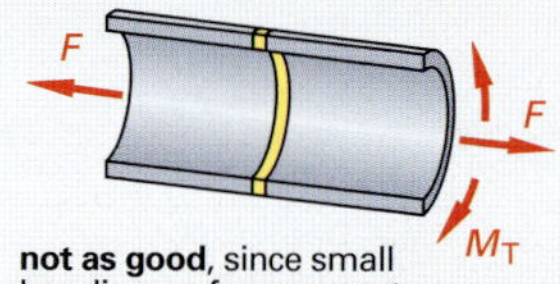

not as good, since small bonding surfaces cannot withstand tensile and shear load

Test methods

Test method standard	Contents
Bending peel test DIN 54461	Tests resistance of bonded joints against peeling forces
Tensile shear test DIN EN 1465	Tests tensile shear strength of high-strength bonded lap joints
Fatigue test DIN EN ISO 9664	Tests fatigue properties of structural adhesives under tensile-shear loads
Tensile test DIN EN 15870	Tests tensile strength of bonded butt joints perpendicular to bonded surface
Roller peel test DIN EN 1464	Tests resistance to peeling forces
Compression shear test DIN EN 15337	Tests shear strength, primarily of anaerobic[1] adhesives

[1] Sets with exclusion of air

PE

Adhesive behavior as a function of temperature and size of bonding surface

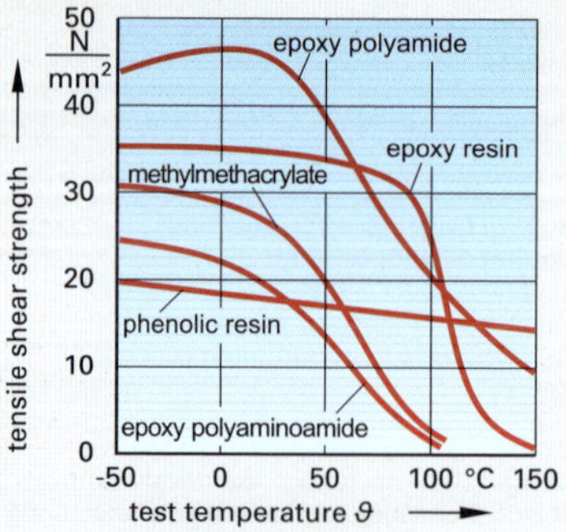

Tensile shear strength of overlap bonded joints

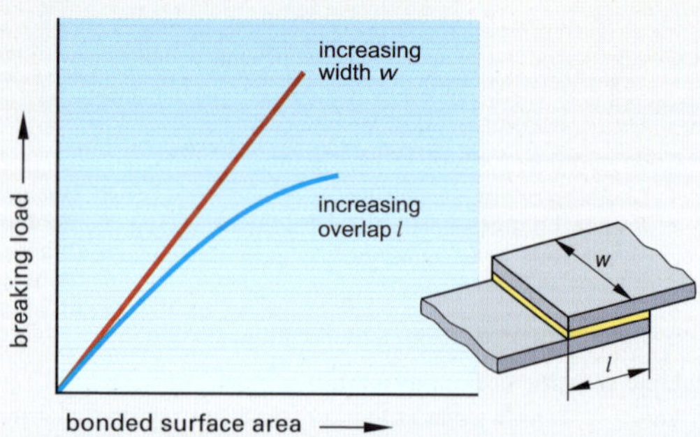

Effect of adhesive joint surface area on breaking load

Safety colors, Prohibitive signs*

Safety colors

cf. DIN 4844-1 (2005-05) and BGV A8[1] (2002-04)

Color	red	yellow	green	blue
Meaning	stop, prohibited	caution! potential danger	safety, first aid	mandatory signs, notices
Contrast color	white	black	white	white
Color of graphic symbol	black	black	white	white
Application examples (see pages 361 and 362)	Stop signs, emergency stop prohibitive signs, fire fighting equipment	Notice of hazards (e.g. fire, explosion, radiation); notice of obstructions (e.g. speed bumps, holes)	Identification of ambulances and emergency exits; first aid and emergency aid stations	Requirement to wear personal protective equipment (PPE); location of a telephone

Prohibitive signs

cf. DIN 4844-2 (2001-02) and BGV A8[1] (2002-04)

 Prohibited	 No smoking	 No fires, open flame or smoking	 Pedestrian access prohibited	 Do not extinguish with water	 Non-potable water
 Access prohibited for unauthorized persons	 Access by forklifts prohibited	 Do not touch	 Do not touch – live voltage	 Do not connect	 No access for persons with pacemaker
 Placement or storage prohibited	 Transport of passengers prohibited	 Walking in this area prohibited	 No spraying with water	 No cell phones	 No food or drink allowed
 No magnetic or electronic data media allowed	 Climbing prohibited for unauthorized persons	 Do not use this device in the bathtub, shower or sink	 Do not reach in	 Operating with long hair prohibited	 Hand-held or manually operated grinding not allowed

[1] BGV A8: accident prevention regulation about safety and health protection marking at workplaces Issued by the German Employers' Liability Insurance Association

*) According to European standards

PE

Warning signs*

Warning signs cf. DIN 4844-2 (2001-02) and BGV A8[1] (2002-04)

 Warning: Hazardous area	 Warning: Combustible materials	 Warning: Explosive substances	 Warning: Toxic substances	 Warning: Corrosive sub-stances	 Warning: Radioactive materials or ionizing radiation
 Warning: Suspended load	 Warning: Forklift traffic	 Danger: High voltage	 Warning: Optical radiation	 Warning: Laser beam radiation	 Warning: Oxidizing substances
 Warning: Non-ionic, electromagnetic radiation	 Warning: Strong magnetic field	 Warning: Danger of tripping	 Warning: Danger of falling	 Warning: Biological hazard	 Warning: Extreme cold
 Warning: Substances hazardous to health or irritants	 Warning: Gas cylinders	 Warning: Hazards due to batteries	 Warning: Explosive atmosphere	 Warning: Milling shaft	 Warning: Crushing hazard
 Warning: Danger of tipping when rolling	 Warning: Automatic start-up	 Warning: Hot surface	 Warning: Risk of hand injury	 Warning: Danger of slipping	 Warning: Moving conveyor on track

[1] BGV A8: accident prevention regulation about safety and health protection marking at workplaces issued by the German Employers' Liability Insurance Association

*) According to European standards

PE

Safety signs*

cf. DIN 4844-2 (2001-02) and BGV A8[1] (2002-04)

Mandatory signs

General mandatory sign	Wear safety glasses	Wear hard hat	Wear ear protection	Wear respirator	Wear safety shoes

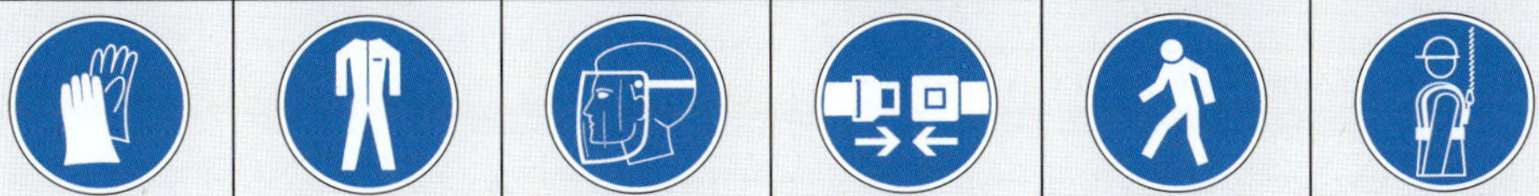

Wear protective gloves	Wear protective clothing	Wear face protection	Use safety belt	For pedestrians	Use safety harness

Use crosswalk	Disc. plug from power bef. opening	Disconnect before working	Wear life preserver	Sound horn	Follow instructions

Escape and rescue signs for escape routes and emergency exits

Direction arrows for First aid stations, escape routes and emergency exits[2]		First aid	Medical stretcher	Emergency shower	Eye rinsing equipment

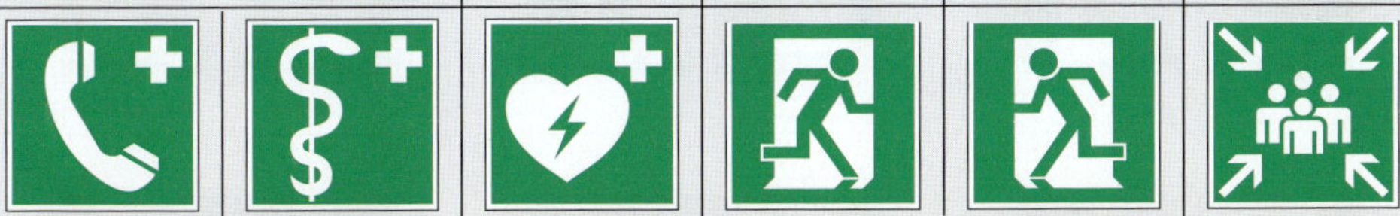

Emergency telephone	Doctor	Defibrillator	Escape route/Emergency exit		Meeting point

Fire protection symbols and additional symbols

Directional arrows		Wall hydrant and fire hose	Ladder	Fire extinguisher	Fire alarm telephone

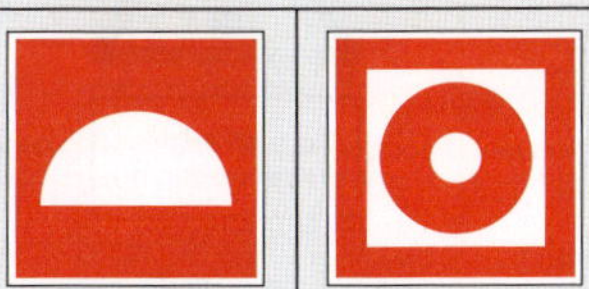

Fire fighting equipment	Manual fire alarm	Work area! Location: Date: Sign may only be removed by: Extra sign which gives more information to supplement the safety sign	High Voltage Danger to life Extra sign which gives more information to supplement the safety sign

[1] BGV A8: accident prevention regulation about safety and health protection signs at workplaces by the German Employers' Liability Insurance Association

*) According to European standards

PE

Safety signs*

cf. DIN 4844-2 (2001-02) and BGV A8[1] (2002-04)

Information signs

Discharge time longer than 1 minute	In case of failure part can have live voltage	Before touching: - discharge - ground - short circuit	5 Safety rules Before beginning work - Employ safety disconnect - Lock out to prevent restart - Check for no voltage - Ground and short circuit - Cover or enclose adjacent parts which have live voltage

Combination signs

Do not connect

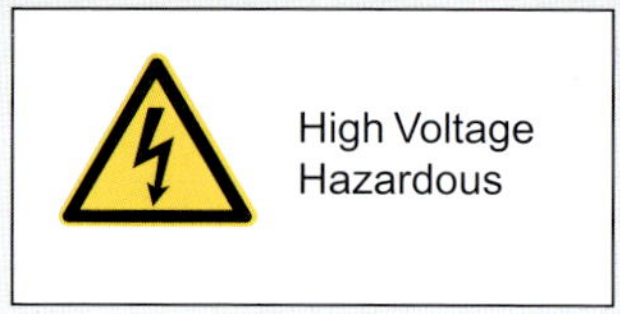

Warning of high voltage

Combination signs for escape routes or emergency exits with corresponding direction indicated by arrows

PE

First aid station

Prohibited! Walking on roof is prohibited.

Fire blanket for fighting fire

Danger of toxic gases

[1] BGV A8: accident prevention regulation about safety and health protection marking at workplaces issued by the German Employers' Liability Insurance Association

*) According to European standards

Danger symbols and description of hazards*

RL 67/548/EEC (2004-04)[1]

Code letter, danger symbol, hazard description	Danger criteria of materials	Code letter, danger symbol, hazard description	Danger criteria of materials	Code letter, danger symbol, hazard description	Danger criteria of materials
T+  **Very toxic**	When consumed in very small amounts leads to death or may cause acute or chronic damage to health. T = toxic	**Xi**  **Irritant**	Contact with skin or mucus membranes can cause inflammation. X = St. Andrew's cross i = irritating	**F**  **Flammable**	Solid material can be easily ignited by a source of ignition. Liquid material with flash point < 21 °C. F = flammable
T  **Toxic**	When consumed in small amounts leads to death or may cause acute or chronic damage to health. T = toxic	**E**  **Danger of explosion**	Risk of explosion by shock, friction, fire or other sources of ignition. E = explosive	**N**  **Environmentally dangerous**	Substances change water, ground, air, climate, animals, plants, etc. in such a way that the environment is endangered. N = noxious (harmful)
Xn  **Harmful to health**	When ingested may result in death or cause acute or chronic harm to health. X = St. Andrew's cross n = noxious	**O**  **Oxidizing**	Substances that substantially increase the risk and severity of a fire, because they produce oxygen. O = oxidizing	**T with R 45**  **Carcinogenic**	Substance may cause cancer from inhaling, swallowing or from contact with the skin. R 45: May cause cancer T = toxic
C  **Corrosive**	Living tissue can be damaged by contact. C = corrosive	**F+**  **Highly flammable**	Liquid substances with flash point < 0 °C and boiling point < 35 °C; gaseous substances, which are flammable in contact with air. F = flammable	**T with R 46**  **Mutagenic substances**	Substances which can have a mutagenic effect on humans. R 46: May cause heritable genetic damage. T = toxic
Xn with R 40  **Limited evidence of mutagenic effect**	Substance which can cause concern due to possible mutagenic effect on humans. However, there is not yet sufficient information available to give conclusive proof. X = St. Andrew's cross n = noxious R 40 = irreversible damage possible (page 199)	**T with R 60, R 61**  **Danger to fertility**	Substances which are known to impair fertility or reproduction. T = toxic R 60 = may impair fertility R 61 = may cause harm to the unborn child	**Xn with R 62, R 63**  **Limited evidence of influence on fertility**	Substances which cause concern due to possible impairment of fertility of humans. X = St. Andrew's cross n = noxious R 62 = possible risk of impaired fertility R 63 = possible risk of harm to unborn child

[1] EU-Directive, Appendix II. The danger symbols have been replaced by the pictograms of the GHS (cf. page 364).
*) According to European standards

PE

Globally Harmonized System (GHS)

CLP[1] Ordinance (EC) no. 1272/2008

Warning signs for hazardous chemicals

Hazard pictogram Code	Meaning of the pictogram	Signal word	Hazard statements, examples		Precautionary statements, examples	
			Code[2]	Wording	Code[3]	Wording
 GHS01	Exploding bomb e. g. explosive substances	Danger	H201	Explosive; mass explosion hazard	P102	Keep out of reach of children
		Warning	H204	Fire or projection hazard	P210	Keep away from heat/ sparks/open flames/hot surfaces. – No smoking
 GHS02	Flame e. g. flammable liquids	Danger	H220	Extremely flammable gas	P251	Pressurized container: Do not pierce or burn, even after use
 GHS03	Flame above a circle e. g. oxidizing solid substances	Danger	H270	May cause or intensify fire; oxidizer	P243	Take precautionary measures against static discharge
 GHS04	Gas cylinder e. g. gas under pressure	Warning	H280	Contains gas under pressure; may explode if heated	P244	Keep pressure regulators free from grease and oil
 GHS05	Corrosive e. g. corrosive to the skin; corrosive to metals	Danger	H318	Causes serious eye damage	P280	Wear protective gloves/ protective clothing/ eye protection/ face protection
 GHS06	Skull and crossbones e. g. severe toxicity	Danger	H300 H311	Fatal if swallowed Toxic in contact with skin	P310	IF SWALLOWED: Immediately call a POISON Center or doctor/physician
 GHS07	Exclamation mark, e. g. severe toxicity; skin irritating	Warning	H319 H317	Causes serious eye irritation May cause an allergic skin reaction	P341	IF INHALED: If breathing is difficult, remove victim to fresh air and keep at rest in position comfortable for breathing
 GHS08	Health hazard e. g. carcinogenicity (cancer-producing effect)	Danger	H370	Causes damage to organs when inhaled	P260	Do not breathe dust/fume/gas/mist/ vapors/spray
 GHS09	Environment e. g. hazardous to the aquatic environment	Warning	H400	Very toxic for aquatic life	P403 + P233	Store in a well-ventilated place. Keep container tightly closed.

[1] CLP = classification, labeling and packaging; [2] Hazard statements, formerly risk phrases;
[3] Precautionary statements, formerly Safety Phrases.

Identification of pipe lines*

cf. DIN 2403 (2007-05)

Area of application and requirements

Area of application: A precise identification marking of pipe lines, indicating the substance being conveyed, is necessary for reasons of safety, fire fighting and proper maintenance and repairs. The identification marking is intended to indicate possible hazards and help to prevent accidents and damage to health.

Requirements concerning identification marking

- Identification marking must be clearly visible and long-lasting.
- Identification can be established by painting, lettering (e.g. via self-adhesive foil strips) or signs.
- Particularly operation-critical and hazardous places should be marked (e.g. beginning and end of branch pipes, wall penetrations, fittings).
- Marking must be repeated at least every 10 m of pipe length.
- Indication of the group and supplemental color (see table below).
- Indication of the flow direction by means of an arrow.
- Indication of the conveyed substance by specifying the name (e.g. water) or the chemical formula (e. g. H_2O).
- With hazardous materials, additional indication of hazard signs (page 363) or warning signs (page 360) if general hazards are implied.

Color assignment according to conveyed substances

Conveyed substance	Group	Group color	RAL	Supplem. color	RAL	Color of lettering	RAL
Water	1	green	6032	–	–	white	9003
Steam	2	red	3001	–	–	white	9003
Air	3	gray	7004	–	–	black	9004
Flammable gases	4	yellow	1003	red	3001	black	9004
Non-flammable gases	5	yellow	1003	black	9004	black	9004
Acids	6	orange	2010	–	–	black	9003
Lyes	7	purple	4008	–	–	white	9003
Flammable liquids and solid materials	8	brown	8002	red	3001	white	9003
Non-flammable liquids and solid materials	9	brown	8002	black	9004	white	9003
Oxygen	0	blue	5005	–	–	white	9003

Identification of special pipe lines

Fire extinguishing lines must be fitted with a red/white/red color marking. The white field contains the graphical symbol of the safety sign „Fire fighting equipment and materials" (cf. page 361) in the color of the extinguishing agent.

Potable water lines must be fitted with a green/white/green color marking. **Non-potable water lines** have a green/blue/green marking. The code letters and their colors are listed in the table below.

Description	Code	Color	Description	Code	Color
Potable water line Potable water line, cold	PW PWC	green	Potable water line, hot, circulating	PWH-C	purple
Potable water line, hot	PWH	red	Non-potable water line	NPW	white

Examples of identification marking

Heating oil

Fire extinguishing unit (water)

Potable water

Compressed air

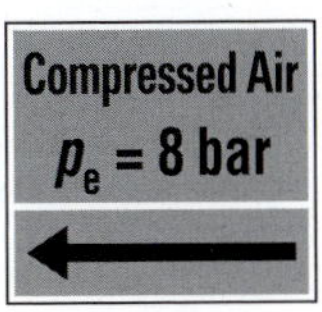

Oxygen (oxidizing, O)

Acetylene (highly flammable, F+)

*) According to European standards

Sound and Noise*

Sonic terms

Term	Explanation
Sound	Sound comes from mechanical vibrations. It propagates in gaseous, liquid and solid bodies.
Frequency	Number of oscillations per second. Unit: 1 Hertz = 1 Hz = 1/s. Pitch increases with frequency. Frequency range of human hearing: 16 Hz–20.000 Hz.
Sound level	Measure of the sound strength (sound energy).
Noise	Undesirable, annoying or painful sound waves; damage depends on strength, duration, frequency and regularity of exposure. For a noise level of 85 dB (A) and higher there is danger of permanent hearing loss.
Decibel (dB)	Standardized unit for sound level.
dB (A)	Since the human ear perceives tones of different heights (frequencies) to have different strengths when they are actually at the same sound levels, noise must be appropriately dampened with filters for certain frequencies. Frequency weighting curve with Filter A compensates for this and indicates the subjective auditory impression. A difference of 3 dB (A) corresponds approximately to a doubling (or halving) of the sound intensity (energy parameter). A difference of 10 dB (A) corresponds to a doubling (or halving) of the subjectively perceived loudness (psychological parameter).

Sound level

Type of sound	dB (A)	Type of sound	dB (A)	Type of sound	dB (A)
Threshold of auditory sensitivity	4	normal speech at distance of 1 m	70	heavy stamping	95–110
Breathing at distance of 30 cm	10	machine tools	75–90	angle grinder	95–115
Soft rustling of leaves	20	loud talking at distance of 1 m	80	car horn at distance of 5 m	100
Whispering	30	welding torch, lathe	85	disco music	100–115
Tearing paper	40	hammer drill, motorcycle	90	hammer and anvil	110
Quiet conversation	50–60	engine test stand, walkman	90–110	jet engine	120–130

Noise protection regulations

cf. Accident Prevention Regulations on "Noise" BGV B3[1] (1997-01)

Accident prevention regulations for noise producing operations

- Requirem. to post signage for noise ranges 90 dB (A) and above.
- Above 85 dB (A) sound protection devices must be available, and they must be used above 90 dB (A).
- If the risk of accidents increases due to noise, appropriate measures must be taken.
- Regular preventative medical checkups are compulsory.
- New operational equipment must conform to the most advanced level of noise reduction.

§ 15 Workplace regulation

Noise limit value for:	max. dB (A)
predominantly mental activities	55
simple, predominantly mechanized activities	70
all other activities (value may be exceeded by 5 dB)	85
break rooms, ready rooms and first-aid rooms	55

Noise harmful to health

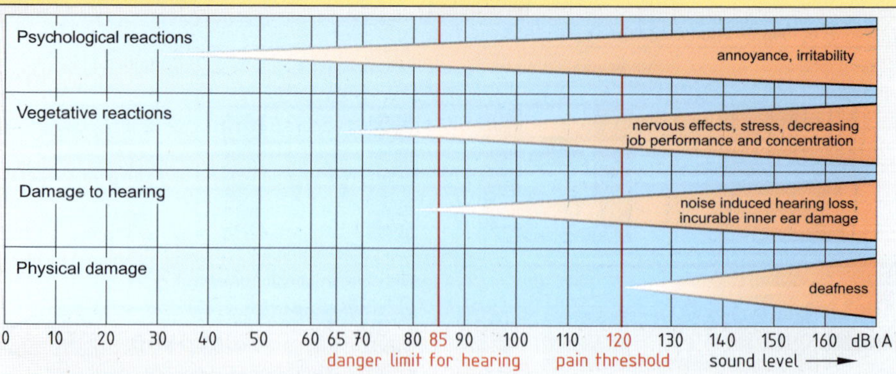

[1] BGV A3: accident prevention regulation about safety and health protection marking at workplaces issued by the German Employers' Liability Insurance Association

*) According to European standards

7 Automation and Information Technology

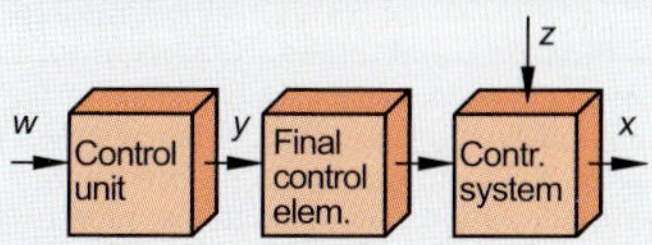

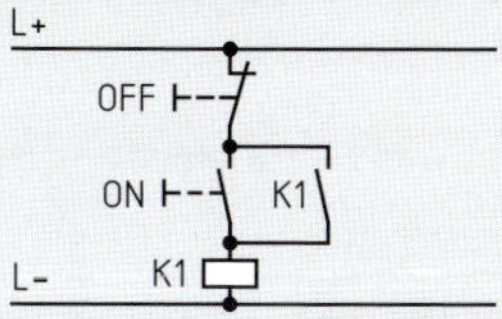

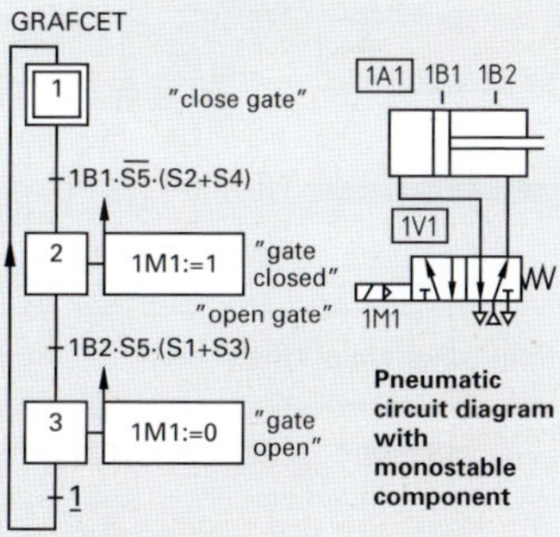

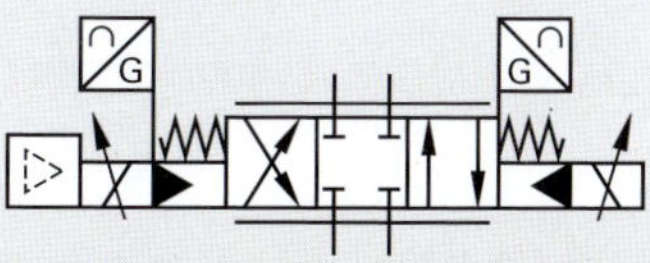

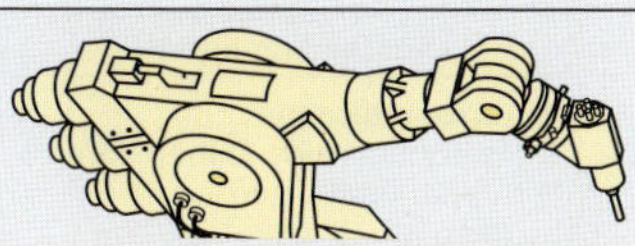

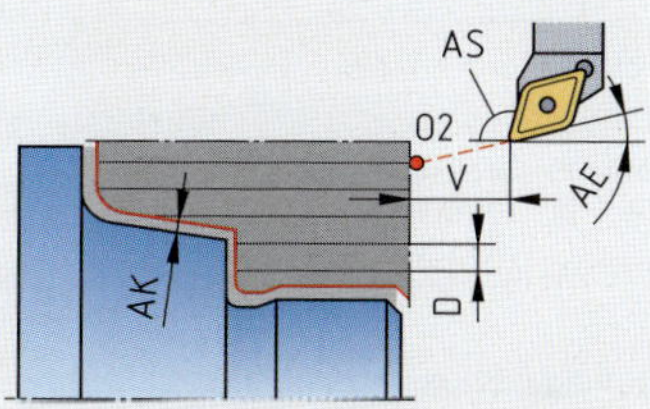

Basic terminology of open loop and closed loop control systems

Basic terminology

cf. DIN 19226-1 to -5 (1994-02), withdrawn

Open loop control	Closed loop control
For open loop control the output variable, such as the temperature in a hardening furnace, is influenced by the input variable, such as the current in the heating coil. The output variable does not have an effect on the input variable. Open loop control has an open action flow.	For closed loop control the controlled variable, such as the actual temp. in an annealing furnace, is continuously monitored and compared to the target temp. (reference variable) and, if there are deviations, adjusted to the reference input variable. Closed loop control has a closed action flow.

Example: Annealing furnace

Schematic presentation

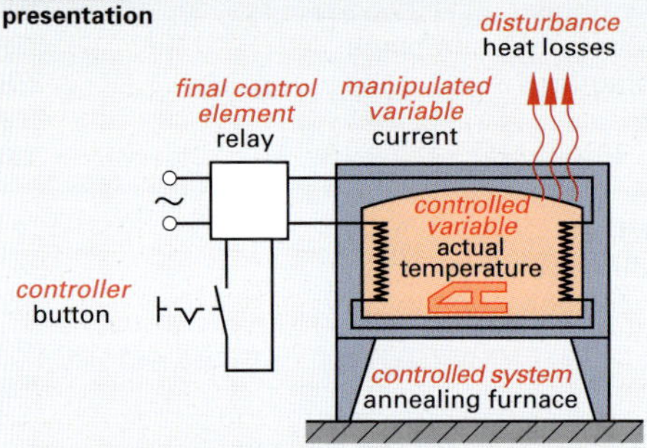

Schematic presentation

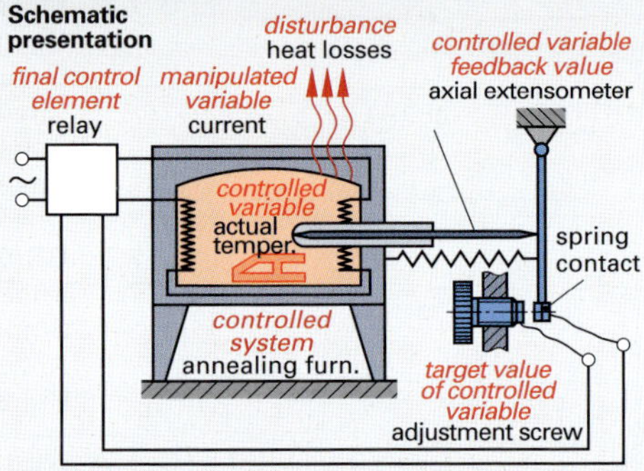

Functional diagram of open loop control system

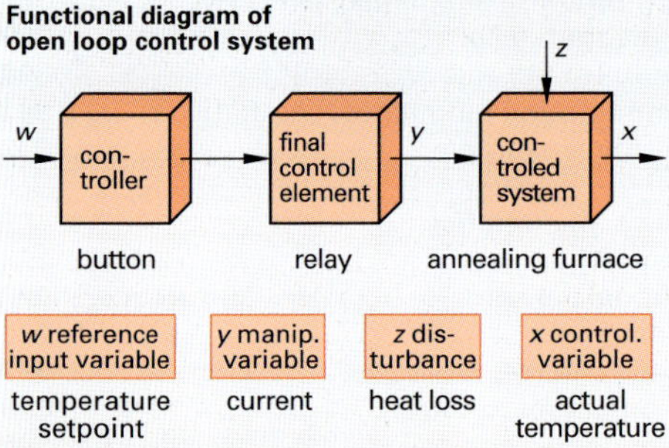

Simplified functional diagram of closed loop control system

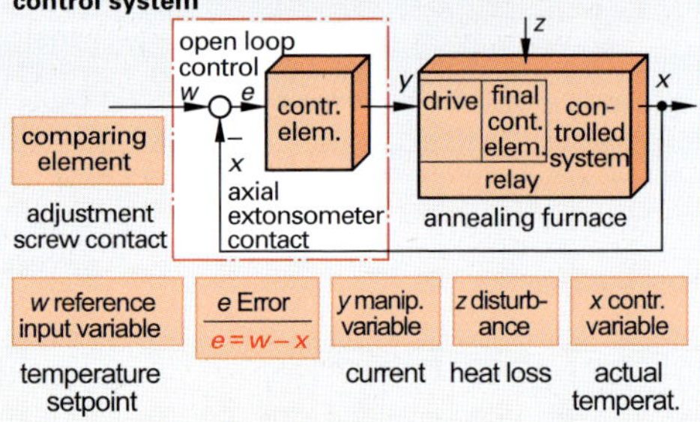

Application-based code letters

cf. DIN 19227-1 (1993-10), withdrawn

Designation example: P D I C

First letters

- D density
- E electrical parameters
- F flow, throughput
- G distance, position, length
- H manual input/intervention
- K time
- L status (e. g. level)
- M humidity
- P pressure
- Q quality parameters
- R radiation parameters
- S speed, rotational speed
- T temperature
- W weight, mass

Supplementary letters

- D difference
- F ratio
- J control point query
- Q sum, integral

Succeeding letters

- A error indication
- C automatic closed loop control
- H upper limit value
- I display
- L lower limit value
- R registration

Example: Differential pressure closed loop control

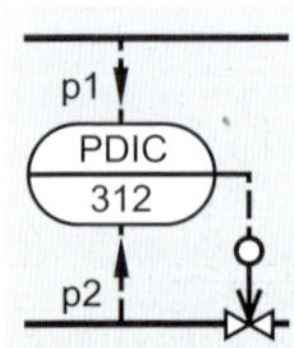

Explanation:
- P pressure
- D difference
- I display
- C automatic closed loop control

In plain language: Pressure differential closed loop control with display of pressure difference

A

Symbols

cf. DIN 19227-1 (1993-10), withdrawn

Location of output & user control

Symbol	Explanation
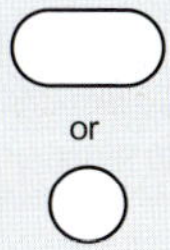	Local, general
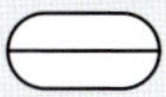	Process control room
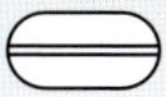	Local control console
	Local, implemented by process control system
	Local, implemented by process computer

Effect on the controlled system

Symbol	Explanation
	Servo motor, general
	Servo motor; the setting for minimal mass flow or flow of energy is set during loss of auxiliary power.
	Servo motor; the setting for maximum mass flow or flow of energy is set during loss of auxiliary power.
	Servo motor; the final control device remains in the most recently acquired setting during loss of auxiliary power.

Measuring point, control point

Symbol	Explanation
	Reference line
	Measuring point, sensor
	Final control element, control point

Example

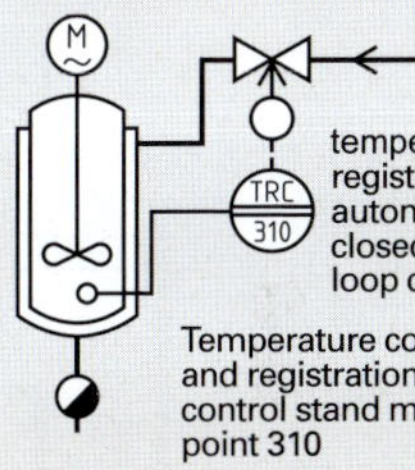

temperature T
registration R
automatic
closed
loop control C

Temperature control and registration at local control stand measuring point 310

Solution based symbols for devices

cf. DIN 19227-2 (1991-02), withdrawn

Sensors

Symbol	Explanation
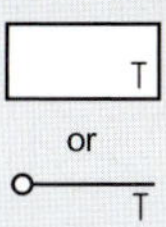	Sensor for temperature, general
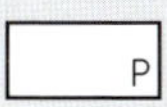	Sensor for pressure
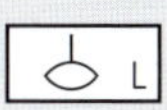	Sensor for level with float
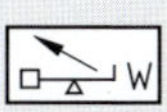	Sensor for weight, scales; indicating

Controllers

Symbol	Explanation
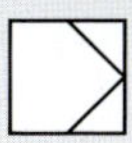	Controller, general
	Two-point controller with switching output and PID behavior
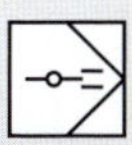	Three-point controller with switching output

Final controlling & user control elements

Symbol	Explanation
	Valve actuator with motor drive
	Valve actuator with solenoid drive
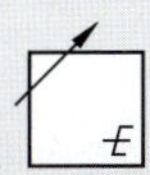	Adjuster for electric signal

Adapters

Symbol	Explanation
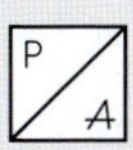	Pressure transducer with pneumatic signal output

Signal designators

Symbol	Explanation
Ɛ	Signal, electrical
A	Signal, pneumatic
∩	Analog signal
#	Digital signal

Output devices

Symbol	Explanation
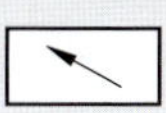	Basic symbol, general display
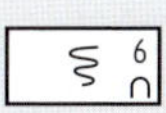	Printer, analog, no. of channels as a numeral
	Monitor

Example: Temperature controller

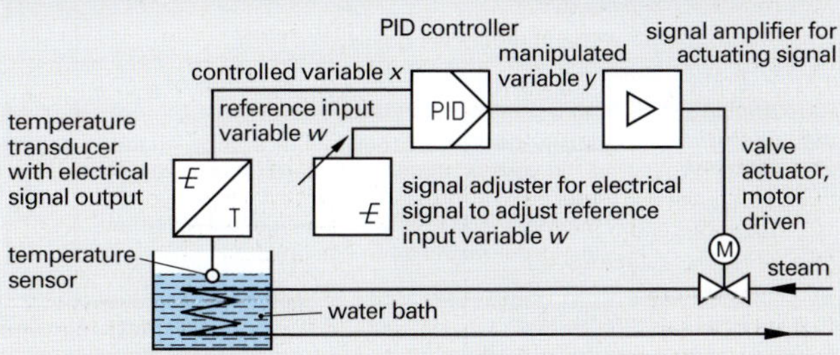

A

Analog controllers

Analog (continuous) controllers

cf. DIN 19225 (1981-12) and DIN 19226-2 (1994-02), both withdrawn

In analog controllers the manipulated variable y may assume any desired value within the control range.

Controller design	Level control example, description	Transition function	Symbol[1] Block representation[2]
P-controllers Proportional controllers Output variable is proportional to input variable. P-controllers have steady-state errors.	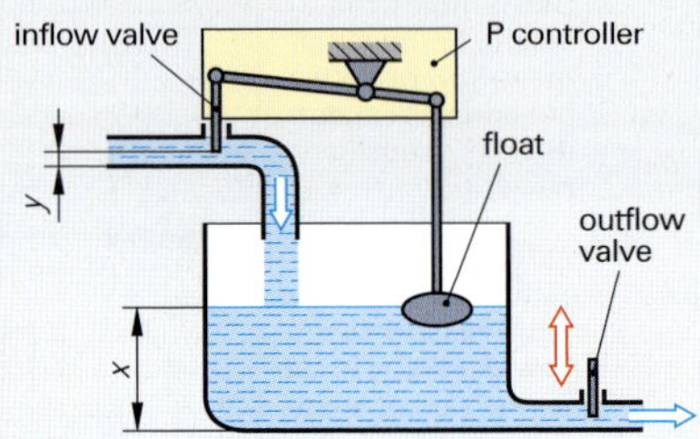	x controlled variable y manipulated variable e error 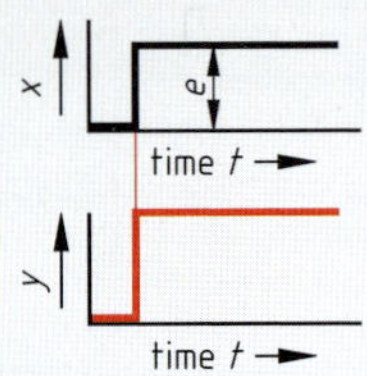	—— step function[3] —— step response[4] 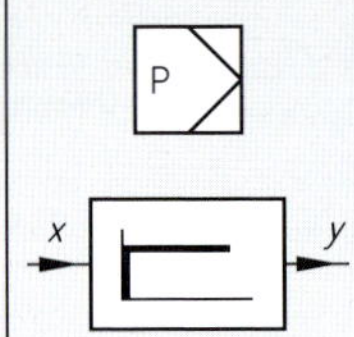
I-controllers Integral controllers I-controllers are slower than P-controllers, but they eliminate all errors.	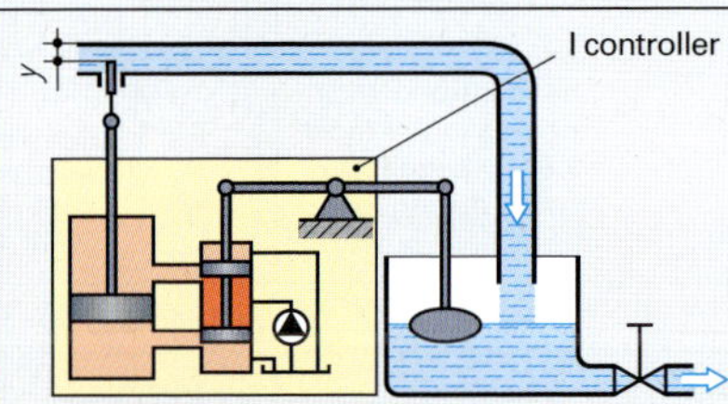	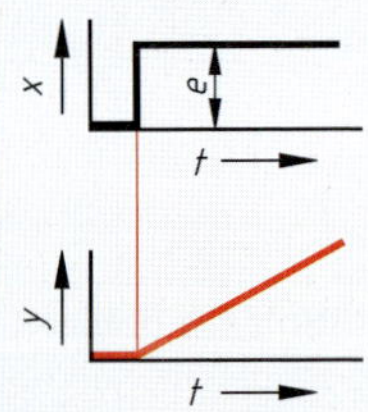	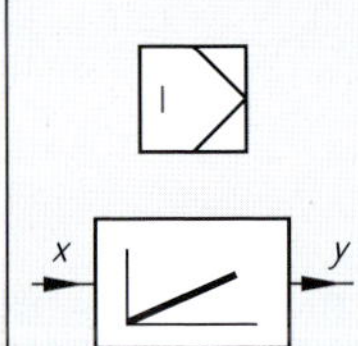
PI-controllers Proportional integral controllers In PI-controllers a P-controller and a I-controller are connected in parallel.	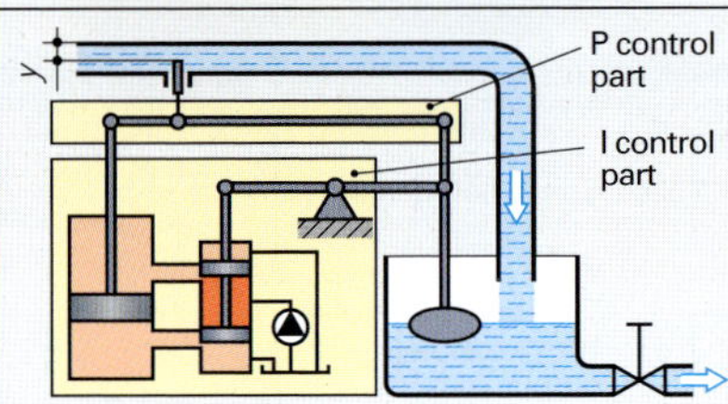	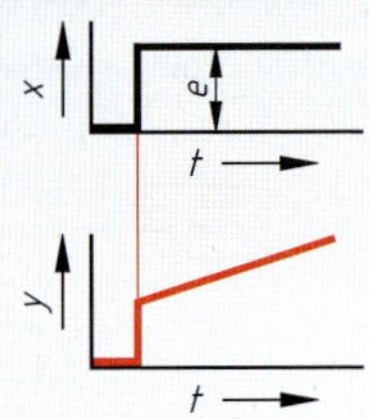	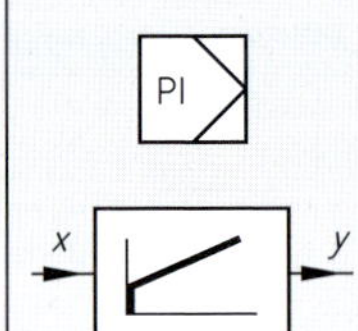
D-controllers Derivative controllers	D-controller systems only occur with P- or PI-controller systems, since pure D-controller behavior with constant error does not provide any manipulated variable and therefore no closed loop control.	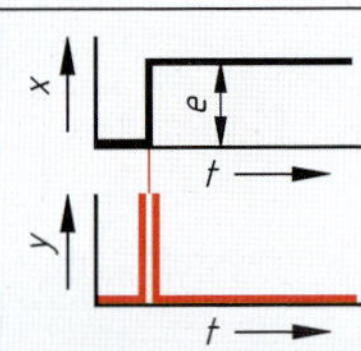	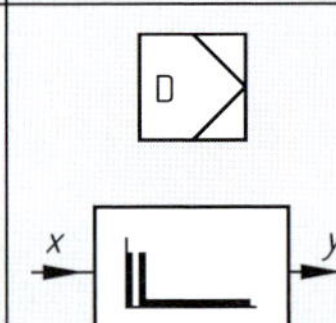
PD-controllers Proportional derivative controllers	PD-controllers are created when a P controller and a D element are connected in parallel. The D part changes the output variable at a rate proportional to the rate of change of the input variable. The P part changes the output variable so that it is proportional to the input variable itself. PD-controllers act quickly.	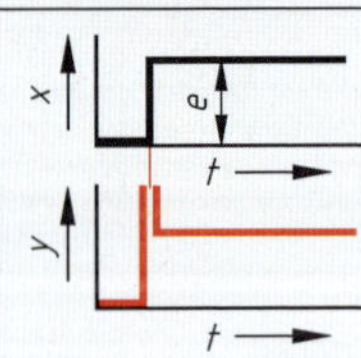	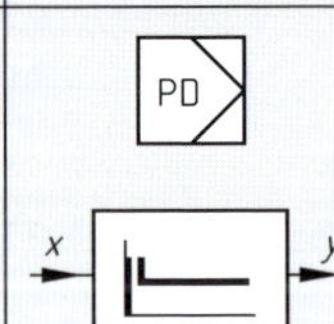
PID-controllers Proportional integral derivative controllers	PID-controllers are created by connecting P, I and D-controllers in parallel. Initially the D part reacts with a large change to the control signal, afterwards this change is reduced to approximately the magnitude of the P element, and finally the effect of the I element causes the response to rise linearly.	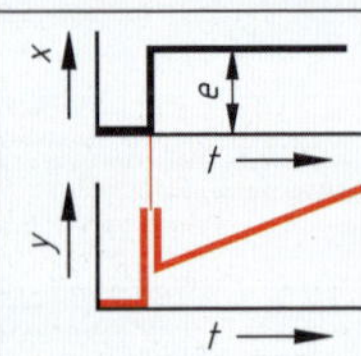	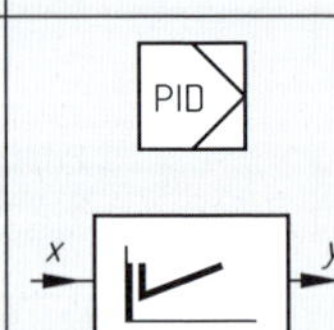

1) Symbol as per DIN 19227-2
2) Block representation as per DIN 19226-2
3) Signal curve at controlled system input
4) Signal curve at controlled system output

A

Discontinuous and digital controllers

Switching (discontinuous) controllers

Switching controllers change the manipulated variable *y* discontinuously by switching in several steps.

Controller design	Example, description	Transition function, switching behavior	Symbol Block representation
Two-point controller	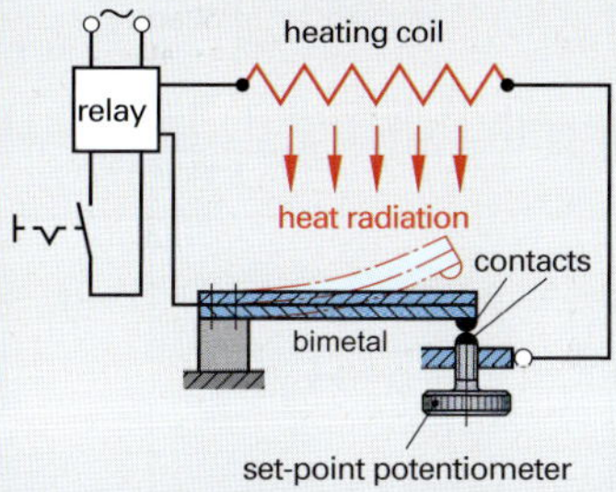	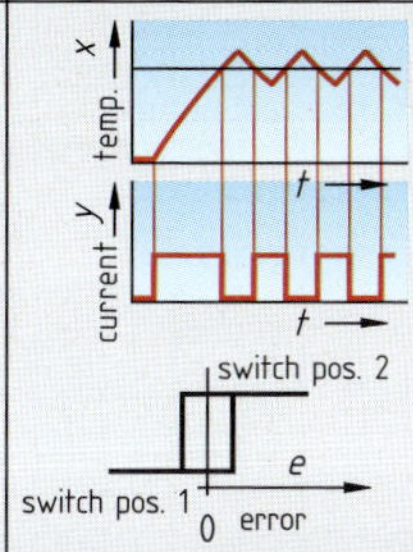	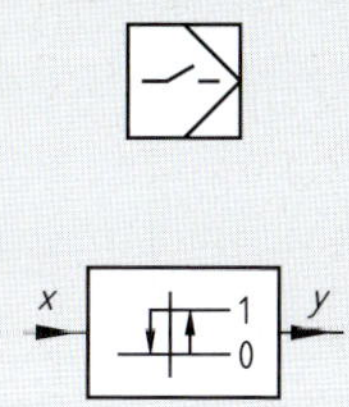
Three-point controller	**Air conditioning system** In an air conditioning system three temperature ranges are assigned three switch positions: – heating ON – heating/cooling OFF – cooling ON	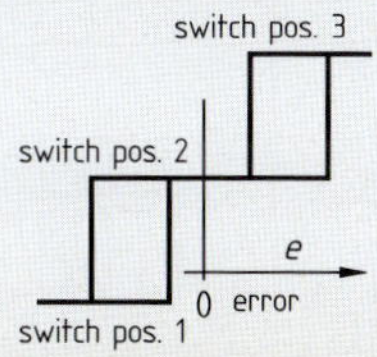	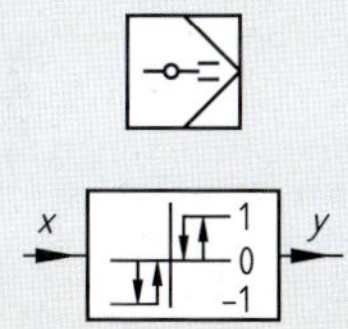

Digital controllers (software controllers)

The operating mode of the digital controller is implemented as a computer program.

Controller design	Example (simplified)	Transient function	Explanation
Computers **Programmable Logic Controllers (PLC)** **Microcontrollers** **Microprocessors**	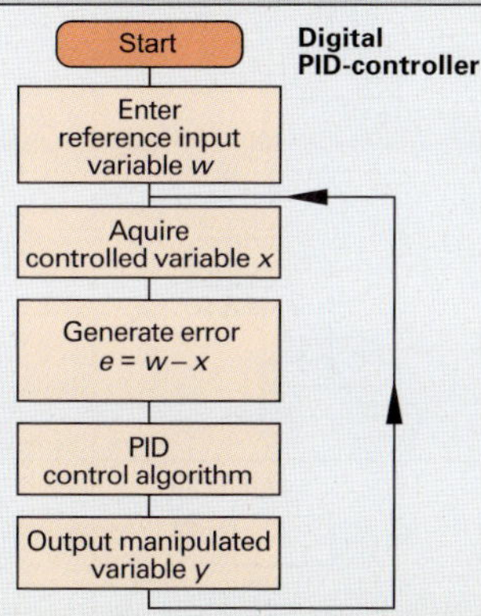	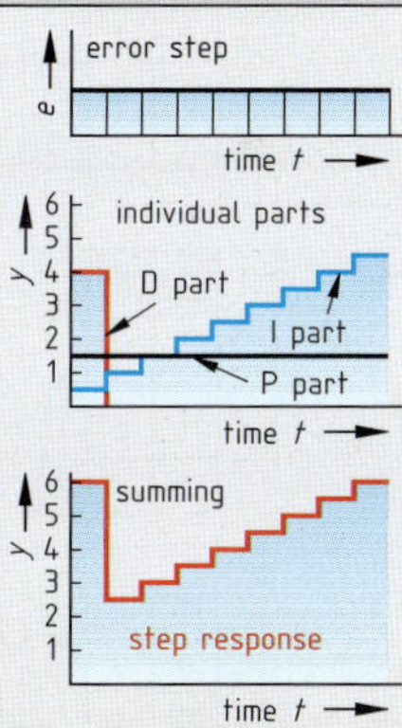	The computer program has the following tasks: – generate error *e* – calculate the manipulated variable *y* based on programmed control algorithms At the step response all P, D and I-parts are summed. Sampling of analog signals and their conversion to digital values and internal program flow causes a time delay of the controlled variable *x* (similar to a T-controlled system).

P-controlled systems with time delay (T part)

Controller design	Example	Transient function	Explanation
P-controlled system with delay 1st order ($P\text{-}T_1$ controlled system)	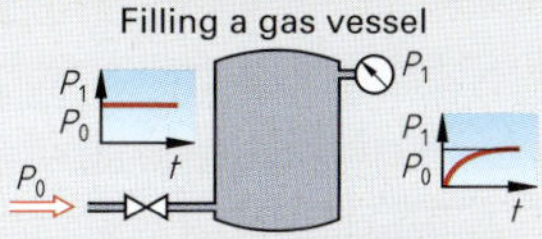	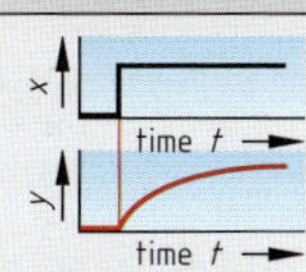	If the pressure vessel is filled by a flow of gas, pressure p_1 in the vessel gradually reaches the pressure of the gas flow.
P-controlled system with delay 2nd order ($P\text{-}T_2$ controlled system)	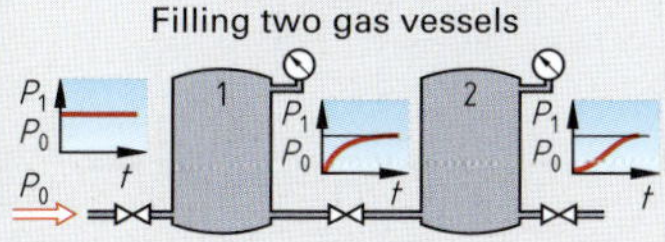	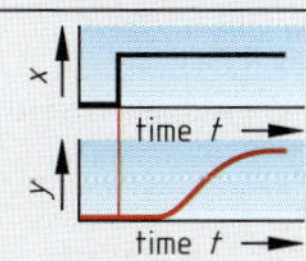	If two vessels are connected in series, pressure p_2 increases in the second vessel slower than pressure p_1 in the first vessel.

Binary logic

cf. DIN EN 60617-12 (1999-04)

Function	Circuit symbols / Logical equation	Function table	Technical implementation: pneumatic	Technical implementation: electric

AND

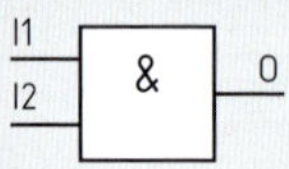

$O = I1 \wedge I2$

I1	I2	O
0	0	0
0	1	0
1	0	0
1	1	1

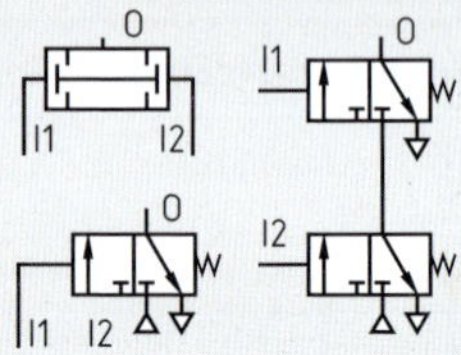

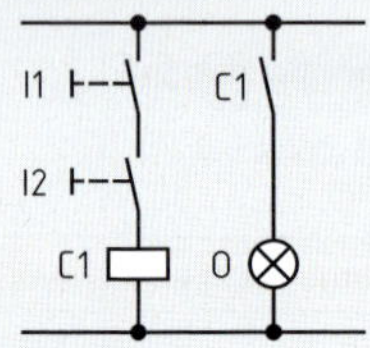

OR

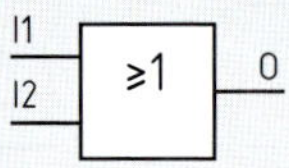

$O = I1 \vee I2$

I1	I2	O
0	0	0
0	1	1
1	0	1
1	1	1

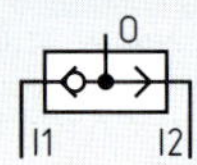

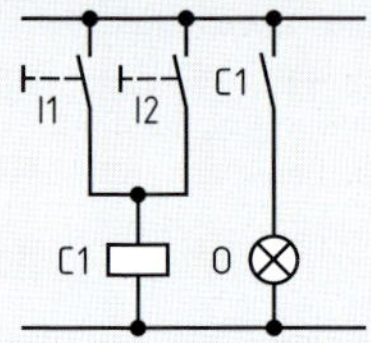

NOT

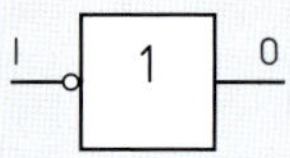

$O = \overline{I}$

I1	O
0	1
1	0

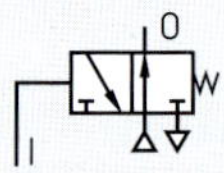

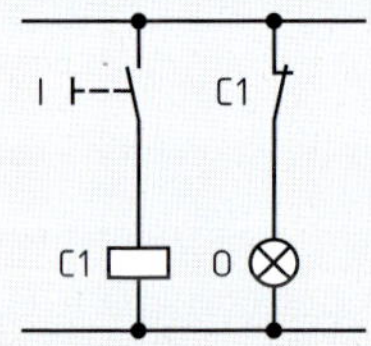

NOT AND (NAND)

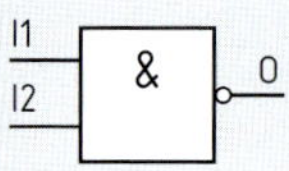

$O = \overline{I1 \wedge I2}$

I1	I2	O
0	0	1
0	1	1
1	0	1
1	1	0

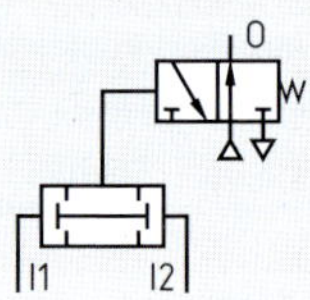

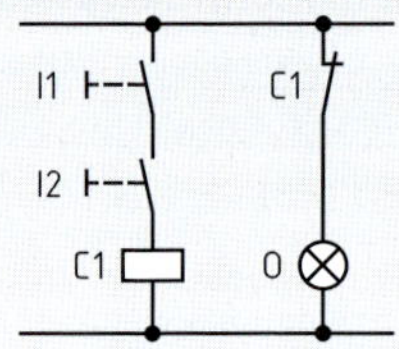

NOT-OR (NOR)

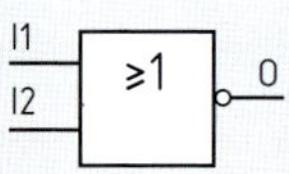

$O = \overline{I1 \vee I2}$

I1	I2	O
0	0	1
0	1	0
1	0	0
1	1	0

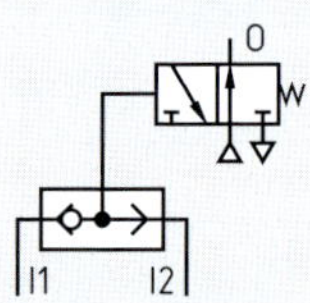

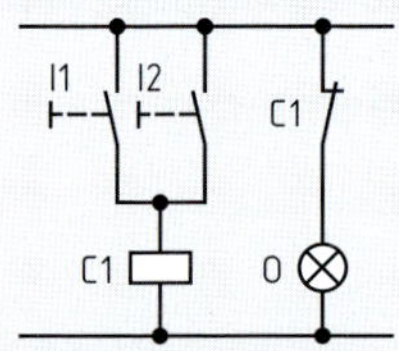

Exclusive OR (XOR)

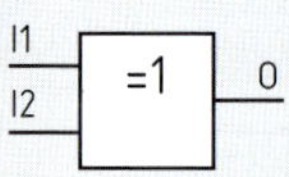

$O = (I1 \wedge \overline{I2}) \vee (\overline{I1} \wedge I2)$

I1	I2	O
0	0	0
0	1	1
1	0	1
1	1	0

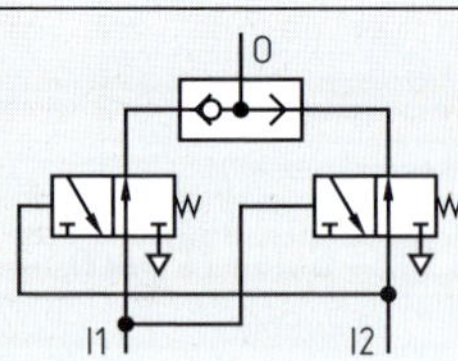

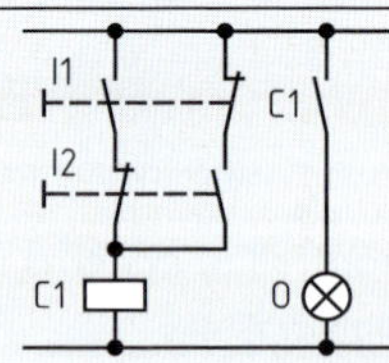

Memory (RS flip-flop)

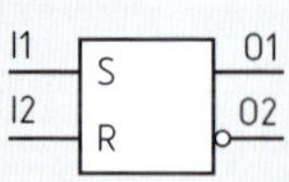

S set
R reset

I1	I2	O1	O2
0	0	•	•
0	1	0	1
1	0	1	0
1	1	□	□

• state unchanged condition
□ indeterminate state

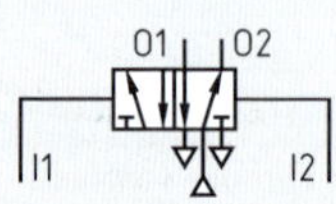

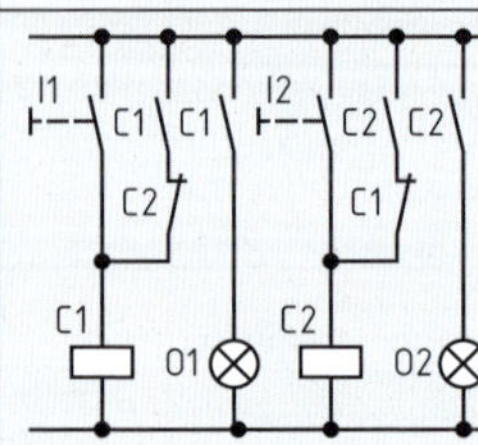

I = inputs O = outputs, e. g. lamps C = relays, contacts

Numbering systems

Decimal system

Base: 10	Numerals (possible numbers): 0, 1, 2, 3, 4, 5, 6, 7, 8, 9			
Decimal number n_{10}	2	0	5_{10}	(say: two zero five to the base 10)
Place value of n_{10}	$10^2 = 100$	$10^1 = 10$	$10^0 = 1$	
Numerical value of n_{10}	$2 \cdot 10^2$	$0 \cdot 10^1$	$5 \cdot 10^0$	
Total of n_{10}	200 +	0 +	5 =	205

Binary system

Base: 2	Numerals (possible numbers): 0, 1				
Binary number n_2	1	0	1	0_2	(say: one zero one zero to the base 2)
Place value of n_{10}	$2^3 = 8$	$2^2 = 4$	$2^1 = 2$	$2^0 = 1$	
Numerical value of n_{10}	$1 \cdot 2^3$ +	$0 \cdot 2^2$ +	$1 \cdot 2^1$ +	$0 \cdot 2^0$	
Total of n_{10}	8 +	0 +	2 +	0 =	10

Hexadecimal system

Base: 16	Numerals (possible numbers): 0, 1, 2, 3, 4, 5, 6, 7, 8, 9, A, B, C, D, E, F Decimal values: 0, 1, 2, 3, 4, 5, 6, 7, 8, 9, 10, 11, 12, 13, 14, 15			
Hexadecimal number n_{16}	A	2	F_{16}	(say: a two f to the base 16)
Place value of n_{10}	$16^2 = 256$	$16^1 = 16$	$16^0 = 1$	
Numerical value of n_{10}	$10 \cdot 16^2$ +	$2 \cdot 16^1$ +	$15 \cdot 16^0$	
Total of n_{10}	2560 +	32 +	15 =	2607

Conversion into other numbering systems

Conversion from binary to decimal (positional notation method)

Binary number n_2	1	0	0	1	1	0	0	1_2
Place values of n_{10}	2^7	2^6	2^5	2^4	2^3	2^2	2^1	2^0

Numerical values of n_{10}: $1 \cdot 2^7 + 0 \cdot 2^6 + 0 \cdot 2^5 + 1 \cdot 2^4 + 1 \cdot 2^3 + 0 \cdot 2^2 + 0 \cdot 2^1 + 1 \cdot 2^0$

Decimal values of n_{10}: $128 + 0 + 0 + 16 + 8 + 0 + 0 + 1 = 153$

Example:

The error monitoring unit of a position control system puts out the following error message in 6-bit binary code via the control bus: 110101_2. Which is the corresponding decimal error code of this message?

Binary number n_2	1	1	0	1	0	1_2
Place values of n_{10}	2^5	2^4	2^3	2^2	2^1	2^0

Numerical values of n_{10}: $1 \cdot 2^5 + 1 \cdot 2^4 + 0 \cdot 2^3 + 1 \cdot 2^2 + 0 \cdot 2^1 + 1 \cdot 2^0$

Decimal values of n_{10}: $32 + 16 + 0 + 4 + 0 + 1 = 53$

Conversion from decimal to binary (decimal remainder method)

Decimal number 192_{10}

192 : 2 = 96	remainder 0 ⇒ binary digit	0
96 : 2 = 48	remainder 0 ⇒	0
48 : 2 = 24	remainder 0 ⇒	0
24 : 2 = 12	remainder 0 ⇒	0
12 : 2 = 6	remainder 0 ⇒	0
6 : 2 = 3	remainder 0 ⇒	0
3 : 2 = 1	remainder 1 ⇒	1
1 : 2 = 0	remainder 1 ⇒	1

1 1 0 0 0 0 0 0_2

Example:

The @ character is the 64th character of the 7-bit ASCII[1)] character set. How is this character binary coded when it is transmitted to the CPU after the key stroke?

Decimal number 64_{10}

64 : 2 = 32	remainder 0 ⇒ binary digit	0
32 : 2 = 16	remainder 0 ⇒	0
16 : 2 = 8	remainder 0 ⇒	0
8 : 2 = 4	remainder 0 ⇒	0
4 : 2 = 2	remainder 0 ⇒	0
2 : 2 = 1	remainder 0 ⇒	0
1 : 2 = 0	remainder 1 ⇒	1

1 0 0 0 0 0 0_2

1) ASCII = **A**merican **S**tandard **C**ode for **I**nformation **I**nterchange, worldwide standard code.

A

Representation in information processing

A structogram or Nassi-Shneiderman diagram (NSD) is a type of diagram that allows the graphical design representation of structured programming. It was developed in 1972/73 by Isaac Nassi and Ben Shneiderman and is incorporated as a standard in DIN 66261. The method divides a complex problem into elementary basic structures such as sequences and controlling structures (selection or loop blocks).

Structogram cf. DIN 66261 (1985-11)	**Description**	**Syntax example in C/C++/C#**	**Program flow chart** cf. DIN 66001 (1983-12)
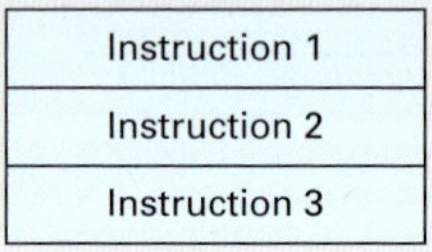	Instruction sequence (sequence block) The instruction sequence may include value assignments, calculations or instructions for the input and output.	*{* *instruction_1;* *instruction_2;* *instruction_3;* *}*	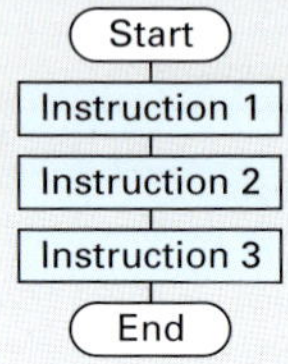
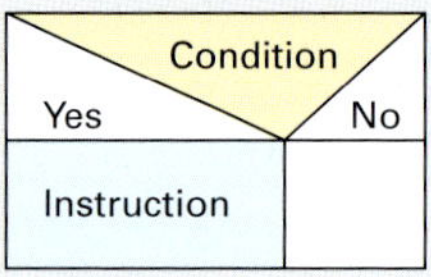	Single-sided branching (selection block) The branching contains only one alternative instruction. The second alternative does not involve any operation.	*if (condition)*[1] *instruction;*	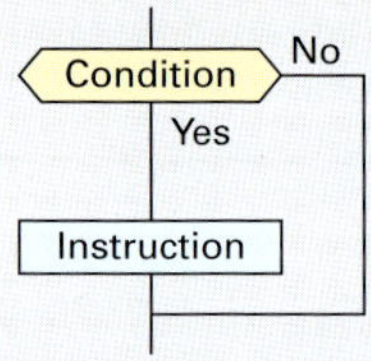
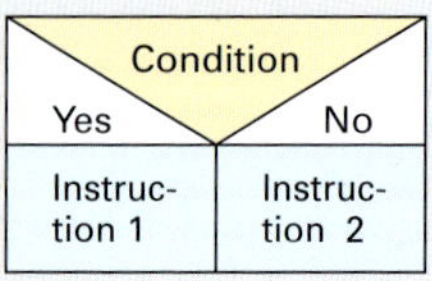	Two-sided branching (selection block) The branching contains two alternatives. Depending on whether yes or no is returned, instruction 1 or 2 is executed.	*if (condition)*[1] *{* *instruction_1;* *}* *else* *{* *instruction_2;* *}*	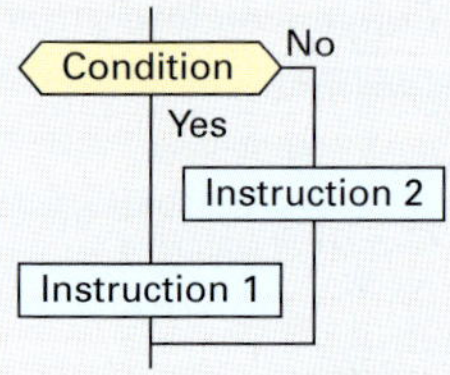
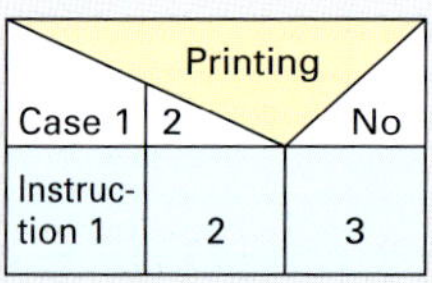	Multiple branching (case selection block) If case 1 and 2 do not apply, instruction 3 is executed according to the *default* setting. After each *break*, the program jumps to the end of the selection block.	*switch (printing)*[2] *{ case 1:* *instruction_1;* *break;* *case 2:* *instruction_2;* *break;* *default:* *instruction_3;* *break;* *}*	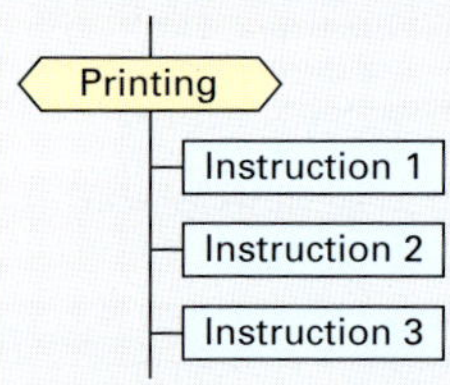
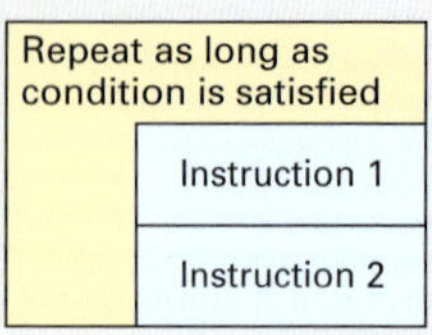	Test first loop block (pre-test loop, while loop) The instructions of the loop block are repeatedly executed as long the condition is satisfied. The condition is tested at the beginning of the loop.	*while (condition)*[1] *{* *instruction_1;* *instruction_2;* *}*	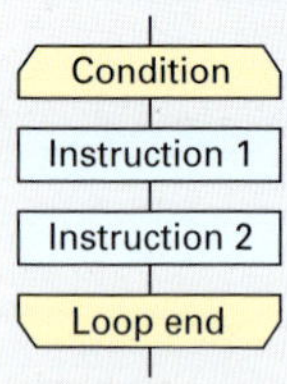
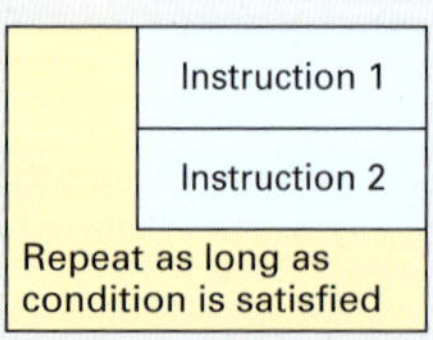	Test last loop block (post-test loop, do loop) The instructions of the loop block are repeatedly executed as long as the condition is satisfied. The condition is tested at the end of the loop.	*do* *{* *instruction_1;* *instruction_2;* *}* *while (condition)*[1]*;*	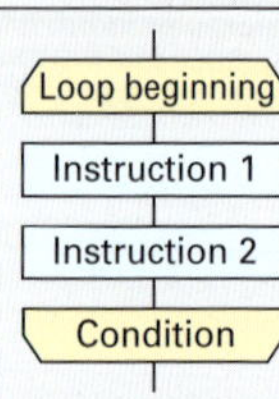

[1] A variable is linked to a value via an operator, e.g. $x = 1$ or $y < 10$
[2] Fixed assignment of a value to a case query or menu option

A

Circuit symbols

cf. DIN EN 60617-1 to -11 (1997-08)

General circuit symbols

- Resistor, general
- Fuse
- Capacitor
- Inductor, coil
- Nonstandard representation
- Permanent magnet
- Lamps, general, optional representation
- Buzzer
- Horn
- Electrolytic component (+ −)
- Converter, transducer

Conductors, connectors and terminals

- Conductor, general
- Conductor, moveable
- Conductor, insulated
- Grounded conductor, PE
- Neutral conductor, PN
- Neutral conductor with protective function PEN
- Junction, optional representation
- Double junction, optional representation
- Connection to ground, optional representation
- Ground
- Ground connector connection

Devices and machines

- Measuring device, machine
- Measuring device, recording
- Transformer, optional representation
- Valve

Semiconductor components

- Semiconductor diode, general
- LED light emitting diode
- PNP transistor
- NPN transistor

Designations

Adjustability

- general
- adjustable
- regulated

Function

- stepped
- continuous

Effect

- thermal
- radiation

Types of current

- DC
- AC with low frequency
- AC with high frequency

Types of connections

- Y connection
- Delta connection
- Y-delta connection

Circuit symbols in wiring system drawings

- Circuit switch a) single-pole b) double-pole (a) b))
- Sensor switch
- Series switch
- Three-way switch, illuminated
- Grounding-type receptacle
- Key button
- Three-pole switch, protective system IP 44 (IP 44)
- Automatic breaker
- Motor circuit breaker
- Ground-fault circuit interrupter

Application examples

- Inductor, continuously adjustable
- Resistor, 5 step variable (5)
- DC-AC converter, regulated
- DC or AC (universal)
- Three-core cable with junction
- Cable with 3 conductors, with ground conductor (G) and 1.5 mm² cross section (3 G 1,5)
- DC motor (M)
- Three-phase motor (M 3~)

A

Circuit symbols

cf. DIN EN 60617-1 to -11 (1997-08)

Relay contacts

Symbol	Meaning
	NO contact, normally open
	NC contact, normally closed
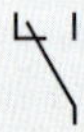	Single pole double throw

Actuation types

Symbol	Meaning
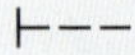	Manual, general
	By pressing
	By pulling
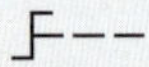	By turning
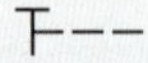	By tilting
	By key
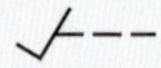	By pedal
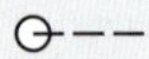	By coil
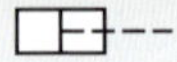	By pressure energy
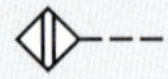	By proximity
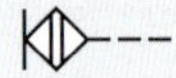	By touching
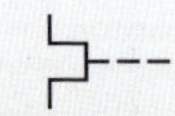	By bimetal (thermal)

Electromech. relays

Symbol	Meaning
	Relay coil, general
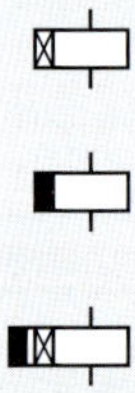	Timer on delay Timer off delay Timer on off delay

Switch behavior

Symbol	Meaning
	Lock, prevents automatic return
a) b) 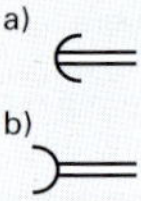	Delayed action (parachute effect) for movement a) to the right b) to the left
	Symbol for "actuated state"

Sensors (Block representation)

Symbol	Meaning
	Capacitive sensor, reacts to proximity of all substances
	Inductive sensor, reacts to proximity of metals
	Magnetic sensor, reacts to close proximity of a magnet (reed switch)
	Optical sensor, reacts to reflection of infrared beam

Examples of switch applications

Symbol	Meaning
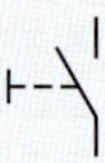	NO contact manually
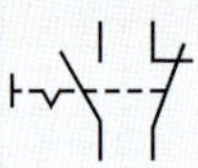	Double pole single throw
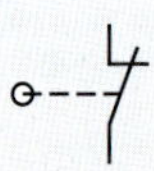	NC contact with roller actuation
a) b) 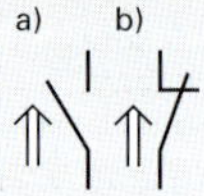	a) NC contact b) NO contact Representation in actuated condition
a) b) 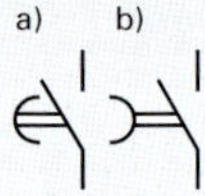	NO contact a) closes b) delayed opening when actuated
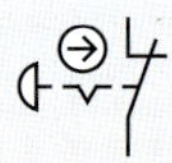	Emergency palm button
	Limit switch, NO contact
	Limit switch, NC contact
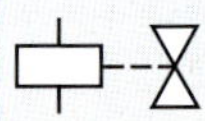	Valve with electromagnetic actuation
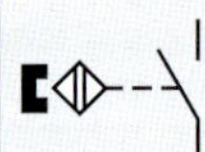	Magnetic proximity switch with NO contact, reacts to proximity of magnetic material.
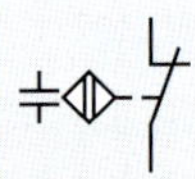	Capacitive proximity switch with NC contact, reacts to proximity of all materials.

Flip-flop elements

RS[1] flip-flop

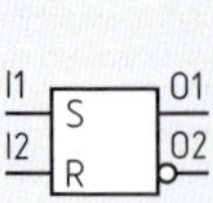

Function table[2]

I1	I2	O1	O2
0	0	●	●
0	1	0	1
1	0	1	0
1	1	□	□

RS flip-flop set dominant

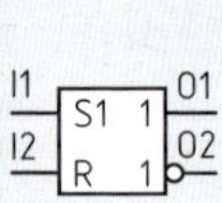

Function table

I1	I2	O1	O2
0	0	●	●
0	1	0	1
1	0	1	0
1	1	1	0

RS flip-flop reset dominant

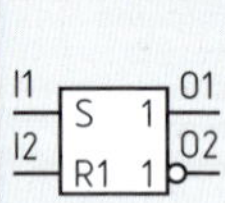

Function table

I1	I2	O1	O2
0	0	●	●
0	1	0	1
1	0	1	0
1	1	0	1

Flip-flops are integrated circuits which store signal conditions.

[1] R = reset
S = set

[2] ● unchanged state
□ indeterminate state

The numeral 1 after an R or S input indicates that the logical state of this input is dominant.

If a signal simultaneously lies at inputs I1 and I2 (I1 = 1 and I2 = 1) the following applies:

Input without the numeral 1 (R for set dominant, S for reset dominant RS flip-flop) is always set to logical state 0.

Delay elements

With rise-delay time

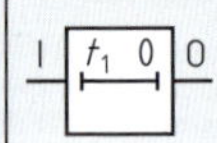

When a signal is applied to input I, output O assumes value 1 after time t_1 elapses.

With turn-off delay

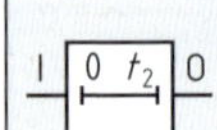

With loss of a signal at input I, output O takes the value 0 after completion of time t_2.

A

Designations in circuit plans*

Designation of devices in circuit diagrams

cf. DIN EN 61346-2 (2005-11)

Example: **S 2 E**

- **S** → **Type of device**
- **2** → **Sequential number**
- **E** → **Device function**

Code letters for type (selection)	
B	Sensor, proximity switch
F	Fuse
K	Switch relay, timed relay
Q	Circuit breaker, contactor
M	Solenoid valve, solenoid
P	Indicator light, horn
R	Resistor
S	Control switch, push-button switch

Code letters for function (not standardized)	
A	Function OFF
B	Direction of movement
E	Function ON
G	Test
K	Jog operation
S	Save, set
R	Clear, reset

Example of circuit diagram

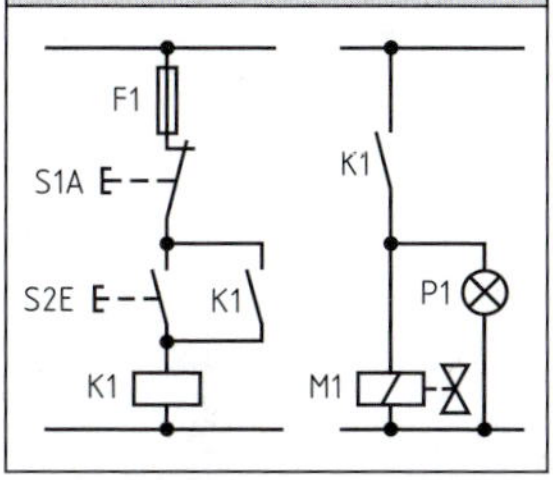

Designation of wires and connections

cf. DIN EN 60446 (2008-02) and DIN EN 60445 (2007-11)

Insulated wires

Type of wire		Code letters	Wire color	Symbols
DC network	positive	L+	black[1]	+
	negative	L–	black[1]	–
	neutral wire	M	light blue	(light blue line)
AC network	Phase conductor 1	L1	black[1]	(line)
	Phase conductor 2	L2	black[1]	(line)
	Phase conductor 3	L3	black[1]	(line)
	neutral wire	N	light blue	(light blue line with mark)
Ground wire		PE	green-yellow	(dashed line with mark) ⏚
PEN wire (neutral wire with ground function, PE + N)		PEN	green-yellow[2]	(dashed line with mark)
Ground		E	black[1]	⏚

Example: Rectifier circuit

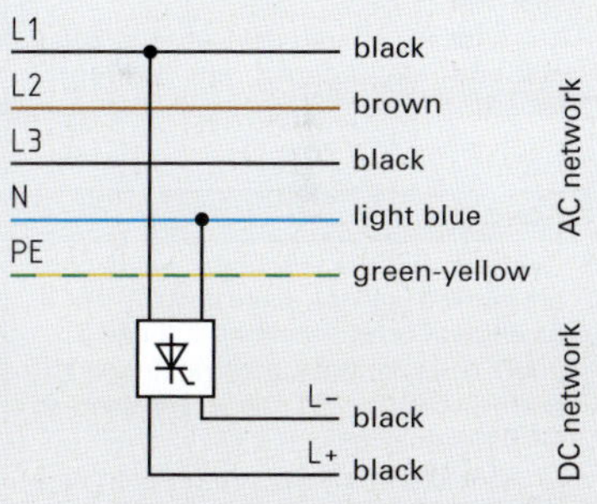

Device connections

Connections for	Designation
Phase conductor 1	U
Phase conductor 2	V
Phase conductor 3	W

Example: Star-connected (squirrel) cage motor

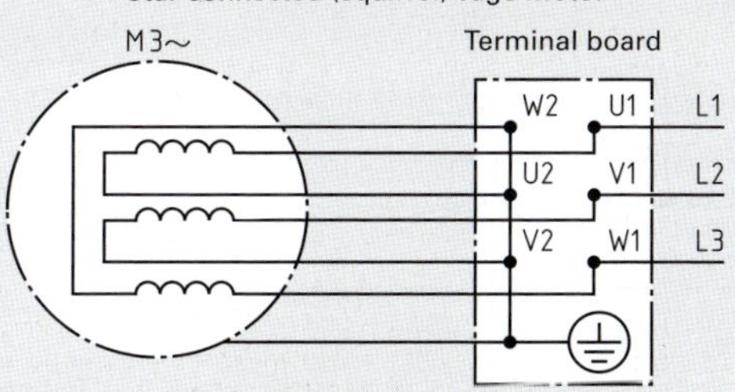

[1] Color is unspecified. Black is recommended, brown to differentiate. Green-yellow may not be used.

[2] PEN-wires have a continuous green-yellow conductor color. To avoid confusion with PE wires, PEN wires are additionally marked with light blue on the ends of the wires, e. g. with a wire clip or adhesive tape.

*) According to European standards

A

Circuit diagrams

cf. DIN EN 61082 (2007-03)

Connector markings on relays

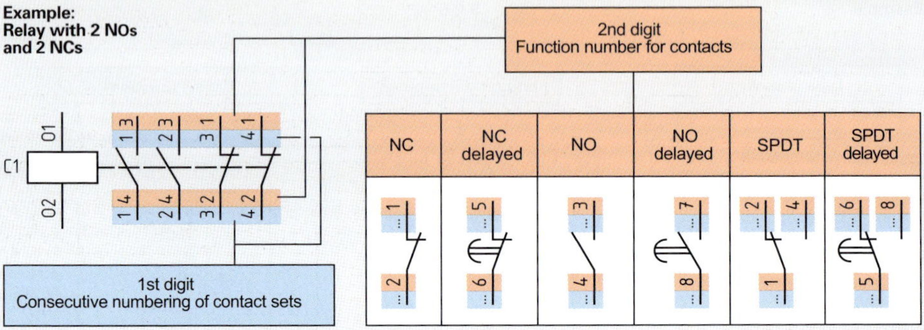

Designing circuit diagrams

Current sections and distribution of electric circuits

- Every electrical device is shown with a vertical current section regardless of the actual spatial arrangement of the elements.
- Current sections are numbered sequentially from left to right.
- The **control circuit** contains devices for signal input and signal processing.
- The **main circuit** contains the necessary final control elements for the working elements.
- The spatially shared devices, e. g. relay coil and relay contact, are not represented.

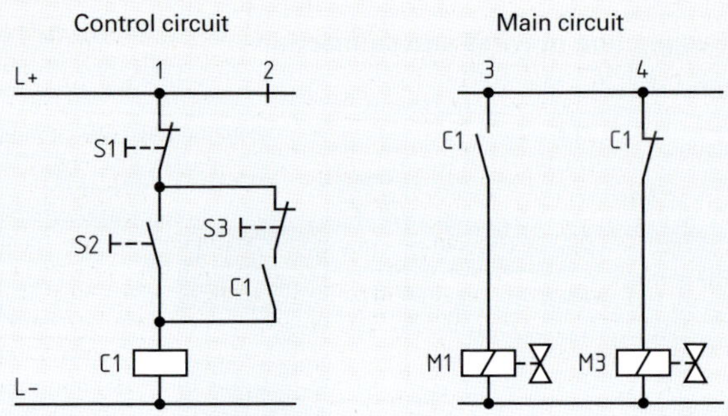

Designation of devices

- Contacts and the associated relay coils are marked with the same code numeral. **Example:** Current sections 1, 2 and 3
- 2 NO contacts belong to relay coil C1, both marked as C1. They are used to latch the relay coil.
- All contacts of a relay are entered as a complete contact set or as a table under the current path of the relay. Both representations indicate the current section on which a contact is located.

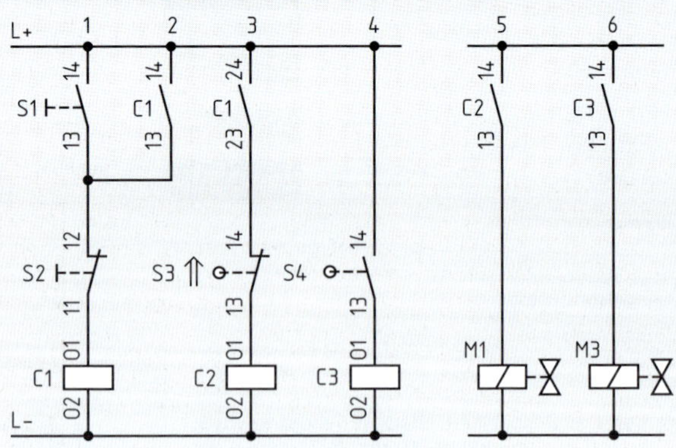

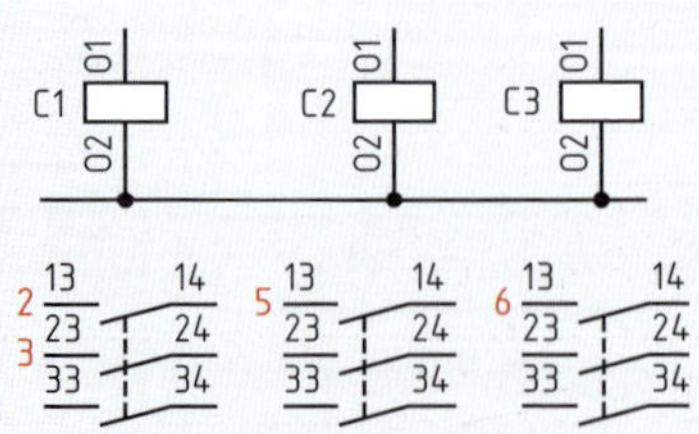

Representation as contact set

Contacts C1	Section
13 – 14	2
23 – 24	3

Contacts C2	Section
13 – 14	5

Contacts C3	Section
13 – 14	6

Representation as table

A

Sensors

Sensors (selection)

Sensors
- Sensors that are sensitive to proximity
  - Inductive sensors
  - Capacitive sensors
  - Photoelectric sensors
  - Ultrasound sensors
  - Magnetic sensors
- Tactile sensors
  - Limit switches

Characteristics of sensors

Sensor type	Symbol	Principle	Advantages	Disadvantages	Object distance
Inductive		Triggers if an object interferes with the alternating magnetic leakage field of the sensor	High degree of protection (IP67), very high switch point precision, dirt tolerant	Only objects with high electrical conductivity, unsuitable where there is greater accumulation of metal chips	1 mm to 150 mm
Capacitive		Triggers if an object interferes with the alternating electric leakage field of the sensor	High degree of protection (IP67), detects all materials; dirt tolerant	Small object distances, larger design than comparable inductive sensors	20 mm to 40 mm
Photoelectric		Triggers if an object returns the infrared field of the sensor	Detects all materials, large distances	Sensitive to dirt, smoke and secondary light, auxiliary power necessary	approx. 2 m
Ultrasound		Evaluates transit times of reflected ultrasonic pulses to determine the distance to an object	Tolerant to dust, dirt and light; detects very small objects at large distances	Slow, use only with standard pressure, not in areas subject to explosion hazards and no high-frequency noise	60 mm to 6 m
Magnetic		A permanent magnet actuates a proximity switch (reed contact) using two contact springs	Suitable in rough environment, high service life, suitable for switches in high frequency circuits	Risk of contact welding; suppresses the current peaks of RC modules	–
Mechanical		Triggered by manual actuation or lever system	Low price, robust, small, unaffected by interference fields, no auxiliary power necessary	Contact chatter, not allowed in food and chemical industries	–

Designation of proximity sensors

cf. DIN EN 60947-5-2 (2008-08)

Example: U 1 A30 A F 2 N

U – Type of detection; 1 – Mechanical mounting conditions; A30 – Design and size; A – Circuit element function; F – Type of output; 2 – Type of connection; N – NAMUR function

Type of detection	Mechanical mounting conditions	Design and size	Circuit element function	Type of output	Type of connection	NAMUR function
I inductive C capacitive U ultrasound D photoelectric diffuse reflected luminous beam M magnetic R photoelectric reflected luminous beam T photoelectric direct luminous beam	1 flush mounting possible 2 flush mounting not possible 3 unspecified	FORM A cylindrical threaded sleeve B smooth cylindrical sleeve C rectangular with square cross-section D square, with rectangular cross-section SIZE (2 digits) for diameter or side length	A NO contact B NC contact C single pole double throw P programmable by user S other	P PNP output, 3 or 4 DC connections N NPN output, 3 or 4 DC connections D 2 DC connections[1] F 2 AC connections[2] U 2 AC or DC connections S other	1 integrated connection line 2 plug connection 3 screw connection 4 … 8 unused 9 other type of connection	N NAMUR[3] function Note: NAMUR sensors are 2 wire sensors that are connected to an external switching amplifier

[1] DC = Direct Current
[2] AC = Alternating Current
[3] NAMUR = **N**ormen**a**rbeitsgemeinschaft für **M**ess- **u**nd **R**egelungstechnik (Standardization Association for Measurement and Control)

A

Safety precautions*

Safety precautions against electrical shock

cf. DIN VDE 0 100-410 (2003-06)

Protection against direct and indirect contact	**Protection against electric shock under normal conditions: against direct contact**	**Protection against electric shock under fault conditions: for indirect contact**
Protection by: – Safety Extra Low Voltage (SELV) – Protective Extra Low Voltage (PELV) – Functional Extra Low Voltage FELV	Protection by: – protective insulation of active parts, e. g. cable – coating as insulation, e. g. housings on electr. devices – distance, e. g. protective hoods, housings of machine screen – barriers, e. g. protective screen, enclosure	Protection by: – automatic disconnect or warning, e. g. residual current protective device – potential equalization – non-conductive areas; e. g. by insulating coverings – protective insulation, e.g. housings encapsulated with insulating material

⇑ Additional protection by residual current circuit breaker GFI's: Ground Fault Interrupter

Effects of alternating current

cf. IEC 60479-1 (1994)

Safety curves for AC 50 Hz from hand to hand or from hand to foot for adults

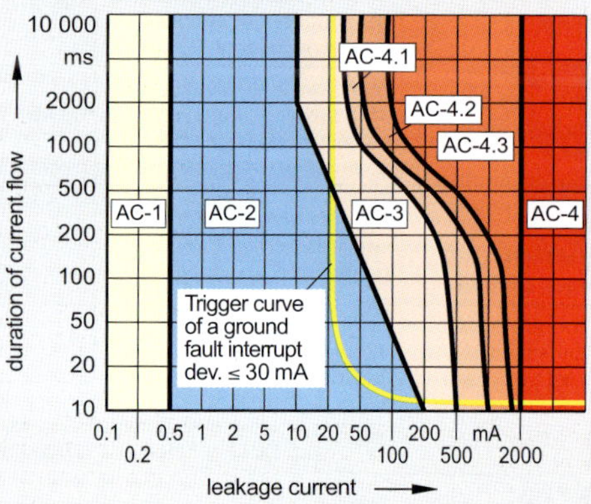

Zone	Physical effects
AC-1	normally no effect
AC-2	normally no damaging physical effects
AC-3	usually no organic damage, difficulty breathing (> 2 s), muscle cramps
AC-4.1	5% probability of ventricular fibrillation
AC-4.2	up to 50% probability of ventricular fibrillation
AC-4.3	over 50% probability of ventricular fibrillation
AC-4	cardiac arrest, cessation of breathing, and extreme burns (increasing with exposure time and current level)

Automatic fuses and wire cross-sectional areas

cf. DIN VDE 0 1000-430 (1991-11)

Rated current of fuse I_n in A	Color code of fuse	Minimum cross-sectional area in mm² for Cu wires by method of installation and number of loaded strands: A1		B1		B2		C	
		2	3	3	3	2	3	2	3
10 (13)	red	1.5	1.5	1.5	1.5	1.5	1.5	1.5	1.5
16	gray	1.5	2.5	1.5	1.5	1.5	1.5	1.5	1.5
20	blue	2.5	2.5	2.5	2.5	2.5	2.5	1.5	2.5
25	yellow	4	4	2.5	4	4	4	2.5	2.5
35	black	6	6	6	6	6	6	4	4
50	white	10	16	10	10	10	10	10	10

Method of installation of cables and insulated wires

cf. DIN VDE 0 298-4 (2003-08)

Method	Illustration	Description
A1	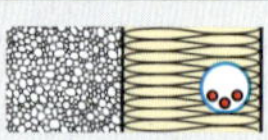	Installation in thermally insulated walls, in electrical conduit
B1	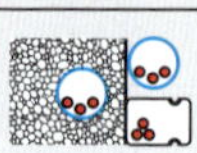	Installation in electrical conduit or in the wall or in cable channels
B2	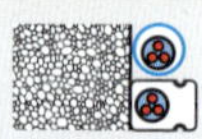	Installation in electrical conduit or in the wall, in cable channels or behind base boards
C	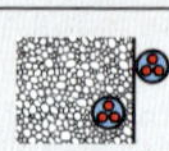	Installation directly on or in the wall

*) According to European standards

Safety precautions*

Protective systems for electrical devices

cf. DIN EN 60529 (2000-09)

Example: IP 3 4 C M

- IP – Protective system designation IP (International Protection)
- 3 – 1st code numeral for protection of device[1] against penetration of solid foreign objects
- 4 – 2nd code number for protection of the device[1] against water with damaging effect
- C – Additional code letters[2]
- M – Supplementary letters

Code no	1st code no. Protection against accidental contact	Protection from foreign objects	Symbol
0	No protection	No protection	
1	Protected against contact by back of the hand	Protected against penetration by foreign objects $d \geq 50$ mm	
2	Protected against contact with finger $d = 12$ mm	Protected against penetration by foreign objects $d \geq 12.5$ mm	
3	Protected against contact with a tool $d = 2.5$ mm	Protected against penetration by foreign objects $d \geq 2.5$ mm	
4	Protected against contact with a wire $d = 1$ mm	Protected against penetration by foreign objects $d \geq 1$ mm	
5	Protected against contact with a wire $d = 1$ mm	Protected from dust	
6	Protected against contact with a wire $d = 1$ mm	Dust proof	

[1] If a code number is not given, the letter X is used in its place, e. g. IP X6 or IP 3X

[2] Is only given if the protection is greater than the 1st code number.

Code no	2nd code number Water protection	Symbol
0	No protection	None
1	Protected against vertical drips	
2	Protected against drips if device is inclined 15°	
3	Protected against water spray impacting device at 60°	
4	Protected against water spray from all directions	
5	Protected against water jets from all directions	
6	Protected against strong water jets from all directions	
7	Protected against temporary submersion in water	
8	Protected against continual submersion in water	...kPa

Additional letters	
A	Protected against contact by back of the hand
B	Protected against contact with finger $d = 12$ mm, 80 mm long
C	Protected against contact with a tool $d = 2.5$ mm, 100 mm long
D	Protected against contact with a wire $d = 1$ mm, 100 mm long

Supplementary letters	
H	Equipment for high voltage
M	Tested on water intake in running machine
S	Tested on water intake on idle machine
W	Suitable for specific weather conditions

Electric equipment for explosive areas

cf. DIN EN 13237 (2003-11)

Example: EEx de II/B T2

- EEx – Symbol for explosion protection
- de – Type of protection
- II/B – Electrical devices group
- T2 – Temperature class

Code	Type of protection
o	oil immersion
p	pressurized enclosure
q	sand filling
d	flameproof enclosure
e	increased safety
i	inherent safety

Group II A	B	C
Risk of explosion by occurrence of the following gases:		
methane, propane, butane, propylene, benzene, toluol, naphthalene, turpentine, petroleum, gasoline, fuel oil, diesel oil, carbon monoxide, methanol, metaldehyde, acetone, acids, chloride	ethylene, acryl nitrite, hydrogen cyanide, dimethylether, propylene oxide, coke oven gas, tetrafluoroethylene	hydrogen, acetylene, carbon bisulphide, ethyl nitrite

Code	Surface temperature
T1	450 °C
T2	300 °C
T3	200 °C
T4	135 °C
T5	100 °C
T6	85 °C

*) According to European standards

Function charts for sequential controls (GRAFCET)

cf. DIN EN 60848 (2002-12)

The function chart in accordance with GRAFCET is a graphical design language for sequential control. However, it does not make any statement about the type of devices used, the direction of lines and the installation of electrical equipment. Only the general representation via symbols is obligatory; dimensions and other details are left to the user.

Important basic terms

GRAFCET	French: **GRA**phe **F**onctionel de **C**ommande **E**tape **T**ransition (say: grafset) English: specification language for function charts of sequential controls
Transition	condition for the transition to the next step
Variable	value that may change
George Boole	British mathematician
Boolean variables	TRUE, FALSE TRUE = logical value 1 FALSE = logical value 0
Initial step	first step
Macro step	condensed representation of a step chain

Basic structure of a GRAFCET

The basic structure of GRAFCETs consists of:

steps, e. g. [2] interactive connections, e. g. |

actions, e. g. [1A1] transitions, e. g. — 1B2

Sequential structure

Steps and transitions are transition conditions that alternate permanently. In linear sequences, only one step is active at a time and can trigger any number of actions. In alternative or parallel branchings, several steps can be active simultaneously.

Example of a GRAFCET (sequence chart)

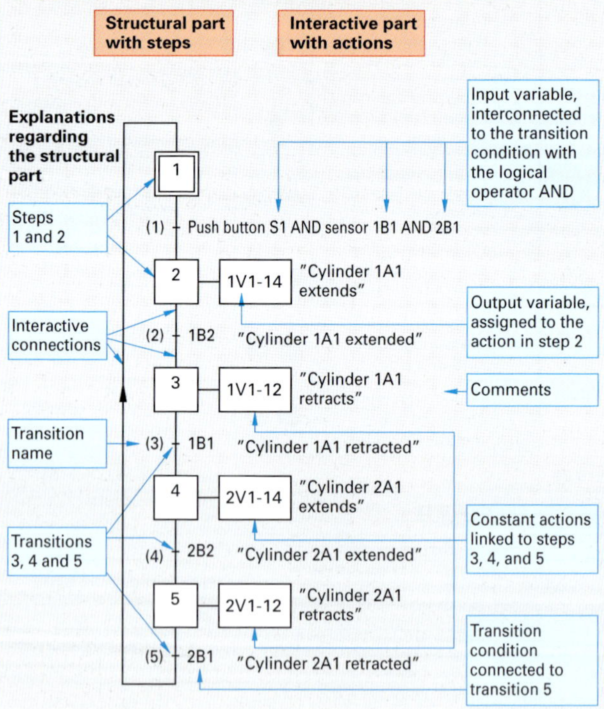

Technology schemes and circuit diagrams are not part of a GRAFCET, but they help considerably to make the task more transparent.

Example: belt conveyor

After activation of push button S1, cylinder 1A1 pushes the workpiece onto a pallet and retracts. Cylinder 2A1 pushes the workpiece subsequently onto a belt conveyor and retracts.

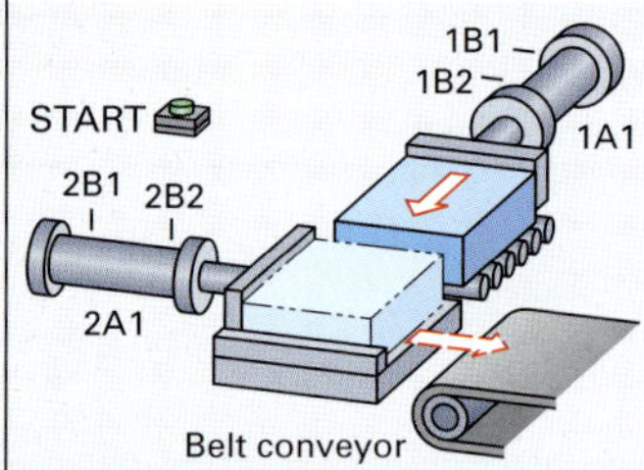

Pneumatic diagram with bistable components

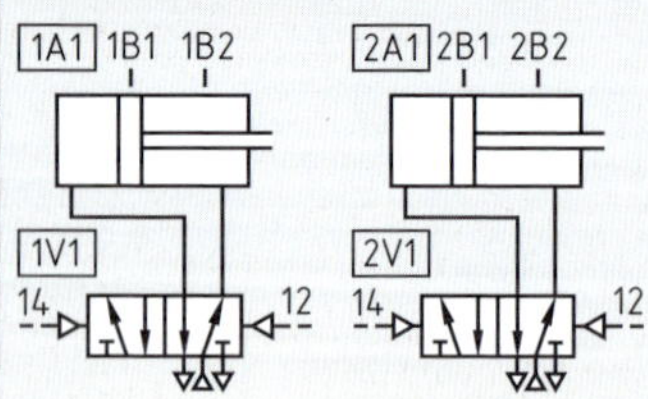

Different representation methods of steps in the interactive part

Explicit and abbreviated step description

Example:

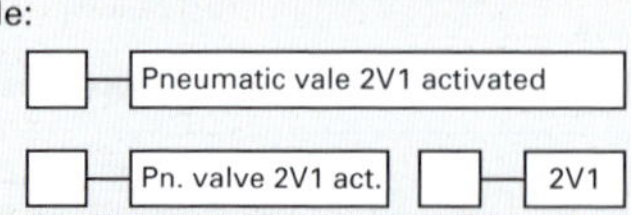

2 actions in the same step

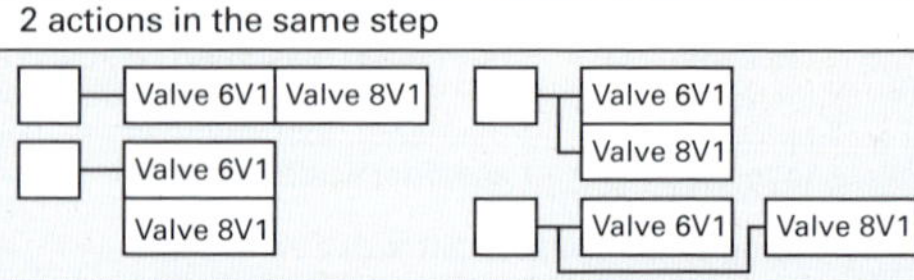

A

Function charts for sequential controls (GRAFCET)

cf. DIN EN 60848 (2002-12)

Example	Explanation

Steps

Initial step

The initial step marks the starting position of a controller. It is the first active step after starting the program.

General step

Step with the step number 2. A step is represented with a square. The step number is marked on top in the middle of the text field. In linear sequences, only one step is active at a time. In parallel sequence charts, several steps may be active simultaneously.

Step variable

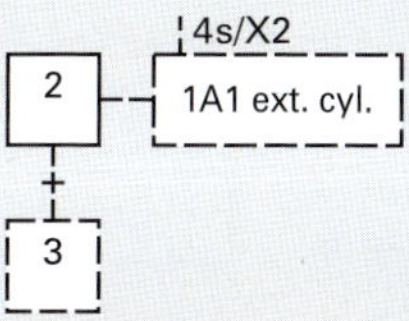

The step variable X is composed of the Boolean variable X and the step name (2).

The step variable can have

- either the value 0 = FALSE (step is not active)
- or the value 1 = TRUE (step is active).

Example: Step 3 is executed 4 seconds after the activation of step 2. This corresponds to a switch-on delay.

Transitions

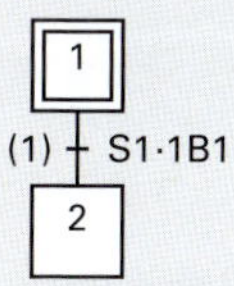

General transition condition

An alphanumerical identification code in brackets can be written on the left.

The transition condition is positioned on the right as a text or a Boolean expression.

Example:
Step 2 becomes active if start button S1 was pressed AND the piston of the lifting cylinder is in the retracted position (1B1).

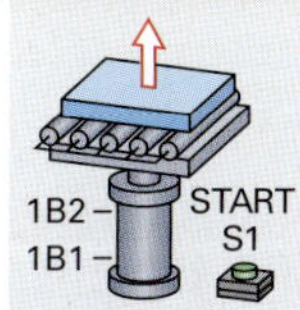

Different representation methods for transition conditions

DIN EN 60848		Boolean representation		Text	Graphical symbol
• corresponds to AND	+ corresponds to OR	∧ corresponds to AND	∨ corresponds to OR		& AND
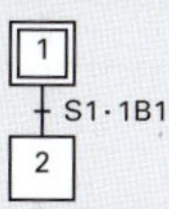	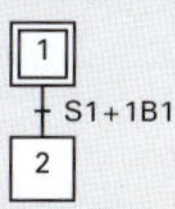	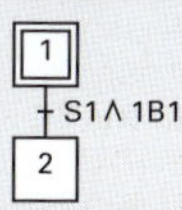	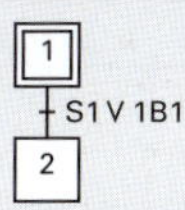	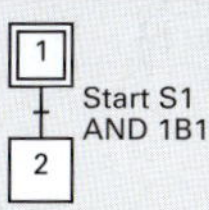	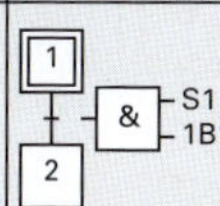

Transitions limited in time

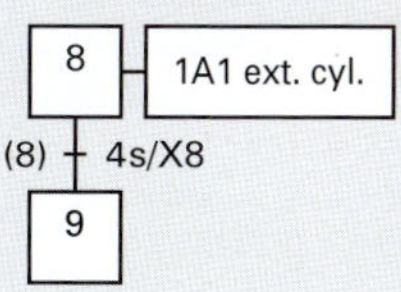

Switch-on delay relating to a step

When step 8 is active, the instruction "1A1 ext. cyl" is executed throughout 4 sec. If step 8 is active for less than 4 sec., the instruction "1A1 ext. cyl." is executed for a shorter period.

Signal-time diagram

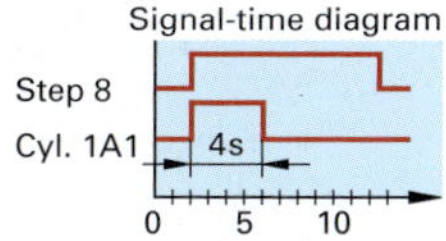

Equivalent representation:

Continuous action limited in time

Transition condition always TRUE

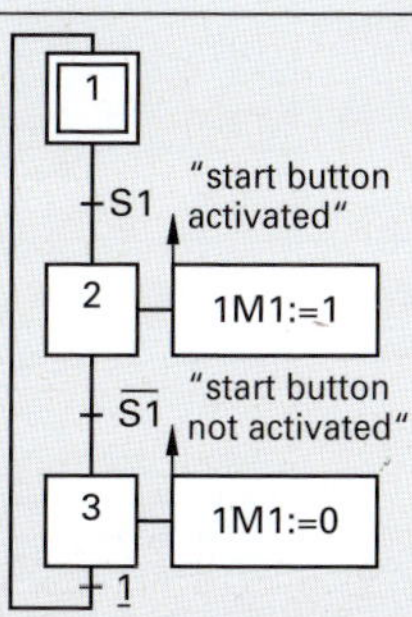

The symbol 1 in this context means that the transition condition is always satisified (TRUE). The result is a transient sequence. The triggering of the transition depends on the activation of the previous step.

Example:
Pressing start button S1 activates step 2 and switches valve solenoid 1M1 on. If the start button is no longer activated, step 3 is launched and switches valve solenoid 1M1 off. Cylinder 1A1 retracts.

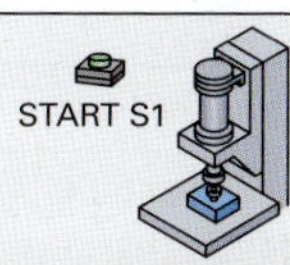

Pneumatic circuit diagram with a monostable component

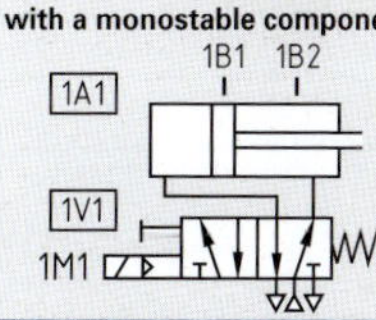

A

Function charts for sequential controls (GRAFCET)

cf. DIN EN 60848 (2002-12)

Example | **Explanation**

Continuous actions

Continuous actions are executed for a definite time.

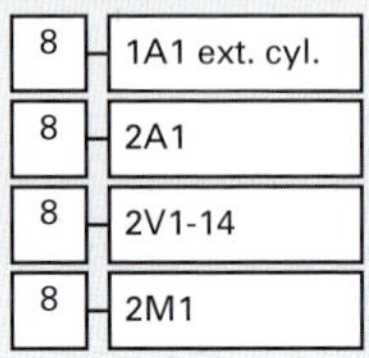

As long as the step is active, the value 1 (TRUE) is assigned to the variable. If it is no longer active, the value 0 (FALSE) is assigned.

Variants for the representation of actions (examples):

1A1 ext. cyl. ⇒ instruction
2A1 ⇒ designation of the drive
2V1-14 ⇒ designation of the valve connection
2M1 ⇒ solenoid in an electro-pneumatic control system

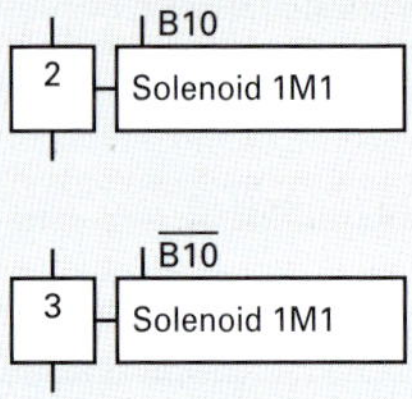

With assignment condition

Example: Press with step 2

If step 2 is active and the assignment condition B10 (bushing available) is satisfied, the value 1 (TRUE) is assigned to the variable 1M1 (solenoid). The solenoid 1M1 switches over and the pneumatic piston 1A1 extends.

A horizontal dash above the variable indicates the negation of a condition.

Pneumatic circuit diagram with monostable component.

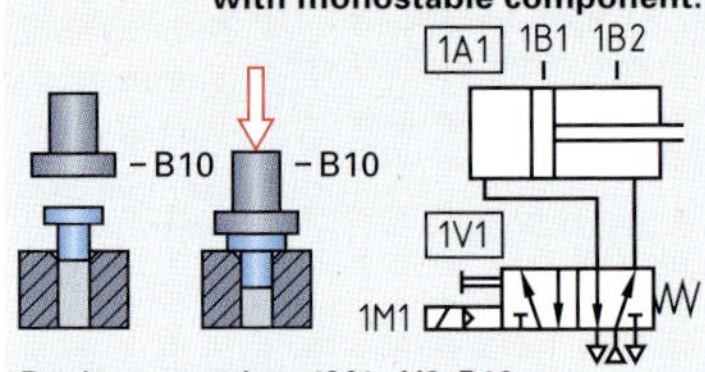

Boolean equation: 1M1=X2·B10

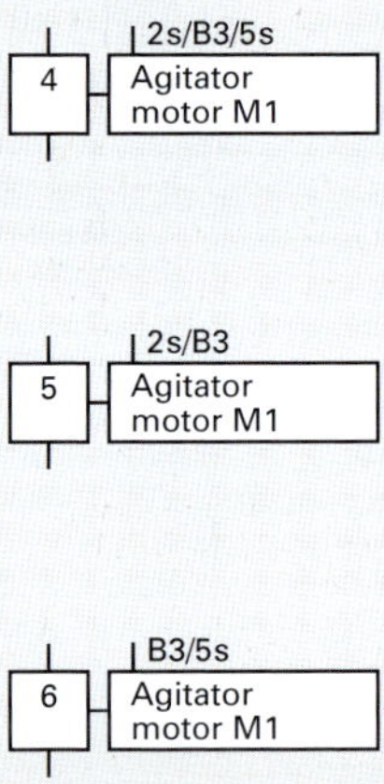

With time-driven assignment condition

Example: Agitator with step 4

2 seconds after the activation of the filling level sensor B3 (rising edge: from 0 to 1), the agitator motor M1 is switched on. This corresponds to a **switch-on delay.**

After deactivation of B3 (falling edge: from 1 to 0), the agitator motor M1 continues running for another 5 seconds. This corresponds to a **switch-off delay.**

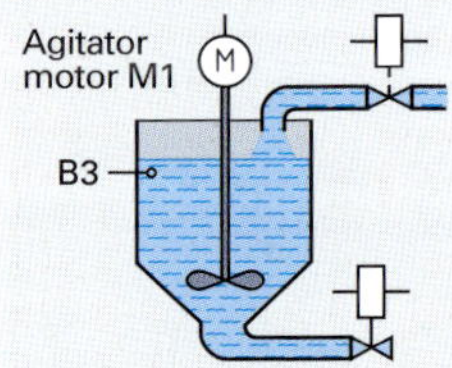

Switch-on and switch-off delay

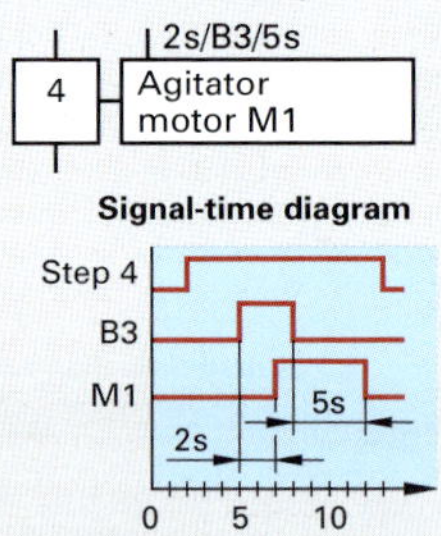

Switch-on delay

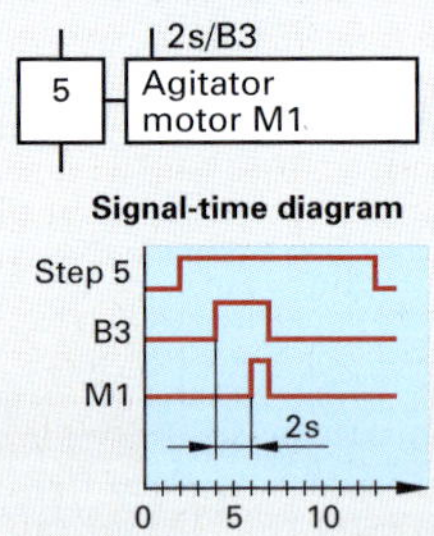

Switch-off delay

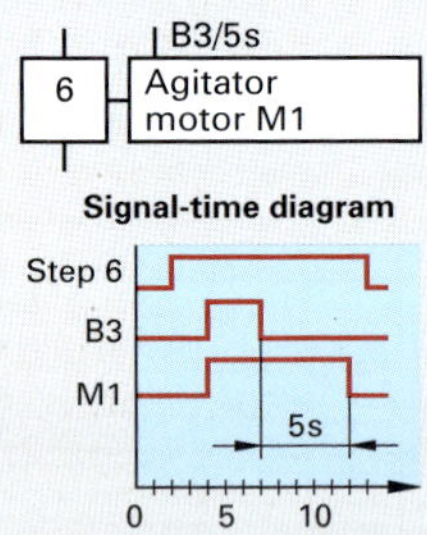

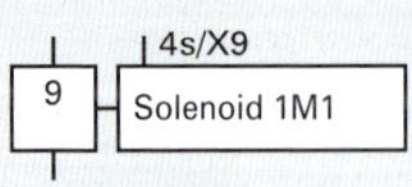

With delay

When step 9 is active, the value 1 (TRUE) is only assigned to the variable 1M1 (solenoid 1M1) after 4 sec. The assignment stops with the deactivation of step 9.

Signal-time diagram

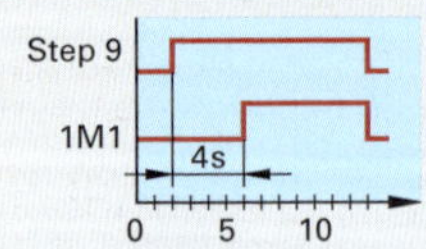

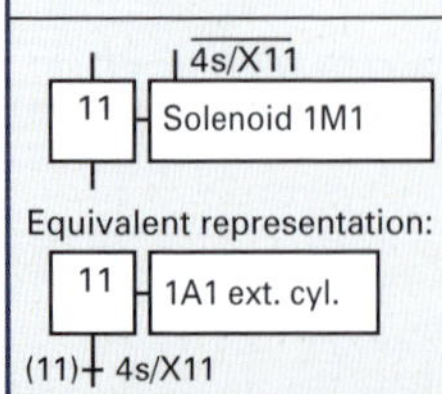

With time limit

When step 11 is active, the action "solenoid 1M1" is executed for a period of 4 sec.
If step 11 is less than 4 sec. active, the action "solenoid 1M1" is executed less long.

This corresponds to a **time limitation.**

Signal-time diagram

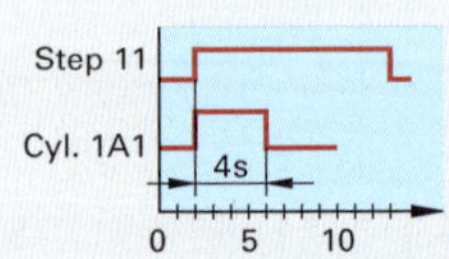

A

Function charts for sequential controls (GRAFCET)

cf. DIN EN 60848 (2002-12)

Example	Explanation

Action with storing effect

If a step becomes active, a value is permanently assigned to the variable of the action. The value remains assigned to this variable beyond the currently active function step until it is overwritten by another action.
Actions with storing effect can be set to the logical value "1" (TRUE) and reset in another step to the logical value "0" (FALSE) after a particular time has elapsed.

Stored with rising edge (activation of the step)

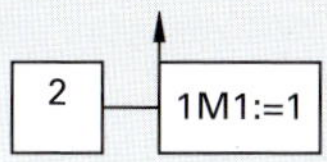

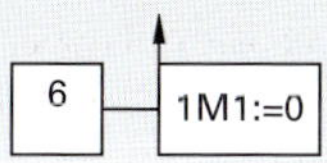

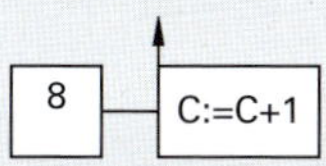

When step 2 is activated, the value 1 is assigned to the valve solenoid 1M1.

When step 6 is activated, the value 0 is assigned to the valve solenoid 1 M1.

The value is stored beyond the active period of the step until it is overwritten or deleted by another action.

When step 8 is activated, the current value of the variable increases incrementally by 1 with each cycle.

Example: factory gate (state: OPEN)

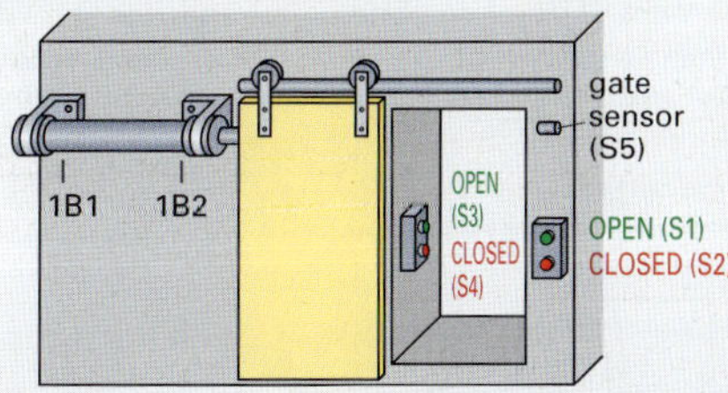

GRAFCET

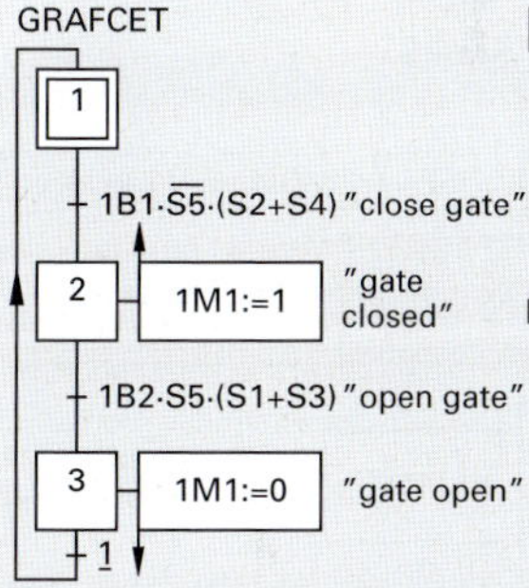

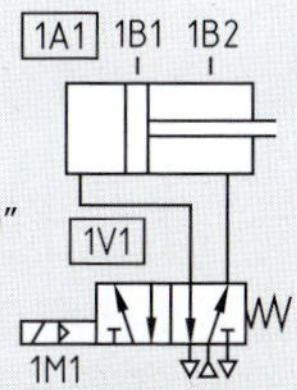

Pneumatic circuit diagram with monostable component

Stored with falling edge (deactivation of the step)

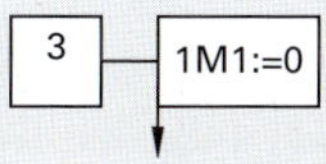

Note: Storing with falling edge can only be realized with a PC or a PLC.

When step 3 is activated, nothing happens.

Only when step 3 is deactivated, the value 0 is assigned to the variable 1M1.

The assignment of this value to 1M1 remains valid until the value is overwritten by another action.

Stored with event

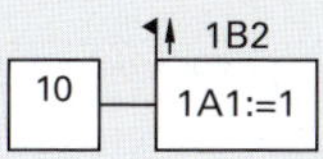

The arrow pointing to the side indicates that the storing action is only executed **when a particular event occurs.**

When step 10 is active and the sensor 1B2 changes from 0 to 1 (rising edge ↑), the value 1 is assigned to the pneumatic cylinder 1A1. The value of the variable 1A1 is stored beyond the activity of function step 10 until it is overwritten.

With delay

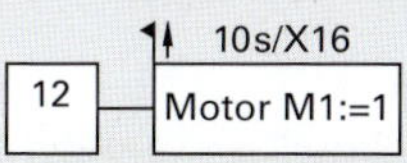

When step 12 is active, the value 1 is assigned to the variable "motor" after 10 sec. have elapsed. The value of the variable M1 is stored beyond the activity of function step 12 until it is overwritten.

Signal-time diagram

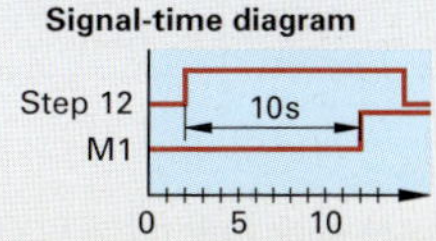

Actuation of cylinders with 5/2 directional control valve (examples)

Monostable (with spring return) **5/2-way pneumatic valve**	**Bistable 5/2-way pulse valve** and actions with continuous effect	**Bistable 5/2-way pulse valve** actions with storing effect
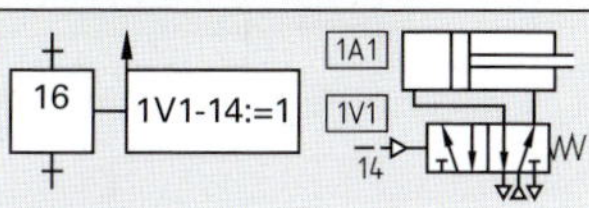	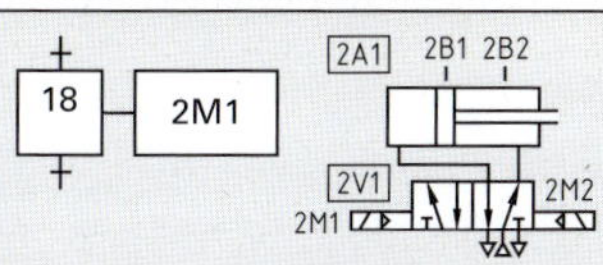	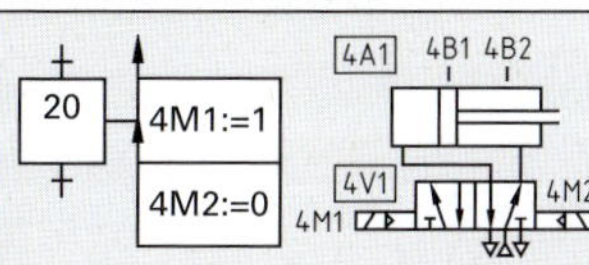
When step 16 is activated, the value 1 is assigned to the compressed air supply connection 14. The value is stored beyond the activity of step 16 until it is overwritten or deleted by another action.	The assignment of the value 1 causes the valve solenoid 2M1 to switch over. Even if the value turns to 0, the valve remains in this switching position. This type of storage is referred to as **mechanical signal storage.**	In order to avoid overlapping of signals, two inverse actions are executed for the same solenoid valve in each case.

A

Function charts for sequential controls (GRAFCET)

cf. DIN EN 60848 (2002-12)

Parallel branching (simultaneous sequence divergence)

Parallel branching allows a simultaneous activation of several partial sequences. If the first function step in the parallel branching becomes active, the partial sequences run independently of each other.

Example: Mixing of two liquids in an agitator

Two different liquids are to be mixed in an agitator vessel. Because of their different viscosities, first liquid 1 flows into the vessel via valve 1. When the liquid level attains filling mark B1, the following happens simultaneously:

1. The motor M1 of the agitator is switched on and
2. valve V1 closes and liquid 2 flows into the tank via valve V2.

Both partial sequences are processed independently of each other.
When the liquid level reaches the filling mark B2, the agitator motor M1 and the valve V2 are switched off again.

GRAFCET
(storing effective actions)

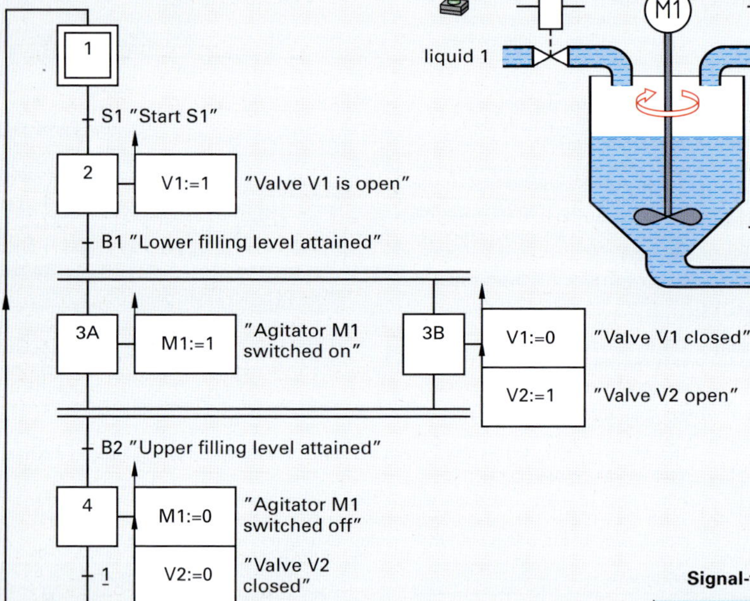

Technological scheme

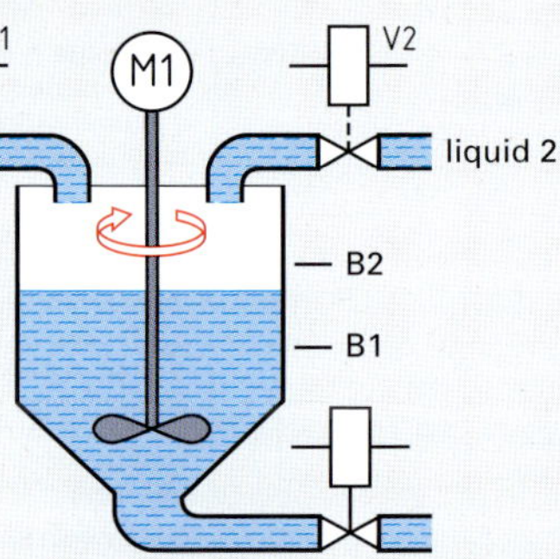

Signal-time diagram

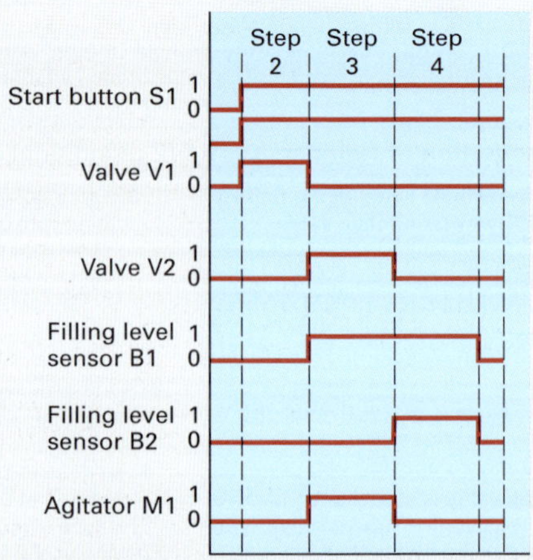

Notes concerning the edges at valve V1:

The rising edge (positive edge) is the change from the state logic 0 to logic 1.
The falling edge (negative edge) is the change from the state logic 1 to logic 0.

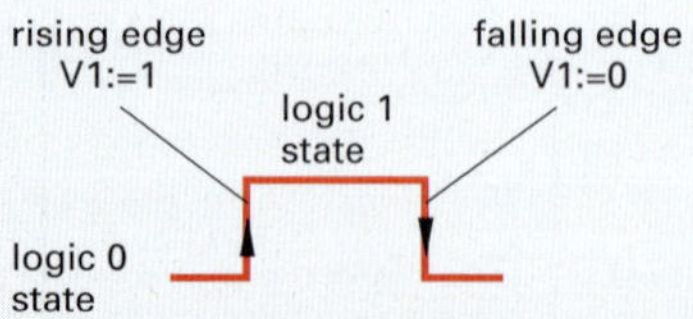

Function charts for sequential controls (GRAFCET)

cf. DIN EN 60848 (2002-12)

Alternative branching (sequence selection divergence)

A sequence can branch into several alternative sequences. In an alternative branching, there must be a transition before and after each step. The transition condition must never be fulfilled together with any other transition condition. After the last active transition, the partial sequences are merged to a common sequence.

Example: Lifting equipment, actions for the selection of thin and thick workpieces

By activating the start button S1, the long-stroke lifting cylinder 1A1 (valve solenoid 1M1) extends and the sensor 1B2 responds. There are two variations:

Variation 1: If a thick workpiece is available, sensor 2B4 responds. The pivoting cylinder 2A1 extends.
Variation 2: If a thin workpiece is available, sensor 2B3 responds. The pivoting cylinder remains in the retracted position.

The transfer cylinder 3A1 pushes the workpiece forward and retracts to its initial position. The long-stroke lifting cylinder 1A1 retracts.

GRAFCET (actions with continuous effects)

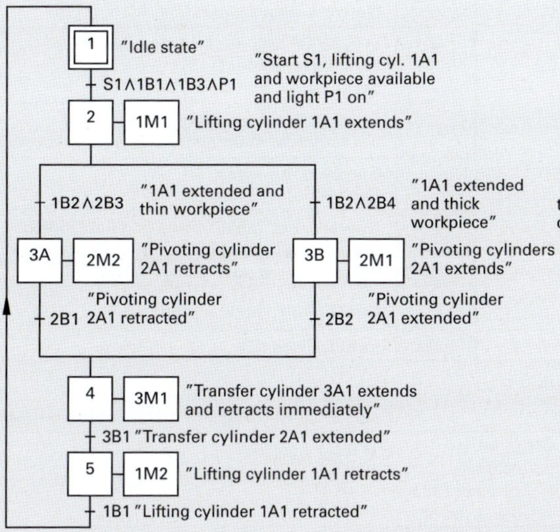

Technological scheme

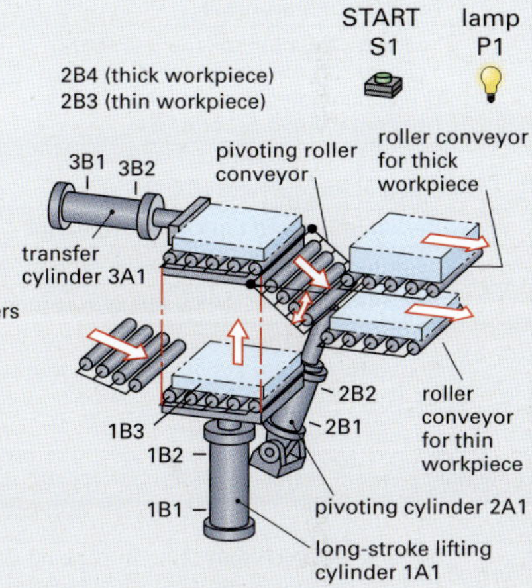

Signal-time diagram

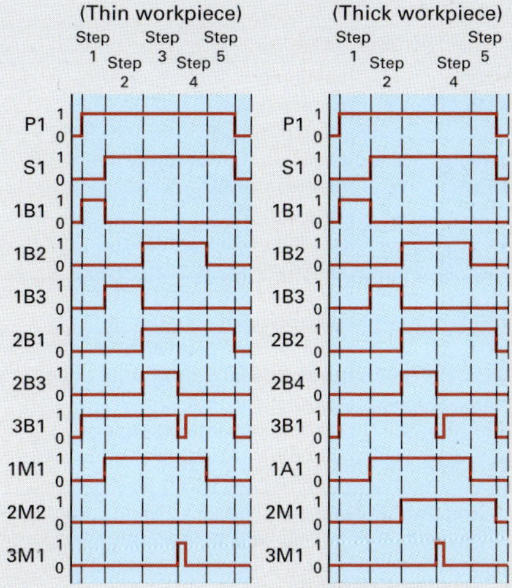

Pneumatic diagram

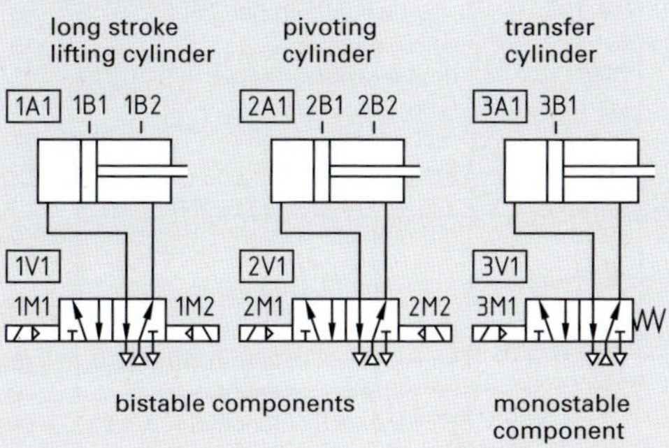

Note: If bistable 5/2-way valves are used to control the cylinder, the pulse valve provides for the storage of the signals.

A

Programming languages

PLC programming languages (overview) cf. DIN EN 61131 (2003-12)

- Text languages
  - Instruction List IL
  - Structured text ST
- Graphic languages
  - Ladder diagram LAD
  - Function block language FBL

Common elements of all PLC languages (selection)

Delimiters (selection) cf. DIN EN 61131 (2003-12)

Symbol	Use
(**)	At beginning and end of comment
+	Leading prefix for decimal numbers Addition operator (ST)
–	Leading prefix for decimal numbers Year-month-day separator Subtraction, negative operator (ST) Horizontal line (LAD and FBL)
: =	Initialization operator Assignment operator (ST)
#	Base number and time literal separator
'	Beginning and end of character strings
$	Beginning of special characters in strings
.	Whole number/fraction separator Separator for hierarchal addresses and structured elements
e *or* E	Real-exponent delimiter

Symbol	Use
:	Step names and variable/type separators Statement label separators (ST) Network label separators (LAD and FBL)
( )	Instruction lists modifier/operator (ST) Function arguments (ST) Delimiter for FBL input lists (ST)
;	Separator for type declaration Separator for statements (ST)
″	Separator for areas Separator for CASE areas (ST)
,	Bulleted lists, initial values and field index separators, operand lists, function argument lists and CASE value lists separators (ST)
%	Direct representation prefix[1)]
| *or* !	Vertical lines (LD)

Individual element variables for storage locations

Variable	Meaning	Variable	Meaning	Example (AWL)
I	storage location input	B	byte size (8 bit)	**ST %QB5**[1)]: Stores current result in byte size in output storage location 5
Q	storage location output	W	word size (16 bit)	
M	storage location tag	D	double word size (32 bit)	
X	(individual) bit size	L	long word size (64 bit)	

Operators

Name	Symbol	Meaning
ADD	+	addition
SUB	–	subtraction
MUL	*	multiplication
DIV	/	division
AND	&	Boolean AND
OR	>= [2)]	Boolean OR
XOR	---- [3)]	Boolean exclusive OR
NOT	---- [3)]	negation
S	---- [3)]	sets Boolean operator to "1"
R	---- [3)]	sets Boolean operator to "0"
GT	>	comparison: greater than
GE	>=	comparison: greater than or equal to
EQ	=	comparison: equal to
NE	< >	comparison: not equal to
LE	<=	comparison: less than or equal to
LT	<	comparison: less than

Elementary data types

Key word	Data type	Bits
BOOL	Boolean	1
SINT	short whole number	8
INT	whole number	16
DINT	double whole number	32
LINT	long whole number	64
REAL	real number	32
LREAL	long real number	64
STRING	variable long number sequence	– [4)]
TIME	duration	– [4)]
DATE	date	– [4)]
BYTE	bit sequence of length 8	8
WORD	bit sequence of length 16	16
DWORD	bit sequence of length 32	32
LWORD	bit sequence of length 64	64

A

1) Directly represented individual element variables have a leading % symbol.
2) This symbol is not allowed as operator in text language.
3) No symbol
4) Manufacturer specific

Programming languages

Ladder diagram (LD)

cf. DIN EN 61131 (2003-12)

A ladder diagram represents the flow in an electromechanical relay system.

Lines and blocks (Symbol | Description)

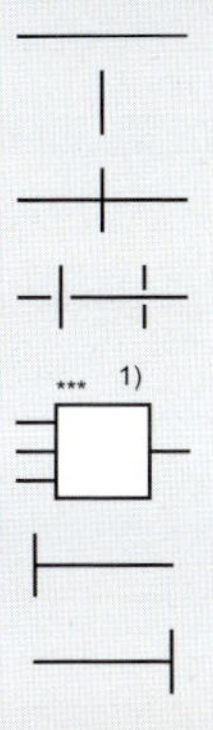

Symbol	Description
	Horizontal line
	Vertical line
	Line junction
	Crossing without connection
*** 1)	Blocks with connection lines
	Left power rail
	Right power rail

Contacts (Symbol | Description)

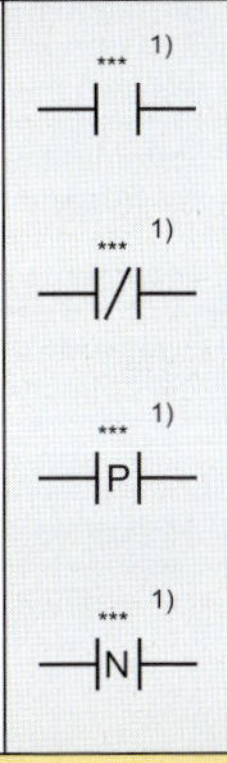

Symbol	Description
*** 1)	NO contact logic condition "1"
*** 1)	NC contact logic condition "0"
*** 1) P	Contact for sensing rising edge, signal from "0" to "1"
*** 1) N	Contact for sensing falling edge, signal from "1" to "0"

Coils (Symbol | Description)

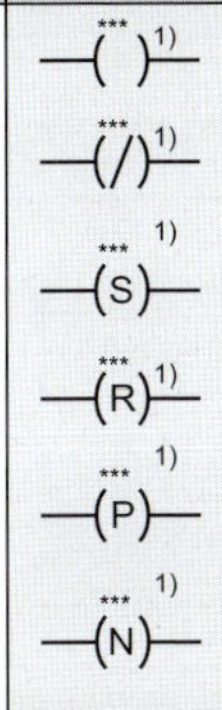

Symbol	Description
*** ( ) 1)	Coil output energize
*** (/) 1)	Coil output deenergize
*** (S) 1)	Latching coil, stores an operation
*** (R) 1)	Unlatching coil
*** (P) 1)	Coil for sensing positive slopes, signal from "0" to "1"
*** (N) 1)	Coil for sensing negative slopes, signal from "0" to "1"

1) component designator

Function block language (FBL)

cf. DIN EN 61131 (2003-12)

Function block language consists of individual function blocks with statistical data. They are useful in implementing frequently recurring functions.

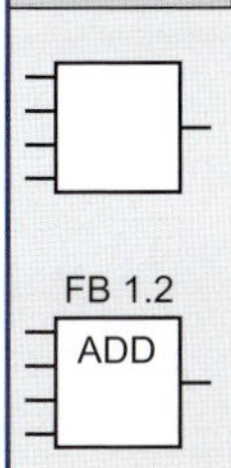

Symbol	Description
	Elements are rectangular or square. Input parameters are placed on the left side and output parameters on the right side.
FB 1.2 ADD	The block's functionality is entered as a name or symbol within the block. The block designator is located above the block.

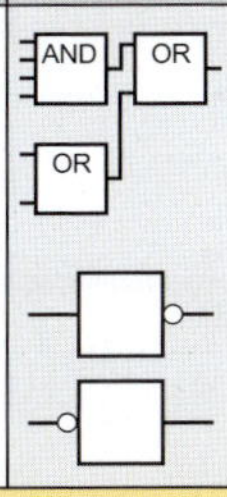

Symbol	Description
AND, OR, OR	Elements must be interconnected by horizontal and vertical signal flow lines.
	Negation of Boolean signals is shown by a circle on the input or output.

Structured text (ST)

cf. DIN EN 61131 (2003-12)

Structured text is a high level language and builds on the syntax of ISO-PASCAL.

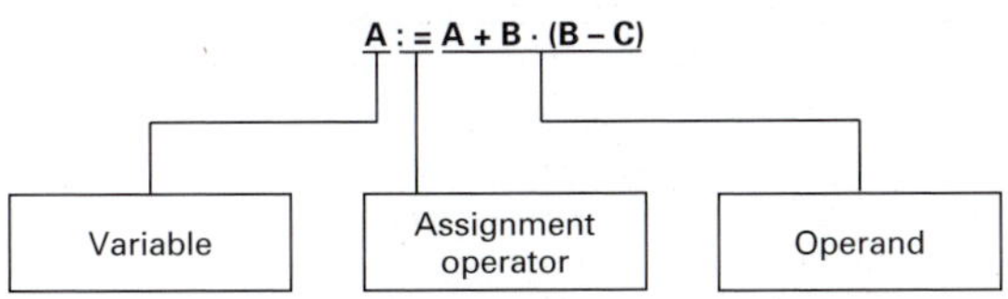

Statement	Type
:=	assignment
IF	conditional statement
CASE	selection statement
FOR	repeat statement
WHILE	repeat statement
REPEAT	repeat statement
EXIT	leaving a repeated statement

Comparison of Function Block Language (FBL) and Structured text (ST)

Function blocks (examples)

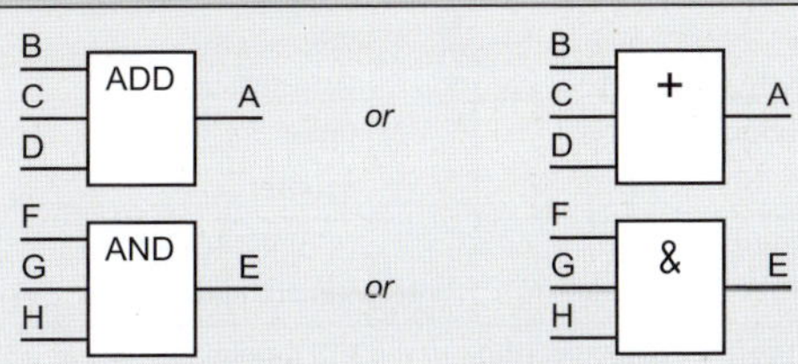

Structured text (examples)

A:= ADD (B, C, D)
or
A:= B + C + D

E:= AND (F, G, H)
or
E:= F & G & H

A

Programming languages

Instruction list (IL)

cf. DIN EN 61131 (2003-12)

Instruction list is a machine-oriented textual programming language, similar to assembly language.

Structure of an instruction

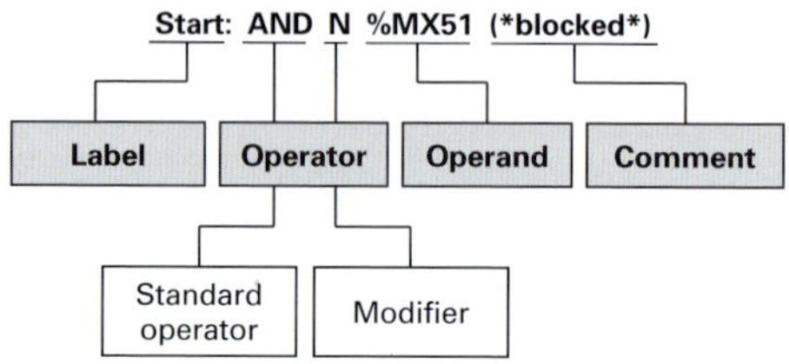

Operator modifiers	
N	Boolean negation of the operand.
C	Statement is only executed if the evaluated result is a Boolean 1.
,	Separates multiple.
(	Evaluation of the operator is deferred until ")" appears.

Standard operators

Operator	Modifier	Meaning
LD	N	setting an operand
ST	N	storing on operand addresses
S	–	sets Boolean operator to 1
R	–	sets Boolean operator back to 0
AND	N,(	Boolean AND
&	N,(	Boolean AND
OR	N,(	Boolean OR
XOR	N,(	Boolean exclusive OR
ADD	(	addition
SUB	(	subtraction
MUL	(	multiplication

Operator	Modifier	Meaning
DIV	(	division
GT	(	comparison: >
GE	(	comparison: >=
EQ	(	comparison: =
NE	(	comparison: < >
LE	(	comparison: <=
LT	(	comparison: <
JMP	C,N	jump to label
CAL	C,N	call of a function block
RET	C,N	jump back
)	–	processing of deferred operations

Information list (IL) according to VDI[1)]

cf. VDI 2880 (1985-09)

Structure of an instruction

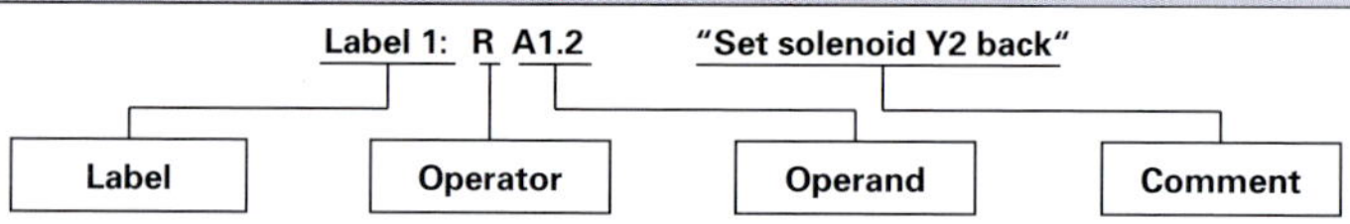

Operators for program organization	
L	load
(	open parenthesis
)	closed parenthesis
NOP	null operation
SP	unconditional jump
SPB	conditional jump
BA	call of a block
BAB	conditional call of a block
BE	block end
"	comment beginning
„	comment end
PE	program end

Operators for signal processing	
U	AND operation
O	OR operation
N	negation
UN	NAND operation
ON	NOR operation
=	assignment
ADD	addition
SUB	subtraction
MUL	multiplication
DIV	division
S	set
R	reset

Operators	
ZV	count forwards
ZR	count backwards
XO	exclusive OR
Operand	
E	input
A	output
M	tag
K	constant
T	timer
Z	counter
P	program block
F	function block

1) In practice, many more PLC controls exist which are programmed according to the VDI guidelines.

A

Programming languages

Comparison of the most commonly used PLC programming languages

Functions as components of programs	Instruction list (IL) according to VDI	Function block language (FBL)	Ladder diagram (LD)
AND with 3 inputs	U E11 U E12 UN E13 = A10	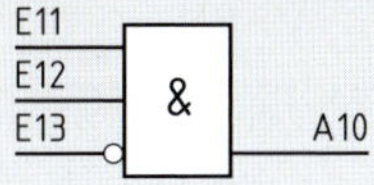	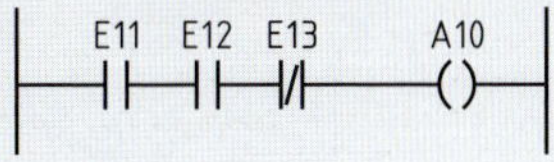
OR with 3 inputs	U E11 O E12 O E13 = A10	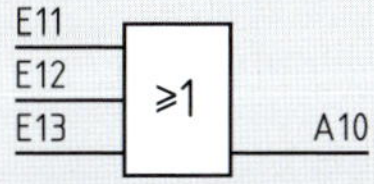	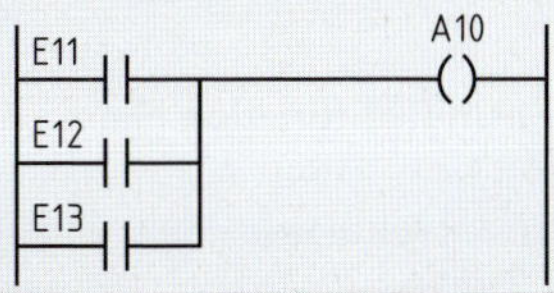
AND before **OR**	U E11 U E12 O U E13 U E14 = A10	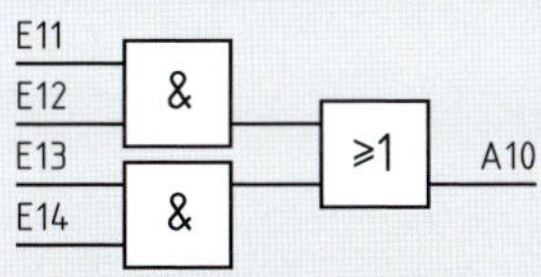	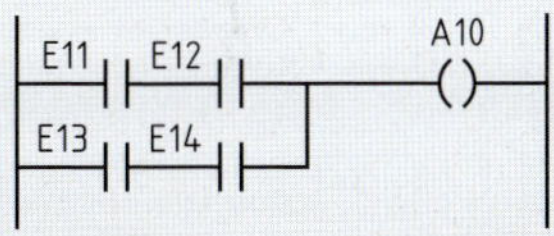
OR before **AND** with intermediate tag	U E11 O E12 = M1 U E13 O E14 U M1 = A10	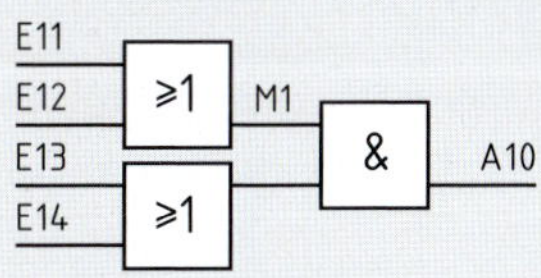	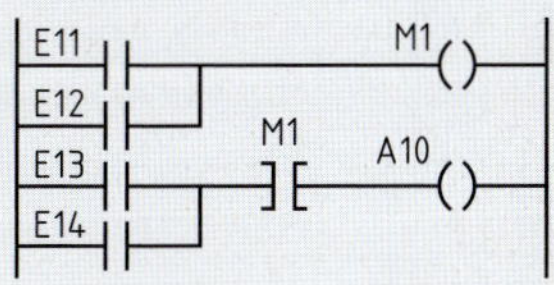
Exclusive **OR** **(XOR)**	U E11 UN E12 O (UN E11 U E12) = A10	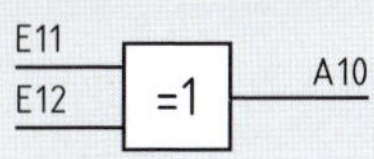	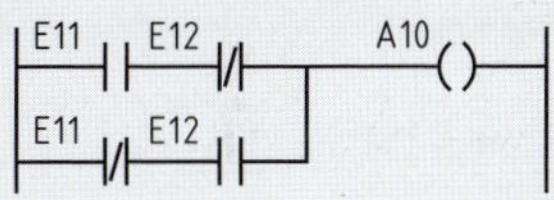
RS flip-flop **Set dominant**	U E12[1)] R A11 U E11 S A11	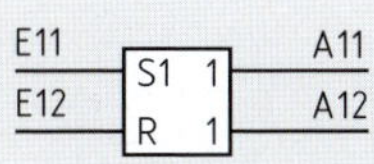	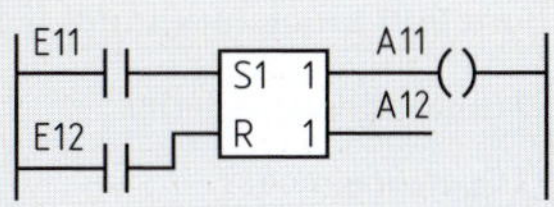
RS flip-flop **Reset dominant**	U E11[1)] S A11 U E12 R A11	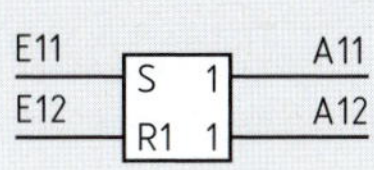	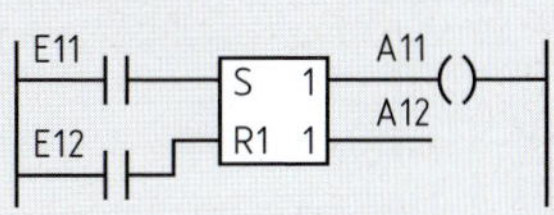
Turn on **delay**	U E11 = T1 U T1 = A10	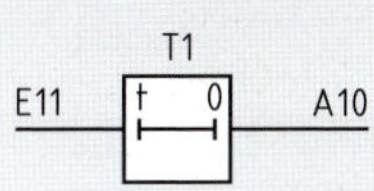	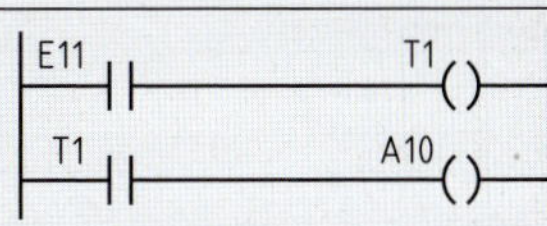
Latch, ON (E 12) dominating	U E12 O A10 UN E11 = A10	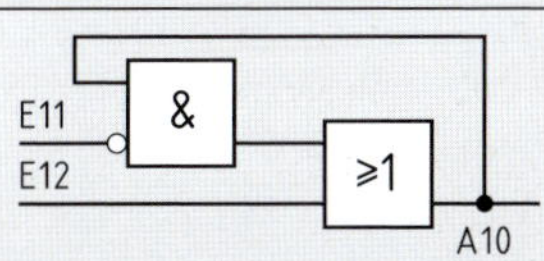	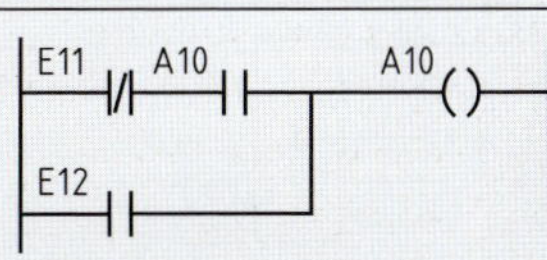

1) The following applies to flip-flops: If S = 1 and R = 1, the last function programmed in the IL dominates.

A

PLC controlled embossing machine tool

Technological scheme

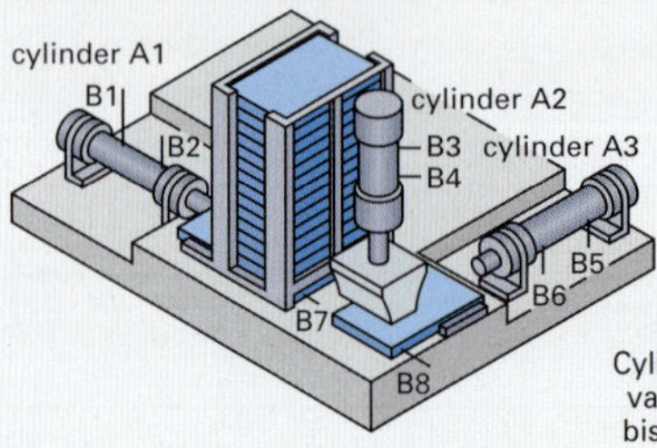

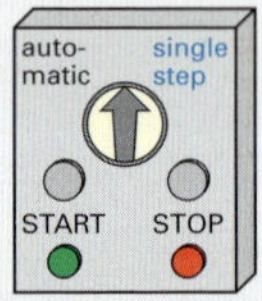

operating panel

Cylinder activation via 5/2-way valve, both sides activated by bistable solenoids, for actions with continuous effect

Description

Workpieces are to be fitted with a workpiece number on an embossing machine tool. The sensor B7 detects whether workpieces are still available in the stacker. The pneumatic cylinder A1 pushes the workpiece out of the stacker into the working position. After this, the embossing cylinder A2 extends and embosses the workpiece. After a delay time of 1 sec., first the embossing cylinder A2 and then the feeding cylinder A1 are retracted. Cylinder A3 serves as an ejector of the embossed workpiece. Sensor B8 detects whether the workpiece was actually ejected.

GRAFCET function chart

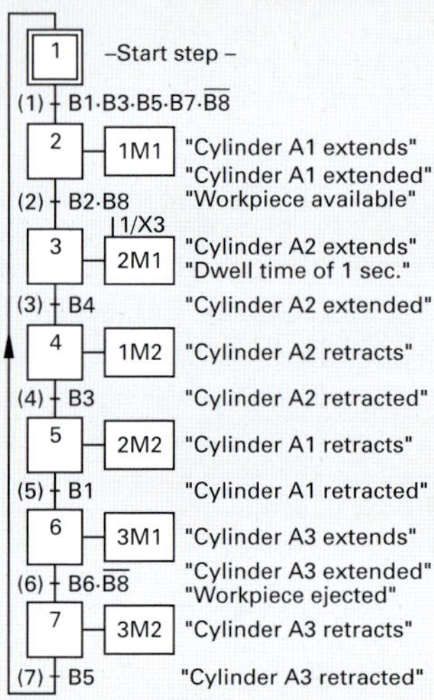

Allocation list

Component and action	Component designation	Address	Remarks
Mode switch AUTOMATIC/STEP	S0/S1	E0.0/E0.1	NO contact/ NC contact
START button	S2	E0.2	NO contact
STOP button	S3	E0.3	NC contact
Proximity switch	B1-B4 B5-B8	E0.4-E0.7 E1.0-E1.3	NO contact
Solenoid valve (with cyl. A1)	1M1 and 1M2	A0.0/A0.1	–
Solenoid valve (with cyl. A2)	2M1 and 2M2	A0.2/A0.3	–
Solenoid valve (with cyl. A3)	3M1 and 3M2	A0.4/A0.5	–

Function block language (FBS)

Organization block OB1

Operating modes

Network 1: function block FB1

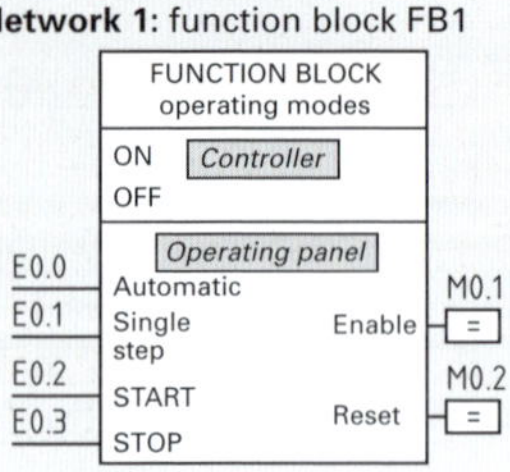

Network 2: idle state

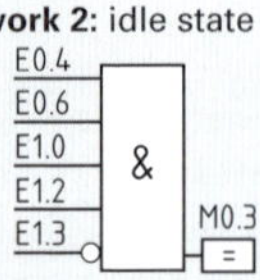

Color marking: step flag in red
Transition in blue

Step chain FC1

Network 3: step 1
Start step

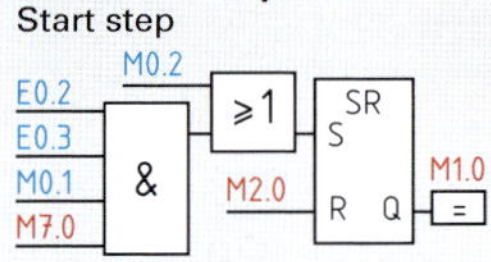

Network 4: step 2
Extend cylinder A1

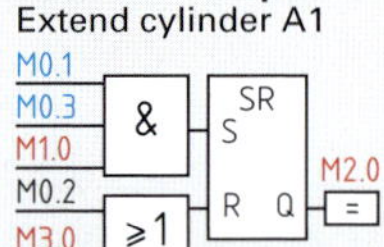

Network 5: step 3
Extend cylinder A2

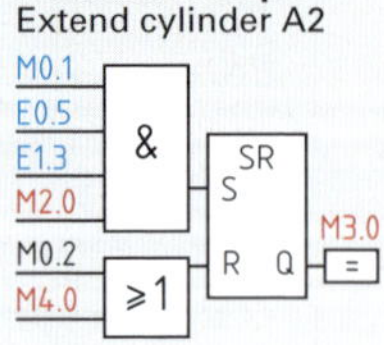

Network 6: step 4
Retract cylinder A2

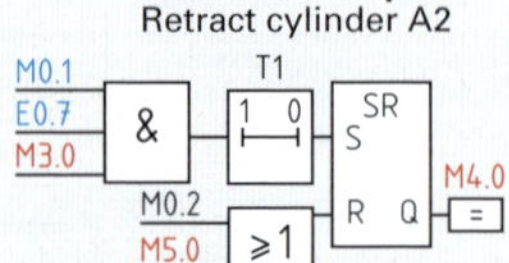

Network 7: step 5
Retract cylinder A1

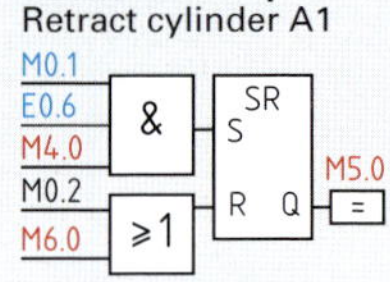

Network 8: step 6
Extend cylinder A3

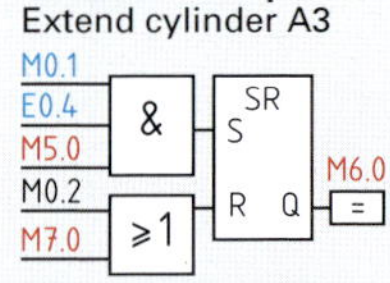

Network 9: step 7
Retract cylinder A3

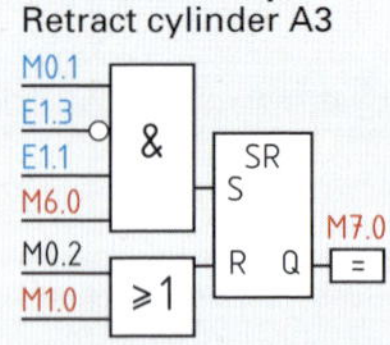

Command output FC2

Networks 10 to 15

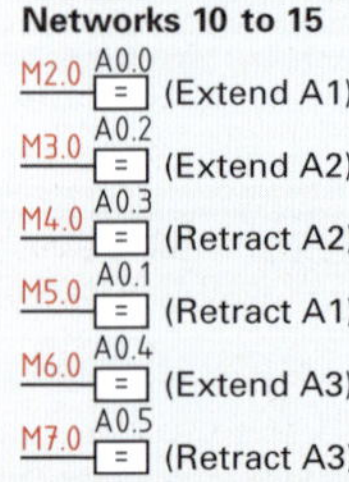

Circuit symbols

cf. DIN ISO 1219-1 (1996-03)

Function elements

Symbol	Designation	Symbol	Designation	Symbol	Designation	Symbol	Designation
	Hydraulic fluid flow Compressed air flow	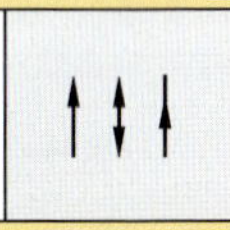	Direction of flow	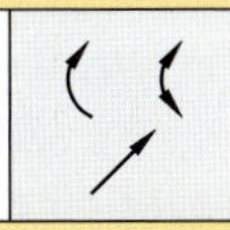	Direction of rotation Adjustability		Spring Flow restriction

Power transmission

Symbol	Designation	Symbol	Designation	Symbol	Designation	Symbol	Designation
	Hydraulic pressure source Pneumatic press. source Working line Control line Leakage current line Enclosure around subassemblies	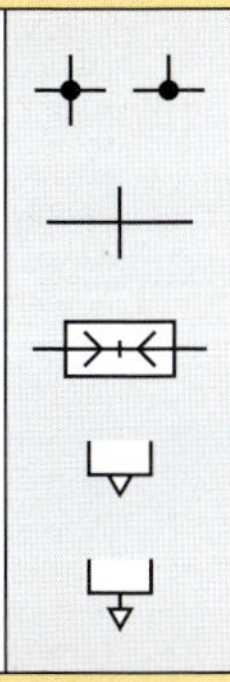	Line junction Line crossing Quick coupling Exhaust without connection Exhaust with connection	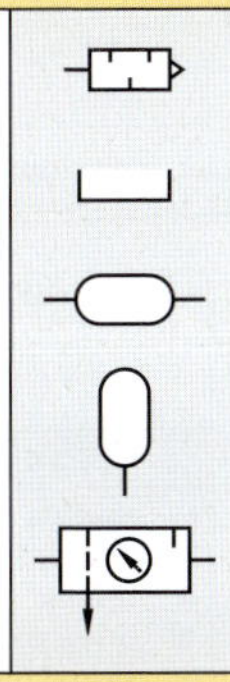	Muffler Tank Air receiver Hydraulic accumulator Service unit (FRL)	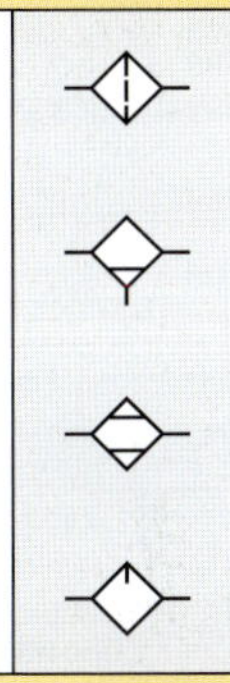	Filter or screen Water separator Air dryer Lubricator

Pumps, compressors, motors

Symbol	Designation	Symbol	Designation	Symbol	Designation	Symbol	Designation
	Fixed displacement hydraulic pump, unidirectional Variable displacement hydraulic pump, bidirectional Compressor, unidirectional	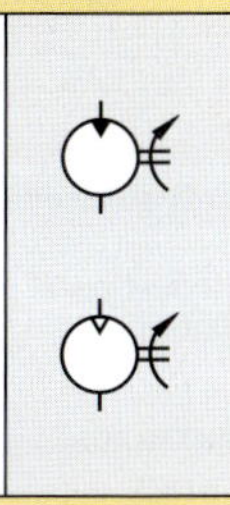	Fixed displacement hydraulic motor, unidirectional Fixed displacement pneumatic-motor, unidirectional	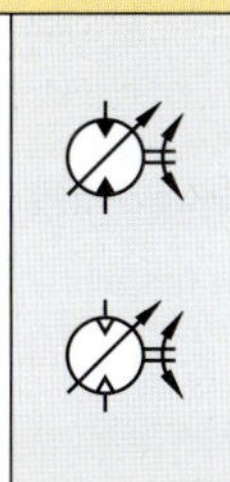	Variable displacement hydraulic motor, bidirectional Variable displacement pneumatic motor, bidirectional	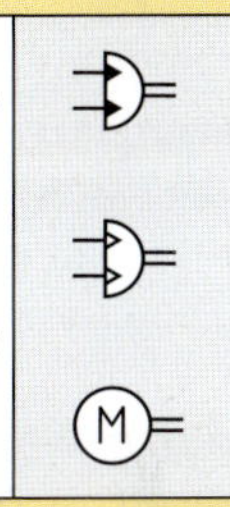	Hydraulic oscillating drive Pneumatic oscillating drive Electric motor

Single-acting cylinders / Double-acting cylinders

Symbol	Designation	Symbol	Designation	Symbol	Designation	Symbol	Designation
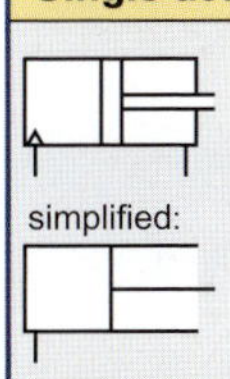 simplified:	Single-acting cylinder, return stroke by undefined power source	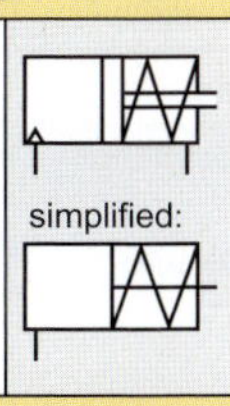 simplified:	Single-acting cylinder, return stroke by integrated spring	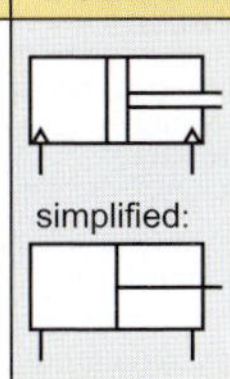 simplified:	Double-acting cylinder with one-sided piston rod	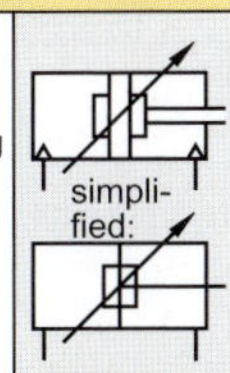 simplified:	Double-acting cylinder with one-sided piston rod and two-sided adjustable end cushion

Check, and/or valves / Pressure valves / Flow control valves

Symbol	Designation	Symbol	Designation	Symbol	Designation	Symbol	Designation
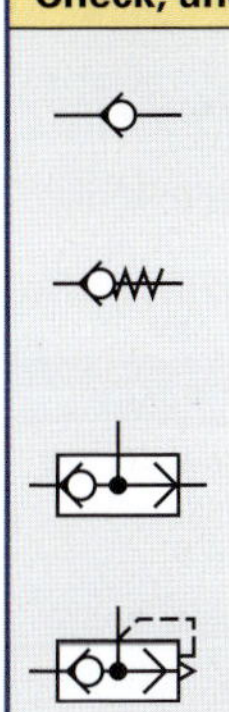	Check valve, unloaded Check valve, spring loaded Shuttle valve (OR function) Quick exhaust valve	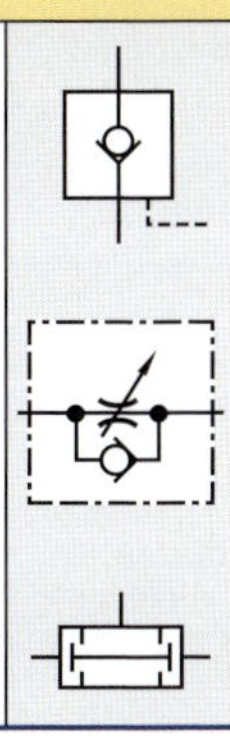	Pilot operated check valve One-way flow control valve Dual-pressure valve (AND function)	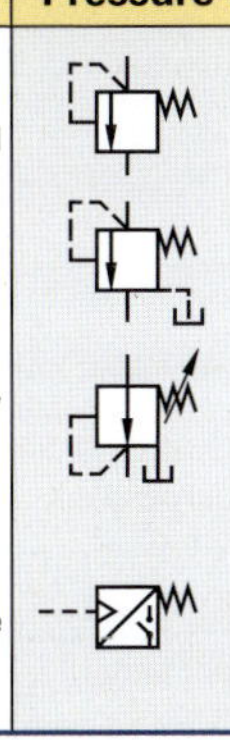	Pressure relief valves Sequence valve 2-way pressure regulator, direct-acting Pressure switch, emits electrical signal for a preset pressure	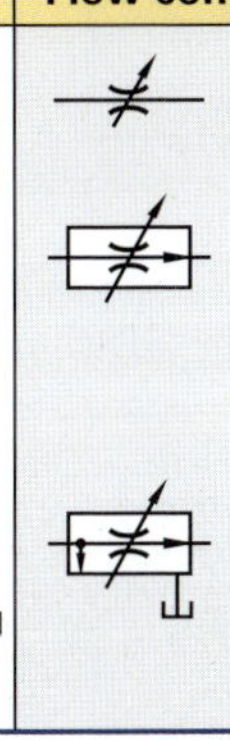	Adjustable throttle valve Adjustable 2-way flow-control valve Adjustable 3-way flow-control valve, relief opening to tank

A

Circuit symbols

Connection designations and codes for directional control valves

cf. DIN ISO 1219-1 (1996-03)
DIN ISO 5599 (2005-12)

Example:
5/2 directional control valve with connection designation

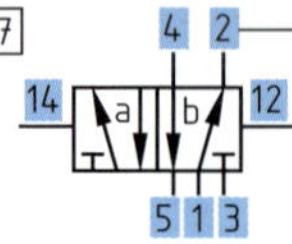

Designator

Code designation

5 / 2 – directional control valve 6 V 7

- 5: Number of connections
- 2: Number of switch positions
- 6: Circuit number
- V: Part designation
- 7: Part number

Switch positions[1)]	
a | b	Valve with 2 positions
a | 0 | b	Valve with 3 positions

[1)] Number of rectangles ≙ Number of positions

Part designation	
P	pumps and compressors
A	drives
M	drive motors
S	signal pick-up
V	valves
Z	all other parts

Connection designations for pneumatic and hydraulic equipment

Connection	as per DIN with numbers	obsolete: with letters[1) 2)]
Inflow, pressure port	1	P
Working ports	2, 4, 6	A, B, C
Vent, drain	3, 5, 7	R, S, T
Leakage oil port	–	L
Control ports[3)]	10, 11, 12, 14	X, Y, Z

[1)] Letters are still frequently used in hydraulic circuit diagrams.
[2)] The sequence of the letters does not necessarily correspond to the number sequence.
[3)] A pulse at control port 12, for example, connects ports 1 and 2.

Designs of directional control valves

2/ directional control valves

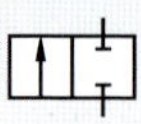
2/2 DCV, normally closed

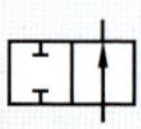
2/2 DCV, normally open

3/ directional control valves

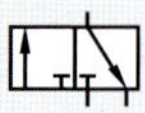
3/2 DCV, normally closed

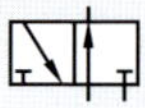
3/2 DCV, normally open

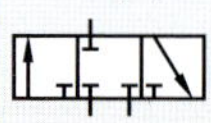
3/3 DCV, NC in middle position

4/ directional control valves

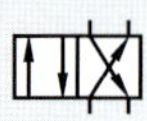
4/2 directional control valve

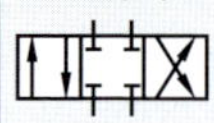
4/3 DCV. NC in middle pos.

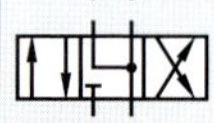
4/3 DCV, with float in middle position

5/ directional control valves

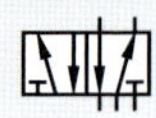
5/2 directional control valve

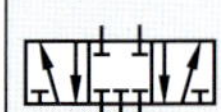
5/3 DCV, NC in middle position

Flow paths

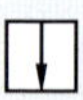
One flow path

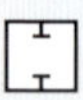
Two closed ports

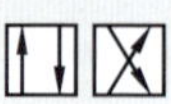
Two flow paths

Two flow paths and one closed port

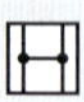
Two interconnected flow paths

One flow path in bypass switch and two closed ports

Actuation of directional control valves

Manually activated

General, no type of actuation indicated

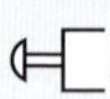
Push button

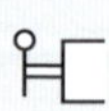
Lever

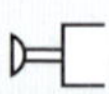
Pull button

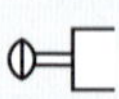
Push and pull button

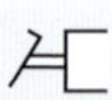
Foot pedal

Mechanical actuation

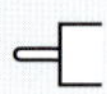
Plunger

Plunger with adjustable stroke limit

Spring

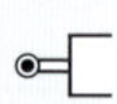
Roller plunger

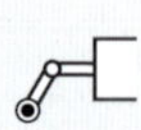
Roller lever, one direction of actuation

Pressure actuation

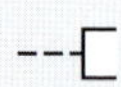
Direct

hydraulic

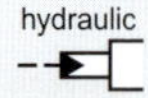

pneumatic

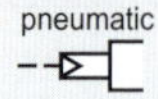
Indirect using pilot valve

Electrical actuation

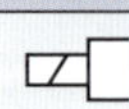
By solenoid

By electric motor

Combined actuation

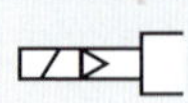
By solenoid and pilot valve

Mechanical components

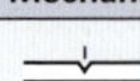
Notch

A

Proportional valves

Basic terminology

Example: Electrically actuated, pilot-operated 4/3-spring centering valve

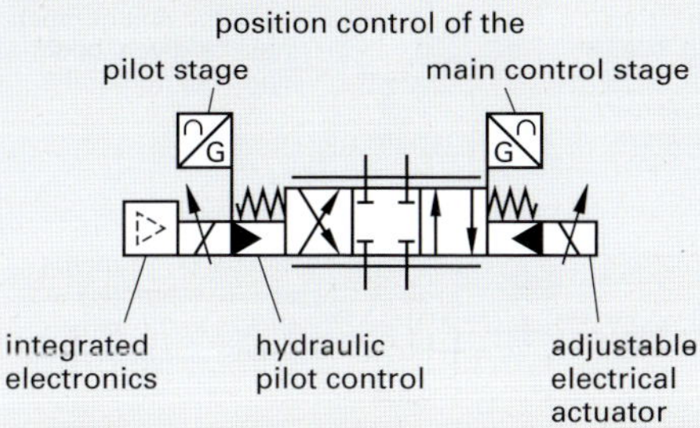

Proportional valves are controlled via a continuously adjustable, electromagnetic actuator. They are mainly used in hydraulic engineering. The output signal, e.g. pressure, flow rate or flow direction, is a quantity proportional to the input signal (current). The pressure, the flow rate and the flow direction allow the quick and precise adjustment of the piston force or velocity of a hydraulic cylinder, for instance, the speed or control of rotation of a hydraulic motor or the extension and retraction of a piston. These parameters can be adjusted automatically via a programmable logical controller (PLC) during operation of the equipment and matched to the automated process.

A proportional valve can also replace one or several valves, e.g. a directional control valve or a flow-control valve.

Circuit symbols (selection) — cf. DIN ISO 1219-1 (2007-12)

Proportional directional valve

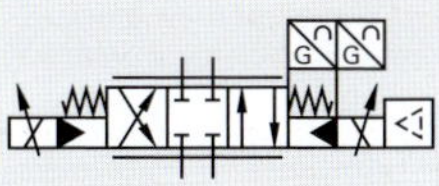	Electro-hydraulic pilot-controlled proportional directional valve with position control of the main control and pilot stage, integrated electronics	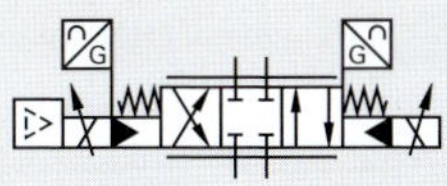	Electro-hydraulic pilot-controlled control valve, with position control of the main control and pilot stage, integrated electronics
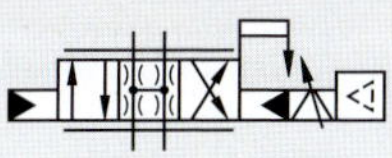	Electro-hydraulic pilot-controlled directional valve, pilot stage acts continuously in both directions, integrated electronics	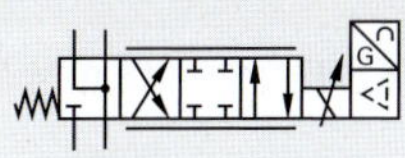	Electro-hydraulic controlled directional valve with priority position in the event of power failure and electrical feedback, integrated electronics
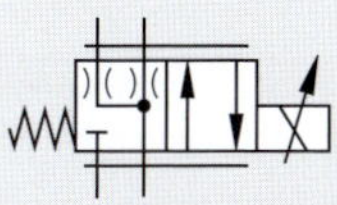	Proportional directional valve, direct actuator	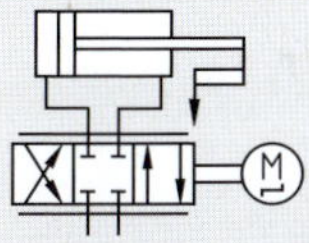	Electro-hydraulic linear drive, comprising a cylinder and a servo valve with step motor, mechanical feedback of the cylinder position.

Proportional pressure valves

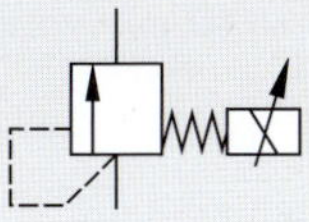	Proportional pressure-relief valves, direct actuator, solenoid acts on the valve cone via a spring	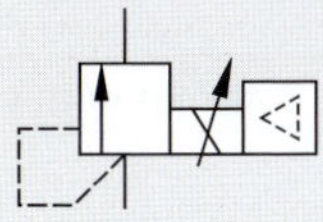	Proportional pressure-relief valve, direct actuator, solenoid acts on valve cone, integrated electronics
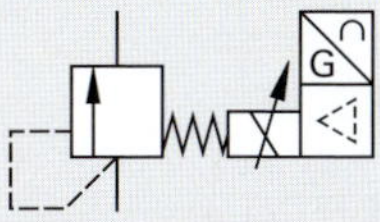	Proportional pressure-relief valve, direct actuator, with solenoid position control, integrated electronics	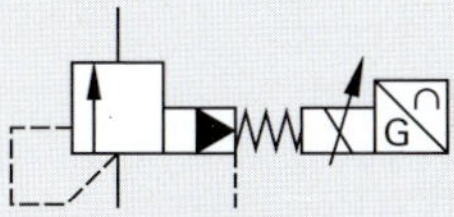	Proportional pressure-relief valve, pilot-controlled, with electrical position detector of the solenoid, with external control oil drain

Proportional flow-control valves

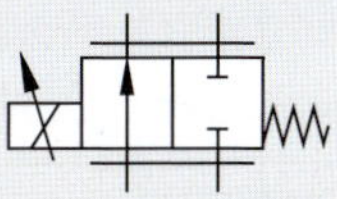	Proportional flow-control valves, direct actuator	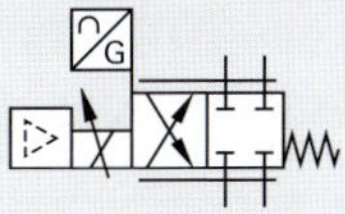	Proportional flow-control valve, direct actuator, with solenoid position control, integrated electronics
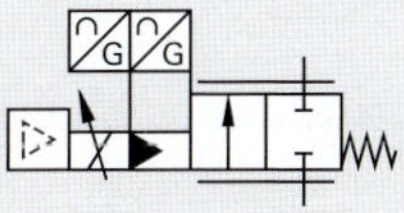	Proportional flow-control valve, pilot-controlled, with position control of the main control and pilot stage, integrated electronics		Flow-control valve, adjustable valve orifice to compensate viscosity variations, adjustment via proportional solenoid

A

Circuit diagrams

cf. DIN ISO 1219-2 (1996-11)

Designing a circuit plan

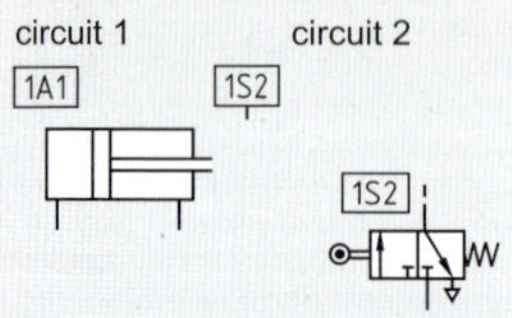

The circuit is subdivided into subcircuits with related control functions.

The actual spatial arrangement of the components is not considered.

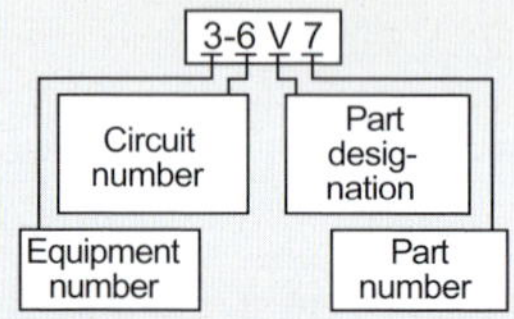

If the circuit diagram is made of several units, the unit number must be given, beginning with numeral 1.

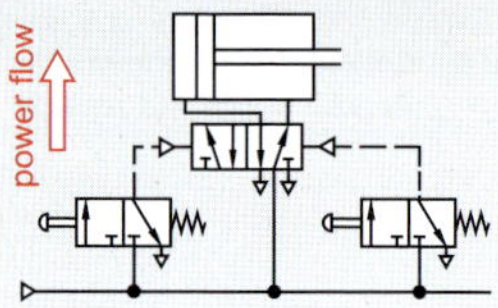

Components are arranged from bottom to top in the direction of power flow and from left to right.

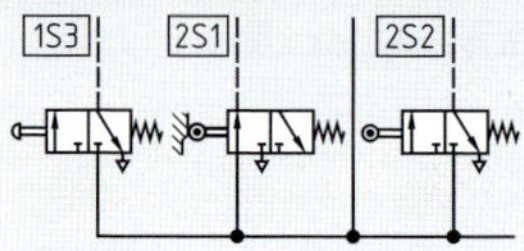

Similar components or subassemblies are shown at the same height within a circuit.

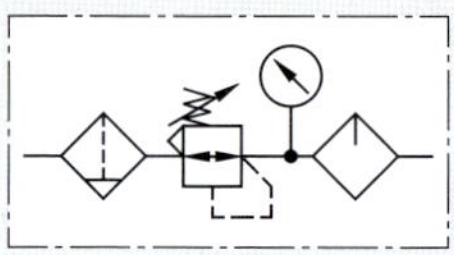

Subassemblies such as throttle check valves or service units (FRL) are enclosed by a dash-dot line.

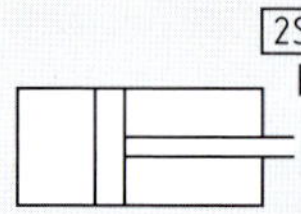

Devices actuated by drives, e. g. limit switches, are represented at their point of activation by a dash and their designator.

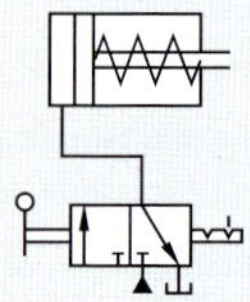

Hydraulic components are shown in their initial positions in the equipment before pressure is applied.

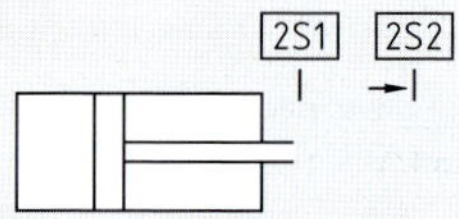

For roller plunger valves operating on one side only, a directional arrow is also placed at the dash.

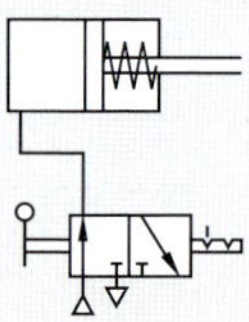

Pneumatic components are shown in their initial positions in the equipment before pressure is applied.

Components of a circuit

Drive elements	Motors, cylinders, valves
Actuators	Valves for controlling drive elements
Control elements	Valves for signal combination
Signal elements	Components used to trigger a switching action
Supply elements	Service unit (FRL), main valve

Example: Pneumatic circuit diagram with two cylinders (lifting device)

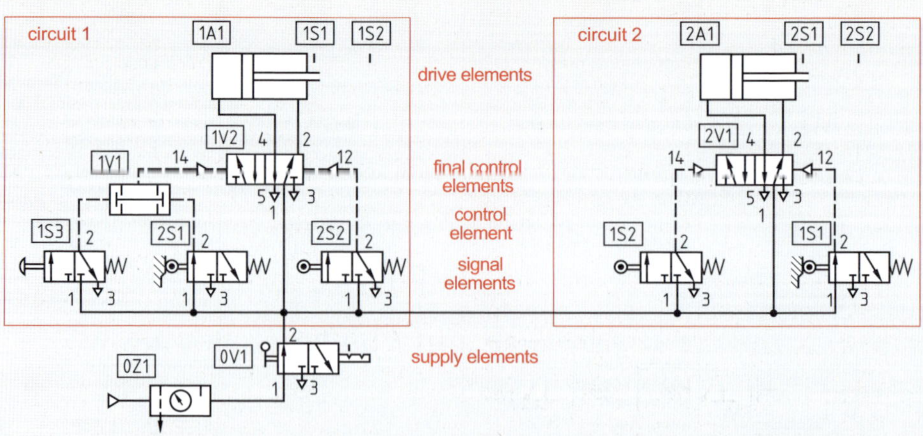

A

Pneumatic control

Sheet metals are manually fed to a bending tool. The tool should bend them to an angle of 90°. The clamping cylinder 1A1 retains the sheet in the working position while the bending cylinder 2A1 with the bending tool extends and forces the sheets to the desired angle. Subsequently, first the bending cylinder 2A1 and then the clamping cylinder 1A1 retract.

Technological scheme

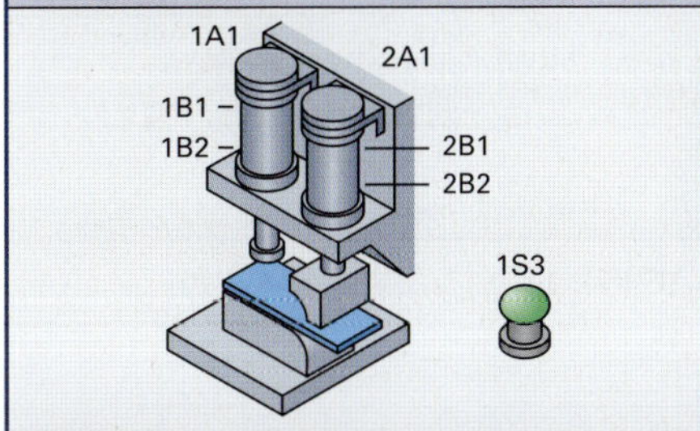

Functional sequence

1A1+	2A1+	2A1–	1A–

Function diagram

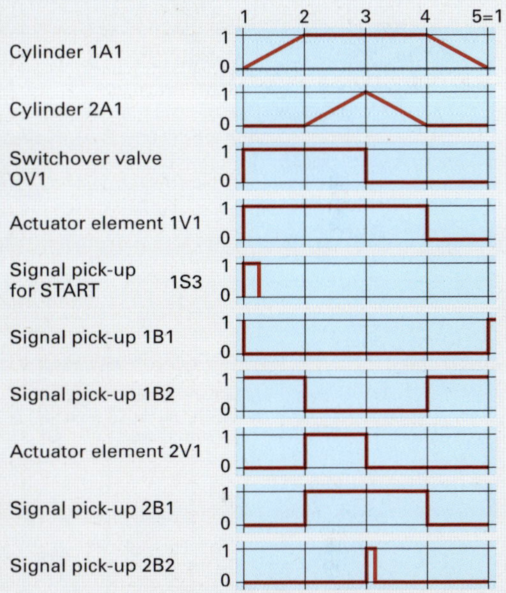

GRAFCET

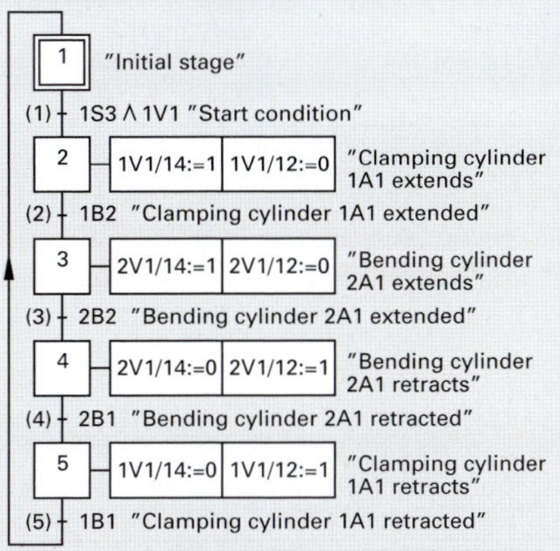

Pneumatic circuit diagram

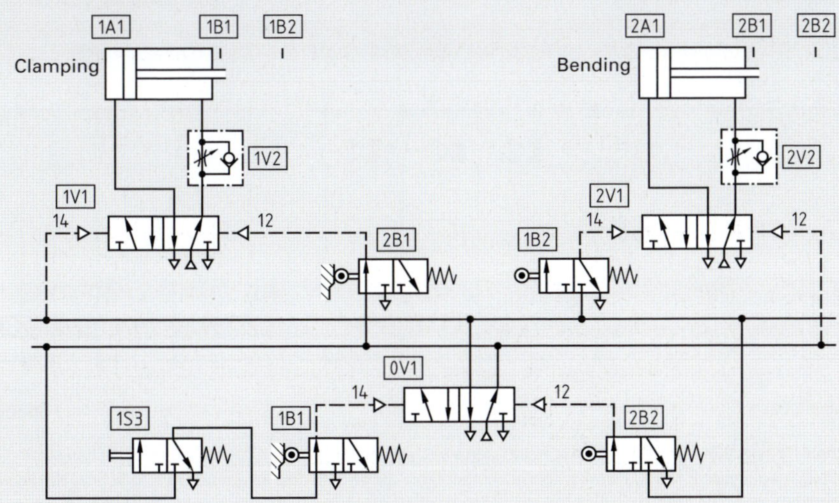

A

Electro-pneumatic controls

Layout

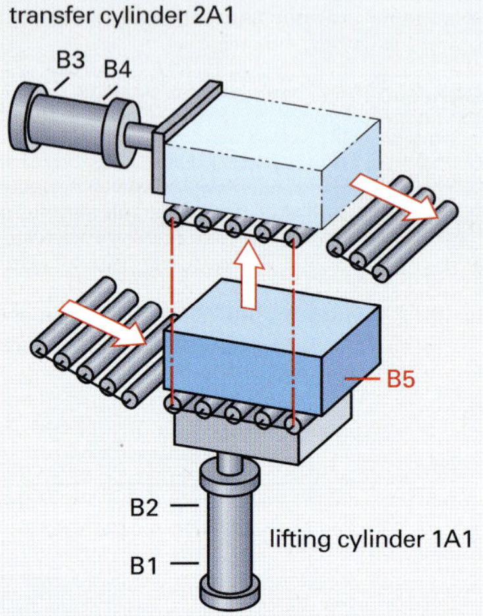

Function diagram

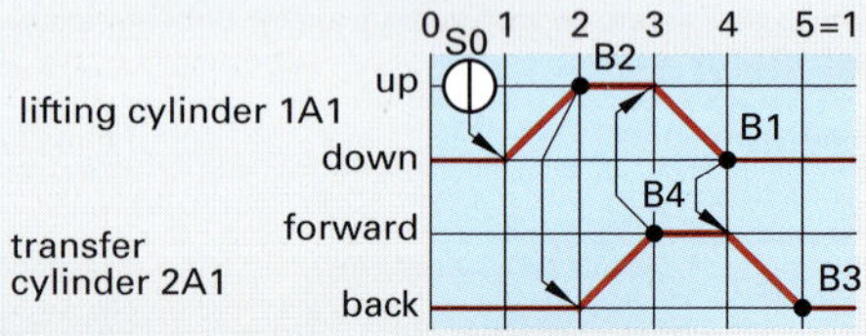

Pneumatic circuit diagram

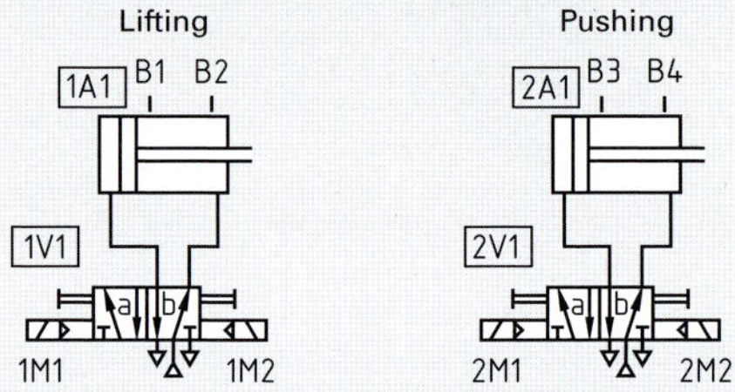

Circuit diagram

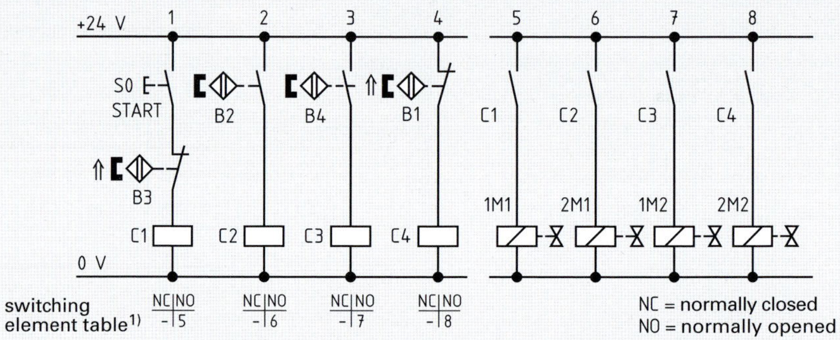

Circuit diagram with the additional functions – magazine query and continuous operation

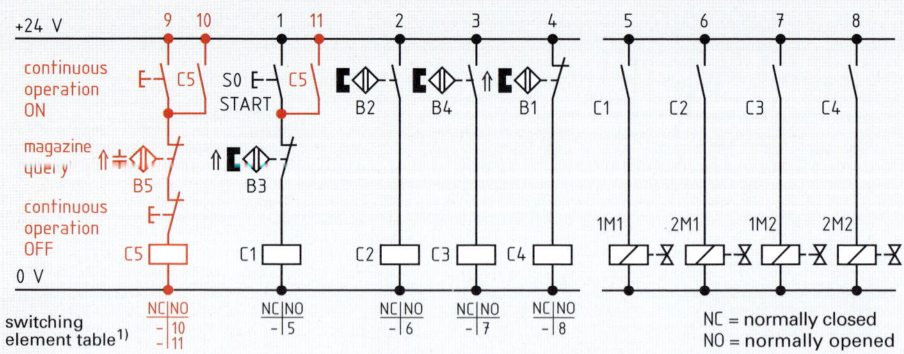

Example for relay K5: Relay K5 has a normally open switch in section 10 and a normally open switch in section 11.

[1] The switching element table is similar to the contact table (page 378) and is often used in practice. However it is not standardized. The table indicates the section in which a NC or NO relay contact can be found.

A

Electro-hydraulic control

A workpiece manufactured from two different materials is to be machined with a reamer. The activation of the start button S1 causes the piston to extend in feed mode and work on the softer material. Sensor 1B2 ensures that the feed motion is slowed down when working on the harder material. When the piston reaches 1B3, the machining of the hard material is finished and the piston retracts to its home position after a delay time of 2 sec.

Technological scheme

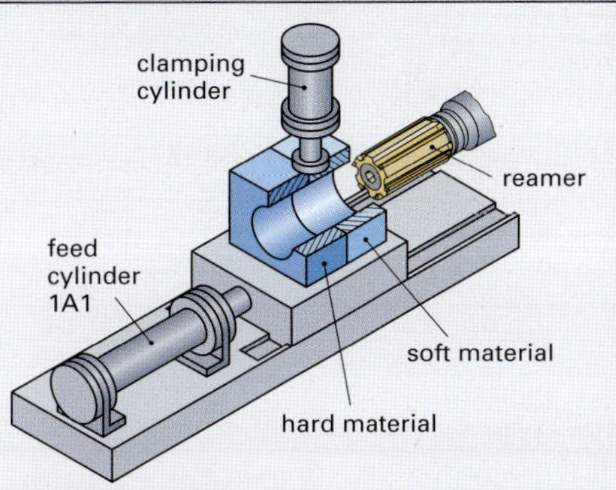

Hydraulic circuit diagram with proportional directional valve

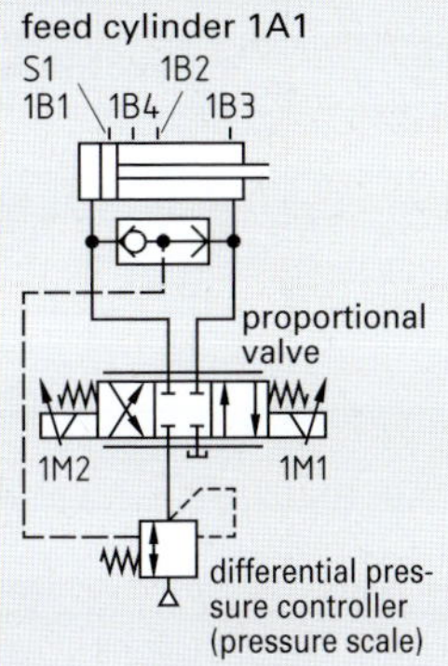

With a proportional directional valve, one can influence the direction, the velocity, the acceleration and the delay of a hydraulic cylinder in the feed unit. The use of such a valve can replace choke valves, flow-control valves, check and changeover valves. A proportional amplifier allows the programming of different ramps for the extending and retracting motion.

GRAFCET

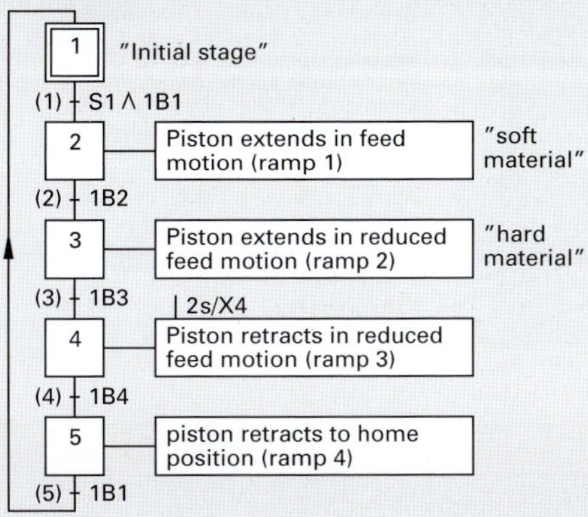

Directional diagram

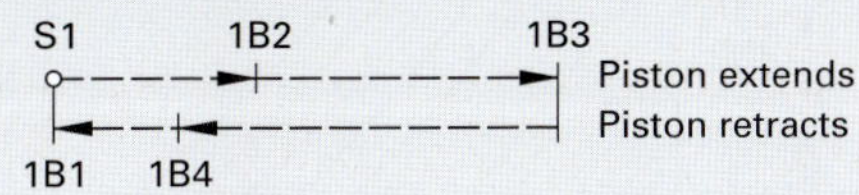

Velocity diagram

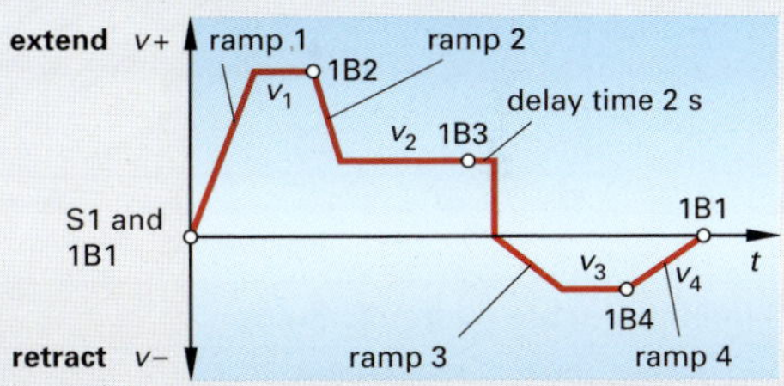

Electro-hydraulic circuit diagram

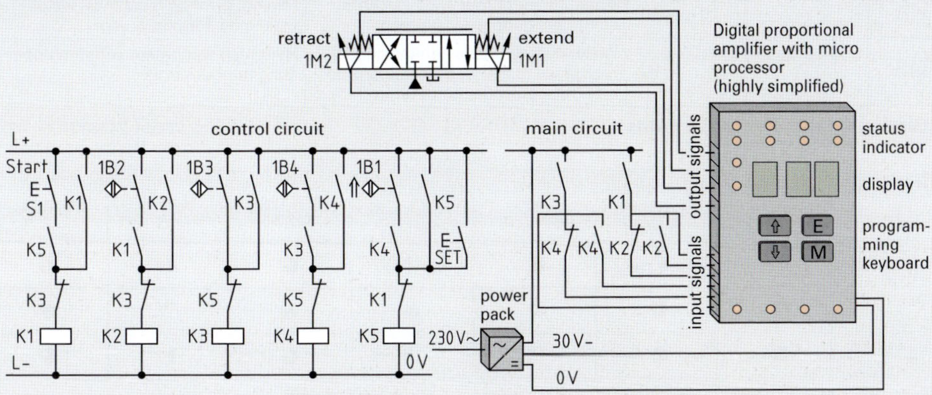

A

Hydraulic fluids

Mineral oil based hydraulic oils

cf. DIN 51524-1 to -3 (2006-04)

Type	Standard	Effect of the ingredients		Applications
HL	DIN 51524-1	Increase in corrosion resistance + Increase in aging resistance	–	Hydraulic units up to 200 bar, with high temperature requirements
HLP	DIN 51524-2		+ Reduction of wear due to scoring in mixed friction area	Hydraulic units with hydro pumps and hydro motors above 200 bar operating pressure and with high temperature requirements
HVLP	DIN 51524-3		+ Reduction of wear due to scoring in mixed friction area + Improvement of viscosity-temperature behavior	

Properties		**HL 10 HLP 10**	**HL 22 HLP 22**	**HL 32 HLP 32**	**HL 46 HLP 46**	**HL 68 HLP 68**	**HL 100 HLP 100**
Kinematic viscosity in mm^2/s	at – 20 °C	600	–	–	–	–	–
	at 0 °C	90	300	420	780	1400	2560
	at 40 °C	9–11	19.8–24.2	28.8–35.2	41.4–50.6	61.2–74.8	90–110
	at 100 °C	2.4	4.1	5.0	6.1	7.8	9.9
Pour point[1] equal to or lower than		30 °C	– 21 °C	–18 °C	–15 °C	–12 °C	–12 °C
Flash point above		125 °C	165 °C	175 °C	185 °C	195 °C	205 °C

[1] The pour point is the temperature at which hydraulic oil still flows under the force of gravity.

⇒ **Hydraulic oil DIN 51524 – HLP 46:** Hydraulic oil of type HLP, kinematic viscosity = 46 mm^2/s at 40 °C

Viscosity-temperature behavior of HL and HLP hydraulic oils

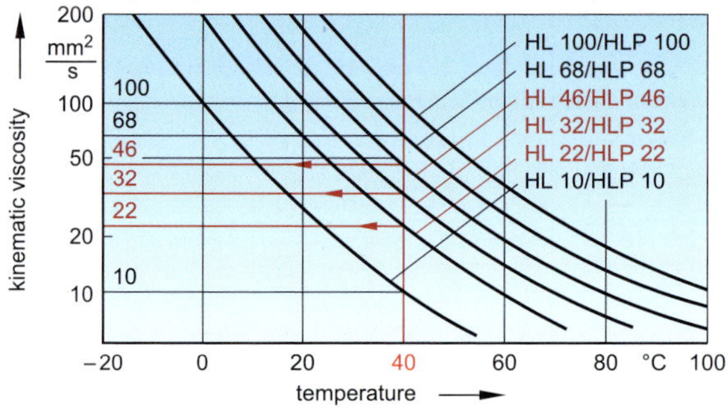

Example of reading from diagram:
A gear pump operates at an average operating temperature of 40°C. During operation the allowable kinematic viscosity of the hydraulic oil is allowed to fluctuate between 20 to 50 mm^2/sec.

According to the diagram there are 6 hydraulic oils to select from:

- HL 22/HLP 22
- HL 32/HLP 32
- HL 46/HLP 46

Non-flammable hydraulic fluids

Type	ISO Viscosity classes	Suitability for temperatures °C	Characteristics	Applications
HFC	15, 22, 32, 46, 68, 100	– 20 to + 60	Aqueous monomer and/or polymer solutions, good wear protection	Mining, printing machines, welding machines, forging presses
HFD		– 20 to + 150	Water free synthetic liquids, good resistance to aging, lubricating property through wide temperature range	Hydraulic equipment with high operating temperatures

Biodegradable hydraulic fluids

cf. VDMA 24569 (1994-03)

A

Hydraulic fluid	Suitability and properties						
	Low temperature flowability	High temperature oxidation stability	Rust protection	Compatibility with inner coatings	Seal compatibility	Cost effectiveness	Fluid life
Unsaturated esters	◐	◐	●	◔	◕	◐	◕
Saturated esters	●	●	●	◕	◕	◔	●
Polyglycol oils	●	●	◔	◔	◕	◐	◕

Suitability: ● very good ◕ good ◐ average ◔ limited/poor

Pneumatic cylinders

Dimensions and piston forces

Piston diameter		12	16	20	25	32	40	50	63	80	100	125	160	200
Piston rod diameter (mm)		6	8	8	10	12	16	20	20	25	25	32	40	40
Coupling thread		M5	M5	G1/8	G1/8	G1/8	G1/8	G1/4	G3/8	G3/8	G1/2	G1/2	G3/4	G3/4
Pushing force[1] at p_e = 6 bar in N	single-act. cyl.[2]	50	96	151	241	375	644	968	1560	2530	4010	–	–	–
	double-act. cyl.	58	106	164	259	422	665	1040	1650	2660	4150	6480	10 600	16 600
Pulling force[1] at p_e = 6 bar in N	double-act. cyl.	54	79	137	216	364	560	870	1480	2400	3890	6060	9960	15 900
Stroke in mm	single-act. cyl.	10, 25, 50					25, 50, 80, 100					–		
	double-act. cyl.	to 160	to 200	to 320	10, 25, 50, 80, 100, 160, 200, 250, 320, 400, 500									

[1] For a cylinder efficiency $\eta = 0.88$ [2] The return force of the spring is considered.

Calculating air consumption

Single-acting cylinder

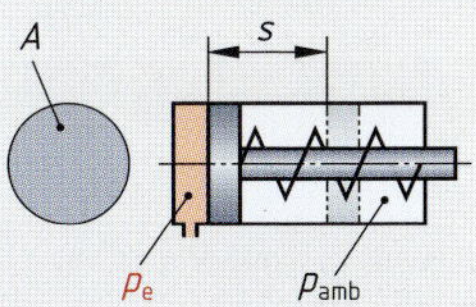

Double-acting cylinder

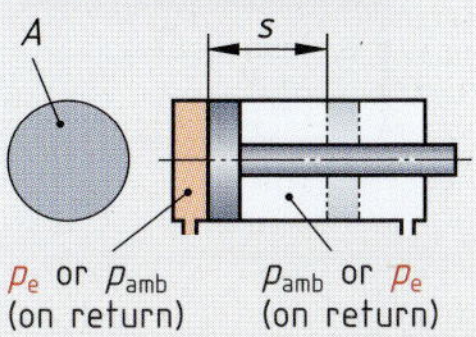

Q air consumption
p_e gage pressure in cylinder
p_{amb} ambient air pressure
n number of strokes
A piston surface area
q specific air consumption per cm piston stroke
s piston stroke

Example:

Single-acting cylinder with d = 50 mm; s = 100 mm; p_e = 6 bar; n = 120/min; p_{amb} = 1 bar; air consumption Q in l/min?

$$\boldsymbol{Q} = A \cdot s \cdot n \cdot \frac{p_e + p_{amb}}{p_{amb}}$$

$$= \frac{\pi \cdot (5\ \text{cm})^2}{4} \cdot 10\ \text{cm} \cdot 120\ \frac{1}{\text{min}} \cdot \frac{(6+1)\ \text{bar}}{1\ \text{bar}}$$

$$= 164\,934\ \frac{\text{cm}^3}{\text{min}} \approx \mathbf{165}\ \frac{\text{l}}{\textbf{min}}$$

Air consumption[1]
Single-acting cylinder

$$Q = A \cdot s \cdot n \cdot \frac{p_e + p_{amb}}{p_{amb}}$$

Air consumption[1]
Double-acting cylinder

$$Q \approx 2 \cdot A \cdot s \cdot n \cdot \frac{p_e + p_{amb}}{p_{amb}}$$

Air consumption determined from the diagram

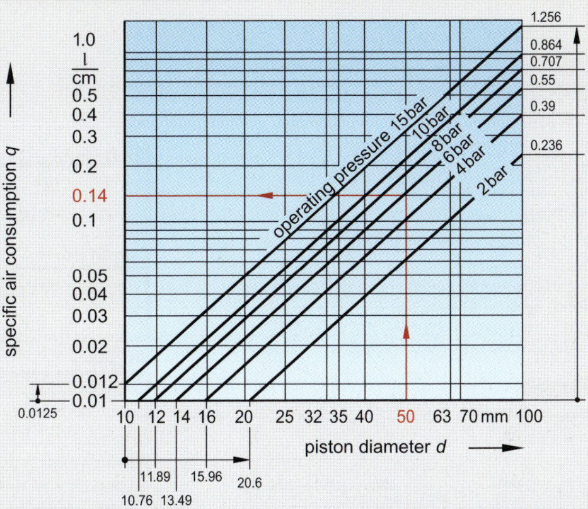

Air consumption[1]
Single-acting cylinder

$$Q = q \cdot s \cdot n$$

Air consumption[1]
Double-acting cylinder

$$Q \approx 2 \cdot q \cdot s \cdot n$$

Example:

Determine the air consumption of a single-acting cylinder of d = 50 mm, s = 100 mm and n = 120/min from the diagram for p_e = 6 bar.
According to the diagram the piston stroke is q = 0.14 l/cm.
$Q = q \cdot s \cdot n =$
= 0.14 l/cm · 10 cm · 120/min
= **168 l/min**

[1] When dead spaces are filled, actual air consumption may be up to 25 % greater. Dead spaces include air lines between the directional control valve and the cylinder and unused space in the end position of the piston. The cross-sectional area of the piston rod is not taken into consideration.

A

Hydraulic cylinders, Hydraulic pumps

Piston forces

Extending

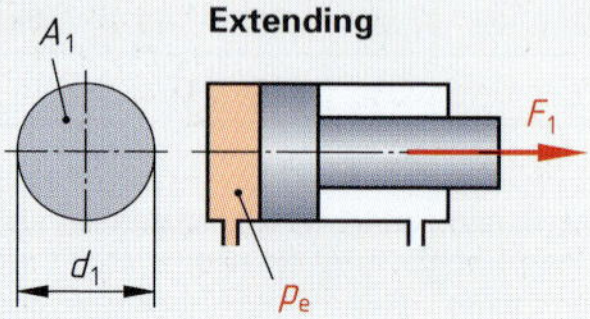

Retracting

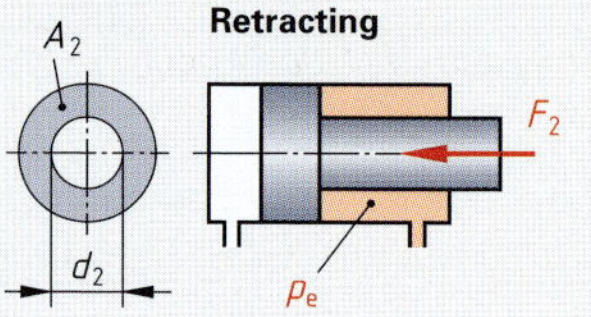

p_e	gage pressure	d_1	piston diameter
A_1, A_2	piston areas	d_2	piston rod diameter
F_1, F_2	effective piston forces	η	efficiency

Effective piston force

$$F = p_e \cdot A \cdot \eta$$

Example:

Hydraulic cylinder with d_1 = 100 mm; d_2 = 70 mm; η = 0.85 and p_e = 60 bar.
What are the effective piston forces?

Extending:

$$\boldsymbol{F_1} = p_e \cdot A_1 \cdot \eta = 600\,\frac{\text{N}}{\text{cm}^2} \cdot \frac{\pi \cdot (10\text{ cm})^2}{4} \cdot 0.85$$

$$= \mathbf{40\,055\ N}$$

Retracting:

$$\boldsymbol{F_2} = p_e \cdot A_2 \cdot \eta$$

$$= 600\,\frac{\text{N}}{\text{cm}^2} \cdot \frac{\pi \cdot [(10\text{ cm})^2 - (7\text{ cm})^2]}{4} \cdot 0.85$$

$$= \mathbf{20\,428\ N}$$

Pressure units

$$1\text{ Pa} = 1\,\frac{\text{N}}{\text{m}^2} = 10^{-5}\text{ bar}$$

$$1\text{ bar} = 10\,\frac{\text{N}}{\text{cm}^2} = 0.1\,\frac{\text{N}}{\text{mm}^2}$$

$$1\text{ mbar} = 100\text{ Pa} = 1\text{ hPa}$$

Piston speeds

Extending

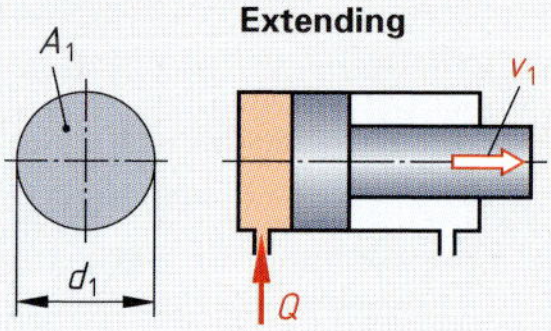

Retracting

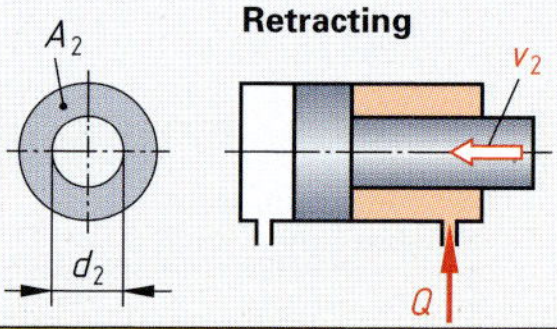

Q	volume flow rate
A_1, A_2	effective piston areas
v_1, v_2	piston speeds

Piston speed

$$v = \frac{Q}{A}$$

Example:

Hydraulic cylinder with d_1 = 50 mm; d_2 = 32 mm and Q = 12 l/min. v_1 = ?; v_2 = ?

Extending:

$$\boldsymbol{v_1} = \frac{Q}{A_1} = \frac{12\,000\text{ cm}^3/\text{min}}{\frac{\pi \cdot (5\text{ cm})^2}{4}} = 611\,\frac{\text{cm}}{\text{min}} = \mathbf{6.11\,\frac{m}{min}}$$

Retracting:

$$\boldsymbol{v_2} = \frac{Q}{A_2} = \frac{12\,000\text{ cm}^3/\text{min}}{\frac{\pi \cdot (5\text{ cm})^2}{4} - \frac{\pi \cdot (3.2\text{ cm})^2}{4}}$$

$$= 1035\,\frac{\text{cm}}{\text{min}} = \mathbf{10.35\,\frac{m}{min}}$$

Power of pumps

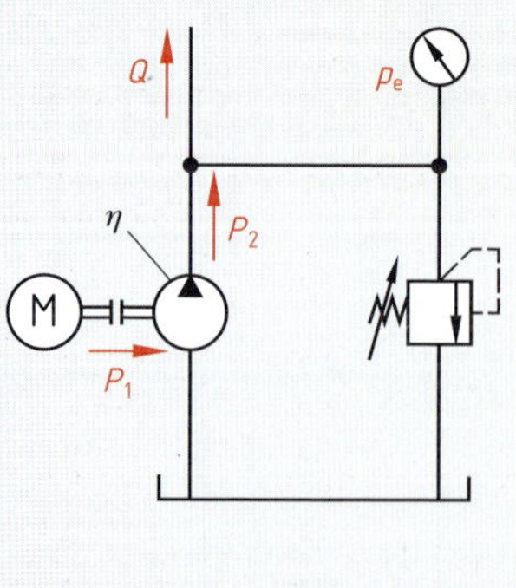

P_1	input power on pump drive shaft
P_2	output power on pump outlet
Q	volume flow rate
p_e	gage pressure
η	efficiency of the pump
M	torque
n	rotational speed
9550	conversion factor
600	conversion factor

Input power

$$P_1 = \frac{M \cdot n}{9550}$$

Output power

$$P_2 = \frac{Q \cdot p_e}{600}$$

Efficiency

$$\eta = \frac{P_2}{P_1}$$

Formulae for input and output power with:
P in kW, M in N · m, n in 1/min, Q in l/min, p_e in bar

Example:

Pump with Q = 40 l/min; p_e = 125 bar; η = 0.84; P_1 = ?; P_2 = ?

$$\boldsymbol{P_2} = \frac{Q \cdot p_e}{600} = \frac{40 \cdot 125}{600}\text{ kW} = \mathbf{8.333\ kW}$$

$$\boldsymbol{P_1} = \frac{P_2}{\eta} = \frac{8.333}{0.84}\text{ kW} = \mathbf{9.920\ kW}$$

A

Tubes

Seamless precision steel tubes for hydraulic and pneumatic lines

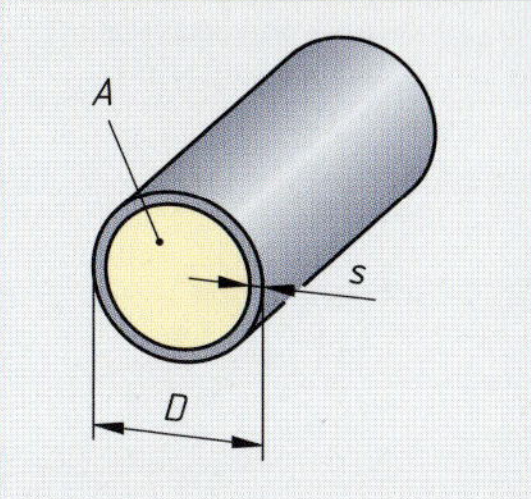

Materials	E235 (St37.4), E355 (St52.4) according to DIN 1630			
Mechanical properties	Material	Tensile strength R_m N/mm²	Yield strength R_e N/mm²	Elongation at fracture *EL* %
	E235	340 to 480	235	25
	E355	490 to 630	355	22
	Good cold workability, surface phosphatized or electroplated and chromed			
Applications	For lines in hydraulic or pneumatic systems at maximal rated pressures up to 500 bar			

Delivery type: Normal manufactured length: 6 m, normalized. Tubes have a surface quality of $Ra \leq 4$ µm.

⇒ **Tube HPL-E235-NBK-20 x 2:** Seamless precision steel tube for hydraulic and pneumatic applications, made of E235, normalized, bright-drawn, outside diameter 20 mm, wall thickness 2 mm

Outside diameter *D* mm	Wall thickness *s* mm	Flow sectional area *A* cm²	Outside diameter *D* mm	Wall thickness *s* mm	Flow sectional area *A* cm²	Outside diameter *D* mm	Wall thickness *s* mm	Flow sectional area *A* cm²
4	0.8	0.05	20	2.0	2.01	38	2.5	8.55
4	1.0	0.01	20	2.5	1.77	38	4.0	7.07
5	0.8	0.10	20	3.0	1.54	38	5.0	6.16
5	1.0	0.07	20	4.0	1.13	38	7.0	4.52
6	1.0	0.13	22	1.0	3.14	38	10.0	2.55
6	1.5	0.07	22	2.0	2.54	42	2.0	11.34
8	1.0	0.28	22	3.0	2.01	42	5.0	8.04
8	1.5	0.20	22	3.5	1.77	42	8.0	5.31
8	2.0	0.13	25	1.5	3.80	50	4.0	13.85
10	1.0	0.50	25	2.5	3.14	50	5.0	12.57
10	1.5	0.39	25	3.0	2.84	50	8.0	9.08
10	2.0	0.28	25	3.5	2.55	50	10.0	7.07
12	1.0	0.79	25	4.5	2.01	50	13.0	4.52
12	1.5	0.64	25	6.0	1.33	55	4.0	17.35
12	2.0	0.50	28	1.5	4.91	55	6.0	14.52
14	1.0	1.13	28	2.0	4.52	55	8.0	11.95
14	1.5	0.95	28	3.0	3.80	55	10.0	9.62
14	2.0	0.79	28	3.5	3.46	60	5.0	19.64
15	1.0	1.33	28	4.0	3.14	60	8.0	15.21
15	1.5	1.13	30	2.0	5.31	60	10.0	12.57
15	2.5	0.79	30	2.5	4.91	60	12.5	9.62
16	1.0	1.54	30	3.0	4.52	70	5.0	28.27
16	2.0	1.13	30	5.0	3.14	70	8.0	22.90
16	3.0	0.79	30	6.0	2.55	70	10.0	19.64
16	3.5	0.64	35	2.5	7.07	70	12.5	15.90
18	1.0	2.01	35	3.5	6.16	80	6.0	36.32
18	1.5	1.77	35	4.0	5.73	80	8.0	32.17
18	2.0	1.54	35	5.0	4.91	80	10.0	28.27
18	3.0	1.13	35	6.0	4.16	80	12.5	23.76

Rated pressure depending on wall thickness

Outside diameter *D* in mm	Rated pressure *p* in bar					
	64	100	160	250	320	400
	Wall thickness *s* in mm					
6	1.0	1.0	1.0	1.0	1.0	1.5
8	1.0	1.0	1.0	1.5	1.5	2.0
10	1.0	1.0	1.0	1.5	1.5	2.0
12	1.0	1.0	1.5	2.0	2.0	2.5
16	1.5	1.5	1.5	2.0	2.5	3.0
20	1.5	1.5	2.0	2.5	3.0	4.0
25	2.0	2.0	2.5	3.0	4.0	5.0
30	2.5	2.5	3.0	4.0	5.0	6.0
38	3.0	3.0	4.0	5.0	6.0	8.0
50	4.0	4.0	5.0	6.0	8.0	10.0

Coordinate systems and axes

cf. DIN EN ISO 9787 (2000-07)

Robot axes

Coordinate system	Robot main axes for positioning		Robot auxiliary axes for orientation
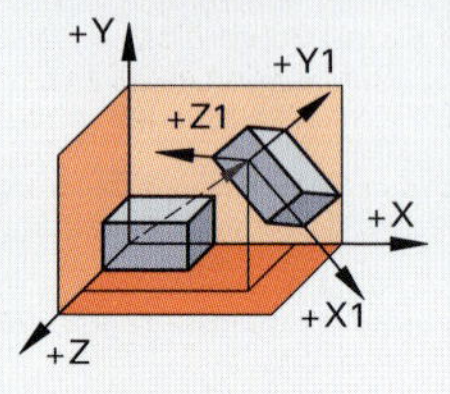	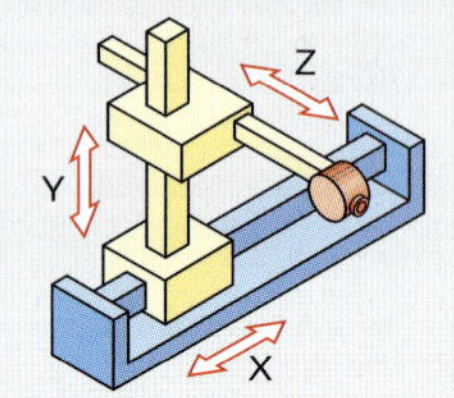	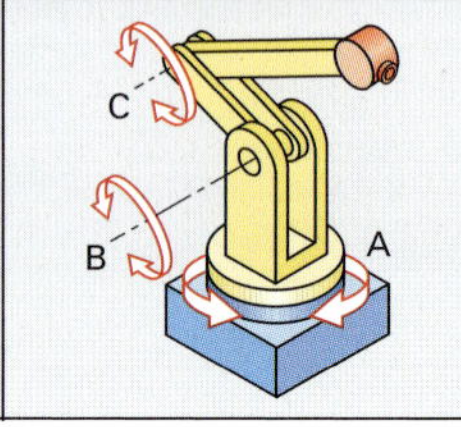	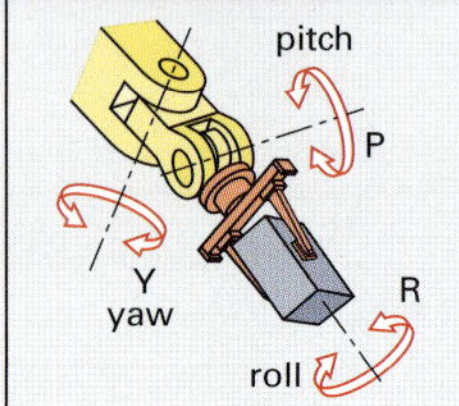
To manipulate workpieces or tools in space, the following are necessary: • 3 degrees of freedom for positioning and • 3 degrees of freedom for orientation	To reach a desired point in space, 3 robot main axes are necessary. **Cartesian robots** 3 translation axes (T axes) designated X, Y and Z	**Articulated arm robots** 3 rotational axes (R-axes) designated A, B and C	3 robot auxiliary axes for spatial orientation • R (roll) • P (pitch) • Y (yaw)

Coordinate systems

cf. DIN EN ISO 9787 (2000-07)

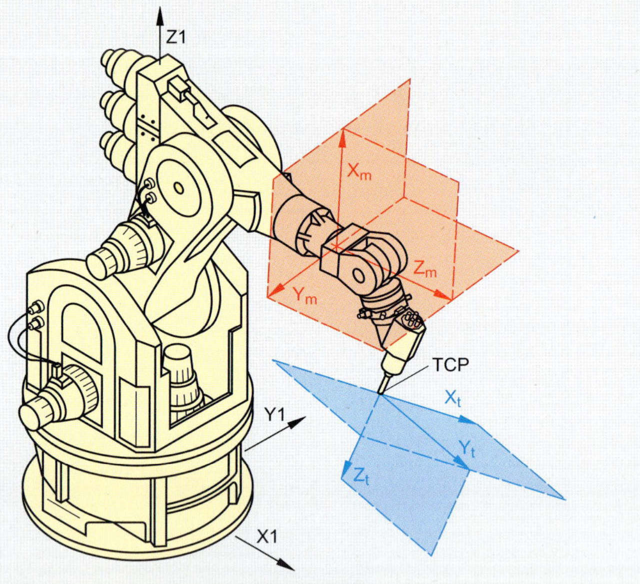

Base coordinate system

The base coordinate system references
- the level mounting surface for the X-Y plane
- the center of the robot for the Z axis

Flange coordinate system

The flange coordinate system references the end surface of the terminating main axis of the robot.

Tool coordinate system

The origin of the tool coordinate system lies at the tool center point *TCP* (*T*ool *C*enter *P*oint).

The speed of the tool center point is referred to as the robot speed and the path of tool travel as the robot trajectory.

Symbols for representing robots (selection)

cf. VDI 2861 (1988-06)

Designation	Symbol	Designation	Symbol	Example RRR robots
Translation axis (T-axis)[1] Translation aligned (telescoping) Translation out of alignment	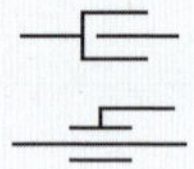	**Rotation axis (R-axis)**[2] Rotation aligned Rotation out of alignment	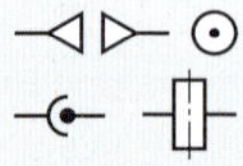	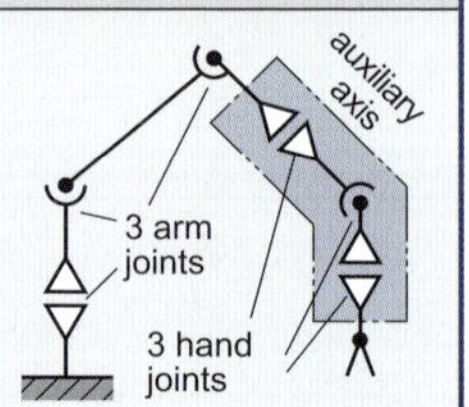 auxiliary axis; 3 arm joints; 3 hand joints
Gripper		Auxiliary axis (e. g. for roll, pitch and yaw)		

[1] Translation = straight line motion [2] Rotation = rotational motion

A

Robot designs

cf. DIN EN ISO 9787 (2000-07)

Mechanical structure[1]	Kinematics[2] and working envelope	Examples of design types	Characteristics, areas of application
Cartesian robots 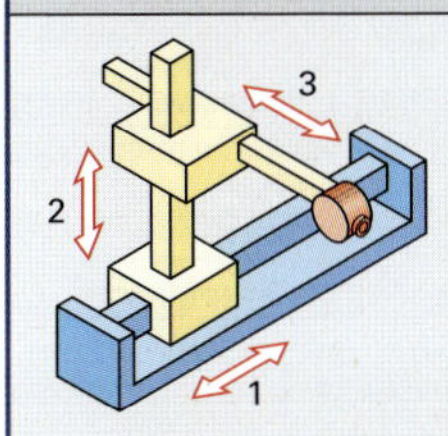	TTT-Kinematics 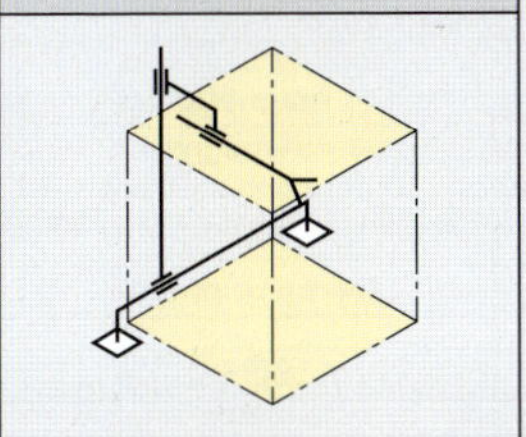	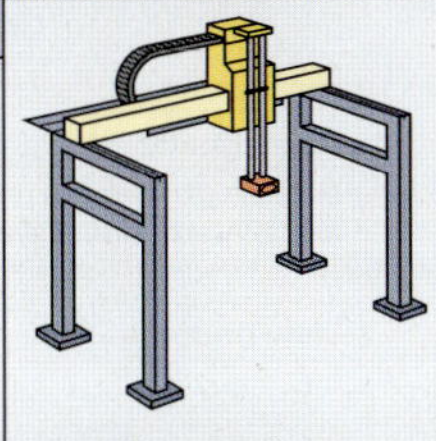 Gantry robot	Main axes: • 3 translational Areas of application: • large working envelope, therefore often in overhead gantry • tool and workpiece feed in production cells • sheet processing with laser beam and water jet cutting • palletizing
Cylindrical robots 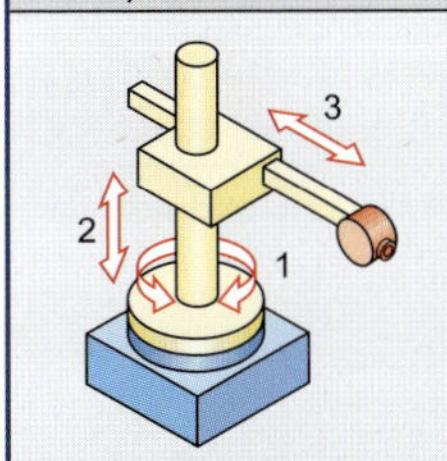	RTT Kinematics 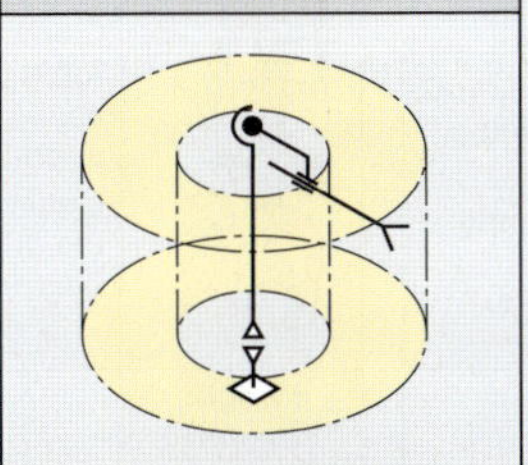	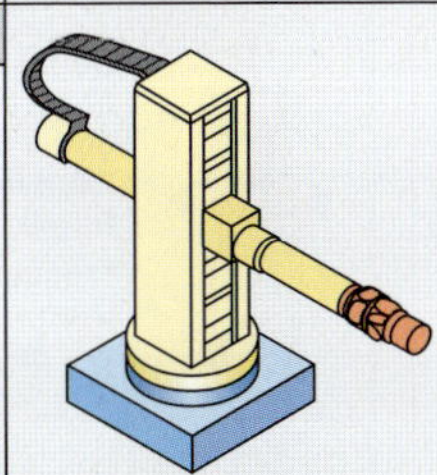 Base robot	Main axes: • 1 rotational • 2 translational Areas of application: • suitable for heavy masses • handling of heavy forged and cast parts • transport of pallets and tool cartridges • pick and place
Polar robot 1 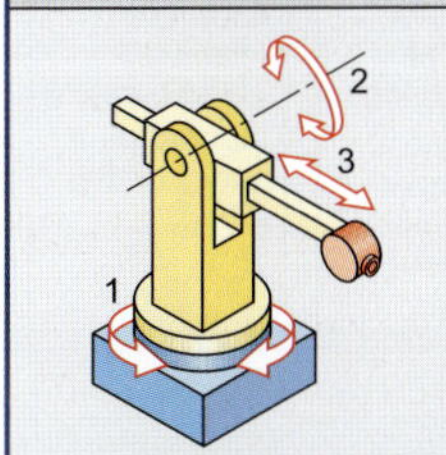	RRT-Kinematics 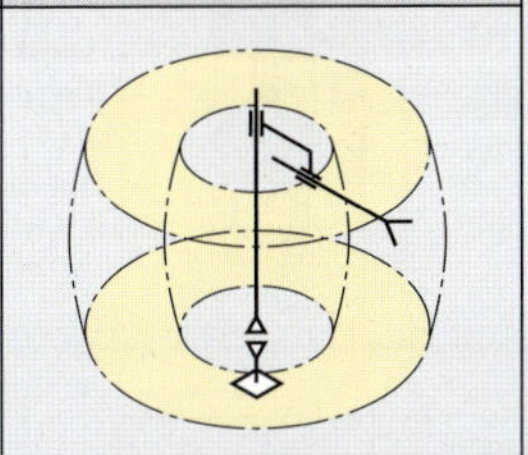	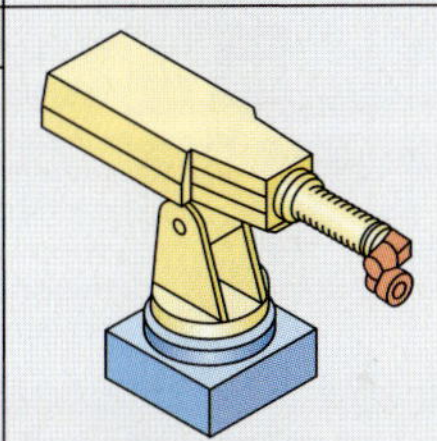 Vertical swivel arm robot	Main axes: • 2 rotational • 1 translational Areas of application: • telescoping type axis 3, consequently deeper working envelope • point and simple path welding, e. g. on car bodies • pick and place with die casting machines
Polar robot 2 Type: SCARA[3] robot 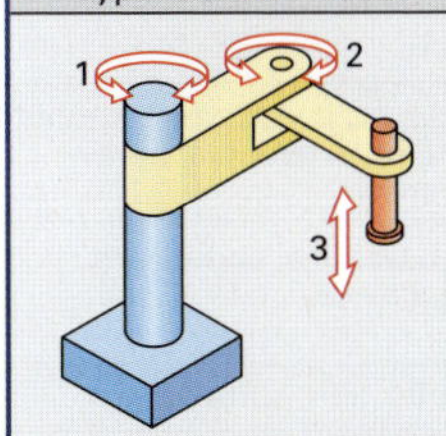	RRT-Kinematics 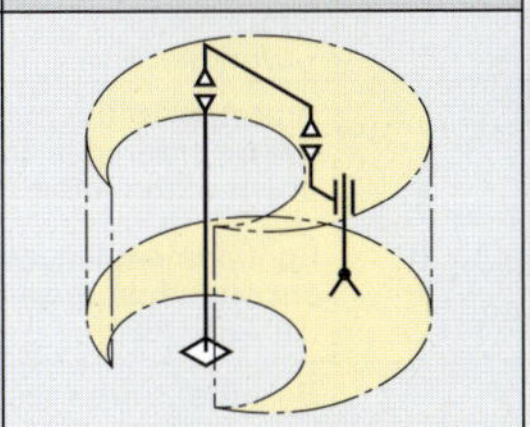	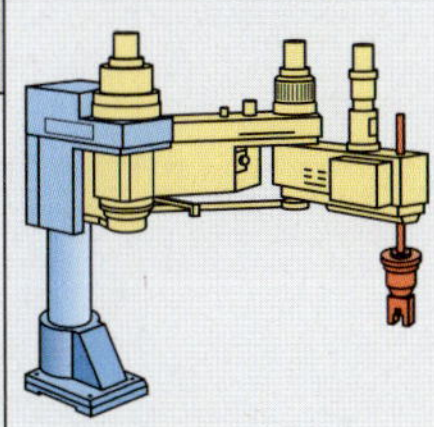 Horizontal swivel arm robot	Main axes: • 2 rotational as horizontal revolute joint • 1 translational Areas of application: • primarily in vertical assembly area • point and simple path welding • pick and place work
Articulated arm robots 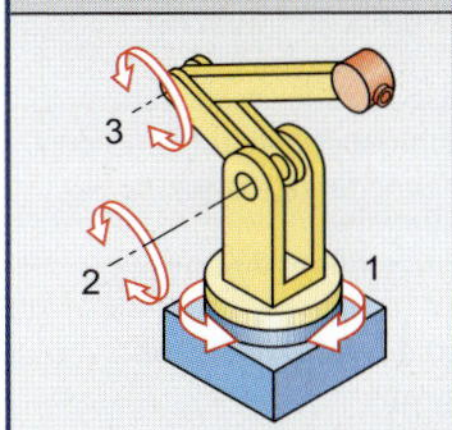	RRR-Kinematics 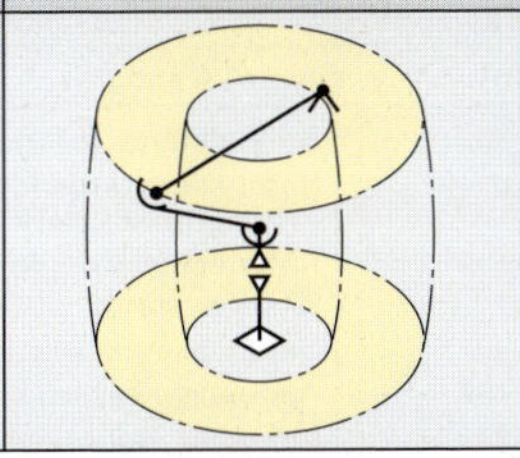	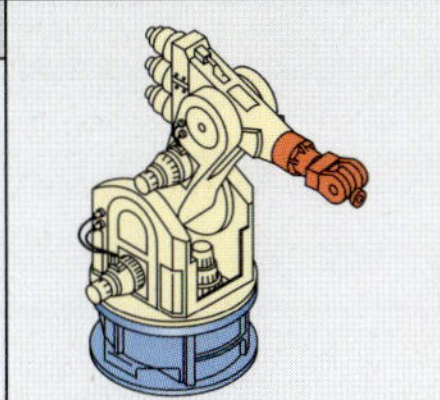 Vertical swivel arm robot	Main axes: • 3 rotational Areas of application: • handling and assembly area • complex path welding • painting work • adhesive bonding • low space requirement yet large working envelope

1) Axes are designated with numbers, where axis 1 is the axis of the first motion.
2) R = rotational axis; T = translational axis (Designations "R" and "T" are not standardized.)
3) SCARA = Selective Compliance Assembly Robot Arm

A

Grippers, Job safety

Gripper

cf. DIN EN ISO 14539 (2002-12) and VDI 2740 (1995-04)

Gripper

- **mechanical**
- **pneumatic**
  - suction gripper
  - articulated finger gripper
- **magnetic**
  - electromagnets
  - permanent magnets
- **adhesive**
  - velcro fastener gripper

Mechanical grippers: **Finger grippers** | **Jaw grippers** | **Clamp grippers** | **Needle grippers**

Finger grippers

Linear grippers	Characteristics
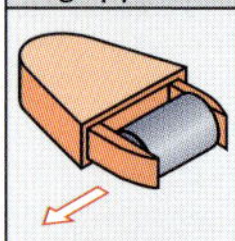	1 degree of movement
Flat gripper	
	3 degrees of movement
Spatial gripper	
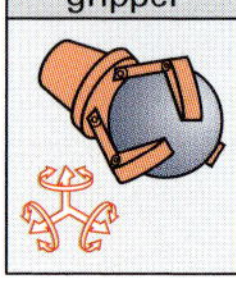	6 degrees of movement

Jaw grippers

Scissors grippers	Characteristics
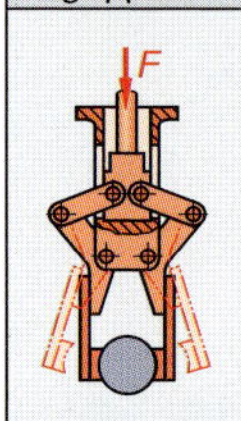	Both gripper fingers turn about an axis fixed in the frame. Frequently used grippers.
Parallel gripper	
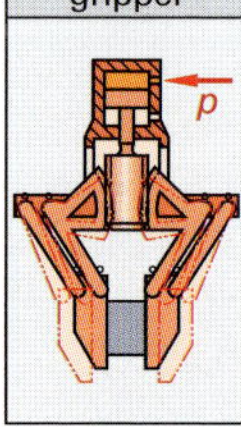	Both gripper fingers are pushed parallel to each other opposite to the gripper housing.

Clamp grippers

Spring loaded	Characteristics
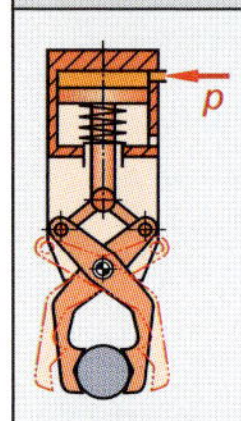	Clamping force is created by a spring. Opening of the gripper by pressure.
Weight loaded	
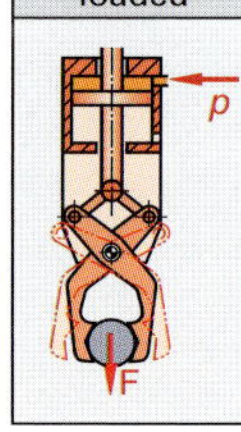	Clamping force created by the own weight of the gripping object. Opening of the gripper by pressure.

Needle grippers

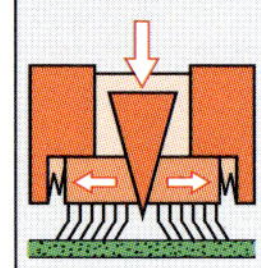

Used in textile industry. Four nail plates are extended by a tapered plug and grip the fabric.

Work safety for handling and robot systems*

cf. DIN EN ISO 10218-1 (2009-07)

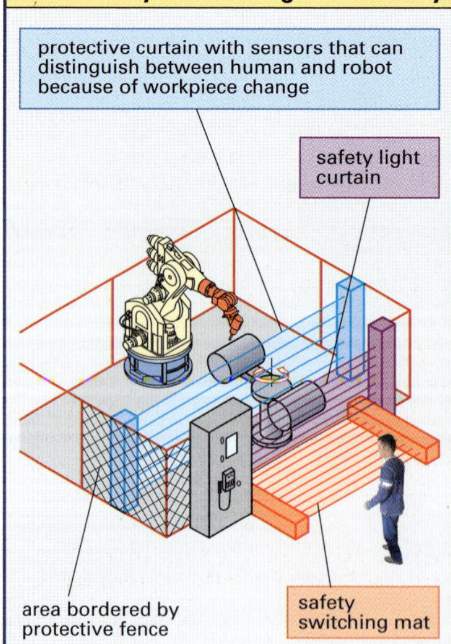

Concepts	Explanations
Maximum space	Area encompassing: • moving parts of robot • tool flange • workpiece
Restricted space	A portion of the maximum space which should not be entered in case of an eventual breakdown of the robot system
Separating safeguards	Containment fences, coverings, permanent encasements, locking devices (DIN EN 1088)
Protective systems with contactless activation	Hazardous area security: light curtains and light barriers Area monitoring: laser scanners Access security: light grills and light barriers
Important safety relevant standards	
DIN EN 292	Safety stand. for machines, basic terminology
DIN EN 61496	Safety standards for machines, contactless activation of safety systems
DIN EN 418	Safety standards for machines, emergency STOP systems
DIN EN 294	Safety around machines, safe distances
DIN EN 457	Acoustical hazard signals
CSA Z 434-03	Industrial Robots and Robot systems
ANSI R 15.06	American Standard for Industrial Robots

*) According to European standards

Coordinate axes

cf. DIN 66217 (1975-12)

Coordinate system

Right hand rule

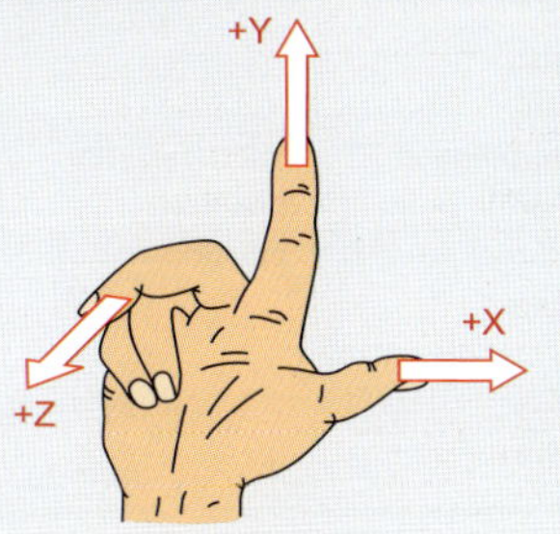

Cartesian coordinate system

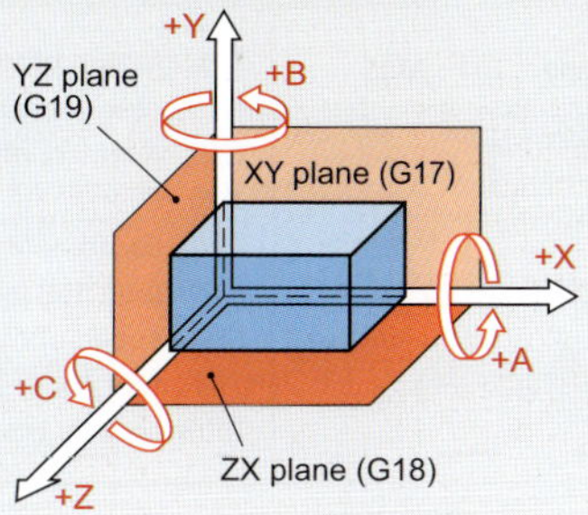

Coordinate axes X, Y and Z are perpendicular to each other.

This arrangement can be represented by thumb, index finger and middle finger of the right hand.

Axes of rotation A, B and C are assigned to coordinate axes X, Y and Z.

When looking down one axis in the positive direction, the positive direction of rotation is clockwise.

Coordinate axes in programming

Vertical milling machine

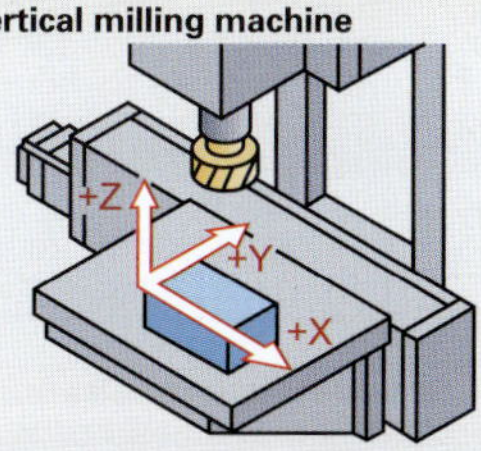

Horizontal milling machine

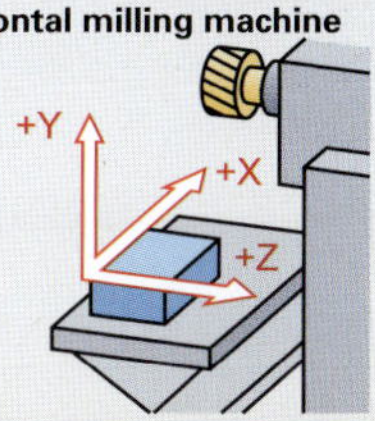

Lathe

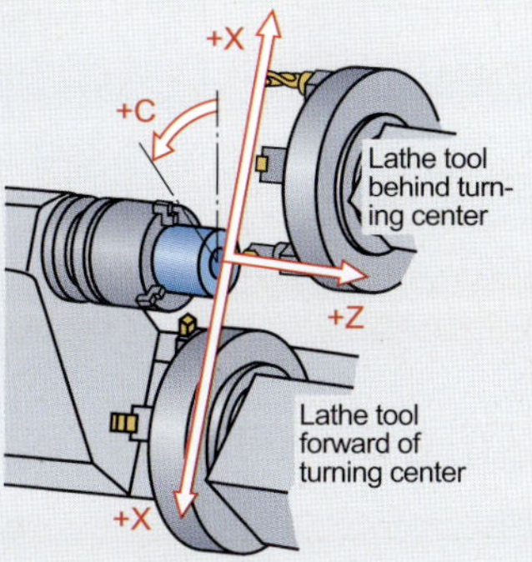

Example:
2-carriage lathe with programmable main spindle

Coordinate axes and the resulting directions of motion are aligned to the main slideways of the CNC machine and are essentially relative to the clamped workpiece with its workpiece zero point.

Positive directions of motion always result in greater coordinate values on the workpiece.

The Z axis always runs in the direction of the main spindle.

To simplify programming it is assumed that the workpiece remains motionless and only the tool moves.

Zero points and reference points

cf. DIN ISO 2806 (1996-04)

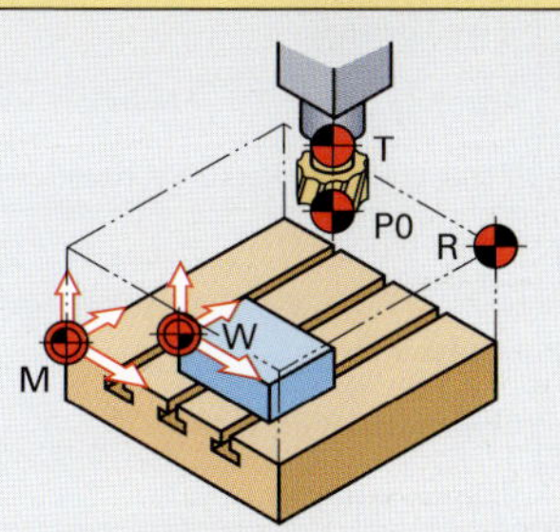

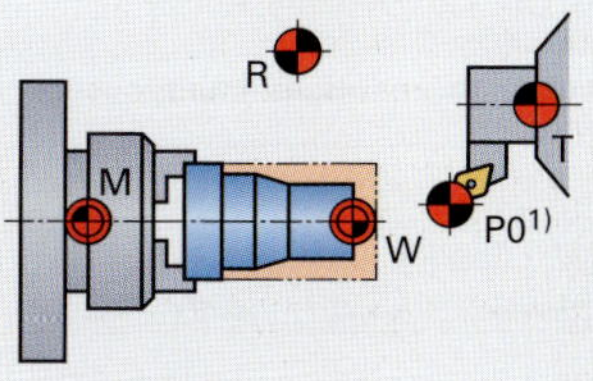

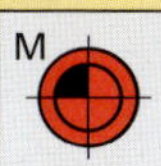

Machine zero point M
Origin of the machine coordinate system and set by the machine manufacturer.

Workpiece zero point W
Origin of the workpiece coordinate system and set by the programmer based on processing aspects.

Tool holder reference point T
Lies central to the limiting face of the tool holder. On milling machines, this is the abutting surface of the tool spindle, on lathes the abutting face of the tool holder on revolver.

Reference point R
Origin of incremental position measurement system with a distance to the machine zero point set by the machine manufacturer.

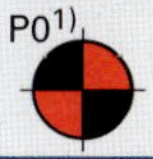

Program zero point P0
Indicates the coordinates of the point at which the tool is located before start of the program.

[1)] not standardized

A

Program structure

Tasks of the control program

Block structure

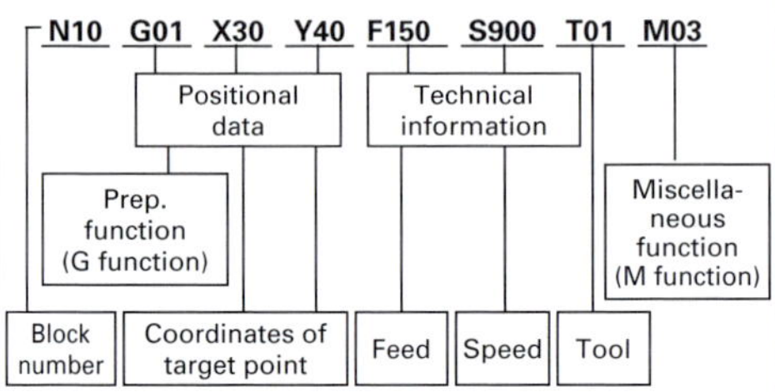

Explanation of words:

N10 block number 10
G01 feed, linear interpolation
X30 coordinate of target point in X direction
Y40 coordinate of target point in Y direction
F150 feed 150 mm/min
S900 speed of main spindle 900/min
T01 tool no. 1
M03 spindle clockwise

Program structure

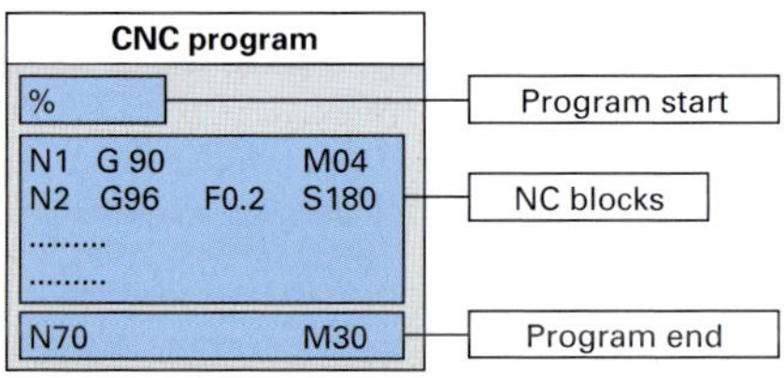

Example:

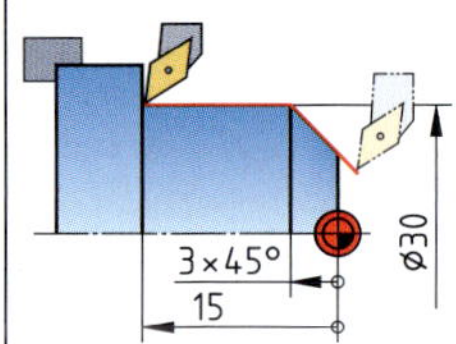

CNC program			
% 01			
N1	G90		M04
N2	G96	F0.2	S180
N3	G00	X20	Z2
N4	G01	X30	Z-3
N5			Z-15
N6	G00	X200	Z200
N7			M30

Preparatory functions

Prep. functions	Effective-ness	Meaning	Prep. functions	Effective-ness	Meaning
G00	●	Positioning at rapid rate	G53	●	Cancel shift
G01	●	Linear interpolation	G54–G59	●	Shift 1––Shift 6
G02	●	Circle interpolation clockwise			
G03	●	Circle interpol. counterclockwise	G74	●	Approach reference point
G04	● (red)	Dwell time predetermined	G80	● (red)	Cancel fixed cycle
G09	● (red)	Exact stop	G81–G89	●	Fixed cycle 1––Fixed cycle 9
G17	●	Plane selection XY			
G18	●	Plane selection ZX	G90	●	Absolute dimensional notation
G19	●	Plane selection YZ	G91	●	Incremental dimensional notation
G33	●	Thread cutting, constant pitch	G94	● (red)	Feed rate in mm/min
G40	●	Cancel tool offset	G95	●	Feed in mm
G41	●	Cutter compensation, left	G96	●	Constant cutting speed
G42	●	Cutter compensation, right	G97	●	Spindle speed in 1/min

● modal: Preparatory functions that remain effective until they are overwritten by a similar type of condition.

● (red) non-modal: Preparatory functions that are only effective in the block in which they are programmed.

Universal miscellaneous functions (m-functions, selection)

cf. DIN 66025-2 (1988-09)

M00	Programmed stop	M04	Spindle counterclockwise	M07	Cooling lubricant ON
M02	Program end	M05	Spindle stop	M09	Cooling lubricant OFF
M03	Spindle clockwise	M06	Tool change	M30	Program end with reset

Tool offset and Cutter compensation

Turning	Milling

Tool offset

Turning

Positional codes[1] for cutting tool point P in relation to center M of cutting radius r_ε

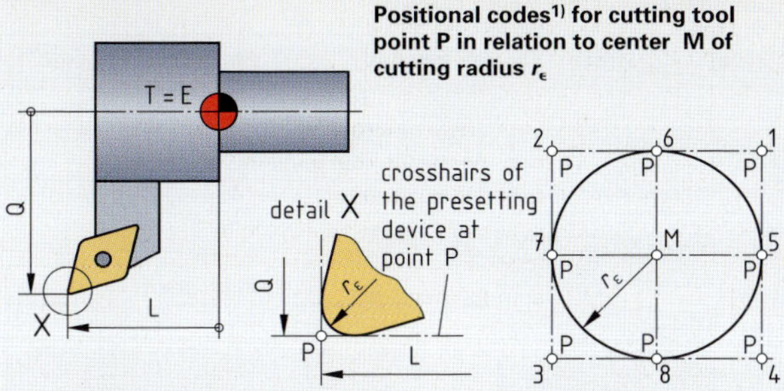

Q transverse offset of X axis
L longitudinal correction of Z axis
r_ε cutting radius
1–8 positional code digits
T tool holder reference point
E tool reference point
M center of cutting radius r_ε
P tool cutting point

[1] not standardized

Milling

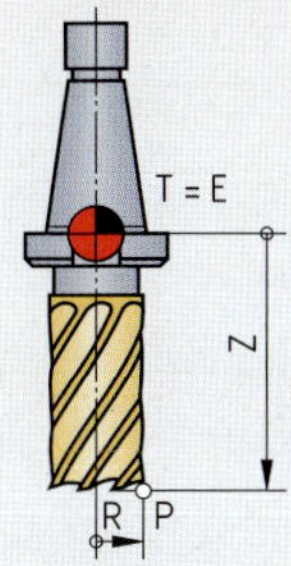

Z tool length
R tool radius
T tool holder reference point
E tool reference point
P tool cutting point

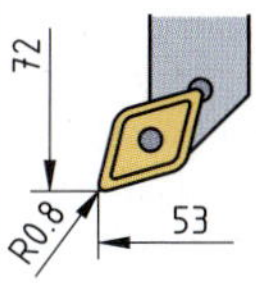

Offset memory	
Q	72
L	53
r_ε	0.8
Positional digit	3

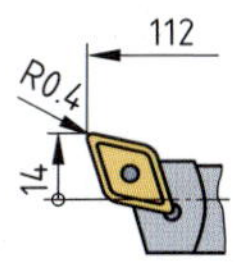

Offset memory	
Q	14
L	112
r_ε	0.4
Positional digit	2

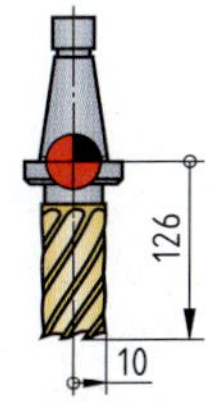

Offset memory	
Z	126
R	10

Cutter compensation

G41	Lathe tool left	G42	Lathe tool right	G41	Milling cutter left

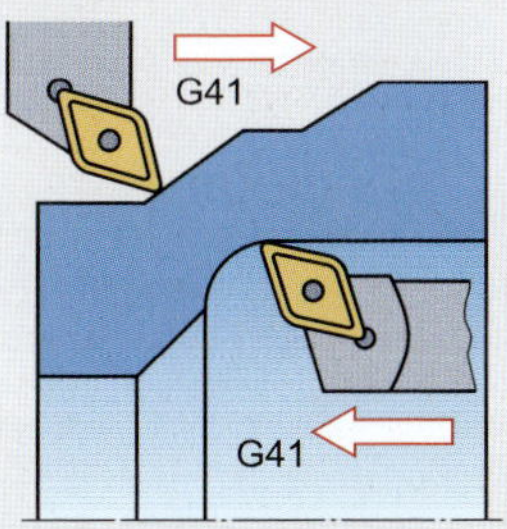

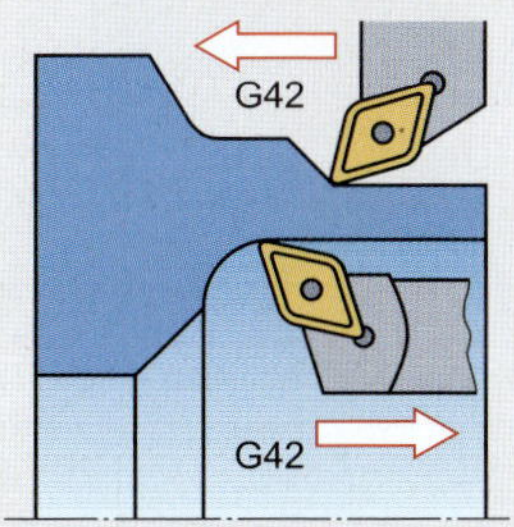

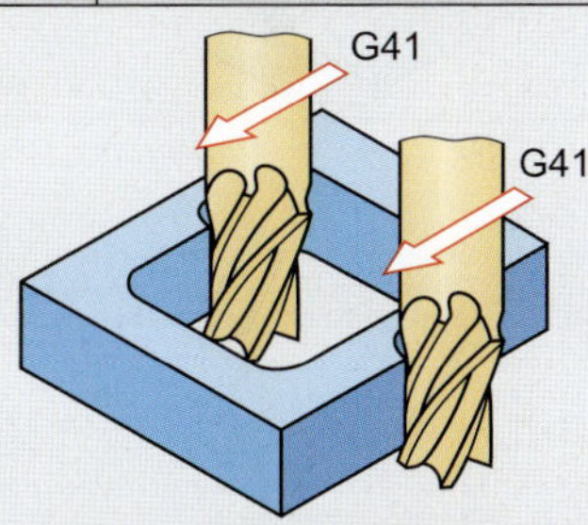

Lathe tool in front of spindle axis

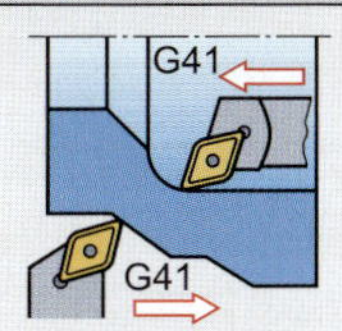

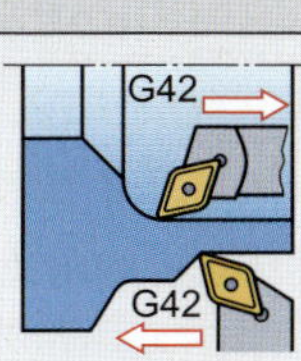

Condition of lathe tool position in front of spindle axis according to DIN 66217: Because of the different perspective in the X-Z plane, the cutter compensation would be opposite for the user looking down on the workpiece and for programming.

G42	Milling cutter right

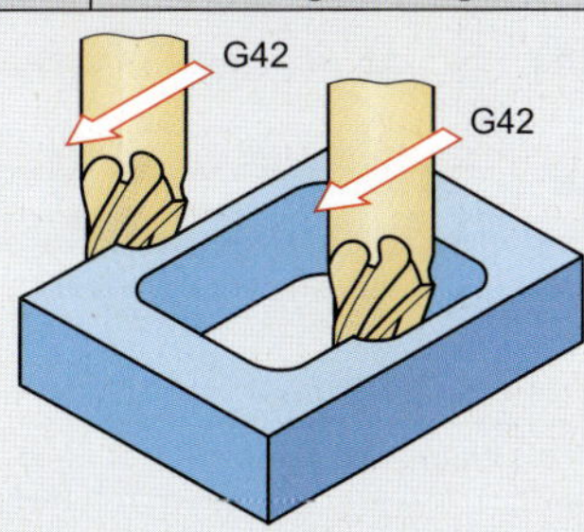

Cutter compensations G41 and G42 may be canceled with function G40.

A

Program structure of CNC machines according to DIN

Machining motion for vertical milling machines

cf. DIN 66025-2 (1983-01)

G01 Linear motion

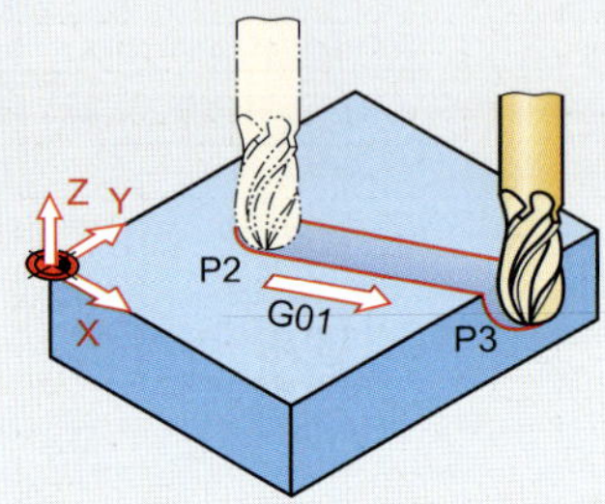

Designation and machining example:

N30 | G01 | X50 | Y19 | Z-8

- G01: Linear interpolation, machining motion in programmed feed
- X50, Y19, Z-8: Coordinates of target point – in X direction, in Y direction, in Z direction

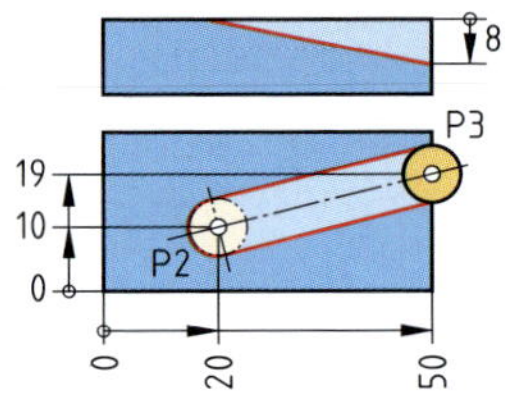

CNC program

N...					
N10	G00	X20	Y10	Z1	(P1)
N20	G01			Z0	(P2)
N30		X50	Y19	Z-8	(P3)
N...					

G02 Clockwise circular movement

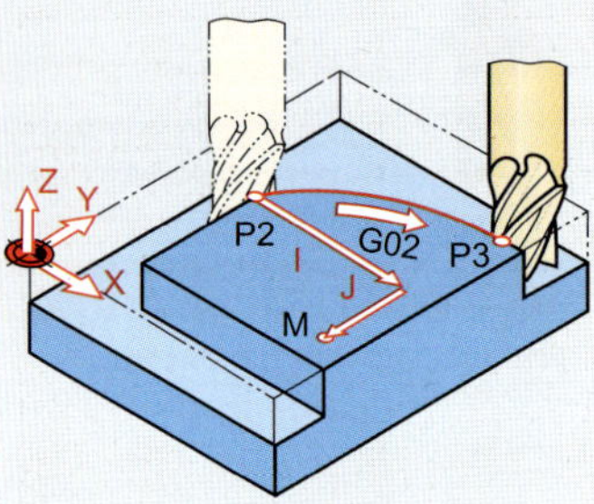

Designation and machining example:

N40 | G02 | X32 | Y38 | I26 | J-10.39

- G02: Clockwise circular interpolation, machining motion in programmed feed
- X32, Y38: Coordinates of circle endpoint – in X direction, in Y direction
- I26, J-10.39: Incremental input of the center point relative to circle starting point – in X direction, in Y direction

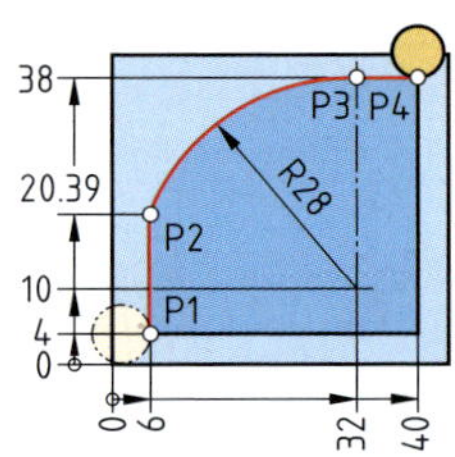

CNC program

N...						
N10	G41					
N20	G01	X6	Y4			(P1)
N30			Y20.39			(P2)
N40	G02	X32	Y38	I26	J -10.39	(P3)
N50	G01	X40				(P4)
N...						

G03 Counterclockwise circular movement

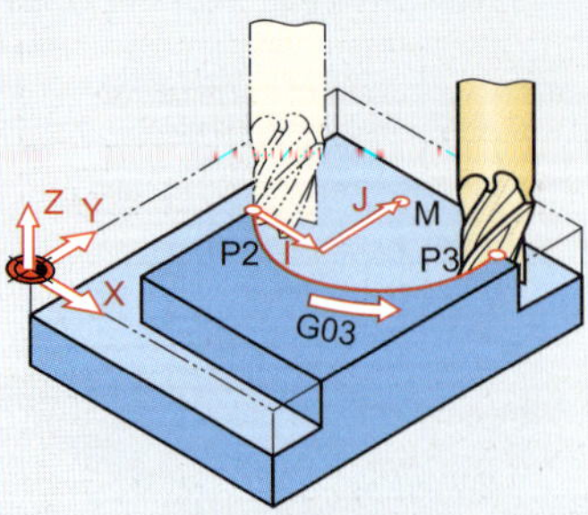

Designation and machining example:

N40 | G03 | X32 | Y38 | I8 | J16.12

- G03: Counterclockwise circle interpolation, machining motion in programmed feed
- X32, Y38: Coordinate of circle endpoint – in X direction, in Y direction
- I8, J16.12: Incremental input of the center point relative to circle starting point – in X direction, in Y direction

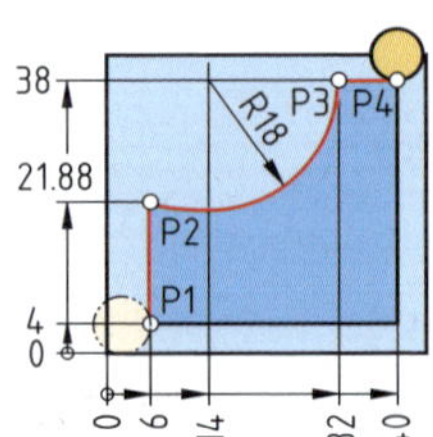

CNC program

N...						
N10	G41					
N20	G01	X6	Y4			(P1)
N30			Y21.88			(P2)
N40	G03	X32	Y38	I8	J16.12	(P3)
N50	G01	X40				(P4)
N...						

Program structure of CNC machines according to DIN

Machining motions of lathes

cf. DIN 66025-2 (1983-01)

G01 Linear movement

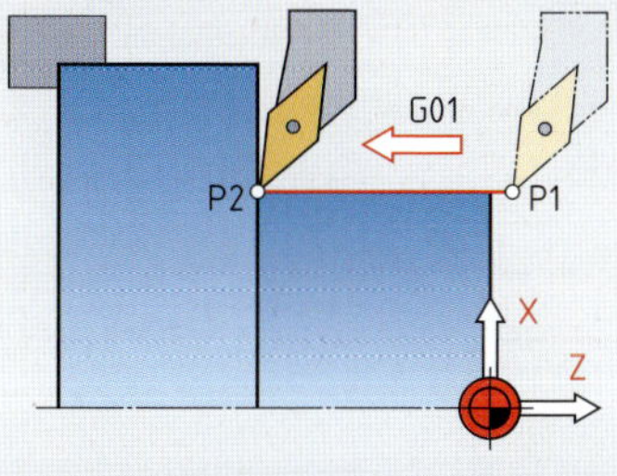

Designation and machining example:

N20	G01	X60	Z-50
	Linear interpolation, machining motion in programmed feed	Coordinates of target point – in X direction	Coordinates of target point – in Z direction

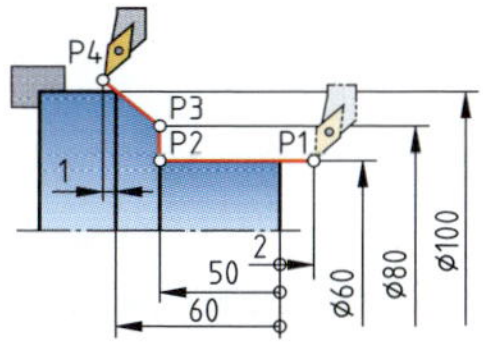

CNC program

N...				
N10	G00	X60	Z2	(P1)
N20	G01		Z-50	(P2)
N30		X 80		(P3)
N40		X102	Z-61	(P4)
N...				

G02 Clockwise circular movement

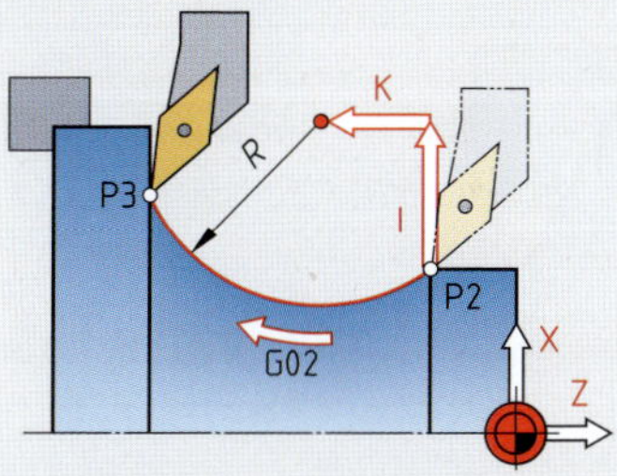

Designation and machining example:

N30	G02	X100	Z-60	I20	K0
	Clockwise circular interpolation, machining motion in programmed feed	Coordinates of circle endpoint – in X direction	Coordinates of circle endpoint – in Z direction	Incremental input of center point relative to circle starting point – in X direction	Incremental input of center point relative to circle starting point – in Z direction

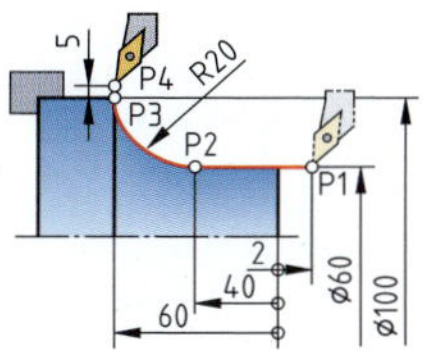

CNC program

N...						
N10	G00	X60	Z2			(P1)
N20	G01		Z-40			(P2)
N30	G02	X100	Z-60	I20	K0	(P3)
N40	G01	X110				(P4)
N...						

G03 Counterclockwise circular movement

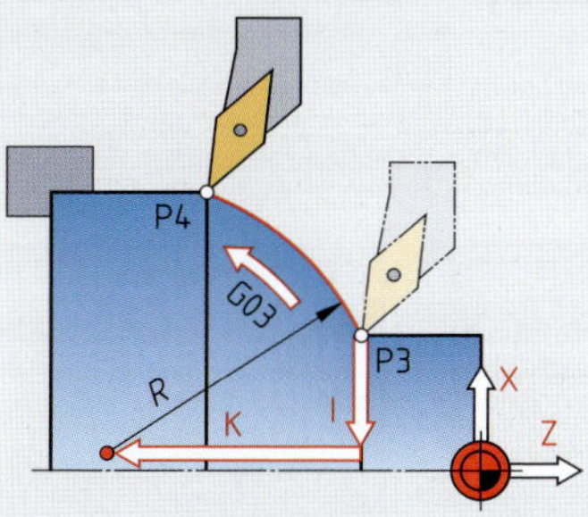

Designation and machining example:

N40	G03	X90	Z-55	I0	K-15
	Counterclockwise circle interpolation, machining motion in programmed feed	Coordinates of circle endpoint – in X direction	Coordinates of circle endpoint – in Z direction	Incremental input of center point relative to circle starting point – in X direction	Incremental input of center point relative to circle starting point – in Z direction

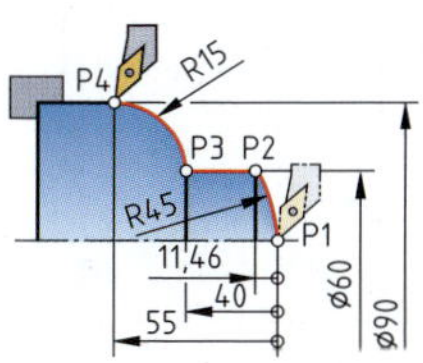

CNC program

N...						
N10	G01	X0	Z0			(P1)
N20	G03	X60	Z-11.46	I0	K-45	(P2)
N30	G01		Z-40			(P3)
N40	G03	X90	Z-55	I0	K-15	(P4)
N...						

A

Program structure of CNC machines according to PAL[1)]

Linear and circular interpolation for lathes and milling machines

Turning	Milling

Incremental programming with XI, YI and ZI coordinates in NC programs with G90

Turning

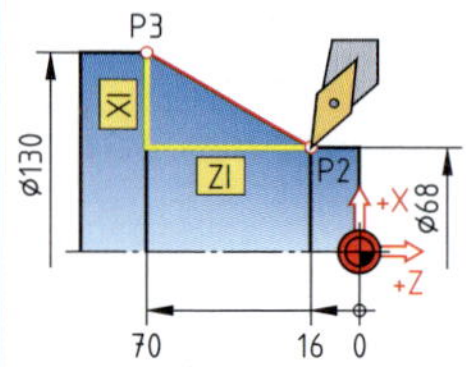

NC program

```
N10...
N15 G90
N20...
N25 G1   X68 Z-16   ;P2
N30 G1  XI31 ZI-54  ;P3
N35...
```

Milling

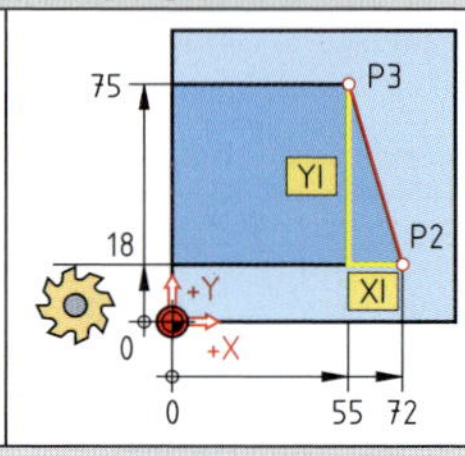

NC program

```
N10...
N15 G42
N20 G0   X...
N25 G1   X72          ;P2
N30 G1  XI-17 YI57    ;P3
N35...
```

Absolute programming with XA, YA and ZA coordinates in NC programs with G91

Turning

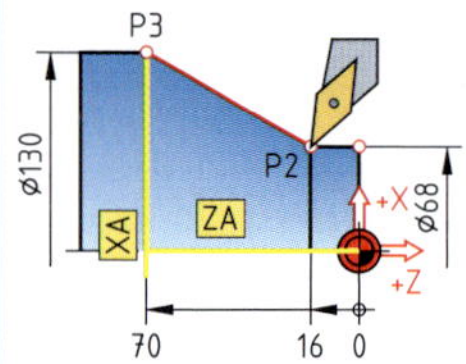

NC program

```
N10...
N15 G90
N20 G0 X68 Z-16       ;P2
N25 G91
N30 G1  XA130 ZA-70   ;P3
N35...
```

Milling

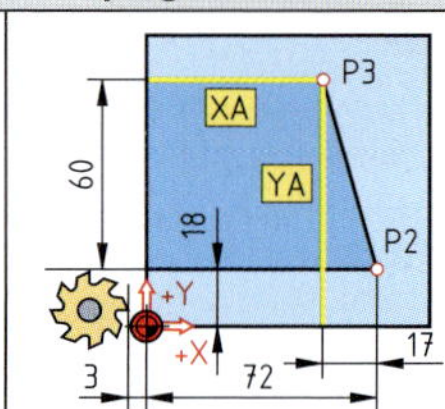

NC program

```
N10...
N15 G42 G0 X-16 Y18
N20 G91
N25 G1   X88          ;P2
N30 G1  XA55 YA78     ;P3
N35...
```

Start angle AS with coordinate value X

Turning

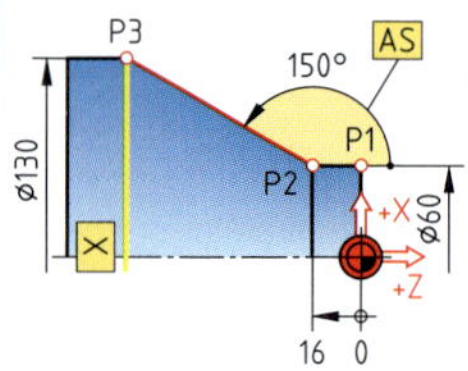

NC program

```
N10...
N15
N20...
N25 G1   X60 Z-16     ;P2
N30     AS150 X130    ;P3
N35...
```

Milling

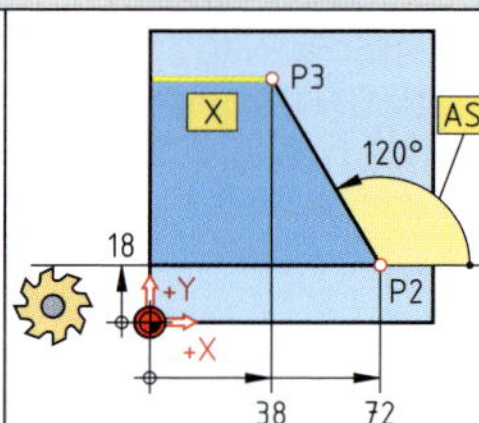

NC program

```
N10...
N15 G42
N20 G0   X...  Y18
N25 G1   X72          ;P2
N30 G1  AS120 X38     ;P3
N35...
```

Circular interpolation with absolute center point coordinates

Turning

Block structure:

```
G90
G1  X..  Z..              ;P2
G2  X..  Z..  IA..  KA..  ;P3
```

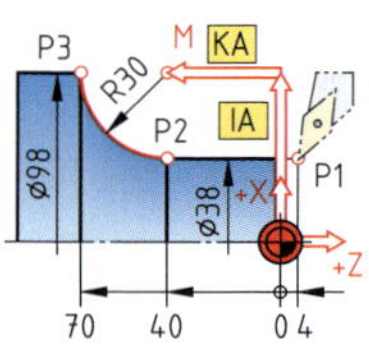

NC program

```
N10 ...
N15 G90
N20 G0   X38  Z4                 ;P1
N25 G1   Z-40                    ;P2
N30 G2 X98 Z-70 IA49 KA-40       ;P3
N35 ...
```

Milling

Block structure:

```
G90
G1  X..  Y..              ;P2
G2  X..  Y..  IA..  JA..  ;P3
```

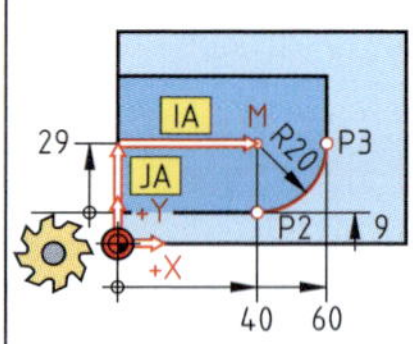

NC program

```
N10 ...
N15 G90 G42
N20 G0   X...  Y9                ;P1
N25 G1   X40                     ;P2
N30 G3 X60 Y29  IA40  JA29       ;P3
N35 ...
```

Transition elements radius RN+ and phase RN–

The radius RN+ and the phase RN– are transition elements between two contour elements (circles, straight lines)

Turning

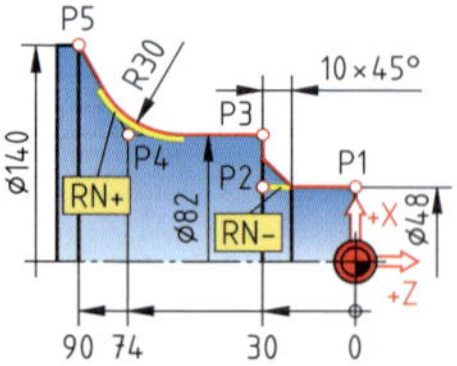

NC program

```
N10...
N15 G90
N20 G0  X48 Z0        ;P1
N25 G1  Z-30  RN–10   ;P2
N30 G1  X82           ;P3
N35 G1  Z-74  RN+30   ;P4
N40 G1  X140 Z-90     ;P5
```

Milling

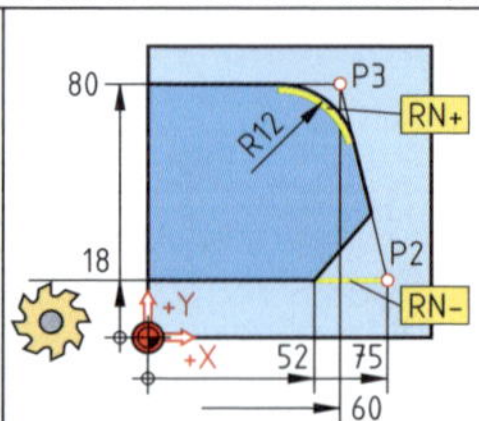

NC program

```
N10...
N15 G42
N20 G0 X... Y18
N25 G1 X75  RN–23     ;P2
N30 G1 X60 Y80 RN+12  ;P3
N35...
```

A

[1)] **P**rüfungs**a**ufgaben- und **L**ehrmittelentwicklungsstelle (PAL) (Institute for the development of training and testing material)

Program structure of CNC machines according to PAL

PAL functions for lathes and milling machines

Programming coordinates and interpolation parameters

XA, YA, ZA	Absolute input of coordinate values relative to the workpiece zero point
XI, YI, ZI	Incremental input of coordinate values relative to the current tool position
IA, JA, KA	Absolute input of the interpolation parameters relative to the workpiece zero point

T-addresses for tool change

T	Tool storage place in the tool revolver or holder
TC	Selection of the number from the offset memory
TR	Incremental tool radius or cutting edge offset in the selected offset memory
TL	Incremental tool length offset in the selected offset memory (milling)
TZ	Incremental tool length offset in Z direction in the selected offset memory (turning)
TX	Incremental diameter offset in X direction in the selected offset memory (turning)

Additional M-functions[1] according to PAL

M13	Clockwise spindle rotation, coolant ON	M17	End of sub program
M14	Counter clockwise spindle rotation, coolant ON	M60	Constant feed
M15	Spindle and coolant OFF	M61	M60 + corner shaping

PAL functions for lathes

G-functions

Types of interpolation

G0	Rapid travel/motion
G1	Linear interpolation with feed rate
G2	Circular interpolation, clockwise
G3	Circular interpolation, counter clockwise
G4	Dwell time
G9	Exact stop
G14	Travel to configured tool change point
G61	Linear interpolation for contour routing
G62	Circular interpolation for contour routing, clockwise
G63	Circular interpolation for contour routing, counter clockwise

Reference points

G50	Cancellation of incremental zero point shift and rotations
G53	Cancellation of all zero point shifts and rotations
G54–G57	Adjustable absolute zero points
G59	Incremental Cartesian zero point shift and rotation

Machining planes and rechucking

G18	Selection of the plane of rotation
G17	Face machining planes
G19	Shell surface/segment surface machining planes
G30	Rechucking/opposed spindle takeover

Dimensions

G70	Setting input to inch units
G71	Setting input to metric units (mm)
G90	Absolute dimensions
G91	Input of incremental dimensions

Cutter compensation

G40	Cancel tool radius offset TRO
G41	Tool radius offset TRO to the left of the programmed contour
G42	Tool radius offset TRO to the right of the programmed contour

Feeds and speeds

G92	Rotational speed limitation
G94	Feed in mm per minute
G95	Feed in mm per revolution
G96	Constant cutting speed
G97	Constant rotational speed

Program features

G22	Call sub program
G23	Repeat program section
G29	Conditional jumps

Cycles

G31	Thread cycle
G32	Tapping cycle
G33	Thread chasing cycle
G80	Completion of a machining cycle contour description
G81	Longitudinal rough-turning cycle
G82	Rough facing cycle
G83	Rough-turning cycle parallel to the contour
G84	Drilling cycle
G85	Undercut cycle
G86	Radial grooving cycle
G87	Radial contour cutting cycle
G88	Axial grooving cycle
G89	Axial contour cutting cycle

A

Program structure of CNC machines according to PAL

G-functions for lathes

G22 Call sub program

Structure of NC block
G22 L [H] [/]

Obligatory addresses:
L number of the sub program

Optional addresses:
H number of repetitions
/ extract level

Main program %900

```
N10 G90..
N15 F.. S.. M4
N20 G0 X42 Z6 ;P1
N25 G22 L911 H2
N30..
N35..
N150 M30
```

jump → / ← return

Sub program L911

```
N10 G91
N15 G0 Z-16
N20 G1 X-6
N25 G1 X6
N30 G0 Z-6
N35 G1 X-6
N40 G1 X6
N45 M17
```

Machining example

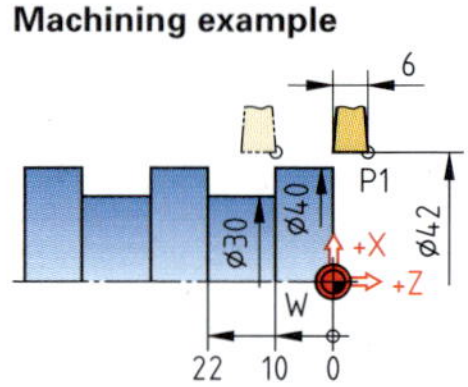

G23 Repeat program section

Structure of NC block
G23 N N [H]

Obligatory addresses:
N start block number of the program section to be repeated
N end block number of the program section to be repeated

Optional addresses:
H number of repetitions

```
N10 ..
N15 G0 X58 Z-15 M4
N20 G91
N25 G1 X-11
N30 G1 X11
N35 G0 Z-16
N40 G23 N20 N35 H2
N45 G90
N50 ...
```

Machining example

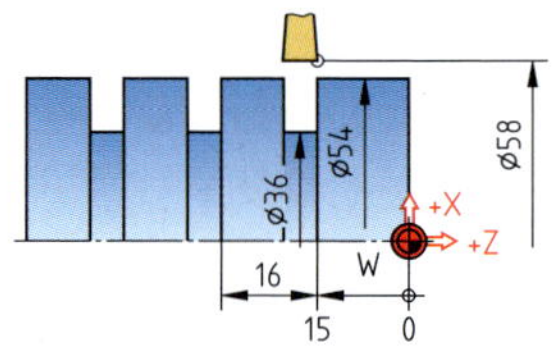

G14 Travel to tool change point

Structure of NC block
G14 [H]

Optional addresses:
H0 travel to tool change point simultaneously in all axes
H1 first X axis, then Z axis H2 first Z axis, then X axis

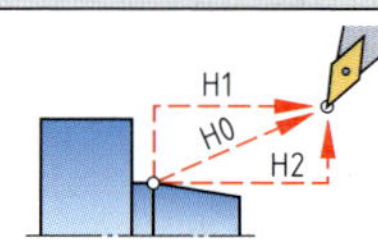

PAL cycles for lathes

G84 Drilling cycle

Structure of NC block
G84 ZI/ZA [D] [V] [VB] [DR] [DM] [R] [DA] [U] [O] [FR] [E]

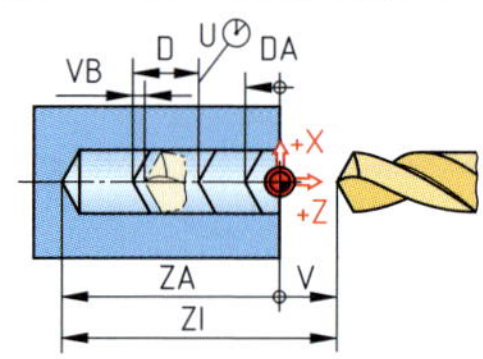

Obligatory addresses:
ZI depth of hole, incremental depth relative to the current tool position
ZA depth of hole, absolute depth

Optional addresses (selection):
D pecking amount
(if D is not specified, pecking depth is equal to the final drilling depth)
V safety distance
VB safety distance to the hole bottom
DR reduction value of the pecking amount
DM minimum infeed
R retract level/distance
DA spot-drilling depth
U dwell time at hole bottom
O dwell time selection
O1 in seconds
O2 in revolutions
FR rapid travel reduction in %
E spot-drilling feed

Machining example

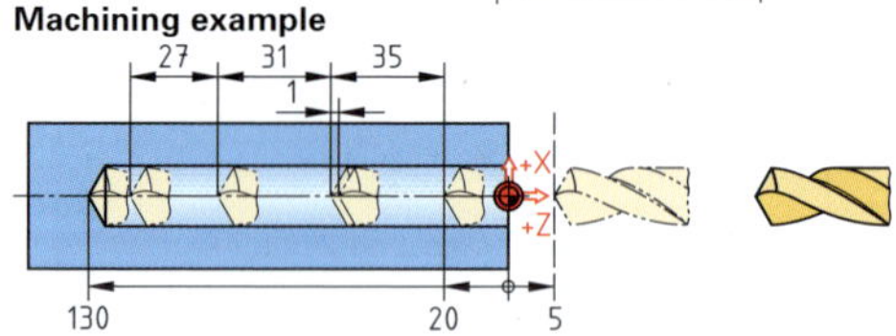


```
N10 G90
N15 G84 Z-130 D30 V5 VB1 DR4 DA20 U0.5
N20 ..
```

G32 Tapping cycle

Structure of NC block
G32 Z/ZI/ZA F

Obligatory addresses:
Z, ZI, ZA thread end point in Z direction
I incremental, A absolute
F pitch of thread

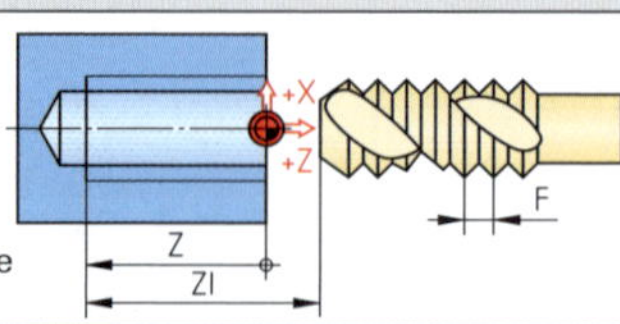

Machining example

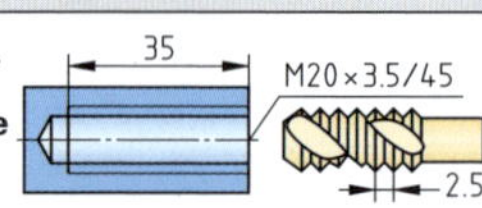


```
N10 G90
N15 G32 Z-35 F2,5 S.. M..
```

A

Program structure of CNC machines according to PAL

PAL cycles for lathes

G31 | Thread cycle

Structure of NC block
G31 Z/ZI/ZA X/XI/XA F D [ZS] [XS] [DA] [DU] [Q] [O] [H]

Obligatory addresses:

Z, ZI, ZA thread end point in Z direction
Z controlled by G90/G91;
I incremental, A absolute

X, XI, ZI thread end point in X direction;
X controlled by G90/G91,
I incremental, A absolute

F thread pitch

D thread depth

Optional addresses [..]:

ZS thread starting point, absolute in Z
XS thread starting point, absolute in X
DA approach
DU overrun
Q number of cuts
O number of idle cycles
H selection of infeed type and residual cuts (RC)
- H1 without offset (radial infeed), RC OFF
- H2 infeed at left flank, RC OFF
- H3 infeed at right flank, RC OFF
- H4 alternating infeed, RC OFF
- H11 without offset (radial infeed), RC ON
- H12 infeed at left flank, RC ON
- H13 infeed at right flank, RC ON
- H14 alternating infeed, RC ON

Residual cuts $^{1}/_{2}$, $^{1}/_{4}$, $^{1}/_{8}$, $^{1}/_{8}$ x (D/Q)

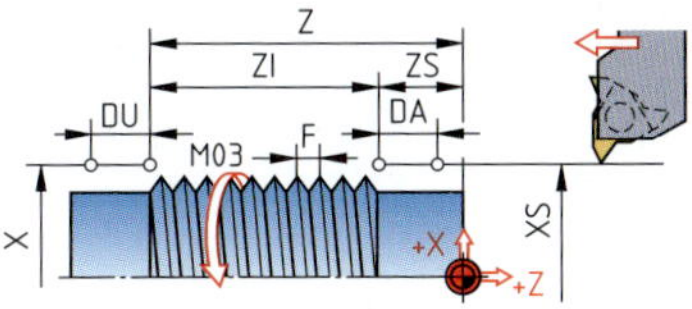

Radial infeed H1/H11 | Flank infeed left H2/H12 | Flank infeed right H3/H13 | Alternating infeed H4/H14

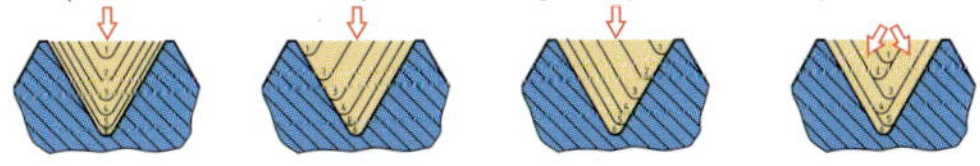

Machining example

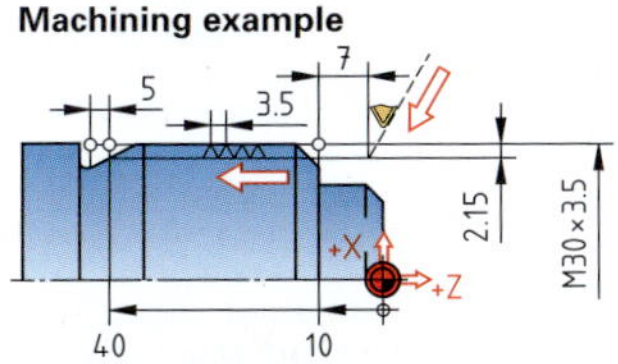


```
N10 G90
N15 G31 Z-40 X30 F3.5 D2.15 ZS-10 XS30 Q12 O13 H14
N20 ..
```

G81 | Longitudinal rough-turning cycle — G82 | Rough facing cycle

Structure of NC block
G81 (or G82) H4 [AK] [AZ] [AX] [AE] [AS] [AV] [O] [Q] [V] [E]
or
G81 (or G82) D [H1/H2/H3/H24]

Obligatory addresses:

D infeed

Optional addresses [..]:

H type of machining
- H1rough machining, removal below 45°
- H2stepwise angle-cutting along the contour
- H3 like H1 with final contour cut
- H4 contour finishing
- H24 rough-machining with H2 and subsequent finishing

AK contour allowance parallel to the contour
AZ contour allowance in Z direction
AX contour allowance in X direction
AE immersion angle (final angle of the tool)
AS emergence angle (lateral adjustment angle of tool)
AV safety angle reduction for AE and AS
O machining starting point
- O1: current tool position
- O2: calculated from contour

Q idle step optimization
- Q1: optimization OFF
- Q2: optimization ON

V safety distance for idle step optimization
- G81: in Z direction
- G82: in X direction

E immersion feed

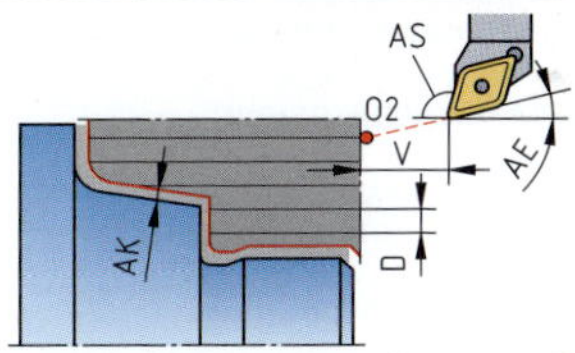

Longitudinal rough turning cycle with G81

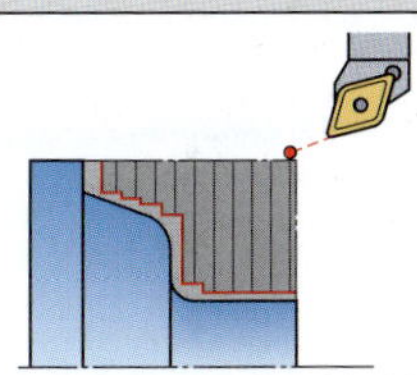

Rough facing cycle with G82

Machining example: longitudinal rough-machining cycle

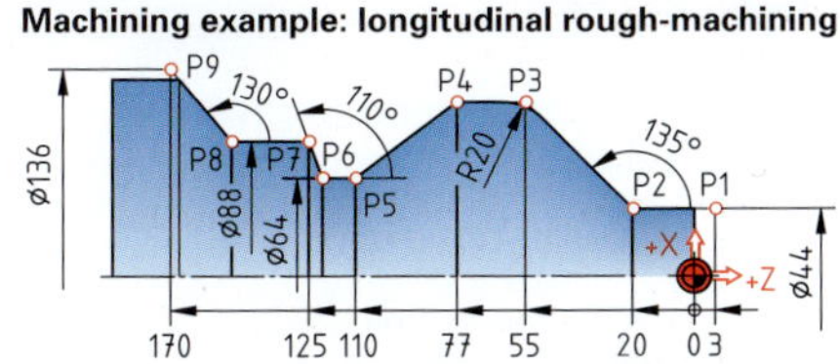


```
N10
N15 G81 D3 H3 E0.15 AZ0.1 AX0.5
N20 X44 Z3                 ;P1
N25 G1 Z-20                ;P2
N30 G1 Z-55 AS135 RN20     ;P3
N35 G1 Z-77 AS180          ;P4
N40 G1 Z-110 X64           ;P5
N45 AS180                  ;P6
N50 AS110 X88 Z-125        ;P7
N55 AS180                  ;P8
N60 AS130 X136 Z-170       ;P9
N65 G80
```

Program structure of CNC machines according to PAL

PAL cycles for lathes

G86	Radial grooving cycle	G88	Axial grooving cycle

Structure NC block
G86 Z/ZI/ZA X/XI/XA ET [EB] [D] [..] (selection)
G88 Z/ZI/ZA X/XI/XA ET [EB] [D] [..] (selection)

Obligatory addresses:
Z, ZI, ZA grooving position in Z direction;
Z controlled by G90/G91,
ZI incremental, ZA absolute
X, XI, XA grooving position in X direction;
X controlled by G90/G91,
XI incremental, XA absolute
ET G86 absolute diameter of grooving depth
G88 absolute grooving depth

Optional addresses [..]:
EB grooving width and position
EB+ grooving in direction Z+ relative to the programmed grooving position P
EB– grooving in direction Z– relative to the programmed grooving position P
D pecking amount (if no value is specified, the pecking depth is equal to the groove depth ET)
AS flank angle of grooving at the starting point relative to the grooving direction (X or Z)
AE flank angle of grooving at the end point relative to the grooving direction (X or Z)
RO rounding or chamfering of upper corners
RO+ rounding
RO– chamfer width
RU rounding or chamfering of lower corners
RU+ rounding
RU– chamfer width
AK contour allowance parallel to the contour
AX contour allowance in X direction (contour offset)
EP setpoint definition for groove cutting (position P)
EP1: setpoint in upper corner of the groove
EP2: setpoint in bottom corner of the groove
H type of processing
H1 roughing cut H14 roughing and finishing
H2 plunge turning H24 plunge turning and finishing
H4 finishing
DB infeed in % of the cutting tool width for grooving
V safety distance above groove
E feed rate into solid material

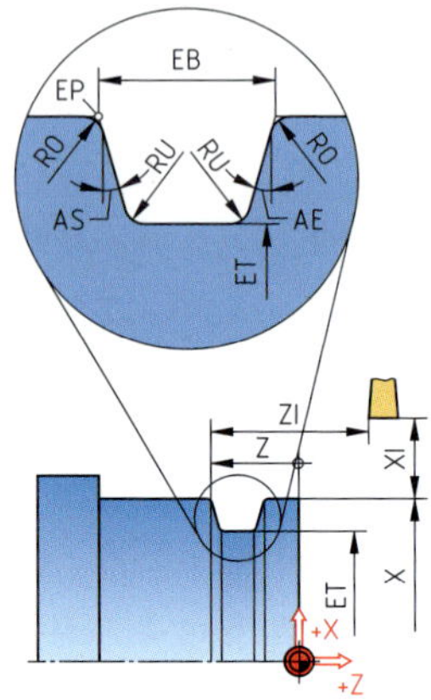

Radial grooving cycle with G86

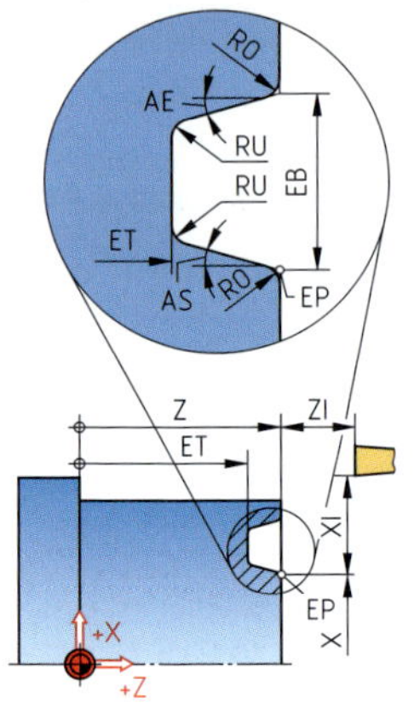

Axial grooving cycle with G88

Machining example: radial grooving cycle with G86:

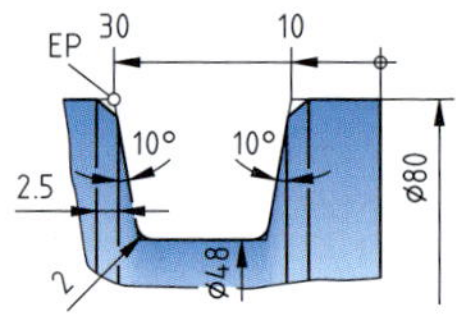


```
N10 G0 X82 Z-32
N35 G86 Z-30 X80 ET48 EB20 D4 AS10 AE10 RO-2.5 RU2 H14
```

G85	Undercut and thread undercut cycle

Structure of NC block
G85 Z/ZI/ZA X/XI/XA I/[I] K[K] [RN] [SX] [H] [E]

Obligatory addresses:
Z, ZI, ZA undercut position in Z direction;
Z controlled by G90/G91,
ZI incremental, ZA absolute
X, XI, XA undercut position in X direction;
X controlled by G90/G91,
XI incremental, XA absolute
I undercut depth; obligatory parameter for DIN 76 (H1)
K undercut length; obligatory parameter for DIN 76 (H1)

Optional addresses [..]:
RN corner radius
SX grinding allowance
E feed rate for plunging
H undercut shape
H1 DIN 76 H2 DIN 509 E H3 DIN 509 F

Thread undercuts acc. to DIN 76

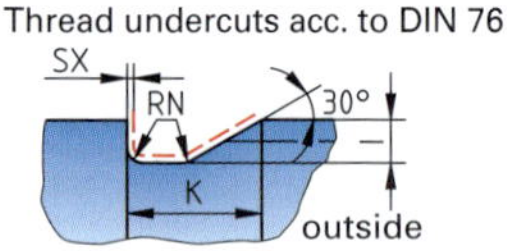

Undercuts acc. to DIN 509

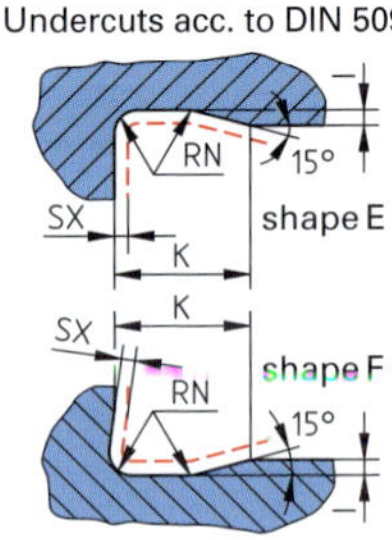

Machining process with DIN 76

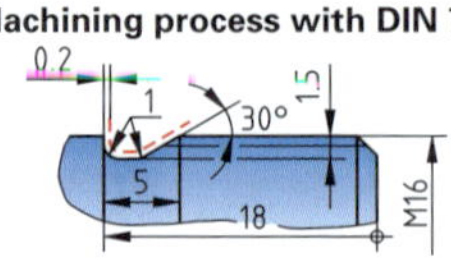


```
N10 G0 ..
N15 G85 ZA-18 XA16 I1.5 K5 RN1 SX0.2 H1 E0.15
```

Further information on p. 84 and p. 87

G80	Completion of a contour description in a rough-machining cycle

Structure of NC block **Optional addresses [..]:** ZA absolute Z-coordinate of the machining limit parallel to the X axis
G80 [ZA] [XA] XA absolute X-coordinate of the machining limit parallel to the Z axis

Program structure of CNC machines according to PAL

PAL functions for milling machines

G-functions

Types of interpolation, contours

G0	Rapid motion
G1	Linear interpolation with feed rate
G2	Circular interpolation, clockwise
G3	Circular interpolation, counter clockwise
G4	Dwell time
G9	Exact stop
G10	Rapid motion in polar coordinates
G11	Linear interpolation with polar coordinates
G12	Circular interpolation with polar coordinates, clockwise
G13	Circular interpolation with polar coordinates, counter clockwise
G45	Linear tangential approach to a contour
G46	Linear tangential retraction from a contour
G47	Tangential approach to a contour in a quarter circle
G48	Tangential retraction from a contour in a quarter circle
G61	Linear interpolation for contour routing
G62	Circular interpolation for contour routing, clockwise
G63	Circular interpolation for contour routing, counter clockwise

Reference points, rotation, mirror images, scaling

G50	Cancellation of the incremental zero point shift and rotations
G53	Cancellation of all zero point shifts and rotations
G54–G57	Adjustable absolut zero points
G58	Incremental zero point shift, polar and rotation
G59	Incremental Cartesian zero point shift and rotation
G66	Mirror image across the X or Y axis, mirror image off
G67	Scaling (enlarging or reducing or cancellation)

Plane selection, dimensions

G17–G19	Plane selection, 2½ D processing
G70	Setting input to inch units
G71	Setting input to metric units (mm)
G90	Input of absolute dimensions
G91	Input of incremental dimensions

Tool offsets

G40	Cancel cutter compensation
G41–G42	Cutter compensation left Cutter compensation right

Feeds and speeds

G94	Feed in mm per minute
G95	Feed in mm per revolution
G96	Constant cutting speed
G97	Constant spindle speed

Program features

G22	Call sub program
G23	Repeat program section
G29	Conditional jumps

Fixed cycles

G34	Start-up of the contour pocket cycle
G35	Rough-machining technology of the contour pocket cycle
G36	Residual material technology of the contour pocket cycle
G37	Finishing technology of the contour pocket cycle
G38	Contour description of the contour pocket cycle
G80	Completion of the G38 cycle
G39	Call contour pocket cycle with material removal either parallel to the contour or in meanders
G72	Rectangular pocket milling cycle
G73	Circular pocket and spigot milling cycle
G74	Slot milling cycle
G75	Circular slot milling cycle
G81	Drilling cycle
G82	Deep drilling cycle with pecking
G83	Deep drilling cycle with pecking and full retraction
G84	Tapping cycle
G85	Reaming cycle
G86	Boring cycle
G87	Plunge milling cycle
G88	Internal thread milling cycle
G89	External thread milling cycle
G76	Multiple cycle call on a straight line (line of holes)
G77	Multiple cycle call on a pitch circle (line of holes)
G78	Cycle call at a particular point (polar coordinates)
G79	Cycle call at a particular point (Cartesian coordinates)

Program structure of CNC machines according to PAL

PAL cycles for milling machines

G1 Linear interpolation with feed rate

Structure of NC block
G1 [X/XI/XA] [Y/YI/YA] [Z/ZI/ZA] [D] [AS] .. (selection)

Obligatory addresses:
X, XI, XA X coordinate of the target point
Y, YI, YA Y coordinate of the target point
Z, ZI, ZA Z coordinate of the target point

Optional addresses [..]:
D length of travel distance
AS ascent angle relative to the X axis
RN transition element to the next contour element
RN+ rounding radius RN– chamfer width
H selection among two solutions via angle criterion
H1 small ascent angle H2 greater ascent angle
TC selection of the offset memory number
TR incremental change of the tool radius value
TL incremental change of the tool length offset

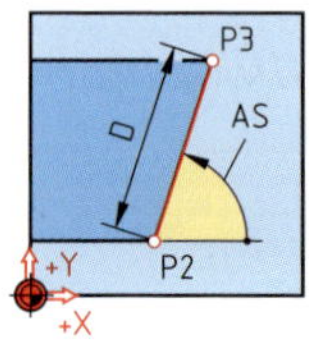

Machining example

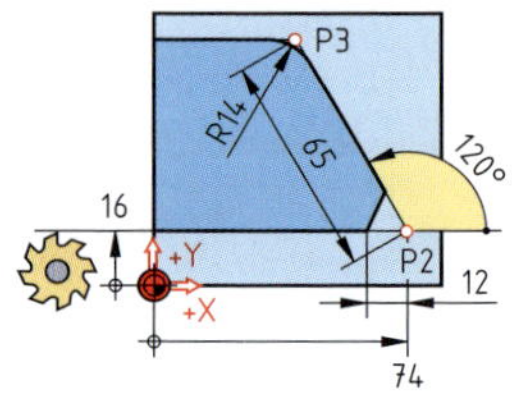


```
N10 ...
N15 G1 X74 Y16 RN-12      ;P2
N20 G1 D65 AS120 RN+14    ;P3
```

G11 Linear interpolation with polar coordinates

Structure of NC block
G11 RP AP/AI [J/JA] [Z/ZI/ZA] [RN] .. (Auswahl)

Obligatory addresses:
RP polar radius
AP polar angle relative to the positive X axis
AI incremental polar angle

Optional addresses [..]:
I, IA X coordinate of the polar center
J, JA Y coordinate of the polar center
Z, ZI, ZA infeed in Z direction
RN transition to the next contour element
RN+ rounding radius RN– chamfer width
TC selection of the offset memory number
TR incremental change of the tool radius value
TL incremental change of the tool length offset

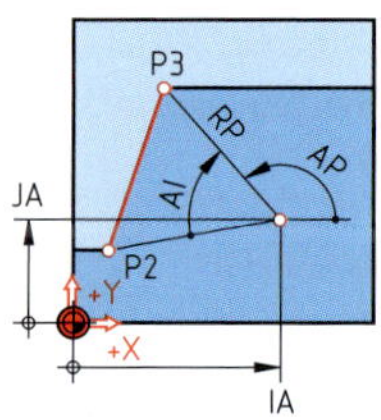

Machining example

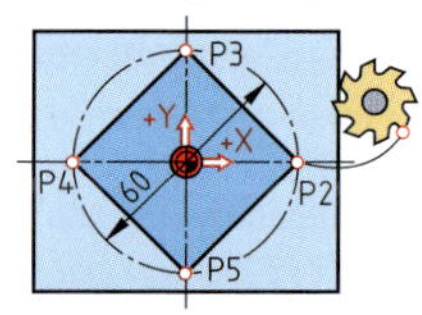


```
N15 G42 G47 R20 X30 Y0 Z-3    ;P2
N20 G11 IA0 JA0 RP30 AP90     ;P3
N25 G11 IA0 JA0 RP30 AP180    ;P4
N30 G11 IA0 JA0 RP30 AP270    ;P5
N35 G11 IA0 JA0 RP30 AP0      ;P2
```

G2/G3 Circular interpolation with Cartesian coordinates

Structure of NC block
G2 [X/XI/XA] [Y/YI/YA] [Z/ZI/ZA] ((I/IA [J/JA]) / ([I/IA] J/JA) / R / A0 [RN] [O] [F] [S] [M]
G3 [X/XI/XA] ....

Optional addresses [...]:
X, XI, XA X coordinate of the target point
Y, YI, YA Y coordinate of the target point
Z, ZI, ZA Z coordinate of the target point
I, IA, J, JA center point coordinates
R radius of arc and
selection of solution via arc length criterion
R+ shorter arc R– longer arc
AO aperture angle
RN transition element
RN+ rounding radius RN– chamfer width
O selection of solution via arc length criterion
O1 shorter arc O2 longer arc

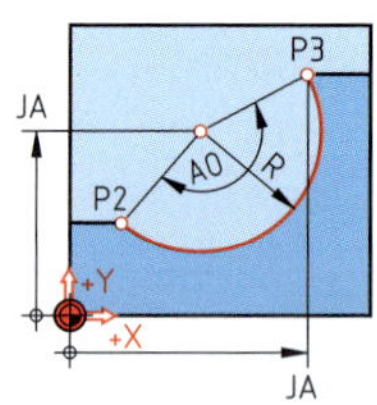

Machining example

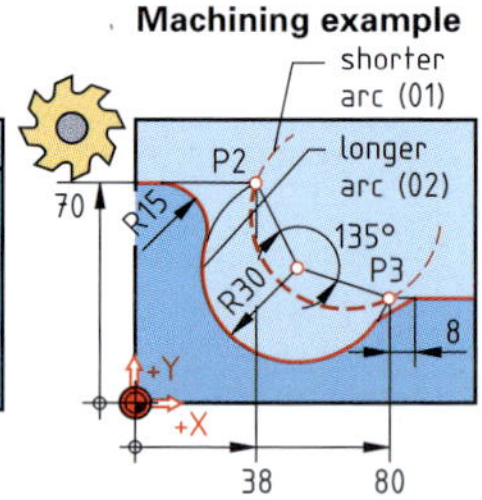


```
N10 ...
N15 G1 X38 Y70 RN+15              ;P2
N20 G3 XA80 R30 AO135 RN-8 O2     ;P3
```

G12/G13 Circular interpolation with polar coordinates

Structure of NC block
G12 AP/AI [I/IA] [J/JA] [Z/ZI/ZA] [RN] [F] [S] [M]
G13 AP/AI [I/IA] [J/JA] [Z/ZI/ZA] [RN] [F] [S] [M]

Obligatory addresses:
AP polar angle of target point
AI incremental polar angle

Optional addresses [...]:
I, IA X coordinate of polar center
J, JA Y coordinate of the polar center
RN+ rounding radius RN– chamfer width

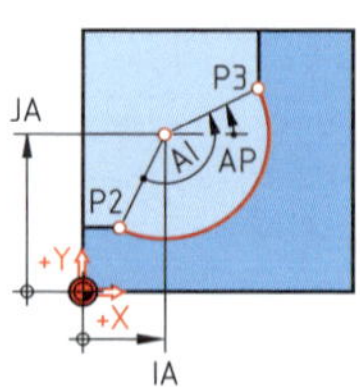

Machining example

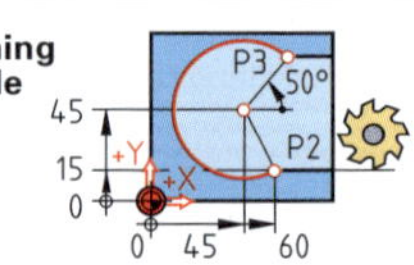


```
N15 G1 X60 Y15              ;P2
N20 G12 IA45 JA45 AP50      ;P3
```

Program structure of CNC machines according to PAL

PAL functions for milling machines

G45 Linear tangential approach to the contour | G46 Linear tangential retraction from the contour

Structure of NC block
G41/G42 G45 D [X/XI/XA] [Y/YI/YA] [Z/ZI/ZA] [W] [E] [F] [S] [M]
G46 G40 D [Z/ZI/ZA] [W] [F] [S] [M]

Obligatory addresses:
with G45: D distance to the first contour point, unsigned
with G46: D length of the retracting motion, unsigned

Optional addresses [..]:
X, XI, XA X coordinate of the first contour point
Y, YI, YA Y coordinate of the first contour point
Z, ZI, ZA with G45: infeed at approach point in the Z axis
with G46: retracting motion at the end point in the Z axis
W absolute position in fast motion in the infeed axis
E feed rate for plunging

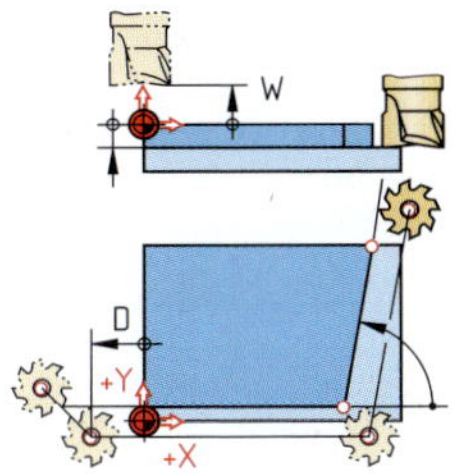

Machining example

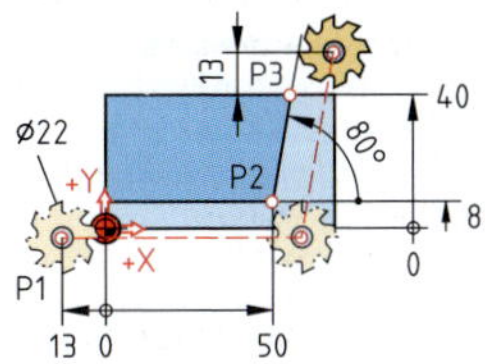


```
N10 ...
N15 G42 G45 X0 Y8 D13    ;P1
N20 G1 X50               ;P2
N25 G1 Y40 AS80          ;P3
N30 G40 G46 D13          ;P4
```

G47 Tangential approach to the contour in a quarter circle | G48 Tangential retraction from the contour in a quarter circle

Structure of NC block
G41/G42 G47 R [X/XI/XA] [Y/YI/YA] [Z/ZI/ZA] (W) [E] [F] [S] [M]
G48 G40 R [Z/ZI/ZA] [W] [F] [S] [M]

Obligatory addresses:
with G47: R radius of the approach motion relative to the center path of the cutter
with G48: R radius of the retracting motion relative to the center path of the cutter

Optional addresses [..]:
X, XI, XA X coordinate of the first contour point
Y, YI, YA Y coordinate of the first contour point
Z, ZI, ZA infeed at the approach point in the Z axis
W absolute position in fast motion in the infeed axis
E feed rate for plunging

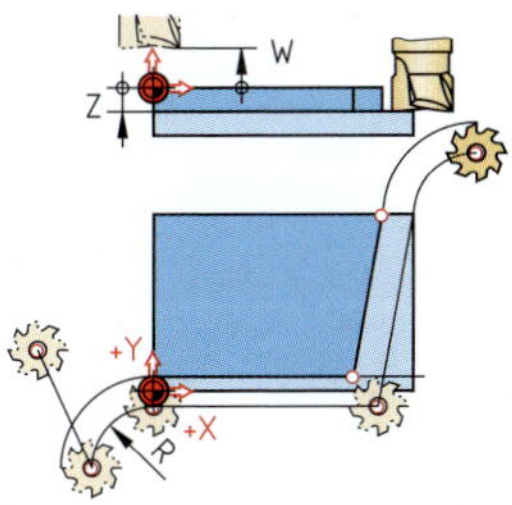

Machining example

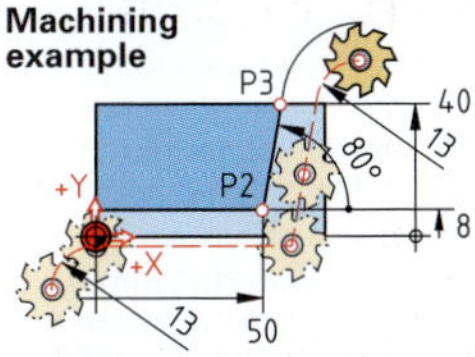


```
N10 ...
N15 G42 G47 X0 Y8 R13    ;P1
N20 G1 X50               ;P2
N25 G1 Y40 AS80          ;P3
N30 G40 G48 R13          ;P4
```

G54–G57 Adjustable absolute zero point shift

Structure of NC block
G54 *or* **G55** *or* **G56** *or* **G57**

Explanatory notes:
The workpiece zero point W is determined by the commands G54 to G57 and has a defined distance to the machine zero point. The operator enters the shift values into the zero point register of the controller before starting the program. The zero point is always specified in absolute coordinates (XA, YA, ZA) relative to the machine zero point.

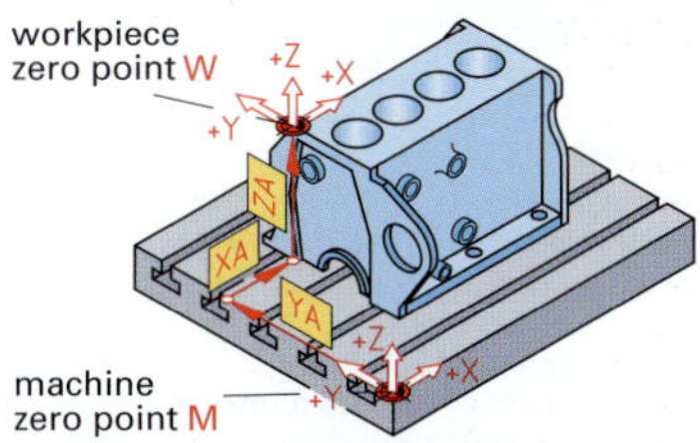


```
N10 ...
N15 G54 ;W
N20
```

G59 Incremental zero point shift and rotation

Structure of NC block
G59 [XA] [YA] [ZA] [AR]

Optional addresses [..]:
XA absolute X coordinate of the new workpiece zero point
YA absolute Y coordinate of the new workpiece zero point
ZA absolute Z coordinate of the new workpiece zero point
AR angle of rotation of the new coordinate system relative to the X axis

Explanatory notes:
If the coordinate system of the workpiece is rotated in its current position, only the angle of rotation is specified:
N… G59 AR–
The zero point shift launched via G54...G57 is reset by:
N… G50

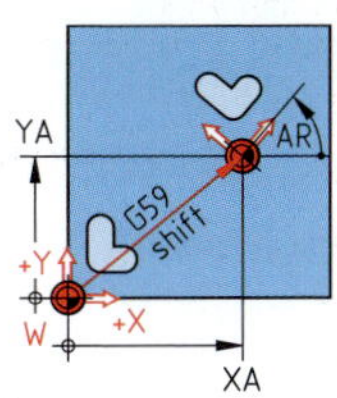

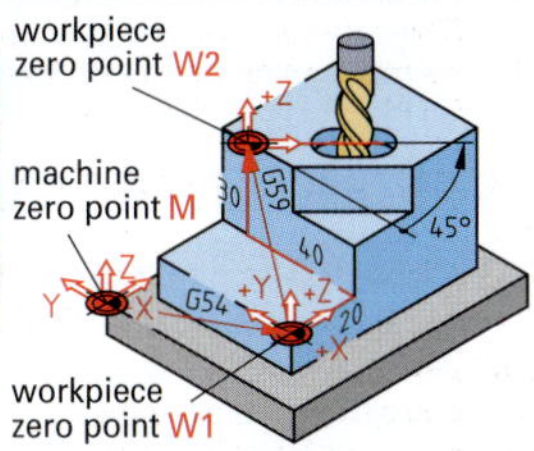


```
N10 ..
N15 G54                        ;W1
N20 G59 X20 Y40 Z30 AR45       ;W2
```

A

Program structure of CNC machines according to PAL

PAL cycles for milling machines

G81 Drilling cycle

Structure of NC block
G81 ZI/ZA V [W] [F] [S] [M]

Obligatory addresses:
ZI depth of bore in the feed axis
ZA absolute depth of bore relative to the coordinate system of the workpiece
V safety distance from the top edge of the hole

Optional addresses [..]:
W retract level relative to the coordinate system of the workpiece

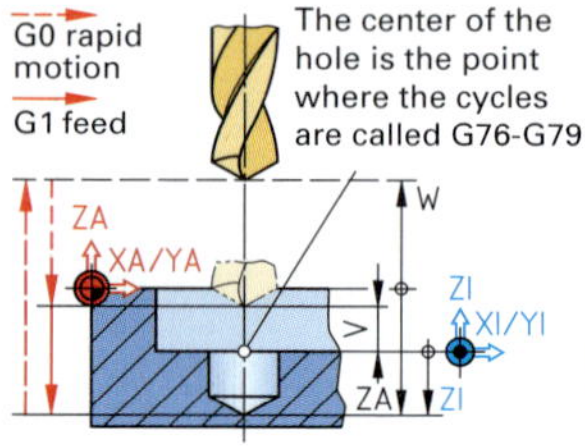

Machining example

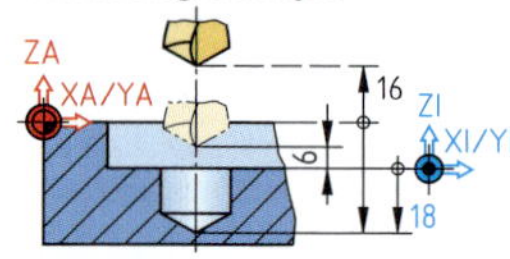


```
N10 ...
N15 G81 ZI-18 V6 W16
N20 G79 X.. Y.. Z.. ;cycle call
```

G82 Deep drilling cycle with pecking

Structure of NC block
G82 ZI/ZA D V [W] [VB] [DR] [DM] [U] [O] [DA] [E] [F] [S] [M]
G83 ZI/ZA D V [W] [VB] [DR] [DM] [U] [O] [DA] [E] [FR] [F] [S] [M]

Obligatory addresses:
ZI/ZA depth of bore in the feed axis
- ZI incremental depth from the top edge of the hole
- ZA absolute depth in workpiece coordinates

D pecking amount
V safety distance above the top edge of the hole

Optional addresses [..]:
W retract level relative to the coordinate system of the workpiece
VB retract distance to the current hole bottom
DR reduction value of the last pecking amount
DM minimum pecking amount (unsigned)
U dwell time at hole bottom (relative to pecking)
O unit of the dwell time
- O1 dwell time in seconds
- O2 dwell time in number of revolutions

DA incremental spot-drilling depth of the first infeed
E spot-drilling feed rate

G83 Deep drilling cycle with pecking and full retraction

G83 has the following features:
– the same addresses as G82
– retracts to the safety distance V for chip removal
and in addition
FR rapid motion reduction in %

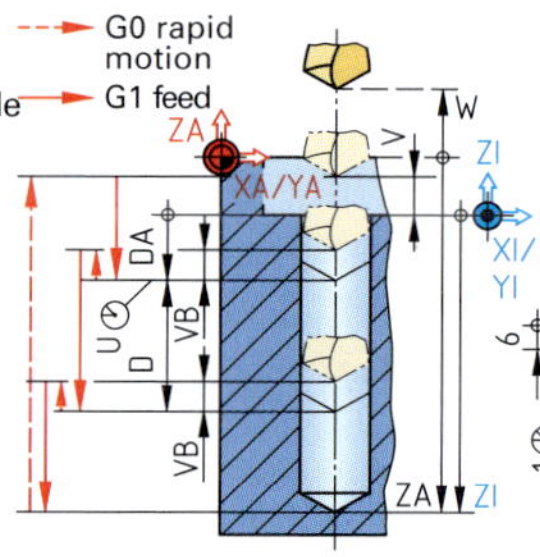

Machining example

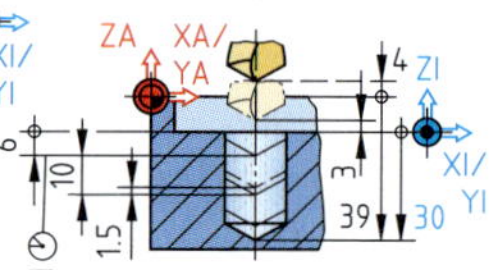


```
N10 ...
N15 G82 ZI-30 D10 V3 W4 VB1.5 DR3 U1 O1 DA6
N20 G79 X.. Y.. Z.. ;cycle call
```

G84 Tapping cycle

Structure of NC block
G84 ZI/ZA F M V [W] [S]

Obligatory addresses:
ZI incremental depth from the top edge of the hole
ZA absolute depth in workpiece coordinates
F thread pitch
M direction of tool rotation for plunging
M3 right-hand thread M4 left-hand thread
V safety distance to the top edge of the hole

Optional addresses [..]:
W retract level relative to the coordinate system of the workpiece

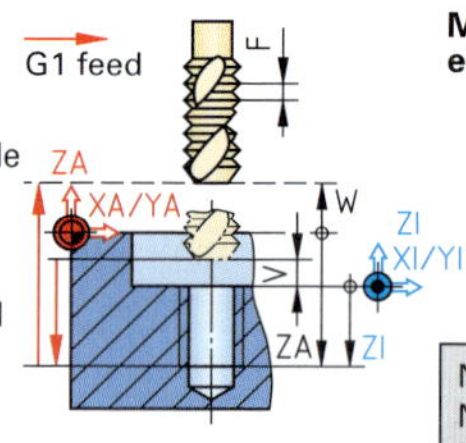

Machining example

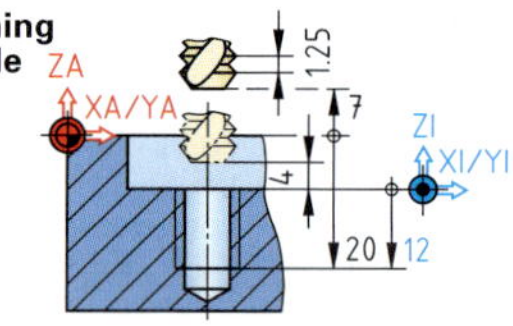


```
N10 ...
N15 G84 ZI-12 F1.25 M3 V4 W7 S800
N20 G79 X.. Y.. Z.. ;cycle call
```

G85 Reaming cycle

Structure of NC block
G85 ZI/ZA [W] [E] [F] [S] [M]

Obligatory addresses:
ZI/ZA drilling depth in the infeed axis
ZI incremental depth from the top edge of the hole
ZA absolute depth in workpiece coordinates
V safety distance from the top edge of the hole

Optional addresses [..]:
W retract level relative to the coordinate system of the workpiece
E feed speed of the retracting motion

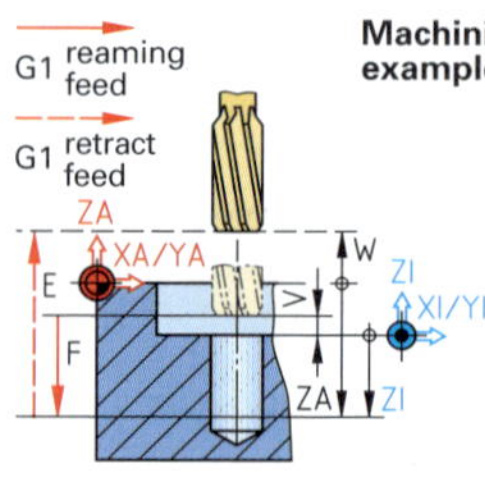

Machining example

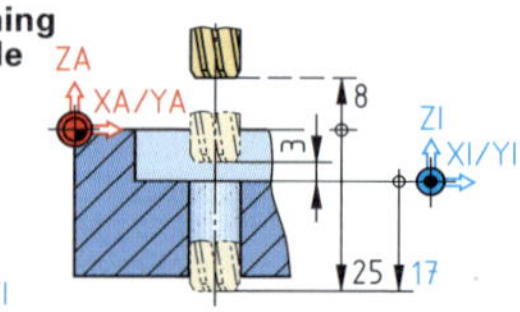


```
N10 ...
N15 G85 ZI-17 V3 W8 E260
G79 X.. Y.. Z.. ;cycle call
```

A

Program structure of CNC machines according to PAL

PAL cycles for milling machines

G86 Boring cycle

Structure of NC block
G86 ZI/ZA V [W] [DR] [F] [S] [M]

Obligatory addresses:

ZI/ZA depth to be bored out
ZI depth of bore in the infeed axis
ZA absolute depth of bore relative to the coordinate system of the workpiece
V safety distance from the top edge of the hole

Optional addresses [..]:

W retract level relative to the coordinate system of the workpiece
DR radial retract distance to the contour

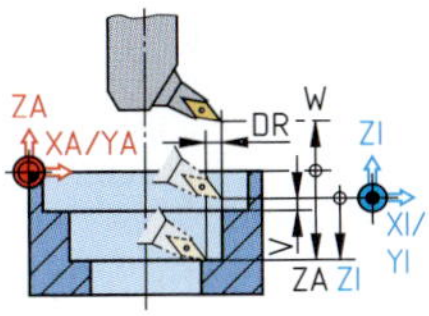

Machining example

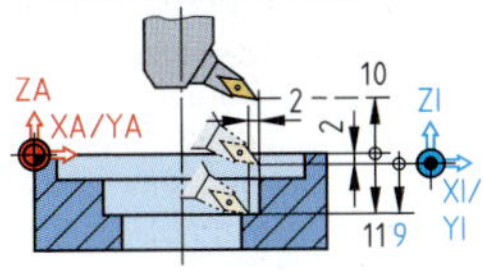

N10 ...
N15 G86 ZI-9 V2 W10 DR2
N20 G79 X.. Y.. Z.. ;cycle call

G87 Plunge milling cycle

Structure of NC block
G87 ZI/ZA R D V [W] [BG] [F] [S] [M]

Obligatory addresses:

ZI/ZA depth of hole to be bored out
ZI incremental depth from the top edge
ZA absolute depth of bore relative to the coordinate system of the workpiece
R radius of the hole to be milled out
D infeed per helical line (pitch of the helical motion)
V safety distance from the top edge of the hole

Optional addresses [..]:

W retract level relative to the coordinate system of the workpiece
BG2 machining, clockwise
BG3 machining, counter clockwise

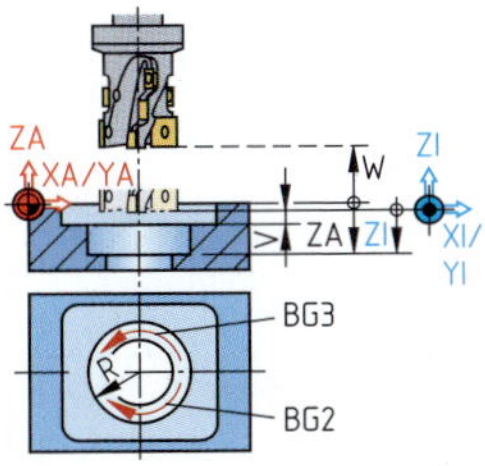

Machining example

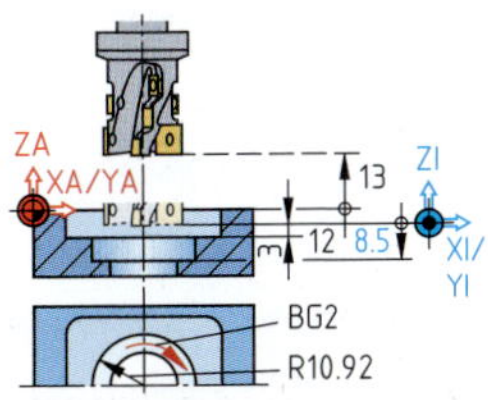

N10 ...
N15 G87 ZI-8,5 R10.92 D3 V3 W13 D3 BG2
N20 G79 X.. Y.. Z.. ;cycle call

G88 Internal thread milling cycle

Structure of NC block
G88 ZI/ZA DN D Q V [W] [BG] [F] [S] [M]

Obligatory addresses:

ZI/ZA depth of thread
ZI incremental depth of thread from the top edge
ZA absolute depth of thread relative to the coordinate system of the workpiece
DN nominal diameter of the internal thread
D thread pitch
Q number of thread grooves of the tool
V safety distance from the top edge of the hole

Optional addresses [..]:

W retract level relative to the coordinate system of the workpiece
BG2 machining, clockwise
BG3 machining, counter clockwise

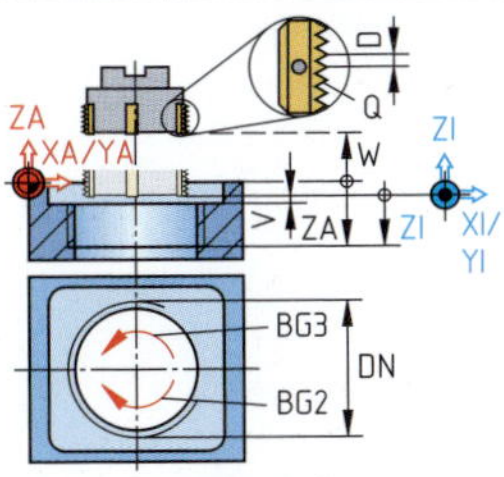

Machining example

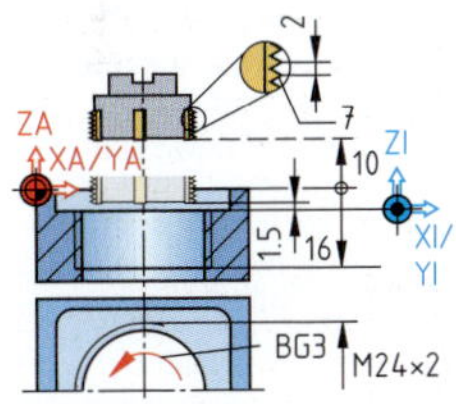

N10 ...
N15 G88 ZA-16 DN24 D2 Q7 V1.5 W10 BG3 F..
N20 G79 X.. Y.. Z.. ;cycle call

G89 External thread milling cycle

Structure of NC block
G89 ZI/ZA DN D Q V [W] (BG] [F] [S] [M]

Obligatory addresses:

ZI incremental depth of thread from the top edge
ZA absolute depth of thread relative to the coordinate system of the workpiece
DN nominal diameter of the external thread
D thread pitch
Q number of thread grooves of the tool
V safety distance to the top edge of the hole

Optional addresses [..]:

W retract level
BG2 machining, clockwise
BG3 machining, counter clockwise

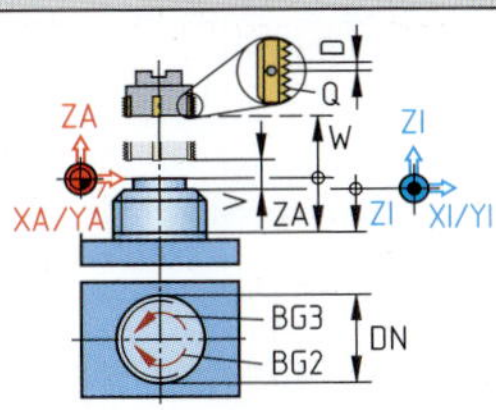

Machining example

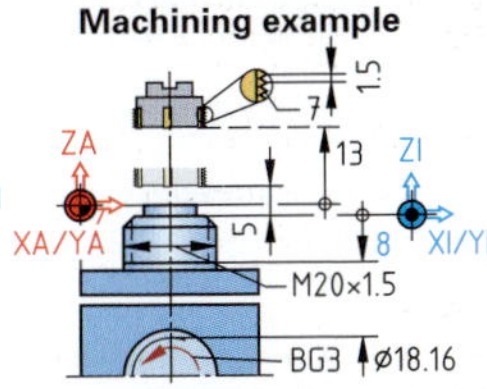

N10 ...
N15 G89 ZI-8 DN18.16 D1.5 Q7 V5 W13 BG3 F..
N20 G79 X.. Y.. Z.. ;cycle call

A

Program structure of CNC machines according to PAL

PAL cycles for milling machines

G72 Rectangular pocket milling cycle

Structure of NC block
G72 ZI/ZA LP BP D V [W] [RN] [AK] [AL] [EP] [DB] [RH] [DH] [O] [Q] [H] [E] [F] [S] [M]

Obligatory addresses:
ZI/ZA depth of the circular pocket in the infeed axis
ZI incremental from the top edge of the pocket
ZA absolute, relative to the coordinate system of the workpiece
LP length of the rectangular pocket in X direction
BP width of the rectangular pocket in Y direction
D maximum depth of cut
V safety distance to the material surface

Optional addresses [..]:
AK pocket wall finish allowance
AL pocket bottom finish allowance
RN corner radius
EP0, EP1, EP2, EP3 definition of the setpoint at cycle call
W retract level, in fast motion
H type of machining
H1 rough machining H4 finishing
H2 face roughing of the rectangular surface
H14 rough-machining and finishing with the same tool
E feed rate for plunging

Machining example

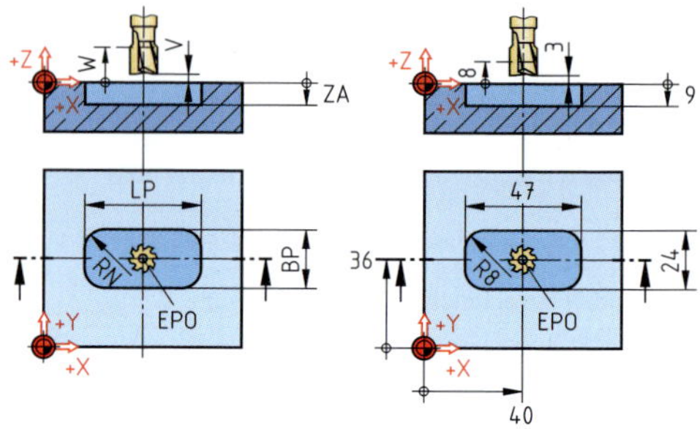


```
N15 G72 ZA-9 LP47 BP24 D4 V3 AK0.4 AL0.5 W8
N20 G79 X40 Y36    ;cycle call for G72
```

G73 Circular pocket and spigot milling cycle

Structure of NC block
G73 ZI/ZA R D V [W] [RZ] [AK] [AL] [DB] [RH] [DH] [O] [Q] [H] [E] [F] [S] [M]

Obligatory addresses:
ZI/ZA depth of circular pocket in the feed axis
ZI incremental from the top edge of the pocket
ZA absolute, relative to the coordinate system of the workpiece
D maximum depth of cut
V safety distance to the material surface

Optional addresses [..]:
RZ radius of the optional spigot
AK pocket wall finish allowance
AL pocket bottom finish allowance
DB cutter path overlap in %
W retract level, in fast motion
H–E as with G72

Machining example

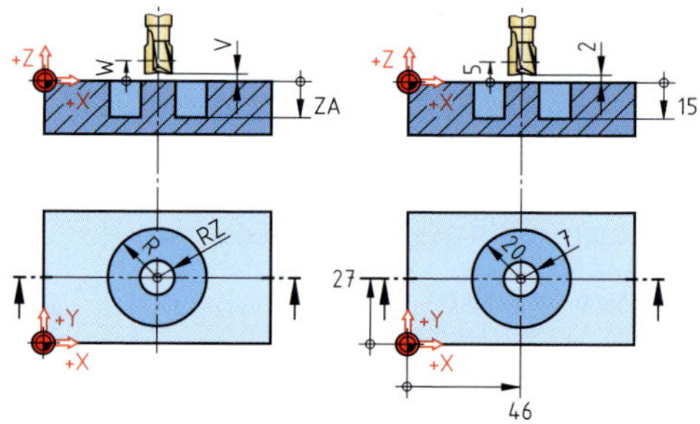


```
N15 G73 ZA-15 R20 D4 V2 AK0.4 AL0.5 W5
N20 G79 X46 Y27    ;cycle call for G73
```

G74 Slot milling cycle (longitudinal slot)

Structure of NC block
G74 ZI/ZA LP BP D V [W] [AK] [AL] [EP] [O] [Q] [H] [E] [F] [S] [M]

Obligatory addresses:
ZI/ZA depth of the slot in the infeed axis
ZI incremental from the top edge of the slot
ZA absolute, relative to the coordinate system of the workpiece
LP slot length
BP slot width
D maximum depth of cut
V safety distance

Optional addresses [..]:
W retract level
AK pocket wall finish allowance
AL pocket bottom finish allowance
EP0, EP1, EP2, EP3 definition of the setpoint at cycle call
O infeed motion
O1 vertical tool immersion
O2 ramping tool immersion
H–E as with G72

Machining example

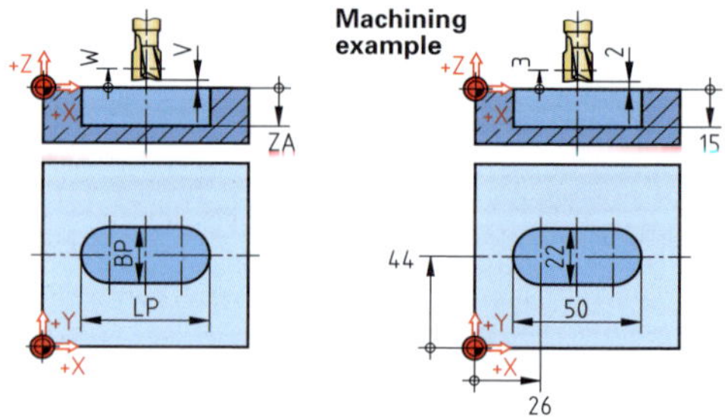


```
N15 G74 ZA-15 LP50 BP22 D3 V2 ;definition of longitudinal slot via G74
N20 G79 X... Y... ;cycle call at a particular point via G79
```

A

Program structure of CNC machines according to PAL

PAL cycles for milling machines

G75 | Slot milling cycle (arc)

Structure of NC block
G75 ZI/ZA BP RP AN/AO AO/AP D V (W] (AK] (AL]
[EP] [O] [Q] [H] [E] [F] [S] [M]

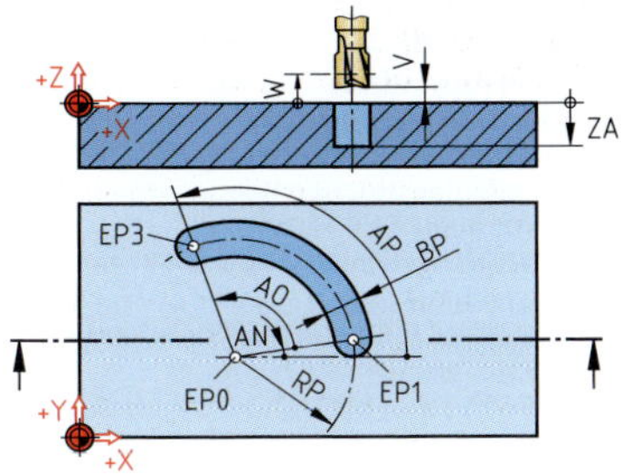

Obligatory addresses:
ZI/ZA slot depth
ZI incremental from the top edge of the slot
ZA absolute depth
BP slot width
RP slot radius
AN polar start angle relative to the positive X axis and the center point of the slot's first end radius
AO polar aperture angle between the center points of the slot's end radii
AP polar final angle relative to the positive X axis and the center point of the slot's second end radius (only 2 of the 3 polar angles need to be defined)
D maximum depth of cut
V safety distance

Optional addresses [..]:
EP definition of the calling point for the slot cycle
EP0 center of the circular slot
EP1 center of the right or top semicircle at the rear end
EP3 center of the left or bottom semicircle at the rear end
W retract level, in fast motion
AK slot wall finish allowance
AL slot bottom finish allowance
Q direction of motion
Q1 climb milling
Q2 conventional milling
H type of machining
H1 rough machining
H4 finishing
H14 rough machining and finishing
E feed rate for plunging

Machining example

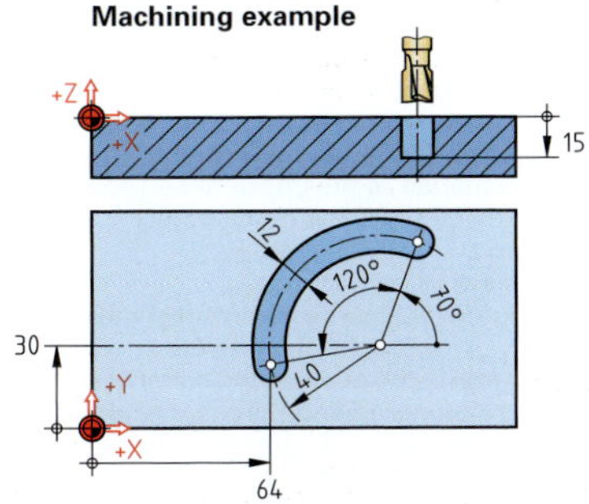

N15 G75 ZA-15 BP12 RP80 AN70 AO120 AK0.3 AL0.5 EP3 D5 V3 W6
N20 G79 X64 Y30 ;cycle call for G75 at EP3

G76 | Cycle call on a straight line (hole line)

Structure of NC block
G76 [X/XI/XA] [Y/YI/YA] [Z/ZI/ZA] AS D O [AR] [W] [H]

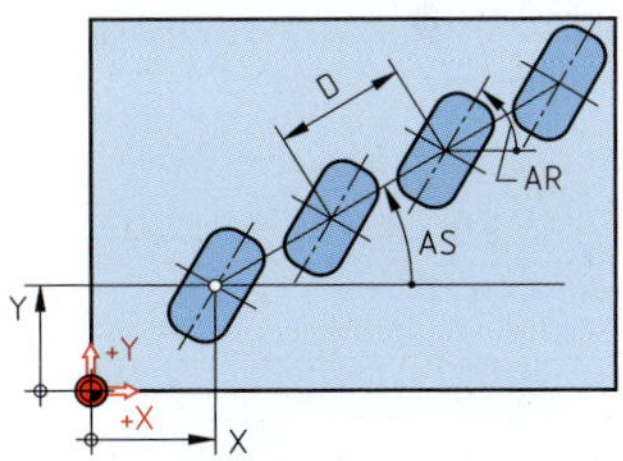

Obligatory addresses:
AS angle of the straight line relative to the first geometry axis
+ counter clockwise
– clockwise
D spacing of the cycle calls on the line
O number of cycle calls on the line

Optional addresses [..]:
X, XI, XA X coordinate of the first point
X absolute or incremental X coordinate (G90, G91)
XI difference in coordinates between the current tool position and the first point on the line
XA absolute coordinate input of the starting point
Y, YI, YA Y coordinate of the first point
Y absolute or incremental Y coordinate (G90, G91)
YI difference in coordinates between the current tool position and the first point on the line
YA absolute coordinate input of the starting point
Z, ZI, ZA Z coordinate of the first point
Z absolute or incremental Z coordinate (G90, G91)
ZI difference in coordinates between the current tool position and the first point on the line
ZA absolute coordinate input of the starting point
AR angle of rotation relative to the positive X axis
W retract level, absolute
H reversing position
H1 tool travels to safety distance between two positions and to the retract level after the last position
H2 tool travels to the retract level between two positions

Machining example

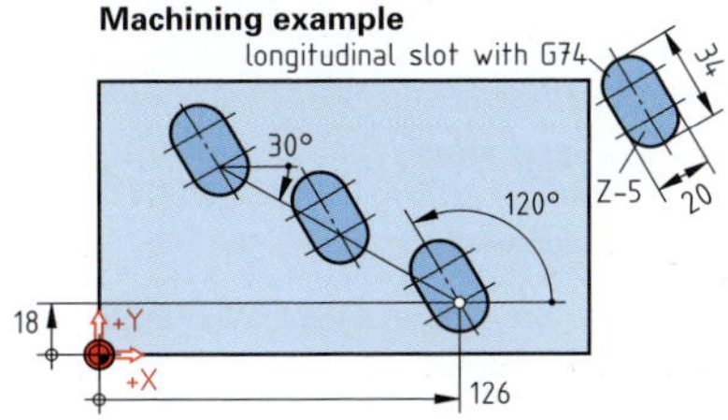

N15 G74 ZA-5 LP34 BP20 .... ;definition of longitudinal slot with G74 N20 G76 X126 Y18 Z0 AR120 D42 O3 AS150 ;cycle call

A

Program structure of CNC machines according to PAL

PAL cycles for milling machines

G77 | Cycle call on a pitch circle (hole circle)

Structure of NC block
G77 [I/IA] [J/JA] [Z/ZI/ZA] R AN/AI AI/AP O [AR] [W] [H] [FP]

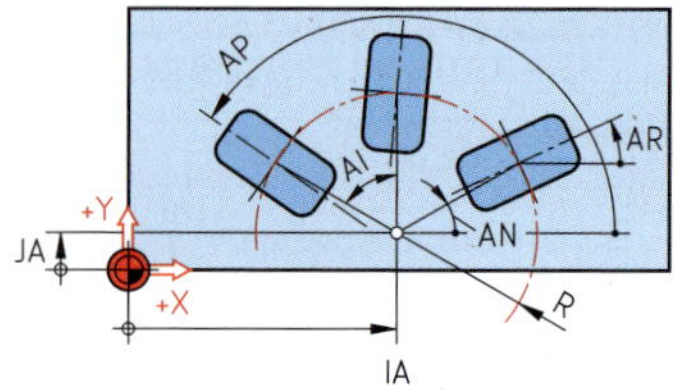

Obligatory addresses:

- R radius of pitch circle
- AN polar angle of first object
- AI constant segment angle
- AP polar angle of last object
- O number of objects on the pitch circle

Optional addresses [..]:

- I difference in X coordinates between the circle center and the starting point
- IA absolute X coordinate of the circle center
- J difference in Y coordinates between the circle center and the starting point
- JA absolute Y coordinate of the circle center
- Z absolute or incremental input via G90/G91
- ZI difference in Z coordinates between the current tool position and the pitch circle center
- ZA absolute coordinate of the target point
- AR angle of rotation in direction of the positive first geometry axis
- Q orientation of the object to be processed
  Q1 forced rotation of the object Q2 fixed orientation of the object
- W retract level, absolute
- H retracting motion
  - H1 the tool travels to the safety distance V after completion of the machining process
  - H2 the tool travels to the retract level W after completion of the machining process
  - H3 like H1, but the tool travels to the next position on the pitch arc
  - FP positioning feed in G94 on the pitch arc at H3

Machining example

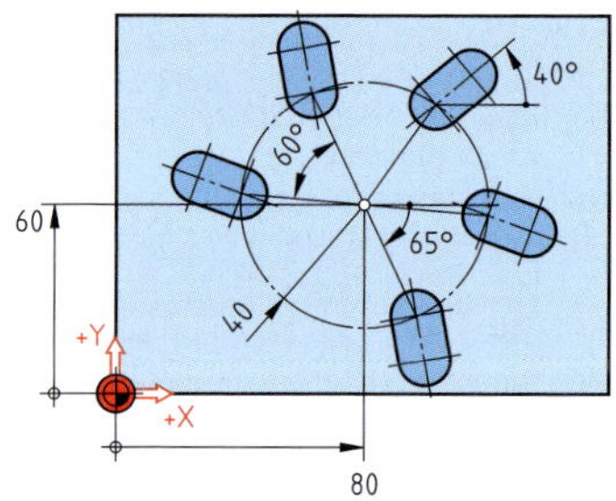


```
N15 G74 ZA-5 LP34 BP20 .... ;longitudinal slot with G74
N20 G77 R40 AN-65 AI60 AR40 O5 IA80 JA60 ;cycle call
```

G78 | Cycle call at a particular point (with polar coordinates)

Structure of NC block
G78 [I/IA] [J/JA] RP AP [Z/ZI/ZA] [AR] [W]

Obligatory addresses:

- I, IA X coordinate of the center of rotation
- J, JA Y coordinate of the center of rotation
- RP radius of the rotation circle
- AP angle of rotation relative to the X axis

Optional addresses [..]:

- Z, ZI, ZA Z coordinate of the top edge
- AR angle of rotation of the object relative to the X axis
- W retract level

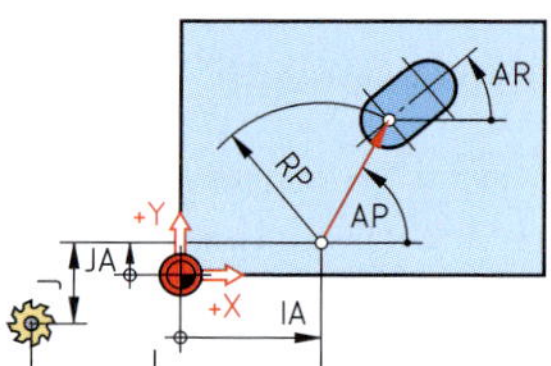

Machining example

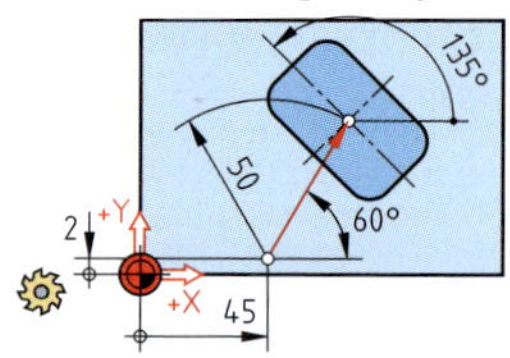


```
N15 G72 ZA.. LP.. BR.. ;rectangular pocket with G72
N20 G78 IA45 JA2 RP50 AP60 AR135 ;cycle call G78
```

G79 | Cycle call at a particular point (with Cartesian coordinates)

Structure of NC block
G79 [X/XI/XA] [Y/YI/YA] [Z/ZI/ZA] [AR] [W]

Optional addresses [..]:

- X, XI, XA X coordinate of the first point
- Y, YI, YA Y coordinate of the first point
- Z, ZI, ZA Z coordinate of the first point
- AR angle of rotation of the object relative to the X axis
- W retract level, absolute in workpiece coordinates

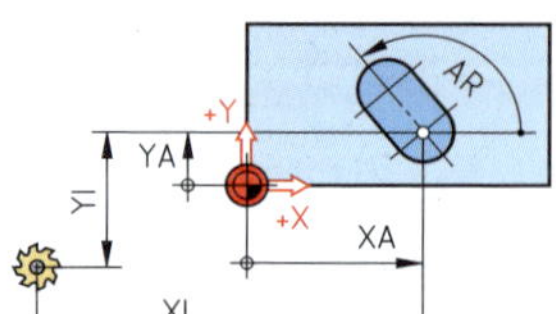

Machining example

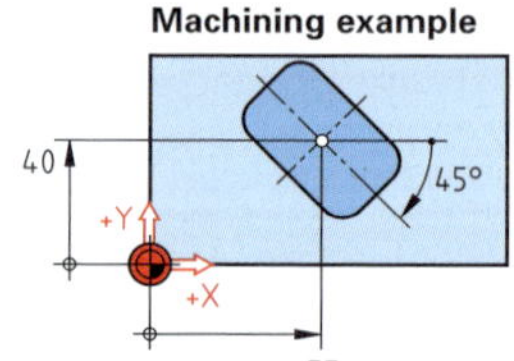


```
N15 G72 ZA.. LP.. BP.. ;rectangular pocket with G72
N20 G79 XA55 YA40 AR-45 ;cycle call G79
```

A

International Material Comparison Chart

Chart I

Germany		USA	U. K.	France	Japan	Sweden
Standard						
DIN, DIN EN	Mat. No.	AISI/SAE	BS	AFNOR	JIS	SS
Structural and machine construction steels						
S185	1.0035	A 283 (A)	1449 15 HR; HS	A 33	–	1300
S235JR	1.0037	1015, A 283	Fe 360 B	E 24-2	STKM 12 A; C	1311
S235JRG1	1.0036	A 283 (C)	Fe 360 B 4360-40 B	–	–	1311, 1312
S235JRG2	1.0038	A550.36	Fe 360 B; 6323-ERW 3; CEW 3	E 24-2 NE	STKM 12A; C	1312
S235J0	1.0114	–	4360-40 C	E 24-3, E 24-4	–	–
S235J2G3	1.0116	A 515 (55)	Fe 360 D 1 FF	E 24-3, E 24-4	–	1312, 1313
S235J2G4	1.0117	1513	A 2	E 36-4	–	–
S275JR	1.0044	1020	Fe 430 B FU	E 28-2	SN 400 B; C; SN 490 B; C	1412
S275J0	1.0143	A 572 (42)	4360-43 C	E 28-3, E 28-4	–	1414-01
S275J2G3	1.0144	A 500 (A; B; D)	Fe 430 D1 FF	E 28-3, E 28-4	SM 400 A; B; C	1411, 1412, 1414
S355JR	1.0045	–	4360-50 B	E 36-2	STK 400	2172
S355J0	1.0553	A 678 (C)	A 3	320-560 M	–	1606
S355J2G3	1.0570	1024; 1524	1449 50/35 HR; HS	E 36-3, E 36-4	STK 500	2132 to 2134, 2174
S355J2G4	1.0577	A 738 (A; C)	Fe 510 D2 FF	A 52 FP	–	2174
S355K2G3	1.0595	A 678 (C)	224-430	–	–	–
S355K2G4	1.0596	A 678 (C)	224-430	–	–	–
E295	1.0050	A 570 (50)	Fe 490-2 FN	A 50-2	SS 490	1550, 2172
E335	1.0060	A 572 (65)	Fe 590-2 FN	A 60-2	SM 570	1650
E360	1.0070	–	Fe 590-2 FN	SM 570	1650	–
Unalloyed quality steels						
S275N	1.0490	A 516 (60)	–	–	–	–
S275M	1.8818	A 715 (7)	–	–	–	–
S355N	1.0545	A 714 (III)	4360-50 E	E 355 R	–	2334-01, 2134-01
S355M	1.8823	A 715 (7)	–	–	–	–
Alloy high grade steels						
S420N	1.8902	A 633 m	–	E 420 R	–	–
S420M	1.8825	–	–	–	–	–
S460N	1.8901	A 633 m	–	E 460 R	–	–
S460M	1.8827	A 734 (B)	–	–	–	–
Quenched and tempered structural steels with higher yield strength						
S460QL	1.8906	–	4360-55 F	S 460 Q, T	SM 520 B, C	2143
S500QL	1.8909	–	–	S 500 T	–	–
S620QL	1.8927	–	–	S 620 T	–	–
S960QL	1.8933	–	–	S 960 T	–	–
Unalloyed steels – Case hardened steels						
C10E	1.1121	1010	040 A 10, 045 M 10	C 10, CX 10	S 9 CK, S 10 C	1265
C10R	1.1207	1011	–	E 355 C	–	–
C15E	1.1141	1015	040 A 15, 080 M 15	XC 12	S 15, S 15 CK	1370
C15R	1.1140	1016	080 A 20	–	–	–
Alloy steels – Case hardened steels						
16MnCr5	1.7131	5115	527 M 17	16 MC 5, 16 Mn Cr 5	–	2173
16MnCrS5	1.7139	5115	620-440	16 MC 5	–	2127
18CrMo4	1.7243	5120/5120 H	527 M 20	20 MC 5	Scr 420 M	2523
18CrMoS4	1.7244	5120/5120 H	527 M 20	20 MC 5	Scr 420 M	2523
20MoCr4	1.7321	K 12220	–	–	–	–
20MoCrS4	1.7323	K 12220	–	–	–	–
15NiCr13	1.5752	3310	655 H 13	12 NC 15	SNC 815 (H)	–
20NiCrMo2-2	1.6523	8620 H	805 H 20	20 NCD 2	SNCM 220 H	2506
20NiCrMoS2-2	1.6526	8620/8620 H	–	20 NCD 2	SNCM 220 M	2506
17NiCrMo6-4	1.6566	–	815 M 17	18 NCD 6	–	2523

International Material Comparison Chart

Chart II

Germany		USA	U. K.	France	Japan	Sweden
Standard						
DIN, DIN EN	Mat. No.	AISI/SAE	BS	AFNOR	JIS	SS
17NiCrMoS6-4	1.6569	4718/47 18 H	–	–	–	–
20MnCr5	1.7147	5120	527 M 20	20 MC 5	SMn C 420 H	–
20MnCrS5	1.7149	5120/5120 H	527 M 20	20 MC 5	Scr 420 M	2523
14NiCrMo13-4	1.6657	9310	832 M 13	16 NCD 13	–	–
18CrNiMo7-8	1.6687	–	–	18 NCD 6	–	–
Unalloyed steels – Quenched and tempered steels						
C22	1.0402	1020	055 M 15	AF 42 C 20	S 20 C, S 22 C	1450
C22E	1.1151	1023	055 M 15	2 C 22, XC 18, XC 25	S 20 C	1450
C25	1.0406	1025	070 M 26	1 C 25	–	–
C25E	1.1158	1025	(070 M 26)	2 C 25, XC 25	S 25 C, S 28 C	1450
C35	1.0501	1035	060 A 35	C 35, 1 C 35	S 35 C, S 35 CM	1572, 1550
C35E	1.1181	1035	080 A 35	C 35	S 35 C	1550, 1572
C45	1.0503	1045	080 A 46	C 45	S 45 C, S 45 CM	1672, 1650
C45E	1.1191	1042, 1045	080 M 46	XC 42 H 1	S 45 C	1672
C60	1.0601	1060	060 A 62	C 60	S 58 C	–
C60E	1.1221	1064	060 A 62, 070 M 60	2 C 60	S 58 C, S 60 CM, S 65 CM	1665, 1678
C30	1.0528	G 10300	080 A 30	XC 32	S 30 C	–
C35	1.0501	1035	060 A 35	–	–	–
C40	1.0511	1040	080 M 40	AF 60 C 40	–	F. 114A
C50	1.0540	G 10500	080 M 50	XC 50	S 50 C	–
C55	1.0535	1055	070 M 55, 5770-50	C 54; 1 C 55	S 55 C, S 55 CM	1655
Alloy steels – Quenched and tempered steels						
38Cr2	1.7003	–	120 M 36	38 C 2, 38 Cr 2	–	–
38CrS2	1.7023	5140	530 A 40	42 C 4	Scr 440 M	2245
46Cr2	1.7006	5045	–	42 C 2, 46 Cr 2	–	–
46CrS2	1.7025	A 768 (95)	–	–	SNB 5	–
34Cr4	1.7033	5132	530 A 32	32 C 4, 34 Cr 4	SCr 430 (H)	–
34CrS4	1.7037	4340/4340 H	818 M 40	35 NCD 6	SNCM 439	–
37Cr4	1.7034	5135	530 A 36	37 Cr4, 38 C 4	Scr 435 (H) (M)	–
37CrS4	1.7038	5135/5135 H	–	38 Cr 4	Scr 435 H	–
25CrMo4	1.7218	4118	708 M 25	25 CD 4	SCM 420	2225
24CrMoS4	1.7213	4130/4130 H	CDS 110	30 CD 4	SCM 430 M	2223-01
41Cr4	1.7035	5140	530 A 40	41 Cr 4, 42 C 4	Scr 440 (H) (M)	–
41CrS4	1.7039	L 1	524 A 14	–	–	2092
34CrMo4	1.7220	4137	708 A 37	35 CD 4	SCM 432	2234
42CrMo4	1.7225	4140	708 M 40	42 CD 4	SCM 440 (H)	2244
50CrMo4	1.7228	4150, 4147	708 A 47	50 Cr Mo 4	SCM 4454 (H)	2512
51CrV4	1.8159	6150	735 A 50	50 CV 4	SUP 10	2230
36CrNiMo4	1.6511	9840	817 M 37	36 CrNiMo 4, 35 NCD 5, 40 NCD 3	–	–
34CrNiMoS4	1.6582	4337, 4240	816 M 40, 817 M 40	34 CrNiMo 8	SNCM 447	2541
30NiCrMo8	1.6580		823 M 30	30 CrNiMo 8	SNCM 431	–
36NiCrMo16	1.6773	5135/5135 H	–	38 Cr 4	Scr 435 M	–
Nitriding steels						
31CrMo12	1.8515	–	722 M 24	30 CD 12	–	2240
34CrAlMo5-10	1.8507	A 355 Cl.D	–	30 CAD 6.12	–	–
40CrAlMo7-10	1.8509	E 7140	905 M 39, En 41 B	40 CAD 6.12	SACM 1, SACM 645	2940
40CrMoV13-9	1.8523	–	897 M 39	–	–	–
Steels for flame and induction hardening						
Cf45	1.1193	1045	060 A 47, 080 M 46	XC 42 H 1 TS	S 45 C, S 45 CM	1672
42Cr4	1.7045	5140	530 A 40	42 C 4 TS	Scr 440	2245
41CrMo4	1.7223	4142	708 M 40, 3111-5/1	42 CD 4 TS	SNB 22, SCM 440	2244
Cf35	1.1183	1035	080 A 35	XC 38 H 1 TS	S 35 C, S 35 CM	1572

S

International Material Comparison Chart

Chart III

Germany		USA	U. K.	France	Japan	Sweden
Standard						
DIN, DIN EN	Mat. No.	AISI/SAE	BS	AFNOR	JIS	SS
Cf53	1.1213	1050	070 M 55	XC 48 H 1 TS	S 50 C, S 50 CM	1674
Cf70	1.1249	–	–	–	–	–
Free cutting steels						
11SMn30	1.0715	1213	230 M 07	S 250	SUM 22	1912
11SMnPb30	1.0718	12 L 13	–	S 250 Pb	SUM 23 L	1914
11SMn37	1.0736	1215	–	S 300	SUM 25	–
11SMnPb37	1.0737	12 L 14	–	S 300 Pb	–	1926
10S20	1.0721	1108, 1109	(210 M 15)	10 F 2	–	–
10SPb20	1.0722	–	–	10 Pb F 2	–	–
35S20	1.0726	1140	212 M 36	35 MF 6	–	1957
46S20	1.0727	1146	En 8 DM	45 MF 4	SUM 43	–
Cold work steels, unalloyed						
C80U	1.1525	W 108	–	C 80 E 2 U, Y_1 80	–	–
C105U	1.1545	W 1	BW 1 A	Y 105	SK 3	1880
Cold work steels, alloy						
45WCrV7	1.2542	S 1	BS 1	45 WCrV 8	S 1	2710
60WCrV8	1.2550	S 1	BS 1	55 WC 20	–	–
100MnCrW4	1.2510	O 1	BO 1	90 MnWCrV 5	SKS 3	–
90MnCrV8	1.2842	O 2	BO 2	90 Mn V 8, 90 MV 8	–	–
X210Cr12	1.2080	P 3	BD 3	Z 200 C 12	SKD 12	2710
102Cr6	1.2067	L 3	(BL 3)	100 Cr 6, Y 100 C 6	SUJ 2	–
45NiCrMo16	1.2767	–	BP 30	Y 35 NCD 16	–	–
X153CrMoV12	1.2379	D 2	BD 2	Z 160 CDV 12	SKD 12	2260
X100CrMoV51	1.2363	A 2	BA 2	Z 100 CDV 5	SKD 12	2260
X40CrMoV51	1.2344	H 13	BH 13	Z 40 CDV 5	SKD 61	2242
X210CrW12	1.2436	D4 (D6)	BD 6	Z 210 CW 12-01	SKD 2	2312
Hot work steels						
55NiCrMoV7	1.2714	–	–	–	SKS 51	–
X37CrMoV5-1	1.2343	H 11	BH 11	Z 38 CDV 5	SKD 6	–
32CrMoV12-28	1.2365	H 10	BH 10	32 CDV 12-28	–	–
High speed steels						
HS6-5-2C	1.3343	M 2	BM 2	HS 6-5	SKH 51	2722
HS6-5-2-5	1.3243	M 35	BM 35	Z 85 WDKCV 06-05-04-02	SKH 55	2723
HS10-4-3-10	1.3207	–	BT 42	HS 10-4-3-10	SKH 57	–
HS2-9-2	1.3348	M 7	–	HS 2-9-2, Z 100 DCWV 09-04-02-02	–	2782
HS2-9-1-8	1.3247	M 42	BM 42	HS 2-9-1-8	SKH 59	2716
S2-9-2-8	1.3249	M 42	BM 34	–	–	–
Stainless steels, austenitic						
X10CrNi18-8	1.4310	301	301 S 21/22	Z 12 CN 18-09	SUS 301	2331
X2CrNi18-9	1.4307	F 304 L	304 L	–	SUS F 304 L	–
X5CrNi18 9	1.4350	304	304 S 31	Z 5 CN 18.09	SUS 304	2332
X2CrNiN19-11	1.4306	304 L	304/305 S 11	Z 2 CN 18-10	SCS 19, SUS 304 L	2352
X2CrNi18-10	1.4311	304 LN	304 S 61	Z 3 CN 18-07 Az	SUS 304 LN	2371
X5CrNi18-10	1.4301	304	304 S 17	Z 5 CN 17-08	SUS 304	2332, 2333
X8CrNiS18-9	1.4305	303	303 S 22/31	Z 8 CNF 18-09	SUS 303	2346
X6CrNiTi18-10	1.4541	321	321 S 31/51	Z 6 CNT 18-10	SUS 321	2337
X4CrNi18-12	1.4303	305/308	305 S 17, 305 S 19	Z 5 CN 18-11 FF	SUS 305 J1, SUS 305	–
X5CrNiMo17-12-2	1.4401	316	316 S 13/17/19	Z 3 CND 17-11-01	SUS 316	2347
X6CrNiMoTi17-12-2	1.4571	316 Ti	320 S 18/31	Z 6 CNDT 17-12	SUS 316 Ti	2350
X2CrNiMo18-14-3	1.4435	316 L	316 S 11/13/14	Z 3 CND 17-12-03/ Z 3 CND 18-14-03	SUS 316 L	2353

International Material Comparison Chart

Chart IV

Germany		USA	U. K.	France	Japan	Sweden
Standard						
DIN, DIN EN	Mat. No.	AISI/SAE	BS	AFNOR	JIS	SS
X2CrNiMoN17-13-3	1.4429	316 LN	326 S 63	Z 3 CND 17-12 Az	(SUS 316 LN)	2375
X2CrNiMoN17-13-5	1.4439	316 L	316 S 11	Z 2 CND 17-12	SUSF 316 L	2348
X1NiCrMoCu25-20-5	1.4539	USN N 08904	–	Z 2 NCDU 25-20	–	2562
Stainless steels, ferritic						
X2CrNi12	1.4003	A 268	–	–	–	–
X6Cr13	1.4000	403	403 S 17	Z 8 C 12, Z 8 C 13 FF	SUS 403	2301
X6Cr17	1.4016	430	430 S 15	Z 8 C 17	SUS 430	2320
X2CrTi12	1.4512	409	409 S 19	Z 3 CT 12	SUH 409	–
X6CrMo17-1	1.4113	434	434 S 17	Z 8 CD 17.01	SUS 434	–
X2CrMoTi18-2	1.4521	443/444	–	–	SUS 444	2326
Stainless steels, martensitic						
X12CrS13	1.4005	416	416 S 21 Z 11 CF13	SUS 416	–	2380
X12Cr13	1.4006	410	410 S 21	Z 10 C 13	SUS 410	2302
X20Cr13	1.4021	420	420 S 37	Z 20 C 13	SUS 420 J 1	2303
X30Cr13	1.4028	420 F	420 S 45	Z 30 C 13	SUS 420 J 2	2304
X46Cr13	1.4034	–	(420 S 45)	Z 44 C 14, Z 38 C 13 M	SUS 420 J2	2304
X39CrMo17-1	1.4122	5925	–	–	–	–
X3CrNiMo13-4	1.4313	CA 6-NM	425 C 11	Z 4 CND 13.4 M	SCS 5, SCS 6	2384
Hot rolled steels for springs						
38Si7	1.5023	–	–	41 Si 7	–	–
46Si7	1.5024	9255	–	51 S 7, 51 Si 7	–	2090
55Cr3	1.7176	5155	525 A 58	55 Cr 3, 55 C 3	SUP 9 (A) (M)	2253
61SiCr7	1.7108	9261, 9262	–	61 SC 7	–	–
51CrV4	1.8159	6150	735 A 50	55 Cr V 4	SUP 10	2230
Cold rolled strip and sheet from soft steels						
DC03	1.0347	A 619	1449 3 CR	E	CR 2	1146
DC04	1.0338	A 620 (1008)	1449 2 CR; 3 CR	ES	SPCE; HR 4	1147
Cast iron with flake graphite (gray iron)						
EN-GJL-100	EN-JL-1010	A 48 20 B	1452 Grade 100	Ft 10 D	G 5501 FC 10	0110-00
EN-GJL-150	EN-JL-1020	A 48 25 B	1452 Grade 150	A 32-101 FGL 150; FT 15 D	G 5501 FC 15	0115-00
EN-GJL-200	EN-JL-1030	A 48 30 B	1452 Grade 220	A 32-101 FGL 200; FT 20 D	G 5501 FC 20	0120-00
EN-GJL-250	EN-JL-1040	A 48 40 B	1452 Grade 250/ 260	A 32-101 FGL 250; FT 25 D	G 5501 FC 25	0125-00
EN-GJL-300	EN-JL-1050	A 48 45 B	1452 Grade 300	A 32-101 FGL 300; FT 30 D	G 5501 FC 30	0130-00
EN-GJL-350	EN-JL-1060	A 48 50 B	1452 Grade 350	A 32-101 FGL 350; FT 35 D	G 5501 FC 35	0135-00
Cast iron with spheroidal (nodular) graphite						
EN-GJS-350-22	EN-JS-1010	–	–	–	–	0717-15
EN-GJS-500-7	EN-JS-1050	A 536 60-45-12	2789 Grade 500/7	A 32-201 FGS 500-7	G 5502 FCD 500	0727-02
EN-GJS-600-3	EN-JS-1060	A 536 80-55-06	2789 Grade 600/3	A 32-201 FGS 600-3	G 5502 FCD 600	0732-03
EN-GJS-700-2	EN-JS-1070	A 536 100-70-03	2789 Grade 700-2	A 32-201 FGS 700-2	G 5502 FCD 700	0737-01
Malleable cast iron						
EN-GJMW-350-4	EN-JM 1010	–	86681 W 35-04	A 32-701 MB 35-7	G 5703 FCMW 330	–
EN-GJMW-400-5	EN-JM 1030	–	6681 W 40-05	A 32-701 MB 40-05	G 5703 FCMW 370	–
EN-GJMW-450-7	EN-JM 1040	–	6681 45-07	A 32-701 MB 450-7	G 5703 FCMWP 440	–
EN-GJMB-350-10	EN-JM 1130	A 47 Grade 22010+32510	310 B 340/12	A 32-702 MN 350-10	G 5703 FCMB 340	0815-00
EN-GJMB-450-6	EN-JM 1140	–	6681 P 45-06	A 32-703 MP 50-5	–	0854-00
EN-GJMB-550-4	EN-JM 1160	–	6681 P 55-04	A 32-703 MP 60-3	G 5703 FCMP 540	0856-00
EN-GJMB-650-2	EN-JM 1180	–	6681 P 65-02	–	–	0862-03
EN-GJMB-700-2	EN-JM 1190	A220 Grade 70003	6681 P 70-02	A 32-703 MP 70-2	G 5703 FCMP 690	0862-03

International Material Comparison Chart

Chart V

Germany		USA	U. K.	France	Japan	Sweden
Standard						
DIN, DIN EN	Mat. No.	AISI/SAE	BS	AFNOR	JIS	SS
Cast steels for general applications						
GS-38	1.0420	–	–	–	SC 360	–
GS-45	1.0446	A 27	–	–	SC 450	–
Cast steels for pressure vessels						
GP240GH	1.0619	A 216 Grade WCC	1504-161 Gr. B	–	–	–
G17CrMo5-5	1.7357	A 217 Grade WC 6	–	–	–	–
Aluminum and wrought aluminum alloys						
			old new	old new		
Al 99.5	1050 A	1050 A	1 B 1050 A	A-5 1050 A	A 1050	4007
Al Mn1	3103	3103	N 3 3103	–	–	4054
Al Mn1Cu	3003	3003	(3103)	A-M 1 3003	A 3003	–
Al Mg1	5005 A	5005 A	N 41 5005	A-G 0.6 5005	A 5005	4106
Al Mg2	5251	5251	N 4 5251	A-G 2 M 5251	–	–
Al Mg3	5754	5754	–	A-G 3 M 5754	–	4125
Al Mg5	5019/5119	5019/5119	–	A-G 5	–	–
Al Mg3Mn	5454	5454	N 51 5454	A-G 3 MC 5454	A 5454	–
Al Mg4.5Mn0.7	5083	5083	N 8 5083	A-G 4.5 MC	A 5083	4140
AlCuPbMgMn	2007	2007	–	A-U 4 PB	–	4335
Al Cu4PbMg	2030	2030	–	–	–	–
Al MgSiPb	6012	6012	–	A-SGPB	–	–
Al Cu4SiMg	2014	2014	H 15 (2014 A)	A-U 4 SG	–	–
Al Cu4MgSi	2017	2017	–	A-U 4 G	A 2017	–
Al Cu4Mg1	2024	2024	2 L 97/9 2024	A-U 4 G1 2024	A 2024	–
Al MgSi	6060	6060	H 9 (6063)	A-GS 6060	A 6063	4103
Al Si1MgMn	6082	6082	H 30 6082	A-SGM 0.7 6082	–	4212
Al Zn4.5Mg1	7020	7020	H 17 7020	A-Z 5 G 7020	(A 7 N 01)	4425
Al Zn5Mg3Cu	7022	7022	–	A-Z 4 GU	–	–
Al Zn5.5MgCu	7075	7075	2 L 95/96 7075	A-Z 5 GU 7075	A 7075	–
Aluminum casting alloys						
AC-AlSi7Mg	AC-42000	A 356	L M 25	A-S 7 g	–	–
Magnesium alloys, Titanium, Titanium alloys						
MgMn2	3.3520	M 1 A	MAG-E-101	G-M 2	–	–
MgAl3Zn	3.5312	AZ 31 B	MAG-E-111	G-A 3 Z 1	–	–
MgAl6Zn	3.5612	AZ 61 A	MAG-E-121	G-A 6 Z 1	–	–
MgAl8Zn	3.5812	AZ 80 A	–	G-A 7 Z 1	–	–
Ti1	3.7025	–	T A 1	–	–	–
Ti2	3.7035	–	T A 2	–	–	–
TiAl6V4	3.7165	–	T A 10-13, 28, 56	–	–	–
TiAlMo4Sn2	3.7185	–	T A 45-51, 57	–	–	–

Index of cited standards and other regulations

[1] These standards have been withdrawn. Replacement standards are given on the referenced page.

S

Index of cited standards and other regulations

[1] These standards have been withdrawn. Replacement standards are given on the referenced page.

S

Index of cited standards and other regulations

1) These standards have been withdrawn. Replacement standards are given on the referenced page.

S

Index of cited standards and other regulations

Index of cited standards and other regulations

S

Subject index

A

B

C

D

E

F

G

H

I, J

K

L

M

N

O

P

Q

R

S

T

U

V

W, X